Classical and Discrete Differential Geometry

This book introduces differential geometry and cutting-edge findings from the discipline by incorporating both classical approaches and modern discrete differential geometry across all facets and applications, including graphics and imaging, physics and networks.

With curvature as the centerpiece, the authors present the development of differential geometry, from curves to surfaces, thence to higher dimensional manifolds; and from smooth structures to metric spaces, weighted manifolds and complexes, and to images, meshes and networks. The first part of the book is a differential geometric, both classical and discrete, study of curves and surfaces in the Euclidean space, while the second part deals with higher dimensional manifolds centering on curvature by exploring the various ways of extending it to higher dimensional objects and more general structures and how to return to lower dimensional constructs. The third part focuses on computational algorithms in algebraic topology and conformal geometry, applicable for surface parameterization, shape registration and structured mesh generation.

The volume will be a useful reference for students of mathematics and computer science, as well as researchers and engineering professionals who are interested in graphics and imaging, complex networks, differential geometry and curvature.

David Xianfeng Gu is a SUNY Empire Innovation Professor of Computer Science and Applied Mathematics at State University of New York at Stony Brook, USA. His research interests focus on generalizing modern geometry theories to discrete settings and applying them in engineering and medical fields and recently on geometric views of optimal transportation theory. He is one of the major founders of an interdisciplinary field, Computational Conformal Geometry.

Emil Saucan is Associate Professor of Applied Mathematics at Braude College of Engineering, Israel. His main research interest is geometry in general (including Geometric Topology), especially Discrete and Metric Differential Geometry and their applications to Imaging and Geometric Design, as well as Geometric Modeling. His recent research focuses on various notions of discrete Ricci curvature and their practical applications.

Classical and Discrete Differential Geometry

Theory, Applications and Algorithms

David Xianfeng Gu and Emil Saucan

CRC Press
Taylor & Francis Group
Boca Raton London New York

CRC Press is an imprint of the
Taylor & Francis Group, an **informa** business

The cover image was created by David Gu to illustrate the Koebe-Poincare's surface uniformization theorem based on discrete surface Ricci flow algorithm.

First edition published 2023
by CRC Press
6000 Broken Sound Parkway NW, Suite 300, Boca Raton, FL 33487-2742

and by CRC Press
4 Park Square, Milton Park, Abingdon, Oxon, OX14 4RN

CRC Press is an imprint of Taylor & Francis Group, LLC

ISBN: 978-1-032-39017-8 (hbk)
ISBN: 978-1-032-39620-0 (pbk)
ISBN: 978-1-003-35057-6 (ebk)

DOI: 10.1201/9781003350576

Typeset in LM Roman
by KnowledgeWorks Global Ltd.

Publisher's note: This book has been prepared from camera-ready copy provided by the authors.

Dedication: E.S. dedicates this book to the memory of Tzvi Har'el, who introduced him to the wondrous world of classical Differential Geometry, and of Robert Brooks, who first revealed to him the power of Ricci curvature and the astounding interplay between smooth and discrete Differential Geometry.

D.G. dedicates this book to Professor Shing-Tung Yau, who taught him that the true meaning of life lies in the pursuit of truth and beauty.

Thanks: E.S. is grateful to his wife for her patience and understanding regarding the almost total suspension of weekends and holidays during the extended gestation period of this book.

D.G. is grateful to his family, collaborators, friends and students for their long-term support.

Contents

Section II Differential Geometry, Computational Aspects

Preface

One can ask, and rightly so, what novelty one more book on Differential Geometry might convey? Indeed, it is hard to be original in writing a Differential Geometry book, given not only the long history of the field and its many illustrious contributors, but also considering that there are already available a number of such classics as those due to do Carmo, Spivak and Klingenberg, to name just a few well-known and widely appreciated ones (and which served us, in the past, as textbooks in the number of courses taught over the years).

Still, we believe our book is highly original, its originality lying in the natural manner in which the classical, smooth Differential Geometry always taught blends in a synergistic manner with the modern Discrete Differential Geometry in all its facets, and applications: Graphics and Imaging, Physics, Networks, etc. We view the two approaches not only as complementing each other in some abstract, vague manner, but rather supporting their reciprocal understanding. (It is common to discretize classical differential geometric notions, and largely we follow the same paradigm. However, it is our belief, supported by certain instances in the book, that discrete notions give a better understanding, in their turn, to the original ones, which are sometimes technical and difficult to grasp). Furthermore, while there are, indeed, very few novel developments (if at all) that one might conceivably contribute to the corpus of classical Differential Geometry, Discrete Differential Geometry is dynamic, vibrant and ever-expanding. In fact, quite a few of the new directions and applications included herein are part of the authors' contributions to the field, a fact that also contributes, we believe, to the novelty and originality of the book.

We should add a few words regarding the content of this book (more technical, guiding ones follow below): It is commonly expected that each book should have a good story, running like a red thread along its pages and that, moreover, each such story needs to be centered around a center character, who focuses our attention and clusters around him all the other characters and drives the main action of the tale. As we have stated elsewhere, we are firm believers in the regretted Robert Brooks' saying that "The fundamental notion of differential geometry is the concept of curvature." We can take this even further and, only slightly exaggerating, maintaining Differential Geometry is nothing but the study of curvature.

Thus the hero of this book is curvature, and the story it presents is its development, from curves to surfaces, thence to higher dimensional manifolds, and from smooth structures, to metric spaces, weighted manifolds and complexes, to images, meshes and networks. It is, we feel, a beautiful story of great importance in mathematics but also in other sciences and various computer-driven applications. Moreover, it seemingly is a never-ending story to which new pages are added at this very moment. We most sincerely hope, therefore, that the reader will also be caught up in this riveting story and the marvels it unveils.

A natural question for the reader to ask upon picking up this book (and for the conscientious author to address) is how one should read it? – In particular, is there a special order in which the chapter should be read? The essential answer is that there is no complicated special ordering of the chapters. The book consists of three parts and a number of appendices.

The first part represents a primer on Elementary Differential Geometry, that is the differential geometric study of curves and surfaces in the Euclidean space, and it might thus be used as a textbook on the subject. It is, however, enhanced by the various discretizations of all the essential notions introduced in the theoretical, "standard" part. In many instances, if not in most of the text, the modern – metric, discrete and applicative aspects are intertwined with the main text. Clearly, the reader can hardly avoid the encounter with the discrete approach and, indeed, we hope he or she will be not able to do so, and not only because this interlacing of the two aspects represents the very spirit of this book, but also because we truly believe this will help in forging a better and deeper understanding of the classical notions. Furthermore, this passing between smooth and discrete and back again represents a significant part of the modern view of Geometry (but not only, as many other facets of Mathematics, and Group Theory, in particular, were influenced and benefited by it). Needless to say, the applicative part clearly represents perhaps the greatest benefit of this book, and we wrote it having in mind readers motivated by the practical, computer-driven uses of Differential Geometry. Its leitmotif – one might even say its vital force – is the story of curvature that motivates and permeates the "story". In consequence, this is the part which is written in the most didactic, even "chatty"– hence, we hope – readable style. We tried to emulate as best as we could be feeling of attending a live class, with wherein a Socratic, maieutic approach is favored; thus questions that drive the development of new ideas and notions are posed (and encouraged). In consequence, the most varied readership can (we hope) follow this part and benefit from it, even if their previous mathematical education does not go beyond the rudiments of Calculus and Linear Algebra (and, in certain places, even less background knowledge is required).

The second part deals with higher dimensional manifolds. However, this is not a proper, ordered and comprehensive introduction to Riemannian Geometry. Instead, the drive is again the notion of curvature, more precisely the various ways one can extend it to higher dimensional objects and to more general structures and, once having achieved this, how to return to lower dimensional constructs and adapt to those, in a discrete manner, the newly defined notions. In consequence, the tone is somewhat less didactic, the pace is slightly more rapid, and less care is taken in presenting all the cumbersome technical preliminaries, but rather proceeds faster toward the heart of the matter, that is the various discretizations in their multiple and powerful applications to a variety of applied fields.

The third part focuses on the computational algorithms in algebraic topology and conformal geometry. These algorithms have been developed in the past twenty years and applied in many engineering and medical fields. The algorithms compute the homology/cohomology groups, the harmonic maps between surfaces and the Riemannian metrics by prescribed curvatures and meromorphic differentials. These methods can be directly used for surface parameterization, shape registration and structured mesh generation. This part briefly introduces the theories and emphases on the algorithmic design; hence, it is more in the style of computer science. Readers can use this part as a recipe for geometric algorithms in practice.

Returning to the original question of the reading order: The order is linear, and the reader who does not wish to continue to the second part, can proceed through Chapters 1-13 in their given order. Those who wish to continue to the second part (and we clearly hope that many will do that and, in truth, encourage them to do so) might have one small exception to this linearity rule, so to say. Namely, while Chapter 5 on the mean and Gauss curvature flows was written having as a background its preceding chapters, and can be useful for readers who do not wish to pursue the generalizations, it can also be skipped in the first reading and, instead, be considered only as preparatory for the Ricci curvature and flow chapter (Chapter 12). The order for the algorithm part is also linear, and the algorithms introduced later depend on those introduced earlier.

The appendices are, in contrast, rather technical, more terse in style, and they cover deep and varied ground that completes and enhances the main body of the book.

David Xianfeng Gu and Emil Saucan

I

Differential Geometry, Classical and Discrete

2

Differential Geometry : Classical and Discrete

CHAPTER 1

Curves

The study of Differential Geometry usually starts with that of curves in the plane and space. The reasons for doing this are multiple and evident: Historically, Differential Geometry evolved, in parallel with Calculus (or rather like a branch of it), beginning thus from Newton; thus, it concerned itself, in the beginning with planar curves. Their study gives the impetus and first directions for the study of surfaces; thus, the understanding of the geometry of curves is necessary for that of surfaces (thence of higher dimensional manifolds as well). Also, on a didactic level, it is traditionally deemed necessary to encourage students by gently introducing them to the subject, via the simplest case, i.e. that of curves. Strangely enough, students do not tend to respond positively to this approach, as they find it tedious and perceive that the time dedicated to this first chapter is far longer than it truly is.

However, we still dedicate a relatively large portion of this book to the study of curves. The first and basic reason for our decision is the fact that understanding the essential notion of curvature in this simple case is not just a technical prerequisite for the study of surfaces, as it rather develops the essential geometric intuition that resides behind many fundamental ideas, such as the very definition of surface curvature due to Gauss (that we shall encounter in the next chapter). Furthermore, and not less important for the spirit in which book was written, even at this incipient stage we can bring the first discretizations of the classical notions, together with a number of application, some of them better known, some of them new, to Analysis but also (and mainly) to Imaging, Wavelets and Complex Networks, among others. Such discretizations are not only interesting in themselves (as the applications mentioned will demonstrate), but they also will constitute the basis for further ones, in higher dimensions and, perhaps more importantly, will underline the methods and ideas behind adaptations of notions developed for smooth geometric objects to the setting of discrete mathematical objects. In consequence, we hope that the reader will appreciate this chapter for its own merits, as well as a portal toward the following chapters.

DOI: 10.1201/9781003350576-1

1.1 CURVES

Before we can discuss about the Differential Geometry of curves, we must first define them properly so that we can have a precise notion of the type of objects we want to study.

The realm in which classical Differential Geometry of curves operates is that of Euclidean plane $\mathbb{R}^2$ and space $\mathbb{R}^3$, but an extension to higher dimensional spaces $\mathbb{R}^n$ is also considered. Thus we shall begin by introducing the essential definitions that are involved in this case. However, since we shall also introduce already in this chapter generalizations of the smooth notions to the setting of metric space, a number of specific definitions will also be added herein.

We begin by noting that curves are geometric objects, i.e. subsets, say, of the plane. However, since we wish to study the differentiable properties of curves, that is to say, we want to be able to apply ideas and techniques of Calculus. To this end, we clearly need to be able to make an appeal to functions. We are therefore conducted to the notion of *parameterized curves*:

Definition 1.1.1 (Parameterized Curves) *Let $I \subseteq \mathbb{R}^3$ (or $\mathbb{R}^2$) be an open interval, and let $c : I \to \mathbb{R}^3$ be a differentiable function. Then c is called a (differentiable) parameterized curve c, while $c(I)$ is called the image (or trace) of the curve. Furthermore, the variable $t \in I$, i.e. such that $c(t) = (x(t), y(t), z(t))$ is called the (eponymous) parameter.*

Note that since we wish to ensure differentiability at every point, here we consider open intervals. Also, if not specified otherwise, we shall denote the unit interval $I = (0, 1)$ (or $I = [0, 1]$ if the need for closed intervals arises). By "differentiable" we mean $\mathcal{C}^\infty$, which means, in fact, that the function is as smooth as needed, while in practice $\mathcal{C}^3$ usually suffices.

Most importantly, note the distinction made between the curve c and its image $c(I)$. While in practice we often refer to the second simply as a "curve", the distinction is essential, since the former is a function while the later is a set (in plane or in space). It is this later geometric object that we wish to study, and to this end, we make appeal to the first notion. Clearly, this might generate confusion, which one tries as best as possible to avoid by using the proper terminology of "parameterized curve". Still this points to the limitations of classical Differential Geometry with its dissociation at times, between the geometric object one wishes to investigate, and a differentiable structure imposed on it largely for convenience reasons. However, passing to the much more general case of curves in metric spaces actually has the advantage of allowing us to concentrate on the geometry of the object itself, without the need for the artifice of parameterization.

Remark 1.1.2 *The (frequent) artificiality of the parametrization is made manifest by the fact that the different parameterizations may give rise to the same image (or rather than the same geometric object admits different parameterizations).*

Example 1.1.3 $c : \mathbb{R} \to \mathbb{R}^2$, $c(t) = (\cos t, \sin t)$ *and* $\gamma : \mathbb{R} \to \mathbb{R}^2$, $\gamma(t) = (\cos 2t, \sin 2t)$ *have the same image: The unit circle* $\mathbb{S}^1$.

This problem is compounded by the fact that not all curves in the plane (for example), not even some simple ones, are differentiable curves.

Counterexample 1.1.4 *The graph of the absolute value function, i.e. the image of curve $c : \mathbb{R} \to \mathbb{R}^2$, $c(t) = (y, |t|)$, is not a parameterized differentiable curve, because the function $t \mapsto |t|$ is not differentiable at $t = 0$.*

On the other hand, even curves with cusps can be represented as the traces of parameterized differentiable curves, such as the one in the example below:

Example 1.1.5 $c : \mathbb{R} \to \mathbb{R}^2,.$ $c(t) = (t^3, t^2)$.

Furthermore, even self-intersecting curves (which are not simple closed ones) can also be obtained as parameterized differentiable curves.

Example 1.1.6 *Descartes folium $c : (-1, \infty) \to \mathbb{R}^2$, $c(t) = \left(\frac{3at}{1+t^3}, \frac{3at^2}{1+t^3} \right)$, $a \neq 0$ has a self-intersection at the point $O = (0, 0)$.*

We do not bring more examples of curves here, as some of the most interesting ones will illustrate important definitions and interesting cases later on.

In the case of metric spaces, one replaces open intervals with the generalizations of closed ones, i.e *continua*, where a continuum is defined as follows:

Definition 1.1.7 *Let (X, d) be a metric space and let $A \subseteq X$, $|A| \geq 2$. Then A is called a (metric) continuum if it is compact and connected.*

The definition above renders the most general extension of curve that we shall study here. However, in many case we shall consider only (*metric*) arcs:

Definition 1.1.8 *Let (X, d) be a metric space and let $\varphi : I \xrightarrow{\sim} X$ be a homeomorphism of the closed (unit) interval. Then $\varphi(I)$ is called a (metric) arc.*

Furthermore, intervals themselves admit a generalization to the setting of general metric spaces. More precisely, we have the following:

Definition 1.1.9 *Let (X, d) be a metric space and let $A \subseteq X$. Then A is called a metric segment if it is isometric to a segment $[a, b] \subset \mathbb{R}$.*

After this digression into the realm of metric curves, we return now to parameterized differentiable curves and bring more definitions that are essential in the sequel.

Definition 1.1.10 *Given the differentiable parameterized curve $c : I \to \mathbb{R}^3$, $\dot{c}(t) = (\dot{x}(t), \dot{y}(t), \dot{z}(t))$ is called the tangent vector (or velocity vector) at c in the point t.*

Clearly, if $\dot{c}(t) \neq 0$, then the line $l(t) = \dot{c}(t)t + c(t)$ is the *tangent line* to c at t. Since in Differential Geometry curves need to have tangents at every point, we are conduced to formulate

Definition 1.1.11 *Let $c : I \to \mathbb{R}^3$, be a parameterized differentiable curve, and let $t \in I$ such that $\dot{c}(t) = 0$. Then t is called a singular point. Otherwise, t is called a regular point. The curve c itself is called regular if all its points are regular, i.e. if $\dot{c}(t) \neq 0$, for all $t \in I$.*

As we know from Calculus, knowledge of $\dot{c}(t)$ allows us to compute the length of the curve in a manner that we formalize as

Definition 1.1.12 *Let $c : I \to \mathbb{R}^3$, be a parameterized differentiable curve. Given $t_0 \in I$, the arc length of c from the point t_0 is defined as*

$$s(t) = \int_{t_0}^{t} ||\dot{c}(t)|| dt \,;$$

where $||\dot{c}(t)|| = \sqrt{\dot{x}^2(t), \dot{y}^2(t), \dot{z}^2(t)}$.

The notation $s(t)$ is not our own, but it rather is the standard one. The use of a standardized notation points out to the fact that this simple definition is quite important in the sequel. Indeed, this is the case, and it is justified by the following simple observation: If $t \equiv s$, then $ds/dt = ||\dot{c}(t)|| = 1$ and, reciprocal, if $||\dot{c}(t)|| = 1$ then

$$s = \int_{t_0}^{t} ||\dot{c}(t)|| dt = t - t_0 \,,$$

that is t is precisely the arc length of c measured from t_0. Clearly, having curves parameterized such at every point the *speed* $||\dot{c}(t)||$ equals 1 is advantageous, a fact that will be made most evident in the sequel. Therefore, it would be most advantageous if we could ensure that any curve can be thus parameterized. This turns out to be the case, at least for regular curves,[1] as guaranteed by

Proposition 1.1.13 *Any regular differentiable curve $c : I \to \mathbb{R}^3$ can be parameterized by arc length.*

From the proposition above allows us to presume that any regular curve is parametrized by arc length (which we shall do, unless otherwise specified).

Exercise 1.1.14 *Prove Proposition 1.1.13.*

1.2 CURVATURE

Remark 1.2.1 *It is a well-known fact of Calculus 101 that the derivative of a function measures rate of change; thence the second derivative measures the rate of change of the (first) derivative. This somewhat physical observation has an important geometric meaning: Since the derivative represents the slope of the tangent (at a given point), the change of the derivative between the points P_1 and P_2 (see Figure 1.1) is represented by the angle $\alpha = \angle(t_1, t_2)$ between the tangents t_1 and t_2 (at the points P_1, P_2, respectively). But $\alpha = \pi - \beta = \angle(n_1, n_2)$ – the angle between the normals at the points P_1, P_2, therefore $\sin \angle(t_1, t_2) = \sin \angle(n_1, n_2)$.*

---

[1]This being one more reason why classical Differential Geometry concentrates on their study.

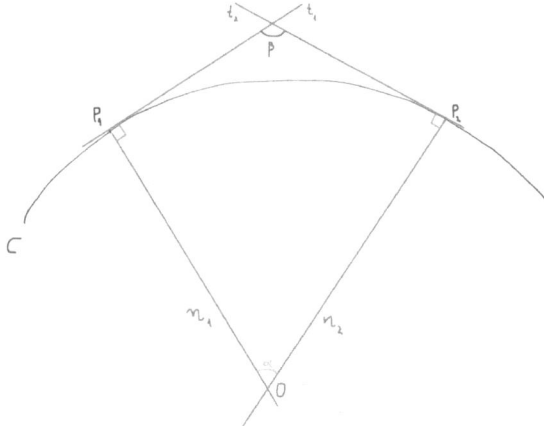

Figure 1.1 The radius of the osculating circle as given by the limit of the angles between normals.

It follows that, if $P_2 \to P_1$, we have

$$\lim_{P_1 \to P_2} \frac{\measuredangle(n_1, n_2)}{P_1 P_2} = \lim_{P_1 \to P_2} \frac{\sin \measuredangle(n_1, n_2)}{P_1 P_2} = \lim_{P_1 \to P_2} \frac{P_1 P_2}{OP_1 \cdot P_1 P_2} = \lim_{P_1 \to P_2} \frac{1}{OP_1} = \frac{1}{R} = \kappa.$$

Remark 1.2.2 *The geometric importance of the observation above is further revealed by the following observation: Since the curve may be presumed to be parametrized by arc-length, i.e. the tangent vectors at each point may be supposed to have length one, they may be viewed as points on the unit circle $\mathbb{S}^1$ (see Figure 1.2, left). Thus the arc $P_1 P_2$ is mapped via this tangential mapping to the corresponding arc $Q_1 Q_2$ on the unit circle (see Figure 1.2, right). Since $\measuredangle(t_1, t_2) = \measuredangle(Q_1 O Q_2)$ and, moreover, $\measuredangle(Q_1 O Q_2)$ equals the length of the arc $Q_1 Q_2$, we conclude (and write quite informally), that*

$$\kappa = \lim \frac{\text{small curve arc}}{\text{its tangent image}}.$$

1.2.1 The Osculating Circle

It is hard to reconcile, at least at first glance to identify the two notions of curvature introduced above, i.e. to identify the simple geometric intuition resting behind the original approach (essentially of Newton[2]) and the modern approach with its handy formulas. This appears to be even more critical if one tries to apply the osculating circle definition to any *concrete computations*, when one seems to be powerless when equipped only with the geometric definition. (Nevertheless, we shall shortly see that modern Mathematics has found a way to circumvent this.) However, we shall show that, indeed, the two definitions coincide. To this end, consider a curve in the plane in parametric form $c : I \to \mathbb{R}^2 ; c(t) = (x(t), y(t))$, and let $t_0 \in I$, such that $\kappa(t_0) \neq 0$ (so we can talk about curvature, at least in the analytic approach). Let us also consider the

---

[2]1665

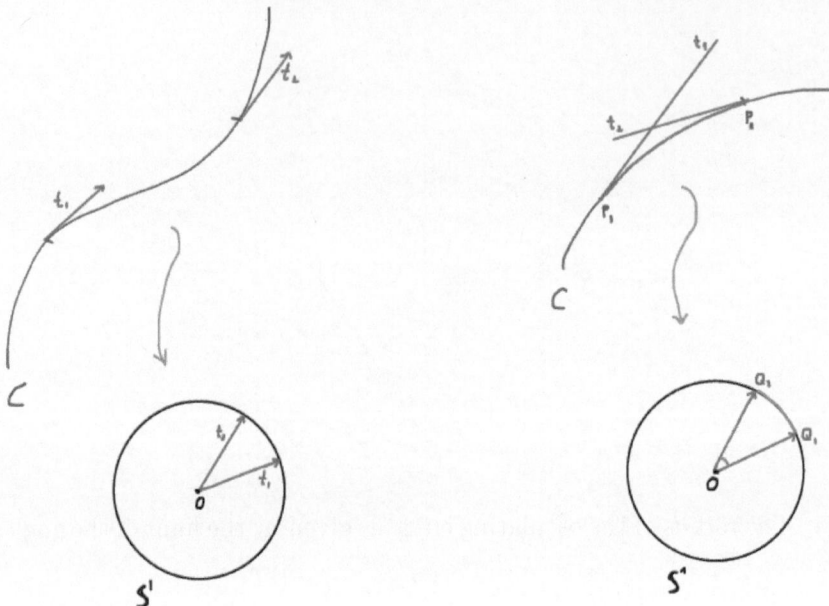

Figure 1.2 Curvature measures the ratio between the length of an infinitesimal arc length and its tangential image.

circle passing through three points P_0, P_1, P_2, where $P_0 = c(t_0), P_1 = c(t_1), P_2 = c(t_2)$, and let the points P_1, P_2 tend to $P_0 = P_0(x_0, y_0)$, that is the osculating circle at P_0) – see Figure 1.3.

Since the equation of a circle can be written in the following form:

$$(\mathcal{C}) : x^2 + y^2 - 2ax - 2by + c = 0,$$

the intersection between a circle and the given curve is given by

$$(c) \bigcap (\mathcal{C}) : x^2(t) + y^2(t) - 2ax(t) - 2by(t) + c = 0. \tag{1.2.1}$$

We make the substitution $t \mapsto t_0 + h$, and we then obtain the following expansion into series:

$$\begin{cases} x(t_0 + h) = x(t_0) + \frac{h}{1!}\dot{x}(t_0) + \frac{h^2}{2!}\ddot{x}(t_0) + \dots \\ y(t_0 + h) = y(t_0) + \frac{h}{1!}\dot{y}(t_0) + \frac{h^2}{2!}\ddot{y}(t_0) + \dots \end{cases}$$

(Not only have we seen that it suffices to restrict to second derivatives, since the circle is a quadric, there is no need (or value) in passing beyond quadratic terms.

From here and from (1.2.1), we obtain

$$x_0^2 + y_0^2 - 2ax_0 - 2by_0 + \kappa + 2\frac{h}{1!}(x_0\dot{x}_0 + y_0\dot{y}_0 - a\dot{x}_0 - b\dot{y}_0) + h^2(\dot{x}_0^2 + \dot{y}_0^2 + x_0\ddot{x}_0 + y_0\ddot{y}_0 - a\ddot{x}_0 - b\ddot{y}_0) + \dots \tag{1.2.2}$$

(Here we put, for simplicity, $x_0 = x_0(t), \dot{x}_0(t)$, etc.)

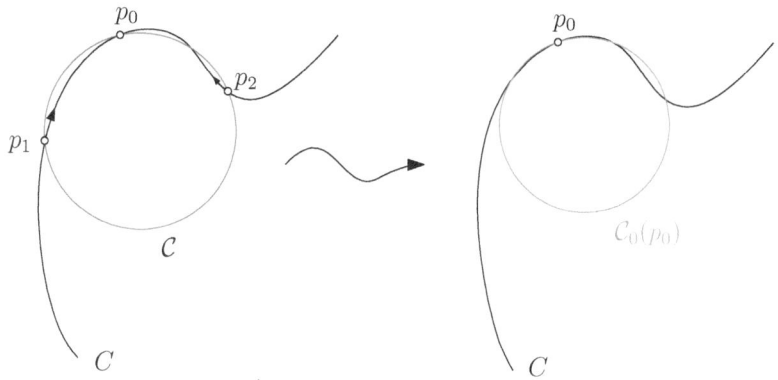

Figure 1.3 P_1, P_2 tend to $P_0 = P_0(x_0, y_0)$, that is the osculating circle at P_0.

Then the circle $\mathcal{C}$ intersects the curve c in three coinciding points iff $h = 0$ is a triple root of (1.2.2), i.e. if and only if

$$\begin{cases} x_0^2 + y_0^2 - 2ax_0 - 2by_0 + c = 0 \\ x_0\dot{x}_0 + y_0\dot{y}_0 - a\dot{x}_0 - b\dot{y}_0 = 0 \\ \dot{x}_0^2 + \dot{y}_0^2 + x_0\ddot{x}_0 + y_0\ddot{y}_0 - a\ddot{x}_0 - b\ddot{y}_0 = 0 \end{cases} \tag{1.2.3}$$

But $\kappa \neq 0 \iff \dot{x}_0\ddot{y}_0 - \dot{y}_0\ddot{x}_0 \neq 0$, that is $\dot{c}, \ddot{c}$ are linearly independent. Therefore (1.2.3) implies that

$$a = x_0 + \dot{y}\frac{\dot{x}_0^2 + \dot{y}_0^2}{\dot{x}_0\ddot{y}_0 - \dot{y}_0{}_0\ddot{x}_0} \; ; \quad b = y_0 - \dot{x}_0\frac{\dot{x}_0^2 + \dot{y}_0^2}{\dot{x}_0\ddot{y}_0 - \dot{y}_0\ddot{x}_0} \; ; \tag{1.2.4}$$

and

$$c = x_0^2 + y_0^2 + \frac{2(x_0\dot{y}_0 - \dot{x}_0 y_0)(\dot{x}_0^2 + \dot{y}_0^2)}{\dot{x}_0\ddot{y}_0 - \dot{y}_0\ddot{x}_0} \; . \tag{1.2.5}$$

From (1.2.4) and (1.2.5) above, we immediately obtain that the radius of the limiting circle.

$$R_O = R_O(t_0) = a^2 + b^2 - c^2 = \frac{1}{\kappa(t_0)}.$$

(Also, (a, b) represent the coordinates of the osculating circle, also called the *center of curvature*.)

$\qquad\qquad\qquad\qquad\qquad\qquad\qquad\qquad\qquad\qquad\qquad\qquad\qquad\qquad \Box$

Exercise 1.2.3 *Complete the computations and explicate the formula for $\kappa(t_0)$. (Compare with Exercise 1.2.8.)*

Remark 1.2.4 *Since the considered limit does not depend on the specific way the points P_1, P_2 are selected, we can choose P_1 such that it will be much closer (orders of magnitude closer) to P_0 than P_2; thus, P_0 and P_1 will coincide before P_2 does. But saying that P_0, P_1 coincide is to say (as we well know from our basic course in*

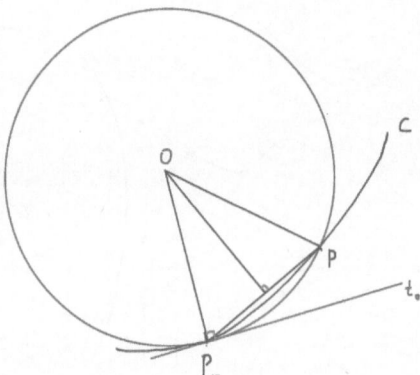

Figure 1.4 The center of the osculating circle, that is the *center of curvature* is the limit of the centers of circles tangent to the curve.

Calculus), that the segment $P_0 P_1$ tends to the tangent t_0 at the point P_1. Therefore, by taking in the limiting process defining the osculating circle, as we did, P_1 to be much closer to P_0 than P_2, we have shown that the osculation circle represents the limit of tangent circles at P_0 – see Figure 1.4. Thus

Remark 1.2.5 *The simplest quadratic curve one encounters in his elementary mathematical studies, e.g. in Algebra or Calculus, is the parabola, whose equation is far simpler than that of the circle. This familiarity and simplicity conduce one, quite naturally ask her/himself whether one can define curvature via the osculating parabola, instead of the osculating circle. This is indeed possible – see, e.g., [159], [309], p. 43. However, for the curious reader, we also offer this as a problem:*

Problem 1.2.6 *Define the osculating parabola (of a curve c at a point p) and study its properties.*

We cannot, in good conscience, proceed further on without bringing the standard formulas for computing the curvature of a planar regular curve, not the least because they represent a standard of any Differential Geometry course but also because they allow us to compute the curvature for a number of classical, essential types of curves. (We shall return to the more geometric definitions of curvature and see how one actually compute curvature in the frame of that paradigm.)

Proposition 1.2.7 *Let $c : I \to \mathbb{R}^2, c(t) = (x(t), y(t)$ be a regular curve. Then*

$$\kappa(t) = \frac{\det(\dot{c}(t), \ddot{c}(t))}{||c(t)||^3} = \frac{||\dot{c}(t) \times \ddot{c}(t)||}{||c(t)||^3} ; \qquad (1.2.6)$$

which in coordinates can be written as

$$\kappa(t) = \frac{\dot{x}(t)\ddot{y}(t) - \ddot{x}(t)\dot{y}(t)}{(\dot{x}^2(t) + \dot{y}^2(t))^{3/2}} . \qquad (1.2.7)$$

Exercise 1.2.8 *Prove the proposition above. (Hint: Look at Formula (1.2.5).)*

Example 1.2.9 *The curvature of the ellipse $\gamma(t) = (c\cos t, b\sin t); a > b > 0$ is given by $\kappa(t) = \dfrac{ab}{\left(a^2\sin^2 t + b^2\cos^2 t\right)^{3/2}}$. In particular, the curvatures at its vertices are $\kappa(0) = \frac{a}{b^2}$ and $\kappa(\frac{\pi}{2}) = \frac{b}{a^2}$. Moreover, if $a = b$, then $\kappa(t) \equiv \frac{1}{a}$, that is we recover the expected curvature of the circle (who is, after all, it's own osculating circle!).*

Exercise 1.2.10 *Verify the computations above.*

Exercise 1.2.11   *1. Show that if the curve c is given in parametric form, i.e. $c(t) = \rho e^{it}$, then $\kappa(t)\dfrac{|\rho^2 + 2\dot\rho^2 - \rho\ddot\rho|}{|\rho^2 + (rho)^2|^{3/2}}$.*

   2. Compute, as a particular case, the curvature of the curve given by $\rho(t) = 1 + 2\cos t$.

Remark 1.2.12 *The formulas above show that the notion of curvature is not defined at points where $\|\dot c\| = 0$. This fits our geometric intuition that at a point of inflection, the very notion of circle of curvature is not definable.*

Remark 1.2.13 *If $\|\dot c\| = 1$, then $\ddot c \perp \dot c$ and, moreover, $|\kappa| = \|\ddot c\|$; thus $|\kappa|$ represents the area of the rectangle of sides $\|\dot c\|$ and $\|\ddot c\|$.*

We can now revisit the notion of center of curvature and reexamine with the help of Formula (1.2.7). More precisely, we shall investigate the *evolute* of a given curve, which is defined as follows:

Definition 1.2.14 *Given a planar (regular) curve c, its evolute $\tilde c$ is defined to be the geometric locus of the centers of curvature of (the points) of c; that is*

$$\tilde c = c + \frac{1}{\kappa}\mathbf{n}. \tag{1.2.8}$$

Remark 1.2.15 *We have the following (immediate, but essential in understanding the definition of evolutes) fact, namely $\tilde{\mathbf{t}} = \mathbf{n}$.*

From the definition of the evolutes, a simple formula for computing the length of an arc (of the evolute) follows easily:

Proposition 1.2.16
$$\ell(\tilde C)\big|_{s_0}^{s_1} = \left| \frac{1}{\kappa(s_0)} - \frac{1}{\kappa(s_1)} \right|.$$

Proof

$$\ell(\tilde C)\big|_{s_0}^{s_1} = \int_{s_0}^{s_1} d(\tilde c) = \int_{s_0}^{s_1} |\dot{\tilde c}|ds = \int_{s_0}^{s_1} \left(\frac{1}{\kappa(s)}\right)' ds = \left| \frac{1}{\kappa(s_0)} - \frac{1}{\kappa(s_1)} \right|$$

□

By using Formula (1.2.7) above, it is to write the expression of $\tilde{c}$ in coordinates: If $c = (x, y)$, then $\tilde{c} = (\tilde{x}, \tilde{y}$, where

$$\tilde{x} = x - \frac{\dot{x}^2 + \dot{y}^2}{\dot{x}\ddot{y} - \ddot{x}\dot{y}}\dot{y}\,;$$

$$\tilde{y} = y + \frac{\dot{x}^2 + \dot{y}^2}{\dot{x}\ddot{y} - \ddot{x}\dot{y}}\dot{x}\,.$$

Exercise 1.2.17 *Prove that, if $y = f(x)$, then*

1. $\kappa = \frac{\ddot{y}}{(1+\dot{y}^2)^{3/2}}$;

2. $\tilde{x} = x - \frac{1+\dot{y}^2}{\ddot{y}}\dot{y}$, $\tilde{y} = y + \frac{1+\dot{y}^2}{\ddot{y}}$.

Exercise 1.2.18 *Show that the involute of the tractrix is the catenary.*[3]

A related notion is that of *involute* (or *evolent*):

Definition 1.2.19 *Let c be a regular (planar) curve, and let $P_0 = c(s_0)$ be a point on c. The involute $\tilde{c}$ is defined as follows:*

$$\tilde{c}_0(s) = c(s) + (s - s_0)\mathbf{t}\,. \tag{1.2.9}$$

Exercise 1.2.20 *Prove that*

$$\tilde{c}_0(s) = c(s) + \frac{\dot{c}}{\sqrt{\dot{c}^2}} \int_s^{s_0} \sqrt{\dot{c}^2}\,dt$$

, *where $s < s_0$.*

Remark 1.2.21 *The notion of evolute admits a simple, mechanical interpretation: If the curve represents a taught string, fixed at $c(s)$, then the trajectory of the point $P_0 = c(s_0)$, when the string is unwound tangentially to its curve, is precisely the evolute.*

As the names suggests, the notions of evolute and involute are essentially dual to each other, with the proviso that, while any curve has only one evolute, it has an infinity of involutes. Still, the following holds:

- Any (regular) curve is the involute of its evolute;

- Any (regular) curve is the evolute of any of its involutes.

Exercise 1.2.22 *Prove the facts above.*

---

[3]Recall their definitions brought in the first section of this chapter.

1.2.2 Menger Curvature

While there exists a tradition of applying the classical notion of curvature to Graphics and GAD, using smooth approximations and such tools as *splines* (see, e.g. [30]), one can object that this amounts, in a certain sense, to cheating. Even without going that far with a purist approach, one might demur for purely mathematical reasons – and rightly so – that the data one encounters in Graphics, Imaging, etc. is discrete, thus no use of derivatives is possible, since they are not even properly definable. The same basic argument can be considered from a Computer Science viewpoint as well, for introducing tools that allow for the application of classical Calculus methods requires further techniques that not only heavily expand the computational overhead, but also – and perhaps more importantly – do not use the data "as is", but rather modify in a manner that might appear acceptable and controllable, but that might also render results that might appear "nice", yet be far from the measurable ground truth.

It turns out that for curvature, a plethora of discretizations exists that allow for an "honest" computational approach, and the present book represents largely an exploration of these ideas. In fact, the first such examples applies already for the discretization for curvature, and goes back in time for almost a century (to an epoch were computers were, at best, a Sci-Fi dream). Note that we used "examples", rather than example, because actually two discretizations of curvature were proposed already in the first half of the 20th century. The first – and the simplest – one was introduced by Menger [221], [222] in 1930, and we concentrate first on it in the present section.[4]

The idea is very simple and intuitive and is based on the notion of osculating circle. The discretization of this classical concept is obtained – in a manner that we shall encounter again later on – by not going to the limit, as there is no notion of infinitesimals in the discrete setting, but rather prescribe a curvature to any (metric) triangle. More precisely, given a triangle T one defines its *Menger curvature* as follows:

Definition 1.2.23 *Given a triangle T, i.e., a metric triple of points with sides of lengths a, b, c, the* Menger curvature κ_M *of T is defined as $\frac{1}{R(T)}$, where $R(T)$ is the radius of the circle circumscribed to the triangle, that is*

$$\kappa_M(T) = \frac{1}{R(T)} = \frac{4\sqrt{p(p-a)(p-b)(p-c)}}{abc}, \qquad (1.2.10)$$

where $p = \frac{(a+b+c)}{2}$ denotes the half-perimeter of T.

Exercise 1.2.24 *Prove that, indeed, $\frac{1}{R(T)} = \frac{4\sqrt{p(p-a)(p-b)(p-c)}}{abc}$.*

The diligent reader has undoubtedly noticed that the formula above is one that holds in the Euclidean plane. Thus, while extremely simple and intuitive, it also

---

[4]It is hardly coincidental that Menger subsequently wrote a rigorous introduction to Calculus [223], where all the concepts and ideas are developed from step functions (for Integral Calculus) and piecewise-linear functions (for Differential Calculus), i.e. from the basic ingredients in Imaging and Graphics.

suffers precisely because of the same reason, of the limitation that it imposes an Euclidean type of geometry on an arbitrarily metric space. However, as we shall see, this doesn't impede the Menger curvature in becoming an easily implementable, flexible notion fitted for the analysis of a variety of discrete phenomena of interest in Computer Science at large. Furthermore, the notion itself can be extended to include other, more general types of geometry, beyond the Euclidean one (see the Remark below).

Remark 1.2.25 *For the more advanced student, who has a knowledge of the Basis of geometry, we mention here that there exist fitting adaptations of Menger's curvature to the cases of Spherical and Hyperbolic Geometry. (Furthermore, we shall return to these geometries in the sequel.) The relevant formulas[5] are:*

$$\kappa_{M,S}(T) = \frac{1}{\tan R(T)} = \frac{\sqrt{\sin p \sin (p-a) \sin (p-b) \sin (p-c)}}{2 \sin \frac{a}{2} \sin \frac{b}{2} \sin \frac{c}{2}}, \qquad (1.2.11)$$

for the spherical case and

$$\kappa_{M,H}(T) = \frac{1}{\tanh R(T)} = \frac{\sqrt{\sinh p \sinh (p-a) \sinh (p-b) \sinh (p-c)}}{2 \sinh \frac{a}{2} \sinh \frac{b}{2} \sinh \frac{c}{2}}. \qquad (1.2.12)$$

for the hyperbolic one, respectively. (Thus, the Menger curvature we defined in the beginning should be called the Euclidean Menger curvature, and thus be denoted, appropriately, by $\kappa_{M,E}$.)

Remark 1.2.26 *In the derivation of Menger curvature we have used two formulas for computing the area of a triangle, both of them "purely metric", i.e. using only the lengths of the triangle's sides. However, by a well-known high-school formula, the area of the triangle can be expressed also in terms of the sinus of an angle, therefore rendering the following alternate formula(s) for the Menger curvature of a triangle $T = T(a, b, c)$ of sides a, b, c and angles α, β, γ:*

$$\kappa_M(T) = \frac{2 \sin \alpha}{bc} = \frac{2 \sin \beta}{ac} = \frac{2 \sin \gamma}{ab}. \qquad (1.2.13)$$

While this formula may seem like a departure from the purely metric framework, we shall see its analogs when we shall later encounter the metrizations of the notion of torsion, which constitutes the subject of the next section. Furthermore, it is an important ingredient in the proof of the following result that shows that the very existence of Menger curvature at every point of a curve guarantees that the curve satisfies an important property of "well-behavior":

Proposition 1.2.27 *If a continuum C in a metric space (X, d) has finite Menger curvature at a point $p \in C$, then C is a rectifiable arc in a neighborhood of p. If $\kappa_M(p)$ exists and it is finite, at every point $p \in C$, then C is a rectifiable arc or a rectifiable simple closed curve.*

---

[5]See, e.g. [177].

Unfortunately, we cannot provide here the proof of the proposition above, which shows how powerful an invariant is Menger curvature, since it makes an appeal to further notions and results that would not fit well within the context of this book.

Remark 1.2.28 *For the advanced reader (as well as for the highly motivated reader who is looking for new avenues to explore) who has taken the time to take a leisurely stroll through the discretizations of curvature we present in the present book, we should mention that Menger curvature to be nothing else than Gromov's $\mathbf{K}_3(\{p, q, r\})$ [142], where p, q, r represent the vertices of a triangle. However, while the general setting of the modern discourse is certainly important, there still is use for the classical, "parochial" Menger curvature, as demonstrated by its many contemporary uses we shall describe below, thus interest and mathematical activity around the "old fashioned" Menger curvature still exists, even though the much more general and modern setting of $\mathbf{K}_n(X)$ is much more alluring. Moreover, it should be noted that, beyond this generality, there rests a lot of uncertainty, as Gromov himself notes [142].*

In this context, we should also mention the very recent development [32]. Here a kind of "comparison Menger" curvature is introduced. Very loosely formulated, whereas in the classical Menger curvature a specific, Euclidean in nature, formula is developed for the circumradius, here it is only compared to that of a triangle of the same sides, in different model (or gauge) surfaces. This is similar (and presumably inspired by) the Alexandrov curvature (which we shall discuss later on).

Furthermore, The factors $(p-a), (p-b), (p-c)$ in the definition above are, in fact, the so-called Gromov products of the vertices of T, which is employed in the definition of Gromov δ-hyperbolicity [142]. Thus Menger curvature may be viewed as a local version of the Gromov hyperbolicity. Given the fact that Gromov hyperbolicity is a global property, hence difficult to compute, it is good to be able to substitute, whenever possible, a local related notion of curvature. Moreover, while Gromov hyperbolicity represents a coarse, global notion of sectional curvature, the Menger curvature which we propose herein is a local, notion of sectional curvature (which we shall introduce formally in Chapter 11), thus fitting better the classical Riemannian paradigm of the development of curvature, first locally, then globally.

To define the Menger curvature at a given point on a curve, one passes to the limit (precisely like in the classical osculating circle definition) and formulate the somewhat cumbersome definition below:

Definition 1.2.29 *Let (M, d) be a metric space and let $p \in M$ be an accumulation point. We say that M has at p Menger curvature $\kappa_M(p)$ iff for any $\varepsilon > 0$, there exists $\delta > 0$, such that for any triple of points p_1, p_2, p_3, satisfying $d(p, p_i) < \delta$, $i = 1, 2, 3$; the following inequality holds: $|\kappa_M(p_1, p_2, p_3) - \kappa_M(p)| < \varepsilon$.*

It is an immediate observation that $\kappa_M \geq 0$ and that $\kappa_m \equiv 0$ on a Euclidean line, as the reader can easily heck. Furthermore, it coincides, by its very definition, with the classical notion, at any point on a smooth curve in the plane.

However, Menger curvature is not necessarily defined at all the points of any curve/of a metric space, as the following simple example demonstrates:

Example 1.2.30 *Let (X, d) the metric space consisting of three rays $\overrightarrow{PX}, \overrightarrow{PY}, \overrightarrow{PZ}$ in $\mathbb{R}^2$ having P as the only common point, endowed with the metric d, where $d(A, B)$ is the usual Euclidean distance d_2 if A, B are on the same ray, and $d(A, B) = d_2(A, P) + d_2(P, B)$ if A, B are on different rays.*

Then $\kappa_M(A) = 0$ for any point $A \in X, A \neq 0$, but $\kappa_M(P)$ is not defined.

Exercise 1.2.31 *Check the assertion above.*

One could object that as we have seen above, in the smooth setting, one only needs two points to converge to a third, fixed one, hence one could simplify accordingly Menger's definition of curvature as follows:

Definition 1.2.32 **(Alt curvature)** *Let (M,d) be a metric space and let $p \in M$ be an accumulation point. Then M is said to have at the point p Alt curvature $\kappa_A(p)$ iff the following limit exists*

$$\kappa_A(p) = \lim_{q,r \to p} K(p, q, r);$$
(1.2.14)

where $K(p, q, r) = 1/R$ and R is the circumradius of the triangle of vertices p, q, r.

However, it turns out that Alt's curvature is, in fact, a more general notion than Menger's curvature. Indeed, while the existence of κ_M obviously implies, precisely as in the case of smooth curve, the existence of κ_A, the reverse implication is false, as demonstrated by

Counterexample 1.2.33 *Let (X, d) be as in Example 1.2.30 above. Then $\kappa_A(P) = 0$ (while, as we have seen, $\kappa_M(P)$ is not defined).*

Exercise 1.2.34 *Verify that, indeed, $\kappa_A(P) = 0$.*

Since the generality in question applies to rather esoteric spaces, and since, on the other hand, Menger's curvature proved to be strong enough to deal (quite efficiently) with such problems as finding estimates (obtained via the Cauchy integral) for the regularity of fractals and the flatness of sets in the plane (see [243]), it's generality seems to suffice for the (interested) mathematical community. Still, at least in the majority practical instances, the two types of curvatures coincide. The most notable – and important – instance is that of the Euclidean plane, where this equivalence (that prompted Alt's idea) is used more than once to find different interpretations of curvature (see, for instance, [309], [163], [105]).

Remark 1.2.35 *Needles to say, Alt curvature suffers from the same imperfection as Menger curvature: since they are both modeled closely after the Euclidean Plane, they convey this Euclidean type of curvature upon the space they are defined on.*

At first sight, the derivation of the spherical and hyperbolic formulas (while quite elementary and natural) seems to take us further away from the purely metric approach we have adopted in this section. However, by using the cosine formula, one can express the area of the triangle as a determinant, e.g.

$$\mathrm{Area}T(a,b,c) = \sqrt{-16D(a,b,c)}\,; \tag{1.2.15}$$

where

$$D(a,b,c) = \begin{vmatrix} 0 & 1 & 1 & 1 \\ 1 & 0 & c^2 & b^2 \\ 1 & c^2 & 0 & a^2 \\ 1 & b^2 & a^2 & 0 \end{vmatrix}. \tag{1.2.16}$$

Therefore, one can write the formula for Menger curvature again in purely metric terms, but this time using a determinant, as

$$\kappa_M(T) = \frac{\sqrt{-D(a,b,c)}}{abc}. \tag{1.2.17}$$

Albeit at this stage the formula above is, at best, a compact form of expressing Menger curvature and, at worst, a simple curiosity, it is a model for further developments of the notion of curvature and related notions, using determinants (see Sections 1.3.5 and Wald below).

1.2.2.1 Applications of Menger Curvature

Despite being somewhat simplistic in imposing a Euclidean structure on the analyzed spaces – or perhaps precisely because of that reason – and Menger curvature has been employed with considerable success to the study of such problems as finding estimates (obtained via the Cauchy integral) for the regularity of fractals and the flatness of sets in the plane (see [243], [302]). As far as practical implementations are concerned, Menger curvature has been used[6] for curve reconstruction ([174]).

Perhaps the most evident (but if applied straightforwardly not the most efficient one) application of Menger curvature is to approximate the *principal curvatures* of surface, in Graphics and Imaging tasks. Since this is a notion that we shall introduce in the next chapter, we defer till then the presentation of some examples.

However, we can still present two more advanced applications of Menger curvature to Imaging and Pattern Recognition. However, these require not simply the original version but rather the derived one of *Menger curvature measure*[7]:

Definition 1.2.36 *Let T be a (metric) triangle. Its* Menger curvature measure $\mu(T)$ *is defined as*

$$\mu(T) = \kappa_M(T) \cdot (\mathrm{diam}(T))^2. \tag{1.2.18}$$

Moreover, if $\mathcal{T}$ is a triangulation (e.g. of a domain in plane or on a surface), its Menger curvature measure is naturally defined as follows:

$$\mu(\mathcal{T}) = \sum_{T \in \mathcal{T}} \kappa_M(T) \cdot (\mathrm{diam}(T))^2. \tag{1.2.19}$$

---

[6]in conjunction with the *traveling salesman* algorithm.

[7]This is, in fact, the manner in which Menger curvature is applied to the study of fractals.

Remark 1.2.37 *The definition above is the basic one. One can generalize it in three manners:*

1. *Replace $\kappa_M(T)$ by $\kappa_M^p(T)$, for some $p > 1$;*

2. *Instead of $(\mathrm{diam}(T))^2$ use any other fitting measure $\nu(T)$ associated to the triangle T;*

3. *Extend the very definition of Menger curvature to include*

The first approach is the one used in the theoretical works mentioned above and it is also fitting to apply, for instance in the analysis of fractal-like textures, which can thus be studied at many/best scales. The second one is ideal in the case on has to deal with some probability measure, but also when more general measures that can be attached to a texture are to be considered. The last one is devised to be used for volumetric data.

While the common model for images is the *stick model* (see, e.g. [254]), where each pixel is viewed as a square, it is easy to subdivide (in more than one manner, actually) each square into triangles (see Figure 1.5), thus obtaining a triangulation. So far we have thus applied the Menger curvature measure to Imaging in two instances. For one, we have employed it (as well as other notions of curvature that we shall introduce in the sequel) to the understanding and classification of textures, especially of the so called *stochastic* textures (these being the "fractal-like" ones alluded above) [27]. (See an example of the computation of the Menger curvature measure in

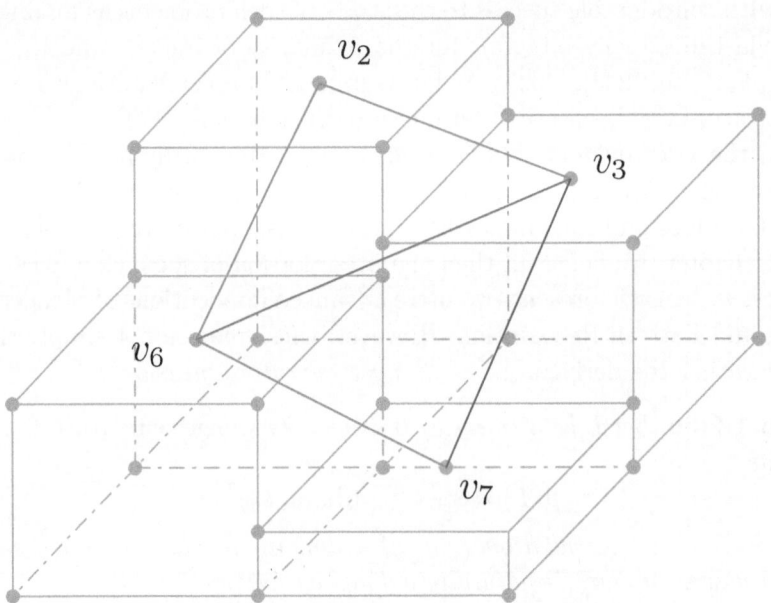

Figure 1.5 Possible triangles choice for the computation of the Menger curvature measure.

Figure 1.6 A standard test image (left), its Menger curvature measure (middle) and the segmentation it renders combined with non-local diffusion (right) [133]. Note that it is a good distinguisher of texture types and, in consequence, an excellent edge detector.

Figure 1.6.) We have also used the Menger curvature measure, in conjunction with *non-local gradients* [128], [127] again to the segmentation of textures [133] (see Figure 1.6 for an illustrative result).

However, we do not need to make appeal to the extended definition of Menger curvature and can apply successfully the original one in the study of Complex Networks [284], [264], [28]. Clearly, in this setting, one should take into account the eventual triangles that appear in network (but we shall return in more detail to this observation later). For some first applications we refer the reader to the references above, however the full force of the idea of using Menger curvature in this setting will become clearer once we shall introduce, in Chapter 10 the notion of Ricci curvature and its discretization. Here we would only like to note that, As defined, the Menger curvature is always positive. This may not be desirable, as we shall discover when passing to surfaces, and it certainly constitutes an impediment when dealing with *oriented* networks (or *directed* graphs). However, for such networks a sign $\varepsilon(T) \in \{-1, 0, +1\}$ is naturally attached to a directed triangle T (Figure 1.7), and the Menger curvature of the directed triangle is then defined, in a straightforward manner as

$$\kappa_{M,O}(T) = \varepsilon(T) \cdot \kappa_M(T) . \tag{1.2.20}$$

Note that we didn't specify here which of the variants of Menger curvature one is to consider. While usually one presumes that the standard, Euclidean notion is appealed to where κ_M could be the Euclidean, the spherical or the hyperbolic version, accordingly to the specific given setting.

Remark 1.2.38 *In truth, in practical applications Alt curvature is used, rather than Menger curvature. However, tradition established that Menge's name be attached to such experiments, and we follow it too. Given the identity of the two notions in the case of "normal" spaces, and Menger's priority in the development of the idea, this is not a serious problem.*

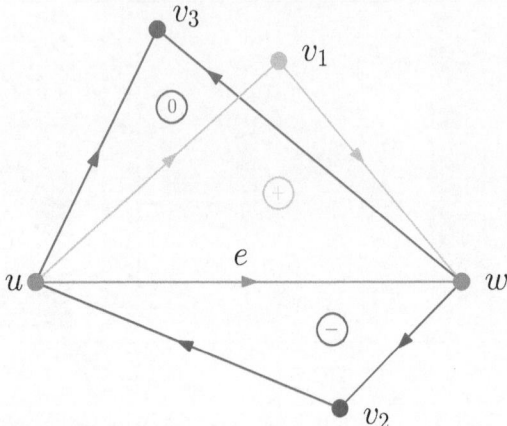

Figure 1.7 Sign convention for directed triangles in a network. While by now means unique, this specific choice of the direction of triangles is by now commonly accepted by many in the Complex Networks community and it is motivated largely by the work of Alon and colleagues on network motifs [227]. More precisely, positively oriented triangles correspond to feed forward loops, negatively oriented ones to feed backward loops, while zero curvature (no contribution) is attached to triangles that belong to neither of these types.

Remark 1.2.39 *In recent years, Hyperbolic geometry has come to be considered better suited to represent the background network geometry as it captures the qualitative aspects of networks of exponential growth such as the World Wide Web, and thus, it is used as the setting for variety of purposes. However, spherical geometry is usually not considered as a model geometry for networks because that geometry has finite diameter, hence finite growth. However, spherical networks naturally arise in at least two important instances. The first one is that of global communication, where the vertices represent relay stations, satellites, sensors or antennas that are distributed over the geo-sphere or over a thin spherical shell that can, and usually is, modeled as a sphere. The second one is that of brain networks, where the cortex neurons are envisioned, due to the spherical topology of the brain, as being distributed on a sphere or, in some cases, again on a very thin (only a few neurons deep) spherical shell, that can also be viewed as essentially spherical.*

1.2.3 Haantjes Curvature

As we have already underlined before, both Menger and Alt curvatures suffer from the same impediment, namely that they impose a Euclidean geometry on the studied space/data. Even if one allows for the Spherical and Hyperbolic versions, one still operates using a constant background geometry. Furthermore, both curvatures can take into account, by their very definitions, only triangles, a fact that represents a serious limitation in many real life applications. Fortunately, a much more flexible notion of metric curvature for 1-dimensional geometric objects is available, namely the so called *Haantjes curvature* or *Finsler-Haantjes curvature*.[8] Furthermore it is connected with other notions (not just of curvature) and, moreover, it represents a simple and direct alternative – at least in certain applications – of more involved and fashionable concepts.

Definition 1.2.40 (*Haantjes curvature*) *Let (M, d) be a metric space and let $c :$ $I = [0, 1] \overset{\sim}{\to} c(I) \subset M$ be a homeomorphism, and let $p, q, r \in c(I)$, $q, r \neq p$. Denote by $\widehat{qr}$ the arc of $c(I)$ between q and r, and by qr line segment from q to r.*

We say that c has Haantjes curvature $\kappa_H(p)$ at the point p iff:

$$\kappa_H^2(p) = 24 \lim_{q, r \to p} \frac{l(\widehat{qr}) - d(q, r)}{\left(l(\widehat{qr}) \right)^3} \; ; \tag{1.2.21}$$

where "$l(\widehat{qr})$" denotes the length – in intrinsic metric induced by d – of $\widehat{qr}$.

Alternatively, since for points/arcs where Haantjes curvature exists, $\frac{l(\widehat{qr})}{d(q,r)} \to 1$, as $d(q, r) \to 0$ (see [175]), κ_H can be defined (see, e.g. [184]) by

$$\kappa_H^2(p) = 24 \lim_{q, r \to p} \frac{l(\widehat{qr}) - d(q, r)}{\left(d(q, r) \right)^3} \; ; \tag{1.2.22}$$

In applications it is this alternative form of the definition of Haantjes curvature that will prove to be more malleable, as we shall illustrate shortly.

Exercise 1.2.41 *Prove Formula (1.2.22).*

In any of its versions, the intuition behind the notion of Haantjes curvature is quite transparent: The longer is the arc as compared to the chord, the more "curved" it is. (The longer the bow is in comparison to its string, the more "bowed", i.e. curved it is.) However, its complicated form is far less intuitive. For now, let us observe that it is proportional to $1/l$ (or $1/d$), which hints to the radius of curvature (and to Menger curvature). Less transparent and definitely more cumbersome is, however, the factor of "24" appearing in the definition. However, the proof of our next theorem will show that the two are interrelated and that the "24" factor arises naturally.

Theorem 1.2.42 *Let $\gamma \in \mathcal{C}^3$ be smooth curve in $\mathbb{R}^3$ and let $p \in \gamma$ be a regular point. Then the metric curvature $\kappa_H(p)$ exists and equals the classical curvature of γ at p.*

Simply put, for smooth curves in the Euclidean plane (or space), Haantjes curvature coincides with the standard (differential) notion, proving that, it represents, indeed, a proper generalization of the classical concept of curvature. We defer the proof for later, when we can employ the so called *Serret-Frènet formulas* (see Section 1.3.1).

In the course of the proof above, we didn't only prove the main assertion, but we also showed that for curves in Euclidean space, Menger and Haantjes curvatures coincide. Moreover, the same holds if one replaces Menger curvature by Alt curvature. It turns out that, in fact, if these notions of metric curvature are applicable, then the following holds:

---

[8] Named after Haantjes [175], who extended to metric spaces an idea introduced by Finsler in his PhD Thesis.

Theorem 1.2.43 (*Haantjes*) *Let γ be a rectifiable arc in a metric space (M, d), and let $p \in \gamma$. If κ_A (κ_M) and κ_H exist, then they are equal.*

Proof We follow the proof in [212], and begin by denoting bt κ any of the curvatures κ_M or κ_A that exists.

Let $q, r \in \gamma$ be two points on the same side of p, such that $d(p, q) = d(q, r) = d$, and let $d(p, r) = a$. Then

$$\kappa^2(p) = \lim_{d \to 0} \frac{a^2(2d + a)(2d - a)}{a^2 d^4} = \lim_{d \to 0} \frac{4}{d^2} \left(1 - \frac{a}{2d}\right) \left(1 + \frac{a}{2d}\right).$$

Given that $\kappa^2(p)$ is defined, it follows that $\lim_{d \to 0} \frac{a}{2d} = 1$, therefore

$$\kappa^2(p) = 4 \lim_{d \to 0} \frac{2d - a}{d^3}.$$

Therefore, from the existence of $\kappa_H(p)$ it follows that

$$\frac{1}{4!}\kappa^2(p) = \lim_{d \to 0} \frac{l(\widehat{pq}) - d}{l(\widehat{pq})^3} = \lim_{d \to 0} \frac{l(\widehat{qr}) - d}{l(\widehat{qr})^3} = \lim_{d \to 0} \frac{l(\widehat{pq}) + l(\widehat{qr}) - a}{[l(\widehat{pq}) + l(\widehat{qr})]^3}.$$

If we simplify the notation by putting $l(\widehat{pq}) = l_1, l(\widehat{qr}) = l_2$, we obtain that

$$\frac{l_1 + l_2 - a}{(l_1 + l_2)^3} = \frac{l_1 - d}{l_1^3} \frac{l_1^3}{(l_1 + l_2)^3} + \frac{l_2 - d}{l_2^3} \frac{l_2^3}{(l_1 + l_2)^3} + \frac{2d - a}{d^3} \frac{d^3}{(l_1 + l_2)^3}.$$

Since

$$\lim_{d \to 0} \frac{l_1}{d} = \lim_{d \to 0} \frac{l_2}{d} = 1,$$

it follows that

$$\kappa^2(p) = 4 \lim_{d \to 0} \frac{2d - a}{d^3} = \kappa^2(p).$$

$\square$

Exercise 1.2.44 *Prove that, indeed,*

$$\lim_{d \to 0} \frac{l_1}{d} = \lim_{d \to 0} \frac{l_2}{d} = 1.$$

As the reader might have observed earlier already, it is important to ensure that the arc γ is rectifiable, since Haantjes curvature is definable only for such arcs. Thus it would appear that it is a more restricted notion of curvature than Menger curvature, which do not necessitate this additional assumption. However, by Proposition 1.2.27 the existence of Menger curvature at the points of a metric arc ensures its rectifiability; thus Haantjes curvature is also applicable.

Remark 1.2.45 *The formulation of the proposition above one cannot but notice the conditional "If" regarding the simultaneous existence of the various curvatures, hence ask her/himself what, if any, are the relationships between them. Not to transform this textbook into a technical one regarding metric curvatures, we restrict ourselves to bringing the essential connections (beyond those already mentioned above) and point to a number of important (counter-)examples:*

1. *The existence of Alt curvature does not imply the existence of Haantjes curvature.*

   Counterexample 1.2.46 *Let $C \subset \mathbb{R}^2$, $C = \{(x, y) \mid y = 0 \text{ if } x = 0; \text{ and } y = x^4 \sin 1/x, \text{ if } x \neq 0\}$. Then $\kappa_A(0)$ exists, while $\kappa_H(0)$ is not defined.*

   Exercise 1.2.47 *Prove the assertion above.*

2. *The existence of $\kappa_M(p)$ implies the existence of $\kappa_H(p)$.*

   Unfortunately, the proof is to involved and lengthy and we are thus forced to refer the reader to, e.g. [47].

   However,

3. *The existence of $\kappa_H(p)$ does not imply the existence of $\kappa_M(p)$.*

   Counterexample 1.2.48 *Let $C = [0, 1]$, with the metric $d(x, y) = t - \frac{1}{3!}t^3 + \frac{1}{4!}t^4 \sin \frac{1}{t}$, where $t = |x - y|$, $x \neq y$. Then (C, d) is a rectifiable metric arc, with $\kappa_H(p)$ existing at every point $p \in C$, while $\kappa_A(p), \kappa_M(p)$ are not defined at any point $p \in C$.*

   Exercise 1.2.49 *Prove the assertion above and show that $\kappa_H \equiv 2$.*

As we have already observed above, while the idea behind Haantjes curvature is quite simple, the notion itself appears somewhat cumbersome. Therefore, it is only natural to try and simplify it – even at the price of discarding a dimensionality condition – so long as the essential geometric motivation is preserved. This motivation, as well as the expression of Haantjes curvature, brings one to consider the classical notion of *excess*:

Definition 1.2.50 (*Excess*) *Given a triangle[9] $T = \triangle(pxq)$ in a metric space (X, d), the excess of T is defined as*

$$e = e(T) = d(p, x) + d(x, q) - d(p, q). \tag{1.2.23}$$

This extremely simple (one might think even "naive") definition represents a notion of metric geometry that has been proven to be flexible and efficient, in many mathematical contexts, not least in the study of the Global Geometry of Manifolds (see, e.g., [144], [145] and the bibliography therein).

A local version of this notion was introduced (seemingly by Otsu [342]), namely the *local excess* (or, more precisely, the *local d-excess*):

$$e_d(X) = \sup_p \sup_{x \in B(p,\rho)} \inf_{q \in S(p,\rho)} (e(\triangle(pxq))), \tag{1.2.24}$$

---

[9]not necessarily geodesic.

where $\rho \leq \mathrm{rad}(X) = \inf_p \sup_q d(p,q)$, (and where $B(p,\rho), S(p,\rho)$ stand – as they usually do – for the ball and respectively sphere of center p and radius ρ).

In addition, global variations of this quantity have also been defined:

$$e(X) = \inf_{(p,q)} \sup_x \left(e(\triangle(pxq)) \right), \qquad (1.2.25)$$

and, the so called *global big excess* (see [342]):

$$E(X) = \sup_q \inf_p \sup_x \left(e(\triangle(pxq)) \right). \qquad (1.2.26)$$

(Of course, in all practical settings, "sup" and "inf" replace "max" and "min" respectively, for all the instances above.)

Evidently, (local) excess and Haantjes curvature are closely related notions, since the geometric "content", so to say, of the notion of local excess being that, for any $x \in B(p,d)$, there exists a (minimal) metric segment γ from p to $S(p,d)$ such that γ is close to x. To be more precise, using a simplified notation and discarding (for convenience/simplicity) the normalizing constant "24", one has the following relation between the two notions:

$$\kappa_H^2(T) = \frac{e}{d^3}, \qquad (1.2.27)$$

where $\rho = \rho(p,q)$, and where by the curvature of a triangle $T = T(pxq)$ we mean the curvature of the path $\widehat{pxq}$. Thus Haantjes curvature can be viewed as a *scaled* version of excess. Keeping this in mind, one can define also a global version of this type of metric curvature, namely by defining, for instance:

$$\kappa_H^2(T) = \frac{e}{\rho^3}, \qquad (1.2.28)$$

or

$$\kappa_H^2(X) = \frac{e(X)}{\mathrm{diam}^3(X)}, \qquad (1.2.29)$$

as preferred.

Of course, one can proceed in the "opposite direction", so to speak, and express the proper (i.e. point-wise) Haantjes curvature via the definition (1.2.24) of local excess, as

$$\kappa_H^2(x) = \lim_{\rho \to 0} e(x). \qquad (1.2.30)$$

1.2.4 Applications of Haantjes Curvature

As expected, based on the experience garnered with Menger/Alt curvature, the most natural (and perhaps the most efficient) applications of Haantjes curvature are to the study of Complex Networks. The first such utilization of this metric curvature was to DNA Microarray Analysis [286]. However, this application makes appeal to ideas that we shall develop in the next chapter, therefore we defer for later its details. Further applications to networks sampling, Deep Learning and Information Geometry also need the reader to be familiar with the notion of *Ricci curvature*, which we shall encounter later on. However, some immediate, yet still novel and powerful applications of Haantjes curvature to Semantic Networks are at hand [90].

We can, however present in some detail, a most unforeseen application, namely to the study of wavelets, more precisely to the understanding of the notion of scale. While perhaps unexpected, the connection between scale and curvature (more specific Haantjes curvature) is quite evident, once one is willing to notice and accept it. Indeed, consider the generic PL wavelet φ in Figure 1.8 below, and let $\widehat{AE}$ be the arc of curve between the points A and E, and let $d(A, E)$ is the length of the line-segment AE. Then

$$l(\widehat{AE}) = a + b + c + d \; ; d(A, E) = e + f. \tag{1.2.31}$$

Using the alternative formula for the Haantjes curvature (1.2.22), the following discretization of the Haantjes curvature is, therefore, natural:

$$\kappa_H^2(\varphi) = 24[(a + b + c + d) - (e + f)]/(e + f)^3. \tag{1.2.32}$$

In addition to the total curvature of the wavelet φ, one can also compute the "local" curvatures of the partial wavelets $\varphi_1 = \widehat{ABC}$ and $\varphi_2 = \widehat{CDE}$, that is the curvatures at the "peaks" B and D:

$$\kappa_H^2(B) = 24(a + b - e)/e^3 , \tag{1.2.33}$$

and

$$\kappa_H^2(D) = 24(c + d - f)/f^3 , \tag{1.2.34}$$

as well as the mean curvature of these peaks:

$$H_H(\widehat{AE}) = [\kappa_H(B) + \kappa_H(D)]/2. \tag{1.2.35}$$

Even though these variations may prove to be useful in certain applications, we believe that the correct approach, in the sense that it best corresponds to the scale of the wavelet, would be to compute the total curvature of φ. Of course, there is nothing special about PL wavelets (except their simplicity) and one can compute the Haantjes curvature for the Haar scaling function and wavelet, as well as for other, smooth families of wavelets (see [109], [16]). The connection scale-curvature

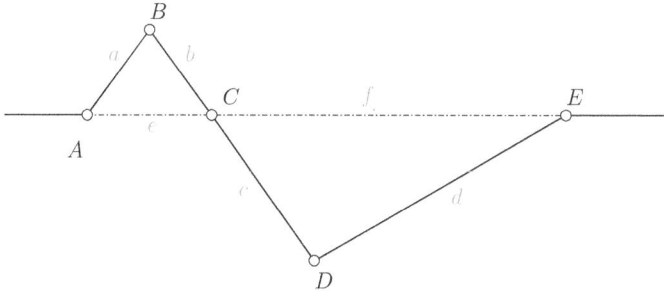

Figure 1.8   A generic PL wavelet.

Figure 1.9   Texture segmentation of an urban landscape image (left): Average Haantjes curvature (middle) and the texture segmentation of image (right), using 7 scales [16].

for a number of families of curves, as well as some general estimates, were given in [109], [16]. (Note also that, via the principal curvatures of a surfaces, one can extend the scale-curvature duality to curvaturelets, ridgelets and shearlets, as well [109].)

A first application of this connection is presented in [16], where the scale-curvature connection is employed to texture analysis and segmentation in images (see, for example, Figure 1.9 as well as [16]). In addition, in both papers mentioned above, an automatic scale detection is suggested (as well as the obvious use of wavelet curvature as edge detector).

More important than any specific application, is the fact that the scale-curvature duality allows for a first bridging, even if only a partial one, of the gap between the two basic, largely non-intersecting, approaches prevalent in Image Processing and related fields: The geometric one, that is closely related to the Graphics community philosophy; and the more classical, Fourier Analysis/Wavelets driven one. In fact, it is quite natural to replace the vaguely defined (but intuitively clear) concept of "scale", to the classical, well defined one of "curvature", by formally defining scale by means of the Haantjes curvature, at least in the purely theoretical setting. This is more relevant in the context of 2-dimensional (as well, of course, as higher dimensional ones), nonseparable signals, where a proper notion of scale is far less intuitive then in the 1-dimensional, classical, case.

Remark 1.2.51 *Before concluding the present section on metric curvatures for curves we should mention that the original, older ones discussed above were, unfortunately, largely forgotten for a long period of time. As already mentioned, Menger's curvature made, rather recently, quite a "revival" in Analysis. Sadly, this cannot be said about the Haantjes curvature (even though, in many practical applications, it is the more adjustable and useful of the two). Still, there is a ray of light, out of this darkness, so to say: Modern (and more "tight") versions of it (and of Menger's curvature) are devised in the modern literature [6], in conjunction with the extensively used notion of Alexandrov space (see below).*

We prefer the classical versions of these types of curvatures, i.e the Menger and Haantjes curvatures, to their more modern counterparts, for a number of (related) reasons:

- *They are simpler and far more intuitive, thence conducive toward applications;*

- *They are more ready to lend themselves to discretization, hence admit easy and direct "semi-discrete" (or "semi-continuous") versions, as the one above (and the ones that we shall introduce in the following chapter);*

- *They are applicable to more general spaces, fact that represents a further incentive in their application in discrete (i.e. Computer Science driven) settings;*

- *While the Alexander-Bishop [6] variants of metric curvatures are more "tight", so to say, they coincide with their classical counterparts on all but the most esoteric spaces;*

- *No apriori knowledge of the global geometry of the ambient space (i.e. Alexandrov curvature) is presumed, nor is it necessary to first determine the curves of constant curvature (see [6]) in order to compute these curvatures; furthermore*

- *They are easy to compute in a direct fashion, at least in the discrete setting (at least among those discrete versions we encountered).*

1.3 TORSION

We have seen that curvature measures the departure from a planar curve of being a straight line. In analogy with it, it is desirable to have at our disposal a quantity that measures the degree (or speed) of departure of a general curve (or *skew* curve) from being planar. In order to define it we need first to better understand curves in space.

Let $c : I \to \mathbb{R}^3$ let be a curve parametrized by arc length, i.e. $||\dot{c}|| \equiv 1$, and let us denote (as usual) $\mathbf{t} = \dot{c}$. Denote (as before) by

$$\mathbf{n} = \frac{\dot{\mathbf{t}}}{|\mathbf{t}|}$$

the *normal vector.* Clearly

$$(*) \qquad \dot{\mathbf{t}}(s) = \kappa(s)\mathbf{n}(s)$$

We also denote

$$\mathbf{b} = \mathbf{t} \times \mathbf{n} \qquad\qquad (1.3.1)$$

the *binormal vector* (to c in $c(s)$).

The triple $\mathbf{t}, \mathbf{n}, \mathbf{b}$ constitutes the so called *Frènet frame* (Figure 1.10).

By their very definitions these vectors satisfy the following relations:

$$\begin{cases} \mathbf{t}^2 = \mathbf{n}^2 = \mathbf{b} = 1 \\ \mathbf{t} \cdot \mathbf{n} = \mathbf{n} \cdot \mathbf{b} = \mathbf{b} \cdot \mathbf{t} = 0 \\ \mathbf{t} = \mathbf{n} \times \mathbf{b} \quad \mathbf{n} = \mathbf{b} \times \mathbf{t} \quad \mathbf{b} = \mathbf{t} \times \mathbf{n} \end{cases} \qquad (1.3.2)$$

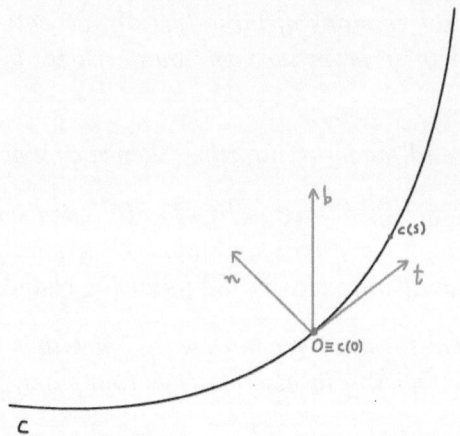

Figure 1.10 The Frènet frame.

By differentiating Formula (1.3.1) we obtain that

$$\dot{\mathbf{b}} = \dot{\mathbf{t}} \times \mathbf{n} + \mathbf{t} \times \dot{\mathbf{n}}.$$

Since $\dot{\mathbf{t}} \times \mathbf{n} = 0$, it follows that

$$\dot{\mathbf{b}} = \mathbf{t} \times \dot{\mathbf{n}} = \mathbf{t} \times \dot{\mathbf{n}}.$$

From here and from the fact that $\dot{\mathbf{b}}(s) \perp \mathbf{b}$, it follows that $\dot{\mathbf{b}}(s) \parallel \mathbf{n}$, i.e. $\dot{\mathbf{b}}(s)$ is proportional to $\mathbf{n}$, a fact which is written as

$$(***) \quad \dot{\mathbf{b}}(s) = -\tau(s)\mathbf{n}(s).$$

We can now formally introduce our next

Definition 1.3.1 *Let* $c : I \to \mathbb{R}^3$ *be a regular curve, such that* $\|\dot{c}\| \equiv 1$ *and such that* $\ddot{c}(s) \neq 0$, *for any* $s \in I$. *Then* $\tau(s)$ *in Formula (***) above is called the torsion of the curve* c *at the point* s.

Remark 1.3.2 *Note that here and above, we have used the notation "s" for the parameter of* c, *instead of the traditional "t". Given that* $\|\dot{c}\| \equiv 1$, *i.e. the curve is parameterized by arc length (cf. the remarks preceding Proposition 1.1.13), this is fully justified.*

Moreover, since (1.3.1) can be written also as $\mathbf{n} = \mathbf{b} \times \mathbf{t}$, we obtain by derivation that

$$\dot{\mathbf{n}} = \dot{\mathbf{b}} \times \mathbf{t} + \mathbf{b} \times \dot{\mathbf{t}},$$

that is

$$(**) \quad \dot{\mathbf{n}}(s) = -\kappa(s)\mathbf{t} + \tau(s)\mathbf{t}.$$

Remark 1.3.3 *We concentrate below a number of observations regarding torsion and its definition.*

The condition that $\ddot{c}(s) \neq 0$, i.e. that $\kappa(s) \neq 0$ is essential. In its absence, it is not clear which plane the curve is departing from – see the counterexample below. (For further details and an in depth analysis, see [309], pp. 25-27.)

1. **Counterexample 1.3.4** *Consider the curve $c : \mathbb{R} \to \mathbb{R}^3$,*

$$
c(t) = \begin{cases} (t, 0, e^{-1/t^2}) \; ; & t > 0 \\ (0, 0, 0) \; ; & t = 0 \\ (t, e^{-1/t^2}, 0) & t < 0 \end{cases}
$$

2. *Note that, contrary to curvature, torsion can take negative values. We shall soon see the geometric interpretation of the sign of τ.*

3. *It is easy to see that (the sign of) τ is invariant under change of orientation. (Check!)*

Before proceeding further, let us bring a classical, yet edifying example:

Example 1.3.5 (The circular helix) *Consider the circular helix, that is the cylindrical curve $c(t) = (a \cos t, a \sin t, bt)$, where $a, b \neq 0$. Then its torsion is $\frac{b}{a^2+b^2}$ (and its curvature is $\frac{|a|}{a^2+b^2}$).*

Exercise 1.3.6 *Verify the computations above.*

Remark 1.3.7 *Note that the pitch of the helix– that is the distance between two consecutive intersections of the helix with a generator of the cylinder (i.e. "the height of one complete helix turn, measured parallel to the axis of the helix") – equals $2\pi b$.*

1.3.1 The Serret-Frènet Formulas

The formulas $(*) - (* * *)$ we obtained above are – for evident reasons – grouped as

$$
\begin{cases} \dot{\mathbf{t}} = \kappa \mathbf{n} & (*) \\ \ddot{\mathbf{n}} = -\kappa \mathbf{t} + \tau \mathbf{b} & (**) \\ \dddot{\mathbf{b}} = -\tau \mathbf{n} & (* * *) \end{cases} \tag{1.3.3}
$$

Formulas 1.3.3 are called the *Serret-Frènet Formulas*, and their importance in the understanding of spatial curves can hardly be underestimated, as it will become evident in the sequel.

Remark 1.3.8 *The vector $\mathbf{d} = \kappa \mathbf{b} + \tau \mathbf{t}$ is called the Darboux vector. It's usefulness resides in the fact that $\dot{v} = \mathbf{d} \times v$, thus differentiation along the curve may be written formally as $\frac{d}{ds} = \mathbf{d} \times$.*

A first application of the formulas (1.3.3) above is the following fundamental fact (which can be viewed as an exercise, so the diligent student might chose to try an solve it before reading the proof):

Corollary 1.3.9 *Let* $c : I \to \mathbb{R}^3$, $||\dot{c}|| = 1$. *Then* $\dot{c}, \ddot{c}, \dddot{c}$ *are linearly independent iff* c *is a plane curve.*

Proof Recall that, given vectors u, v, w, their *mixed* (or *triple*) *product* $u \cdot (v \times w)$ represents the signed volume of the parallelepiped constructed on the three vectors. Therefore, u, v, w are linearly independent $\Leftrightarrow u \cdot (v \times w) = 0$.

Since (a) $\dot{c} = \mathbf{t}$, it follows from $(*)$ that (b) $\ddot{c} = \kappa \mathbf{n}$, thence $\dddot{c} = (\kappa \mathbf{n})\dot{} = \dot{\kappa} \mathbf{n} + \kappa \dot{\mathbf{n}}$. From the last expression and $(**)$ it follows, (after a couple of manipulations that the reader can provide for he/himself) that (c) $\dddot{c} = \dot{\kappa} \mathbf{n} + \kappa(\tau \mathbf{b} - \kappa \mathbf{t})$.

Then, by using formulas (a), (b) and (c) that we just derived, we obtain that

$$\dot{c} \cdot (\ddot{c} \times \dddot{c}) = \kappa \mathbf{b} \cdot \dot{\kappa} \mathbf{n} + \kappa^2 \tau \mathbf{t} \cdot (\mathbf{n} \times \mathbf{b}) - \kappa^2 \mathbf{t} \cdot (\mathbf{n} \times \mathbf{t})$$

From the definition of $\mathbf{t}, \mathbf{n}, \mathbf{t}$ it follows that the last expression equals $0 + \kappa^2 \tau \mathbf{t} \cdot \mathbf{t} - 0 = \kappa^2 \tau$.

Therefore, $\dot{c} \cdot (\ddot{c} \times \dddot{c}) = 0 \Leftrightarrow \kappa^2 \tau = 0$, that is either $\kappa = 0$, i.e. c is straight line, hence planar, or $\tau = 0$' that is c is contained in a plane.

$\square$

Exercise 1.3.10 *Prove the following analogous formulas for the triple products of the remaining elements of the Frènet triple*

1. $\dot{\mathbf{t}} \cdot (\ddot{\mathbf{t}} \times \dddot{\mathbf{t}}) = \kappa^3(\kappa\dot{\tau} - \dot{\kappa}\tau) = \kappa^5 \frac{d}{ds}\left(\frac{\tau}{\kappa}\right)$;

2. $\dot{\mathbf{b}} \cdot (\ddot{\mathbf{b}} \times \dddot{\mathbf{b}}) = \tau^3(\dot{\kappa}\tau - \kappa\dot{\tau}) = \tau^5 \frac{d}{ds}\left(\frac{\kappa}{\tau}\right)$.

A result similar in spirit to Corollary 1.3.9, that is, again, both simple to prove but extremely important, is the following:

Corollary 1.3.11 *Let* $c : I \to \mathbb{R}^3$, *such that* $||\dot{c}|| \equiv 1$. *Then*

1. $\kappa \equiv 0$ *iff* c *is (part of) a straight line.*

2. $\tau \equiv 0$ *iff* c *is planar (i.e. it is included in a plane).*

Again, we suggest the student to view this as a (rather straightforward) exercise, before reading the proof below.

Proof

1. If $\kappa = 0$, then, by $(*)$ it follows that $\dot{\mathbf{t}} = 0 \Leftrightarrow \ddot{c} = 0$, thence it follows that $\dot{c} = \alpha = $ const., where α is a constant vector. From here it immediately follows that $c(s) = \alpha s + \beta$, where β is a constant vector.

2. If $\tau = 0$, then, by $(***)$ it follows that $\dot{\mathbf{b}} = 0$, thence $\mathbf{b} = \lambda = (\lambda_1, \lambda_2, \lambda_3) = $ const. By scalar multiplying this last equality by $\dot{c}$ we obtain that $\mathbf{b} \cdot \dot{c} = \lambda \cdot \dot{c} \Leftrightarrow \mathbf{b} \cdot \mathbf{t} = \lambda \cdot \dot{c}$, therefore, by (1.3.2) that $= \lambda \cdot \dot{c} = 0$. By integrating this last expression we obtain that $\lambda \cdot c = a$, where a is a constant scalar. This represents, as it is well-known from a Linear Algebra course, the equation of a plane, which can be written in coordinates as $\lambda_1 x(s) + \lambda_2 y(s) + \lambda_3 z(s) = a$. Thus, c is, indeed, a planar curve.

This concludes the proof.

□

The following simple, yet geometrically interesting results (since they characterize some classical, important types of curves), can be viewed as exercises; as such the reader is strongly encouraged to try and solve them before reading their proofs.

Proposition 1.3.12 *Let c be a curve such that all the tangents to c pass trough a fixed point. Then c is (included in) a straight line.*

Proof Let p be the fixed point. Then $c(s) = p - \lambda \dot{c}(s)$, $(\dot{c} \neq 0)$. By differentiating

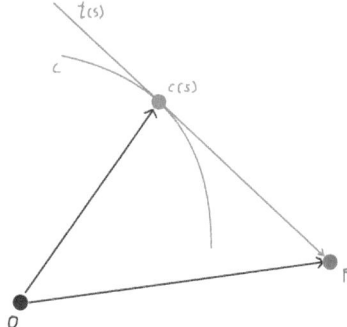

the equation above we obtain, using the first of the Serret-Frènet formulas, that

$$(1 + \dot{\lambda})\dot{c} = -\lambda \kappa \mathbf{n} \Leftrightarrow ((1 + \dot{\lambda})\dot{c}) \cdot \mathbf{n} = -\lambda \kappa \mathbf{n} \cdot \mathbf{n}.$$

Since $\dot{c} \cdot \mathbf{n} = 0$ and $\mathbf{n} \cdot \mathbf{n} = 1$, we obtain that $\kappa \equiv 0$, i.e. c is (part of) a line.

□

Proposition 1.3.13 *Let c be a curve such that all the (principal) normals to c pass trough a fixed point. Then c is (included in) a circle.*

Proof Again, let p denote the fixed point. Then the condition of statement can be written as

$$c(s) + \lambda(s)\mathbf{n}(s) = p$$

$$\Leftrightarrow \dot{c} = \dot{\lambda}\mathbf{n} + \lambda\dot{\mathbf{n}} = 0 \Leftrightarrow \mathbf{t} + \dot{\lambda}\mathbf{n} + \lambda(-\kappa\mathbf{t} + \tau\mathbf{b}) = 0 \Leftrightarrow \mathbf{t}(1 - \lambda\kappa) + \dot{\lambda}\mathbf{n} - \tau\lambda\mathbf{b} = 0$$

By the orthogonality of **t**, **n**, **b** it follows that $1 - \kappa\lambda = 0$, hence (i) $\lambda = 1/\kappa$. Furthermore, by the same argument, we have that $\dot{\lambda} = 0$, thence (ii) $\lambda = \text{const}$. From (i) and (ii) above it follows that (iii) $\kappa = \text{const}$. Moreover, again by the orthogonality of **t**, **n**, **b**, we have that $\tau\lambda = 0$, hence, since $\lambda \neq 0$, $\tau = 0$, therefore (iv) c is included in a plane. From (iii) and (iv) above we immediately reach the desired conclusion.

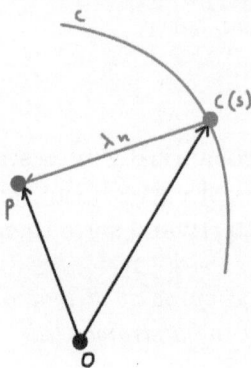

$\square$

The next result represents a characterization of spherical curves:

Proposition 1.3.14 *Let $c : I \to \mathbb{R}^3$ be a parameterized curve, such that $\tau(s) \neq 0, \kappa'(s) \neq 0, \forall s \in I$. Then c is a spherical curve $\equiv R^2 + (R')^2 T^2 = \text{const.}$, where we used the following notation: $R = 1/\kappa$, $T = 1/\tau$.*

The proof is technical and rather lengthy. Moreover, it also is of the same type as shorter proofs that we have already given. Therefore, we omit it and concentrate on the more novel and applicative matters. However, we leave it as an exercise for the highly motivated reader.

Exercise 1.3.15 *Prove the proposition above.*

Problem 1.3.16 *The "$\Rightarrow$" of the proof above admits a simple, geometrical argumentation. – Provide it!*

We can also bring the following characterization of helixes:

Proposition 1.3.17 *Let $c : I \to \mathbb{R}^3$ be a (parameterized) curve in general position, such that $\kappa(t)\tau(t) \neq 0, \forall t \in I$. Then the following conditions are equivalent:*

1. *The tangents to the curve form a constant angle with fixed direction.*

2. *The principal normals to the curve are perpendicular to a fixed direction.*

3. *The binormals to the curve are perpendicular to a fixed direction.*

4. *$\kappa(t) = \lambda\tau(t), \forall t \in I$, where $\lambda = \text{const.}$*

The proof is similar in spirit to the ones above, but we include it nevertheless for didactic reasons:

Proof We begin by noting that, as usual, we can suppose that $||\dot{c}|| \equiv 1$.

(1) $\Rightarrow$ (2) Let $\vec{l}$ the unit vector giving the fixed direction, and let $\alpha = \sphericalangle(\mathbf{t}, \vec{l})$. Then (1) $\Leftrightarrow \mathbf{t} \cdot \vec{l} = \cos \alpha$. By differentiating this last expression we obtain $\dot{\mathbf{t}} \cdot \vec{l} = 0$, and therefore, from $(*)$ it follows that $\kappa \mathbf{n} \cdot \vec{l} = 0$, thus $\mathbf{n} \cdot \vec{l} = 0$, i.e. $\mathbf{n} \perp \vec{l}$.

(2) $\Rightarrow$ (1) $\mathbf{n} \perp \vec{l} \Leftrightarrow \mathbf{n} \cdot \vec{l} = 0$, therefore from $(*)$ we obtain $\kappa \mathbf{n} \cdot \vec{l} = 0$, thence $\dot{\mathbf{t}} \cdot \vec{l} = 0$. It follows that $\mathbf{t} \cdot \vec{l} = \text{const.} = \cos \alpha$, hence (given that $||\mathbf{t}|| = ||\dot{c}|| \equiv$) $\sphericalangle(\mathbf{t}, \vec{l}) = \alpha = \text{const.}$

(2) $\Leftrightarrow$ (3) $\mathbf{n} \perp \vec{l} \Leftrightarrow \mathbf{n} \cdot \vec{l} = 0 \Leftrightarrow -\tau \mathbf{n} \cdot \vec{l} = 0 \overset{(***)}{\Longleftrightarrow} \dot{\mathbf{b}} \cdot \vec{l} = 0 \Leftrightarrow \mathbf{b} \cdot \vec{l} = \text{const.} = \cos \beta$, where $\beta = \sphericalangle(\mathbf{b}, \vec{l})$.

(2) $\Rightarrow$ (4) $\mathbf{n} \perp \vec{l} \Leftrightarrow \mathbf{n} \cdot \vec{l} = 0$. From here and from the euivalence between assertions (1) and (2) it follows that (a) $\mathbf{t} \cdot \vec{l} = \cos \alpha$ and (b) $\mathbf{b} \cdot \vec{l} = \cos \beta$.

On the other hand, by differentiating the expression $\mathbf{n} \cdot \vec{l} = 0$, we obtain $\dot{\mathbf{n}} \cdot \vec{l} + \mathbf{n} \cdot \dot{\vec{l}} = 0$, thence, by $(**)$ we obtain $-\kappa \mathbf{t} \cdot \vec{l} + \tau \mathbf{b} \cdot \vec{l} + \mathbf{n} \cdot 0 = 0$. By applying expression (b) above, we obtain that (c) $-\kappa \cos \alpha + \tau \cos \beta = 0$.

Since $\mathbf{b} \perp \mathbf{t}$ it follows that $(\cos \alpha, \cos \beta) \neq (0,0)$, therefore one can divide (c), for example, by τ, thus $\kappa/\tau = const.$

(4) $\Rightarrow$ (1) By multiplying the third Serret-Frènet equation by λ and adding the result to the first one, we obtain that $\dot{t} + \lambda \dot{b} = \kappa \mathbf{n} - \lambda \tau \mathbf{n}$. Thus, since, by (4) $\kappa = \lambda \tau$, it follows $\dot{\mathbf{t}} + \lambda \dot{\mathbf{b}} = 0$. From here, by integration, we obtain that $\mathbf{t} + \lambda \mathbf{b} = \vec{v}$, where $\vec{v}$ is a constant vector. But the last relation is equivalent to $\mathbf{n} \cdot (\mathbf{t} + \lambda \mathbf{b}) = \mathbf{n} \cdot \vec{v}$. Since $\mathbf{n} \cdot \mathbf{t} + \lambda \mathbf{n} \cdot \mathbf{b} = 0$, it follows that $\mathbf{n} \cdot \vec{v}$. By taking $\vec{v} = \vec{l}$ the desired result follows.

$\square$

Exercise 1.3.18 *Check that the properties above characterize the circular helix.*

The Serret-Frènet Equations also allow us to define and characterize another special class of curves, namely the so called *Bertrand curves*:

Definition 1.3.19 *Let $c : I \to \mathbb{R}^3$ be a parametrized curve, such that $\kappa(t)\tau(t) \neq 0, \forall t \in I$. c is called a Bertrand curve if there exists a (parametrized) curve $\gamma : J \to \mathbb{R}^3$ such that the principal normals to c are also principal normals of γ (see Figure 1.11). In this case γ is called a Bertrand mate of c and c, γ are called Bertrand mates.*

The first basic result states that the only "interesting" Bertrand curves are spatial ones, and that the condition $\kappa(t)\tau(t) \neq 0, \forall t \in I$ is not a formal one:

Lemma 1.3.20 *Any planar curve is a Bertrand curve.*

While this could constitute an excellent simple exercise, we also give its proof below (while, again, recommending the reader to solve it first).

Proof If c is a straight line, then the result is trivial.

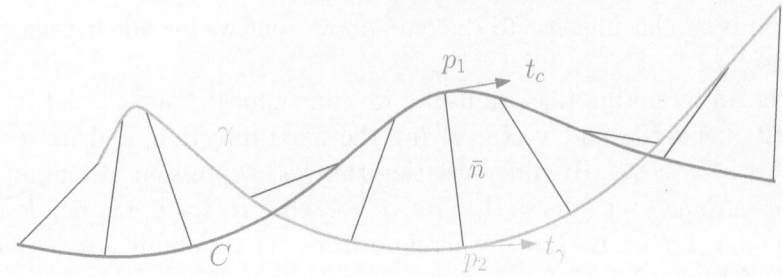

Figure 1.11 Betrand mates.

If c is not a straight line, it follows that $\kappa \neq 0$. Then γ – the locus of the curvature centers of c is well defined an it is also a curve in the plane of c. Furthermore, by its definition,

$$\gamma(s) = c(s) + \frac{1}{\kappa}\mathbf{n}.$$

By differentiating the formula above and via the Serret-Frènet equations we obtain

$$\dot\gamma = -\frac{\dot\kappa}{\kappa^2}\mathbf{n};$$

i.e. c has as a principal normal the tangents to γ. Analogously, this holds for any curve Γ perpendicular to the tangents to γ, that is to sy c, γ have the same principal normal.

$\square$

We have the following characterization of Bertrand mates:

Proposition 1.3.21 *Let c be a Bertrand curve, and let κ, τ denote its curvature and torsion, respectively. Then the following hold:*

1. *$\mathrm{dist}(p_1, p_2) = $ const., where p_1, p_2 are like in Figure 1.11;*

2. *$\angle(t_c, t_\gamma) = $ const.;*

3. *There exists constants a, b, c such that $a\kappa(s) + b\tau(s) + c = 0$, for any $s \in I$.*

Since the proof is technical and rather lengthy and, moreover, it is of the same type as shorter proofs that we have already given, we leave it as an exercise to the diligent reader.

Exercise 1.3.22 *Prove Proposition 1.3.21.*

We add below a few other problems on Bertrand curves:

Problem 1.3.23 *Let c be such that $\kappa_c \equiv $ const. Then, if $\tau_c \not\equiv $ const., the Bertrand mate γ_c of c represents the locus of the curvature centers of c and, moreover, $\kappa_c\gamma_c = \kappa_c$.*

Problem 1.3.24 *A helix has an infinity of Bertrand mates.*

Problem 1.3.25 *Let c, γ be Bertrand mates. Prove that $\tau_c(s) \cdot \tau_c\gamma(s) \equiv$ const. > 0.*

Hint: Write the formula $a\kappa + b\tau + c = 0$ of Proposition 1.3.21 (iv) in trigonometric form.

Problem 1.3.26 *Let c be a planar curve, and let Γ_1, Γ_2 be involutes of c. Then Γ_1, Γ_2 constitute a Bertrand pair.*

Problem 1.3.27 *Prove that c is a Bertrand pair $\Longleftrightarrow$ There exists a function $v : I \to \mathbb{R}^3$ such that (i) $\|v(t)\| \equiv 0$, and (ii) $\|\dot{v}(t)\| \equiv 1$, and such that it satisfies the equation*

$$c(t) = a \int v(t)dt + b \int v(t) \times \dot{v}(t)dt.$$

1.3.2 Haantjes Curvature Revisited

We are now ready to return to the proof of Theorem 1.2.42. However, while the proof of the theorem, as originally stated, does not require additional definitions, and can be generalized to more general spaces[10], it is also quite lengthy and technical, therefore we do not bring it here. Instead, we prove it under more relaxed conditions, that will allow us to make appeal to the Serret-Frènet formulas. More precisely, (following [47]) we shall prove, using quite elementary tools, the following:

Theorem 1.3.28 *Let $c \subset \mathbb{R}^3; c(s) = (x_1(s), x_2(s), x_3(s))$ be a curve parametrized by arc length, such that $x_i \in \mathcal{C}^3, i = 1, 2, 3$. Then κ_H exists and $\kappa_H \equiv \kappa$ – the classical curvature of c.*

Proof The change of notation for the components/coordinates of c adopted in the statement of the theorem allows us to simplify the proof, at least in so far as the length of the computations involved is concerned.

We begin by selecting a mobile coordinate system who's origin O is at p – an arbitrary point on c, and who's axes x_1, x_2, x_3 are along the tangent, normal and binormal at p, respectively. Also, let $q \neq r \neq p \in c$, and denote $l(\widehat{pq}) = u, l(\widehat{qr}) = v$. (If one wishes to taken orientation into account, then choose the points such that $v > 0$.)

Then, by standard expansion into Maclaurin series, we can write

$$x_i(u + v) - x_i(u) = vx_i'(u) + \frac{1}{2}v^2x_i''(u) + \frac{1}{6}v^3x_i'''(u + \xi_iv);$$

where $0 < \xi_i, i = 1, 2, 3$. Then

$$d(q, r) = \sqrt{\sum_{i=1}^{3}(x_i(u + v) - x_i(u))^2} = v\sqrt{\sum_{i=1}^{3}\left(x_i'(u)\frac{1}{2}v^2x_i''(u) + \frac{1}{6}v^3x_i'''(u + \xi_iv)\right)^2},$$

---

[10]Haantjes, cf. [47].

where d denotes the usual Euclidean metric in the plane. Therefore,

$$\frac{l(\widehat{qr}) - d(q,r)}{l^3(\widehat{qr})} = \frac{v - d(q,r)}{v^3} = \frac{1 - \sqrt{\sum_{i=1}^3 \left(x_i'(u)\frac{1}{2}v^2 x_i''(u) + \frac{1}{6}v^3 x_i'''(u + \xi_i v)\right)^2}}{v^2}.$$

After some technical manipulations of the right hand side, and noting that $q, r \to p \Leftrightarrow v \to 0$, we obtain that

$$\lim_{q,r \to p} \frac{l(\widehat{qr}) - d(q,r)}{l^3(\widehat{qr})} = -\frac{1}{6}\sum_{i=1}^3 x_i'(0)x_i'''(0) - \frac{1}{8}\sum_{i=1}^3 (x_i''(0))^2.$$

Give our choice of axis at $O \equiv p$, we have that $x_1'(0) = 1, x_2'(0) = x_3'(0) = 0$. By using the Serret-Frènet formulas we obtain that $x_1'''(0) = -\kappa^2$ and $\sum_{i=1}^3 (x_i''(0))^2 = \kappa^2$. Therefore, we obtain that

$$\lim_{q,r \to p} \frac{l(\widehat{qr}) - d(q,r)}{l^3(\widehat{qr})} = \frac{1}{6}\kappa^2 - \frac{1}{8}\kappa^2 = \frac{1}{24}\kappa^2,$$

that is $\kappa_H(p) = \kappa$.

□

Remark 1.3.29 *Given that, by its very definition, $\kappa_M(p) = \kappa$, the theorem above could have been formulated in terms of Menger curvature, instead of Haantjes curvature.*

Exercise 1.3.30 *Supply the technical details omitted in the proof above.*

1.3.3 The Local Canonical Form

It is most natural to try and express the curve not in the abstract, absolute coordinates of the ambient space, but rather into local coordinates given by the *moving frame* $\mathbf{t}, \mathbf{n}, \mathbf{b}$. (Intuitively, this would like the feeling of a person in a roller-coaster ride who is trying to find his bearing.)

To this end, let us suppose, as it often was the case before, that $c : I \to \mathbb{R}^3$ is a curve parametrized by arc length, such that $\dot{c}, \ddot{c} \neq 0$; and let us suppose that the base point of reference $p \in c$ is $p = c(0)$ (see Figure 1.12). Then, by developing into Maclaurin series, one can write

$$c(s) = c(0) + \dot{c}(0)s + \frac{\ddot{c}(0)}{2!}s^2 + \frac{\dddot{c}(0)}{3!}s^3 + \omega(s^4);$$

where $\omega(s^4) \xrightarrow[s \to 0]{} 0$.

Since $\dot{c}(0) = \mathbf{t}, \ddot{c}(0) = \mathbf{n}$ and $\dddot{c} = \dot{\kappa}\mathbf{n} + \kappa(\tau\mathbf{b} - \kappa\mathbf{t})$ (see the proof of Corollary 1.3.9 above), we obtain that

$$c(s) - c(0) = \left(s - \frac{\kappa^2 s^3}{3!}\right)\mathbf{t} + \left(\frac{s^2 \kappa}{2} + \frac{s^3 \dot{\kappa}}{3!}\right)\mathbf{n} - \frac{s^3}{3!}\kappa\tau\mathbf{b} + \omega(s^4).$$

(Here $\kappa = \kappa(0)$, etc.)

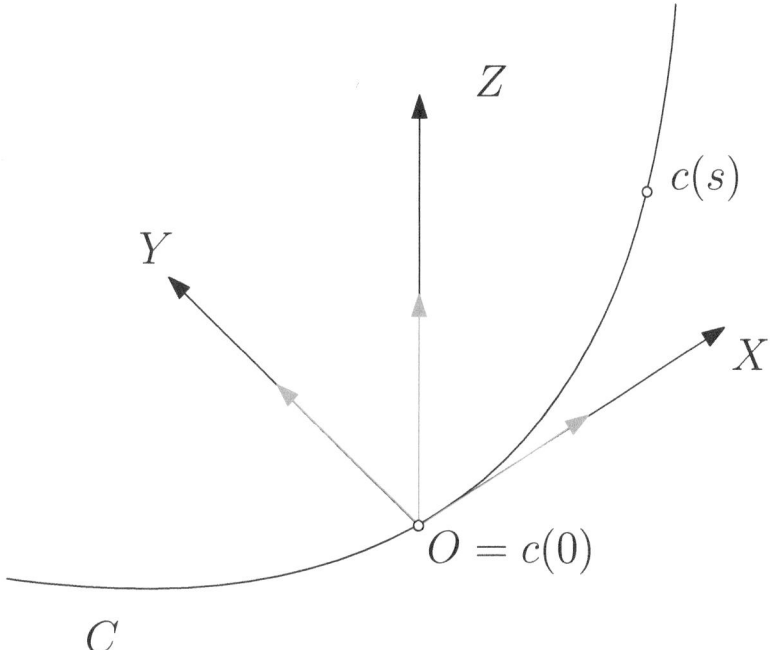

Figure 1.12 The Frènet frame.

Writing, in the coordinates system considered above, $c(s) = (x(s), y(s), z(s))$, one obtains the following system:

$$\begin{cases} x(s) = s - \frac{\kappa^2 s^3}{3!} + \omega(x) \\ y(s) = \frac{s^2 \kappa}{2} + \frac{s^3 \dot{\kappa}}{3!} + \omega(y) \\ z(s) = \frac{s^3}{3!}\kappa\tau + \omega(z) \end{cases} \qquad (1.3.4)$$

where $\omega(s^4) = \omega(x)\mathbf{t} + \omega(y)\mathbf{n} + \omega(z)\mathbf{b}$. Thus

$$\begin{cases} x(s) = s + \alpha(s^3) \\ y(s) = \frac{s^2 \kappa}{2} + \beta(s^4) \\ z(s) = \frac{s^3}{3!}\kappa\tau + \gamma(s^4) \end{cases}$$

By simple algebraic manipulations we obtain the following connections between the coordinate functions (in the moving coordinate frame):

$$\begin{cases} y = \frac{\kappa}{2}x^2 \text{ (up to order 2 terms in the osculating plane)} \\ z = \frac{\kappa\tau}{6}x^3 \text{ (up to order 3 terms in the rectifying plane)} \\ z^2 = \frac{2}{9}\frac{\tau^2}{\kappa}y^3 \text{ (up to order 3 terms in the normal plane)} \end{cases} \qquad (1.3.5)$$

In other words, the projections of the curve on the osculating, rectifying and normal planes have (up to higher order terms) the forms depicted in Figure 1.13. One can discern, by looking at the shapes, hence properties, of the projections above, that, as already noted above, the projection on the osculating plane approximates the curve

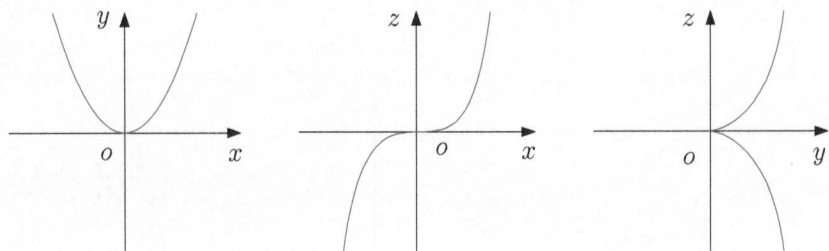

Figure 1.13 The projection of the curve c on the moving frame coordinate planes: Osculating plane (left), rectifying plane (middle) and normal plane (right).

in the best manner, followed. in descending order, by projection on the rectifying and normal planes. (Indeed, the projection onto the normal plane is not even differentiable at 0, as it has a cusp.)

Beyond this simple observation, the local canonical form offers some immediate consequences:

The first such conclusion is nothing less than the geometric interpretation of torsion (Figure 1.14):

1. It follows from the third equation of (1.3.4) that (for small enough s) if $\tau > 0$, then z grows, as s grows; and if $\tau < 0$, then z decreases, as s grows.

A similar fact can be stated about curvature, more precisely:

2. There exists a segment J, $0 \in J \subset I$ such that $\Rightarrow y(s) \geq 0$, with "=" holding iff $s = 0$ – see Figure 1.15. (Indeed, recall that for space curves $\kappa > 0$.)

The third conclusion (which reminds us of a similar fact regrading planar curves) is geometrically significant and we have seen its metric analog already in Section 1.3.5.

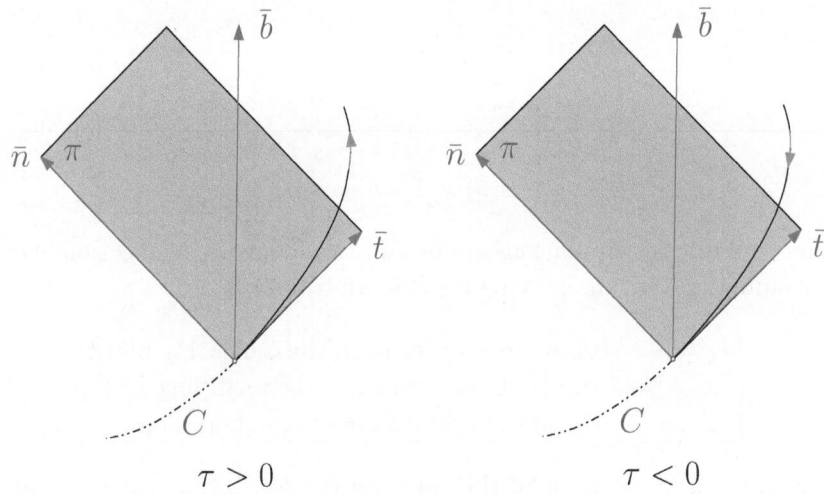

Figure 1.14 The geometric interpretation of the sign of τ.

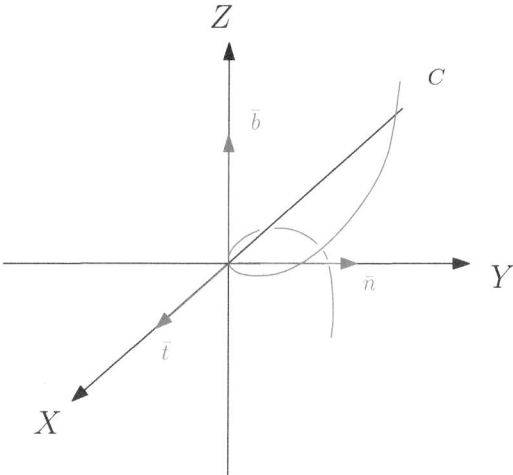

Figure 1.15 The geometric interpretation of the sign of κ.

3. Considers the planes $\pi_h = \langle \mathbf{t}, c(s+h) \rangle$. Then the osculating plane at p, is the limiting position of the planes π_h, that is $\pi(p) = \lim_{h \to 0} \pi_h$ (see Figure 1.16). Since we proved a similar result to the last assertion in the metric setting, and since the proof is also available in cite [105], we leave the proof as an exercise to the reader.

Exercise 1.3.31 *Prove Assertion 3 above.*

Remark 1.3.32 *There are some (unexpected, at times) differences between planar and spatial curves. However, we do not dwell here in this kind of pathologies, not to deviate too much from our road map of our book.*

1.3.4 Existence and Uniqueness Theorem

The Serret-Frènet formulas not only allow us to prove such results on spatial curves as we have seen above, but also encapsulate a deeper significance, as by using them one can show that the curvature and torsion function uniquely determine the curve (up to isometry). More precisely, we have the following:

Theorem 1.3.33 (The Existence and Uniqueness Theorem of Curves[11])
Given differentiable functions $\kappa(s) > 0$ and $\tau(s), s \in I$, there exists a regular parametrized curve $c : I \to \mathbb{R}^3$, such that s represents its arc length, and $\kappa(s), \tau(s)$ are the curvature, resp. torsion of c. Moreover, c is unique up to Euclidean isometry.

Recall that an *Euclidean isometry* (or *rigid motion*) can be written as an orthogonal linear map of positive determinant O and a translation T of vector $\vec{\mathbf{v}}$.

---

[11]a.k.a. The Fundamental Theorem of Curves of the Local Theory of Curves

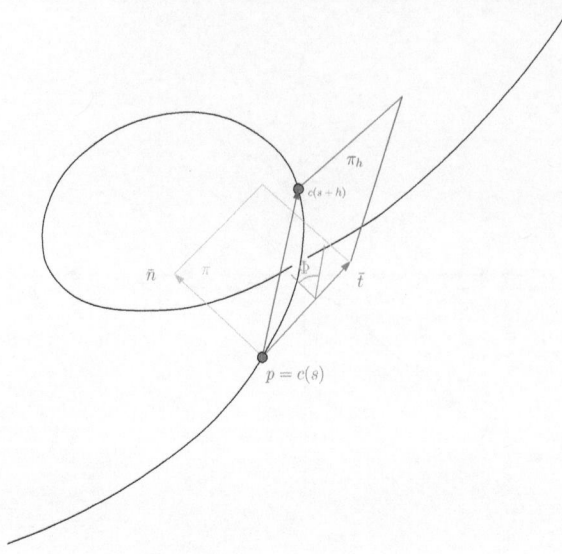

Figure 1.16 The geometric interpretation of torsion.

The full proof of the Existence and Uniqueness Theorem requires making appeal – not surprisingly, given its name – to the Existence and Uniqueness Theorem for Ordinary Differential Equations. Therefore, we do not bring the proof here, since our focus is placed on discretizations, rather than on the classical theory. Furthermore, a full, rigorous proof is available, for instance in the appendix to Chapter 4 of [105]. Furthermore, he also gives, in Section 5 of Chapter 1, an elementary proof of the uniqueness part of the theorem, which we invite the reader to peruse.

We would also like to mention already here that, an analogous theorem for surfaces also exists, which we shall formulate at the proper place in the text. For now, we would only like to mention that its proof, in turn, makes appeal to the theory of Partial Differential Equations. Thus, in a sense, it is a pity we cannot explore more the proofs this technical theorems, since, although less geometrically intuitive, they shed light on the deep connections between Differential Geometry and Differential Equations, which we shall yet encounter later on in the book.

1.3.5 Metric Torsion

Unfortunately, no simple, intuitive approach to the metrization of the notation of torsion (akin to that based on the osculating circle leading to Menger curvature) is available. However, this does not mean that there exists no such geometrically motivated definition. It is based on the following result, akin to that detailed in Remark 1.2.1.[12]

---

[12]and which could have been presented as at the beginning of this section, albeit in less natural manner.

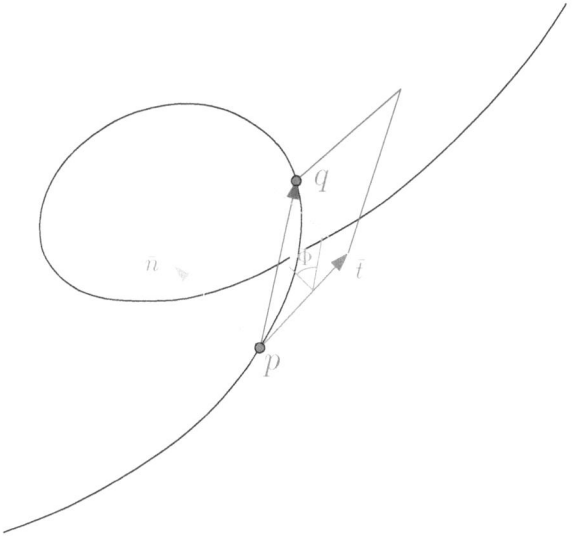

Figure 1.17 Geometric interpretation of torsion's magnitude.

Theorem 1.3.34 *Let ϕ be the angle between the osculating plane to the curve c at the point p, and the plane $\pi(\mathbf{t}, pq)$, where $q \in c, q \neq p$ (see Figure 1.17). Then*

$$\frac{1}{|\tau(p)|} = 3 \lim_{q \to p} \frac{\sin \phi}{pq} \qquad (1.3.6)$$

Sketch of Proof Let us select a coordinate system with the origin at p and axes having versors (i.e. direction vectors of norm 1) $\mathbf{t}, \mathbf{n}, \mathbf{b}$, and denote by $\vec{q}$ the position vector of points on the curve c, relative to this coordinate system, i.e. $\vec{q} = (x(s), y(s), z(s))$. Then the plane $\pi(\mathbf{t}, pq)$ has normal vector $\vec{n} = (0, -y, -z)$, thence

$$\sin \phi = \frac{|z|}{\sqrt{y^2 + z^2}}.$$

On the other hand, given that by the classical limit $\lim |s|/pq = 1$, we obtain that

$$\lim_{q \to p} \frac{\sin \phi}{pq} = \lim_{q \to p} \frac{\sin \phi}{|s|} = \lim_{q \to p} \frac{|z|}{|s|\sqrt{y^2 + z^2}}.$$

Then, by the canonical development into series of y and z and some algebraic manipulations we obtain the desired result.

□

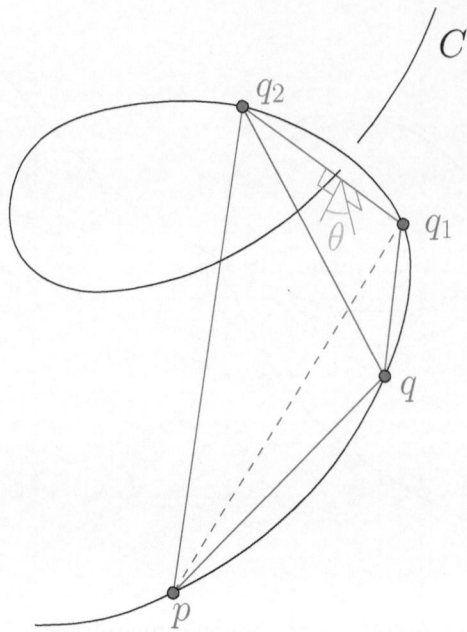

Figure 1.18 The osculating tetrahedron.

Exercise 1.3.35 *Provide the missing details in the proof above.*

Let us also consider points $q_1, q_2 \in c$, as well as the planes $\pi(p, q_1, q_2)$ and $\pi(q, q_1, q_2)$, and let us denote by θ the smallest of the two dihedral angles determined by $\pi(p, q_1, q_2)$ and $\pi(q, q_1, q_2)$. Then

$$\frac{\sin \phi}{pq} = \lim_{q_1, q_2 \to p} \frac{\sin \theta}{pq},$$

as the reader is invited to check. Therefore,

$$\frac{1}{|\tau(p)|} = 3 \lim_{q \to p} \lim_{q_1, q_2 \to q} \frac{\sin \measuredangle(\pi(p, q_1, q_2), \pi(q, q_1, q_2))}{pq}. \tag{1.3.7}$$

(See Figure 1.18)

As the reader surely guessed by now, the metrization of torsion is obtained by expressing the right-side of Formula (1.3.7) in purely metric terms, using its geometric meaning as a dihedral angle in the tetrahedron $T = T(p, q, q_1, q_2)$ – see Figure 1.18. Again, as already seen for Menger's curvature, this is done via the determinant

$$D(p, q, q_1, q_2) = \begin{vmatrix} 0 & 1 & 1 & 1 & 1 \\ 1 & 0 & pq^2 & pq_1^2 & pq_2^2 \\ 1 & pq^2 & 0 & qq_1^2 & qq_2^2 \\ 1 & pq_1^2 & qq_1^2 & 0 & q_1q_2^2 \\ 1 & pq_2^2 & qq_2^2 & q_1q_2^2 & 0 \end{vmatrix}. \tag{1.3.8}$$

By elementary row (or column) operations and by making appeal to the law of cosines, one obtains that

$$D(p, q, q_1, q_2) = 8pq^2 \cdot pq_1^2 \cdot pq_2^2 \cdot \Delta \qquad (1.3.9)$$

where

$$\Delta = \begin{vmatrix} 1 & \cos \angle pq_1 q & \cos \angle pq_1 q_2 \\ \cos \angle pq_1 q & 1 & \cos \angle qq_1 q_2 \\ \cos \angle pq_1 q_2 & \cos \angle qq_1 q_2 & 1 \end{vmatrix} . \qquad (1.3.10)$$

(See Figure 1.19 for the geometric meaning of the entries of the determinant above.)

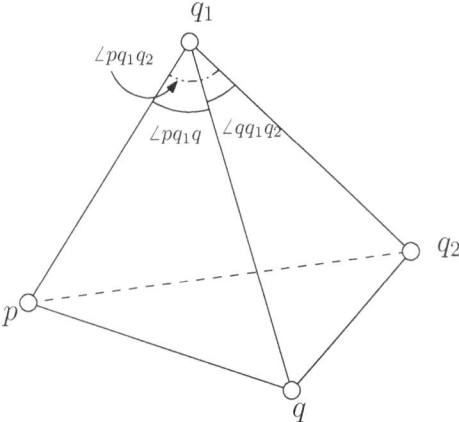

Figure 1.19 The geometric meaning of the entries of Δ.

Exercise 1.3.36 *Supply the full proof of Formula (1.3.10).*

Further manipulations of the determinant above, and application of the spherical law of cosines, renders

$$\Delta = \sin^2 \angle(p, q_1, q_2) \cdot \sin^2 \angle(q, q_1, q_2) \cdot \sin^2 \angle(q_1, q_2; p, q) ; \qquad (1.3.11)$$

where $\sphericalangle(q_1, q_2; p, q)$ represents the dihedral angle between the faces $T(p, q_1, q_2)$ and $T(q, q_1, q_2)$ of the tetrahedron $T(p, q, q_1, q_2)$ – see Figure 1.19.

Exercise 1.3.37 *Complete the proof of Formula (1.3.11).*

Since, by a high-school formula,

$$\mathrm{Area}(T(p, q_1, q_2)) = \frac{1}{2}pq_1 \sin \angle(p, q_1, q_2), \mathrm{Area}(T(q, q_1, q_2)) = \frac{1}{2}qq_2 \sin \angle(q, q_1, q_2);$$

and, moreover (using the notation in (1.2.16)),

$$-16\mathrm{Area}^2(T(p, q_1, q_2)) = D(p, q_1, q_2), -16\mathrm{Area}^2(T(q, q_1, q_2)) = D(q, q_1, q_2);$$

we obtain from Formula (1.3.9) that

$$\sin \sphericalangle(q_1, q_2; p, q) = q_1 q_2 \sqrt{\frac{18D(p, q, q_1, q_2)}{D(p, q_1, q_2)D(q, q_1, q_2)}} . \qquad (1.3.12)$$

In consequence, we attained the desired metric form of torsion of curves in $\mathbb{R}^3$, namely

$$\frac{1}{|\tau(p)|} = 3 \lim_{q \to p} \lim_{q_1, q_2 \to q} \frac{p_2 p_3}{p_1 p_4} \sqrt{\frac{18 D(p_1, p_2, p_3, p_4)}{D(p_1, p_2, p_3) D(p_2, p_3, p_4)}} \,. \qquad (1.3.13)$$

Note that, since we assumed that $\tau(p)$ is defined, the triples considered above are not collinear, it follows that $D(p_1, p_2, p_3), D(p_2, p_3, p_4) < 0$ and $D(p_1, p_2, p_3, p_4) \geq 0$, therefore the right-hand side of the formula above is well defined.

Exercise 1.3.38 *Prove the inequalities in the observation above.*

We can now proceed and present Blumenthal's metrization of torsion, devised, cf. [212], in 1939:

Let $p_1, p, p_3, p_4 \subset \mathbb{R}^3$ points in general position. We define *Blumenthal metric torsion* of the quadruple p_1, p, p_3, p_4 by

$$\tau_B^2(p_1, p, p_3, p_4) = \frac{18 D(p_1, p_2, p_3, p_4)}{D(p_1, p_2, p_3) D(p_1, p_3, p_4) D(p_1, p_2, p_4) D(p_2, p_3, p_4)} \,. \qquad (1.3.14)$$

To define the Blumenthal metric curvature at an accumulation point p of a metric space, we pass, as usual for is by now, to the limit:

$$\tau_B(p) = \lim_{p_1, p_2, p_3 \to p} \frac{\sqrt{18 |D(p, p_1, p_2, p_3)|}}{\sqrt[4]{D(p_1, p_2, p_3) D(p_1, p_3, p_4) D(p_1, p_2, p_4) D(p_2, p_3, p_4)}} \,. \qquad (1.3.15)$$

It turns out that this is a "correct" metrization of torsion and, more precisely we have the following:

Theorem 1.3.39 *Let $c \subset \mathbb{R}^3$ be a regular smooth curve and let $p \in c$. Then $\tau_B(p)$ exists and $\tau_B(p) = |1/\tau(p)|$.*

Exercise 1.3.40 *Prove Theorem 1.3.39 above by showing that*

$$\tau_B(p_1, p_2; p_3, p_4) = 3 \lim_{p_2, p_3, p_4 \to p_1} \frac{\sin \angle(p_1, p_2; p_3, p_4)}{p_3 p_4} \,. \qquad (1.3.16)$$

(Hint: Use the geometric interpretation of the determinants $D(p_1, p_2, p_3)$, etc. and their connection with Menger curvature (see Formula (1.2.17) above)

In the very same spirit as from passing from the Alt curvature to the Menger curvature, one is conduced to the following *strong form of Blumenthal's metric torsion*:

$$\tau_B^*(p) = \lim_{p_i \to p} \frac{\sqrt{18 |D(p_1, p_2, p_3, p_4)|}}{\sqrt[4]{D(p_1, p_2, p_3) D(p_1, p_3, p_4) D(p_1, p_2, p_4) D(p_2, p_3, p_4)}} \,; i = 1, 2, 3, 4 \,. \qquad (1.3.17)$$

The relationship between the two types of Blumenthal metric torsion are summarized in the exercise below:

Exercise 1.3.41 *Prove that*

1. *The existence of $\tau_B(p)$ does not imply the existence of $\tau_B^*(p)$.*

2. *$\tau_B^*(p)$ may be defined, even though $\tau_B(p)$ is not.*

One can show (following J. W. Sawyer, cf. [212]) that $\tau_B^*(p)$ exists at any point of a regular smooth curve in $\mathbb{R}^3$, and therefore it equals, by Theorem 1.3.39 it equals $1/|\tau(p)|$. Moreover, it can be shown that if $p \in c$, where c is a continuum in $\mathbb{R}^3$, such that $\kappa_M(p) \neq 0$ and $\tau_B^*(p)$ exists, then

$$p_1, p_2; p_3, p_4 = 3 \lim_{p_i \to p} \frac{\sin \angle(p_1, p_2; p_3, p_4)}{p_3 p_4}, i = 1, 2, 3, 4; \qquad (1.3.18)$$

where the points p_1, p_2, p_3, p_4 are in general position.

Using formula 1.3.18 above one can prove the following result:

Theorem 1.3.42 *If p is a point on metric continuum such that $\tau_B^*(p)$ is defined, and such that $\kappa_M(p) \neq 0$, then there exists a plane $\pi(p)$ such that $\pi(p_1, p_2, p_3) \to \pi(p)$, that is c has an osculating plane at p, in the generalized sense (which coincides with the classical one for smooth curves in Euclidean space).*

Even though the proof is not long, we do not bring it here , so not to transform this book into a technical volume on Metric Geometry. However, we leave it as an exercise to the reader.

Exercise 1.3.43 *Prove the theorem above.*

Furthermore, metric torsion and curvature also satisfy the following property that shows that they do represent, indeed, proper generalizations of the classical notions, as hoped for:

Theorem 1.3.44 *Let $c \subset \mathbb{R}^3$ a continuum such that, for any $p \in C$, the following conditions hold: (a) $\kappa_M(p) > 0$; and (b) $\tau_B^*(p) > 0$.*
Then c is contained in a plane.

(For a proof the interested reader is referred to [212].)
In fact, if a metric curve for which κ_M, τ_B^* are defined and positive at each point p satisfies all the essential properties of a regular curve, such as rectifiability, existence of tangents, etc. (see [212]).

Remark 1.3.45 *Blumenthal's metric definitions of torsion were not the first one. In fact, just a year before Blumenthal, G. Alexits [126] proposed the following definition of metric torsion, which is derived much more directly from Formula (1.3.13):*

Definition 1.3.46 *Let (X, d) be a metric space and let p_1, p_2, p_3, p_4 an ordered (metric) quadruple, such that $D(p_1, p_2, p_3), D(p_2, p_3, p_4) \neq 0$. The Alexits metric torsion of the ordered quadruple is defined as*

$$\tau_A(p_1, p_2, p_3, p_4) = \frac{p_2 p_3}{p_1 p_4} \sqrt{\frac{18|D(p_1, p_2, p_3, p_4)|}{D(p_1, p_2, p_3) D(p_2, p_3, p_4)}}. \qquad (1.3.19)$$

If $p \in X$ is an accumulation point, then the torsion $\tau_A(p)$ of X is defined as

$$\tau_A(p) = \lim_{q \to p} \lim_{q_1, q_2 \to q} \tau_A(p, q, q_1, q_2) ; \qquad (1.3.20)$$

with the necessary provisos that (a) The double limit exists; and (b) that the right limit does not depend on the rder of the points q, q_1, q_2.

The definition above, even it is directly derived from the metric form of torsion in Euclidean space, and even thought it has the same geometric motivation as the previous definition of torsion in metric spaces, Alexits' one suffers from two major disadvantages, that cause it being supplanted by Blumenthal's one: It necessitates an iterated limit and, furthermore, it depends on the ordering of the quadruple.

However, by the proof of Theorem 1.3.39 above, the two definitions coincide for regular points on smooth regular curves in $\mathbb{R}^3$, and both of them equal $1/\tau$.

At this point the reader is certainly asking herself or himself whether there are no applications of metric torsion (after all, such applications were presented in Section 1.2.3 and 1.2.3 above, and more will be presented in the following chapters). While applications to Graphics, for instance, can be envisioned, clearly the ones with the highest potential are those regarding networks, especially for the study of the spatial structure of molecules, proteins and genes, i.e. in Biomedicine, Genetics, etc. Unfortunately, we have no published results so far to illustrate these ideas.

1.3.5.1 The Metric Existence and Uniqueness Theorem of Curves

Given that the metric curvatures and torsions were developed to mimic the classical notion and, moreover, since they coincide with the classical notions in the case of smooth curves in Euclidean plane and space, it is only natural to ask oneself whether an existence and uniqueness theorem might be formulated in terms of metric curvature and torsion. This is not just a question raised by simple curiosity; it is, in fact, much deeper, since it probes how well the metric notions in question generalize the classical ones. In other words, whether they are separate discretizations or if they do, indeed, integrate into a metric theory of spatial curves. The answer is again, positive and, again, it is due to Gadum [173]. More precisely, we have the following metric generalization of the Fundamental Theorem (of the local theory of curves):

Theorem 1.3.47 *Let $c, \gamma \subset \mathbb{R}^3$ be two smooth metric arcs such that (a) κ_M and τ_B^* are defined and positive at all the points of the two metric arcs.*

Then the c, γ are isometric iff there exists a bijective application $\Phi : c \to \gamma$ that preserves (1) arc length; (2) the metric curvature κ_M and (3) the metric torsion τ_B^*.

The proof is, unfortunately, rather technical thus the interested reader should see the original paper of Gadum as well as Blumenthal's [212].

We conclude this section with an exercise illustrating again the parallels between classical and metric curve theory.

Exercise 1.3.48 *Let* $c, \gamma \subset \mathbb{R}^3$ *be two smooth metric arcs such that (a)* κ_M *and* τ_B^* *are defined and positive at all the points of the two metric arcs., and let* $p \in c$.

Then $\kappa_M(p)$ *is equal to the curvature of* p *of the normal projection of* c *on the osculating plane of* c *at* p.[13]

1.4   HIGHER DIMENSIONAL CURVES

The differential geometry of curves we presented here extends, rather automatically, to curves in $\mathbb{R}^n$, for $n > 3$. However, it is not only less intuitive, it also is of far less use in most practical applications. Furthermore, the theory of metric curvatures that we have introduced above transcends, in generality, this simple dimensionality extension. As such, we do not dwell on it in any detail, but only bring here the most basic facts, for the sake of completeness. (For more details see, e.g. [329].)

By mimicking the construction of a Frènet frame for curves in $\mathbb{R}^3$, one constructs a (distinguished) Frènet frame in the n dimensional case as follows: First define

$$e_1(t) = \frac{\dot{c}(t)}{||\dot{c}(t)||} \, ; \tag{1.4.1}$$

and

$$e_j(t) = \frac{\tilde{e}_j(t)}{||\tilde{e}_j(t)||} \, 2 \le j \le n \, ; \tag{1.4.2}$$

where $\tilde{e}_j(t)$ is defined inductively as follows:

$$\tilde{e}_j(t) = -\sum_{k=1}^{j-1} \left(c^{(j)}(t) \cdot e_k(t) \right) e_k(t) + c^{(j)}(t) \, . \tag{1.4.3}$$

A moving frame $\{e_i(t)\}_1^n$ associated to a curve $c(t)$ in $\mathbb{R}^n$ satisfies the following equations:

$$\dot{c}(t) = \sum_i \alpha_i(t) e_i(t)$$

$$\dot{e}_i(t) = \sum_j \omega_{ij}(t) e_j(t)$$

where

$$(\star) \quad \omega_{ij}(t) = \dot{e}_i(t) \cdot e_j(t) = -\omega_{ij}(t) \, .$$

If, furthermore, $\{e_i(t)\}_1^n$ is the distinguished Frènet frame we just defined, it also satisfies the fitting Serret-Frènet equations, namely

$$(\star\star_1) \quad \alpha_1 = ||\dot{c}(t)|| \, , \; \alpha_1 = 0 \, , \; i > 1 \, ;$$

and

$$(\star\star_2) \quad \omega_{ij}(t) = 0 \, , \; j > i+1 \, .$$

It is important to notice how the differential equations above behave under changes of variables. To this effect, we have

---

[13]Recall that the existence of the osculating plane is guaranteed by 1.3.42.

Proposition 1.4.1 *The coefficients ω_{ij} are invariant under isometries. Furthermore, if the isometry is orientation preserving, so are $\frac{\omega_{ij}(t)}{||\dot{c}(t)||}$.*

The importance of the last fact reveals itself in light of our next definition:

Definition 1.4.2 *Let $c :\to \mathbb{R}^n$ be a curve admitting a distinguished Frènet frame. Then*

$$\kappa_i(t) = \frac{\omega_{ij}(t)}{||\dot{c}(t)||} \,; i = 1, 2, \ldots, n-1. \tag{1.4.4}$$

is called the i-th curvature of c.

We have the following:

Proposition 1.4.3 $\kappa_i(t) > 0$ *for $1 \leq i \leq n-2$.*

In particular, if $n = 3$, there are only two curvature functions: $\kappa_1 \equiv \kappa$ – the (classical) curvature, which, as we know, is always positive; and $\kappa_2 \equiv \tau$ – the (classical) curvature, which, of course, can be both positive and negative. (If $n = 2$, then there is only one curvature function, $\kappa_1 \equiv \kappa$, which is always positive). Thus, out of the $n - 1$ curvature functions, the first $n - 2$ are positive, and can be thus viewed as proper curvatures, while the $(n-1)$-curvature function, which can take both positive and negative values, corresponds to the classical torsion of curvatures in $\mathbb{R}^3$.

Exercise 1.4.4 *We have proceeded in the declared spirit of this book, from simple to complicate and from intuition to formulae. However, it is quite possible to go from general to the particular (and this is indeed done in many texts, e.g. in the classical [329]). – Recover the classical theory of curves in plane and space from the formulae introduced in this section.*

Exercise 1.4.5 *Consider the curve $c : \mathbb{R} \to \mathbb{R}^4$, $c(t) = (\cos t, \sin t, t, t)$.*

1. Write the distinguished Frènet frame of c.

2. Compute the curvatures of c.

Remark 1.4.6 *Interesting enough, even though the definitions extend, as we have seen, immediately, not all classical types of curves have their immediate higher dimensional analogs. In particular, there are no Bertrand curves in $\mathbb{R}^3$, $n \geq 4$ (see [219]).*

CHAPTER 2

Surfaces: Gauss Curvature – First Definition

We now concentrate on the differential geometry of surfaces, which not only represents the central part of any introductory course in Differential Geometry, it also provides most of the essential theoretical tools essential in Graphics, CAGD and Imaging. Thus, this and the following chapters represents the very core of this book.[1]

2.1 SURFACES

The title of this section is rather vague. Partly, this is due to our general motivation as presented in the brief introduction to this chapter. However, we deliberately chose not to be more specific in the title, in order to try and explain why we only consider a quite specific class of surfaces. Since this is a (basic) course in Differential Geometry, one needs to be able to operate with differential operators, in a manner that resembles – and stems from – Calculus. Thus our basic modus operandi, at least in the basic, incipient stages is largely that encountered in Calculus 2.[2] That means that we shall, at least in the beginning, concentrate only on the *local theory* of surfaces, the locality being imposed, as the reader might have guest, by the need to be able to consider differentials and derived notions. This has as a consequence the fact that the "our" surfaces are not (for the most part) the ones one might expect, i.e. not the ones encountered in a introductory Algebraic Topology course. This might be a disappointment to some of the readers. However, our experience sadly testifies to the fact the opposite is the common case: Many students in a find it hard to transcend the bounds of local, technical apparatus of Calculus and view surfaces as from the *global* , topological viewpoint. Thus, beyond the need for locality, we need to deal with surfaces in a manner that allows for derivatives computation, i.e. that are smooth enough or, in technical term specific to the field *regular*. Furthermore, while a basic topological definition might have sufficed in other settings, for practical computations we will need our surfaces to be precisely like those in Calculus 2,

---

[1]and of any course on the subject, as generations of students know only to well.

[2]Indeed, many of the notions relevant to us, including some of those discussed at length in our previous chapter, can be encountered in some Calculus 2 books.

DOI: 10.1201/9781003350576-2

namely parametrized surfaces. Furthermore, yet again like in a Calculus course, we essentially concentrate on surfaces in Euclidean Space $\mathbb{R}$. Having explained our goals and motivation, we can proceed to the technical part, which we shall try and keep to a bare minimum so that we can concentrate on the geometric ideas as soon as possible. (For further details the reader can consult, e.g. [105], [168] or [240].)

Definition 2.1.1 *A set $S \subset \mathbb{R}^3$ is called a regular surface if, for any $p \in S$, there exists a neighborhood V of p in $\mathbb{R}^3$, and an open set $U \in \mathbb{R}^2$ together with a surjection $f : U \to V \cap S$, such that*

1. *$f \in \mathcal{C}^3$;*

2. *f is a homeomorphism;*

3. *f is regular at any point $q \in U$, i.e. $df_q : \mathbb{R}^2 \to \mathbb{R}^3$ is injective.*

Then f is called a parameterization or f^{-1} is called a local coordinate system[3] (which is also continuous), and $V \cap S$ is called a coordinate neighborhood.

Remark 2.1.2 *The main subject of Differential Geometry is, as we already know, curvature, which can be seen as generalization or geometrization of the second derivative; thus, it is not surprising that class $\mathcal{C}^2$ suffices for most results in the sequel. However, it would be false to presume this is always the case, and we shall point out in due time where the higher differentiability class is absolutely need.*

Remark 2.1.3 *Condition (3) is equivalent[4] to the condition that the Jacobian matrix $J_f(q)$ has rank 2,) for any $q \in U$. This can be expressed in a more geometric manner as $\partial f / \partial u \times \partial f / \partial u \neq \bar{0}$. In other words, S admits a tangent plane at any point; thus, it does nowhere reduce to a point or a curve (so it can't be expressed solely in terms of u or v).*

Remark 2.1.4 *Condition (1) ensures that S has no self-intersections, while Condition (2) prevents q to be the vertex of cone (see also our previous remark).*

Example 2.1.5 *The simplest examples are*

1. *A plane;*

2. *An open disk;*

   The first nontrivial example is given by

3. *The sphere $\mathbb{S}^2$.*

   By the compactness of the sphere, it follows that it is not possible to use only one coordinate neighborhood, thus that at least to such open sets are needed to parametrize $\mathbb{S}^2$.

---

[3]Sometimes the later name is attached, especially in Differential Topology, to f^{-1}
[4]as we again have learned in a Calculus 2 course.

(a) *The simplest way is to consider the stereographic projection from the North Pole $N = N(0,0,1)$ and from the South Pole $S = S(0,0,1-)$. Recall that (standard) stereographic projection from the North Pole, $\Pi_N : \mathbb{S}^2 \to \mathbb{R}^2$, is given by*

$$(x,y,z) \in \mathbb{S}^2 \longmapsto (u,v) \in \mathbb{R}^2 \,;\; u = \frac{x}{1-z}, y = \frac{y}{1-z}\,.$$

Exercise 2.1.6 *(i) Prove the formula above.*
(ii) Find the corresponding formula for Π_S.

Exercise 2.1.7 *There is another version of the stereographic projection in which N is taken to be $(0,0,2)$ and $S = (0,0,0)$. Determine the formulas of Π_N and Π_S for this variant of the stereographic projection.*

(b) *Another method of finding a parametrization of the sphere, again with two coordinate neighborhoods, is the one encountered in Calculus 2, of geographical (and astronomical) inspiration, namely taking as one parametrization the mapping $\phi : (0,\pi) \times (0,2\pi) \to \mathbb{S}^2$, given by*

$$\phi(u,v) = (\sin v \cos v, \sin u \sin v, \cos u)\,.$$

(Question to the reader: What is the second natural parametrization?)

(c) *The easiest to use in practice way to parameterize the sphere, in the sense that it is natural from its expression as $x^2 + y^2 + z^2$ uses six – not just two – coordinate neighborhoods (thus emphasizing the fact that their number is not canonical), by using the idea of the so called "Magdeburg Spheres" from Physics. For example, $U_1 = x^2 + y^2 < 1$ and $f_1(x,y) = (x, y, \sqrt{1-x^2-y^2})\,.$*

Exercise 2.1.8 *Write the other parametrizations $f_i : U_i \to \mathbb{S}^2$, $i = 2, \ldots, 6$.*

Clearly, the definition of a surface we brought above is not always an easy one to operate with in general. Therefore, one might hope to be able to view surfaces as graphs of functions (again, as one does in Calculus). This is, indeed, the case, and its proof is left as an exercise.

Exercise 2.1.9 *Let $f : U \to U = \text{int } U \subset \mathbb{R}^2$ be a C^3 function. Then its graph is a surface.*

While this nomenclature is less commonly used, such surfaces are called *simple surfaces* and f is called a *Monge parametrization*. Note that any surface in $\mathbb{R}^3$ can be obtained by "gluing" (or "stitching") simple surfaces.

Another canonical and useful manner to obtain surfaces (again taught in Calculus courses) is to express them in terms of *regular values* of functions, where these are defined as follows:

Definition 2.1.10 *Let $F : U = \operatorname{int} U \subset \mathbb{R}^3 \to \mathbb{R}^3$ be a diffeomorphism. Then $p \in U$ is called a critical point (of or for F) if $d_p F : \mathbb{R}^3 \to \mathbb{R}^3$ is not surjective. In this case, $F(p)$ is called a critical value]. Otherwise, p is called a regular value.*

Remark 2.1.11 *Clearly the definition can be extended for mappings from $\mathbb{R}^m$ to $\mathbb{R}^n$, for any m, n. To understand the inspiration for this theorem, the reader is invited to solve the following simple*

Exercise 2.1.12 *What are the critical points and values and the regular points in the case $m = n = 1$?*

Remark 2.1.13 *It turns there are "very few" critical values; thus "most" values are regular values. (This is a intuitive formulation of the classical Sard Theorem – or Morse-Sard Theorem; see, e.g., [225].)*

This last remark is extremely important in view of the following:

Theorem 2.1.14 *Let $f : U = \operatorname{int} U \subset \mathbb{R}^3 \to \mathbb{R}$ be a $\mathcal{C}^3$ function, and let $a \in f(U)$ be a regular value. Then $f^{-1}(a)$ is a regular surface (in $\mathbb{R}^3$).*

The proposition above is extremely useful in practice, since it allows to easily produce regular surfaces, as the following examples demonstrate:

Example 2.1.15   1. *The ellipsoid $\frac{x^2}{a^2} + \frac{y^2}{b^2} + \frac{z^2}{c^2} = 1$ can be viewed as $f^{-1}(0)$, for $f(x, y, z) = \frac{x^2}{a^2} + \frac{y^2}{b^2} + \frac{z^2}{c^2} - 1$.*

   2. *The hyperboloid of two sheets $\frac{x^2}{a^2} - \frac{y^2}{b^2} - \frac{z^2}{c^2} = 1$ can be obtained as $f^{-1}(0)$, for $f(x, y, z) = \frac{x^2}{a^2} - \frac{y^2}{b^2} - \frac{z^2}{c^2} - 1$.*

> **Remark 2.1.16** *This last example demonstrates that a surface is not necessarily connected. However, connected surfaces can be characterized by the property in the elementary exercise below:*
>
> **Exercise 2.1.17** *Prove that $f : S \subset \mathbb{R}^3 \to \mathbb{R}$ is a continuous function such that $f(p)f(q) > 0$, for any $p, q \in S$, where S is a connected surface, then $\operatorname{sign} f$ is constant.*
>
> **Exercise 2.1.18** *Check that 0 is a regular value in both the examples above.*

   3. *Let $\mathbb{R}^3 \to \mathbb{R}$, $f(x, y, z) = z^2 + (\sqrt{x^2 + y^2} - a)^2$. Then r^2 is a regular value for any $r < a$, thence the torus of revolution $z^2 = z^2 - (\sqrt{x^2 + y^2} - a)^2$ is a regular surface.*

> **Exercise 2.1.19** *Check the assertion above.*

There also exists a local converse of Theorem 2.1.14:

Proposition 2.1.20 *Let $S \subset \mathbb{R}^3$ be a regular surface and let $p \in S$. Then there exists a neighborhood $U \subset S$ of p, such that U is the graph of a function of one of the following types:*

$$z = f(x, y) \,, \; y = g(x, z) \,, \; x = h(y, z) \,.$$

(As the reader might have noticed, this result is one encountered usually in Calculus 2.)

Exercise 2.1.21 *Prove that the (one-sheeted) cone $z = \sqrt{x^2 + y^2}$, $(x, y) \in \mathbb{R}^2$, is not a surface.*

The proposition above can be used to reduce the number the conditions in our definition of surfaces. More precisely, the reader is invited to solve the following

Exercise 2.1.22   *1. Show that Condition (2) in Definition 2.1.1 is redundant.*

2. Using (1) above prove that the torus of revolution $f(u, v) = ((R + r\cos u)\cos v,$ $(R + r\cos u)\sin v, r\sin u)$ is a surface.

To conclude this primer on the notion of surface, as employed in Differential Geometry, we bring the following one, which expresses surfaces in terms of mappings into, rather then subset of $\mathbb{R}^3$:

Definition 2.1.23 *A parametrized surface $f : U = \operatorname{int} U \subset \mathbb{R}^2 \to \mathbb{R}^3$ is a $\mathcal{C}^3$ mapping. f is called regular if its differential $d_p f : \mathbb{R}^2 \to \mathbb{R}^3$ is injective for all $p \in U$. A point $q \in U$, such that $d_q f$ is not regular is called a singular point (of/for f).*

It is precisely this definition, which generalizes our definition of curves and, in turn, will be generalized itself later to yield the notion of manifold.

Exercise 2.1.24 *Let $c : I \to \mathbb{R}^3$ be a non-planar curve, such that $\kappa_c(t) \neq 0$, for all $t \in I$. Show that*

$$f(t, v) = c(t) + v\dot{c}(t) \,, t \in I, v \in \mathbb{R} \,, \tag{2.1.1}$$

is a parametrized surface that is regular for $v \neq 0$ (called the tangent surface of c).

Remark 2.1.25 *The example in the exercise above shows that a regular parametrized surface can self-intersect.*

We wind up this succinct overview of the basic facts regarding surfaces[5] with the following result which shows that we can actually consider, at least locally, regular parametrized surfaces instead of surfaces as defined before:

Proposition 2.1.26 *Let $f : U = \operatorname{int} U \subset \mathbb{R}^2 \to \mathbb{R}^3$ be a regular parametrized surface, and let $p \in U$. Then there exists a neighborhood $V \in U$ of p, such that $f(V) \subset \mathbb{R}^3$ is a surface.*

---

[5]We restricted here only to the bare minimum. Other ideas and definitions, such as the *tangent plane* will be introduced when necessary. (The reader is also asked to recall that this specific notion was also introduced in Calculus 2.)

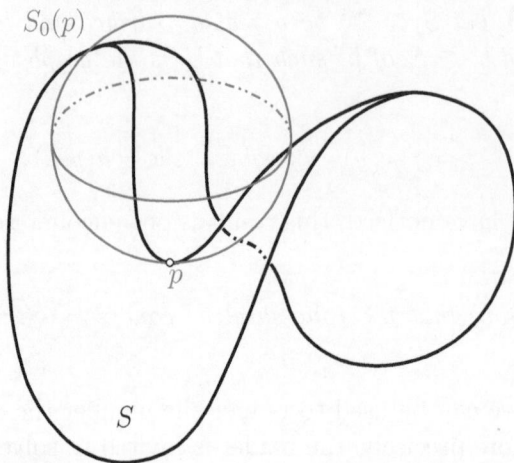

Figure 2.1 The osculating sphere and saddle points on surfaces.

2.2 GAUSS CURVATURE – FIRST DEFINITION

When passing from planar curves to surfaces in Euclidean space, that is to say to their most natural generalization, it is very tempting to try and follow the same route, namely to define curvature via the *osculating sphere* (at each point of the surface). However, while this approach would certainly still be valid for convex surfaces, one cannot avoid but noticing that it is not applicable when the considered surfaces has *saddle points* (see Figure 2.1), as these points represent (at least in a sense) the analog of inflexion points for curves, for which we have seen that then osculating circle (hence curvature) is not even defined (definable?).

Exercise 2.2.1 *Properly define the osculating sphere at a point on a surface, determine its center and radius and write its equation.*

Remark 2.2.2 *Precisely like in the curve's case, one might be naturally conduced to the idea that it might be possible to define surface curvature by approximating them with paraboloids, rather then with surfaces. This turns, indeed, to be a successful strategy, for surfaces as well, not just for planar curves. We do not elaborate here along this line, even though we shall briefly return to it in the sequel, and refer the reader to [159], [310], pp. 42-43. Nevertheless, we also offer it is a challenging problem to the motivated reader.*

Problem 2.2.3 *Define the osculating paraboloid and explore its properties.*

One is thus conducted to the next most natural approach, namely to make appeal to a well-known (by now) concept, namely that of curvature of (planar) curves. To do this, one "cuts" the surface with *normal planes* Π (see Figure 2.2), each such "cut" producing a curve $C = S^2 \cap \Pi$, having curvature $k_C = k_C(p)$ at the point p. (Note that $\Pi = < N_p, t_c >$, where N_p denotes the normal vector to S at p and t_C represent

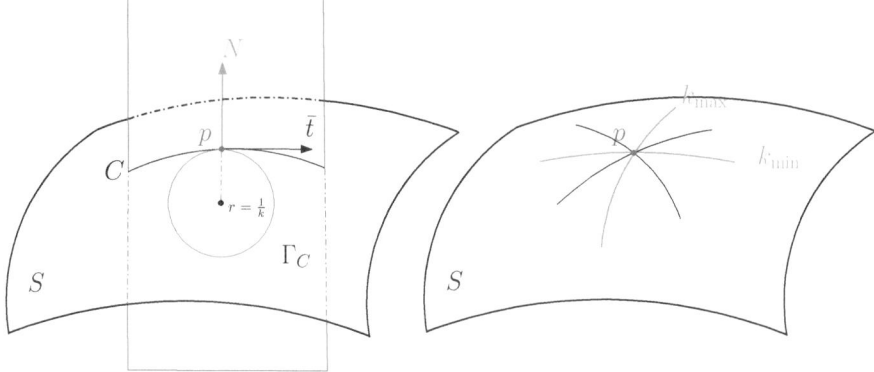

Figure 2.2 Principal curvatures on a smooth surface.

the tangent vector to C at p; hence the importance of "cutting" the surface only with normal planes.) Having adopted this approach we can now present[6] "what they knew about surfaces before Gauss".

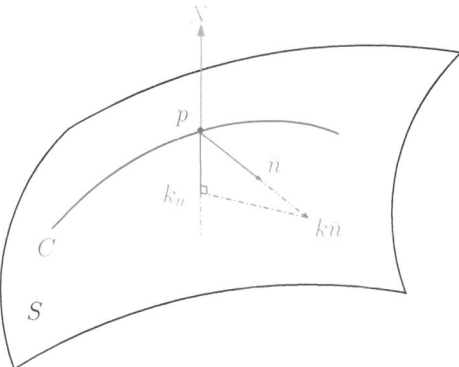

Figure 2.3 Normal curvature.

The first such result (at least from a logical, if not historical) viewpoint relates shows how to compute the curvature of a curve obtained by sectioning the surface by a non-normal plane; thus showing that the choice of *normal sections* is, indeed, not truly restrictive. But first, let us introduce a new definition that will allow us to formulate the theorem in question in a simple form.

Definition 2.2.4 *Let C be a regular curve passing through a point p on a surface S, and let $k = k_C(p)$ denote its curvature at p. Denote by θ the angle between the normal N to the surface at P, i.e. the angle given by $\cos(\theta) = n \cdot N$, where n represents the normal vector to C at p. The length of projection of the vector kn on N, i.e. $k\cos(\theta)$ is denoted by k_n and is called the normal curvature (of C at the point p) (See Figure 2.3.).*

---

[6]using Spivak's formulation [309].

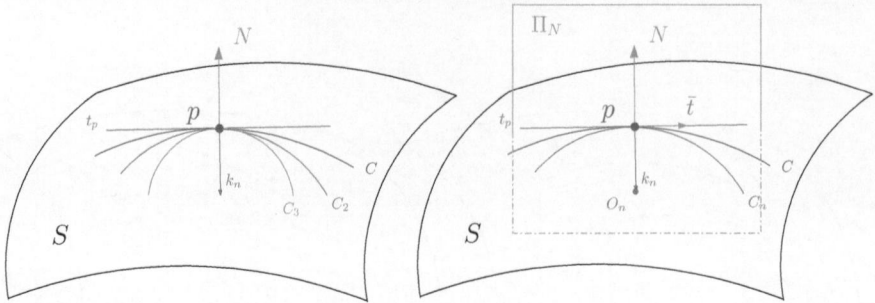

Figure 2.4 Meusnier's Theorem: All curves on the surface S, having the same tangent vector at p, have the same normal curvature $k_n = k_n(p)$ (left); The normal curvature k_n is the curvature of the curve obtained by sectioning the surface S with the normal plane through p (right).

As the notation clearly suggests, the definition of normal curvature is independent of the choice of the curve C, and we can state this fact as

Theorem 2.2.5 (Meusnier, 1976) *All curves on a surface S passing through a given point $p \in S$, and having the same tangent vector at p, have the same normal curvature $k_n = k_n(p)$.*

In other words, if Π_φ is a plane intersecting the surface S at the point p, such that $\angle(\Pi_\varphi, \Pi) = \varphi$, then the curvature (at p) k_φ of the curve $\gamma = S \cap \Pi_\varphi$ is related to the normal curvature via the following simple relation:

$$k_n = k_\varphi \cdot \cos\varphi .$$

The geometric interpretation of Meusnier's Theorem is that the osculating circles of the plane sections (See Figure 2.4.) though the same tangent to the surface are contained in a sphere – see Figure 2.4 (and, for some different aspects, [163]).

Another geometric insight into Meusnier's Theorem is given in the following exercise:

Exercise 2.2.6 *Let c be a curve on a surface S. Then its center of curvature at a point p is the normal projection onto the osculating plane at the surface at S of the center of curvature of c_n – the normal section at p.*

It follows that one can talk about the normal curvature in (or along) a given direction at a point on a surface, and study only the curvature of normal sections, a fact that will prove important in the sequel.

Note also that, from its definition, it follows immediately that the normal curvature at a point p of a curve c on a surface S, does not depend on the orientation of C (since the curvature of planar curves does not), but it does depend on the orientation of S (given that it is defined by means of the normal at p).

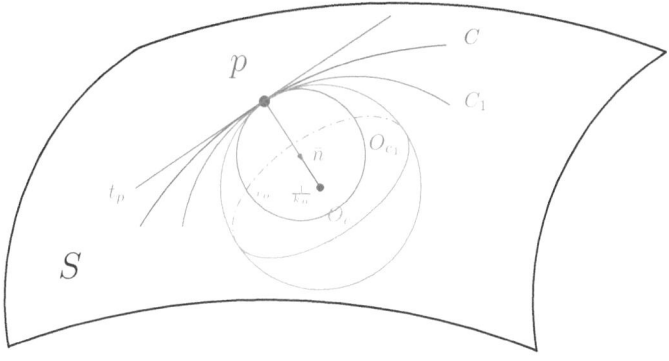

Figure 2.5 The geometric interpretation of Meusnier's Theorem.

Of special import are the extremal values of k_n. It turns out that not only are the minimal and maximal values $k_{\min}(p), k_{\max}(p)$ of k_n attained, they occur, in fact, in orthogonal directions – let they be given by the orthonormal basis $\{e_1, e_2\}$ of $T_p(S)$. This warrants a formal definition:

Definition 2.2.7 *The minimal and maximal values $k_{\min}(p), k_{\max}(p)$ of normal curvature are called the principal curvatures (at p). The corresponding directions given by e_1, e_2 are called the principal directions (at p).*

Remark 2.2.8 *Traditionally the principal directions are denoted as $k_1 = k_{\min}(p)$, $k_2 = k_{\max}(p)$; thus the associated principal directions are numbered fittingly.*

Example 2.2.9 *A number of important, if immediate, examples must be listed:*

1. *Obviously, for any point in the plane $k_{\min}(p) = k_{\max}(p) = 0$.*

2. *The principal curvatures are clearly equal also at any point of the sphere and equal to $1/R$, where R is the radius of the sphere (since they are the reciprocal of the osculating radius of a great circle).*

3. *On a cylinder, the principal curvatures are $1/R$, where R is the radius of the base circle and 0 – the curvature of the generatrix. Thus all other sections (which are ellipses) have curvatures between 0 and $1/R$. (The reader is encouraged to provide for him/herself the accompanying drawing.)*

4. *The principal curvatures of the saddle surface (hyperbolic paraboloid) $z = x^2 - y^2$ at $O = (0, 0)$ are opposite to each other. (We shall compute them later, as an exercise. Again, reader is encouraged to provide for him/herself the accompanying drawing.)*

Having obtained the possible information regarding the curvature of one curve, passing through a given point on a surface, one can study the curvatures of all the members of the family of curves on S passing through p.[7] It turns out that one can even compute the normal curvature (at any point) of any curve on a surface, given that the principal curvatures (at the given point) are known. More precisely, we have the following, quite classical theorem:

Theorem 2.2.10 (Euler, 1760) *Let $p \in S$ be such that $k_{\min}(p) \neq k_{\max}(p)$, and let $\mathbf{v} \in \vec{\mathbf{T}}_{\mathbf{p}}(\mathbf{S})$ such that $\angle(\vec{\mathbf{v}}, \mathbf{e_1}) = \theta$. Then*

$$k_n = k_{\min} \cos^2 \theta + k_{\max} \sin^2 \theta. \qquad (2.2.1)$$

A proof, in the spirit of Euler's original one, can be found in [309]. We shall present one, more modern in nature, later on, once we introduce the contemporary mathematical tool of the two fundamental forms (see next section).

Remark 2.2.11 *1. One can consider the angle φ with e_2, rather than with e_1, given that $\varphi = \pi/2 - \theta$, hence $\sin \varphi = \cos \theta, \sin \theta = \cos \varphi$.*

2. One can choose the vector $-\vec{\mathbf{v}}$, which will accordingly affect the signs of the principal curvatures. (Check!)

The following natural observation we prefer to leave as an exercise to the reader:

Exercise 2.2.12 *What can be said if $k_{\min}(p) = k_{\max}(p)$?*

The principal curvatures, in conjunction, define the curvature at a point of the surface. More precisely, we have

Definition 2.2.13 *The product $K = k_{\min} k_{\max}$ is called the Gauss curvature (or Gaussian curvature) of the surface (at a point p).*

While this definition is quite natural (and it seemed so in Gauss' own time [130]), it is not less (or even perhaps even more so) logical to consider the arithmetic, rather than (the square of) the geometric mean.

Definition 2.2.14 *$H = (k_{\min} + k_{\max})/2$ is called the mean curvature of the surface (at a point p).*

Clearly the second definition is simpler to handle, thus one would expect both notions to be at least equally important. However, we shall see that Gaussian curvature is more "powerful", so to say. This is not to say that mean curvature is not an extremely important notion, with a plethora of practical applications We shall make all this much more precise in the sequel, but for now let us content with the fact that Gaussian curvature is the one that allows us to classify the points on a surface.

Definition 2.2.15 *A p be a point on a (smooth) surface $S \subset \mathbb{R}^3$ is called*

1. Elliptic, if $K(p) > 0 \Leftrightarrow k_{\min} \cdot k_{\mathrm{Maz}} > 0$;

---

[7]As it so often happens in science, the logical development and the historical concatenation of discoveries do not coincide, and this result predates that of Meusnier (albeit by just a few years).

2. Hyperbolic, *if* $K(p) < 0 \Leftrightarrow k_{\min} \cdot k_{\mathrm{Maz}} < 0$;

3. Parabolic, *if* $K(p) = 0$, *but not both principal curvatures are 0*;

4. Planar, *if both principal curvatures are 0*.

It should be not too surprising, after going through a Calculus 2 course, that the four types of points encode the basic possible shapes of surfaces in a vicinity of a point. More precisely, in the neigborhood of a point p a surface is

1. An elliptic paraboloid, if p is an elliptic point;

2. A hyperbolic paraboloid if p is a hyperbolic point;

3. A parabolic cylinder, if p is a parabolic point;

4. In this last case, the situation is more complicated: Clearly a plane satisfies it, but so does, for instance, the surface generated by the rotation of the graph of $z = y^4$ around the Oz axis, as well as the "monkey saddle" $z = x^3 - 3xy^2$.

The classical quadratic surfaces above do not appear by pure chance. Instead, they are the result of a way of circumventing, at least partially our failure to understand curvature via the osculating sphere. The idea is to use, instead, the *osculating paraboloid* which approximates the surface up to order 2, at the point p. It is defined, using our freshly gained knowledge of principal curvatures, as

$$\psi(u, v) = \frac{1}{2}(k_{\min}u^2 + k_{\max}v^2). \tag{2.2.2}$$

(Here we identify p with $0 = (0, 0) \in \mathbb{R}^2$.)

By restricting this expression to the tangent plane at p and considering the *asymptotic directions* (i.e. having director vectors equal to the *principal vectors*), we obtain the so called *Dupin indicatrix*

$$k_{\min}u^2 + k_{\max}v^2 = \pm 1 \tag{2.2.3}$$

It is, as expected, an ellipse, if p is an elliptic point and a hyperbola if p is a hyperbolic point., while for a planar point it degenerates into a pair of parallel lines. (A full technical treatment can be found in [310].) We shall encounter the osculating paraboloid again shortly, after introducing the *second fundamental form*.

Lines of curvature   There are two types of "interesting" types of curves on surfaces which we can introduce quite early in the course. (The definition of a third one, much more important, will have to be postponed for a while.) The first such kind of curves is quite natural to define:

Definition 2.2.16 *Let* $c \subset S$ *a regular, connected curve, such that* $\mathbf{t}(p)$ – *the unit tangent to* c *at* p – *is a principal direction for any* $p \in c$ *is called a* line of curvature *of* S.

Lines of curvature admit the following characterization:

Theorem 2.2.17 (O. Rodrigues, 1813) *A regular, connected curve c on a surface S is a line of curvature if and only iff*

$$N'(t) = \alpha(t)\tilde{c}'(t) ;$$

where $\tilde{c}(t)$ is any parametrization of c, $N'(t) = N \circ \tilde{c}$, and where $\alpha(t)$ is a differentiable function (of t). Moreover, in this case, $-\alpha(t)$ represents the principal curvature in the direction of $\tilde{c}'(t)$.

The proof is quite simple and short, therefore we leave it to the reader.

Exercise 2.2.18 *Prove Theorem 2.2.17.*

Before passing to the next class of special curves on a surface, let us note that in the plane and on the sphere all directions at any point are principal ones, whereas this special situation certainly does not happen at any point of the cylinder or the parabolic hyperboloid, for instance. Thus points where all directions are principal directions are clearly special, thus deserving a definition of their own:

Definition 2.2.19 *Let $p \in S$, such that $k_{min}(p) = k_{max}(p)$. Then p is called an umbilical (or an umbillic[8]) point.*

Example 2.2.20   *1. As we have seen, all the points of the sphere and the plane are umbilics.*

   2. The point $O = (0,0,0)$ is an umbilical point of the paraboloid $z = x^2 + y^2$.

   3. The point $O = (0,0,0)$ is an umbilical point of the monkey saddle $z = x^3 - 3xy^2$. (In this case $k_{min}(O) = k_{max}(O) = 0$.)

   4. On the other hand, the torus of revolution has no umbilics.

The natural question arises whether there exist other surfaces, apart from the plane and the sphere, on which all point are umbilics? The answer is "No" – this property is characteristic for the plane and the sphere:

Theorem 2.2.21 *Let S be a connected surface such that all its points are umbilics. Then S is a subset of a plane or of a sphere.*

We do not bring here the proof; the interested reader can find it [105].

Having seen that there are few surfaces composed solely from umbilical points, one can still hope to easily find many such points on a variety of surfaces. Indeed, as we have seen, the ellipsoid has four umbilics, while the monkey's saddle has one umbilical point, for which $k_{min}(p) = k_{max}(p) = 0$. (Thus one says that p is a *planar umbilical point*.) On the other hand, the torus of rotation has no umbilical points. Still, one might hope that (closed) surfaces other than the torus resemble more the ellipsoid and do have umbilics. This is, indeed, the case as we can state the following

---

[8]From the Latin word for "navel".

Proposition 2.2.22 *Let S be a smooth surface in $\mathbb{R}^3$. Then S has at least one umblical point.*

The proof is not hard, but it makes appeal to some notions that we have not introduced yet. One is entitled to believe that, due to symmetry reasons, and motivated by the case of the ellipsoid, if not all, then, at least surfaces homeomorphic to the sphere have at least two umbilical points. Most surprisingly, this is not a settled matter, and at this point in time it still remains a conjecture:

Conjecture 2.2.23 (Caratheodory) *If $S \subset \mathbb{R}^3$ is a compact surface, homeomorphic to the sphere, then S has at least two umbilics.*

While the conjecture has been proved for real analytic surfaces, the proof is very involved and, moreover, the conjecture has not yet been settled in its generality. (For more details and bibliography, see, e.g. [37], [39].)

Asymptotic lines The second type of remarkable lines on surfaces that we can present already are the so called *asymptotic lines* (or *curves*):

Definition 2.2.24 *Let $c : I \to S$ be a curve. If $\vec{n}(s) \perp \vec{N}(s)$, for all $s \in I$, then c is called an* asymptotic curve *(or line).*

In other words, asymptotic lines are characterized by the property of being perpendicular to their spherical image. Equivalently, $\mathbf{n}$ is always contained in the tangent plane which coincides with the osculating plane[9].

Exercise 2.2.25 *Show that c is an asymptotic line $\Longleftrightarrow k_n \equiv 0$.*

As expected, straight lines are asymptotic lines. More precisely, we have

Proposition 2.2.26 *Let $l \subset S$ be a straight line. Then*

1. l is an asymptotic line.

2. $K|_l \equiv 0 \Leftrightarrow N|_l \equiv 0$.

We leave the proof as a straightforward exercise for the reader – See Figure 2.7 and 2.8.

Exercise 2.2.27 *Prove Proposition 2.2.26.*

As an immediate consequence of the result above we remark that the asymptotic lines of the hyperbolic paraboloid $S : x^2 - y^2 = 2z$ are its rectilinear generators, i.e. the curves $\{x = \text{const.}\}$ and $\{y = \text{const.}\}$ (see Figure 2.6).

The example above conforms with our intuition, as well as with Proposition 2.2.26 above, that straight lines – and asymptotic lines in general – tend "fast" to infinity (see Figure 2.2). It is less intuitive, perhaps that the upper and lower meridians of torus of revolution are also asymptotic lines, since they are, of course, circles, thus far from intuitively being lines. Yet they – and their normals – are contained in the tangent plane to the torus. In fact, one can view them as being lines "at infinity" since they are contained, entirely, in tangent plane (see Figure 2.6).

---

[9]when it is defined.

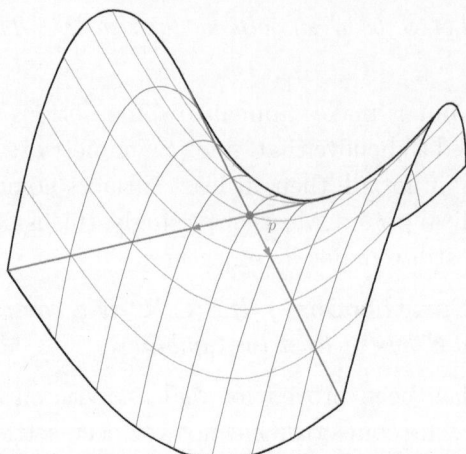

Figure 2.6 Asymptotic lines and directions on the parabolic hyperboloid.

Remark 2.2.28 Asymptotic directions, *that is directions which are tangent vectors to asymptotic lines, exist only at points p such that $K(p) \leq 0$.*

Exercise 2.2.29 *Prove the (simple) assertion above.*

The classical case of smooth surfaces, i.e. of class $\mathcal{C}^k, k \geq 2$, deserves, of course, special attention: Let $U = int(U)$ be an open set and let $f : U \subseteq \mathbb{R}^2 \to \mathbb{R}^3$ be a smooth function (i.e. $f \in \mathcal{C}^k, k \geq 2$). Then the expression of K in local coordinates is:

$$K = \frac{eg - f^2}{EG - F^2} ; \qquad (2.2.4)$$

where

$$E = f_u \cdot f_u , \ F = f_u \cdot f_v , \ G = f_v \cdot f_v ; \qquad (2.2.5)$$

and

$$e = \frac{det(f_u, f_v, F_{uu})}{\sqrt{EG - F^2}} , \ f = \frac{det(f_u, f_v, F_{uv})}{\sqrt{EG - F^2}} , \ g = \frac{det(f_u, f_v, F_{vv})}{\sqrt{EG - F^2}} ; \qquad (2.2.6)$$

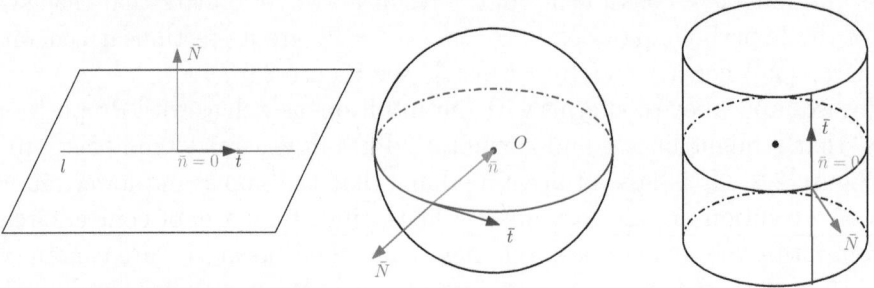

Figure 2.7 Left: Straight lines as asymptotic lines; they tend fast to infinity. Right: Big circles on the sphere do not tend at all to infinity.

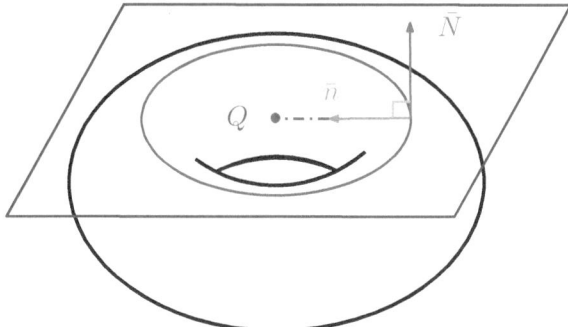

Figure 2.8 The upper meridian of the torus is an asymptotic line, since its normal **n** is perpendicular to the normal to the surface $\vec{N}$. (The same holds, of course, for the lower meridian.)

where $f_u = \partial f / \partial u$, etc. and "·" denotes the scalar product.

$$I_f = \begin{pmatrix} E & F \\ F & G \end{pmatrix} \text{ and } II_f = \begin{pmatrix} e & f \\ f & g \end{pmatrix} \text{ are called } \textit{the first}, \text{ respectively } \textit{the second}$$

fundamental form of S. The more modern approach, which allows us to obtain a series of important results (albeit at the cost of lack of adaptability to discrete settings) represents the focus of the next section.

2.3 THE FUNDAMENTAL FORMS

We now bring the more modern (and formal) approach to the essential geometric ideas above.

2.3.1 The First Fundamental Form

The *first fundamental form* of a surface encodes the manner in which a surface $S \subset \mathbb{R}^3$ inherits the (natural) inner product of the ambient space $\mathbb{R}^3$; thus the manner in which infinitesimal lengths are defined on the surface, and thus provides us with road-map to computing lengths of curves and angles between them, as well as areas of domains on the surface, in an intrinsic manner, i.e. without making appeal each time to the metric in the surrounding $\mathbb{R}^3$.

More precisely, given a parameterized surface $S = f(D), f : D \subseteq \mathbb{R}^2 \to \mathbb{R}^3$, and a point $p \in S$, consider $w_1, w_2 \in T_p(S)$. Then $w_1, w_2 \in \mathbb{R}^3$, therefore the inner product $w_1 \cdot w_2$ is well defined, and we use this simple observation to define the *first fundamental form* by *lifting* the inner product on $\mathbb{R}^2$ to $T_p(S)$

Definition 2.3.1 *Let S and p be as above. The first fundamental form of S (at p) $I_p : T_p(S) \to \mathbb{R}$ is defined as $I_p(w) = w \cdot w = ||w||^2 \geq 0$ (See Figure 2.9.).*

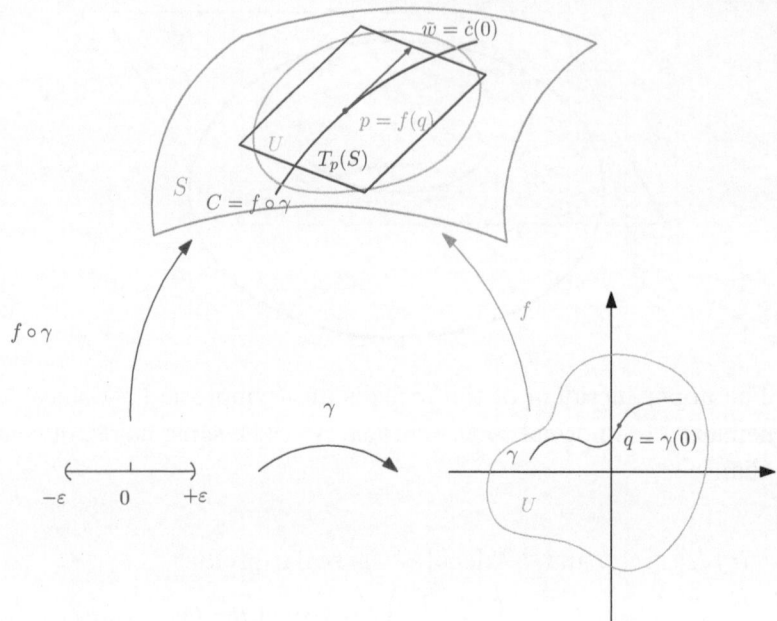

Figure 2.9 The first fundamental form on a surface.

In matricial form the first fundamental form is written as $I_p = I = (g_{ij})$, this being the form which we shall use mainly later on or, as already introduced above as

$$I = \begin{pmatrix} E & F \\ F & G \end{pmatrix}.$$

Remark 2.3.2 *While I is a bilinear form, $I_p : T_P(S)^2 \times T_P(S^2 \to \mathbb{R}$, it is symmetric (given defined via the internal product of $\mathbb{R}^3$), thus not a differential form (which has to be, by definition, antisymmetric.)*

We wish to express I_p in terms of the (canonical) base $< \frac{\partial f}{\partial u}, \frac{\partial f}{\partial v} >$: Given $w \in T_p(S)$, we can write (by its very definition) w as the tangent vector at p to a curve $c \in S$ passing through p, that is $w = \dot{c}(0) = (,)$ Then:

$$I_p(\dot{c}(0)) = (\dot{c}(0) \cdot \dot{c}(0))_p = \left(\frac{\partial f}{\partial u}\dot{u} + \frac{\partial f}{\partial v}\dot{v}\right) \cdot \left(\frac{\partial f}{\partial u}\dot{u} + \frac{\partial f}{\partial v}\dot{v}\right)$$

$$= \left(\frac{\partial f}{\partial u} \cdot \frac{\partial f}{\partial u}\right)(\dot{u})^2 + 2\left(\frac{\partial f}{\partial u} \cdot \frac{\partial f}{\partial v}\right)\dot{u}\dot{v} + \left(\frac{\partial f}{\partial v} \cdot \frac{\partial f}{\partial v}\right)(\dot{v})^2.$$

$$= E(\dot{u})^2 + 2F\dot{u}\dot{v} + G(\dot{v})^2. \tag{2.3.1}$$

We can now list the following formulary:

A. *Curve length* Given a curve $c : I \to S$, it is a classical Calculus fact that

$$\ell(c(t)) = \int_0^t ||\dot{c}(t)||dt = \int_0^t \sqrt{I(c(t))}dt. \tag{2.3.2}$$

By denoting $c(t) = f(u(t), v(t))$, we obtain

$$\ell(c(t)) = \int_0^t \sqrt{E(\dot{u})^2 + 2F\dot{u}\dot{v} + E(\dot{u})^2(\dot{v})^2}dt \tag{2.3.3}$$

The length element of a curve on a surface $d\ell(t)$ is usually denoted simply by ds and from the formula above we obtain that

$$ds^2 = Edu^2 + 2Fdudv + Gdv^2, \tag{2.3.4}$$

which represents a notation that will become very important in the sequel, since it is consecrated in *Riemannian Geometry*.

B. *Angles* Since the angle between the curves c and γ at an intersection point $p_0 = c(t_0) = \gamma(t_0)$ is given by

$$\cos(\theta) = \frac{\dot{c}(t_0) \cdot \dot{\gamma}(t_0)}{||\dot{c}(t_0)|| \cdot ||\dot{\gamma}(t_0)||}, \tag{2.3.5}$$

we obtain, with the same notation as above, that

$$\cos(\theta) = \frac{\frac{\partial f}{\partial u} \cdot \frac{\partial f}{\partial v}}{||\frac{\partial f}{\partial u}|| \cdot ||\frac{\partial f}{\partial v}||} = \frac{F}{\sqrt{EG}} \tag{2.3.6}$$

Remark 2.3.3 *Note that from Formula (2.3.6) above it follows immediately that the coordinate curves are orthogonal iff $F = F(u, v) \equiv 0$.*

Example 2.3.4 *The* loxodromes *(from the Greek "loxos' – slanting or crosswise), or "rhumb lines" of the sphere, are the spherical curves that make a constant angle with all the meridians. Since such curves appear in the Mercator projection as straight lines that intersect the images of the meridians, which are also straight lines, at constant angles (by the conformality of the Mercator projection), their role in navigation, especially in the Age of Discoveries can be hardly underestimated. Using the following standard parametrization of the sphere $f : V \times V \to \mathbb{R}^3, V = (0, \pi), f(\theta, \varphi) = (\sin\theta\cos\varphi, \sin\theta\sin\varphi, \cos\theta)$, we obtain $f_\theta(w) = (\cos\theta\cos\varphi, \cos\theta\sin\varphi, -\sin\theta)$, and $f_\varphi(w) = (-\sin\theta\sin\varphi, \sin\theta\cos\varphi, 0)$; where $w = (\theta, \varphi)$. Therefore, by (2.2.5) we obtain $E = 1, F = 0, G\sin^2\theta$. (Check!) Thence, by (2.3.4), $I_f = a^2 + b^2\sin^2\theta$.*

We can now determine the curves that make a constant angle β with the meridians, i.e the curves $\{\varphi = \text{const.}\}$, that is the loxodromes (See Figure 2.10.). To this end, let us suppose that such a curve is given by the parametrization

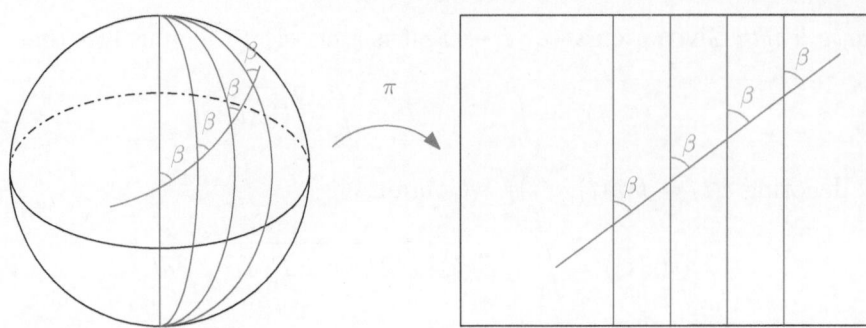

Figure 2.10 The meridians and loxodromes of the sphere are mapped by the Mercator projection onto straight lines intersecting at a constant angle, since the Mercator projection is conformal.

$\gamma(t) = (f \circ c)(t)$, where $c(t) = (\theta(t), \varphi(t))$ is a planar curve. Then, by (2.3.5), $\cos \beta = \frac{f_\theta \cdot \gamma'(t)}{|f_\theta||\gamma'(t)|}$. But, in the basis (f_θ, f_φ), we have $\gamma'(t) = \theta' f_\theta + \varphi' f_\varphi$ and, of course, $f_\theta = 1 \cdot f_t theta + 0 \cdot f_\varphi$. Therefore $\cos \beta = \frac{\theta'}{\sqrt{(\theta')^2 + (\varphi')^2 \sin^2 \theta}}$, thence, by some standard algebraic manipulations the defining condition of the loxodromes becomes $(\theta')^2 \tan^2 \beta - (\varphi')^2 \sin^2 \theta = 0 \Leftrightarrow \frac{\theta'}{\sin \theta} = \pm \frac{\varphi'}{\tan \beta}$. By integrating the last equation we obtain the sought for expression, namely

$$\ln\left(\tan \frac{\theta}{2}\right) = \pm(\varphi + c)\cot \beta;$$

where the constant c is determined by the initial point.

C. *Areas* Precisely like we have seen with regard to lengths, the area for formula is also, essentially, the one from Calculus 2, namely

$$\text{Area}(R) = \iint_Q \left\|\frac{\partial f}{\partial u} \times \frac{\partial f}{\partial v}\right\| du dv = \iint_Q \sqrt{EG - F^2} du dv \qquad (2.3.7)$$

2.3.1.1 *Examples*

1. *Surfaces of Revolution* The generic case of surfaces of revolution[10] is important not only as a relatively simple, while general, example, but mainly because it includes many of the surfaces of huge import in geometry and Topology. These are surfaces generated by the rotation, around the Oz axis of a curve $c : I \to \mathbb{R}, c(t) = (\varphi(t), 0, \psi(t))$, where $\dot\varphi^2(t) + \dot\psi^2(t) > 0$ and $\varphi(t) \neq 0$ for any $t \in I$ – See Figure 2.11. In other words, these are surfaces parametrized by a function $f(u, v) = (\varphi(v) \cos(u), \varphi(v) \sin(u), \psi(v)), u \in (0, 2\pi), v \in (a, b), \varphi(v) \neq 0$. Then, for $S = f(u, v)$, we have $E = \varphi^2, F = 0, G = \dot\varphi^2 + \dot\psi^2$. (Note that if c is parameterized by arc length – as commonly supposed – then $G = 1$.) In particular, for

---

[10]"revolution", not "rotation", as sometimes called, due to the analogy with the dynamics of the solar system.

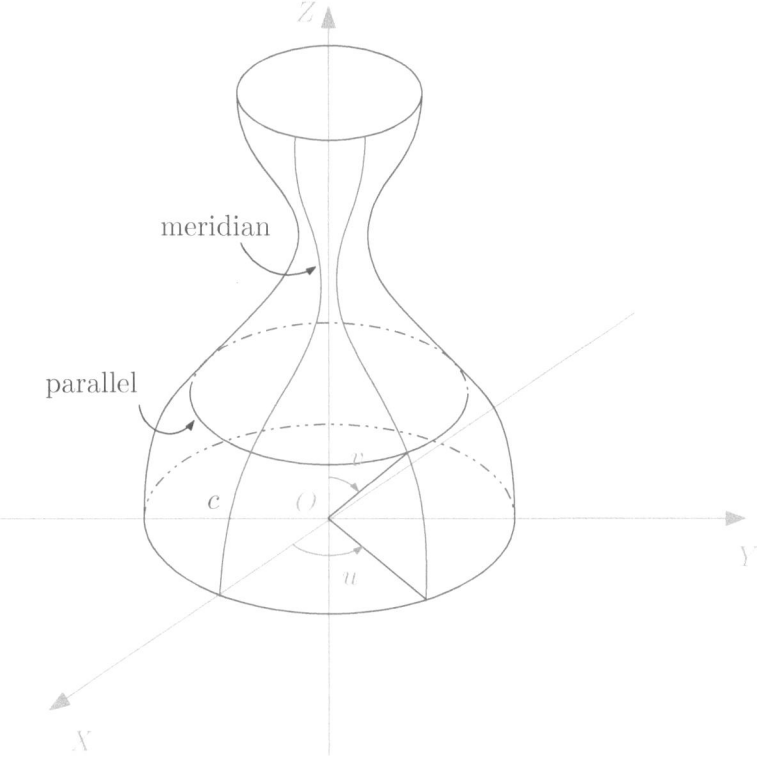

Figure 2.11 A typical surface of revolution.

(a) $\varphi(v) = \sin v, \psi(t) = \cos(v)$, i.e. for the *unit sphere*, $E = \sin^2, F = 0, G = 1$.

(b) $\varphi(v) = R + r\cos(v), \psi(t) = r\sin(v), 0 < r < R$, i.e. for the *torus (of revolution)*, $E = r^2, F = 0, G = (R + r\cos(v))^2$. In consequence, the area of the torus is, cf. Formula (2.3.7), $4rR\pi^2$ (this being a particular case of the classical *Pappus (first) Theorem* [11]) – See Figure 2.12.

(c) $\varphi(v) = r\sin v, \psi(v) = r\ln|\tan(v/2)| + \cos(v)$, that is if the generating (profile) curve is the tractrix (see Chapter 1 above), the resulting surface is the famous *pseudosphere*, to which we shall shortly return. In this case, $E = r^2\cot^2(x), F = 0, G = r^2\sin^2(x)$.

Exercise 2.3.5 *Verify all the computations in the examples above.*

Exercise 2.3.6 *Let $f : \mathbb{R}^2 \to \mathbb{R}^3, f(u, v) = (u\cos v, u\sin v, 2u^2)$, and the planar curves $c_i : \mathbb{R} \to \mathbb{R}^2, i = 1, 2, 3; c_1(t) = (t^2, t), c_2(t) = (-t^2, t), c_3(t) = (t, 1)$.*

(a) Compute the first fundamental form of $S = f(\mathbb{R})$.

---

[11]Recall that Pappus first Theorem states that the area of the surface of revolution generated by the rotation of a curve c around an external axis equals the product of the arc length with the distance traveled by the *centroid* of c, i.e. the arithmetic mean of the points of c, that is their center of mass.

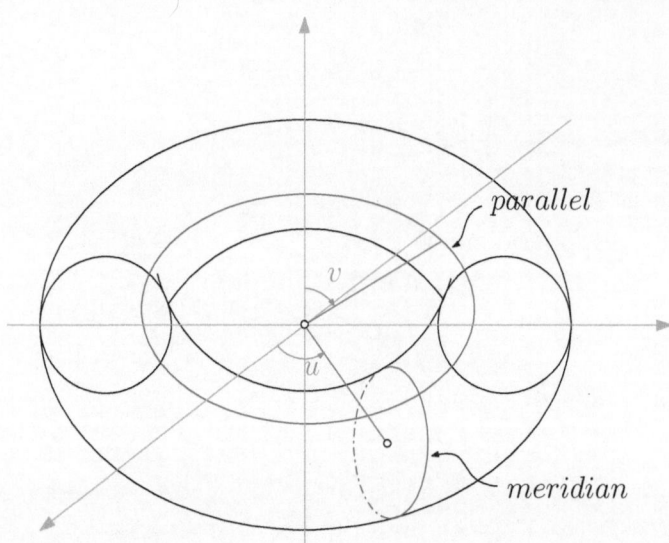

Figure 2.12 The torus of revolution. Note that in Topology, the names of the meridians and parallels are interchanged.

 (b) *Compute the lengths of the sides of the triangle T determined on S by the curves $\gamma_i = f \circ c_i, i = 1, 2, 3$;*
  and

 (c) *The angles of T.*

2. We conclude with an important example that is not a surface of revolution, namely the *helicoid*, which is the surface generated by lines perpendicular to the Oz axis and passing through the points of the (circular) helix (see Chapter 1). Its parametrization is easily seen to be $f(u, v) = (v \cos u, v \sin u, bu)$, where $b \neq 0$. Then $E = b^2 + v^2, F = 0, G = 1$. (Check!)

2.3.1.2 *The Second Fundamental Form*

Besides the first fundamental for, the other essential in ingredient in defining the curvature of a surface is the *second fundamental form*. Its definition rests upon the intuitive notion of *Gauss map*, that attaches to each point on S its unit normal (See Figure 2.13.). More formally, we have

Definition 2.3.7 (*Gauss map*) *Let $S \subset \mathbb{R}^3$ be an oriented surface, and let N denote its outward normal (i.e. such that $(f_u \times f_v) \cdot N > 0$). The Gauss (or normal) map of S is defined as*

$$N(p) = \frac{f_u \times f_v}{\|f_u \times f_v\|}(p) \tag{2.3.8}$$

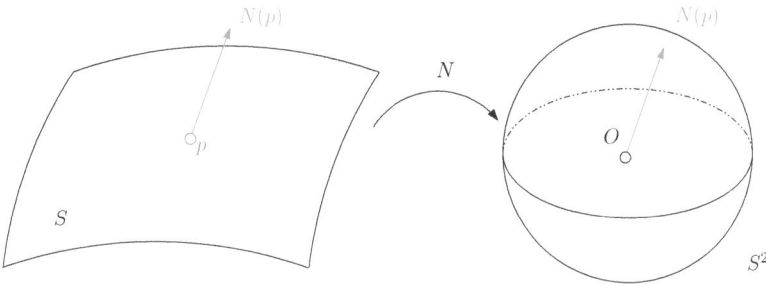

Figure 2.13 The Gauss map.

It is to verify that the Gauss map is differentiable. Moreover, given that $T_p(S) \equiv T_N(p)(\mathbb{S}^2)$ as vector spaces, one can view the differential dN_p as a linear map defined on $T_p(S)$ with values on itself.

If $c : I \to S$ is a parametrized curve, then

$$\dot{N}(0) = dN_p(\dot{c}(0))$$

Thus dN_p measures the rate of change of N along the curve c (at $t_0 = 0$), that is it measures the rate of change (in a neighborhood of p.) of N along c. (This is analogy the with(similar to) the role of curvature for curves, however in this case N is not a scalar.)

Exercise 2.3.8 *Find the image $N(c)$ of the curve c on a surface S and $N(S)$, when*

1. *A plane.*

2. *$S = \mathbb{S}^2$ and c is (a) A great circle; (b) The 60% parallel.*

3. *S is a circular cylinder and c is (a) A meridian; (b) A parallel.*

4. *S is a torus of revolution and c is (a) A meridian; (b) A parallel.*

5. *S is the cone $z = \sqrt{x^2 + y^2}$ and c is (a) A generator; (b) The circle of height $z_0 = 1$.*

6. *S is the hyperbolic paraboloid $z = y^2 - x^2$ and c is a small circle around the point $p = (0, 0, 0)$.*

7. *S is the paraboloid $z = x^2 + y^2$ and $c = S \cap \{y = 0\}$.*

We are now ready to define the *second fundamental form*:

Definition 2.3.9 *The bilinear form $II_p : T_p(S) \times T_P(S) \to \mathbb{R}$, given by*

$$II_p(\mathbf{v}, \mathbf{v}) = -dN(\mathbf{v}) \cdot \mathbf{v} \qquad (2.3.9)$$

is called the second fundamental form of the surface S.

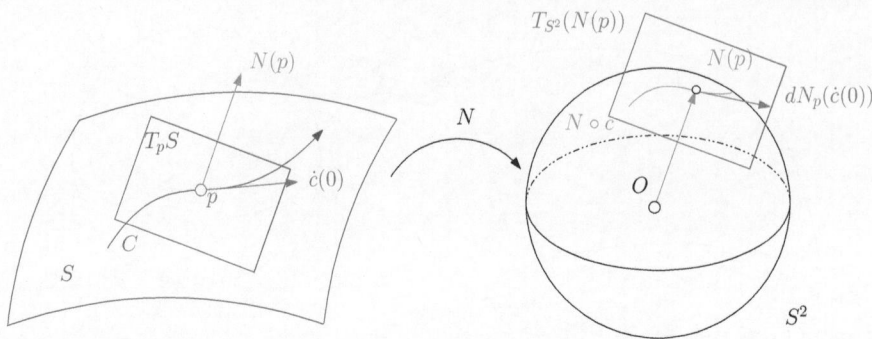

Figure 2.14 The Weingarten map.

We can reinterpret Meusnier's Theorem in terms of the second fundamental form as:

Proposition 2.3.10

$$II_c(s)(\dot{c}(s), \dot{c}(s)) = k_c(s)N(s) \cdot n(s) \tag{2.3.10}$$

i.e.

$$II(\dot{c}, \dot{c}) = k_n(c). \tag{2.3.11}$$

In local coordinates, the second fundamental form is written as

$$II_c(s)(\dot{c}(s), \dot{c}(s)) = e(\dot{u})^2 + 2fF\dot{u}\dot{v} + g(\dot{v})^2. \tag{2.3.12}$$

We leave the proof as an exercise to the reader

Exercise 2.3.11 *Prove Proposition 2.3.12. (See also Exercise 2.3.23 below.)*

Example 2.3.12

Let us return to the important case of surfaces of revolution (see Example 2.3.1.1). In the generic case, it is not hard to verify that $e = -\varphi\dot{\psi}, f = 0, g = \dot{\psi}\ddot{\varphi} - \dot{\varphi}\ddot{\psi}$. Therefore, it follows that

$$K = -\frac{\ddot{\varphi}}{\varphi}. \tag{2.3.13}$$

Exercise 2.3.13 *Verify the formulas above.*

In particular, for

1. *For the unit sphere, $e = \sin^2(v), f = 0, g = \sin(v)(\sin(v) + \cos(v))$, and $K = -\frac{-\sin(v)}{\sin(v)} = +1$, as expected.*

2. *For the torus of revolution, $K \sim \frac{\cos(v)}{R + r\cos(v)}$. An immediate consequence of this formula is the fact that the torus has no umbilics. Another is that $K = 0$, if $v = \pi/2$ or $3\pi/2$; $K > 0$, for $v \in (0, \pi/2) \cap (3\pi/2, 2\pi)$; and $K < 0$ for $v \in (\pi/2, 3\pi/2)$. Thus, on the exterior of the torus curvature is positive, on it's interior ("close" to the hole) it is negative, and it is zero on the upper and lower parallels.*

3. *For the pseudosphere, $e = -r \cot x, f = 0, g = r \sin x \cos x$, therefore, using the computations for the First Fundamental form we obtained in 2.3.1.1, we see that the pseudosphere's Gauss curvature is $K \equiv -1$.*

Remark 2.3.14 *This result has prompted some to offer the pseudosphere as a model of Hyperbolic Geometry, a misconception that was promoted along time by many a successful book on Mathematics popularization. While, it is indeed tempting to view the pseudosphere as a "materialization" into Euclidean space of a sphere of radius i (since $K \equiv -1 = i^2$), this is an interpretation that holds only locally. In truth, the pseudosphere fails to fulfill two of the fundamentals geometric conditions of a plane: (a) It is not a* complete *surface, since it has a singular curve at $r = 0$; and (b) it is not* simply connected, *given that it has the topology of cylinder. In fact, not only does the pseudosphere provide an example of the realization in Euclidean space of the Hyperbolic plane, such an embedding is in fact not possible, due to a famous theorem of Hilbert, from 1901 that states that no such smooth embedding exists i $\mathbb{R}^3$. (In fact, the minimal embedding dimension is $n = 5$.)*

The power of the discrete approach to Differential Geometry is illustrated also here: While, as we have seen, no model of a true Hyperbolic plane is possible in Euclidean space, a PL paper, made of gluing 7 equilateral triangles at each vertex is easy to produce. Since the discrete (defect) curvature at each vertex is $2\pi - 7\pi/3 = -\pi/3 \approx -1$, this is, indeed, a good – as well as simple – model of the Hyperbolic plane. In fact, a paper model for the pseudosphere is also quite easy to build – see [322] and Figure 2.15.

Figure 2.15 A paper model of the Hyperbolic plane, constructed using the instructions (based on an idea of Bill Thurston [322]) given in [255]. Note that even on a 4-four neighborhood of a vertex the surface almost intersects itself, suggesting what is the main obstruction to the embedding of the Hyperbolic plane into $\mathbb{R}^3$.

As an example of handling the second fundamental form beyond its use in computing Gaussian curvature for specific surfaces, we bring here the (promised) proof of Euler's Formula:

Proof Let $\mathbf{v} \in T_p(S), ||\mathbf{v} = 1$, and let e_1, e_2 be an orthonormal basis of $T_p(S)$. Then $\mathbf{v} = e_1 \cos \theta + e_2 \sin \theta$. Therefore,

$$k_n = II_p(\mathbf{v}, \mathbf{v}) = -dN_p(\mathbf{v}) \cdot \mathbf{v} = -dN_p(e_1 \cos \theta + e_2 \sin \theta) \cdot (e_1 \cos \theta + e_2 \sin \theta).$$

Recalling that the eigenvalues of $-dN_p$ are $k_{\min}$ and k_{Max}, it follows that

$$k_n = (k_{\min}e_1 \cos \theta + k_{\text{Max}}e_2 \sin \theta) \cdot (e_1 \cos \theta + e_2 \sin \theta) = k_{\min} \cos^2 \theta + k_{\text{Max}} \sin^2 \theta.$$

□

Problem 2.3.15 *Determine all the surfaces of revolution for whom $K \equiv$ const. (Hint: For instance, to imagine surfaces of constant positive curvature apart from the sphere, think of spindles and of wedding rings.*[12]

Problem 2.3.16 *Determine all the surfaces S^2 for whom $|\nabla K| \equiv 1$.*

The determinant of the second fundamental form, or simply put, the Gaussian curvature, proves to be an important tool in understanding the (local) geometry of surfaces, even at a very basic level, and we have the alternative classification of points in terms of dN_p, which is less intuitive than the one formulated in terms of Gaussian curvature, but perhaps easier to explore in the planar case: A p be a point on a (smooth) surface $S \subset \mathbb{R}^3$ is called

1. *Elliptic*, if $\det(dN_p) > 0$;

2. *Hyperbolic*, if $\det(dN_p) < 0$;

3. *Parabolic*, if $\det(dN_p) = 0$, and $dN_p \neq 0$;

4. *Planar*, if $dN_p = 0$.

As we have seen, $dN_p : T_p(S) \to T_p(S)$ (or $-dN_p$, as it appears in the definition of the second fundamental form), is it itself a measure of curvature, called the *Weingarten map* or, sometimes, most suggestively, the *shape operator*.

The intuition behind this definition is quite straightforward: – See Figure 2.14. The amount of change of N_p in the direction $\mathbf{v}$ measures the variation of the tangent planes to the surface in that direction, thus gauging the (directional) departure of S from being a plane, i.e. of its degree of curving.

---

[12]While we certainly hope the student will at least attempt to solve this not overly difficult problem by his/herself, we note that a full solution can be found in [310].

Remark 2.3.17 *Both the "+" and "−" signs can (and, alternatively, are) be used to define the shape operator. While the "+" would seem more intuitive, we have seen that using the "−" one is natural in explaining the geometric signification of the second fundamental form via Meusnier's Theorem. Therefore, this will be convention that we shall use.*

Remark 2.3.18 *It is customary to discuss the Weingarten map closer, or even before, the definition of the second fundamental form. However, while we discussed it here, for completeness sake, we preferred to conform to the geometric vein adopted in this book and concentrate first on the more intuitive aspects.*

Exercise 2.3.19 *Find the shape operator of the following surfaces: (a) A plane; (b) The unit sphere; (c) A circular cylinder; (d) The saddle surface $z = xy$.*

The following algebraic fact regarding dN_p (or $-dN_p$) is quite significant from a geometric viewpoint as well:

Proposition 2.3.20 $dN_p : T_p(S) \to T_p(S)$ *is a self-adjoint linear map.*

(Recall that a linear mapping $L : V \to W$ is *self-adjoint* if $L(\langle \mathbf{v}, \mathbf{w} \rangle.)$[13]
Since we wish to concentrate on the main, geometric ideas we do not bring here the proof, but rather encourage the reader to supply it:

Proposition 2.3.21 *Prove Proposition 2.3.20.*

The fact that dN_p (or $-dN_p$) is self-adjoint is important due to the following fact:

Theorem 2.3.22 *Let $L : V \to W$ be a self-adjoint linear map. Then there exists an orthonormal basis $\{e_1, e_2\}$ of V such that e_1, e_2 are eigenvectors of V.*

(For a proof of this result, as well as for other complementary material on Linear Algebra, see e.g. [197].)
The importance for us of the result above resides in the fact that, for $L = -dN_p$, the eigenvalues corresponding to e_1, e_2, respectively, are k_{min}, k_{Max}. This fact is geometrically intuitive, yet the theorem above and Proposition 2.3.20 render a formal proof of this fact and, moreover, they show that the maximal and minimal sectional curvatures occur in orthogonal directions. Furthermore, since the eigenvalues of a linear operator satisfy its *characteristic polynomial*, it follows that the following equation holds: $k^2 + 2(k_{min} + k_{Max})k + k_{min} \cdot k_{Max} = 0$, that is

$$k^2 + 2Hk + K = 0 ; \qquad (2.3.14)$$

where H, K are, of course, the mean and Gauss curvature. Thus, we have algebraically recuperated (or even could have defined) the two curvatures of a surface. Given that any linear mapping can be written as a matrix (which, in this case is diagonal), it follows that K and JH can be written as

$$K(p) = \det A_{dN_p} ; H(p) = \frac{1}{2}\text{trace} A_{dN_p} ;$$

---

[13]which, in the case at hand is simply expressed as $L(\mathbf{v}) \cdot \mathbf{w}$

where A_{dN_p} is the (2×2) matrix of $-dN_p$.

Furthermore, to any self-adjoint mapping $L : V \to V$ corresponds a quadratic form Q in V, namely $Q\mathbf{v} = L\mathbf{v} \cdot \mathbf{v}$. Then the eigenvalues λ_1, λ_2 corresponding to e_1, e_2 above, are the minimum, respective maximum, of Q on the unit circle in V. In our case $L = -dN_p$ and Q is nothing else than the second fundamental form II_p, thus indeed the $\lambda_1 = k_{min}$, $\lambda_2 = k_{Max}$ are obtained on orthogonal directions in the unit circle of $T_p(S)$.

Exercise 2.3.23 *Let $S = f(U)$, $f : U \subset \mathbb{R}^2 \to \mathbb{R}^3$ be a parametrized surface.*

1. *Prove that*

$$\begin{cases} N_u = a_{11} f_u + a_{21} f_v \\ N_v = a_{12} f_u + a_{22} f_v \end{cases} \tag{2.3.15}$$

(Equations (2.3.15) are called the equations of Weingarten.*)*

2. *Show that*

$$\begin{pmatrix} a_{11} & a_{21} \\ a_{12} & a_{22} \end{pmatrix} = - \begin{pmatrix} e & f \\ f & g \end{pmatrix} \begin{pmatrix} E & F \\ F & G \end{pmatrix}^{-1},$$

and compute a_{ij}; $1 \le i, j \le 2$.

The coefficients e, f, g of the second fundamental form play an analogous role to the curvature and torsion of a curve, at least in the sense that they allow us to understand how well surfaces are (locally) approximated up to order two. We begin by recalling that any surface can be expressed as the graph of a function $f(x, y, z) = (x, y, \varphi(x, y))$. Then, we have the following simple formulas for K and H:

$$K = \frac{\varphi_{xx} - \varphi_{xy}^2}{\sqrt{1 + \varphi_x^2 + \varphi_y^2}}; \tag{2.3.16}$$

$$H = \frac{}{\sqrt{1 + \varphi_x^2 + \varphi_y^2}}. \tag{2.3.17}$$

No less important (and perhaps more, at least at this stage) is the fact that in this case

$$II_p(x, y) = \varphi_{xx}(0, 0)x^2 + 2\varphi_{xy}(0, 0)xy + \varphi_{yy}(0, 0)y^2. \tag{2.3.18}$$

that is the second fundamental form is nothing but the Hessian of φ. This strengthens and concertizes our rather vague observation that curvature *is* second curvature (in some avatar or another).

Exercise 2.3.24 *Prove Formulas (2.3.16), (2.3.17) and (2.3.18).*

Beyond this rather philosophical observation, Formula (2.3.16) allows us to show that, indeed, e, f, g play an analogous role to the κ and τ play for curves. This will allow us to *en passant* rediscover the osculating paraboloid and Dupin indicatrix (as promised). To this end, let $\varepsilon > 0$ sufficiently small, such that the curve $C = \{(x, y) \in T_pS \,|\, \varphi(x, y) = \varepsilon\}$ be a regular curve. Then, if p is not planar, C is the Dupin indicatrix.

Proof 2.1 *Indeed, in this case,* $f = \varphi_{xy}(0,0) = 0$, $k_{min} = \varphi_{xx}(0,0)$, $k_{Max} = \varphi_{yy}(0,0)$. *Then by developing into Maclaurin series around* $(0,0)$ *and taking into account that* $\varphi_x(0,0) = \varphi_y(0,0) = 0$, *we obtain that*

$$\varphi(x,y) = \frac{1}{2}(\varphi_{xx}^2(0,0)x^2 + 2\varphi_{xy}(0,0)xy + \varphi_{yy}^2(0,0)y^2) + R_2$$

$$= \frac{1}{2}(k_{min}x^2 + k_{Max}y^2) + R_2\,;$$

that is $C = \{k_{min}x^2 + k_{Max}y^2 + 2R_2 = 2\varepsilon\}$. *Thence, if* p *is not planar, then we can consider* $C' = \{k_{min}x^2 + k_{Max}y^2 = 2\varepsilon\}$ *as the order 1 approximation of* C. *Then, if we make the change of variables* $x = \alpha\sqrt{2\varepsilon}$, $y = \beta\sqrt{2\varepsilon}$, *and the curve has the expression* $C'' = k_{min}\alpha^2 + k_{Max}\beta^2 = 1$, *i.e. it represents the restriction of the osculating paraboloid to the tangent plane, that is the Dupin indicatrix.*

Remark 2.3.25 *If* p *is a planar point, then* $k_{min}\alpha^2 + k_{Max}\beta^2$ *gives the tangent plane* $T_p(S)$.

Exercise 2.3.26 *Supply the missing details in the proof above.*

Exercise 2.3.27 *Let* $p \in S$. *Show that if*

1. p *is elliptic, then there exists a neighbourhood* V *in* S *of* p, *such that* V *is contained in one of the half-spaces determined by* $T_p(S)$;

2. p *is hyperbolic, then any neighborhood* V *in* S *of* p, *intersects both of the half-spaces determined by* $T_p(S)$.

While for any practical purposes, at least in a first course in Differential Geometry, the essential first two fundamental forms suffice, it is interesting to note that a *third fundamental form* also exists, $III_p : T_p(S) \times T_p(S) \to \mathbb{R}$, which is defined as

$$III_p(p)(\mathbf{v},\mathbf{w}) = -dN_p(\mathbf{v}) \cdot (-dN_p(\mathbf{w})) = dN_p(\mathbf{v}) \cdot dN_p(\mathbf{w})\,;\mathbf{v},\mathbf{w} \in T_p(S). \qquad (2.3.19)$$

Exercise 2.3.28 *Prove that*

$$III_p(p)(\mathbf{v},\mathbf{w}) = (dN_p(\mathbf{v}))^2 \cdot \mathbf{w}\,. \qquad (2.3.20)$$

While the third fundamental form has its uses, it is, in fact, expressible in terms of the first and second fundamental forms (and the Gaussian and mean curvatures), and we have

$$III - 2H \cdot II + K \cdot I = 0. \qquad (2.3.21)$$

Exercise 2.3.29 *Prove Formula (2.3.21) above. (Hint: Recall that* $-dN_p$ *satisfies its characteristic polynomial.)*

Clearly, the importance of fundamental forms of higher order is quite modest. However, it might be instructive to show that, for instance, a *forth fundamental form*, can be defined (in a manner suggested by Formula (2.3.20)) as follows:

$$IV_p(p)(\mathbf{v},\mathbf{w}) = (dN_p(\mathbf{v}))^3 \cdot \mathbf{w}\,. \qquad (2.3.22)$$

2.3.1.3 Distinguished Curves Revisited

Having at our disposal the language of fundamental forms, provides us with the tools that can also make the study of lines of curvature and asymptotic curves not just more technical, but also more concrete and efficient.

The Differential Equation of the Lines of Curvature   Recall that c is a line of curvature iff

$$dN(\dot{c}(t)) = \lambda(t)\dot{c}(t).$$

By expressing dN in terms of the coefficients of the fundamental forms we obtain that the equation above is equivalent to

$$
\begin{cases}
\frac{fF-eG}{\sqrt{\Delta}}\dot{u} + \frac{gF-fG}{\sqrt{\Delta}}\dot{v} = \lambda\dot{u} \\[2mm]
\frac{eF-fE}{\sqrt{\Delta}}\dot{u} + \frac{fF-gE}{\sqrt{\Delta}}\dot{v} = \lambda\dot{v}
\end{cases}
$$

(Here, as before, $\Delta = EG - F^2$.) By eliminating λ from both equations we obtain

$$(fF - eG)(\dot{u})^2 + (gF - fG)\dot{u}\dot{v} + (gF - fG)(\dot{v})^2 = 0, \tag{2.3.23}$$

which can be written in determinant (and easily mnemonic) form as

$$
\begin{vmatrix}
(\dot{v})^2 & -\dot{u}\dot{v} & (\dot{u})^2 \\
E & F & G \\
e & f & g
\end{vmatrix}
= 0. \tag{2.3.24}
$$

From here the following lemma follows easily:

Lemma 2.3.30 *The parametric curves are lines of curvature iff*

$$F = f = 0$$

Exercise 2.3.31 *Prove Lemma 2.3.30*

The following important particular case follows:

Corollary 2.3.32 *The parametric curves of a surface of revolution are lines of curvature.*

Thus, for example, the parallels and meridians of the torus of revolution are lines of curvature, as we already noted above.

The Differential Equation of the Asymptotic Lines Since c is a line of curvature iff $II() \equiv 0$, the differential equations of the lines of curvature is

$$e(\dot{u})^2 + 2f\dot{u}\dot{v} + g(\dot{u})^2 = 0. \qquad (2.3.25)$$

From here we immediately obtain

Corollary 2.3.33 *If $eg - f^2 < 0$ (i.e. at hyperbolic points) the coordinate curves are asymptotic lines iff*

$$e = g = 0.$$

Exercise 2.3.34 *Determine the asymptotic lines of the surface given by $f : \mathbb{R}^2 \to \mathbb{R}^3$, $f(x, y) = (x + y, x - y, 2xy)$. (See also the example of the parabolic hyperboloid we gave before.)*

Remark 2.3.35 *The apparent simplicity of Equations (2.3.24) and (2.3.25) represents a habitual trap for the CS or EE students who try to determine the special lines on a generic surface, by solving the equations in question, a task that is practically impossible (except in such simple cases as above) by using only the customary tricks such students are learning in their prevalent ODE courses.*

Lines of Curvature Revisited We now concentrate on the lines of curvatures and explore them using the new tools introduced above. One can prove a number of nice results regarding lines of curvature that are not only interesting in themselves, but have interesting applications, some of them quite far-reaching. We begin with

Theorem 2.3.36 (*Bonnet*) *Let c be a curve on the surface S. Then c is a line of curvature of S if Σ_c is planar, where Σ_c denotes the surface formed by the normals to S along c.*

(Recall that S is called *flat* if $K \equiv 0$.)

Before bringing the proof of this theorem, we need to introduce a new definition, namely

Definition 2.3.37 *A surface S is said to be* ruled *if it is given by a parametrization of the form $f(s, t) = c(s) + t\delta(s)$, where c, δ are curves. The curve c is called the* directrix *of the surface.*

Example 2.3.38 *Beside the trivial example of the plane, we mention here the following well-known surfaces:*

1. *Cylinders (not necessarily right circular ones);*

2. *Cones (again, not only the right, circular ones);*

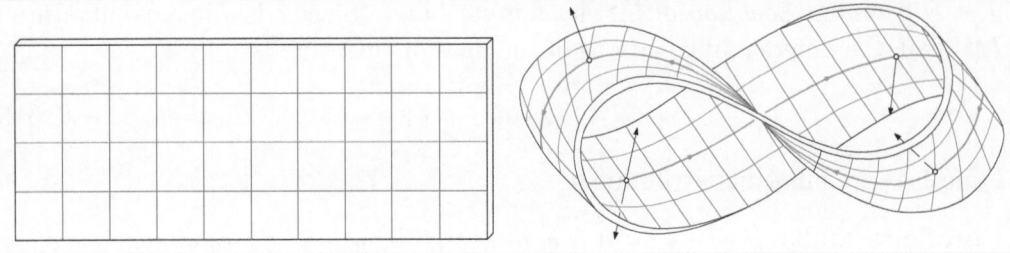

Figure 2.16 The Möbius strip as a ruled surface.

3. *Helicoids;*

4. *Catenoids;*

5. *The Möbius strip.*

Using the formulary developed above, one easily obtains the following formula for the Gaussian curvature of a ruled surface (See Figure 2.16.), whose proof we leave as an undemanding exercise for the reader

$$K = -\frac{[\dot{c} \cdot (\delta \times \dot{\delta})]^2}{||(\dot{c} + t\dot{\delta}) \times \delta||^2} \cdot \tag{2.3.26}$$

Exercise 2.3.39 *(a) Prove Formula (2.3.26) above.*
(b) Compute the Gaussian curvature of the helicoid.

We can now proceed and prove Bonnet's theorem:
Proof We begin by observing that Σ_c is a ruled surface, more precisely that $\Sigma_c = c(s) + tN(c(s))$. Then from (2.3.26) follows that Σ_c is flat $\Leftrightarrow \dot{c} \cdot (\delta \times \dot{\delta}) = 0 \Leftrightarrow \dot{c}(s) \cdot \left(N(c(s)) \times \frac{dN(c(s))}{ds}\right) \Leftrightarrow \frac{dN(c(s))}{ds} \sim \dot{c}(s)$, i.e. if c is a line o curvature.

$\square$

Bonnet's theorem has an important (even if somewhat particular) consequence:

Corollary 2.3.40 *Let S be a surface of revolution. Then its meridians and parallels are lines of curvature.*

This is immediate once we notice that the corresponding ruled surfaces are planes and cones (see Figure 2.17 below).

It turns ot that the property of being a line of curvature on a surface is strong enough to have implications for a second surface, is the given curve is the intersection of the two surfaces. More precisely, we have the following not very well-known theorem:

Theorem 2.3.41 *(Joachimsthal) Let S_1, S_2 be two surfaces and let $c = S_1 \cap S_2$, such that C is a line of curvature in S_1. Then c is also a line of curvature in S_2 iff S_1 and S_2 intersect at a constant angle along c.*

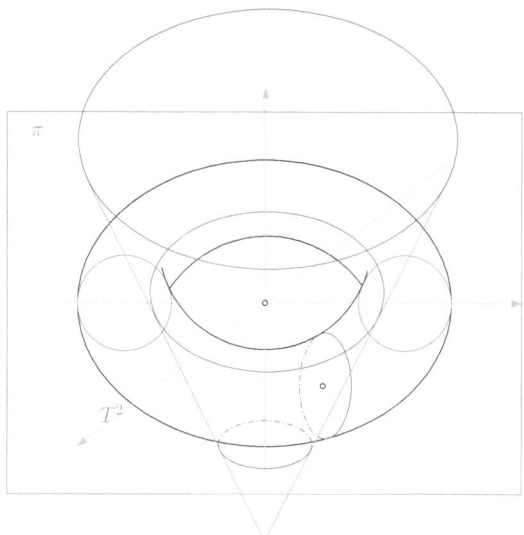

Figure 2.17 The ruled surfaces corresponding to the meridians and parallels of the torus of revolution.

As a consequence we have obtain yet another proof of the fact that the meridians and parallels of the torus are lines of curvature – See Figure 2.17. (The reader is invited to supply the details if he/she feels this is not an immediate consequence from his/hers viewpoint.)

Proof The condition in the statement of the theorem is formally expressed as $\angle(N_{S_1}(c(s)), N_{S_2}(c(s)))$. By differentiation we obtain that

$$\frac{d}{ds}\left(N_{S_1} \cdot N_{S_1} N_{S_2}\right) = \frac{dN_{S_1}}{ds} \cdot N_{S_1} + N_{S_1} \cdot \frac{dN_{S_2}}{ds}.$$

Since c is a line of curvature on S_1

□

Exercise 2.3.42 Let $c = S_1 \cap S_2$. Prove that the curvature of c is given by

$$k_c = \frac{1}{\sin\varphi}\sqrt{k_{n1}^2 + k_{n2}^2 - 2k_{n1}k_{n2}\cos\varphi} \tag{2.3.27}$$

where k_{n1}, k_{n2} denote the curvatures of the normal sections of S_1, S_2, respectively and $\varphi = \angle(N_1, N_2)$.

Moreover, lines of curvature can characterize intersections even of triples of surfaces. More precisely, we have the following theorem:

Theorem 2.3.43 *(Dupin) The intersection curves (lines) of a triply orthogonal family are lines of curvatures on the surfaces.*

Here *triply orthogonal families* are formally defined as follows:

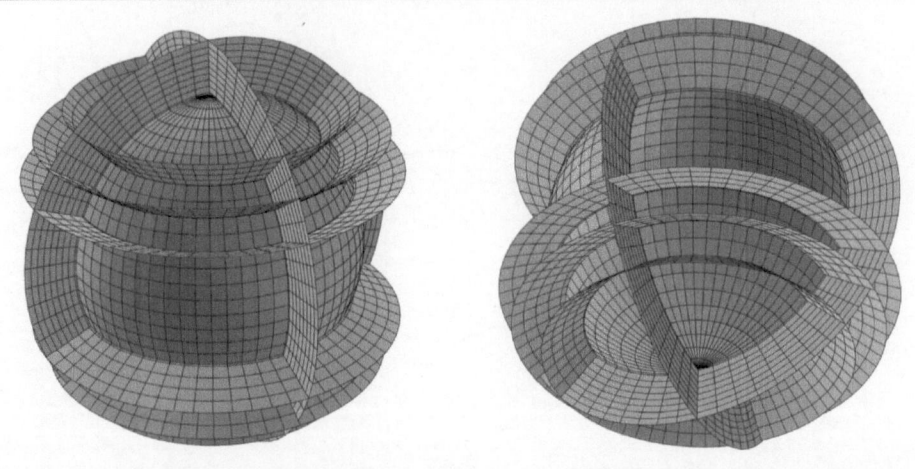

Figure 2.18 A triply orthogonal family of surfaces: The spheres centered at the origin; $\mathcal{F}_2$ – the planes through Oz; $\mathcal{F}_3$ – the cones generated by lines through O and making a constant angle with the axis Oz (after [310]).

Definition 2.3.44 *A triply orthogonal family of surfaces is a diffeomorphism* $F :$ $W = \mathrm{int}W \subset \mathbb{R}^3 \to \mathbb{R}^3$, *such that* $dF(u, v, w) : T_{(u,v,w)}\mathbb{R}^3 \to T_{F(u,v,w)}\mathbb{R}^3$ *is a bijection; and such that*

$$F_u \cdot F_v = F_u \cdot F_w = F_v \cdot F_w = 0. \tag{2.3.28}$$

In other words, for any $p = (u_0, v_0, w_0) \in \mathbb{R}^3$, the correspondences The basic examples are the following:

Example 2.3.45   *1. The planes parallel to the coordinate planes;*

   2. Circular cylinders with Oz as axis, as well as planes through Oz and also the planes parallel to the plane xOz;

   3. The three families of surfaces consisting of $\mathcal{F}_1$ – the spheres centered at the origin; $\mathcal{F}_2$ – the planes through Oz; $\mathcal{F}_3$ – the cones generated by lines through O and making a constant angle with the axis Oz. (See Figure 2.18.)

   4. The family $\mathcal{F}_\lambda$, where $\mathcal{F}_\lambda = \frac{x^2}{a^2 - \lambda^2} + \frac{y^2}{b^2 - \lambda^2} + \frac{z^2}{c^2 - \lambda^2} ; 0 < a^2 < b^2 < c^2$. The family consists of ellipsoids for $\lambda < a^2$; of hyperboloids with one sheet, for $a^2 < \lambda < b^2$; and hyperboloids with two sheets for $b^2 < \lambda < c^2$. (See Figures 2.18, 2.19, and 2.20.)

   ***Exercise 2.3.46** Prove that the surfaces in the example above indeed represent a triply orthogonal family.*

It is precisely this last example, augmented by the observation that any ellipsoid can be realized as one of the surfaces in a triply orthogonal family like the one in the last example, that motivated Dupin's Theorem.

We now return and prove Dupin's Theorem:

Proof By differentiating Equation (2.3.28) by u, v and w we obtain

$$\begin{cases} F_{uv} \cdot F_w + F_{wu} \cdot F_v = 0 \\ F_{vw} \cdot F_u + F_{uv} \cdot F_w = 0 \\ F_{wu} \cdot F_v + F_{vw} \cdot F_u = 0 \end{cases}$$

By subtracting the first equation from the third we obtain $F_{vw} \cdot F_u - F_{wu} \cdot F_v = 0$ From here, and from the last equation in the system above it follows that $F_{vw} \cdot F_u = F_{wu} \cdot F_v = 0$. From these equalities, by applying again (2.3.28), we obtain that $F_u, F_v, F_{uv} \perp F_w$, hence that F_u, F_v, F_{uv} are coplanar. Therefore, we have that $F_u \cdot (F_v \times F_{uv}) = \det(F_u, F_v, F_{uv}) = 0$.

Since the diagonal element M of the second fundamental form is given by $\frac{\det(F_u, F_v, F_{uv})}{\sqrt{EG - F^2}}$, we have shown that, for $w = \text{const.} = w_0$, we have that $M = M(u, v) = 0$. It follows that the curves $\{u = \text{const.}\}$ and $\{v = \text{const.}\}$ are perpendicular, therefore $F = F_u \cdot F_v = 0$. It follows that he curves $\{u = \text{const.}\}$ and $\{v = \text{const.}\}$ are lines of curvature on the surface $w = \text{const.}$ Analogously one proves that the similar fact holds for the surfaces $u = \text{const.}$ and $v = \text{const.}$

$\square$

Exercise 2.3.47 *In the proof above we made appeal to the following fact:*

Lemma 2.3.48 $\{u = \text{const.}\} \perp \{v = \text{const.}\} \iff M = F = 0$.

Prove Lemma 2.3.48 above.

Remark 2.3.49 *The reciprocal of Dupin's Theorem does not hold, as the following counterexample shows:*

Counterexample 2.3.50 *Non-circular cylinders with Oz as axis, as well as planes through Oz and also the planes parallel to the plane xOz.*

However, a partial converse result does hold. More precisely, we have the following:

Theorem 2.3.51 (Darboux) *If two families of surfaces are orthogonal and, moreover, their curves of intersections are lines of curvature, then there exists a third family of surfaces orthogonal two the first two families*

We do not bring the proof and refer the reader to [310].

Remark 2.3.52 *Dupin's Theorem might appear at best as a mildly interesting curiosity; after all triply orthogonal families of surfaces are rather exotic objects. However, this is not the case, as Dupin's Theorem proves to be an essential ingredient in the proof of Liouville's Theorem on the characterization of conformal mappings in space (see Chapter 19).*

Asymptotic Lines Revisited While, as we have seen, using the differential equation of asymptotic lines might be far less useful than one might have expected, one can still characterize asymptotic lines and, moreover, in a much more geometric manner than their differential equation might have had.

The first step in this direction is given by

Theorem 2.3.53 (Beltrami-Enneper Theorem, the weak version) *Let $c \subset S$ be an asymptotic line, and let $p = c(0)$, such that $k(0) \neq 0$. Then*

$$|\tau(0)| = \sqrt{-k(0)} \,. \tag{2.3.29}$$

(The reader will note that, since $p = c(0)$ admits an asymptotic line, $k(0) \leq 0$.)

Remark 2.3.54 *The result above is called "the weak version" because it provides no insight on the sign of $\tau(0)$. (However, as its name also suggests, we shall be able to prove a strong version shorty.)*

Proof As usual, we may presume that c is parameterized by arc length. Furthermore, we can naturally set $N = \mathbf{t} \times \mathbf{n}$. Since, by the second of the Serret-Frènet Formulas, $-\dot{\mathbf{b}} = \mathbf{t} \times \mathbf{n}$, the matrix of the linear mapping $-dN : T_p(S) \to T_p(S)$ is, in the basis $\{\mathbf{t}(0), \mathbf{n}(0)\}$,

$$M = \begin{pmatrix} 0 & \tau(0) \\ \tau(0) & 0 \end{pmatrix}.$$

. Since $K(p) = \det(-M)$, it follows that $K(p) = -\tau^2(0)$. (Recall that, since c is an asymptotic line, $K|_c \leq 0$.)

□

To attain the strong version of the Beltrami-Enneper we need to introduce yet another concept, namely that *Darboux frame*:

Definition 2.3.55 *Let $c : I \to \mathbb{R}^3$ be a parametrized curve. Then the* Darboux *frame* $\mathbf{t}, \mathbf{u}, \mathbf{v}$ *is defined as follows:*

$$\mathbf{t}(s) = \dot{\mathbf{c}}(s)$$

$$\mathbf{v}(s) = \mathbf{t}(s) \times \mathbf{u}(s)$$

where $\mathbf{u}(s)$ is defined in the following manner:

$$(i) \qquad ||\mathbf{u}|| = 1$$

$$(ii) \qquad \mathbf{u} \perp \dot{\mathbf{c}}(s)$$

$$(iii) \qquad \text{The frame } \langle \dot{\mathbf{c}}(s), \mathbf{u}(s) \rangle \text{ is positively oriented}$$

$$(iv) \ \dot{\mathbf{c}}(s) \times \mathbf{u}(s) = N(s)$$

While, in contrast with the Frènet frame, defining the Darboux frame requires the curve c to be embedded in a surface, thus appears to be more restricted, it is also definable even when $k(s) = 0$, thus is applicable for more general curves than the more common Frènet frame. It's efficiency, so to say, for the study of curves on surfaces becomes evident by considering the natural adaptation of the Serret-Frènet Formulas to the Darboux frame:

$$\begin{cases} \dot{\mathbf{t}} = \qquad\quad k_g\mathbf{u} + k_n\mathbf{v} & (*) \\ \dot{\mathbf{u}} = -k_g\mathbf{t} \qquad\quad + \tau_g\mathbf{v} & (**) \\ \dot{\mathbf{v}} = -k_n\mathbf{t} - \tau_g\mathbf{u} & (***) \end{cases} \qquad (2.3.30)$$

The factor k_n appearing in formulas $(*)$ and $(**)$ is the normal curvature that we already studied, while k_g and τ_g are called the *geodesic curvature* and *geodesic torsion*, respectively (See Figure 2.21.)

Exercise 2.3.56 *Prove Formulas (2.3.30) above and show that, indeed, the factor k_n appearing therein is, indeed, the normal curvature of c.*

Exercise 2.3.57 *Let $c \subset S$ be a curve. Prove that the following statements are equivalent:*

1. *c is an asymptotic curve.*

2. *$k = \pm k_g$.*

Exercise 2.3.58 *Let $c \subset S$ be a curve. Prove that the following statements are equivalent:*

1. *c is a line of curvature.*

2. *$\tau_g = 0$.*

It is easy to check that, given a unit vector $\mathbf{x} \in T_p(S)$ the following formulas hold:

$$\begin{cases} k_n = -dN(\mathbf{x}) \cdot \mathbf{x} & (*) \\ \tau_g = -dN(\mathbf{x}) \cdot \mathbf{x}^\perp & (**) \end{cases} \qquad (2.3.31)$$

where $\mathbf{x}^\perp \in T_p(S), \mathbf{x}^\perp \perp \mathbf{x}, ||\mathbf{x}^\perp|| = 1$ and the frame $\langle \mathbf{x}, \mathbf{x}^\perp \rangle$ is positively oriented.

Exercise 2.3.59 *Prove Formulas (2.3.31).*

As might be expected from the similar form of the formulas for k_n and τ_g in (2.3.31) above, a result similar to Euler's formula can be proved for the geodesic torsion. More precisely, we have

Proposition 2.3.60 *Let e_1, e_2 be the principal directions at $p = c(0) \in S$, with corresponding principal curvatures k_1 and k_2, and let $\theta = \angle(e_1, \mathbf{x})$, where $\mathbf{x} \in T_p(S), ||\mathbf{x}|| = 1$. Then*

$$\tau_g(\mathbf{x}) = (k_2 - k_1)\cos\theta\sin\theta \qquad (2.3.32)$$

The proof is quite straightforward and we leave it to the reader:

Exercise 2.3.61 *Prove Formula (2.3.32) above.*

Using Formulas (2.3.31) one can easily prove the following results relating the torsion of an asymptotic line on a surface (an intrinsic notion) to its geodesic torsion (an extrinsic notion):

Proposition 2.3.62 *Let $c \in S$ be an asymptotic line with curvature k and torsion τ, respectively. If $k \neq 0$, then $\tau = \tau_g$.*

This represents an excellent opportunity for the reader to exercise working with Formulas (2.3.31):

Exercise 2.3.63 *Prove Proposition 2.3.62.*

We are now equipped with the necessary tools that allow us to formulate and prove the strong version of The Beltrami-Enneper Theorem:

Theorem 2.3.64 *Let $c \in S$ be an asymptotic line with curvature k, such that $c(0) = p$ and $k(0) \neq 0$. Then*

$$|\tau(0)| = \sqrt{-k(0)} \,. \tag{2.3.33}$$

Furthermore, if $K(p) < 0$ and both asymptotic lines at p, c_1 and c_2 have non-zero curvatures at p, then $\tau_{c_1} = -\tau_{c_2}$.

Proof If c is an asymptotic line, then $\tau = \tau_g$, by Proposition 2.3.62. Therefore, by Formula (2.3.32)

$$\tau_g(0) = (k_2 - k_1) \cos\theta \sin\theta$$

where τ is as above. Moreover, $\mathbf{x}$ is an asymptotic vector, therefore by Euler's formula

$$k_{min} \cos^2\theta + k_{min} \sin^2\theta = 0 \,.$$

From the two formulas above we immediately obtain that

$$k_{Max} = -\kappa_{min} \frac{\cos^2\theta}{\sin^2\theta} \,.$$

From this last formula and the first one, we have

$$\tau(0) = -k_{min} \left(\frac{\cos^2\theta}{\sin^2\theta} + 1 \right) \cos\theta\sin\theta = -k_{min} \frac{\cos\theta}{\sin\theta} \,.$$

However, from the third formula obtained and from the fact that $K(p) = \kappa_{min} k_{Max}$ we obtain that $K(p) = -\kappa_{min}^2 \frac{\cos^2\theta}{\sin^2\theta}$, i.e.

$$K(p) = -\tau^2(0) \,.$$

as desired. Moreover, the first formula we derived shows that $\tau(0)$ changes sign when we replace θ by $-\theta$, i.e. when passing from one asymptotic line to the other,

$\square$

2.4   SOME IMPLEMENTATION ASPECTS

We cannot conclude this chapter without mentioning, however briefly, the role the notions introduced above play in CAD, Graphics, Imaging, Vision, Geometric Modeling and Manufacturing an the manners such notions are computed (or rather, approximated) in practice. While the approaches listed below are both older and different in spirit (being based essentially, on approximations) from the ones we promulgate in this book (and which rest mainly on the idea of discretization), they are never the less important, and not just historically.

Let us begin with the observation that for such implementations, polygonal meshes constitute the basic representations of geometry that are employed. Analysis of such data sets is extremely important in many applications such as reconstruction, segmentation and recognition or even non-photorealistic rendering. In this context, curvature analysis plays a major role. As an example, curvature analysis of 3D scanned data sets were shown quite a long time ago to be one of the best approaches to segmenting the data [345].

To begin with, let us note that in [157, 311], the principal curvatures and principal directions of a triangulated surface are estimated at each vertex by a least square fitting of an osculating paraboloid to the vertex and its neighbors. In these papers references use quadratic approximation methods where the approximated surface is obtained by solving an over-determined system of linear equations. In [217], circular cross sections, near the examined vertex, are fitted to the surface. Then, the principal curvatures are computed using Meusnier's and Euler's theorems. In [320], the principal curvatures are estimated with the aid of an eigenvalues/vectors analysis of 3×3 symmetric matrices that are defined via an integral formulation. Clearly, these can ultimately be used to approximate Gauss and mean curvature.

Another method for the computation of Gaussian curvature based on the asymptotic analysis of the paraboloid fitting scheme is given in [220]. Also, in [182], the fact that $\frac{1}{2\pi} \int_0^{2\pi} k_n(\varphi) d\varphi = H$ and $\frac{1}{2\pi} \int_0^{2\pi} k_n(\varphi)^2 d\varphi = \frac{3}{2}H^2 - \frac{1}{2}K$, where $k_n(\varphi)$ is the normal curvature in direction φ, is used to estimate K and H, via discrete approximations of these integrals around a vertex.

While we restrict here only to this modicum of methods, shall return to the problem of approximating surface curvature in the next chapters and we shall examine it in detail, from a theoretical, as well as practical viewpoint. However, let us note that more details and comparisons regarding the above mentioned approaches can be found in [319], [33].

Moreover, not just computation of Gaussian and mean curvature are for interest in the applicative fields, but also the extraction of other geometric features of interest, such as the determination of umbilics and of lines of curvature [216]. The extremalities of curvature are also of interest [347]. More details an all the directions presented above can be found in the monograph [244].

We wouldn't like to conclude this chapter with this enumerative note, however important it might be to bring these studies to the reader's attention. Instead, we would like to emphasize the importance of curvature in Imaging, Vision, CAD, and related fields, since abrupt changes in its magnitude indicate the presence of *edges* and *contours*. We do not expatiate on this subject since it is intuitive and, more importantly, we already addressed with specific applications in our chapter on curves and we shall again return to it in the sequel.

Instead we rather consider here the sometimes forgotten subject of projections and their curvature. The importance of this topic, in Imaging, Vision, Graphics and GAGD should be evident: After all, all (2-dimensional) images (photographs in particular) are nothing but the projection of 3-dimensional object on a 2-dimensional screen, so they might be viewed as the shades of the object on the said "wall" (plane). Given that axonometric drawings is a more than classical tool, indeed of paramount importance, in Technical (Machine drawing in particular) and Architecture, it is rather strange that the main result below is rather new. (A reason for this might be not just a certain degree of conservatism, but also the fact that people tended to concentrate on straight lines designs.) This result is due to Koenderink, and the reader is warmly invited to read his inspiring book [194]. To be able to enunciate it, we need the following:

Definition 2.4.1 *Let S be a surface and let consider the orthogonal projection pr_Π of S onto a plane Π. The contour C_Π of the projection $S_\Pi = \mathrm{pr}_\Pi(S)$ is called the apparent contour (of S on Π). The points of S projected onto the apparent contour constitute the* rim *of S as viewed on (or by) Π.*

We are now ready to state Koenderink's result:

Theorem 2.4.2 ("Shape from contour") *Let S be a surface and consider the orthogonal projection pr_Π of S onto a plane Π. Then the Gaussian curvature K of S satisfies the following equality:*

$$K = k_c k_n \, ; \tag{2.4.1}$$

where k_c denotes the curvature of the resulting contour and k_n the normal curvature in the projection direction (See Figure 2.22.)

Proof Let $P \in S$, such that $\mathrm{pr}_\Pi \in C_\Pi$. We can chose a coordinate system with origin at P, and such that Ox is in the projection direction, Oy is parallel to the tangent to C_Π and Oz is parallel to the normal to C_Π. Then S can be expressed locally as

$$z = f(x, y) = \frac{1}{2}(ax^2 + 2bxy + cz^2) + O_3(x, y) \, .$$

On the other hand, the apparent contour C_Π satisfies the following condition:

$$\vec{n_C} \times e_1^C = \frac{(-f_x, -f_y, 1)}{\sqrt{1 + f_x^2 + f_y^2}} \times (1, 0, 0) \, ,$$

which locally at the origin has the form

$$ax + by + O_3(x, y) = 0 \Longleftrightarrow x = -\frac{1}{a}by + O(y^3).$$

By introducing this last expression in the local form of the surface, we obtain that C_{II} has the local form near the origin:

$$z = \frac{ac - b^2}{2a}y^2 + O(y^3).$$

Therefore, it follows immediately that

$$k_c = \frac{ac - b^2}{a}.$$

On the other hand, the intersection of S with the normal plane $\{y = 0\}$ has the following local expression at the origin:

$$z = \frac{1}{2}ax^2 + O(x^3)$$

whose curvature at the origin is a. Finally, let us note that $LK = ac - b^2$.

□

Exercise 2.4.3 *Verify the missing computations in the proof above.*

Instead of orthogonal projection onto a plane, one might consider the central projection onto the unit sphere $\mathbb{S}^2$ from a point O. (Think, for instance, on the way images are produced in a planetarium.) There is another theorem of Koendering corresponding to this case:

Theorem 2.4.4 *Let S be a surface, and consider the central projection π_O of S from O onto the unit sphere $\mathbb{S}^2_O$ of center O. Let P be a point on the rim of S on $\mathbb{S}^2_O$, and denote $d = \|\overrightarrow{OP}\|$. Then the Gaussian curvature K of S satisfies the following equality:*

$$K = \frac{1}{d}k_c k_n; \tag{2.4.2}$$

where k_n is the curvature of the normal section at P of S by the plane $\Psi = <\overrightarrow{OP}, \vec{\mathbf{n}}>$, and where k_g is the geodesic curvature of the apparent contour at $\pi_O(P)$.

Since this second theorem is less important than the first one, certainly in the setting of this book, we leave its proof as a (non-trivial) exercise to the reader.

Exercise 2.4.5 *Illustrate and prove Theorem 2.4.4.*

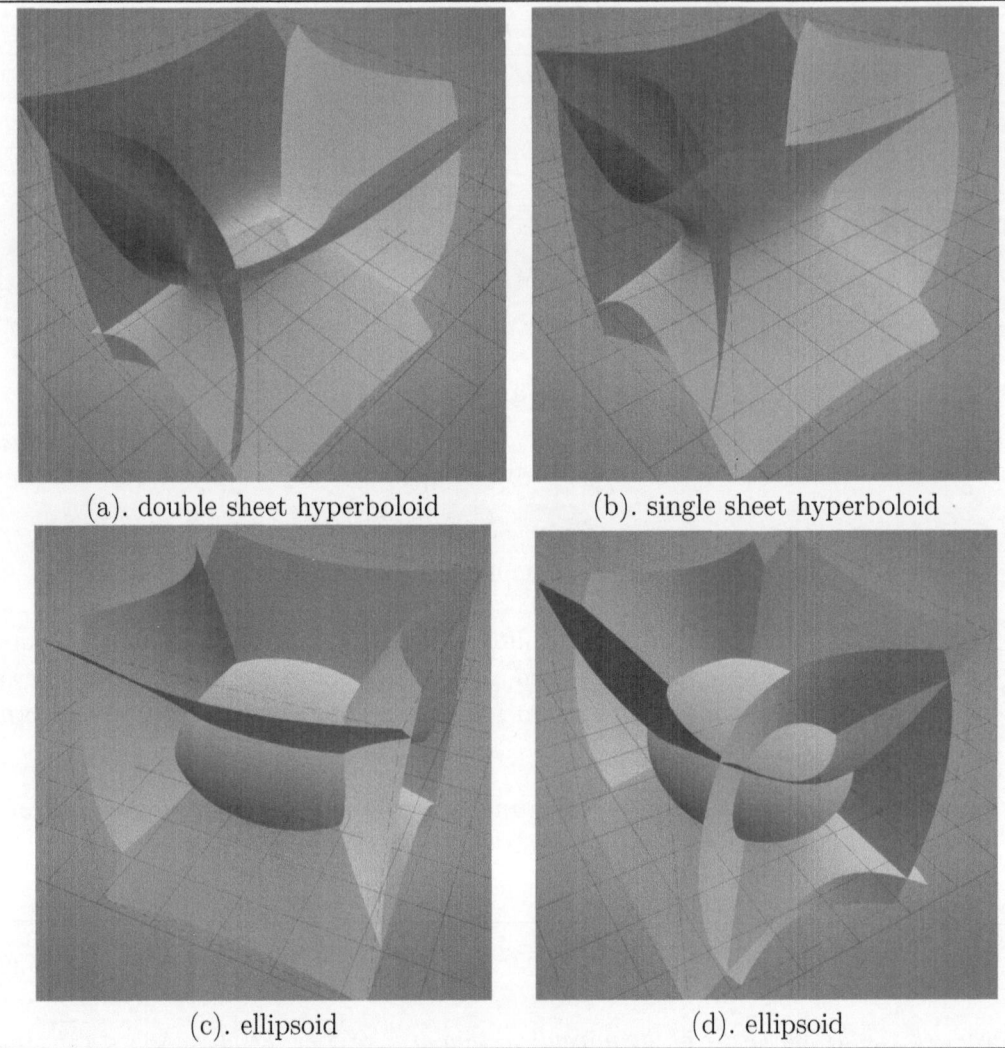

(a). double sheet hyperboloid   (b). single sheet hyperboloid

(c). ellipsoid   (d). ellipsoid

Figure 2.19   The triply orthogonal family in Example 2.3.45 (4) (after [310]), confocal quadratics.

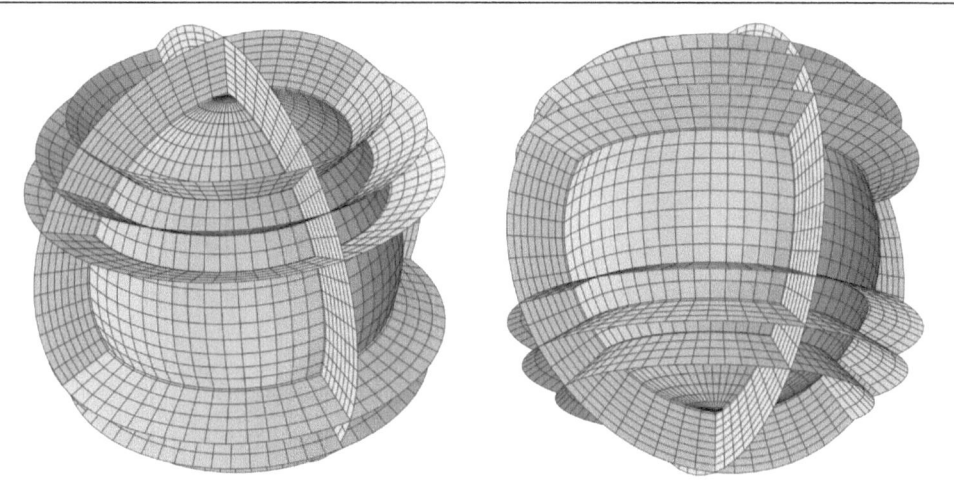

Figure 2.20 A triply orthogonal family of surfaces: The spheres centered at the origin; $\mathcal{F}_2$ – the planes through Oz; $\mathcal{F}_3$ – the cones generated by lines through O and making a constant angle with the axis Oz (after [310]).

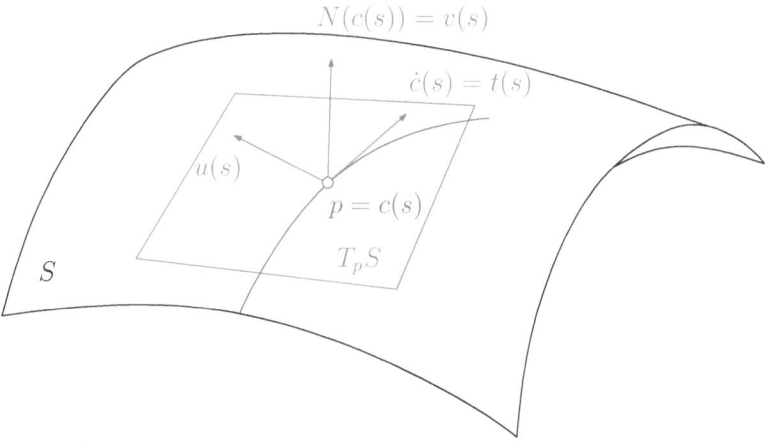

Figure 2.21 The Darboux frame.

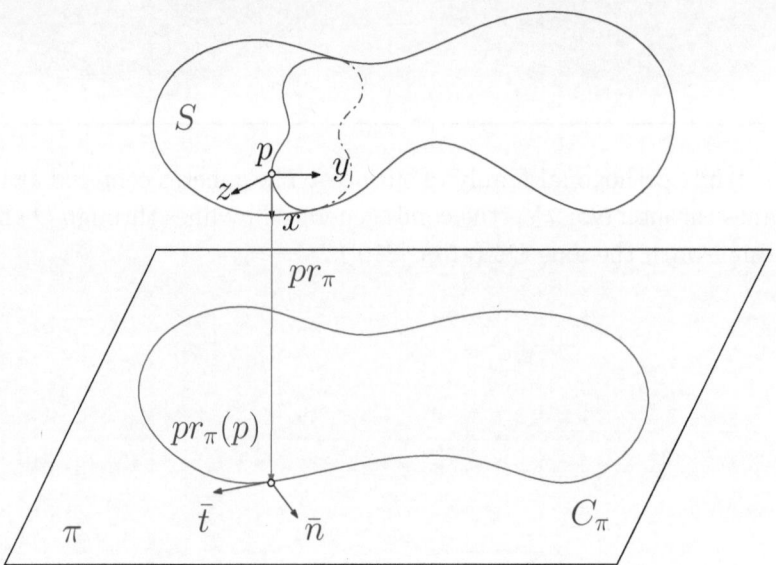

Figure 2.22 Shape from contour.

CHAPTER 3

Metrization of Gauss Curvature

Before even beginning this section, we should make quite clear that, when it comes to applications to Graphics and CAD, the best discrete type of curvature, by far, that one has at his/her disposal is the defect one, that we have introduce already when discussing Theorema Egregium. We shall encounter it again when we shall introduce the *combinatorial Ricci flow*. However, other discretizations of curvature are certainly of interest from a theoretical viewpoint and, furthermore, might prove themselves quite efficient in other, more general contexts.

Indeed, after the experience we had with metric approximations of the curvature (and torsion) of curves it is only natural to have developed similar hopes of success with metric analogs, or at least approximations, of Gauss and mean curvature. In fact, the immediate approach would be make use of Menger/Alt and Haantjes curvatures to this end.

3.1 METRIC APPROXIMATION OF SECTIONAL CURVATURES

The idea of using Menger/Alt and Haantjes curvatures to approximate Gauss and mean curvatures is a most natural one if one recalls the definitions of these surface curvatures based on principal curvatures. This is a well established approach in the fields above, see [319], [203] and the bibliography therein. Certainly, given that for a discrete surface, such as encountered in Images and Graphics there are only a finite number of natural directions to consider, the determination of the principal directions and the approximation of their respective curvatures in this directions is a rather straightforward task. On the flip side of things, the reduced number of directions might not suffice to efficiently "guess" the principal curvatures. In graphics, where there exist a large number of spatial degrees of liberty in improving the data there are, however, means of improving the result – see [319], [285]. In Imaging, where the surface corresponding to an image is rigidly anchored at the grid points, such degrees of freedom are not readily available. Still, some encouraging results are available, see Figure 3.1. Moreover, one can increase the number of directions in which a digital surface is explored, by taking larger (but not too large, to avoid extensive averaging

DOI: 10.1201/9781003350576-3

Figure 3.1 The "Cameraman" test image (left) and its Menger curvature (right) [133]. Note that for triangles, Menger and Haantjes curvatures essentially coincide, thus for images the too approaches render very similar results, at least for the standard 3×3 masks.

effects), e.g. 5×5 or 7×7 – see Figures 3.1 and 3.2. We have already illustrated, in Chapter 1, the application of this approach to the discretization of mean (and Gauss curvature), in tandem with the use of wavelets, to the segmentation of images.

Computing principal curvatures, mainly κ_{Max}, using either Haantjes or Menger curvature has yet another use in Imaging, more precisely in Medical Imaging, where the flattening with controlled distortion of (noisy, in many cases) images is an important task in certain applications. To this end, we have shown – using a theoretical result of Gehring and Väisälä [131] – that "patches" of triangular (polygonal) meshes can be mapped *quasi-conformally* and, indeed, *quasi-isometrically* on the plane, as long as the normal n to the given PL surface S does not deviate, on any simply connected region (patch) $U \subseteq S$, "too much" from an initial value n_0 (see [287]). Note that quasi-conformal mappings, are, indeed, as their name implies "almost preserving" angles (for the proof in the 2-dimensional case, see [5]. Recall also that quasi-isometric mappings are defined as follows:

Definition 3.1.1 *Let $D \subset \mathbb{R}^n$ be a domain. A homeomorphism $f : D \to \mathbb{R}^n$ is called a* quasi-isometry *(or a* bi-lipschitz mapping*), if there exists $1 \leq C < \infty$ such that*

$$\frac{1}{C}|p_1 - p_2| \leq |f(p_1) - f(p_2)| \leq C|p_1 - p_2|, \text{ for all } p_1, p_2 \in D; \tag{3.1.1}$$

where "$| \cdot |$" denotes the standard (Euclidean) metric on $\mathbb{R}^n$.

The coefficient $C(f) = \min\{C \mid f$ is a quasi–isometry$\}$ is called the minimal distortion *of f (in D).*

In our adaptation of Gehring and Väisälä's theoretical result (in whose presentation above we have restricted ourselves solely to the setting relevant here), from

Figure 3.2 Mean Haantjes-Finsler curvature of an image with respect to different window sizes [16]. From top left in clockwise order: Original image, 3×3, 7×7 and 5×5 window size.

smooth surfaces to triangular meshes, we had to make a choice of the normal and of a tangent plane at a vertex of the triangulation; which we did by choosing (as it is common in Graphics) n_0 to be the mean of the normals to the faces adjacent to the considered vertex. The choice of the starting point (which determines the size of the patch U) was shown to be best done by considering the Gaussian curvature, more precisely its (classical by now, see e.g. [163], [149]) discretization as *angular defect* – See the example in Figure 3.3. (For more details, including precise algorithms and examples, see [287].)

Remark 3.1.2 *This somewhat empiric approach can be made rigorous by making appeal to results of Semmes (see [142], Appendix B and the relevant bibliography within). Technically put, the deviation of the normal to a surface S can be controlled by the $\|\cdot\|$ norm:*

$$\|\mathbf{n}\|_* = \sup_{x \in S, R > 0} \frac{1}{|B(x,R) \cap S|} \int_{|B(x,R) \cap S|} |n(y) - n_{x,R}| dy, \qquad (3.1.2)$$

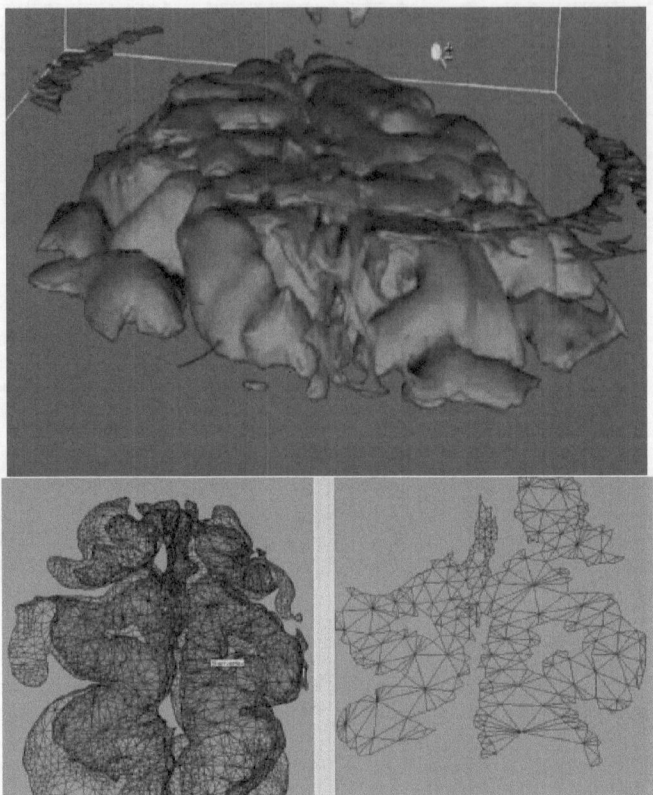

Figure 3.3 Flattening of 3D Cerebral Cortex MRI. Here the resolution is 15110 triangles and the chosen angle of 10° produces a dilatation of 1.1763 [287]. The cortical region selected for flattening is shown in the top image, while a resulting (non-simply connected) patch is illustrated below (before flattening, left and after flattening, right). Notice the position of the starting vertex for the flattening algorithm is also indicated.

where $n_{x,R}$ is the normal mean on $B(x,R) \cap S$, that is

$$n_{x,R} = \frac{1}{|B(x,R) \cap S|} \int_{B(x,R) \cap S} n(y) dy. \tag{3.1.3}$$

The geometric condition we seek is expressed using $||\mathbf{n}||_*$ via

Definition 3.1.3 S^d *is called a* chord-arc surface with small constant *(CASSC) if*
(i) $||\mathbf{n}||_* \leq \varepsilon$,
and
(ii)

$$\sup_{x \in S, R > 0} \sup_{y \in B(x,R) \cap S} \frac{1}{R} | < x - y, n_{x,R} > | \leq \varepsilon_1. \tag{3.1.4}$$

Indeed, condition (3.1.4) of the preceding definition takes place iff for any $x \in S$ and for all $R > 0$, the set $B(x,R) \cap S$ remains close to the hyperplane through x normal to

$n_{x,R}$. Formally, there exist a hyperplane $H = T_{n_{x,R}}$ such that, for any $y \in B(x,R) \cap S$ the following holds:

$$d(y, T_{n_{x,R}}) \leq \varepsilon_2 R. \tag{3.1.5}$$

In metric terms we can formulate the following equivalent definition:

Definition 3.1.4 *The surface S is called ε-flat if*

$$l(\gamma) \leq (1 + \varepsilon)|x - y|, \tag{3.1.6}$$

and

$$(1 - \varepsilon')\text{Area}_{\text{Eucl}}(B^2(R)) \leq |B(x,R) \cap S| \leq (1 + \varepsilon')\text{Area}_{\text{Eucl}}(B^2(R)), \tag{3.1.7}$$

R being as above.

Note that a surface being ε-*flat* represents a stronger statement than the simple quasi-isometry condition (3.1.6), due to the low area distortion expressed by condition (3.1.7). Also, note that all the definitions and conditions above can be trivially extended to higher dimensions (only "Area" being replaced by "Vol", of course).

As expected, the deviation of the normal (hence flatness) can controlled by curvature, more precisely if S has "*small curvature in the mean*", i.e. if

$$\int_S (k_{Max}(x))^2 dx = ||k_{Max}||_{L^2} < \delta, \tag{3.1.8}$$

for small δ, then S is ε-flat.

Let us make a few remarks regarding the small curvature in the mean condition above:

1. As above one can extend, *mutatis mutandis*, this condition to any dimension.

2. Condition (3.1.8) is, in fact, stronger that ε-flatness, because it requires higher order derivatives.

3. Condition (3.1.8) is scale invariant.

Thus, for a surface to have small curvature in the mean does not just imply mere ε-flatness; in fact condition (3.1.8) implies (see [324]) that there exists a $(1 + \eta)$-bilipschitz parameterization of M, where $\eta = \eta(\delta)$, and where $\eta \to 0$, when $\delta \to 0$.

We can thus conclude that the use Haantjes or Menger metric curvatures provides us with a powerful tool for flattening and bilipschitz parametrization of noisy and "rough" data.

3.2 WALD CURVATURE

The idea of using Menger and Haantjes curvatures to emulate the classical principal curvatures is, albeit, natural but also of rather limited accuracy, as we have seen. Furthermore, it is somewhat naive, in the sense that it reflects the understanding of surface curvature prior to Gauss's seminal work. Thus, emulating Gauss' definition of surface curvature in the metric context is a logically necessary step. This stride was taken first by Wald [330], [331]. His idea was to go back to Gauss' original method of defining surface curvature by comparison to a standard, model surface (i.e. the unit sphere in $\mathbb{R}^3$), while extending it to general gauge surfaces, rather than restrict himself to the unit sphere. Moreover, instead of comparing infinitesimal areas (which would be an impossible task in general metric space not endowed with a measure), he compared quadrangles. More precisely, his starting point was the following definition:

Definition 3.2.1 (*Metric quadruple*) *Let (M, d) be a metric space, and let $Q = \{p_1, ..., p_4\} \subset M$, together with the mutual distances: $d_{ij} = d_{ji} = d(p_i, p_j)$; $1 \leq i, j \leq 4$. The set Q together with the set of distances $\{d_{ij}\}_{1 \leq i,j \leq 4}$ is called a* metric quadruple.

Remark 3.2.2 *The reader has undoubtedly already recognized that the definition above conducts toward Gromov's $\mathbf{K}_4(X)$ [142], similar to the way that Menger curvature identifies with $\mathbf{K}_3$ (see Remark 1.2.28). Indeed, we can view, in a sense, this chapter as representing an extended overview of and discussion on $\mathbf{K}_4(X)$.*

Remark 3.2.3 *The following slightly more abstract definition can be also considered, one that does not make appeal to the ambient space: a metric quadruple being a 4 point metric space, i.e. $Q = (\{p_1, ..., p_4\}, \{d_{ij}\})$, where the distances d_{ij} verify the axioms for a metric. However, this comes at a price, as we shall shortly see in Remark 3.2.5.*

Before being able to pass to the next definition we need to introduce some additional notation: S_κ denotes *the complete, simply connected surface of constant Gauss curvature κ* (or *space form*), i.e. $S_\kappa \equiv \mathbb{R}^2$, if $\kappa = 0$; $S_\kappa \equiv \mathbb{S}^2_{\sqrt{\kappa}}$, if $\kappa > 0$; and $S_\kappa \equiv \mathbb{H}^2_{\sqrt{-\kappa}}$, if $\kappa < 0$. Here $S_\kappa \equiv \mathbb{S}^2_{\sqrt{\kappa}}$ denotes the sphere of radius $R = 1/\sqrt{\kappa}$, and $S_\kappa \equiv \mathbb{H}^2_{\sqrt{-\kappa}}$ stands for the hyperbolic plane of curvature $\sqrt{-\kappa}$, as represented by the Poincaré model of the plane disk of radius $R = 1/\sqrt{-\kappa}$.

Definition 3.2.4 *The* embedding curvature $\kappa(Q)$ *of the metric quadruple Q is defined to be the curvature κ of the gauge surface S_κ into which Q can be isometrically embedded – if such a surface exists (See Figure 3.4).*

Remark 3.2.5 *Even though the basic idea of embedding curvature is, in truth, quite intuitive, care is needed if trying to employ it directly, since there are a number of issues that arise (as we have anticipated in Remark 3.2.3 above):*

1. *If one uses the second (abstract) definition of the metric curvature of quadruples, then the very existence of $\kappa(Q)$ is not assured, as it is shown by the following*

   Counterexample 3.2.6 *The metric quadruple of lengths*

   $$d_{12} = d_{13} = d_{14} = 1; \; d_{23} = d_{24} = d_{34} = 2$$

   admits no embedding curvature.

   Exercise 3.2.7 *Prove the assertion in Counterexample 3.2.6 above.*

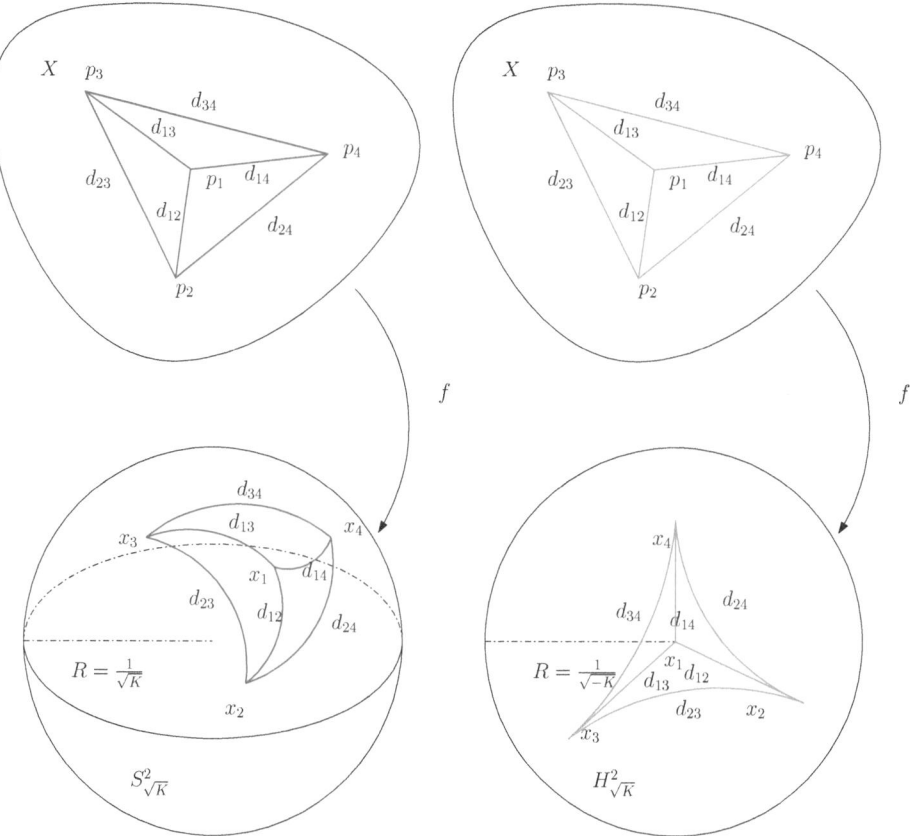

Figure 3.4 Isometric embedding of a metric quadruple in a gauge surface: $\mathbb{S}^2_{\sqrt{\kappa}}$ (left) and $\mathbb{H}^2_{\sqrt{\kappa}}$ (right).

2. *Any linear quadruple is embeddable, apart from the Euclidean plane, in all hyperbolic planes (i.e. of any strictly negative curvature), as well as in infinitely many spheres (whose radii are sufficiently large for the quadruple to be realized upon them).*

3. *Moreover, even if a quadruple has an embedding curvature, it still may be not unique (even if Q is not linear); as it is illustrated by the following examples:*

Example 3.2.8 *(a) For each $\kappa > 0$, each neighborhood of any point $p \in S_\kappa$ contains a non-degenerate quadruple that is also isometrically embeddable in $\mathbb{R}^2$. (For the proof see [212], pp. 372-373).*

(b) The quadruple Q of distances $d_{13} = d_{14} = d_{23} = d_{24} = \pi$, $d_{12} = d_{34} = 3\pi/2$ admits exactly two embedding curvatures: $\kappa_1 = \frac{1}{2}$ and $\kappa_2 \in \left(\frac{1}{4}, \frac{4}{9}\right)$. The proof can be found in [47], however we suggest it as a more challenging exercise exercise for the committed reader.

Exercise 3.2.9 *Prove the assertion above.*

We can now define the *Wald curvature* [330],[331] (or, more precisely, its modification due to Berestovskii [34]):

Definition 3.2.10 *Let (X, d) be a metric space. An open set $U \subset X$ is called a region of curvature $\geq \kappa$ iff any metric quadruple can be isometrically embedded in S_m, for some $m \geq k$. A metric space (X, d) is said to have Wald-Berestovskii curvature $\geq \kappa$ iff any $x \in X$ is contained in a region $U = U(x)$ of curvature $\geq \kappa$.*

Remark 3.2.11 *Clearly, in the context of polyhedral surfaces, which represent, as we have already noted, the geometric medium Graphics, the natural choice for the set U required in Definition 3.2.10 is the star of a given vertex v, that is, the set $\{e_{vj}\}_j$ of edges incident to v. Therefore, for such surfaces, the set of metric quadruples containing the vertex v is finite.*

Equipped with this quite simple and intuitive choice for U (and in analogy with Alexandrov spaces – see also Appendix A below) it is quite natural to consider, for PL surfaces, the following definition of the Wald curvature $K(v)$ at the vertex v:

$$K_W(v) = \min_{v_i, v_j, v_k \in \mathrm{Lk}(v)} K_W^{ijk}(v),$$

where $K_W^{ijk}(v) = \kappa(v; v_i, v_j, v_k)$, and where $\mathrm{Lk}(v)$ denotes the link of the vertex v – see Footnote 11 below.[1] (See Figure 3.5 for a concrete instance.) Note that here we consider the (intrinsic) PL distance between vertices.

The embedding curvature at a point can now be defined naturally as a limit. However, we need first yet another preparatory definition:

Definition 3.2.12 *(M, d) be a metric space, let $p \in M$ and let N be a neighborhood of p. Then N is called* linear *iff N is contained in a metric segment.*

Definition 3.2.13 *Let (M, d) be a metric space, and let $p \in M$ be an accumulation point. Then M has (embedding) Wald curvature $\kappa_W(p)$ at the point p iff*

1. *Every neighborhood of p is non-linear;*

2. *For any $\varepsilon > 0$, there exists $\delta > 0$ such that if $Q = \{p_1, ..., p_4\} \subset M$ and if $d(p, p_i) < \delta, i = 1, ..., 4$; then $|\kappa(Q) - \kappa_W(p)| < \varepsilon$.*

Remark 3.2.14 *Even if the basic idea of embedding curvature is, in fact, quite natural, one should take care when trying to employ it directly, since there are a number of "surprises" that arise:*

1. *If one uses the second (abstract) definition of the metric curvature of quadruples, then the very existence of $\kappa(Q)$ is not assured, as it is shown by the following*

---

[1]Recall that the *link* $\mathrm{lk}(v)$ of a vertex v is the set of all the faces of $\overline{\mathrm{St}}(v)$ that are not incident to v. Here $\overline{\mathrm{St}}(v)$ denotes the *closed star* of v, i.e. the smallest subcomplex (of the given simplicial complex K) that contains $\mathrm{St}(v)$, namely $\overline{\mathrm{St}}(v) = \{\sigma \in \mathrm{St}(v)\} \cup \{\theta \mid \theta \leqslant \sigma\}$, where $\mathrm{St}(v)$ denotes the *star* of v, that is the set of all simplices that have v as a face, i.e $\mathrm{St}(v) = \{\sigma \in K \mid v \leqslant \sigma\}$.

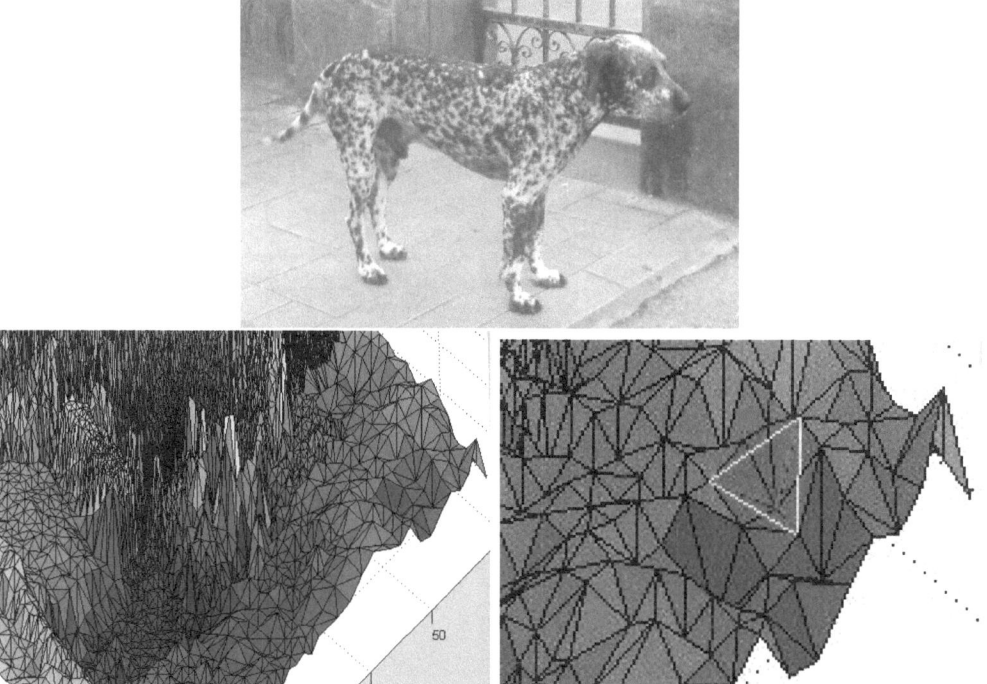

Figure 3.5 Detail of the triangulation (below, left) corresponding to a natural image (top); and the computation of the Wald curvature at a vertex (below, right) [279]. Notice that only the quadruples generated by edges incident to the vertex (red edges) are considered. Note that distances between adjacent vertices (yellow edges) should be considered, even though they are not among the edges of the triangulation.

Counterexample 3.2.15 *The metric quadruple of lengths*

$$d_{12} = d_{13} = d_{14} = 1; \ d_{23} = d_{24} = d_{34} = 2$$

admits no embedding curvature.

Exercise 3.2.16 *Prove the assertion in the counterexample above.*

2. *Any linear quadruple is embeddable, apart from the Euclidean plane, in all hyperbolic planes (i.e. of any strictly negative curvature), as well as in infinitely many spheres (whose radii are sufficiently large for the quadruple to be realized upon them).*

3. *Moreover, even if a quadruple has an embedding curvature, it still may be not unique (even if Q is not linear); as it is illustrated by the following examples:*

Example 3.2.17 *(a) For each $\kappa > 0$, each neighborhood of any point $p \in S_\kappa$ contains a non-degenerate quadruple that is also isometrically embeddable in $\mathbb{R}^2$. (For the proof see [212], pp. 372-373).*

> (b) *The quadruple Q of distances $d_{13} = d_{14} = d_{23} = d_{24} = \pi$, $d_{12} = d_{34} = 3\pi/2$ admits exactly two embedding curvatures: $\kappa_1 = \frac{1}{2}$ and $\kappa_2 \in \left(\frac{1}{4}, \frac{4}{9}\right)$.*

As before, even though the proof can be found in [47], we prefer to leave the proof of the last assertion to the reader.

Exercise 3.2.18 *Prove the assertion above.*

Fortunately, for "nice" metric spaces – that is to say spaces that are locally sufficiently "plane like" – the embedding curvature exists and it is unique (see, e.g, [212] and, for a briefer but more easily accessible presentation, [271]).

Moreover – and this represents a fact that is very important for some of our own goals, as will be made clear in our chapter dedicated to the metric surface Ricci flow, namely Chapter 12 – this embedding curvature coincides with the classical Gaussian curvature. Indeed, one has the following result due to Wald:

Theorem 3.2.19 (Wald [331]) *Let $S \subset \mathbb{R}^3$, $S \in \mathcal{C}^m$, $m \geq 2$ be a smooth surface. Then, given $p \in S$, $\kappa_W(p)$ exists and $\kappa_W(p) = K(p)$, where $K(p)$ denotes the Gaussian curvature at p.*

Remark 3.2.20 *In the theorem above the metric considered in the computation of Wald curvature is the intrinsic one of the surface. (Indeed, the reciprocal Theorem 3.2.21 below is formulated, at least prima facie, for a much more general class of metric spaces than mere smooth surfaces embedded in Euclidean 3-space.) However, in applications Euclidean (extrinsic) distances are used instead. However, this does not represent a theoretical obstruction (only, perhaps, a practical one) – see, for instance, [278], Section 4.3 and the references therein.[2]*

Moreover, Wald also has shown that the following partial reciprocal theorem also holds:

Theorem 3.2.21 Let M be a compact and convex metric space. If $\kappa_W(p)$ exists, for all $p \in M$, then M is a smooth surface and $\kappa_W(p) = K(p)$, for all $p \in M$.

Remark 3.2.22 *Obviously, here the metric considered is the abstract one of the given metric space, that is proven to coincide with the intrinsic one of a smooth surface.*

The results above, in conjunction, show that Wald curvature represents, indeed, a proper metrization of the classical (smooth) notion, and not just a mathematical "divertissement", lacking any significant geometric content.

We continue with a definition whose full significance will become more clear sooner, rather than later, when its role in the actual computation of Wald curvature will be revealed:

Definition 3.2.23 *A metric quadruple $Q = Q(p_1, p_2, p_3, p_4)$, of distances $d_{ij} = dist(p_i, p_j)$, $i = 1, ..., 4$, is called* semi-dependent *(or a sd-quad, for brevity), there exist 3 indices, e.g. 1,2,3, such that: $d_{12} + d_{23} = d_{13}$.*

---

[2]The literature on the subject being too vast to even begin and enumerate it here.

Remark 3.2.24 *The condition in the definition above implies, in fact, that the three points in question lie on a common metric segment i.e. a subset of a given metric space that is isometric to a segment in $\mathbb{R}$ (see [212], p. 246).*

Perhaps the main advantages of sd-quads stems from in the following fact:

Proposition 3.2.25 *An sd-quad admits at most one embedding curvature.*

The proof follows easily from the following result (for a proof of whom we refer to [212], [47]).

Proposition 3.2.26 *Let $Q_1 = \{p_1, q_1, r_1, s_1\}$, $Q_2 = \{p_2, q_2, r_2, s_2\}$ be non-linear and non-degenerate quadruples in $\mathcal{S}_{\kappa_1}$, $\mathcal{S}_{\kappa_2}$, respectively. If $\triangle(p_1, q_1, r_1) \cong \triangle(p_2, q_2, r_2)$ and $\kappa_1 < \kappa_2$, then:*

1. *$p_1 s_1 = p_2 s_2$, $q_1 s_1 = q_2 s_2 \implies r_1 s_1 > r_2 s_2$;*

2. *$r_1 s_1 = r_2 s_2$, $q_1 s_1 = q_2 s_2 \implies p_1 s_1 > p_2 s_2$;*

3. *$p_1 s_1 = p_2 s_2$, $r_1 s_1 = r_2 s_2 \implies q_1 s_1 < q_2 s_2$.*

Exercise 3.2.27 *Prove Proposition 3.2.25.*

In fact, there also exists a classification criterion – due to Berestovskii [34], see also [252], Theorem 18 – for embedding curvature possibilities in the general case:

Theorem 3.2.28 *Let M, Q be as above. Then one and only one of the following assertion holds:*

1. *Q is linear.*

2. *Q has exactly one embedding curvature.*

3. *Q can be isometrically embedded in some $\mathcal{S}_\kappa^m$, $m \geq 2$; where $\kappa \in [\kappa_1, \kappa_2]$ or $(-\infty, \kappa_0]$, where $\mathcal{S}_\kappa^m \equiv \mathbb{R}^m$, if $\kappa = 0$; $\mathcal{S}_\kappa^m \equiv \mathbb{S}_{\sqrt{\kappa}}^m$, if $\kappa > 0$; and $\mathcal{S}_\kappa^m \equiv \mathbb{H}_{\sqrt{-\kappa}}^m$, if $\kappa < 0$. Moreover, $\kappa \in \{\kappa_0, \kappa_1, \kappa_2\}$. represent the only possible values of planar embedding curvatures, i.e. such that $m = 2$. (Here $\mathbb{S}_{\sqrt{\kappa}}^m$ denotes the m-dimensional sphere of radius $R = 1/\sqrt{\kappa}$, and $\mathbb{H}_{\sqrt{-\kappa}}^m$ stands for the m-dimensional hyperbolic space of curvature $\sqrt{-\kappa}$, as represented by the Poincaré model of the ball of radius $R = 1/\sqrt{-\kappa}$).*

4. *There exist no m and k such that Q can be isometrically embedded in $\mathcal{S}_\kappa^m$.*

Remark 3.2.29 [3]*[A Local-to-Global Result] Since the definition of Wald curvature is rather abstract, the question whether it is actually computable (and preferably in a computationally efficient manner) naturally arises. However, before passing to the matter of the actual computation of Wald curvature, we include here a result who's*

---

[3]This observation can be skipped upon first reading.

full importance and relevance will become, we hope, clearer later on, but which can also appreciated on its own intrinsic merit. More precisely, we bring the fitting version of the Toponogov Comparison Theorem (or Alexandrov-Toponogov Comparison Theorem).[4]

Theorem 3.2.30 (Toponogov's Comparison Theorem for Wald Curvature)
Let (X, d) be an inner metric space of curvature $\geq k$. Then the entire X is a region of Wald curvature $\geq k$.

Since the proof is somewhat lengthy and technical we do not bring it here – see [250], as well as [252].

3.2.1 Computation of Wald Curvature I: The Exact Formula

Since the definition of Wald curvature is rather abstract, the question whether it is actually computable (and preferably in a computationally efficient manner) naturally arises. As it turns out, a non-negligible part of the attractiveness of Wald curvature does not reside in its simplicity and intuitiveness, but also that it comes endowed, so to say, with a simple formula for its actual computation. (This is in stark contrast with the *Alexandrov comparison curvature* at least in its usual presentation – but we shall elaborate later on this subject.) More precisely, we have the following formula:

$$\kappa(Q) = \begin{cases} 0 & \text{if } D(Q) = 0\,; \\ \kappa, \ \kappa < 0 & \text{if } det(\cosh\sqrt{-\kappa} \cdot d_{ij}) = 0\,; \\ \kappa, \ \kappa > 0 & \text{if } det(\cos\sqrt{\kappa} \cdot d_{ij}) \text{ and } \sqrt{\kappa} \cdot d_{ij} \leq \pi \\ & \quad \text{and all the principal minors of order 3 are } \geq 0\,; \end{cases} \tag{3.2.1}$$

where $d_{ij} = d(x_i, x_j), 1 \leq i, j \leq 4$, $(\cosh\sqrt{-\kappa} \cdot d_{ij})$ is a shorthand for $(\cosh\sqrt{-\kappa} \cdot d_{ij})_{1 \leq i,j \leq n}$, etc., and $D(Q)$ denotes the so called *Cayley-Menger determinant*:

$$D(x_1, x_2, x_3, x_4) = \begin{vmatrix} 0 & 1 & 1 & 1 & 1 \\ 1 & 0 & d_{12}^2 & d_{13}^2 & d_{14}^2 \\ 1 & d_{12}^2 & 0 & d_{23}^2 & d_{24}^2 \\ 1 & d_{13}^2 & d_{23}^2 & 0 & d_{34}^2 \\ 1 & d_{14}^2 & d_{24}^2 & d_{34}^2 & 0 \end{vmatrix}. \tag{3.2.2}$$

There is, in fact, nothing mysterious about the formula above. Indeed, it has a very simple geometric meaning ensuing from the following fact:

$$D(p_1, p_2, p_3, p_4) = 8\big(Vol(p_1, p_2, p_3, p_4)\big)^2, \tag{3.2.3}$$

where $Vol(p_1, p_2, p_3, p_4)$ denotes the (un-oriented) volume of the parallelepiped determined by the vertices $p_1, ..., p_4$ (and with edges $\overrightarrow{p_1 p_2}, \overrightarrow{p_1 p_3}, \overrightarrow{p_1 p_4}$).[5] From here immediately follows that

---

[4]For the formulation and proof of this classical result, see, e.g. [63].

[5]As a historical note, it is perhaps worthwhile to recall that Formula 3.2.3 above was proved by Cayley in his very first mathematical paper [350] (published while he was still begrudgingly making his living as a lawyer!...)

Proposition 3.2.31 *The points $p_1, ..., p_4$ are the vertices of a non-degenerate simplex in $\mathbb{R}^3$ iff $D(p_1, p_2, p_3, p_4) \neq 0$:*

Evidently, this means also that the opposite assertion also holds, namely that a simplex of vertices $p_1, ..., p_4$ is degenerate, i.e. isometrically embeddable in the plane $\mathbb{R}^2 \equiv S_0$.

It is not hard deduce that the expressions appearing in Formula (7.0.3) for the cases where $\kappa \neq 0$ represent the equivalents of $D(Q)$ in the hyperbolic, respective spherical cases, using the well-known fact, that, in the spherical (resp. hyperbolic) metric, the distances d_{ij} are replaced by $\cos d_{ij}$ (resp. $\cosh d_{ij}$). However, the proof of this fact, as well for the analogous formulas and results in higher dimension are far too encumbering for this restricted exposition, so we have no choice but to advise the reader to consult [212].

Remark 3.2.32 *A stronger result along these lines also exists. We formulate it – for convenience and practicality – for the case $n = 3$, only. However, it is readily generalized to any dimension. (For proofs and further details, see [212].)*

Theorem 3.2.33 *Let $d_{ij} > 0, 1 \leq i, j \leq 4, i \neq j$. Then there exists a simplex $\tau = T(p_1, ..., p_4) \subseteq \mathbb{R}^3$, such that $d(x_i, x_j) = d_{ij}, i \neq j$; iff $D(p_i, p_j) < 0, (\forall) \{i, j\} \subset \{1, ..., 4\}$ and $D(p_i, p_j, p_k) > 0, (\forall) \{i, j, k\} \subset \{1, ..., 4\}$; where, for instance,*

$$D(p_1, p_2) = \begin{vmatrix} 0 & 1 & 1 \\ 1 & 0 & d_{12}^2 \\ 1 & d_{12}^2 & 0 \end{vmatrix}$$

and

$$D(p_1, p_2, p_3) = \begin{vmatrix} 0 & 1 & 1 & 1 \\ 1 & 0 & d_{12}^2 & d_{13}^2 \\ 1 & d_{12}^2 & 0 & d_{23}^2 \\ 1 & d_{13}^2 & d_{23}^2 & 0 \end{vmatrix} ;$$

etc.

It turns out that one can further relax the conditions of the previous theorem to obtain

Proposition 3.2.34 *Let $d_{ij} > 0, 1 \leq i, j \leq 4, i \neq j$. Then there exists a simplex $\tau = \tau(p_1, ..., p_4) \subseteq \mathbb{R}^3$, such that $d(x_i, x_j) = d_{ij}, i \neq j$; iff $D(p_1, p_2, p_3, p_4) \neq 0$ and $\text{sign}(D(p_1, p_2, p_3, p_4)) = +1$.*

Proof 3.1(Sketch) *Sufficient to show (by using standard operations on determinants) that:*

$$D(p_1, p_2, p_3)D(p_1, ..., \hat{p}_i, ..., p_4) = M_{i4}^2 + D(p_1, ..., \hat{p}_i, ..., p_4)D(p_1, p_2, p_3, p_4) ;$$

where M_{i4} is the cofactor (in D) of d_{i4}^2, and were we used the notation: $\{p_1, ..., \hat{p}_i, ..., p_4\} = \{p_1, p_2, p_3, p_4\} \setminus \{p_i\}$.

Exercise 3.2.35 *Complete the proof of Proposition 3.2.34.*

3.2.2 Computation of Wald Curvature II: An Approximation

Unfortunately, using Formula (3.2.1) for the actual computation of $\kappa(Q)$ is anything but simple, since the equations involved are – apart from the Euclidean case – transcendental, therefore not solvable, in general, using elementary methods. Moreover, they tend to display a numerical instability when solved with computer assisted methods. (See [296], [285] for a more detailed comments and some numerical results.)

Note that Formula (3.2.1) implies that, in practice, a renormalization might be necessary for some of the vertices of positive Wald-Besetkovskii curvature, which represents yet another impediment in it use.

Fortunate enough, there exists a good approximation result, due to Robinson. Not only does his result give a rational formula for approximating $\kappa(Q)$ and provide good error estimates, it also solves one other problem inherent in the use of the Wald curvature, namely the possible lack of uniqueness of the computed curvature. The way to circumvent this difficulty and the other pitfalls of Formula (3.2.1) is to make appeal to the simpler geometric configuration of sd-quads:

Theorem 3.2.36 ([262]) Given the metric semi-dependent quadruple $Q = Q(p_1, p_2, p_3, p_4)$, of distances $d_{ij} = d(p_i, p_j)$, $i, j = 1, ..., 4$; the embedding curvature $\kappa(Q)$ admits a rational approximation given by:

$$K(Q) = \frac{6(\cos \angle_0 2 + \cos \angle_0 2')}{d_{24}(d_{12} \sin^2(\angle_0 2) + d_{23} \sin^2(\angle_0 2'))} \tag{3.2.4}$$

where: $\angle_0 2 = \angle(p_1 p_2 p_4)$, $\angle_0 2' = \angle(p_3 p_2 p_4)$ represent the angles of the Euclidian triangles of sides d_{12}, d_{14}, d_{24} and d_{23}, d_{24}, d_{34}, respectively.

Moreover the *absolute error* R satisfies the following inequality:

$$|R| = |R(Q)| = |\kappa(Q) - K(Q)| < 4\kappa^2(Q)\mathrm{diam}^2(Q)/\lambda(Q), \tag{3.2.5}$$

where $\lambda(Q) = d_{24}(d_{12} \sin \angle_0 2 + d_{23} \sin \angle_0 2')/S^2$, and where $S = Max\{p, p'\}$; $2p = d_{12} + d_{14} + d_{24}$, $2p' = d_{32} + d_{34} + d_{24}$.

Remark 3.2.37 *(a) The function $\lambda = \lambda(Q)$ is continuous and 0-homogenous as a function of the d_{ij}-s. Moreover: $\lambda(Q) \geq 0$ and $\lambda(Q) = 0 \Leftrightarrow \sin \angle_0 2 = \sin \angle_0 2' = 0$, i.e. iff Q is linear. [Therefore, for sd-quads $\lambda(Q) > 0$ and, moreover, $\lambda(Q) \to 0 \Rightarrow Q \to linearity.]*

(b) Since $\lambda(Q) \neq 0$ it follows that: $K(Q) \in \mathbb{R}$ for any quadrangle Q. In addition: $sign(k(Q)) = sign(K(Q))$.

(c) If Q is any sd-quad, then $\kappa^2(Q)\mathrm{diam}^2(Q)/\lambda(Q) < \infty$. Moreover, if $\lambda(Q) \not\gg 0$,[6] then $\kappa^2(Q)\mathrm{diam}^2(Q)/\lambda(Q) \not\gg 0$ i.e. $|R|$ is small if Q is not close to linearity.
In this case $|R(Q)| \sim diam^2(Q)$ (for any given Q).

The basic idea of the proof of Theorem 3.2.36 is to basically calque, in a general metric setting, the original way of defining Gaussian curvature – in this case,

---

[6]i.e. for not very small values of $\lambda(Q)$

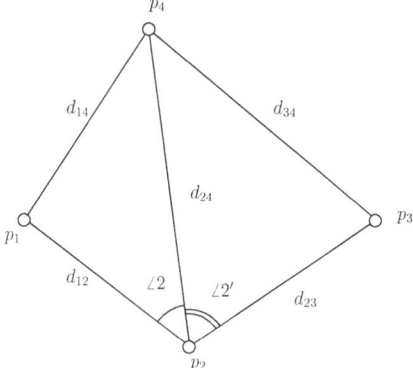

Figure 3.6 Euclidean triangles corresponding to an sd-quad [280]

rather than accounting for the area distortion, one measures the curvature by the amount of "bending" one has to apply to a general planar quadruple so that it may be "straightened" to a triangle $\triangle(p_1p_3p_4)$, with p_2 lying on the edge p_1p_3 – i.e. isometrically embedded as a *sd-quad* – in some S_κ. The proof does not necessitate additional special techniques, however, it is rather convoluted. Therefore, while for the sake of completion we do bring it below, we also would like to assure that it can be skipped without any detriment to the understanding of the remainder of the book.

Proof Consider two planar (i.e. embedded in $R^2 \equiv S_0$) triangles $\triangle p_1 p_2 p_4$ and $\triangle p_2 p_3 p_4$, and denote by $\triangle p_{1,k} p_{2,k} p_{4,k}$ and $\triangle p_{2,k} p_{3,k} p_{4,k}$ their respective isometric embeddings into S_k. Then $p_{i,k} p_{j,k}$ will denote the geodesic (of S_k) through $p_{i,k}$ and $p_{j,k}$. Also, let $\angle_k 2$ and $\angle_k 2'$ denote, respectively, the following angles of $\triangle p_{1,k} p_{2,k} p_{4,k}$ and $\triangle p_{2,k} p_{3,k} p_{4,k}$: $\angle_k 2 = \angle p_{1,k} p_{2,k} p_{4,k}$ and $\angle_k 2' = \angle p_{2,k} p_{3,k} p_{4,k}$ (see Figure 3.6).

But $\angle_k 2$ and $\angle_k 2'$ are strictly increasing as functions of k. Therefore the equation

$$\angle_k 2 + \angle_k 2' = \pi \tag{3.2.6}$$

has at most one solution k^*, i.e. k^* represents the unique value for which the points p_1, p_2, p_4 are on a geodesic in S_k (for instance on $p_1 p_4$).

But that means that k^* is precisely the embedding curvature, i.e. $k^* = \kappa(Q)$, where $Q = Q(p_1, p_2, p_3, p_4)$.

Equation (3.2.6) is equivalent to

$$\cos^2 \frac{\angle_{k^*} 2}{2} + \cos^2 \frac{\angle_{k^*} 2'}{2} = 1$$

The basic idea being the comparison between metric triangles with equal sides, embedded in S_0 and S_k, respectively, it is natural to consider instead of the previous equation, the following equality:

$$\theta(k, 2) \cdot \cos^2 \frac{\angle_0 2}{2} + \theta(k, 2') \cdot \cos^2 \frac{\angle_0 2'}{2} = 1 \tag{3.2.7}$$

where we denote:

$$\theta(k, 2) = \frac{\cos^2 \frac{\angle_{k^*} 2}{2}}{\cos^2 \frac{\angle_0 2}{2}} ; \quad \theta(k, 2') = \frac{\cos^2 \frac{\angle_{k^*} 2'}{2}}{\cos^2 \frac{\angle_0 2'}{2}} .$$

Since we want to approximate $\kappa(Q)$ by $K(Q)$ we shall resort – naturally – to expansion into MacLaurin series. We are able to do this because of the existence of the following classical formulas:

$$\cos^2 \frac{\measuredangle_k 2}{2} = \frac{\sin(p\sqrt{k}) \cdot \sin(d\sqrt{k})}{\sin(d_{12}\sqrt{k}) \cdot \sin(d_{24}\sqrt{k})} ; \; k > 0 ;$$

$$\cos^2 \frac{\measuredangle_k 2}{2} = \frac{\sinh(p\sqrt{k}) \cdot \sinh(d\sqrt{k})}{\sinh(d_{12}\sqrt{k}) \cdot \sinh(d_{24}\sqrt{k})} ; \; k < 0 ;$$

and, of course

$$\cos^2 \frac{\measuredangle_0 2}{2} = \frac{pd}{d_{12}d_{24}} ;$$

were $d = p - d_{14} = (d_{12} + d_{24} - d_{14})/2$ (and the analogous formulas for $\cos^2 \frac{\measuredangle_{k'} 2}{2}$).

By using the development into series of $f_1(x) = \frac{\sin \sqrt{x}}{\sqrt{x}}$ and $f_2(x) = \frac{\sinh \sqrt{x}}{\sqrt{x}}$; one (easily) gets the desired expansion for $\theta(k, 2)$:

$$\theta(k, 2) = 1 + \frac{1}{6}kd_{12}d_{24}\big(\cos(\measuredangle_0 2) - 1\big) + r ; \qquad (3.2.8)$$

where: $|r| < \frac{3}{8}k^2 p^4$, for $|kp^2| < 1/16$.

By applying (3.2.8) to (3.2.7), we receive:

$$\left[1 + \frac{1}{6}k^* d_{12}d_{24}\big(\cos(\measuredangle_0 2) - 1\big) + r\right]\cos^2 \frac{\measuredangle_0 2}{2} + \qquad (3.2.9)$$

$$\left[1 + \frac{1}{6}k^* d_{23}d_{24}\big(\cos(\measuredangle_0 2') - 1\big) + r'\right]\cos^2 \frac{\measuredangle_0 2'}{2} = 1 ;$$

for: $|r| + |r'| < \frac{3}{4}(k^*)^2 (Max\{p, p'\})^4 = \frac{3}{4}(k^*)^2 S^4$.

By solving the linear equation (in variable k^*) (3.2.9) and using some elementary trigonometric transformation one has:

$$k^* = \frac{6(\cos \measuredangle_0 2 + \cos \measuredangle_0 2')}{d_{24}(d_{12}\sin^2(\measuredangle_0 2) + d_{23}\sin^2(\measuredangle_0 2'))} + R ;$$

where:

$$|R| < \frac{12(|r| + |r'|)}{d_{24}(d_{12}\sin^2(\measuredangle_0 2) + d_{23}\sin^2(\measuredangle_0 2'))} < \frac{9(k^*)^2 \max\{p, p'\}}{d_{24}(d_{12}\sin^2(\measuredangle_0 2) + d_{23}\sin^2(\measuredangle_0 2'))} .$$

But $k^* \equiv \kappa(Q)$, so we get the desired formula (3.2.4).

To prove the correctness of the bound (3.2.5) one has only to observe that:

$$S = Max\{p, p'\} < 2diam(Q), \; (diam(Q) = \max_{1 \le i < j \le 4}\{d_{ij}\}),$$

and perform the necessary arithmetic manipulations.

□

Since the Gaussian curvature $k_G(p)$ at a point p is given by:

$$k_G(p) = \lim_{n \to 0} \kappa(Q_n) ;$$

where $Q_n \to Q = \Box p_1 p p_3 p_4$; $diam(Q_n) \to 0$, from *Remark 4.6.(c)* we immediately infer that the following theorem holds:

Theorem 3.2.38 *Let S be a differentiable surface. Then, for any point $p \in S$:*

$$k_G(p) = \lim_{n \to 0} K(Q_n);$$

for any sequence $\{Q_n\}$ of sd-quads that satisfy the following condition:

$$Q_n \to Q = \square p_1 p p_3 p_4; \; diam(Q_n) \to 0.$$

In other words, while the result above is quite expected – but nevertheless quite necessary if one aims for practical implementations – it shows that the ideas and results above do satisfy the necessary convergence properties required in applications.

Remark 3.2.39 *The convergence result provided in Theorem 3.2.38 is not just in the sense of measures and errors of different signs do not simply cancel each other. Indeed, $\text{sign}(\kappa(Q)) = \text{sign}(K(Q))$, for any metric quadruple Q.*

Remark 3.2.40 *It turns out [262] that the approximation formula (which looks quite daunting and forbidding) can be simplified in special case, for which "nicer" formulas cab be obtained:*

1. *If $d_{12} = d_{32}$, then*

$$K(Q) = \frac{12}{d_{13} \cdot d_{24}} \cdot \frac{\cos \angle_0 2 + \cos \angle_0 2'}{\sin^2 \angle_0 2 + \sin^2 \angle_0 2'}; \tag{3.2.10}$$

(here we have of course: $d_{13} = 2d_{12} = 2d_{32}$); or, expressed as a function of distances alone:

$$K(Q) = 12 \frac{2d_{12}^2 + 2d_{24}^2 - d_{14}^2 - d_{13}^2}{8d_{12}^2 d_{24}^2 - (d_{12}^2 + d_{24}^2 - d_{14}^2)^2 - (d_{12}^2 + d_{24}^2 - d_{34}^2)^2} \tag{3.2.11}$$

2. *If $d_{12} = d_{32} = d_{24}$ and if the following condition also holds:*

3. *$\angle_0 2' = \pi/2$; i.e. if $d_{34}^2 = d_{12}^2 + d_{24}^2$ or, considering (2), also: $d_{34}^2 = 2d_{12}^2$ then*

$$K(Q) = \frac{6 \cos \angle_0 2}{d_{12}(1 + \sin^2 \angle_0 2)} = \frac{2d_{12}^2 - d_{14}^2}{4d_{12}^4 + 4d_{14}^2 d_{12}^2 - d_{14}^4}. \tag{3.2.12}$$

Furthermore, when approximating the curvature at a point p on a surface, that is in a typical instance in applications, one can restrict him/herself to a 1-parameter family of metric quadruples $Q(\rho)$, by considering a geodesic γ (see Chapter 6) through the point $p = p_2$, construct on γ the points p_1 and p_3 such that $d(p, p_1) = d(p, p_3) = \rho$, as well as the point p_4 such that $d(p, p_4) = \rho$ and, furthermore, such that $\triangle p p_3 p_4$ is congruent with a Euclidean right triangle [262].

The question whether Formula (3.2.4) (or any of its variations mentioned above) is truly efficient in applications arises naturally. The following example, due also to Robinson, indicates that, at least in some cases, the actual computed error is far smaller then the theoretical one provided by Formula (3.2.5).

Example 3.2.41 ([262]) *Let Q_0 be the quadruple of distances $d_{12} = d_{23} = d_{24} = 0.15, d_{14} = d_{34}$ and of embedding curvature $\kappa = \kappa(Q_0) = 1$. Then $\kappa S^2 < 1/16$ and $K(Q_0) \approx 1.0030280$, whereas the error computed using formula (3.2.5) is $|R| < 0.45$.*

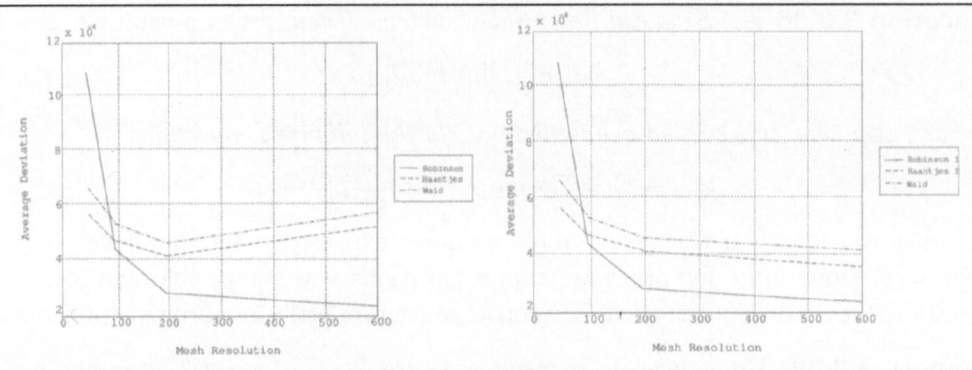

Figure 3.7 Left: Approximation errors with respect to mesh resolution (as number of triangles) [285]. Right: Improved approximation errors with respect to mesh resolution (as number of triangles). Note that marked improvement on the performance of the methods resulting from the addition of even one supplementary direction.

3.2.3 Applications of Wald Curvature

Clearly, as already suggested by the example considered in Figure 3.5 above, the main test cases for applications of Wald curvature that present themselves are in Imaging and Graphics. Much to our chagrin, there is preciously little that we can present the reader in terms of experimental results. A toy experiment, comparing the performance of Wald curvature and its approximation due to Robinson with Haantjes curvature (employed like in Section 3.1 above) of the torus of revolution, computed on the standard square grid in the parametric plane, as well as its improvement by adding all the diagonals in one direction is shown in Figure 3.7. A further implementation on a classical natural test image, displaying very modest results, is shown in Figure 3.8. The results above demonstrate that Wald and Robinson curvatures have limited applications to classical Imaging but paradoxically also show that in the context of Graphics, where more directions (quadruples) are available and, moreover, where various mesh improvement methods are available, they might prove quite useful.

However, for classical Imaging and Pattern Recognition ends, one can make appeal to the related *Wald curvature measure*, akin to the Menger curvature measure we have already encountered:

$$\mu_W(v) = K_W(v) \cdot \text{Area}(St(v));$$ (3.2.13)

where $St(v)$ denotes the star of the vertex v. Evidently, the applications of the Wald measure would be similar to those of the Menger measure, with the potential for uses in more general settings (e.g. cloud of points).

However, it is probable that the most impact Wald curvature might have is in such instances where polyhedral complexes that are not surfaces/manifolds arise, such as in Complex Networks and Biomathematics, or in tandem with such methods as Persistent Homology.

Figure 3.8 Mean curvature, computed using the Wald curvature of the grayscale "Lena" [285].

Moreover, Wald curvature is doubly advantageous from a theoretical viewpoint: On the one hand, it allows us to define a metric Ricci curvature for 3- and higher-dimensional polyhedral manifolds, as well as related Ricci flow (see Chapter 12). On the other hand it allows us to make the connection between the older, yet computable in an applicative context metric curvatures and the more modern Comparison Geometry using *Alexandrov curvature* (see Appendix Alexandrov spaces).

3.2.4   Wald Curvature Revisited

We return to the very definition of Wald embedding curvature and show that it can be characterized in terms of the angles of the model triangles. To this end, we need first some further preliminaries:

Let $Q = \{x_1, x_2, x_3, x_4\}$ be a *metric quadruple* and let $V_\kappa(x_i)$ be defined as follows:

$$V_\kappa(x_i) = \alpha_\kappa(x_i; x_j, x_l) + \alpha_\kappa(x_i; x_j, x_m) + \alpha_\kappa(x_i; x_l, x_m) \qquad (3.2.14)$$

where $x_i, x_j, x_l, x_m \in Q$ are distinct, and κ is any number (see Figure 3.9).

We can now state the sought characterization of Wald-Berestovskii in terms of angle sum:

Proposition 3.2.42 ([252], Theorem 23) *Let (X, d) be a metric space and let $U \in X$ be an open set. U is a region of curvature $\geq \kappa$ iff $V_\kappa(x) \leq 2\pi$, for any metric quadruple $\{x, y, z, t\} \subset U$.*

The presence of angles in result above raises two issues regarding the role of angles in metric geometry (which, after all, purportedly makes appeal only to distances between pairs of points):

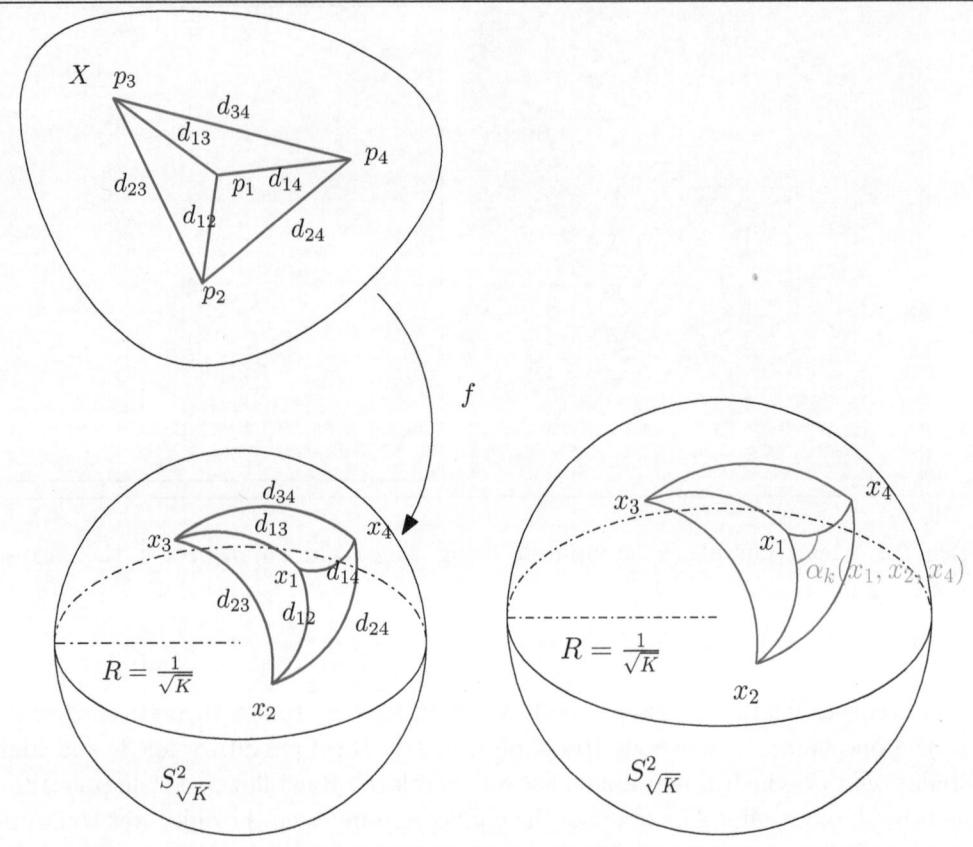

Figure 3.9 The angles $\alpha_\kappa(x_i, x_j, x_l)$ (right), induced by the isometric embedding of a metric quadruple in $\mathbb{S}^2_{\sqrt{\kappa}}$ (left).

1. From a more theoretical viewpoint, Proposition 3.2.42 shows that a "purely metric" (i.e. angle independent) approach is not truly possible for, even if "pure metric"' formulas can be devised, the angles are also always "hidden in the substrate".[7] We might also recall that one can express (at least in the Euclidean case) the Cayley-Menger determinant in terms of (face or, alternatively, dihedral) angles incident to a vertex (see Section Metric Torsion).

2. From the practical (applications oriented) point of view, the proposition above shows that, in fact, the metric approach to curvature is essentially equivalent to the combinatorial (angle-based) one, as far as polyhedral surfaces (in $\mathbb{R}^3$) are considered.[8] Therefore, as far as approximations of smooth surfaces in $\mathbb{R}^3$ are concerned, both approaches converge to the classical Gauss curvature.)

---

[7]We shall encounter again this phenomenon when discussing the discretization of curvature, via thick triangulations and metric curvatures, in the context of the so called *Regge Calculus*.

[8]The metric approach is, to be sure, far more general and applicable to a very large class of metric spaces (thence applied to fields other than Graphics and Imaging).

Recall that, in the context of polyhedral surfaces, the natural choice for the set U required in Definition 3.2.10 is the *star* of a given vertex v, that is, the set $\{e_{vj}\}_j$ of edges incident to v. Therefore, for such surfaces, the set of metric quadruples containing the vertex v is finite.

Equipped with this quite simple and intuitive choice for U (and in analogy with Alexandrov spaces – see also Section 5.1 below) it is quite natural to consider, for PL surfaces, the following definition of the *discrete* (PL, or "finite scale") Wald curvature $K_W(v)$ at the vertex v:

$$K_W(v) = \min_{v_i, v_j, v_k \in \mathrm{Lk}(v)} K_W^{ijk}(v), \qquad (3.2.15)$$

where $K_W^{ijk}(v) = \kappa(v; v_i, v_j, v_k)$, and where $\mathrm{Lk}(v)$ denotes the *link* of the vertex v.[9] Note that here we consider the (intrinsic) PL distance between vertices.

This definition, in combination with Proposition 3.2.42 the general problem of the existence of isometric embeddings of generic metric metric spaces into gauge spaces. While in its full generality this is, of course, an unattainable goal, one would still be interested in the much more restricted, but important in the applied setting (Graphics, Imaging, Mathematical Modeling, Networking etc.), problem of isometric embedding of PL surfaces in $\mathbb{R}^3$. A partial result in this direction is a criterion for the local isometric embedding of polyhedral surfaces in $\mathbb{R}^3$, resemblant to the classical Gauss fundamental (compatibility) equations in the classical differential geometry of surfaces (and which we will discuss at length in Chapter 8), that we proved in [277]:

Given a vertex v, with metric curvature $K_W(v)$, the following system of inequalities should hold:

$$\begin{cases} \max A_0(v) \leq 2\pi; \\ \alpha_0(v; v_j, v_l) \leq \alpha_0(v; v_j, v_p) + \alpha_0(v; v_l, v_p), & \text{for all } v_j, v_l, v_p \sim v; \\ V_\kappa(v) \leq 2\pi; \end{cases} \qquad (3.2.16)$$

Here

$$A_0 = \max_i V_0; \qquad (3.2.17)$$

"$\sim$" denotes incidence, i.e. the existence of a connecting edge $e_i = vv_j$ and, of course, $V_\kappa(v) = \alpha_\kappa(v; v_j, v_l) + \alpha_\kappa(v; v_j, v_p) + \alpha_\kappa(v; v_l, v_p)$, where $v_j, v_l, v_p \sim v$, etc.

Returning to the analogy with the Gauss compatibility equation, the first two inequalities represent the (extrinsic) embedding condition, while the third one represents the intrinsic curvature (of the PL manifold) at the vertex v.

---

[9] Recall that the *link* $\mathrm{lk}(v)$ of a vertex v is the set of all the faces of $\overline{\mathrm{St}}(v)$ that are not incident to v. Here $\overline{\mathrm{St}}(v)$ denotes the *closed star* of v, i.e. the smallest subcomplex (of the given simplicial complex K) that contains $\mathrm{St}(v)$, namely $\overline{\mathrm{St}}(v) = \{\sigma \in \mathrm{St}(v)\} \cup \{\theta \mid \theta \leqslant \sigma\}$, where $\mathrm{St}(v)$ denotes the *star* of v, that is the set of all simplices that have v as a face, i.e $\mathrm{St}(v) = \{\sigma \in K \mid v \leqslant \sigma\}$.

CHAPTER 4

Gauss Curvature and Theorema Egregium

4.1 THEOREMA EGREGIUM

Looking at Equation $(*)^1$ one one would be conducted, inevitably, to presume that Gauss curvature depends on the way the surface is *embedded* ("placed" or "drawn", so to say) in (Euclidean) space, given that it depends on the second fundamental form, hence on the normal mapping, thence clearly on its position in space. In other words, *prima facie*, Gauss curvature would depend on its position in space, thus being what it is called a *extrinsic property*. It is therefore quite a surprising fact that, in truth, Gauss curvature is an *intrinsic property*, i.e. it depends solely on the (geometric properties of the) surface itself, and not on its specific (but clearly aleatory) positioning (*embedding*) in space. Precisely how surprising a fact this is demonstrated by the fact that Gauss himself was quite surprised when making this discovery hence he named it "Theorema Egregium", i.e. in Latin, "The Remarkable"' (or "Excellent") "Theorem". We state it here formally for the record:

Theorem 4.1.1 (Gauss' Theorema Egregium) *Gauss curvature is an intrinsic property.*

The way one would prove this in the classical (smooth) setting – as Gauss has indeed done – is to demonstrate an atrocious formula for K that expresses it solely in terms of the first fundamental form, thus showing it to be, indeed, intrinsic (since shows that K depends only on the metric properties of the surface and not on its position in space). The most common such formula is the one below:[2]

---

[1] $(*)$ $K = \frac{\det(II)}{\det(I)} = \frac{eg-f^2}{EG-F^2}$.

[2] It was first proven by Brioschi in 1852 and, independently by R. Baltzer (see [317], p. 112 and [43], p. 94) Gauss owns original proof of 1827 [130] is much earlier, but his version of the formula, albeit equivalent, is more complicated and much harder to arrive at.

DOI: 10.1201/9781003350576-4

$$K(EG-F^2)^2 = \begin{vmatrix} -\frac{1}{2}G_{uu} + F_{uv} - \frac{1}{2}E_{vv} & \frac{1}{2}E_u & F_u - \frac{1}{2}E_v \\ F_v - \frac{1}{2}G_u & E & F \\ \frac{1}{2}G_v & F & G \end{vmatrix} - \begin{vmatrix} 0 & \frac{1}{2}E_v & \frac{1}{2}G_u \\ \frac{1}{2}E_v & E & F \\ \frac{1}{2}G_u & F & G \end{vmatrix}.$$

(4.1.1)

We shall spare he reader the ordeal of going through the details of the proof, and we rather refer the interested ones to, e.g. [309], pp. 129-131.

Remark 4.1.2 *As already mentioned before, there exist (many) other formulas[3] for K that show that is intrinsic. Arguably the most elegant one is the following one due to Frobenius (see [43] p. 117 or [47], p. 365):*

$$K = -\frac{1}{4(EG-F^2)^2} \begin{vmatrix} E & E_u & E_v \\ F & F_u & F_v \\ G & G_u & G_v \end{vmatrix} - \frac{1}{\sqrt{EG-F^2}} \left(\frac{\partial}{\partial v} \frac{E_v - F_u}{\sqrt{EG-F^2}} - \frac{\partial}{\partial u} \frac{F_v - G_u}{\sqrt{EG-F^2}} \right);$$

(4.1.2)

While unfortunately irrational, it is much simpler and more symmetric (a fact always attractive to the mathematical eye, so to say. However, its charm does not resides solely on these facts and it presents an advantage also from the computational point of view: Given that it only contains only order 1 derivatives of E, F, G (thus only order 3 derivatives of the given parametrization) it is numerically more stable than the Brioschi formula, that contains order 2 derivatives of E, F, G (thus order 4 derivatives of the parametrization), therefore amply compensating for the presence of the square-root (see [113], [306]).

Exercise 4.1.3 *Prove Frobenius' formula.*

As we have stressed above, the classical proof is tedious and not very intuitive. However, the discrete approach has a clear advantage in this case, providing not just a simple, elementary proof, but also revealing the geometric insight behind Gauss' result. Fittingly, the very basic idea is much more geometric: One has to go back to the normal (Gauss) mapping. In fact, this shouldn't be too surprising to the reader, given that we have already seen in the previous chapter the relation between curvature and the variation of the normals in the curves setting (see Theorem 2.2.5). While the ideas that we'll introduce below still need a highly non-trivial understanding of surfaces geometry, it is seemingly possible, or even probable, that Gauss was well aware of the curve case and had in mind when developing the total[4] curvature of surfaces

Given a point p on a surface S, consider a small curve γ on S containing p in its interior. Its image by the normal mapping determines a corresponding small curve on the unit sphere $\mathbb{S}^2$ – see Figure 4.1. Let A, A' be the interior (domains) determined by γ and $N(\gamma)$, respectively.

---

[3]A particularly short and symmetric one one – but, unfortunately, using Christoffel symbols – is due to Liouville (1851) – see [317], p. 114 or [96], p. 367.

[4]better known to know, of course, as Gauss curvature

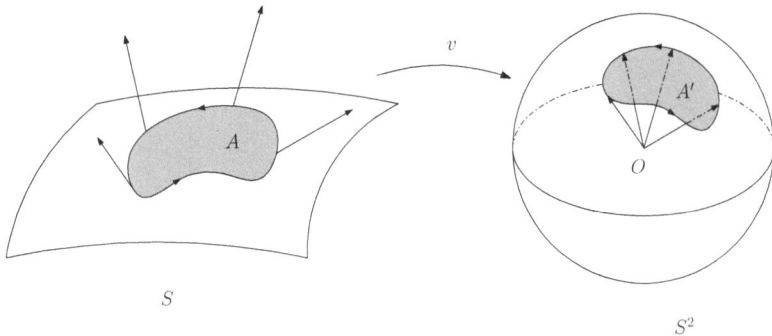

Figure 4.1 A small region around a point p on the surface S is mapped on a small region on the unit sphere, around the normal image $N(p)$ of p.

Denote $K(\gamma) = \frac{\text{Area}(A)}{\text{Area}(A')}$, and let γ collapse to p, such that $\text{diam}\gamma \to 0$. Then (obviously), both $\text{Area}(A)$ and $\text{Area}(A')$ tend to 0.

Question 4.1.4 *Why is it important that* $\text{diam}\gamma \to 0$ *and it is not sufficient to request only that* $\text{Area}(A) \to 0$*? Hint: Recall some Calculus 2 arguments.*

Now define

$$K(p) = \lim_{\text{diam}\gamma \to 0} K_\gamma . \tag{4.1.3}$$

We still have to provide in the definition for the possible negative K. It is here where the seemingly technical condition on the curve γ to be positively oriented intervenes. More precisely, one has to notice that the Gauss map preserves the sign of (small enough) curves around points with positive curvature, but reverses it for those around points of negative curvature – see Figure 4.2.

This is, in fact, Gauss' original definition of curvature – He next *proved* that this definition of curvature can be also expressed as the product of the principal curvatures! The fact that text-books (even ours!...) define Gauss curvature as a product of the maximal and minimal sectional curvatures has two reasons. The first one is hereditary/historical: This is (as we have seen) the first, "embryonic" approach to the Differential Geometry of surfaces before Gauss, and, traditionalism aside, it still appears to be intuitive. The second resides in the fact that, given the modern language and tools of Calculus and Linear Algebra, it is more direct or "economical" to give first this definition. (Recall, however, that these tools – at least in their contemporary versions – were not fully available in Gauss' time.) The problem with this tradition is not only that it lacks geometric intuition and depth, it is also so strongly engraved, by accumulated generations of identical teaching and learning, in students' consciousness (and even subconsciousness) that it prevents them, in sadly, to many cases, from accepting the alternative, geometric approach. This retreat in the comfort of familiar Calculus symbolism and computations is especially sad in the case of students in such fields as Graphics and Imaging, where the discrete, geometric methods

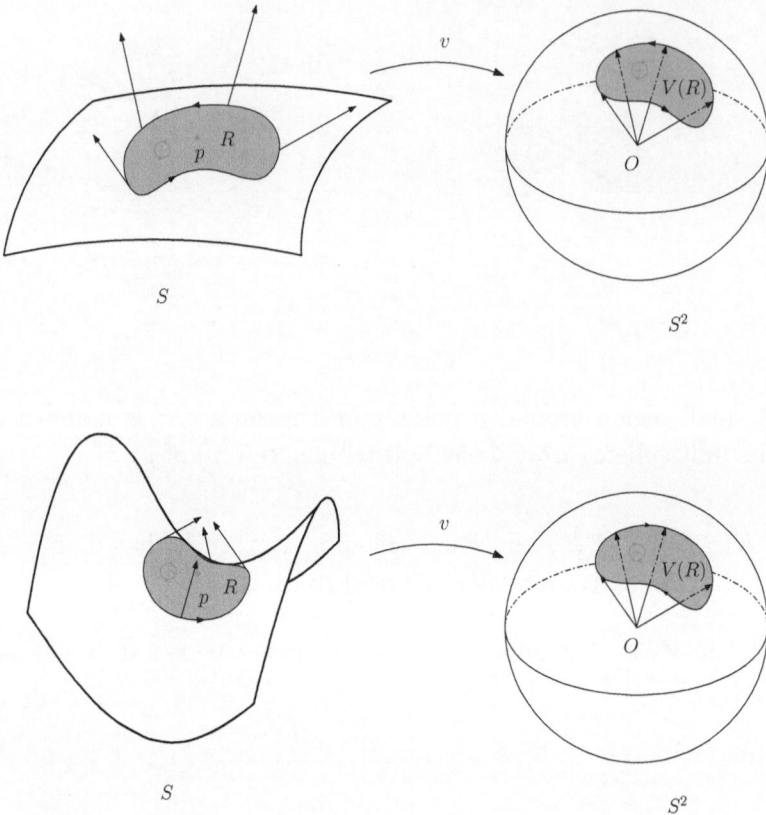

Figure 4.2 Orientation of infinitesimal curves is preserved at points of positive curvature (a), and reversed at points of negative curvature (b).

are – as we shall emphasize and illustrate again and again – essential. (In fact, it has been our firm belief for quite some time that the discrete geometric approach it not only more useful, it also far more simple and intuitive and, as such it should be incorporated in curricula [270].)[5]

Before proceeding further we should verify that this definition truly makes sense, by verifying that we obtain the right (expected) answer when computing K, in this

---

[5]It might help future engineers to appreciate Gauss' definition if they knew the personal history behind Gauss' discovery; It seems that at a certain point in his life, Gauss underwent what might be called a mathematically middle-life crisis: After achieving not-mean success in other fields than Geometry, he decided to abandon Mathematics and he went to work for a while in a "hight-tech industry" of his time, namely in topography. It is quite possible that the intuition he developed in land mensuration conducted him to his seminal work in Differential Geometry. (Incidentally, there exists a mathematically educational, so to say, anecdote (see [343]?) inspired, seemingly, by this period in Gauss life: As the story goes, Gauss was invited to a contest by a topographer, the goal of the contest being to see which of them too would draw more right angled triangles in a short, given period of time (say 30" or 1'). The topographer started drawing as fast as possible such triangles using his instrument (square rule). Gauss, instead, drew first a semicircle over a segment than started to connect points on it with the extremities of the segment. Needles to say that the anecdote ends by making clear who had won...)

manner, for some standard, elementary surfaces. Indeed, if $S \equiv \mathbb{R}^2$, then all the normals to any curve in the plane are perpendicular to the plane, hence the normal image of any curve reduces to a point, therefore the Gauss curvature of the plane is identically 0, as expected. Likewise, if $S \equiv \mathbb{S}^2$ then the normal image of any curve its the very curve itself, hence, for the (unit) sphere $K \equiv 1$, again corresponding to the previous computations.

Exercise 4.1.5 *Verify, using the same type argument, that*
(i) $K(\text{cylinder}) \equiv 0$;
(ii) $K(\text{cone}) \equiv 0$.
(Here, as before, we consider right, circular cones and cylinders.)
Hint: Consider the images of these images under the normal map, as discussed above.

It still remains to show that this new definition of Gauss curvature coincides with the previous ("technical") one. We bring here the following somewhat informal proof:

Proof 4.1 *We have that $K = [N \, N_u \, N_v]/\sqrt{g}$. Since $N_u \times N_v \parallel N$, it follows that*

$$|N_u \times N_v| = |K|\sqrt{g}. \tag{4.1.4}$$

On the other hand, $\text{Area}(A_1) = \int \int |\frac{\partial f}{\partial u} du \times \frac{\partial f}{\partial v} dv| = \int \int \sqrt{g} du dv$. Moreover, $\text{Area}(A_2) = \int \int |N_u du \times N_v dv|$, hence, by (4.1.4) it follows that $\text{Area}(A_2) = \int \int |K|\sqrt{g} du dv$. Therefore,

$$\lim_{\text{diam}\gamma \to 0} \frac{\text{Area}(A_1)}{\text{Area}(A_2)} = K.$$

Exercise 4.1.6 *Fully formalize the proof above.*

The route to the elementary proof we shall give bellow is to consider "*bendings*"[6], i.e transformations (of the surface) that leave lengths of curves (and angles of intersections of curves) invariant. In this suggestive language, Theorema Egregium is formulated as:

Gauss curvature is invariant under bendings.

The proof we bring here follows closely that of Hilbert and Cohn-Vossen [163] (a popularization work of real depth who, incidentally, pioneered, as far as we know, this discrete approach, at least in this context).

Proof 4.2 *Imagine the surface to be composed of a (very large) number of extremely thin – but rigid – plates, connected at the edges by hinges around which they can swivel – see Figure 4.3.*
If the number of plates around each vertex is at least 4, than the "umbrella" around each vertex can change shape by "vibrating" along the hinges.

---

[6]or, in modern language, "local isometries"

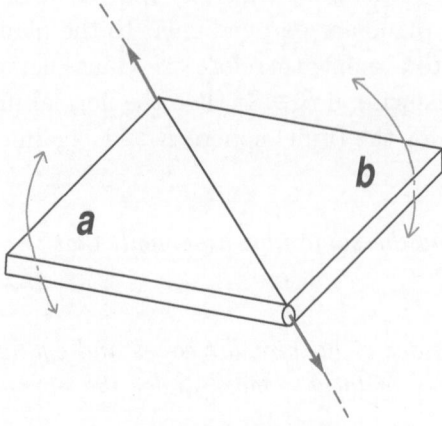

Figure 4.3 A hinge between two triangular rigid plates.

Question 4.1.7 *Why is the condition that the number of plates around each vertex be* > 3 *necessary? Is this a "reasonable"/natural condition?*

However, all this change in the shape (relative position) of the elements of each "umbrella" leave both lengths of curves and angles between intersecting curves invariant, that is to say they are bendings. Here we presume, without loss of generality, that all the curves intersect only in the interiors of the plates, and on the edges.

Question 4.1.8 *Why is this presumption a legitimate one?*

Next we consider unit normals to all the faces (triangles) adjacent (abutting) a vertex v one for each such triangle. Of course, one has an infinity of possibilities for the choice of the base point for the normals, but, since they are all parallel, we can chose, for each as a base point for each triangle in such a manner that the perpendiculars from these base points to each edge fall in the same point (see Figure 4.4, left). Let us denote by P the (non-planar polygon formed by these triangles, and by $\mathcal{P}$ its spherical image (see Figure 4.4, right), that is the spherical polygon of vertices k', l', m', n' *(and edges great circle arcs connecting the relevant pairs of vertices).*

We next prove that Area($\mathcal{P}$) *is invariant under bending. To this end we need the following lemma (which we shall prove slightly later):*

Lemma 4.1.9 Area($\mathcal{P}$) = Area(interior angles of $\mathcal{P}$)*, or, simply put, that the area of* $\mathcal{P}$ *depends solely on the sum of its (interior) angles.*

Therefore, suffices to prove that the sum of the (interior) angles of $\mathcal{P}$ *is invariant under bending. Since* $\alpha = \pi - \varphi$ *for each such angle* α*, (see Figure 4.4), the assertion follows immediately.*

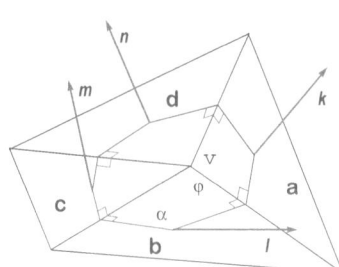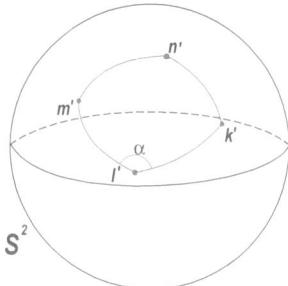

Figure 4.4 The spherical polygon corresponding, under the normal map, to the polygon determined by the normals to the faces around a vertex of triangular surface.

By considering finer and finer approximations of the given surface by such triangular plates, i.e. with ever increasing number of sides, passing to the limit as the number of triangles tends to infinity, the desired results follows.

Before taking a second look at the deep meaning of the proof above, we give the proof of Lemma 4.1.9 (which, by the way, is of no mean importance itself):

Proof 4.3 (Proof of Lemma 4.1.9) *We first determine the area of a lunule[7], ie. half of the area bounded by two great circles – see Figure 4.5. By examining it, it is easy to see that*

$$\frac{\mathrm{Area}(L)}{\mathrm{Area}(\mathbb{S}_{\mathbb{R}}{}^2)} = \frac{\alpha}{2\pi}\mathrm{Area}(\mathbb{S}_{\mathbb{R}}{}^2)\,,$$

where $\mathbb{S}_{\mathbb{R}}{}^2)$ is the sphere of radius R, therefore

$$\mathrm{Area}(L) = 2\alpha R^2\,.$$

We next consider the spherical triangle S, of angles α, β, γ (and sides a, b, c, respectively). Then

$$S + S_1 + S_2 + S_3 = 2\pi R^2$$

(since the left hand represents the area of a hemisphere.)
On the other hand, by examining Figure 4.6, it is easy to check that:

$$S + S_1 = 2\alpha R^2$$

$$S + S_2 = 2\beta R^2$$

$$S + S_3 = 2\gamma R^2$$

By summing the last three equalities we obtain

$$2(\alpha + \beta + \gamma)R^2 = 3S + (S_1 + S_2 + S_3) = 2S + (S + S_1 + S_2 + S_3) = 2S + 2\pi R^2$$

---

[7]i.e., in Latin, a "little moon"

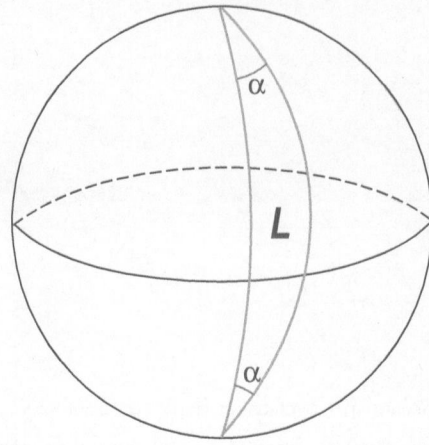

Figure 4.5 A lunule of angle α.

(By the first equality in the series.)
   It follows that

$$S = (\alpha + \beta + \gamma - \pi)R^2 \,. \qquad (4.1.5)$$

(of course, the formula is even simpler for $R = 1$, i.e. for the unis sphere.)
   Before continuing with our proof, let us make the following

Remark 4.1.10 *Since area is a positive function, from (4.1.5) we obtain the immediate – but, nevertheless, important – corollary that the (interior) angle sum of a spherical triangle is strictly greater than π.*

   To finish the proof of the lemma, we have only to recall that, if $\mathcal{P}$ has n sides and angles $\alpha_1, \ldots, \alpha_n$, then it is decomposable in n triangles, $S_1, \ldots, S_n$. Then, by applying (4.1.5) for each and every of these triangles, we obtain

$$\mathrm{Area}(\mathcal{P}) = [(\alpha_1 + \cdots + \alpha_n) - (n-2)\pi]\, R^2 \,.$$

Remark 4.1.11 *While the possibility of approximating arbitrarily well a surface by flat (Euclidean) triangles, and, indeed, dividing a surface in "curvilinear" triangles is a very intuitive statement, it necessitates, however, a quite non-trivial proof, at least in the general case. How non-trivial this proof might be, is underlined by the fact that Hilbert and Cohn-Vossen only claim that their proof of Theorema Egregium holds just for convex surfaces. We shall encounter these triangulations in "full force" further on, in the context of the Gauss-Bonnet Theorem and, instead of merely referring the reader to the Topology of Surfaces literature, we shall bring a quite simple and concise Differential Geometric demonstration for general surfaces, befitted for the case of smooth (at least $\mathcal{C}^2$) surfaces.*

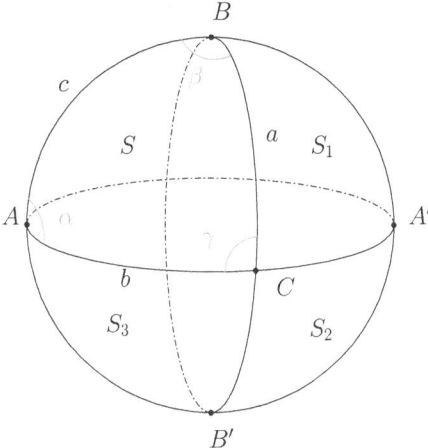

Figure 4.6 Computing the area of a spherical triangle of angles α, β, γ.

We can now return at the proof of the Theorema Egregium itself and try to understand its deeper significance. The main inside that the proof provides is that the Gauss curvature of a triangulated[8] or polyhedral surface at a vertex is a function of the sum angles of the faces incident with that vertex. In the *combinatorial* case, i.e. when all the edges are given length one and all triangles (faces) aria also equal to 1, the contribution of Area(P) in the proof above can, therefore, be take to be always 1, thus we obtain the following simple and appealing

Definition 4.1.12 (Combinatorial curvature) *Let $\mathcal{S}$ be a PL (polyhedral) surface and let v a vertex of $\mathcal{S}$. Then the* combinatorial *(Gauss) curvature (of $\mathcal{S}$ at v) is*

$$K(v) = 2\pi - \sum_i \alpha_i \,, \tag{4.1.6}$$

where the sum is taken over all the adjacent angles at v.

This *defect*[9] definition of curvature is not only elementary, it is also quite intuitive: Six equilateral (regular) triangles group perfectly in the plane (a fact that is commonly exploited and seen in floor tilings); the curvature of the plane is, of course 0 and, indeed, for any vertex we have $K(v) = 2\pi \sum_1^6 \pi/3 = 0$. If less then 6 triangles (but at least 3!) meet at a vertex, then the defect will be positive, hence the combinatorial curvature of the vertex will be positive too, corresponding to the intuition that a pyramid has – as a very bad approximation of an ellipsoid – positive curvature.

---

[8]technically formulated a *PL piecewise linear surface* or, more precisely, a *piecewise flat surface*

[9]$2\pi - \sum_i \alpha_i$ is called the   (of $\mathcal{S}$ at v) and it (obviously) measures the "distance" of a small neighborhood of v from being a planar disk.

When more then 6 faces meet at a vertex, defect, hence curvature, will be negative, a fact expected from such a (piecewise flat) saddle surface.

Remark 4.1.13 *Since $2\pi - 7\pi/3 = -\pi/3 \approx -1.046$, this definition of curvature allows us to construct – using an idea of Bill Thurston – simple and suggestive models of the Hyperbolic Plane: Start with a regular triangular grid in the plane, then cut-and-paste triangles such that, around each vertex, there will be precisely 7 triangles. From the computation above it follows that we obtain a piecewise flat surface, where all the curvature is concentrated at the vertices (and identically equal to -1, up to some small error), thus rendering a PL model of the Hyperbolic Plane, as promised. (For details on the construction and figures, see [255].)*

The combinatorial definition of curvature is not only simple and intuitive, it is also quite powerful, as we shall see, for instance in Section [Discrete Ricci Flow]. However, its importance transcends the fields of Discrete Differential Geometry and Graphics. Most importantly, its very combinatorial nature render it an extremely important tool in such domains where basic combinatorial objects, such as graphs, arise naturally, notably in Group Theory (via, for instance, the so called *Cayley graph*)– see, for instance, [211] and [117].[10]

For Graphics, CAGD, etc., ends for example (but also in Finite Element implementations) one would like to approximate the curvature of smooth surfaces, via the one of some polyhedral approximation. One of the best methods is based on the combinatorial curvature, but with a certain modification(adaptation), namely the following weighted variation:

$$K(v) = \frac{2\pi - \sum_i \alpha_i}{\frac{1}{3}\mathrm{Area}(P)}. \qquad (4.1.7)$$

(Here the same convention for the summation was used as above.)

Exercise 4.1.14 *Justify the method described above. In particular, explain the choice of $1/3$.*

It turns out that this simple approach gives a quite good approximation method, with nice convergence properties (in terms of size of the *mash*, i.e the maximal diameter of the faces of the triangular mesh used to approximate the surface) – see [319].

In fact, [319] brings quite a number of diverse approaches to the curvature approximation problem, especially for the computation of K. These methods, based on approximations/discretizations of the principal curvatures, various differential operators and tensors are less simple and certainly not as far-reaching as combinatorial curvature. We do not, therefore, detail them here further, but rather leave we contend ourselves of leaving them as exercises for the reader, as already sketched in the previous section. and further refer him to [319] and the bibliography within. Instead we shall concentrate, in the next section on a much more general approach based

---

[10]For obvious reasons we are prohibited from exploring this direction of study, even in the slightest detail, in the present text.

on metric curvature, akin to those introduced in Section 2 of Chapter 1 – but much more flexible and powerful.

We shall return to the problem of approximating curvature of surfaces, but first let us take a second, in depth, look at the Hilbert and Cohn-Vossen proof of Theorema Egregium that we presented above. Banchoff's approach, however, is quite different as it makes appeal to topological ideas, more precisely to the so called *Morse Theory* (for a full yet highly readable introduction into the field, see, the instantly classical [225][11]). We shall present here only the polyhedral setting, but we shall also point out to the reader, by means of a number of exercises, the original, smooth definitions and results.

The basic notion we need is that of *critical point of a height function*. To obtain the intuition necessary to develop the polyhedral ideas, let us first review the classical notions which spurred them: Let S^2 be a smooth, closed surface in $\mathbb{R}^3$ and let $\vec{v}$ be a and arbitrary direction in $\mathbb{R}^3$ (i.e. a unit vector, or a point on the unit sphere $\mathbb{S}^2$). We define the *height function h* as being the projection (function) of $\mathbb{R}^3$ on the line l determined by $\vec{v}$ – see Figure 4.7 for the standard example, to which we'll refer again.[12] A point $p \in S^2$ is called a *critical point* for h if the tangent plane to S^2 at p is perpendicular to l, otherwise it is called an *ordinary point*. All the points of the torus in Figure 4.7 are ordinary, except the maximum, the minimum and the two saddle points, that is a total of only four critical points.

To each critical point a numerical value is attached, namely we set $i(p, l) = +1$ if m is a local minimum or maximum, and $i(p, l) = -1$ if m is a saddle point. To this formal (and seemingly arbitrary) assignment we wish to find a more geometrical characterization.[13] The main observation is the following one: If p is an ordinary point, then the tangent plane (to S) at p is not "horizontal" (parallel to l), therefore it meets a "small" (infinitesimal) circle (on S) around p in precisely two points – see Figure 4.8. In contrast, the intersection of the tangent plane with such a circle at maximum or minimum point is void, whereas at a saddle point it will intersect an infinitesimal circle in four distinct points (Figure 4.8).

Based on the observation above one can then formally define the index in the following combinatorial manner:

$$i(p, l) = 1 - \frac{1}{2}|\{T_N(p) \cap C_\varepsilon(p)\}|; \qquad (4.1.8)$$

where $T_N(p)$ is the plane through p normal to l and $C_\varepsilon(p)$ denotes an infinitesimal circle centered at p; in other words

$$i(p, l) = 1 - \frac{1}{2}(\#\text{points in which the plane through } p \qquad (4.1.9)$$

$$\text{perpendicular to } T \text{ meets a "small circle" about } p \text{ on } M^2).$$

---

[11] Indeed, it is considered by some "the best Math book ever written" !...

[12] While in the simple case at hand the name "height function" is quite suggestive and concrete, in a more general context the "height" in question need not be related to a physical/geometric height. For instance, in classical Morse theory, the height is the energy functional of paths (having as critical points the geodesics) [225]; while in Alexandrov spaces height is given by the distance function (from a base point) [246].

[13] We shall indicate in the relevant exercises how the index emergences in the smooth case.

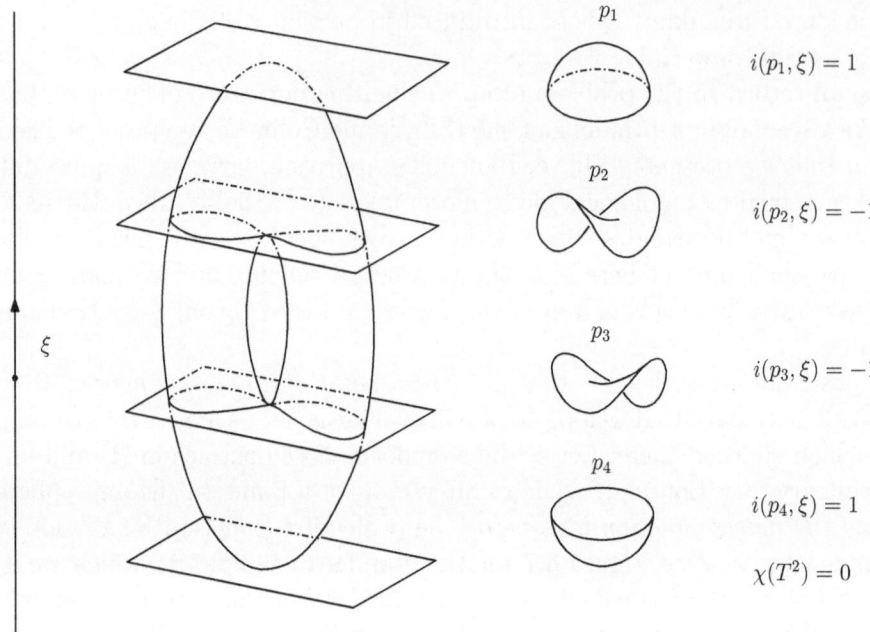

$$i(p_1, \xi) = 1$$

$$i(p_2, \xi) = -1$$

$$i(p_3, \xi) = -1$$

$$i(p_4, \xi) = 1$$

$$\chi(T^2) = 0$$

Figure 4.7 Height function for the torus, critical points and their indices (after [26])

While very simple, this definition is not quite what one would expect from a notion on smooth surfaces, both because of the vagueness of the notion of "small circle", and because, in practice, it would be quite difficult to determine the required number of intersections on a general ("not very smooth" surface, and for a general direction l ($\vec{v}$)). On the other hand, its form is almost what one would request from a definition befitting polyhedral surfaces.

To justify this assertion, let us first note that, for polyhedral surfaces, the *star* $St(p)$ of a vertex p, i.e. the set of all simplices incident to p (that is the edges and faces (including their edges and vertices containing p) plays the role of a "small" disk neighborhood centered at p, while the *link* $Lk(p)$, i.e. the polygon representing the boundary of $St(p)$ represents the polyhedral analog of a "small circle" around p. Observe also that a point is *ordinary* for the height function h if the plane perpendicular to l that passes through p divides $St(p)$ into two pieces. Any interior point of face of an edge is, therefore, an ordinary one for any direction *general* for the given polyhedral surface, i.e. such that $h(p) \neq h(q)$, for any two distinct vertices of S^2. Moreover, given that M^2 has only a finite number of edges, it follows that the number of non-general directions is finite, thus our analysis is not limited by using general directions, given the fact that they are the rule, rather then one of the finite number of exceptional cases. Furthermore, this represents the precise polyhedral equivalent of a classical result, namely that almost any direction $\vec{v} \in \mathbb{S}^2$, the associated height function has only a finite number of critical points, thus almost all directions (up to

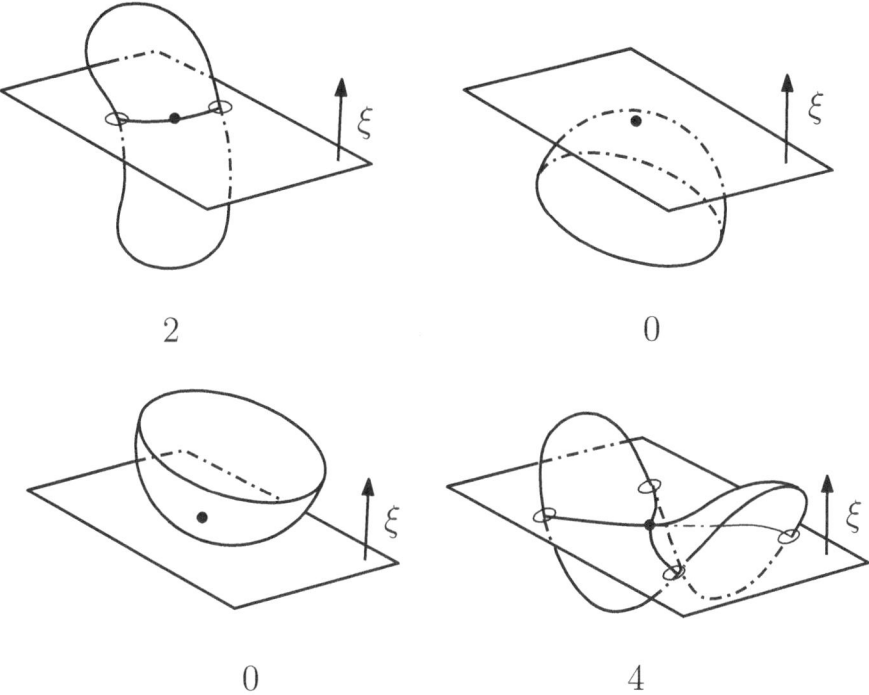

2

0

0

4

Figure 4.8 A geometric approach to the definition of the index (after [26]): $i(p, l) = 1 - \frac{1}{2}(\#\text{points in which the plane through } p \text{ perpendicular}$ to T meets a "small circle" about p on M^2).

a set of zero measure) is general (see [225]). In contrast, vertices represent critical points of all of the types arising for smooth surfaces – see Figure 4.9.

In fact, while for smooth surfaces the only possible critical points are maxima, minima and non-degenerate saddle points ([225]) – see Figure 4.8, right, on polyhedral surfaces *degenerate* critical points can also arise, like the so-called *monkey saddle*[14] – see Figure 4.10. For example, in Figure 4.10, while any hight function (direction) close to l has only non-degenerate critical points, on the polyhedral surface for any such direction there are precisely six triangles with p as a middle, thus for each such direction k, $i(p, k) = -2$, thus p cannot be a non-degenerate critical point.

Note that we said "almost"; however we can improve this definition of the index to become purely combinatorial, by making the following observation: We still can count intersections with "small disks" – only better, because now "small disk" has a precise meaning – the star of a vertex. From this observation easily follows that the number of times the plane through p perpendicular to a triangle T (with vertex p) meets $\text{Lk}(p)$ is then equal to $\#T$ in $\text{St}(p)$, such that one of the vertices of T lies above the plane and the other lies below. In such a case v is called *the middle vertex* of T for l. (See Figure 4.11.)

---

[14]Named thus because it represents the shape necessary for a monkey to be able to comfortably ride on: It needs **three** depressions ("flaps"): The usual two for the legs and an extra one fo the tail.

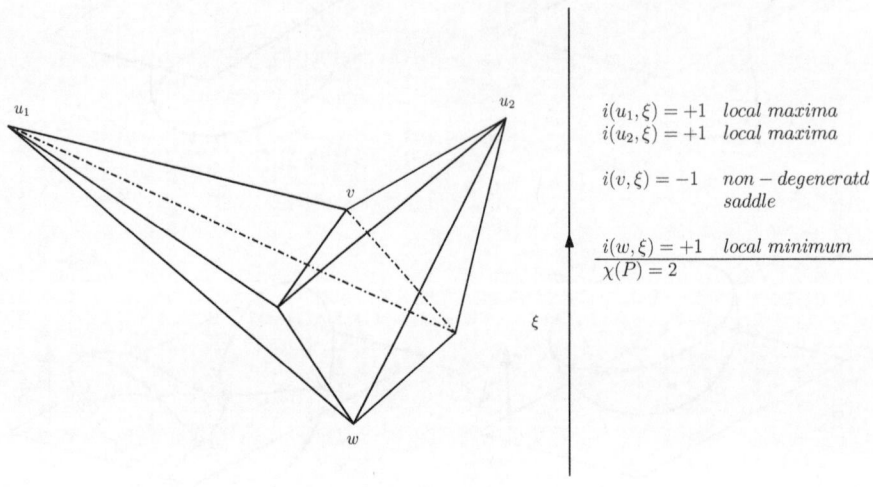

$$i(u_1, \xi) = +1 \quad local\ maxima$$
$$i(u_2, \xi) = +1 \quad local\ maxima$$

$$i(v, \xi) = -1 \quad non-degeneratd\ saddle$$

$$i(w, \xi) = +1 \quad local\ minimum$$
$$\chi(P) = 2$$

Figure 4.9 Critical points on a polyhedral surface (after [26]).

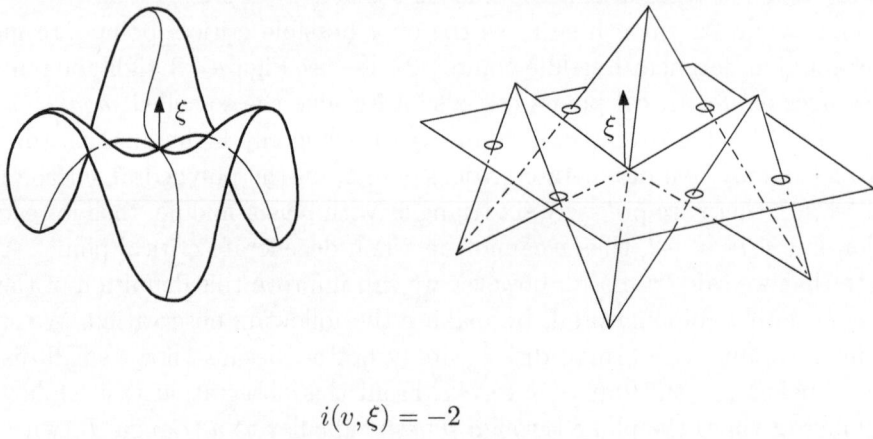

$$i(v, \xi) = -2$$

Figure 4.10 The monkey saddle: Left: A smooth surface; Right: A polyhedral surface. (After [26])

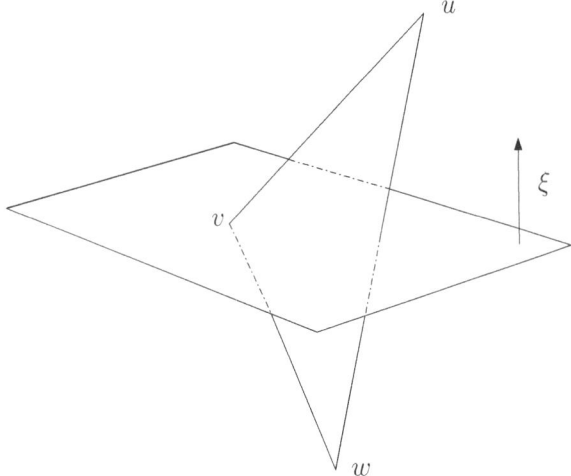

Figure 4.11 The middle vertex of a triangle (after [26]). v is the middle vertex of Δ for ξ.

We are thus conducted to formulate the desired combinatorial definition of the index:

$$i(p,l) = i(p,l) = 1 - \frac{1}{2}(\#T \text{ s.t.} p \text{ is a middle for } l). \qquad (4.1.10)$$

Using the above formula of the index, we can proceed to the proof of the Theorema Egregium. However, in order to do this, we have still to correlate between combinatorial form of the Gauss curvature at a vertex on a polyhedral surface and the vertex's index (w.r.t. a general direction). To this end, let us first return to the smooth case and observe that, with the notations of Figure 4.1, the area of A' is given by its characteristic function[15] $\chi_{A'}$, namely $\text{Area}(A') = \int_{A'} \chi_{A'} dA$. Furthermore, the height function w.r.t. a given direction l has a critical point at $p \in A'$ if and only if the direction vector of l, $\vec{v} \in A'$. Note that to the same direction correspond to points on the unit sphere, p and its antipodal point $p*$. Then, since we can chose A such that A' does not contain antipodal points, we obtain

$$K(A') = \frac{1}{2} \int_{\mathbb{S}^2} \sum_{p \in A'} i(p,l). \qquad (4.1.11)$$

Note that we presumed that the normal mapping ν is injective on A. However, the formula above can be extended to the general case, but this requires a more delicate analysis, that is not required in the polyhedral case, given the fact that the role of A is played by $\text{St}(p)$ and, moreover, the vertex p is the only critical point in $\text{St}(p)$.

---

[15]Recall that the *characteristic function* of a set X is defined as $\chi_X(x) = 1$ if $x \in X$, and 0 otherwise.

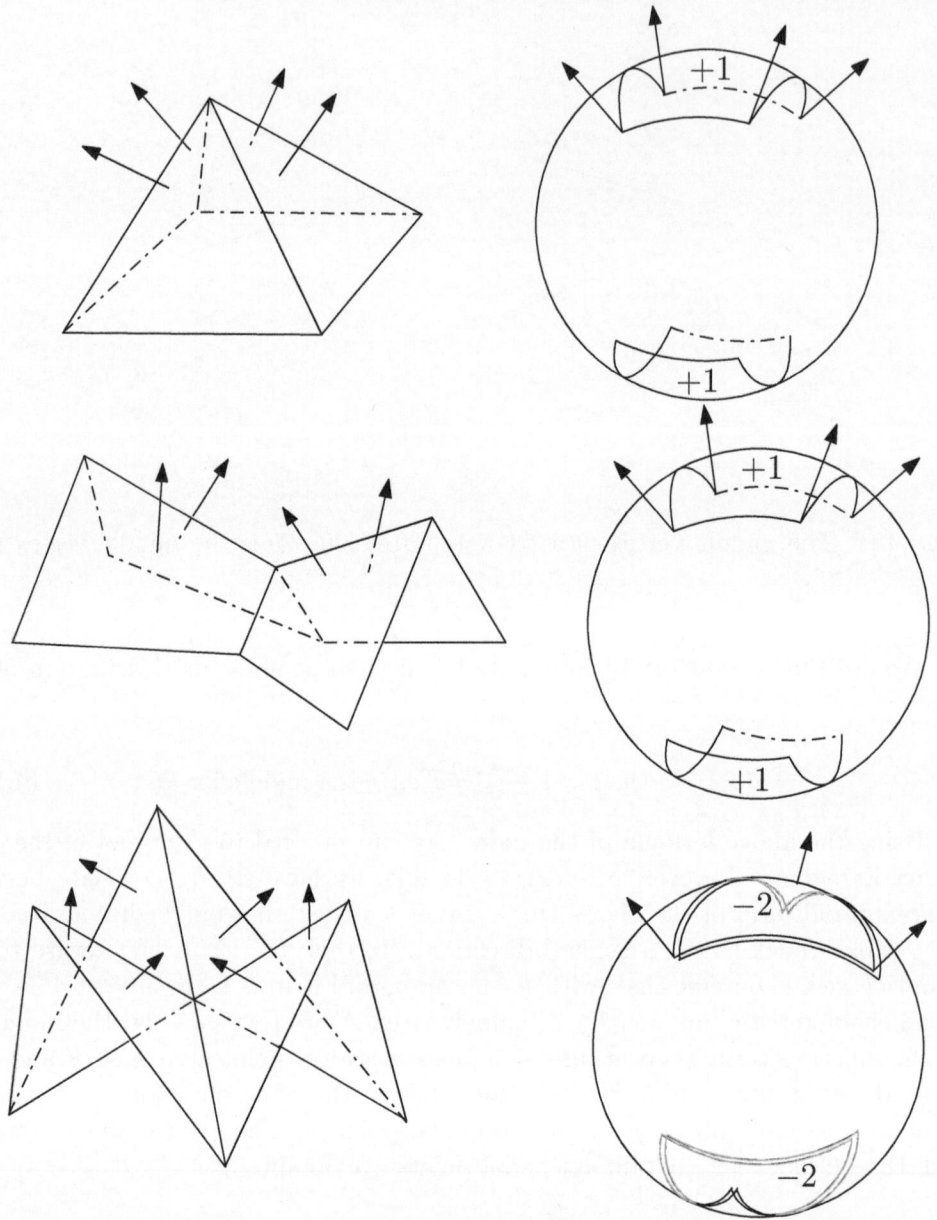

Figure 4.12 Gaussian curvature and index on a polyhedral surface (after [26]).

Therefore, we can use, in the polyhedral case as well, Formula (4.1.11) above (see Figure 4.12) and, moreover,

$$K(p) = \frac{1}{2} \int_{\mathbb{S}^2} i(p,l) dA. \tag{4.1.12}$$

We can now bring the

Alternative proof of Theorema Egregium We start by introducing yet another notation: Let $m = m(p, T, l)$ be the function that takes the value 1, if p is a middle

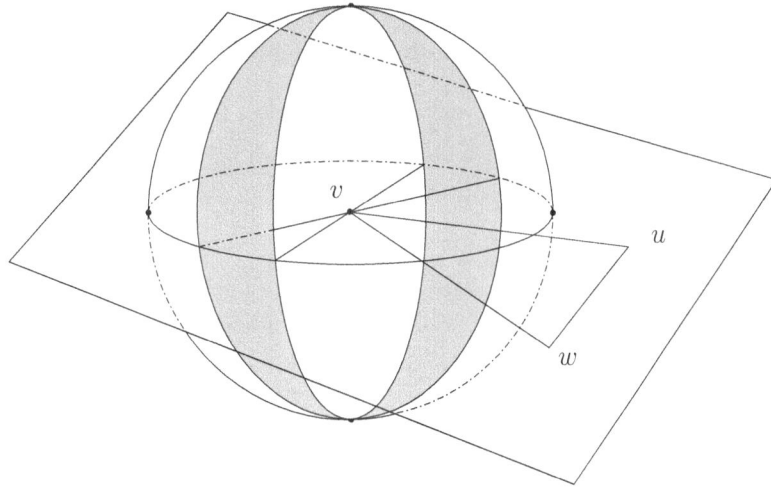

Figure 4.13 Triangle middle vertex and double lunule on the unit sphere.(After [26])

vertex for T, relative to l, and 0 otherwise. Then, $i(p, l) = m(p, T, l)$, by its definition. Therefore, it follows from (4.1.12) that

$$K(p) = \frac{1}{2} \int_{\mathbb{S}^2} \left(1 - \frac{1}{2} \sum_{T \in S^2} m(p, T, l) \right) dA = \frac{1}{2} \int_{\mathbb{S}^2} dA - \frac{1}{4} \int_{\mathbb{S}^2} m(p, T, l) dA.$$

The first term in the right hand equation represents half of the area of the unit sphere, therefore it is equal to 2π, thus we only have to estimate the second term. Since any vector in $\vec{v}$ in $\mathbb{R}^3$ can be uniquely expressed as a sum $\vec{v} = \vec{u} + \vec{v}$, where $\vec{u} \in \Pi(T)$ – the plane determined by the triangle T, and $\vec{v} \perp \Pi(T)$, it follows that $m(p, T, l) = 1$ if and only if

$$\vec{v} \cdot \vec{pq} > 0 > \vec{v} \cdot \vec{pr} \text{ or } \vec{v} \cdot \vec{pq} < 0 < \vec{v} \cdot \vec{pr};$$

that is if and only if $\vec{v}$ belongs to the double lunule of angle[16] $\alpha = \sphericalangle(\vec{pq}, \vec{pr})$ on the unit sphere of center p, perpendicular to the equatorial plane Π_T (see Figure 4.13).

Since the area of such a double lunule is 4α (see the proof of Lemma 4.1.9) it follows that the second term is equal to $\sum_T(\alpha)$, where the sum is taken over all the triangles incident with p. Thus, we discover yet again[17], by following this alternative route of proof, that $K(p) = 2\pi - \sum_T(\alpha)$, thus K is intrinsic. This concludes our alternative proof.

$\square$

Exercise 4.1.15 *This exercise introduces to the reader, via a graduated sequence of questions, the smooth (classical) counterpart of the tools developed in Banchoff's proof (as promised).*

---

[16]We are perhaps surprisingly encountering again this geometric object!...
[17]unsurprisingly, by now

1. Let S^2 be a smooth surface, $f : S^2 \to \mathbb{R}$ a differentiable function, and let $p \in S^2$ be a critical point of p. Denote by $H_p f$ the Hessian of f of at p.

   (a) Show that, if f is the height function of S^2 relative to $T_p(S^2$ – the tangent plane to S^2 at p, i.e. $f(q) = \overrightarrow{(q - p)} \cdot N(p)$, $q \in S^2$., then p is a critical point of f and that $H_p f$ is well defined.

   (b) Prove that if $\vec{v}$, $|\vec{v}| = 1$ then $H_p f(\vec{v}) = k_{n,\vec{v}}$ – the normal curvature of S^2, at p, in the direction $\vec{v}$.

   (c) Prove that $H_p f$ (where f is as above) is precisely the second fundamental form of S^2 at p.

2. Let S^2, f and p be as above. We say that p is a non-degenerate critical point for f if the self-adjoint linear mapping associated to the quadratic form $H_p f$ is nonsingular, and degenerate otherwise. Furthermore, f is called a Morse function if has only non-degenerate critical points.

   (a) Let h_{p_0} be the distance function from S^2 to a fixed point $p_0 \notin S^2$, i.e. $h_{p_0}(p) = \sqrt{\overrightarrow{(p - p_0)} \cdot \overrightarrow{(p - p_0)}}$, $p \in S^2$. Prove that $p \in S^2$ is a critical point for h iff the line $l = \overleftrightarrow{pp_0} \perp S^2$.

   (b) Let p a critical point of h_{p_0}, and let $C \subset S^2$ a curve parameterized by arc length, trough p_0 (i.e. $C(0) = p$), having as (unit) tangent vector at p, $\dot{C}(0) = \vec{v} \in T_p(S^2)$. Show that

   $$H_p h_{p_0}(\vec{v}) = \frac{1}{h_{p_0}} - k_n \; ;$$

   where k_n is, as above, the normal curvature of S^2 in the direction $\vec{v}$. Furthermore, show that p is a degenerate critical point of h_{p_0} iff $h_{p_0} = 1/k_1$ or $h_{p_0} = 1/k_2$, where k_1, k_2 are the principal curvatures at p.

   (c) Prove that the set $X = \{r \in \mathbb{R} \mid h_{p_0} \text{ is a Morse function}\}$ is dense in $\mathbb{R}^3$. (In other words, Morse functions on smooth, regular surfaces are "generic".)

Remark 4.1.16 *As Banchoff [26] points out, his approach represents an (extensive) extension of the one in Pólya's paper [231] concerning polyhedral disks in $\mathbb{R}^3$.*

Remark 4.1.17 *The method above applies, in fact, to any dimension $n \geq 2$ and allows for the extension dimensions higher than 3 of the Theorema Egregium [25]*

Before concluding this section, let us note that both approaches used above (namely that of Hilbert–Cohn-Vossen, as well that Banchoff's one based on the index) can be adapted to obtain the polyhedral version of *Gauss-Bonnet Theorem* – hence, by passing to the limit, of its classical version. However, we defer these proofs for the appropriate section. Moreover, (as the reader might have guessed) we shall be able to provide yet another proof of Theorema Egregium in the course of the proof of the Gauss-Bonnet Theorem.

4.1.1 The Tube Formula and Approximation of Surface Curvatures

Returning to the problem of approximating curvature of surfaces, we should note that, while there are many successful ways of computing (or, rather, approximating) Gauss curvature, the same cannot be said about mean curvature for methods that are quite efficient for determining K render only modest results for the computation of H. This is hardly surprising, given that mean curvature is, as we already mentioned, and in contrast with Gauss curvature, extrinsic (therefore purely metric, for instance are difficult, if not impossible, to apply). Partly to illustrate this phenomenon, but also for the sake of the method itself and for the benefit of encountering the ideas and theories behind it we bring below what we believe it is the simplest, computationally method (yet hardly trivial at a theoretical level).

This method of approximating the mean curvature is based upon the so called *Tube Formula*. While sometimes included (usually in the exercise section or as appendix/facultative reading) in some text-books, it is still largely ignored. As we hope we shall demonstrate or, at least, begin making a case for wider popularization and use, this is quite unfortunate, and for two main reasons. One is the purely scientific one: The tube formula is essential in an important type of extension of the notion of curvature to a large class of non-smooth ("wild") spaces. We shall dwell in more detail in this below and further on in the book. (There are some more technical, so to say "local" benefits – for these one can consult [137], [104].) The second reason is didactic in nature: As the reader will discover shortly, understanding the tube formula, necessitates getting acquainted with some beautiful ideas and important facts.

Let us start with some necessary definitions and notation: Let $S \subset \mathbb{R}^3$ be an orientable surface and let N_p denote, as usual, the unit normal of S at p. For each $p \in S$ consider, in the direction of N_p, the open symmetric interval of length $2\varepsilon_p$, I_{p,ε_p}, where ε_p is chosen to be small enough such that $I_{p,\varepsilon_p} \cap I_{q,\varepsilon_q} = \emptyset$, for any $p, q \in S$ such that $||p-q|| > \xi \in \mathbb{R}_+$. Then, $\mathrm{Tub}_\varepsilon(S) = \bigcup_{p\in S} I_{p,\varepsilon_p}$ is an open set that contains S and such that for any point $x \in \mathrm{Tub}_\varepsilon(S)$, there exists a unique normal line to S through x. $\mathrm{Tub}_\varepsilon(S)$ is called a *tubular neighborhood* of S or just a *tube* – Figure 4.7. The two surfaces $S_{\pm\varepsilon} = S \pm \varepsilon \bar{N}$ are called the *offset surfaces* of S with offset distance ε. We shall consider sets of constant *offset* ε.

Remarkably, the existence of tubular neighborhoods is assured both locally, for any regular, orientable surface (see [104], Proposition 1, p. 111), and globally for regular, compact, orientable surfaces (see [104], Proposition 3, p. 113) (See Figure 4.14). It is also important to underline the fact that, since ε_p depends upon p, $Tub_\varepsilon(S)$

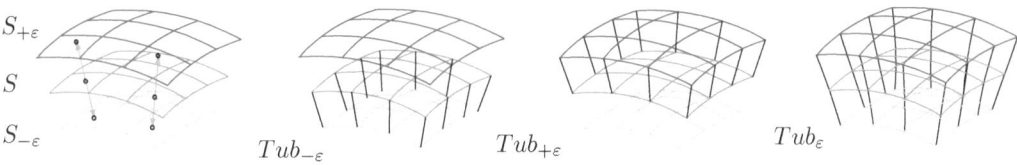

Figure 4.14 Tubular neighborhood of S.

does not, in general, coincide with the *ε-neigborhood* of S, i.e., with the set $N_\varepsilon(S) = \{x \in \mathbb{R}^3 \mid dist(x, S) = \varepsilon\}$.

We also note a number of facts about the existence and regularity of the offset surface S_ε: If S is convex, then $S_{\pm\varepsilon}$ are piecewise $C^{1,1}$ surfaces (i.e., they admit parameterizations with continuous and bounded derivatives), for all $\varepsilon > 0$. Also, if S is a smooth enough surface with a boundary (that is, at least piecewise C^2), then $S_{\pm\varepsilon}$ are piecewise C^2 surfaces, for all small enough ε (see [120], p. 1025). Moreover, for any compact set $S \subset \mathbb{R}^3$, $S_{\pm\varepsilon}$ are Lipschitz surfaces for almost any ε (see [167]). This is extremely pertinent to our setting of Differential Geometry on non-smooth sets, since it allows the computation of the mean curvature for local triangulations, not only for smooth surfaces (see also the discussion further below).

We are now ready to bring the Tube Formula itself.

Theorem 4.1.18 (The Tube Formula) *Let $S \subset \mathbb{R}^3$ be a compact orientable surface. Then*

$$Vol(Tub_\varepsilon(S)) = 2\,\varepsilon\,Area(S) + \frac{2\varepsilon^3}{3}\int_S K dA\,. \qquad (4.1.13)$$

While hardly original (we follow closely [137]) we bring the proof for the sake of completeness.

Proof 4.4 *Let $S = f(U)$ be a local parametrization of the surface. Define $f_\delta : U \to \mathbb{R}^3$, $f_\delta(u, v) = f(u, v) + \delta\bar{N}(u, v)$. Then, for a sufficiently small $|\delta|$, f_ε is injective, for all $|\varepsilon| < |\delta|$. Thus, one can choose ε such that $f_{\pm\varepsilon}$ represents a parametrization of $S_{\pm\varepsilon}$.*

Then:

$$\begin{cases} \frac{\partial f_\varepsilon}{\partial u} = \frac{\partial f}{\partial u} + \varepsilon\frac{\partial\bar{N}}{\partial u}, \\[2ex] \frac{\partial f_\varepsilon}{\partial v} = \frac{\partial f}{\partial v} + \varepsilon\frac{\partial\bar{N}}{\partial v}. \end{cases} \qquad (4.1.14)$$

However, $II_S(\bar{w}) = -d\bar{N} \cdot \bar{w}$, where II_S denotes the second fundamental form of S (see [104], p. 141). Thus, one can express the partial derivatives of f_ε as:

$$\begin{cases} II_S\frac{\partial f}{\partial u} = -\frac{\partial\bar{N}}{\partial u}\,, \\[2ex] II_S\frac{\partial f}{\partial v} = -\frac{\partial\bar{N}}{\partial v}\,. \end{cases}$$

Therefore (2.2) becomes:

$$\begin{cases} \frac{\partial f_\varepsilon}{\partial u} = (1 - \varepsilon II_S)\frac{\partial f}{\partial u}\,, \\[2ex] \frac{\partial f_\varepsilon}{\partial v} = (1 - \varepsilon II_S)\frac{\partial f}{\partial v}\,, \end{cases} \qquad (4.1.15)$$

and

$$\frac{\partial f_\varepsilon}{\partial u} \times \frac{\partial f_\varepsilon}{\partial v} = det(1 - \varepsilon II_S)\left(\frac{\partial f}{\partial u} \times \frac{\partial f}{\partial v}\right).$$

Moreover, $det(1 + \varepsilon I_S) > 0$, for small enough ε, thus

$$\left\|\frac{\partial f_\varepsilon}{\partial u} \times \frac{\partial f_\varepsilon}{\partial v}\right\| = det(1 - \varepsilon II_S)\left\|\frac{\partial f}{\partial u} \times \frac{\partial f}{\partial v}\right\|.$$

From the classical formula expressing the principal curvatures of S in terms of H and K (see [104], p. 212, [240], pp. 208-9):

$$det(1 - \varepsilon II_S) = 1 - 2\varepsilon H + \varepsilon^2 K,$$

it follows (by the well-known differential geometry expression for area – see, e.g., [104], 2-8.) that:

$$Area(f_\varepsilon(U)) = \int_U \left\|\frac{\partial f_\varepsilon}{\partial u} \times \frac{\partial f_\varepsilon}{\partial v}\right\| dudv$$

$$= \int_U (1 - 2\varepsilon H + \varepsilon^2 K)\left\|\frac{\partial f}{\partial u} \times \frac{\partial f}{\partial v}\right\| dudv$$

$$= Area(f(U)) - 2\varepsilon\int_{f(U)} HdA + \varepsilon^2\int_{f(U)} KdA.$$

Thus, by summation over the local parameterizations composing S,

$$Area(S_{\pm\varepsilon}) = Area(S) \mp 2\varepsilon\int_S HdA + \varepsilon^2\int_S KdA. \qquad (4.1.16)$$

Therefore, $Area(S_{+\varepsilon}) + Area(S_{-\varepsilon}) = 2Area(S) + 2\varepsilon^2\int_S KdA$, and follows that:

$$Vol(Tub_\varepsilon(S)) = \int_{-\varepsilon}^{+\varepsilon} Area(S_t)dt = 2\varepsilon\, Area(S) + \frac{2\varepsilon^3}{3}\int_S KdA,$$

whence (2.1) follows immediately.

Unfortunately, the Tube Formula cannot be employed directly to compute the mean curvature. (Explain why!) Moreover, in the case of triangulated surfaces computing K by means of the Tube Formula reduces to approximating $K(p)$ by the angle defect at the point p (see [92], p. 9 and Remark 4.1.20 below); i.e., by approximating $K(p)$ by $K(p) = 2\pi - \sum_1^n \psi_j$, where ψ_j denote the angles of the triangles incident with p (see Figure 4.4) – that is, by applying a well-known method, based upon the Local Gauss-Bonnet Theorem (see [53], [176]). Fortunately, (4.1.13) also yields a much more useful (yet, sadly, usually overlooked) formula, which we will refer to henceforward as the *Half Tube Formula*:

Theorem 4.1.19 (Half Tube Formula) *Let $S \subset \mathbb{R}^3$ be a compact orientable surface. Then*

$$Vol(Tub_{\pm\varepsilon}(S)) = \varepsilon Area(S) \mp \varepsilon^2\int_S HdA + \frac{\varepsilon^3}{3}\int_S KdA. \qquad (4.1.17)$$

Remark 4.1.20 *A similar formula is developed in [92], p. 4, for full tubes, without, however, noticing that the term containing H vanishes. Indeed, our method is closely related to that of [92]. However, they stem from somewhat different considerations. Note that, as already mentioned above, only the use of the half tubes formula allows the computation of the mean curvature. Moreover, the method we propose has the hardly unnoticeable advantage of being much simpler to implement as well as far more intuitive*[18]*than the one proposed in [92].*

Before we can present the Half-Tube Formula based algorithm for the computation of mean curvature, we should emphasize that it (and the Tube Formula) represent generalizations of the classical Steiner-Minkowski Theorem ([35] 12.3.5, [137] Theorems 10.1 and 10.2.) for compact, convex polyhedra with non-empty interiors of dimension $n \geq 2$. Again, both for its relevance here and its elegance and importance in general, we bring below the said theorem (however only in the relevant for us cases, $n = 2$ and $n = 3$).

Theorem 4.1.21 (Steiner-Minkowski) *Let $P \subset \mathbb{R}^n$, $n = 2, 3$ be a compact, convex polyhedron and let $N_\varepsilon(P) = \{x \in \mathbb{R}^n \mid dist(x, P) \leq \varepsilon\}$, $n = 2, 3$.*

1. *If $n = 2$, then,*

$$Area(N_\varepsilon(P)) = Area(P) + \varepsilon Length(\partial P) + \pi \varepsilon^2 , \qquad (4.1.18)$$

   where ∂P denotes the boundary of P.

2. *If $n = 3$, then,*

$$Vol(N_\varepsilon(P)) = Vol(P) + \varepsilon Area(\partial_2 P) + C\varepsilon^2 Length(\partial_1 P) + \frac{4\pi\varepsilon^3}{3} , \quad (4.1.19)$$

   where $\partial_2 P$ denotes the boundary faces of P, $\partial_1 P$ denotes the boundary edges of P and the last term contains the 0-dimensional volume contribution of the boundary vertices of P (by convention: $Vol(\partial_0 P) = |V_P|$ – the number of vertices of P), and where $C = C(P)$ is a scalar value that encapsulates $\iint_S H$ and that essentially depends on the dihedral angles of P.

To gain some insight – that will serve us well in developing the sought for algorithm – we begin by analyzing the 2-dimensional case first. $N_\varepsilon(P)$ naturally decomposes into three components: P itself, a union of rectangles of height ε constructed upon the sides of P (see Figure 4.15, left) and the union of circular sectors of radius ε (see Figure 4.15, left (a)). The total area of the rectangles is:

$$\varepsilon Length(e_1) + \ldots + \varepsilon Length(e_n) = \varepsilon Length(\partial P) ,$$

where $e_1, \ldots, e_n$ denote the boundary edges of P. The angles ϕ_n of the circular sectors are given by: $\phi_i = 2\pi - \pi/2 - \pi/2 - \alpha_i = \pi - \alpha_i$, where α_i is the respective internal angle of P (see Figure 4.15, left (a)). However, $\alpha_1 + \ldots + \alpha_n = (n-2)\pi$, thus

---

[18]albeit less far reaching

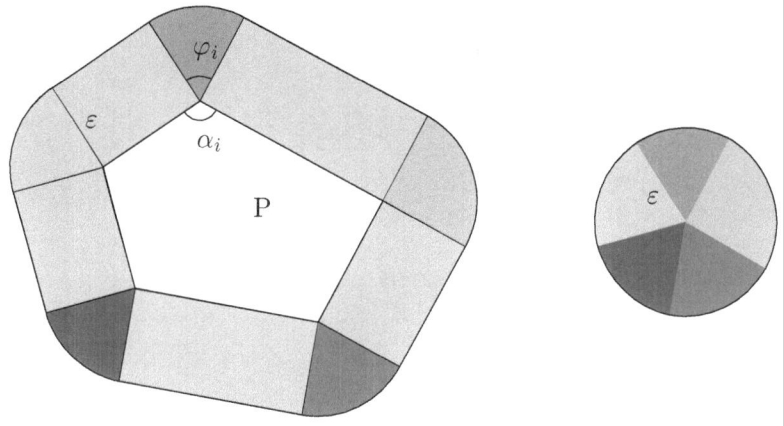

Figure 4.15 Tubular neighborhood of a convex polygon.

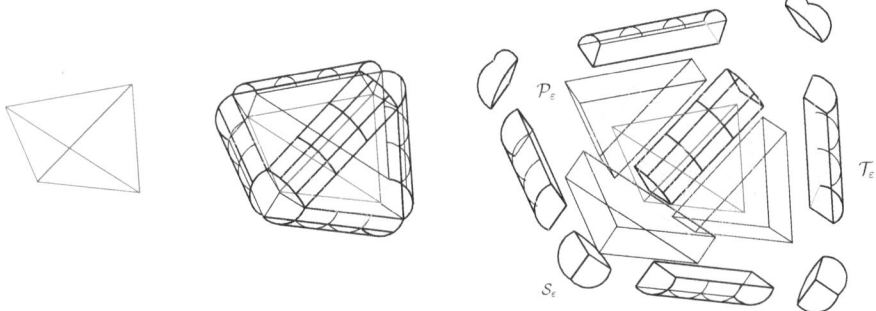

Figure 4.16 Tubular neighborhood of convex polyhedron. Note the right parallelepipeds $\mathcal{P}_\varepsilon$ built upon the faces, the orthogonal products $\mathcal{T}_\varepsilon$ of circular sectors and edges, and spherical sectors $\mathcal{S}_\varepsilon$ associated with the vertices.

$\phi_1 + \ldots + \phi_n = 2\pi$. That is, the circular sectors combine to form a disk of radius ε (see Figure 4.15, right (b)).

In the case of $n = 3$, $N_\varepsilon(P)$ decomposes into the following four components: P, right parallelepipeds $\mathcal{P}_\varepsilon$ of height ε built upon the faces, the orthogonal products $\mathcal{T}_\varepsilon$ of circular sectors and edges, and spherical sectors $\mathcal{S}_\varepsilon$ associated with the vertices of P and whose union is a ball of radius ε (see Figure 4.16). Each of the geometrical objects above gives rise to the terms of (3.2) containing the fitting power of ε.

One cannot fail to notice the similarity between Formulas (4.1.16) and (4.1.19), more precisely the correspondence between the terms pertaining to H in both of the equations. Beyond its integral expression in (4.1.16), due to the limiting process possible on a smooth surface, the main difference resides in the "$\pm$" sign, that appears due to discarding the overly restrictive convexity condition.

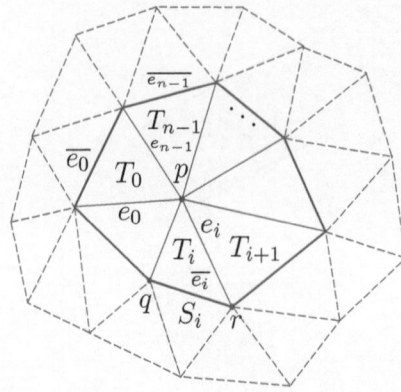

Figure 4.17 The first $(\text{Ring}_1(p)$, solid) and second $(\text{Ring}_2(p)$, dashed) ring around vertex p.

Also, it is important to underline that while in the computation of $H(p)$, the dihedral angles of the edges through p, $e_1, \ldots, e_n$ (see Figure 4.17) are computed, this is done only *in the sense of measures*, this holding for the areas involved in the expression of $K(p)$ (see [125], [92]). By convergence in the "sense of measures" we mean the following: given a patch U (i.e., several *stars*[19] of a certain vertex) on the triangulated surface S, each element in U (i.e., vertices, edges, etc.) makes a certain contribution to H (K). Even if these contributions may not individually be precise, they do add to the correct answer *on the average*, when the density of the mesh increases. It is important to note that this averaging effect is not a local phenomenon.

To this end one should regard, for instance, $\text{Area}(T_i)$ as a weight associated with the triangle T_i and uniformly distributed among its vertices p, q, r. The same uniform distribution is to be considered with respect to the weights naturally associated with edges. Therefore, e.g. the measure of the dihedral angle associated with the edge e_1, is to be equally distributed among the vertices adjacent to it, i.e. p and q.

The same type of uniform distribution is to be considered when computing the contribution of $\text{Area}(T_i)$, for example in the computation of $K(p)$: T_i will contribute $\frac{1}{3}\text{Area}(T_i)$ to the computation of (each of) $K(p), K(q), K(r)$.[20]

However, the edge $\overline{e_i} = qr$ also contributes to $H(p)$, since it is an element of T_i, which is adjacent to p. (This may be counterintuitive to the classical approach of computing curvatures of curves *through* p, but one should remember that these "elementary" edge-curvatures are to be viewed as measures!) Since the boundary edge $\overline{e_i}$ is common to T_i and the second-ring triangle S_i, it's contribution to each of the T_i triangles is half of the associated dihedral angle. Analogous considerations are to be applied in computing the contribution of the boundary vertices (e.g., q), etc.

---

[19]For this and other combinatorial/PL topology notions see, e.g. [229].
[20]This represents a first application of the principle (and "quiz") mentioned above.

Let $\text{Ring}_i(p)$ be the i'th ring around p and denote by $|e_i|$ the length of edge e_i. Then, the formula employed for computing the $H(p)$ follows:

$$H(p) = \frac{1}{\text{Area}(\text{Ring}_1(p))} \left[\frac{1}{2} \sum_{i=0}^{n-1} \varphi(T_i, T_{(i+1) \bmod n}) |e_i| + \frac{1}{4} \sum_{i=0}^{n-1} \varphi(T_i, S_i) |\overline{e_i}| \right],$$

where $S_i \in \text{Ring}_2(p)$ shares edge $\overline{e_i}$ with $T_i \in \text{Ring}_1(p)$, and $\varphi(T_i, T_{i+1})$ denotes the dihedral angles between adjacent triangles T_i and T_{i+1}.

Algorithm 1 condenses this process:

Algorithm 1 Estimates the mean curvature at vertex p

$RingArea \Leftarrow 0$;   ▷ The area of the first ring around vertex p
$ContribSum \Leftarrow 0$;   ▷ The sum of the contributions of the triangles from the first ring around vertex p
$n \Leftarrow |Ring_1(p)|$;   ▷ Number of triangles in the first ring around vertex p
for $i \leftarrow 0$ to $n-1$ **do**
    $RingArea \mathrel{+}= Area(T_i)$;
    $ContribSum \mathrel{+}= \varphi(T_i, T_{(i+1) \bmod n}) |e_i| + \frac{1}{2} \varphi(T_i, S_i) |\overline{e_i}|$;
end for
Return $\frac{ContribSum}{2RingArea}$;

If the user requires the mean curvature for many vertices of the model, this algorithm can be implemented more efficiently as follows: First calculate the contribution of each edge. For each edge, e_i, that is adjacent to triangles T_i and T_{i+1}, we assign an attribute called $EdgeContrib$ that is equal to $\frac{1}{4} \varphi(T_i, T_{i+1}) |e|$. Then, for each triangle we assign an attribute called $TriangleContrib$ that is the sum of all $EdgeContrib$ of its' edges. Finally, we can compute the mean curvature at vertex p using Algorithm 2.

Algorithm 2 Estimates the mean curvature at vertex p. Efficient version.

$RingArea \Leftarrow 0$;   ▷ The area of the first ring around vertex p
$ContribSum \Leftarrow 0$;   ▷ The sum of the contributions of the triangles from the first ring around vertex p
$n \Leftarrow |Ring_1(p)|$;   ▷ Number of triangles in the first ring around vertex p
for $i \leftarrow 0$ to $n-1$ **do**
    $RingArea \mathrel{+}= Area(T_i)$;
    $ContribSum \mathrel{+}= T_i.TriangleContrib$;
end for
Return $\frac{ContribSum}{2RingArea}$;

As already mentioned above, the Half-Tube Formula method produces, in general, good results. We do not discuss here convergence rates, etc., since this would take us too far afield in a quite technical direction. (The interested reader can consult [202].) However, we should note that the algorithm works especially well for surfaces of negative and mixed Gauss curvature. This is due, apparently, to the averaging effect of the tubes (more generally, *normal cycles* ([92], [98], [125]), which represent

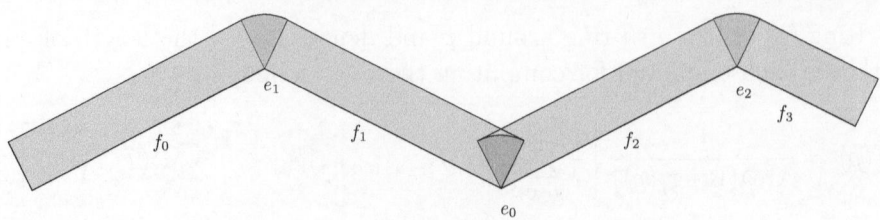

Figure 4.18 Normal Cycle at Negative Curvature Vertex [202].

a generalization of Steiner's approach on convex polyhedra to a much larger class of geometrical objects, in particular to smooth and piecewise-linear manifolds in any dimension above edges adjacent to vertices of negative curvature (see Figure 4.18).

An immediate practical application of our method of approximating mean curvature is for the computation for of *Willmore (elastic) energy* of triangulated surfaces, where the Willmore $W(S)$ energy of be a smooth, compact surface S in $\mathbb{R}^3$ [21] is defined as:

$$W(S) = \int_S H^2 dA - \int_S K dA. \tag{4.1.20}$$

Remark 4.1.22 *From the very definitions of H and K it follows immediately that*

$$W(S) = \frac{1}{4} \int_S (k_1 - k_2)^2 dA.$$

Since, as we shall prove later on,[22] $\int_S K dA$ is a constant that depends only on the *genus* of the closed surface S, it follows that, for all practical purposes, suffices to compute

$$W_0(S) = \int_S H^2 dA. \tag{4.1.21}$$

(Ours represents an approach to the computation of W on triangular meshes alternative to the one proposed in [48].)

It turns out that the Willmore energy is a *conformal invariant* (see [340]), i.e. it is invariant under *conformal mappings* (i.e. *angle preserving mappings*), comparing the Willmore energy under before and after deformation (of a given triangular mesh) would represent a measure of the departure of this deformation from being conformal. A first important application of this fact (and of or algorithm) is in Medical Imaging – see [287]. Furthermore, its computation, as well as of the related *Willmore flow*

$$\frac{dS}{dt} = -\nabla W(S) \,; \tag{4.1.22}$$

---

[21]This holds in general for more surfaces, only isometrically *immersed* $\mathbb{R}^3$. (Again, we shall discuss later on this notion.)

[22]in the Chapter/Section dedicated to the Gauss-Bonnet Theorem

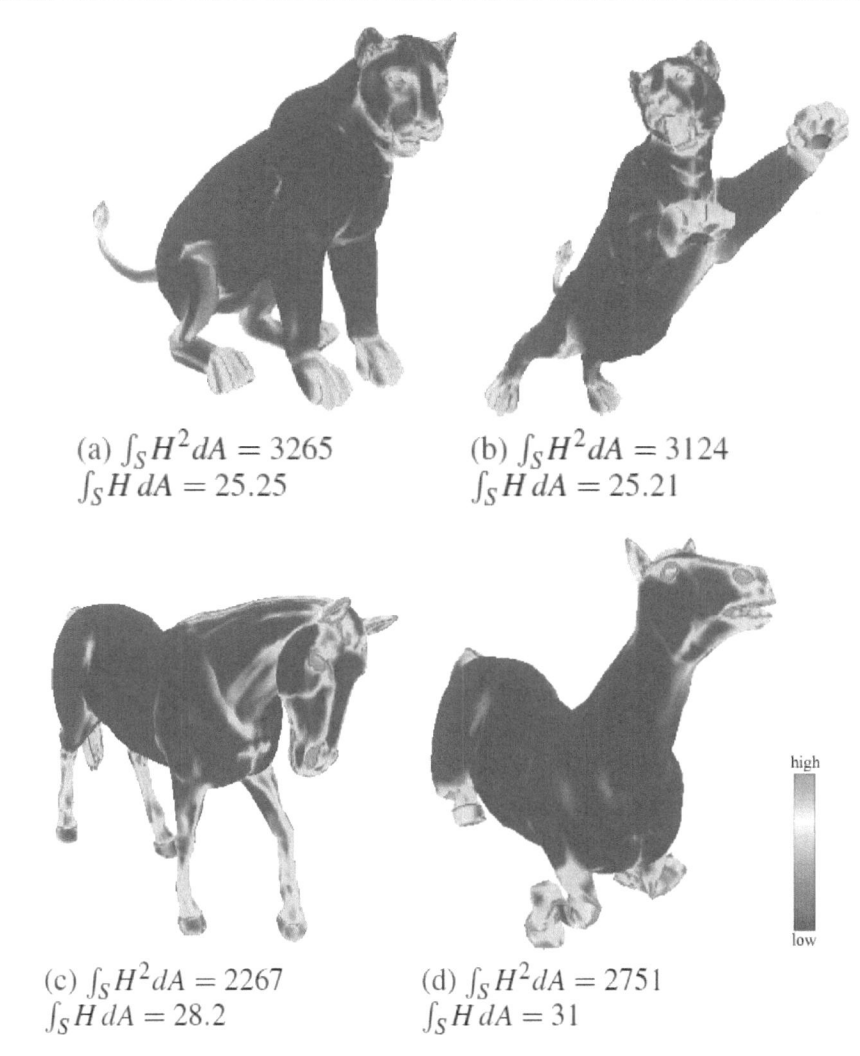

(a) $\int_S H^2 dA = 3265$
$\int_S H\, dA = 25.25$

(b) $\int_S H^2 dA = 3124$
$\int_S H\, dA = 25.21$

(c) $\int_S H^2 dA = 2267$
$\int_S H\, dA = 28.2$

(d) $\int_S H^2 dA = 2751$
$\int_S H\, dA = 31$

high

low

Figure 4.19 Above: $\Delta W = 4.3\%$, $\Delta \int H = 0.1\%$. Below: $\Delta W = 21\%$, $\Delta \int H \simeq 10\%$. It follows that the deformation the "Panther" is almost isometric (and conformal), while the one of the "Horse" fails on both accounts ([232]).

where dS denotes the differential of f, $S = f(u, v)$, is also relevant in a variety of fields and their applications, such as in Conformal Geometry, Variational Surface Modeling, and Thin Structures (see, e.g., [49]) and the bibliography therein).

While the conformal invariance of the Willmore energy – thus of $\int_S H^2 dA$ – is a classical and well-known fact, is a much more recent and surprising result [181] that the integral $\int_S H$ is, in fact, invariant under bending. Therefore, the computation of this simple integral is useful useful in applications involving *PL isometric embeddings* (see, e.g. [60], [276]) and as measure of the departure of deformations from being isometric (see [287], [232]) – See Figure 4.19 for an illustration.

Before concluding the discussion regarding the Tube Formula, its role in the computation of curvature, thus and its applications in Graphics, etc., we have to make a number of more theoretical – but of consequence in possible applications – remarks:

Both the Steiner-Minkowski Theorem and the Tube (and Half-Tube) Formula extend not only to arbitrary convex sets in $\mathbb{R}^n$ (see [35]), but also to open subsets with compact closure and smooth boundary in any Riemannian manifold (see [137], pp. 10-11 and 243-248). The most straightforward generalization is the following theorem (see, e.g. [40], Theorem 6.9.9.):

Theorem 4.1.23 (Weyl) *If X is an k-dimensional submanifold of $\mathbb{R}^n$, then:*

$$Vol(Tub_\varepsilon(X)) = \sum_{i=0}^{\lfloor k/2 \rfloor} c_{2i}\varepsilon^{n-k+2i} \; ;$$

where

$$c_{2i} = \frac{1}{n-k+2i} \int_X K_{2i} dV_{2i} \, ,$$

and where K_{2i} are polynomials of degree i in the curvature tensor R of X, the so-called Weyl curvatures (see [40], [75]).

In particular, we have $c_0 = vol(X)vol(\mathbb{B}^{n-k}(0,\varepsilon))$ and $c_2 = \frac{1}{2}\int_X \kappa \, dX$, where κ denotes the *scalar curvature*, which we shall define later, but which represents on possible, canonical generalization to higher dimension of Gauss curvature. (For the proof o the result mentioned above, see [40], 6.9, [137], Lemma 4.2.) It follows, that in the case of special interest for us, i.e., $n = 3$ and $k = 2$, we have: $Vol(Tub_\varepsilon(X)) = c_0\varepsilon + c_2\varepsilon^3$, where $\kappa \equiv K$ – the Gauss curvature of the surface X, i.e., we recover Formula (4.1.13).

Moreover, via the normal cycle the results above also extend to piecewise linear manifolds (piecewise linearly and isometrically) embedded in $\mathbb{R}^n$ (see [76], [122], [92]).[23] Recently these theorems were generalized to general closed sets in $\mathbb{R}^n$ (see [170]).

This last fact is particularly important for practical, computational purposes, since it allows us to apply the Half Tube Formula not only locally, that is by computing the measures associated with the triangles belonging to the ring of a given vertex p, but also to extend it to the next ring, and, potentially, to more consecutive rings. Moreover, since the Half Tube Formula involves K also, it allows us estimate K by computing over more than one ring, thus permitting the extension of the angle deficiency method to more than one ring. However, since by (4.1.17), the computation of K involves ε^3, one expects difficulties to arise, due to numerical instability.

We close this chapter with the following

Problem 4.1.24 *Does the common assumption regarding the uniform distribution of the area hold also for "bad" triangulations (i.e., containing "thin" triangles, or, in*

---

[23]We shall return later, in more technical detail as well as many applications, to the notion just mentioned.

other words, having low aspect ration), and how much is it affected by the quality of the mesh? In view of the seminal results of [75] regarding the good approximations of curvature measures for smooth manifolds by piecewise linear (PL) manifolds whose simplices are "fat" (i.e., satisfy a certain non-degeneracy condition imposed upon the dihedral angles in all dimensions), one is inclined to conclude that "fat" triangles should be considered if one wants to make use of the uniform distribution of the area assumption, in which case the original question should be replaced by the more technical one: "What is the lower admissible boundary for the "fatness" of the triangles if a uniform distribution is to be considered?" Moreover, one would also like to address the more practical problem of properly weighing the areas of the triangles adjacent to a vertex (instead than using the common "1/3" factor that includes, implicitly, the hidden assumption of (almost) uniformity of the triangles, i.e. similar size and good (i.e large) aspect ratio.

4.2 NORMAL CYCLE

The theory of normal cycles unifies the concepts of curvatures for both smooth and polyhedral surfaces.

Let S be a smooth manifold of dimension at least two embedded in a Euclidean space $\mathbb{R}^k$. A differential 2-form ω on S associates with every point $p \in S$ a skew-symmetric bilinear form on T_pS, denoted as ω_p. Given two tangential vector fields f and g, a differential 2-form can be constructed as follows

$$(f \wedge g)_p(u, v) = \begin{vmatrix} f_p \cdot u & g_p \cdot u \\ f_p \cdot v & g_p \cdot v \end{vmatrix}$$

for all $p \in S$ and $(u, v) \in T_pS$, $f_p \cdot u$ is the inner product of $\mathbb{R}^k$. If A is a linear transformation of the plane spanned by u and v, then $(f \wedge g)_p(Au, Av) = det(A)(f \wedge g)_p(u, v)$. In particular, $(f \wedge g)_p(u_1, v_2) = (f \wedge g)_p(u_2, v_2)$ for any two orthonormal frames (u_1, v_1) and (u_2, v_2) for the plane T_pS. Similarly, we have $(f_1 \wedge g_1)_p(u, v) = (f_2 \wedge g_2)_p(u, v)$ for any couple of orthonormal frames (f_1, g_1) and (f_2, g_2) spanning the same oriented plane. For example, suppose n is the normal field of the surface, (e_1, e_2, n) is an orthonormal frame field, where for each point $(e_1, e_2)_p \in T_pS$, then $e_1 \wedge e_2$ is the area form of the surface.

Suppose $\varphi : S_1 \rightarrow S_2$ is a diffeomorphism between two surfaces, ω is a 2-differential form on S_2, the pullback of ω by φ, denoted as $\varphi^*\omega$ is given by:

$$\varphi^*\omega_p(u, v) = \omega_{\varphi(p)}(D_p\varphi(u), D_p\varphi(v))$$

for all $p \in S_1$ and $u, v \in T_pS_1$. The change of variable formula relates the integral of a 2-differential from with the one of its pullback,

$$\int_{S_1} \varphi^*\omega = \int_{\varphi(S_1)} \omega.$$

Integral 2-currents generalize oriented surfaces. They can be defined as linear combinations of oriented surfaces with integral coefficients. Integration of 2-differential forms is extended to integral 2-currents by linearity:

$$\int_{n_1 S_1 + n_2 S_2} \omega = n_1 \int_{S_1} \omega + n_2 \int_{S_2} \omega.$$

The surface U that is setwise the same as S but with reverse orientation thus corresponds to the current $-S$.

Now set $\mathcal{S} = \mathbb{R}^3 \times \mathbb{S}^2$, $\mathcal{S}$ is a subset of $\mathbb{R}^3 \times \mathbb{R}^3$. The first factor of the product space is called the *point space* E_p, and the second one the *normal space* E_n. Each element in $\mathcal{S}$ can be thought of as a point in the space together with a unit normal vector. If u is a 3-vector, u^n will denote the vector $(0, u) \in E_p \times E_n$, and u^p the vector $(u, 0) \in E_p \times E_n$.

Invariant 2-forms Suppose g is a rigid motion of $\mathbb{R}^3$, we can extend it to $\mathcal{S}$: set $\hat{g}(p, n) = (g(p), \bar{g}(n))$, where $\bar{g}$ is the rotation associated with g. A 2-form ω defined on $\mathcal{S}$ is *invariant under rigid motions* if $\hat{g}^* \omega = \omega$ for all rigid motion g. Let $(x_1, x_2, x_3, y_1, y_2, y_3)$ be the canonical coordinates of $\mathbb{R}^3 \times \mathbb{R}^3$. The vector space Ω of invariant 2-form on $\mathcal{S}$ has dimension 4, which is spanned by

$$
\begin{aligned}
\omega_0 &= y_1 dx_2 \wedge dx_3 &+& \quad y_2 dx_3 \wedge dx_1 &+& \quad y_3 dx_1 \wedge dx_2 \\
\omega_1 &= y_1 dy_2 \wedge dy_3 &+& \quad y_2 dy_3 \wedge dy_1 &+& \quad y_3 dy_1 \wedge dy_2 \\
\omega_2 &= y_1 (dx_2 \wedge dy_3 &+& \quad dy_2 \wedge dx_3) \\
&+ \; y_2 (dx_3 \wedge dy_1 &+& \quad dy_3 \wedge dx_1) \\
&+ \; y_3 (dx_1 \wedge dy_2 &+& \quad dy_1 \wedge dx_2)
\end{aligned}
$$

and

$$
\begin{aligned}
\omega_3 &= (y_2^2 + y_3^2) dx_1 \wedge dy_1 &-& \quad y_1 y_2 dx_1 \wedge dy_2 &-& \quad y_1 y_3 dx_1 \wedge dy_3 \\
&+ \; (y_3^2 + y_1^2) dx_2 \wedge dy_2 &-& \quad y_2 y_1 dx_2 \wedge dy_1 &-& \quad y_2 y_3 dx_2 \wedge dy_3 \\
&+ \; (y_1^2 + y_2^2) dx_3 \wedge dy_3 &-& \quad y_3 y_1 dx_1 \wedge dy_3 &-& \quad y_3 y_2 dx_3 \wedge dy_2
\end{aligned}
$$

Smooth Surface Suppose M is a surface embedded in $\mathbb{R}^3$, which is the boundary of some compact set $V \subset \mathbb{R}^3$. Suppose M is smooth, the curvature measures are defined as follows:

Definition 4.2.1 (Curvature Measure) *The Gaussian curvature measure of M, ϕ_V^G is the function that associates with every (Borel) set $B \subset \mathbb{R}^3$ the quantity:*

$$\phi_V^G(B) := \int_{B \cap M} K(p) dp,$$

where $K(p)$ is the Gaussian curvature of M at point p. Similarly, we define the mean curvature measure ϕ_V^H by:

$$\phi_V^H(B) := \int_{B \cap M} H(p) dp,$$

$H(p)$ being the mean curvature of M at point p. The area measure ϕ_V^A by

$$\phi_V^A(B) := \int_{B \cap M} dp,$$

where dp is the area element of M.

The concept of normal cycle is a generalization of the unit normal bundle of the surface. Intuitively, normal cycles are a way to unfold offsets in a higher dimensional space.

Definition 4.2.2 (Normal Cycle) *The normal cycle $N(V)$ of V is the current associated with the set:*

$$ST^{\perp}V = \{(p, n(p))|p \in M\} \subset E_p \times E_n$$

endowed with the orientation induced by the one of M.

The *Gauss map* is a diffeomorphism from a surface to its normal cycle:

$$i : M \mapsto ST^{\perp}V, \quad p \mapsto (p, n(p)).$$

The integration of the invariant 2-forms on the normal cycle gives curvature measures.

Lemma 4.2.3 *For all Borel set $B \subset \mathbb{R}^3$, we have*

$$\int_{i(B\cap M)} \omega_0 = \phi_V^A(B), \int_{i(B\cap M)} \omega_1 = \phi_V^G(B), \int_{i(B\cap M)} \omega_2 = \phi_V^H(B), \int_{i(B\cap M)} \omega_3 = 0.$$

In the following, we give an equivalent definition of the invariant 2-forms, which is more convenient for the proof.

Definition 4.2.4 *Let $(p, n) \in \mathcal{S}$ and $u, v \in \mathbb{R}^3$ such that (u, v, n) is a direct othonormal frame of $\mathbb{R}^3$. We set:*

$$
\begin{aligned}
\omega^A_{(p,n)} &= u^p \wedge v^p \\
\omega^G_{(p,n)} &= u^n \wedge v^n \\
\omega^H_{(p,n)} &= u^p \wedge v^n + u^n \wedge v^p
\end{aligned}
\tag{4.2.1}
$$

It is easy to show that $\omega_0 = \omega^A$, $\omega_1 = \omega^G$ and $\omega_2 = \omega^H$. We give the proof of the lemma 4.2.3.

Proof 4.5 *By the change of variable formula, we have*

$$\int_{i(B\cap M)} \omega^A = \int_{(B\cap M)} i^*\omega^A.$$

Let (u, v) be a direct orthonormal frame of $T_q M$, where $q \in M$. By definition, we have

$$D_q i(w) = w^p + D_q n(w)^n.$$

Expressing $\omega^A_{i(q)}$ in the frame (u, v, n_q), $(u, v) \in T_q M$,

$$(i^*\omega^A)_q(u, v) = \begin{vmatrix} u^p \cdot (u^p + D_q n(u)^n) & v^p \cdot (u^p + D_q n(u)^n) \\ u^p \cdot (v^p + D_q n(v)^n) & v^p \cdot (v^p + D_q n(v)^n) \end{vmatrix} = \begin{vmatrix} u^p \cdot u^p & v^p \cdot u^p \\ u^p \cdot v^p & v^p \cdot v^p \end{vmatrix} = 1.$$

Similarly,

$$(i^*\omega^G)_q(u,v) = \begin{vmatrix} u^n \cdot (u^p + D_q n(u)^n) & v^n \cdot (u^p + D_q n(u)^n) \\ u^n \cdot (v^p + D_q n(v)^n) & v^n \cdot (v^p + D_q n(v)^n) \end{vmatrix}$$

$$= \begin{vmatrix} u^n \cdot D_q n(u)^n & v^n \cdot D_q n(u)^n \\ u^n \cdot D_q n(v)^n & v^n \cdot D_q n(v)^n \end{vmatrix} = K(q),$$

furthermore, we choose $(u,v) \in T_q M$ as the principle directions, $D_q n(u) = k_1(q)u$ and $D_q n(v) = k_2(q)v$,

$$(i^*\omega^H)_q(u,v) = \begin{vmatrix} u^p \cdot (u^p + D_q n(u)^n) & v^n \cdot (u^p + D_q n(u)^n) \\ u^p \cdot (v^p + D_q n(v)^n) & v^n \cdot (v^p + D_q n(v)^n) \end{vmatrix}$$

$$+ \begin{vmatrix} u^n \cdot (u^p + D_q n(u)^n) & v^p \cdot (u^p + D_q n(u)^n) \\ u^n \cdot (v^p + D_q n(v)^n) & v^p \cdot (v^p + D_q n(v)^n) \end{vmatrix}$$

$$= \begin{vmatrix} u^p \cdot u^p & k_1 v^n \cdot u^n \\ u^p \cdot v^p & k_2 v^n \cdot v^n \end{vmatrix} + \begin{vmatrix} k_1 u^n \cdot u^n & v^p \cdot u^p \\ k_2 u^n \cdot v^n & v^p \cdot v^p \end{vmatrix} = k_1(q) + k_2(q).$$

Polyhedral Surface When M is not smooth but convex, the normal cycle can be generalized by replacing normal vectors by *normal cones*.

Definition 4.2.5 (Normal Cone) *Suppose V is convex, the normal cone $NC_V(p)$ of a point $p \in V$ is the set of unit vectors v such that :*

$$NC_V(p) := \{v | v \in \mathbb{S}^2, s.t. \ \forall q \in V \ (q-p) \cdot v \le 0\}.$$

Definition 4.2.6 (Normal Cycle) *Suppose V is convex, the normal cycle of $M = \partial V$ is the current associated with the set*

$$N(V) := \{(p,n) | p \in \partial V \ n \in NC_V(p)\}$$

endowed with the orientation included by the one of ∂V.

In particular, when V is convex and smooth, this definition $N(V)$ agrees with the previous definition $ST^\perp V$.

As shown in Figure 4.20, the normal cycle of a simplex is the unfolded version of offsets of the simplex, and can be decomposed into spherical parts S_ε, clindric parts T_ε and planar parts T_ε. The difference is that these parts are now live in $E_p \times E_n$. We will say that a subset A of $E_p \times E_n$ lives above a subset $B \subset \mathbb{R}^3$ if the projection of A on the point space is included in B. The spherical, cylindric and planar parts live above vertices, edges and faces of the polyhedron respectively. The normal cycle has the *additivity* property:

Proposition 4.2.7 (Normal Cycle Additivity) *let V_1 and V_2 be two convex sets in $\mathbb{R}^3$ such that $V_1 \cup V_2$ is convex. Then:*

$$N(V_1 \cap V_2) + N(V_1 \cup V_2) = N(V_1) + N(V_2).$$

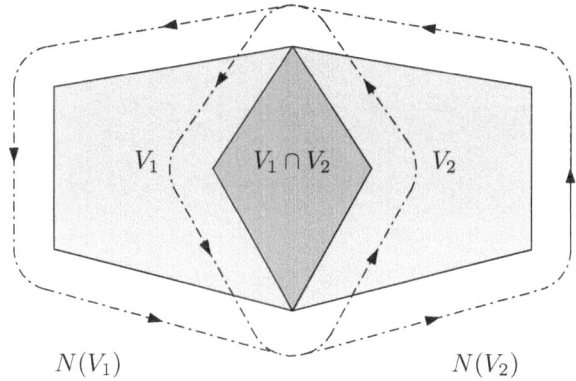

Figure 4.20 Additivity of normal cycles.

Proof 4.6 *It is sufficient to show that the multiplicities of any point (p, n) in $N(V_1 \cap V_2) + N(V_1 \cup V_2)$ and $N(V_1) + N(V_2)$ agree. If p doesn't belong to $\partial V_1 \cap \partial V_2$, this obvious. If p lies in $\partial V_1 \cap \partial V_2$, one concludes easily by the facts:*

$$NC_{V_1 \cap V_2}(p) = NC_{V_1}(p) \cup NC_{V_2}(p)$$

and

$$NC_{V_1 \cup V_2}(p) = NC_{V_1}(p) \cap NC_{V_2}(p).$$

By the additivity property, we can define the normal cycle of polyhedra. Suppose V is a polyhedron, we triangulate it into tetrahedra t_i, $i = 1, 2, \ldots, n$. The normal cycle can be defined by those of t_i's by the *inclusion-exclusion* formula:

$$N(V) := \sum_{k=1}^{\infty} (-1)^{k+1} \sum_{1 \le i_1 < \cdots < i_k \le n} N \left(\bigcap_{j=1}^{k} t_{i_j} \right).$$

Curvature measures for a polyhedron V are defined by integration of corresponding invariant forms on the normal cycle $N(V)$, just like the smooth case. It is sufficient to compute the integral of these forms on the spherical, cylindric and planar parts of the normal cycle.

Suppose $q \in M$ is in interior of a face of the polyhedron, $NC_V(q)$ is in the planar part. The tangent plane $T_q M$ is spanned by two vectors $(u, v) \in T_q M$, then

$$D_q i(u) = u^p + D_q n(u)^n = u^p, \quad D_q i(v) = v^p + D_q n(v)^n = v^p.$$

Hence

$$(i^* \omega^G)_q(u, v) = \begin{vmatrix} u^n \cdot D_q n(u)^n & v^n \cdot D_q n(u)^n \\ u^n \cdot D_q n(v)^n & v^n \cdot D_q n(v)^n \end{vmatrix} = 0$$

and

$$(i^*\omega^H)_q(u,v) = \begin{vmatrix} u^p \cdot (u^p + D_q n(u)^n) & v^n \cdot (u^p + D_q n(u)^n) \\ u^p \cdot (v^p + D_q n(v)^n) & v^n \cdot (v^p + D_q n(v)^n) \end{vmatrix}$$

$$+ \begin{vmatrix} u^n \cdot (u^p + D_q n(u)^n) & v^p \cdot (u^p + D_q n(u)^n) \\ u^n \cdot (v^p + D_q n(v)^n) & v^p \cdot (v^p + D_q n(v)^n) \end{vmatrix}$$

$$= \begin{vmatrix} u^p \cdot u^p & v^n \cdot u^p \\ u^p \cdot v^p & v^n \cdot v^p \end{vmatrix} + \begin{vmatrix} u^n \cdot u^p & v^p \cdot u^p \\ u^n \cdot v^p & v^p \cdot v^p \end{vmatrix} = 0.$$

Therefore, planar parts do not contribute to the curvature measures ϕ_V^H and ϕ_V^G.

The tangent plane to a cylindric part at a point (p,n) is spanned by u^p and v^n, where u is a vector parallel to the corresponding edge and v is orthogonal to u.

$$\omega_{(p,n)}^G(u^p,v^n) = \begin{vmatrix} u^n \cdot u^p & v^n \cdot u^p \\ u^n \cdot v^n & v^n \cdot v_n \end{vmatrix} = 0,$$

and

$$\omega_{(p,n)}^H(u^p,v^n) = \begin{vmatrix} u^p \cdot u^p & v^n \cdot u^p \\ u^p \cdot v^n & v^n \cdot v^n \end{vmatrix} + \begin{vmatrix} u^n \cdot u^p & v^p \cdot u^p \\ u^n \cdot v^n & v^p \cdot v^n \end{vmatrix} = 1.$$

The integral of ω^H over a subset of a cylindric part equals the area of this subset.

A tangent plane to a spherical part is spanned by two vectors of E^n, (u^n,v^n) are orthonormal.

$$\omega_{(p,n)}^H(u^n,v^n) = \begin{vmatrix} u^p \cdot u^n & v^n \cdot u^n \\ u^p \cdot v^n & v^n \cdot v^n \end{vmatrix} + \begin{vmatrix} u^n \cdot u^n & v^p \cdot u^n \\ u^n \cdot v^n & v^p \cdot v^n \end{vmatrix} = 0,$$

and

$$\omega_{(p,n)}^G(u^n,v^n) = \begin{vmatrix} u^n \cdot u^n & v^n \cdot u^n \\ u^n \cdot v^n & v^n \cdot v_n \end{vmatrix} = 1.$$

Integrating ω^G yields the area of the spherical part.

This shows the curvature measure $\phi_V^H(B)$ of a subet $B \subset \mathbb{R}^3$ is the sum of the areas of cylindric parts of $N(V)$ lying above B, weighted by their multiplicites. $\phi_V^G(B)$ is obtained by summing the areas of spherical parts lying above B weighted by their multiplicites. By direct computation, we can get the following definitions.

Definition 4.2.8 (Discrete Curvature Measure) *Assume that V is a polyhedron with vertex set P and edge set E. The discrete Gaussian curvature measure ϕ_V^G of $M = \partial V$ is the function that associates with every Borel set $B \subset \mathbb{R}^3$ the quantity:*

$$\phi_V^G(B) = \sum_{p \in B \cap P} k(p)$$

where $k(p)$ is the angle defect of M at point p, that is 2π minus the sum of angles between consecutive edges incident on p. Similarly, we define the discrete mean curvature measure ϕ_V^H by:

$$\phi_V^H(B) = \sum_{e \in E} l(e \cap B)\beta(e),$$

where $l(e)$ is the length of the edge, $|\beta(e)|$ the angle between the normals to the triangles of M incident on e. The sign of $\beta(e)$ is chosen to be positive if e is convex and negative if it is concave.

Approximation In practice, smooth surfaces are approximated by polyhedra. It is important to approximate the smooth curvature measures by discrete ones with prescribed precision. The approximation accuracy depends on the choice of the samples on M and the triangulation.

Suppose M is a smooth surface, $\mathcal{P}$ denotes a finite subset of M and T is the Delaunay triangulation of $\mathcal{P}$ restricted to V.

Definition 4.2.9 *$\mathcal{P}$ is said to be a ε-sample of M if for all point $p \in M$, the ball $B(p, \varepsilon\, lfs(p))$ centered on p and with radius $lfs(p)$ meets $\mathcal{P}$. $lfs(p)$ is the local feature size, denotes the distance function to the medial axis of $\mathbb{R}^3 \setminus M$.*

The following theorem shows the approximation precision of curvature measures:

Theorem 4.2.10 *Let $\mathcal{P}$ be an ε-sample of M with ε small enough and $\mathcal{P}$ be locally uniform. If B is the relative interior of a union of triangles of ∂T, then*

$$|\phi_T^G(B) - \phi_V^G(\pi(B))| \leq K\varepsilon$$
$$|\phi_T^H(B) - \phi_V^H(\pi(B))| \leq K\varepsilon$$

where $K = O(area(B) + length(B))$, π denotes the projection on M.

The detailed proof can be found in [93]. A more practical method is to conformally map the surface onto a planar domain, then compute the triangulation using planar Delaunay refinement method [204].

CHAPTER 5

The Mean and Gauss Curvature Flows

The reader undoubtedly noted that, after defining it, we scarcely mentioned mean curvature anymore, and she (or he) might wonder why this has been the case. An evident explanation resides is due the intrinsic nature of Gaussian curvature (as opposed to the extrinsic one of mean curvature), a property that allowed us to prove a number of deep, essential theorems of which the first and foremost is the Gauss-Bonnet Theorem. This doesn't mean, however, that mean curvature is not important. In fact, it is essential in the study of the so called *minimal surfaces* (known to many essentially as "soap bubbles"), which are defined by the condition $H \equiv 0$. Unfortunately, the study of minimal surfaces[1], constitutes a huge sub-field in itself of Differential Geometry (and of Geometric Analysis as well). Therefore, we cannot do it justice in any manner in this book, are thus sadly forced to relegate it to a possible future edition.

However, there is yet another application where mean curvature is essential and which, while also vast, allows us to present it succinctly here. Thus is even more important for us, since this allows us to prepare the ground for the introduction, later on, of an essential aspect of modern – smooth as well as discrete Differential Geometry – namely the *Ricci flow*, which we shall discuss at some length in Chapter 12. The application in question is the one of "*snakes*" (a.k.a. *active contours*) [241], [267], [338], [321], which is essentially identical, in dimension one, with the *Beltrami* or *Laplace-Beltrami flow* [304] (to cite only the foundational paper out of a most extensive bibliography). We shall therefore present the mean curvature flow, but we begin with simpler – and motivational – one dimensional case.

5.1 CURVE SHORTENING FLOW

Let us begin with a Physics thought experiment:[2] Imagine a splash of water on a metal plate, which is heated by a candle flame placed under the drop. The heat will propagate on the plate via the *heat kernel*, given by the Laplacian of the temperature function, a fact underlines the role of the Laplacian. On the physical/practical

---

[1] and their *constant mean curvature surfaces*, give by the (obvious) condition $H \equiv$ const.

[2] It can be turned – with the required precaution – into a practical experiment on a home stove.

DOI: 10.1201/9781003350576-5

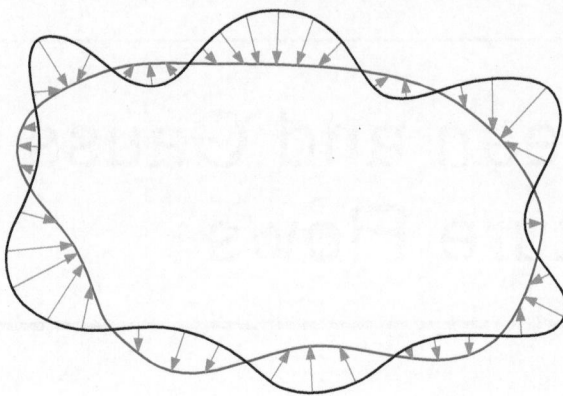

Figure 5.1 The curve shortening flow.

side, what shall happen is the fact that the water will evaporate, thus the drop will contract. In the process it will become smoother and rounder and, just before completely evaporating, it will become a very small, perfectly round droplet. It is this very familiar process that we model below from the viewpoint of curvature.

We begin from the observation that the heat flow is in fact equivalent to the *curvature flow* or *curve-shortening flow*:

$$\frac{d\text{length}}{dt} = -\int \kappa^2 ds\,; \tag{5.1.1}$$

where as always, ds denotes arc length. In other words, given a simple closed smooth[3] curve in the plane, we let each point of the curve move in normal direction with velocity[4] equal to the curvature at that point (see Figure 5.2). We can now return and present (rather informally[5] some properties of the curvature flow. We begin with the one promised by its very name:

Fact 5.1.1 *The curvature flow is shortening.*

This is a natural idea: The regions of higher curvature, i.e. the "bumpier", hence of longer length, are "adjusted" more, since the speed of evolution is given by the curvature (see Figure 5.2).

Remark 5.1.2 *The same equation holds for the evolution of curves on surfaces (since no term depending on the geometry of the surface appears in the equation above).*

Fact 5.1.3 *The curvature flow is smoothing.*

Indeed, even if the initial curve is only $\mathcal{C}^2$, it instantaneously becomes $\mathcal{C}^\infty$ (in fact even real analytic). This is a consequence of the fact that we are face with a *parabolic* PDE.

---

[3]i.e. at least $\mathcal{C}^2$

[4]meaning direction, too!

[5]For the technical details see [129], [139] [140].

Figure 5.2 The curve shortening flow: Regions of higher curvature have to be "adjusted" more.

Remark 5.1.4 *The time of smoothness may be short. Indeed, singularities may – and will – develop.*

Fact 5.1.5 *The curvature flow is collision-free.*

That simply means that two initially disjoint curves remain disjoint. Technically, this is a consequence of the maximum principle for parabolic differential equations. However, we present here instead a simple geometric argument:
Sketch of Proof First, note that is sufficient to study the case when the curves are one inside the other.

Next, observe that at the first time of contact t_0 the curves must be tangential. Then, the curvature of the inner curve is grater then of the exterior one (see Figure 5.3). It follows that the inner curve is moving faster than the exterior one.

Hence, at some $t_0 - \varepsilon$ the curves should have intersected, in contradiction to the choice of t_0.

□

Fact 5.1.6 *Under the curvature flow embedded curves remain embedded.*

That is to say that planar curves with no inflection points will develop no new inflections under the curvature flow. Moreover, given a curve on a surface, it will remain on the surface while evolving by the curvature flow (see Figure 5.4).

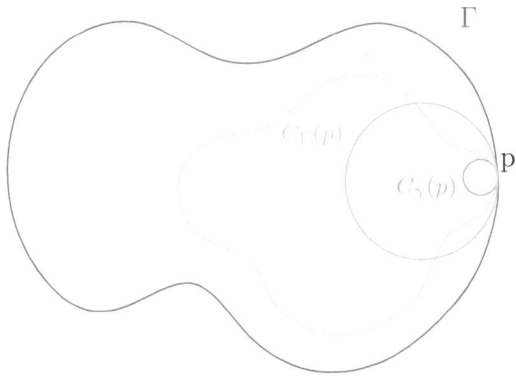

Figure 5.3 Inner curves have higher curvature.

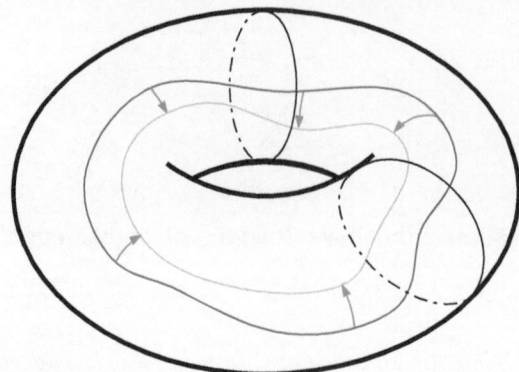

Figure 5.4 The curve shortening flow acting on a loop embed on a surface.

Fact 5.1.7 *Under the curvature flow curves have a finite lifespan.*

Sketch of Proof Every (simple) closed planar curve is contained in the interior of a circle, hence it evolves faster then the circle (by the same argument as above). Since the circle collapses in fine time, therefore so does the curve.

□

Fact 5.1.8 *Under the curvature flow convex curves shrink to round points.*

Formally, this represents the following deep theorem:

Theorem 5.1.9 (Gage-Hamilton, 1986) *Under the curvature flow a convex curve remains convex and shrinks to a point. Moreover, it becomes asymptotically circular, i.e. if the evolving curve is rescaled such that the enclosed area is constant, then the rescaled curve converges to a circle.*

Note that this (hard to prove) theorem justifies our intuition regarding the evolution of a liquid drop on a hot plate.

Fact 5.1.10 *Under the curvature flow embedded curves become convex.*

This is theorem of Greyson [139].

Corollary 5.1.11 *Under the curvature flow embedded curves shrink to round points*

This corollary again proves right our physical intuition.

Remark 5.1.12 *One can trivially generalize the problems above to curves in man-ifolds of any dimension. Moreover, one can consider, instead of the right hand in Equation (5.1.1) any functional of the type*

$$\int_\gamma f(k_g)ds \qquad (5.1.2)$$

where γ is a curve in an m-dimensional manifold of $\mathbb{R}^n$, f is a real function and k_g, ds represent the same notions as before. (One might look at the curvature of γ instead of its geodesic curvature.) Flows based on such functionals can be used for curve straightening (cf. Chapter 6), and they seem to be a strong tool in determining closed geodesics in manifolds [207].

5.2 MEAN CURVATURE FLOW

We now pass to the natural (free) translation into a geometric language of the Laplace-Beltrami flow, namely to the mean curvature flow. It turns out that, happily enough, almost all the properties of the curvature flow of planar curves extend to the case of mean curvature flow of surfaces. We briefly overview these properties below:

Fact 5.2.1 *Under the mean curvature flow surfaces become smoother.*[6]

Fact 5.2.2 *Under the mean curvature flow area decreases.*

Remark 5.2.3 *The mean curvature flow may be regarded as gradient flow for the area functional.*

Fact 5.2.4 *Under the mean curvature flow disjoint curves remain disjoint.*

Fact 5.2.5 *Under the mean curvature flow embedded surfaces remain embedded.*

Fact 5.2.6 *Under the mean curvature flow surfaces have finite lifespans.*

Fact 5.2.7 *Under the mean curvature flow the analog of the Gage-Hamilton theorem holds.*

More precisely, we have the following

Theorem 5.2.8 (Huisken, 1984) *Under the mean curvature flow, compact, convex surfaces shrink to round points.*

However, the analog of Grayson's theorem is false – see the counterexample below:

Remark 5.2.9 *All these results automatically extend to higher dimensions, the role of surfaces in $\mathbb{R}^3$ being taken by n-dimensional hypersurfaces in $\mathbb{R}^{n+1}$ (see Chapter 10). Consequently, the two principals curvatures at each point of surface in Euclidean 3-space are replaced, in this case, by n principal curvatures $k_1, \ldots, k_n$. Thus it is natural to define mean curvature of hypersurfaces as*

$$H = \frac{1}{n}(k_1 + \cdots + k_n).$$

However, hypersurfaces represent a very particular case and, moreover, the geometric meaning of the mean curvature is rather limited. Thus the proper generalization of the notion of mean curvature flow is not by simpling adding more terms (i.e. principals curvatures), but rather at searching for the proper geometric generalization of the idea of heat flow. It turns out the right extension to dimension three and higher is that of Ricci flow, which we shall discuss in some length in a dedicated chapter (See also Figure 5.5).

---

[6]for a short time

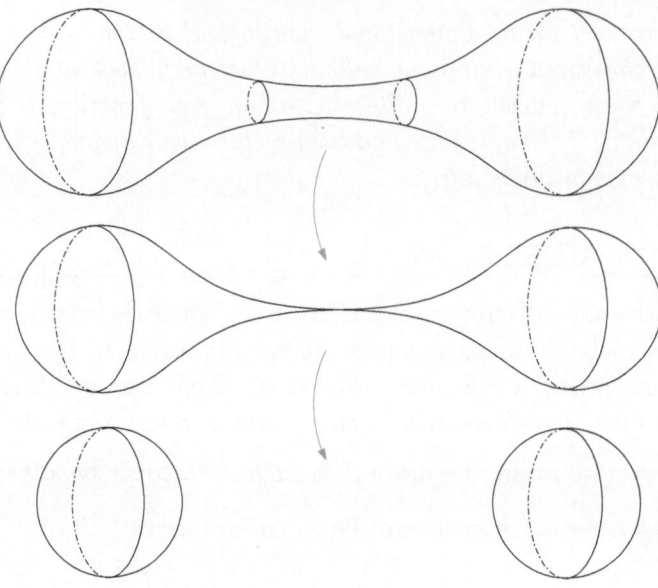

Figure 5.5 A counterexample showing that the surface analog of Greyson's Theorem does not hold: A "dumbbell" surface evolves, under the mean curvature flow, toward two disjoint spheres.

Let us conclude this section with the observation that, precisely like the notion of curve shortening flow which it generalizes, the mean curvature flow also has a clear applications in practice (albeit perhaps fewer or less known) – see, for instance [89] and the bibliography cited there.

5.3 GAUSS CURVATURE FLOW

Less well known (but perhaps easier to see as a direct generalization of the planar curve shortening flow) is the *Gauss curvature flow*. It seemingly was motivated by an observation by Hilbert and Cohn-Vossen [163][7] that pebbles on a beach are most often ellipsoidal in shape (and not spherical as one might have presumed. This conjecture – for it was only a conjecture – was repeated by Rogers in 1976 [198], who was clearly unaware of the 1974 paper by Firey [121] that showed that as times tends to infinity, pebbles become spherical and, moreover, if they become ellipsoidal, it is because they have preferential direction for abrasion.[8] (For more details on the history of the problem, see [39]. See also the quite recent paper [164] where the problem is attacked, with modern tools, from a more experimental viewpoint.)

It is time we formalized the problem. First, from a physical viewpoint, we are seeking to model the process of "changing shape of a tumbling stone subjected to collisions from all directions with uniform frequency" [11]. Note that this problem is

---

[7]This coming to show again how deep and influential this "popularization" book really was and still is.

[8]Formally stated, they are *anisotopic*.

not only a nice puzzle,but it is clearly relevant in Geology. Moreover – and this is in truth the reason this problem has garnered interest beyond its mathematical confines – it is important (and it was more so till not long ago) in the manufacturing of round stone or metal balls (like ball bearings) [39], [164].

Mathematically speaking, one is conduced to study the evolution of the evolution of convex surfaces by Guss curvature, that is under the *Gauss curvature flow*, that is, as the reader clearly expects by now, their evolution in the direction of the normal to the surface, at rate proportional to Gauss curvature K. More formally, given a surface S_0 given by an embedding $\mathbf{x}$ of a surface S, then the evolving surface at time t, $S_t = \mathbf{x}(p, t)$ is given by

$$\frac{\partial \mathbf{x}}{\partial t} = -K(p, t)N(p, t) \tag{5.3.1}$$

where $K(p, t), N(p, t)$ denote the Gauss curvature and normal, respectively, at the point p, at time t. (Formally, one should also add, of course, the initial condition $\mathbf{x}(p, 0) = S_0(p)$.)

Firey [121] proved, besides the limiting result we already mentioned, local existence and conjectured existence at any time t, under some some existence and regularity of solutions conditions and also the symmetry, relative to the origin, of the initial surface. Further progress was made by several authors (see [11]), but the problem was settled only in 1999 by Andrews [11], more precisely he proved the following[9]

Theorem 5.3.1 ([11]) *If S_0 is a compact, smooth, strictly convex surface in $\mathbb{R}^3$, then there exists a unique, smooth solution S_t of the Gauss curvature flow (5.3.1), for $t \in (0, T]$, where*

$$T = \frac{\mathrm{Vol}(S_0)}{4\pi}.$$

Moreover, the surfaces S_t are strictly convex, and

$$\lim_{t \to T} S_t = q \in \mathbb{R}^3.$$

Furthermore, by rescaling at q on obtains smooth convergence to a sphere. More precisely, the embedding $\tilde{\mathbf{x}} \in C^\infty$, where

$$\tilde{\mathbf{x}}(p, t) = \frac{\mathbf{x}(p, t) - q}{\sqrt[3]{3(T - t)}}$$

converges to $\tilde{\mathbf{x}}_T(p)$, and $\tilde{\mathbf{x}}_T(S) = \mathbb{S}^2$.

In other words, while the pebbles would erode to points, they become round balls just before the collapse.

Before concluding this section le us make a couple of remarks:

Remark 5.3.2 *The surfaces S_t are described by the family of curves given by the trajectories $p \longmapsto \mathbf{x}(p, t)$ orthogonal to the surfaces S_t.*

---

[9]Additionally, Andrews also proved a similar result for non-smooth initial surfaces (Theorem 2, [11]).

Remark 5.3.3 *Regarding the ellipsoids vs. spheres shape of the resulting pebbles:*

1. The deviation from sphericity is given by

$$(k_{\text{Max}} - k_{\text{min}})^2 = 4(H^2 - K).$$

2. If in the defining Equation (5.3.1) one replaces K by $\sqrt[4]{K}$, then the limiting shapes will indeed be ellipsoids [12].

The reader interested to study in depth the geometric evolution equations presented in this chapter and related ones, is invited to study the quite new book dedicated to the subject [13].

CHAPTER 6

Geodesics

Geodesics are the analog, for surfaces (and higher-dimensional manifolds) of the straight lines in the good old Euclidean geometry. This is such an innate notion in all of us, not least due to the years of schooling, that we tend to forget that straight lines in the plane have to *different* properties: They both *minimize distances* (between pairs of points) and also represent the *straightest* curves in the plane. While the first concept of geodesics is ubiquitous in Computer Science, with all its ramifications and applications, being essentially identified in the common conception with the famous Dijkstra algorithm, geodesics as straightest curves are generally forgotten, even though it is this avatar of theirs that human beings use in day to day life.

Evidently, such an essential mathematical concept has its applications and interpretations in Physics. To the point, the definition of geodesics as shortest curves has, as physical analog, the *Maupertuis-Jacobi principle of least action*, while their role as straightest curves is mirrored, in Physics, by the *Maupertuis-Gauss-Hertz principle of least constraint* (or *stationary action principle*). In fact, there is yet another characterization of geodesics, which is commonly forgotten in applications, but which is extremely important in Geometry, namely that of geodesics as *frontal* lines which is also best understood beginning from the following intuitive explanation: Imagine a sageway navigating a terrain (i.e. moving on a surface). It's wheels are rigidly connected by the axle, so they travel, essentially the same distance (and at equal speeds). Then the driver, which is placed at the middle of the axle, travels approximately on a geodesic (which is, of course, his desire – to navigate between origin and destination on the shortest possible route). This can be made more formal, by considering a pair of points p, q at ε distance one from the other, that are both moving perpendicularly to the segment $[p, q]$, by a (small distance) δ. Then the curve described by the midpoint m of the considered segment is an approximation of a geodesic, which can be made as good as desired by letting ε (and δ) tend to 0. (See Figure 6.1.) We shall return, in much more detail, to this definition of geodesics, in the sequel. (Also, for a generalization of the notion of geodesic, inspired by bicycle movement, see [19].)

6.1 COVARIANT DERIVATIVE

The essential technical ingredient needed in the study of geodesics is the *covariant derivative*:

DOI: 10.1201/9781003350576-6

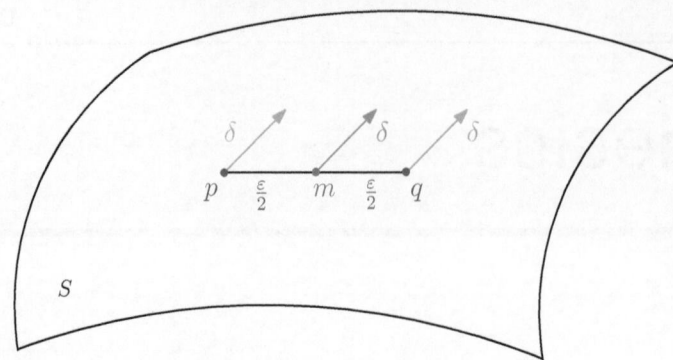

Figure 6.1 Schematic rendering of the geodesics as frontal curves. Here the distance between the initial points p and q is small (ε), as is the distance they travel (δ). The trajectory of the midpoint m is the (local) geodesic c.

Definition 6.1.1 *Let $p \in U = \text{int}\,U \subset S$, let $\mathbf{w}$ be the restriction to U of a vector field on S, $\mathbf{x} \in T_p(S)$, and let $c : (-\varepsilon, +\varepsilon) \to U$, such that $c(0) = p$ and $\dot{c}(0) = \mathbf{x}$. The projection on $T_p(S)$ of $\frac{d\mathbf{w}}{dt}$, i.e. $\nabla_{\mathbf{x}}\mathbf{w}(p)$,*

$$\text{pr}_{T_p(S)} \frac{d\mathbf{w}}{dt} := \frac{\nabla \mathbf{w}}{dt} \tag{6.1.1}$$

is called the covariant derivative *of $\mathbf{w}$, relative to $\mathbf{x}$, at p. (Also denoted by $\frac{D\mathbf{w}}{dt}$ or $D_y\mathbf{w}$.)*

Remark 6.1.2 *The covariant derivative has a quite simple physical intuition – See Figure 6.2: It represents the tangential component of the acceleration $\ddot{\mathbf{c}}$ (i.e. it is "the acceleration at $c(t)$ as seen from the surface S" – see Figure 6.3).*

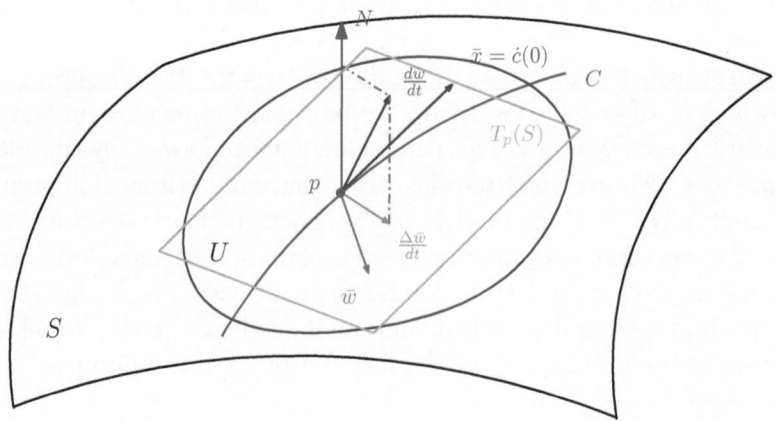

Figure 6.2 The covariant derivative at p of $\mathbf{w}$, relative to $\mathbf{u}$.

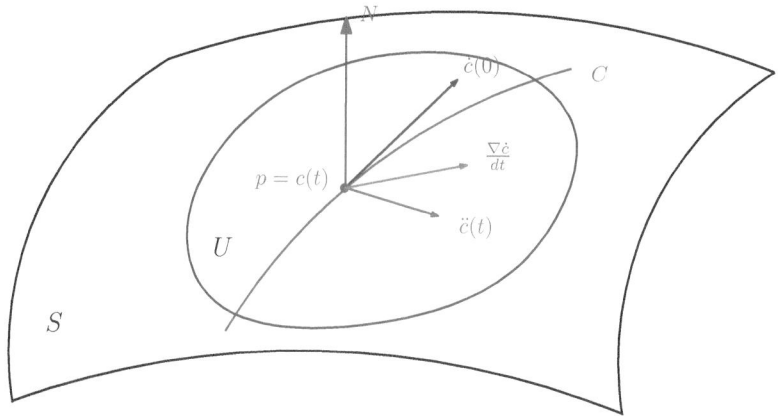

Figure 6.3 The physical interpretation of the covariant derivative.

One can produce a (quite technical) expression for $\nabla_{\mathbf{x}}\mathbf{w}(p)$ by writing $\mathbf{w}$ in the standard basis of $T_p(S)$, namely $\mathbf{w}(t) = a(t)f_u + b(t)f_v$ (given that $c(t) = f(u(t), v(t)) \in S$. Then

$$\frac{d\mathbf{w}}{dt} = a(f_{uu}\dot{\mathbf{u}} + f_{uv}\dot{\mathbf{v}}) + b(f_{vu}\dot{\mathbf{u}} + f_{vv}\dot{\mathbf{v}}) + \dot{a}f_u + \dot{b}f_v .$$

Thus, by (6.1.1), we obtain the following expression for the covariant derivative of $\mathbf{w}$:

$$\frac{\nabla\mathbf{w}}{dt} = (\dot{a}+\Gamma^1_{11}a\dot{\mathbf{u}}+\Gamma^1_{12}a\dot{\mathbf{v}}+\Gamma^1_{12}b\dot{\mathbf{u}}+\Gamma^1_{22}v\dot{\mathbf{v}})f_u+(\dot{b}+\Gamma^2_{11}a\dot{\mathbf{u}}+\Gamma^2_{12}a\dot{\mathbf{v}}+\Gamma^2_{12}b\dot{\mathbf{u}}+\Gamma^2_{22}v\dot{\mathbf{v}})f_v .$$

$$(6.1.2)$$

Remark 6.1.3 *Since (6.1.2) depends only on $\mathbf{x}$ and on the Christoffel symbols, i.e. on the first fundamental form of the surface, it follows that the covariant derivative is an intrinsic notion.*

Remark 6.1.4 *If S is a plane, the covariant derivative is nothing but the usual directional derivative. Indeed, in this case we can chose a parametrization such that $E = G = 1, F = 0$, thus $\Gamma^k_{ij} = 0$, for all triples i, j, k, thus $\frac{\nabla\mathbf{w}}{dt} = \dot{a}f_u + \dot{b}f_v$.*

Note, however, than on a surface different from the plane, parallel vector fields might look very "un-parallel" in the Euclidean sense. The simplest, and most eloquent example is that of the vectors tangent to a meridian of the (unit) sphere: If we consider (as we normally do), the meridian (i.e. a great circle of the sphere) to be parametrized by unit length, then the tangent vectors are orthogonal to the normals to the surface (and meridian), thus this represents a parallel vector field – See Figure 6.4.

Example 6.1.5 *Prove that the covariant derivative is the only operator $\mathcal{D} : V \to V$, where is the set of $\mathcal{C}^{\infty}$ vector fields along a curve c, satisfying the following properties*

1. $\frac{\mathcal{D}(V+W)}{dt} = \frac{\mathcal{D}(V)}{dt} + \frac{\mathcal{D}(W)}{dt}$, for any $V, W \in \mathscr{V}$;

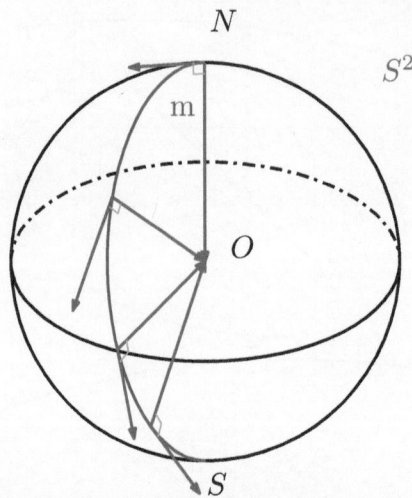

Figure 6.4   Parallel Transport along a meridian of the sphere.

2. $\frac{\mathcal{D}(fV)}{dt} = \frac{df}{dt}V + f\frac{\mathcal{D}(V)}{dt}$, *for any* $V \in \mathscr{V}$ *and any* $f \in \mathcal{C}^\infty([a,b])$;

3. *If* V *is the restriction of a* $\mathcal{C}^\infty$ *vector field* U *to the neighborhood of a point* $p = c(t)$, *then*

$$\frac{\mathcal{D}(V)}{dt} = \nabla_{\frac{dc}{dt}} U \,.$$

To define geodesics, we first need to define *parallel vector fields*, using the covariant derivative that we just introduced:

Definition 6.1.6 *Let* $c : I \to S$ *be a parametrized curve and let* **w** *be a vector field along c.* **w** *is called (*Levi-Civita*) parallel iff* $\frac{\nabla \mathbf{w}}{dt} \equiv 0$.

In the case the surface is a Euclidean plane, the notion of parallelism reduces to that of a constant vector field (along the given curve c).

The following simple result, constitutes a direct generalization of the situation in the plane.

Lemma 6.1.7 *Let* $\mathbf{v}, \mathbf{w}$ *be parallel vector fields along* $c : I \to S$. *Then* $\mathbf{v}(t) \cdot \mathbf{w}(t) \equiv$ const.. *(In particular, if* $\|\mathbf{v}\| =$ const., $\|\mathbf{w}\| =$ const., *then* $\measuredangle(\mathbf{v}, \mathbf{w})$ *is also constant.)*

Proof By the definition of parallelism we have:

$$\frac{\nabla \mathbf{v}}{dt} = 0 \iff \frac{d\mathbf{w}}{dt} \perp T_{c(t)}, \text{ thence } \mathbf{v}(t) \cdot \dot{\mathbf{w}}(t) = 0 \,;$$

and

$$\frac{\nabla \mathbf{w}}{dt} = 0 \iff \frac{d\mathbf{v}}{dt} \perp T_{c(t)}, \text{ thence } \dot{\mathbf{v}}(t) \cdot \mathbf{w}(t) = 0 \,.$$

Therefore, $\mathbf{v}(t) \cdot \dot{\mathbf{w}}(t) + \dot{\mathbf{v}}(t) \cdot \mathbf{w}(t) = 0 \iff (\mathbf{v}(t) \cdot \mathbf{w}(t))' = 0$, thus $\mathbf{v}(t) \cdot \mathbf{w}(t) \equiv$ const..

□

It turns out that the parallel transport along a given curve is determined by the initial vector, more precisely we have the following:

Proposition 6.1.8 *Let $c : I \to S$ be a parametrized curve, and let $\mathbf{v}_0 \in T_{p_0}(S)$, $p_0 = c(t_0)$. Then there exist a unique parallel field $\mathbf{v}(t)$ along c, such that $\mathbf{v}(t_0) = \mathbf{v}_0$.*

Exercise 6.1.9 *Prove the proposition above. (Hint: Use the Existence and Uniqueness Theorem for Ordinary Differential Equations.)*

Proposition 6.1.8 allows us to speak of *the parallel transport of a vector $\mathbf{v}_0$ along a curve c*, as being the parallel field determined by the initial vector $\mathbf{v}_0$.

The next result, is not only simple to prove, it also is quite useful, as we shall shortly show:

Lemma 6.1.10 *Let S_1, S_2 be two tangent curves and along $c = S_1 \cap S_2$. Then the parallel transport along c in S_1 and S_2 coincide.*

This is a straightforward exercise for the reader:

Exercise 6.1.11 *Prove Lemma 6.1.10.*

An immediate demonstration of the usefulness of the lemma is the following

Example 6.1.12 *1. One can compute the parallel transport along a parallel of the (unit) sphere by considering the tangent cone to the sphere, along the chosen parallel and developing the cone onto the plane – see Figure 6.5. Supply the details and compute the parallel transport.*

  2. Can one apply a similar method to compute the parallel transport along a great circle of the sphere? If it is possible, compute it.

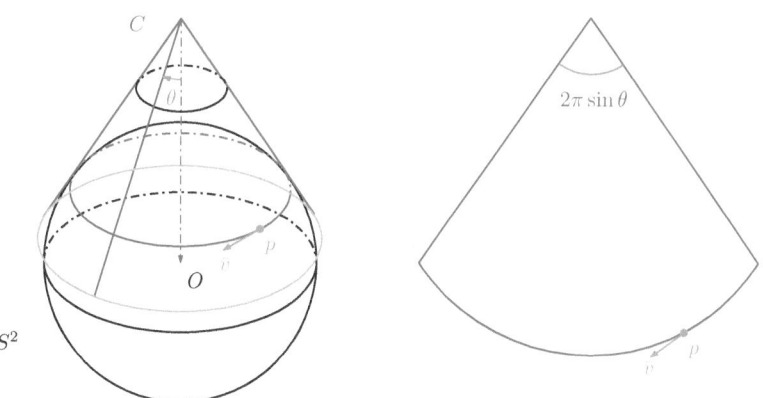

Figure 6.5 Computing the parallel transport along a parallel of the sphere.

6.2   GEODESICS

We are now ready to formally define the notion of geodesic:

Definition 6.2.1 *Let S be a smooth surface and let $c : I \to S$ be a regular, connected curve. c is called a* geodesic *iff, for any $p \in c(I)$, there exists a neigborhood V of p where the following holds:*

$$\frac{\nabla \dot{c}(t)}{dt} = 0 \, . \tag{6.2.1}$$

Remark 6.2.2 *Evidently, if $l \subset S$ is a straight line, then l is a geodesic.*

Remark 6.2.3 *$c \subset S$ is a geodesic $\Longleftrightarrow N \| \ddot{c} = kn \Longleftrightarrow \vec{n}(p) \| N(p)$, for any $p \in S$.*

Example 6.2.4   *1. The great circle on a sphere are geodesics (see the example in the previous section).*

  2. Let C be the standard circular cylinder of radius 1. One can use the canonical parametrization $f(u, v) = (\cos u, \sin u, v)$ and apply the characterizing Equation (6.2.1) to determine the geodesics of the cylinder. This is an exercise which we suggest to the reader. However, there is a much simpler – and intuitive – way of seeing this:

> ***Exercise 6.2.5*** *Compute the geodesic of the circular cylinder using the defining differential equation (6.2.1).*

  While the exercise above is certainly a worthy one, there is a far simpler (and more instructive) way of determining the geodesics of the cylinder. Indeed, intuitively, the cylinder can be constructed by bending a rectangular sheet of paper. Abstractly, it is obtained by identifying the opposite, parallel sides of an infinite strip. Since bending is an (local) isometric transformation, the straight lines that are the geodesics in the plane will be mapped into the geodesics of the cylinder (and this operation is reversible). Therefore, the geodesic of the cylinder are the parallels, the meridians and the helices (skew lines) on the cylinder. Since the whole plane can be filled by an infinity of such strips, an immediate consequence is the (somewhat surprising in the beginning) fact that, in general, there are an infinity of geodesics connecting two points on the cylinder. (The exception being points on the meridians and the parallels.) – See Figure 6.6.

  3. It would be tempting to use the same method to determine the geodesics of the torus. Indeed, the torus is obtained by identifying the opposite sides of the unit square by Euclidean translations, which are isometries. Thus, the geodesics of the plane (restricted to the square) are mapped to the geodesics of the torus. (See Figure 6.7.) However, the torus described is the so called flat torus and, since it was obtained by identifications which are Euclidean isometries, its curvature should be identical 0. However, this is not possible due to the following:

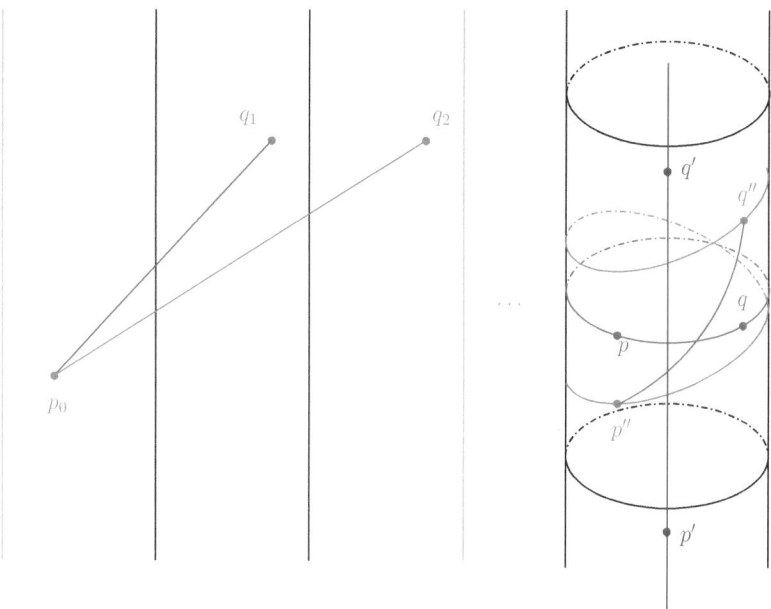

Figure 6.6 The geodesics of the cylinder.

Fact 6.2.6 *Any compact surface in $\mathbb{R}^3$ has at least one point of positive curvature.*

The proof of this result (which could has as well be included) is left to the reader.

Exercise 6.2.7 *Prove Fact 6.2.6. (Hint: Begin from the definition of compact sets in $\mathbb{R}^3$.)*

We conclude the list of first examples with an exercise of a surface constructed in a manner similar to the previous two ones.

Exercise 6.2.8 *Determine the geodesics of the cone. (Hint: Consider also Example 5.1.12 above.)*

Proposition 6.2.9 *Let $c \subset S$ be a geodesic. Then*

1. If c is a line of curvature, then c is planar.

2. If c is an asymptotic line, then c is a straight line.

Proof We make appeal to the Darboux Equations ([Gauss0], 2.3.30):

$$\begin{cases} \dot{\mathbf{t}} = \quad\quad k_g\mathbf{u} + k_n\mathbf{v} & (*) \\ \dot{\mathbf{u}} = -k_g\mathbf{t} \quad\quad + \tau_g\mathbf{v} & (**) \\ \dot{\mathbf{v}} = -k_n\mathbf{t} - \tau_g\mathbf{u} & (***) \end{cases}$$

Furthermore, we need the following, simple yet essential

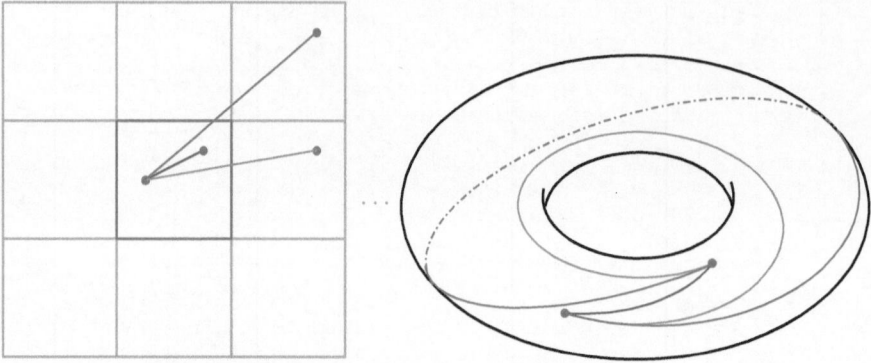

Figure 6.7 The geodesic on the torus are not unique.

Lemma 6.2.10 *Let $c \subset S$ be a regular curve. Then*

1. *c is a geodesic $\Longleftrightarrow k_g \equiv 0$.*

2. *c is a line of curvature $\Longleftrightarrow \tau_g \equiv 0$.*

Proof

1. *c is a geodesic $\Longleftrightarrow \ddot{c} \perp N \Longleftrightarrow \mathbf{n} \perp N$ (for $k \neq 0$).*

   By the first of the Darboux Equations $\dot{\mathbf{t}} = k_g\mathbf{u} + k_n\mathbf{v}$, i.e. $\ddot{c} = \pm k_g\mathbf{n} \pm k_n N$. By scalar multiplication of this last expression with $\mathbf{n}$ the desired conclusion follows immediately.

2. *$\dot{v}(0) = dN$ and, by the third of the Darboux Equations, $\dot{\mathbf{v}} = -k_n\mathbf{t} - \tau_g\mathbf{u}$.*

   On the other hand, $k_n = -dN \cdot t$. However, by Theorem 2.2.16 c is a line of curvature iff $-dN = k\mathbf{t}$.

   Combining the facts above, it follows immediately that c is a line of curvature iff $\tau_g \equiv 0$.

This concludes the proof of the lemma.

□

We are now ready to proceed with the proof of the proposition.

1. From the lemma above it follows that, on the one hand, $\tau_g \equiv 0$ and, on the other hand, $k_g \equiv 0$. Therefore, the second of the Darboux Equations becomes $\dot{\mathbf{u}} = 0$, thus $\mathbf{u} = \mathbf{u}_0 = $ const. Since $\mathbf{t} \cdot \mathbf{u} = 0$, by denoting $\mathbf{t} = (\dot{x}, \dot{y}, \dot{z})$ and $\mathbf{u}_0 = (u_0^1, u_0^2, u_0^3)$, we have that $u_0^1\dot{x}, u_0^2\dot{y}, u_0^3\dot{z} = 0$, thence $u_0^1x, u_0^2y, u_0^3z = c_0 = $ const., i.e. $c = (c, y, z)$ is a plane curve.

2. From the lemma above it follows that, on the one hand, $k_g \equiv 0$ and, on the other hand, since c is an asymptotic line, $k_n \equiv 0$. Therefore $\dot{\mathbf{t}}$, by the first of the Darboux, Equations, hence $\mathbf{t} = $ const., thence $c(s) = as + b$, i.e. c is a straight line.

This concludes the proof of the proposition.

□

It is time we turned to the geometric interpretation of the geodesic curvature k_g, which is rather straightforward, namely $\frac{\nabla \dot{c}(t)}{dt} = k_g(N \times \dot{c})$ (see Figure 6.8 below). This formula gives us another definition (or interpretation) of geodesic curvature, namely as the absolute value of the covariant derivative, i.e.

$$k_g = \left\| \frac{\nabla \dot{c}(t)}{dt} \right\|. \tag{6.2.2}$$

Thus geodesics are, among all the curves on a surface passing to a point on that surface, the ones with the smallest (geodesic) curvature. The formula above also demonstrates that the notion of geodesics as straightest lines is an extrinsic one, since it depends on the normal, i.e. on the manner the surface is realized (embedded) in Euclidean space. One can interpret this fact in terms of osculating plane as follows: A curve $c \subset S$ is a geodesic of S if the osculating plane of the curve contains the normal N to the surface at every point of the curve. This condition can be expressed in terms of tangents, by requiring that the angle between tangents at points close to each other is as small as possible if the curve is to continue to be contained in S.

Exercise 6.2.11 *Justify the assertion above.*

Since this is equivalent to the condition that geodesic curvature is minimal, we recover the previous characterization of geodesics. The osculating plane condition is, seemingly, the one used in practice to determine geodesics on the Earth surface [163], a fact that is ingrained in the very name of "geodesic", that is derived from the Ancient Greek word for Earth – "geos/γεω".

Moreover, one can strengthen this observation, by observing that, as we know, $\ddot{c} = k\mathbf{n}$ (See Figure 6.9.), thus

$$k^2 = k_n^2 + k_g^2. \tag{6.2.3}$$

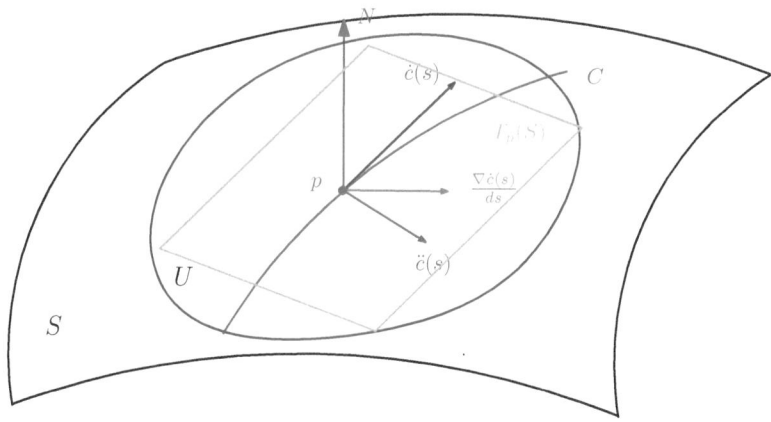

Figure 6.8 The geometric interpretation of the geodesic curvature. The tangential component of $\ddot{c}$ is also called the *geodesic curvature vector*.

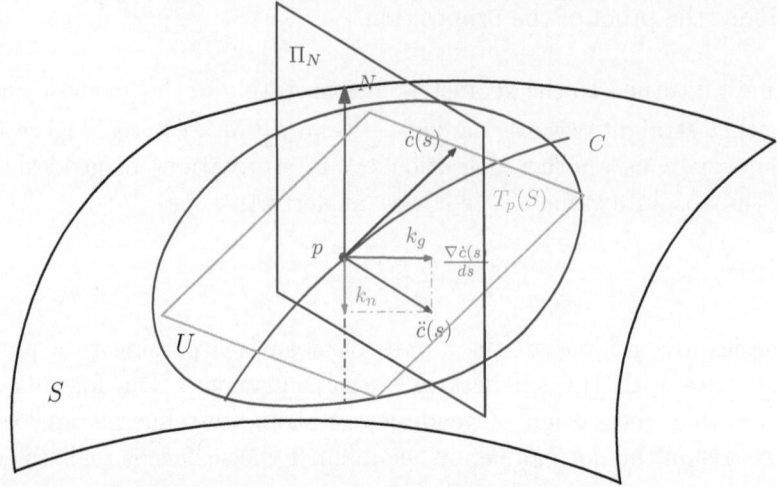

Figure 6.9 The normal and geodesic components of the curvature of curves on a surface.

This is quite a practical fact, as the reader can convince her/himself by solving the following exercise

Exercise 6.2.12 *Compute the geodesic curvature of the φ parallel on the unit sphere. (Particular case $\varphi = 66°$.)*

In the sequel we shall need the formula in the exercise below.

Exercise 6.2.13 *Prove that*

$$k_g = f_s \cdot (f_{ss} \times N) \ . \tag{6.2.4}$$

We concentrate next on the geometric interpretation of τ_g. We begin by observing that, from the second of the Darboux Equations, it follows that

$$\tau_g = \mathbf{v} \times \dot{\mathbf{u}} \ .$$

Next, consider the oriented angle φ between $\mathbf{n}$ and $\mathbf{v} = N$. Then

$$\mathbf{n} = \mathbf{u} \sin \varphi + \mathbf{v} \cos \varphi \ ,$$

$$\mathbf{b} = -\mathbf{u} \cos \varphi + \mathbf{v} \sin \varphi \ .$$

From the two equalities above we can extract $\mathbf{u}, \mathbf{v}$:

$$\mathbf{u} = \mathbf{n} \sin \varphi - \mathbf{b} \cos \varphi \ ,$$

$$\mathbf{v} = \mathbf{n} \cos \varphi + \mathbf{b} \sin \varphi \ .$$

By inserting these formulas in the expression of τ_g above we obtain that

$$\tau_g = (\mathbf{n} \cos \varphi + \mathbf{b} \sin \varphi) \times \frac{d}{ds} (\mathbf{n} \sin \varphi - \mathbf{b} \cos \varphi) \ .$$

But, by the Serret-Frenet formulas

$$\frac{d\mathbf{n}}{ds} = \dot{\mathbf{n}} = -\kappa\mathbf{t} + \tau\,,$$

$$\frac{d\mathbf{b}}{ds} = \dot{\mathbf{b}} = -\tau\mathbf{n}\,.$$

Thence it follows that

$$\tau_g = \tau + \frac{d\varphi}{ds}\,. \tag{6.2.5}$$

Thus geodesic torsion measures the deviation from the curve torsion due to its being embedded in a surface, where the deviation is measured by the rate of change of the angle between the curve and surface normals.

In particular, we have the following

Proposition 6.2.14 *Let $p \in S$ and let $\mathbf{x} \in T_p(S), \|\mathbf{x}\| = 1$. Then $\tau_g(\mathbf{x}) = \tau(0) -$ the geodesic curvature of a curve $c \subset S$, such that $\dot{c}(0) = \mathbf{x}$.*

The proof is simple and short, thus we leave it to the reader.

Exercise 6.2.15 *Prove the proposition above.*

Remark 6.2.16 *There exists a kinematic interpretation of the parallel transport (cf. [37]): Given a surface S, a curve $c \subset S$ and in initial point $p \in c$, fix the tangent plane $T = T_p(S)$ to S at p. The surface S, considered as a rigid body, is rolled on this plane. The points of contact of S on T trace a curve c_T on T. Then $k_{c_T} = k_c$ and, moreover, to any parallel vector field on S, along c, corresponds, along c_T, a field of constant vectors in T (See Figure 6.10).*

We now return to the study of geodesics on surfaces with

Lemma 6.2.17 *Let $S \subset \mathbb{R}^3$ be a connected surface and let $c : I \to S$ be a geodesic, such that*

1. *c is planar;*

2. *$k(c(s)) \neq 0$, for all s.*

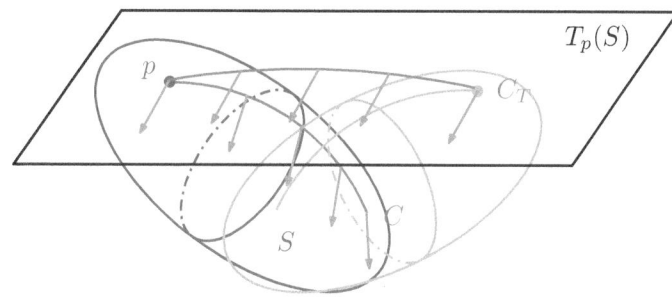

Figure 6.10 The kinematic interpretation of the parallel transport.

Then c is a line of curvature.

Proof Let Π be a plane containing c. Then $\ddot{c}$ exists and is non zero, by the second condition in the statement of the lemma and $\ddot{c} \subset \Pi$. Since c is a geodesics, $\mathbf{n}\|N$, therefore $N \subset \Pi$. It follows that $\frac{dN(c(s))}{ds} \subset \Pi$, thus $\frac{dN(c(s))}{ds} = \lambda\dot{c}$, i.e. c is a line o curvature.

$\square$

We conclude this section with the following

Theorem 6.2.18 *Let $S \subset \mathbb{R}^3$ be a connected surface, such that any geodesic of S is contained either in a plane or in a sphere. The S itself is contained either in a plane or in a sphere.*

Exercise 6.2.19 *Prove the theorem above. (Hint: Prove that an point $p \in S$ is an umbilic.)*

The following exercise is also an essential ingredient in the proof of the theorem above (as well as being an interesting fact in itself):

Exercise 6.2.20 *Let $c \subset S \subset \mathbb{R}^3$, $\|\dot{c}\| \equiv 1$, and let $p = c(0)$. Prove that the curvature of c, $k_c \neq 0$ and $\mathbf{n}$ is not perpendicular to N.*

Since we do not wish to transform this text into yet another book in classical Differential Geometry, we shall omit the (quite technical) proofs of some of the results below[1], and rather concentrate on the more geometrically intuitive aspects with a view on the discretizations and their applications.

Let us first note that, if $\mathbf{v}(t)$ is a unit vector field, then

$$\frac{\nabla \dot{c}(t)}{dt} = \lambda(N \times \mathbf{v}(t));$$

where $\lambda = \lambda(t)$ is called the *algebraic value of the covariant derivative* and it is denoted by $\left[\frac{\nabla \dot{c}(t)}{dt}\right]$. In particular, for $\mathbf{v}(t) = \dot{c}(t)$, the algebraic value of the covariant derivative is nothing else but the geodesic curvature k_g.

The first result involving the algebraic value of the covariant derivative is the following preparatory

Lemma 6.2.21 *Let $c: I \to$, and let $\mathbf{v}, \mathbf{w}$ be unit differentiable vector fields along c. Then*

$$\left[\frac{\nabla \dot{c}(t)}{dt}\right] - \left[\frac{\nabla \dot{c}(t)}{dt}\right] = \frac{d\varphi}{dt}$$

φ *denoting the angle* $\angle(\mathbf{v}, \mathbf{w})$.

Exercise 6.2.22 *Prove the lemma above.*

---

[1] The interested reader is referred, e.g. to [105], [168].

Proposition 6.2.23 *Let $f(u,v)$ be an orthogonal parametrization of an oriented surface S, let $c \subset S$ be a curve, and let $\mathbf{v}$ be a differentiable vector field along c. Then*

$$\left[\frac{\nabla \dot{c}(t)}{dt} \right] = \frac{1}{2\sqrt{EG}} \left(G_u \frac{dv}{dt} - E_v \frac{du}{dt} \right) + \frac{d\varphi}{dt} . \tag{6.2.6}$$

As an immediate consequence of the proposition above we can obtain yet another, independent of the theory of ODEs, proof of Proposition 6.1.8 on the existence and uniqueness of parallel transport, which we leave as an exercise to the reader.

Exercise 6.2.24 *Provide a proof of Proposition 6.1.8, using Proposition 6.2.23 above.*

As another corollary of Proposition 6.2.23 one can show that a results somewhat similar to results for normal curvature (chapter 2, Exercise 2.21) and geodesic curvature (chapter 2, Proposition 2.58), holds for the geodesic curvature of the intersection of two surfaces, more precisely one can prove the following result, sometimes called *Liouville's formula*:

Proposition 6.2.25 *Let $c \subset S$ be (locally) parametrized by arc length, and let $f(u,v)$ be an orthogonal parametrization of S, and let $\varphi(t)$ denote the (oriented) angle between f_u and $\dot{c}(t)$. Then*

$$k_g = k_{g,1} \cos \varphi + k_{g,2} \sin \varphi + \frac{d\varphi}{dt} ; \tag{6.2.7}$$

where $k_{g,1}, k_{g,1}$ are the geodesic curvatures of the coordinate curves $\{v = \text{const.}\}$, $\{v = \text{const.}\}$, respectively.

Since the formula above, however interesting it might be, is not essential in the sequel, we leave its proof as an exercise.

Exercise 6.2.26 *Prove Proposition 6.2.25.*

It is only natural to look for a differential equation for the geodesics of a surface, given that such equations were indeed obtained in Chapter 2 for the other types of important curves on surfaces, namely for lines of curvature and asymptotic lines. Such a formula is indeed possible, however with the caveat that it is not a single equation, but rather the following pair of *differential equations of the geodesics*:

$$\begin{cases} u'' + \Gamma_{11}^1 (u')^2 + 2\Gamma_{12}^1 u'v' + \Gamma_{22}^1 (v')^2 = 0 \\ v'' + \Gamma_{11}^2 (u')^2 + 2\Gamma_{12}^2 u'v' + \Gamma_{22}^2 (v')^2 = 0 \end{cases} \tag{6.2.8}$$

The proof is immediate, therefore we leave it as yet another exercise for the reader.

Exercise 6.2.27 *Prove that geodesics are characterized by the System (6.2.8).*

The system of equations above, which have as coefficients the Christoffel symbols, should be daunting enough to preclude their lighthearted use in practice. Sadly enough, engineering students do try to use them to solve exam problems, under dire time constraints, and with only with the modicum of understanding of ODEs a first such course endows them with. To demonstrate how difficult solving the System (6.2.8) we suggest the following exercise (whose full solution can be found in [105] and [310]).

Exercise 6.2.28 *Let S be a surface of revolution.*

1. *Show that the meridians of S are geodesics.*

2. *Determine a necessary condition for a parallel of S to be a geodesic.*

3. *Prove the following result, known as Clairaut's Theorem or Clairaut's relation:*

   Proposition 6.2.29 *Let γ be a geodesic on a surface of revolution. Show that, if γ intersects a parallel of S, then*

   $$r \cos \theta = C = \text{const.} \qquad (6.2.9)$$

   where r is the radius of the parallel at the point of intersection and $\theta \in [0, \pi/2]$ is the angle between γ and .

   Conversely, if a curve c parametrized by arc length satisfies the equation above, and it is not a parallel, then c is a geodesics.

4. *Using Clairaut's Theorem, describe the global behavior of the geodesic of the torus of revolution (thus completing the analysis of Example 3 in Exercise 6.2.5).*

5. *Let $p = p(x, y, z), z > 0$ be a point on the hyperboloid of revolution $x^2 + y^2 - z^2 = 1$. Show that there exists a geodesic starting from p that approaches asymptotically (but never intersects) the parallel $x^2 + y^2 = 0$.*

Also, we shall show in the last section of this chapter how the problem of determining the geodesics of a surface can be addressed in practice.

Having said that, it is not true, however, that System (6.2.8) is of no import. Indeed, an immediate consequence of these equations is the essential existence and uniqueness result for geodesics:

Proposition 6.2.30 *Let $p \in S$ be a point, and consider a tangent direction $\mathbf{v} \in T_p(S)$. Then there exists exactly one geodesic through p, in the direction $\mathbf{v}$.*

Proof The equations in (6.2.8) are ordinary differential equations of order 2,thus, by the existence and uniqueness theorem for this type of equation, System (6.2.8) has a unique solution u, v, satisfying initial conditions $u(t_0) = u_0, v(t_0) = v_0, u'(t_0) = u_1, v'(t_0) = v_1$. Thus, through each point (u_0, v_0) and tangent to a given direction $(u'(t_0), v'(t_0))$ passes precisely one geodesic. Moreover, the existence theorem assures the geodesic is differentiable. Moreover, it is also regular by Lemma 6.1.7, since parallel transport preserves the lengths of tangent vectors, thus $||c||^2 = (u'(t))^2 + (v'(t))^2 \neq 0$ for all t, since it holds for the initial point p.

$\square$

We have yet to address the role of geodesics as shortest curves, and it is high time we should do this, but in order to remediate this lacuna we have to bring yet another couple of definitions. The ingredient needed is a new local coordinates system, more natural for dealing with geodesics, which we construct as follows:

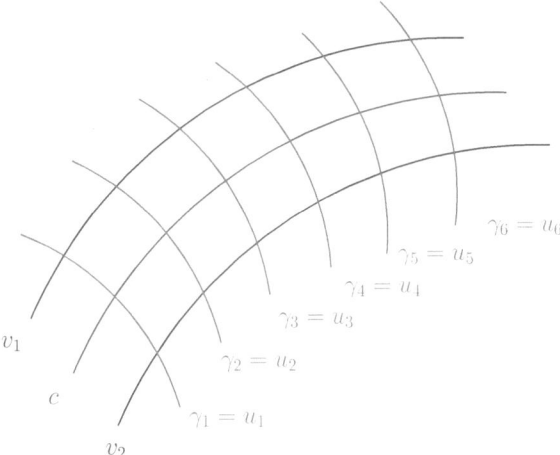

v_1

c

v_2

$\gamma_1 = u_1$

$\gamma_2 = u_2$

$\gamma_3 = u_3$

$\gamma_4 = u_4$

$\gamma_5 = u_5$

$\gamma_6 = u_6$

Figure 6.11 Geodesic coordinates on surface.

Let c be a curve on the surface $S = f(u, v)$. We construct geodesics γ_i orthogonal to c. We define these geodesics as the u parametric curves, while as v parametric curves we chose their orthogonal trajectories (see Figure 6.11). Clearly this construction is purely local, thus the computations and discussion below are restricted to a neighborhood where this construction is possible.[2]

The nomenclature for the families of curves introduced above is quite suggestive:

Definition 6.2.31 *The u and v curves in the construction above are called a* field of geodesics *and* geodesic parallels, *respectively. The new coordinates are called* geodesic *(or* Fermi*) coordinates (See Figure 6.11).*

While we shall dedicate a separate section to the computation of geodesics in various applied, discrete settings, we wish to emphasize that even such smaller aspects of the theory, like geodesic coordinates have their applications in practice. The most striking one is in architecture, when smooth surfaces of varying curvature and nontrivial topology are to be constructed, using standard constructions elements which are rectangular or more general quadrilateral plates or beams made of rigid (or almost rigid) materials, such as concrete, glass metal and wood). – See Figure 6.12 for a few examples. The pioneering work in this direction (and related ones) is that of H. Pottman (see, e.g. [332]).

We can take advantage of the newly defined coordinates as follows: First notice that, since the new parametric curves are orthogonal, we have $F = 0$. Next, let us parametrize the u curves by arc length, which we denote (as usual) by s. Then

$$f_s = f_u \frac{du}{ds} = \frac{f_u}{\sqrt{E}} \, ; f_{ss} = f_u \frac{du}{ds} = \frac{f_s u}{\sqrt{E}} = \frac{f_{uu}}{E} - \frac{E_u f_u}{E^2} \, .$$

Therefore, from he definition of the Christoffel symbols we have that $\Gamma^2_{11} = -\frac{E_v}{2G}$,

---

[2]One can formalize somewhat more this construction (see, e.g. [105]), yet we prefer the more natural approach, for the same reason as before in the text.

Figure 6.12 Geodesic parallels in practice. They are used in the planning of curved (and higher genus) surfaces when the construction itself involves only rigid materials as concrete (above, left – Mumbai airport), glass and metal (above, right and below, left – Seoul) and wood (below, right – the Klein bottle model).

therefore, from the Gauss Formulas and Exercise 6.2.13 it follows that along the u parametric curves we have

$$k_g = -\frac{E_v}{2E\sqrt{G}} = -\frac{(\sqrt{E})_v}{\sqrt{EG}} = 0\,.$$

Therefore, it follows that E is a function depending only on u, thus we can define a new parameter $\tilde{u} = \int \sqrt{E}du$. Then, in the coordinates $(\tilde{u}, v)$ the first fundamental of S (in its length element incarnation) is

$$ds^2 = d\tilde{u}^2 + G(\tilde{u}, v)dv^2,\,. \tag{6.2.10}$$

(We have used here the new parameter $\tilde{u}$, to avoid confusion at this stage, but it usual to denote it also by u, not to encumber the notation.)

Note that we have omitted here the full computations, due to the same motivation mentioned earlier. However, they are suggested as an exercise to the reader.

Exercise 6.2.32 *Complete the computations above.*

Developing Formula (6.2.10) was necessary to prove the following result regarding the frontal geodesics, which we neglected so far:

Proposition 6.2.33 *The geodesics of a field of geodesics are cut by any pair of geodesics parallels in arcs of equal length, and, conversely, if two v curves cut equal length on their respective u curves, the later are geodesics.*

Proof From Formula (6.2.10) it immediately follows that

$$s = \int_{u_0}^{u_1} du = u_1 - u_0 \, .$$

(Note that we reverted to the simpler notation.) Since the u curves are geodesics, the proof first part of the lemma is concluded.

To prove the opposite implication, we only have to note that if any two v curves, say $\{u = u_0\}$ and $\{u = u_1\}$ determine equal lengths on their orthogonal trajectories, then this length

$$\ell = \int_{u_0}^{u_1} \sqrt{du^2 + dv^2}$$

does not depend upon v, hence that $D dv^2 = 0$. Therefore, thus given that $G \neq 0$, the trajectories are u curves, hence geodesics.

□

It turns out that the forward geodesics we just introduced are also needed to deal with the geodesics as shortest curves, at least locally. More precisely, we have the following

Proposition 6.2.34 *Geodesics are length minimizing of all the curves connecting two curves in a geodesic field.*

Proof Let $p_1 = p_1(u_1, v_1), p_2 = p_2(u_2, v_2)$ the two end points (in geodesic coordinates), and let $c = c(u, v)$ be a curve of endpoints p_0, p_1. Then

$$\ell(c) = \int_{u_0}^{u_1} \sqrt{1 + G\left(\frac{du}{dv}\right)^2} \, du \geq \int_{u_0}^{u_1} du = u_1 - u_0 \, .$$

Thus, by Proposition 6.2.33 the minimal length is attained by the geodesic of endpoints p_0 and p_1.

□

Since geodesics permit us, by their very definition, to construct a "usual" (i.e. as close to the Euclidean Geometry, as possible), it is no wonder than can supply us with a number of beautiful and deep results, yet with a marked geometric flavor. However, to obtain them by fully exploit geodesics, we need to introduce yet another type of geodesic coordinates namely the so called *geodesic polar coordinates*. These are intuitively quite simple to define: Given a point on a surface S, we consider all the *radial geodesics* through p, as well as their orthogonal trajectories, which are called (befittingly – see Figure 6.13) – *geodesic circles*. In a manner similar to the one we

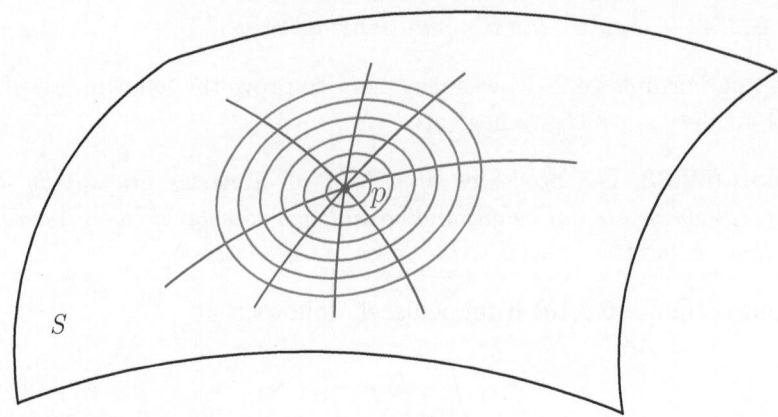

Figure 6.13 Geodesic polar coordinates on surface: Radial geodesics in light blue, geodesic circles in violet.

employed for the field of geodesics and the geodesic parallels we obtain that

$$ds^2 = dt^2 + G(r, \phi)d\phi^2 \,, \tag{6.2.11}$$

where, similar to the planar case r represents the arc length along geodesics through p, and π denotes the angle between a fixed direction through p and a geodesic. There are at least two strategies of proving this:

A. Proceed along the same lines as in the proof of Formula (6.2.10) (that is use the differential equations of the geodesics and Christoffel symbols). Since the computations are quite similar to the ones we already performed, we leave this proof as an exercise for the reader.

Exercise 6.2.35 *Prove Formula (6.2.10) using the differential equations of the geodesics.*

B. Prove that $f_r \cdot f_\phi = 0$.

Exercise 6.2.36 *Justify approach B. above and provide the details of the proof.*

Remark 6.2.37 *The geometric interpretation of the fact that, by Formula (6.2.10), the coefficient F of the first fundamental form equals 0, is of course, that the families of geodesic circles and radial geodesics are orthogonal to each other. This fact is deemed historically sufficiently important to be called the Gauss Lemma. Indeed, it amounts to the fact that "polar coordinates are geodesic coordinates based on a geodesic circle".[3]*

The formal way to do proceed is to define the so called exponential map – See Figure 6.14. The idea behind this notion is a fundamental one in Calculus, as well as in any extension or application of it, namely to study such object as graphs of

---

[3][329]

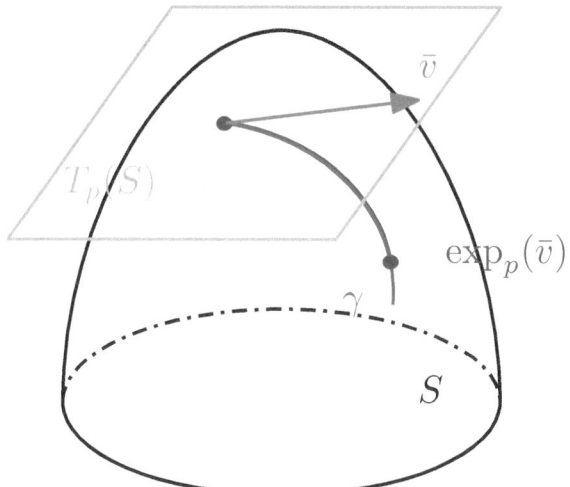

Figure 6.14 The exponential map.

function by linearizing them, locally. In the case of the exponential map, one trans-
forms a small ball on S, centered at point p, to a ball on small ball $T_p(S)$, where,
as usual, p is identified with the origin of $T_{(p)}(S) \cong \mathbb{R}^2$. It is important to note that
the object in S is a ball in the intrinsic metric of S, thus far from being an Eu-
clidean ball – See Figure 6.15. This is intuitively quite clear, if you think of trying
to paste a planar disk on a surface with bumps and creases. and, the more "bumpy"
a surface is, the less Euclidean will be the ball on the surface. But "bumpyness" is
measured by curvature, therefore we gain an intuitive understanding between the ex-
ponential map and curvature. As the reader has probably noted, why the geometric
process would be to map a ball on the surface onto the tangent plane at its center,
is easier in practice to proceed in the opposite direction, as when hinting to the con-
nection between curvature and the exponential map. Thus we define the exponential
map $\exp_p : B(0; \varepsilon) \subset T_{(p)}(S) \to \beta(p; \varepsilon) \subset S$. Geometrically, the definition is quite
simple: Given a vector $\mathbf{v} \in T_p(S)$, one measures, along the geodesic γ through p, in
the direction $\mathbf{v}$ (which exists by one the previous results), a segment of length $\|\mathbf{v}\|$
(see Figure 6.14). Technically, one defines

$$\begin{cases} \exp_p(\mathbf{v}) = \gamma(1, \mathbf{v}) & \mathbf{v} \neq \mathbf{0}; \\ \exp_p(\mathbf{v}) = p & \mathbf{v} = \mathbf{0}. \end{cases} \tag{6.2.12}$$

where $\gamma(1, \mathbf{v}) = \gamma(\|\mathbf{v}\|, \mathbf{v}/\|\mathbf{v}\|)$, for $\mathbf{v} \neq \mathbf{0}$.

Exercise 6.2.38 *Compute $\exp_p(\mathbf{v})$ if*

1. p is the North Pole on the unit sphere and $\|\mathbf{v}\| = 17\pi$;

2. p is a point of a cone of apex u, and $\|\mathbf{v}\| = d(p, u)$.

Exercise 6.2.39 *Sketch geodesic balls of increasing radii on (a) The (unit) sphere;*
(b) A circular cylinder; (c) The hyperbolic paraboloid; and compare them.

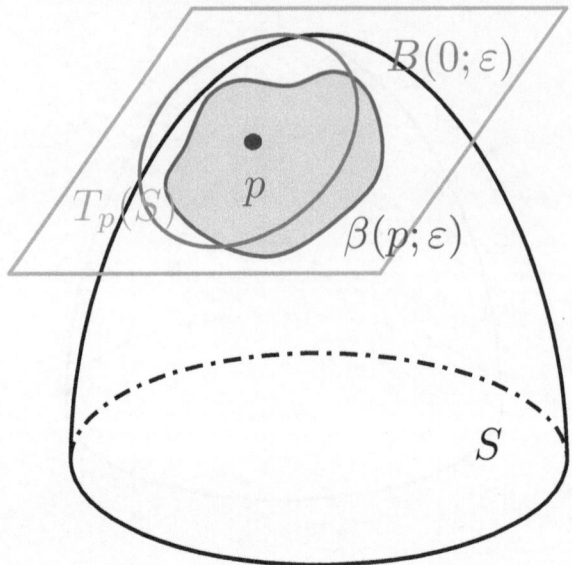

Figure 6.15 The image under exponential map of a small ball $B(0; \varepsilon) \subset T_{(p)}(S) \cong \mathbb{R}^2$, where $p \equiv p$, is small ball $\beta(p; \varepsilon)$ centered at $p \in S$. Note that $\beta(p; \varepsilon)$ is, in general, far from being an Euclidean ball.

Remark 6.2.40 *To make the definition above technically correct one has to prove that it is always possible for small enough* $\|\mathbf{v}\|$.

Exercise 6.2.41 *Formulate and prove the required technical lemma.*

It is most important, however, to note that definition above, based on laying down on a geodesic through p a segment of a given length is problematic, since this is not always possible. Indeed, recall that geodesics are given by a pair of second order ODEs, thus their existence is ensured, but only only locally. This has a crucial importance in the understanding of the geometry of surfaces, but we shall return to this problem later. For now, let us note that, by its very definition, $\exp_p$ is defined (more generally) on *start-like*[4] domains in $T_p(S)$.

However, for small enough ε the exponential map is not only defined, but also has good properties, that ensure it can be used as wished. More precisely, we have the following:

Proposition 6.2.42 *For any* $p \in S$, *there exists* $\varepsilon > 0$ *such that*

1. *$\exp_p$ is defined and differentiable in $B(0; \varepsilon) \subset T_{(p)}(S)$ and, moreover,*

2. *There exists a neighborhood $V \subset B(0; \varepsilon)$ such that $\exp_p : V \to \exp_p(V)$ is a diffeomorphism.*

---

[4]A set $X \subset \mathbb{R}^n$ is called *star-like* if there exists $x_0 \in X$, such that, $[x, x_0] \subseteq X$, for any $x \in X$.

While, for what we believe are clear reasons, we do not bring the proof of this fact, we nevertheless invite the more advanced readers to supply one of their own.

Exercise 6.2.43 *Prove the proposition above.*

The result above allows one to prove the important fact below:

Proposition 6.2.44 *The intrinsic distance is a (proper) metric on S.*

where the *intrinsic distance* (which in view of the proposition about, we shall call it henceforth the *intrinsic metric*) is defined as follows:

Definition 6.2.45 *Let $p, q \in S$. Then* intrinsic distance *(on S) between p and q is defined as*

$$d(p, q) = \inf_{c} \{\ell(c) \mid c \text{ is a curve on S with from } p \text{ to } q\}. \qquad (6.2.13)$$

...

Moreover, one can also demonstrate that, for ε as above, $B(p; \varepsilon$ is *(geodesically) convex*, that is that the following proposition holds:

Proposition 6.2.46 *For any $, q \in B(p; \varepsilon$, there exists one and only geodesic segment of ends p and q, more precisely the geodesic of velocity $\exp_q^{-1}$.*

Furthermore, we can also bring the following characterization of geodesics:

Proposition 6.2.47 *A curve $c = c(s) \subset S$, locally minimizes distances if and only if there exists a smooth mapping $\tau : [s_0, s_1] \to [0, 1]$, having $d\tau/ds \geq 0$ such that $c(s) = \gamma(\tau(s))$, where γ is a geodesic.*

In other words, curves that minimize distances between two points are necessarily geodesics.

We do not prove either of the results above (wishing to preserve, as much as possible the character of this book) but we shall (at least implicitly) make appeal to their intuitive geometric content. (Proofs can be found, for instance in [105].)

It is convenient to have a name for the neighborhood $W = \exp_p(V)$, which is called a *normal neighborhood* (of $p \in S$). We can now return to the initial motivation that we had in introducing the exponential map: Given that it is a diffeomorphism of V, it can be used to define coordinates on W and, indeed, the geodesic polar coordinates, are the image of ordinary polar coordinates in the plane $T_p(S)$, while the image of rectangular coordinates in the same plane are the so called *normal coordinates*, which we define below.

The introduction of yet another type of coordinates is due, at least partly to a disadvantage of the geodesic polar coordinates, namely that, precisely like in the case of the standard polar coordinates in the plane, they are not regular at p (the origin). The desired new type of coordinates, that have the advantage of being regular (i.e. the Jacobian of the transformation $(x, y) \mapsto (u, v)$, where (u, v) are the standard coordinates, is non-zero) in a full neigborhood of the origin are precisely the *normal coordinates*:

$$\begin{cases} x = r \cos \phi \\ y = r \sin \phi \end{cases} \qquad (6.2.14)$$

The coefficients of the first fundamental form in this system of coordinates (at p) are

$$E = G = 1, F = 0. \tag{6.2.15}$$

Furthermore, by choosing u, v to be the arc lengths of orthogonal parametric curves through p, we also ensure that (at p) we have $u = v = 0$.

An immediate application of the normal coordinates can be found in the following

Exercise 6.2.48 *Prove that, in Formula (6.2.10) also holds that*
$\lim_{\rho \to 0} G(\rho, \phi) = 0$; $\lim_{\rho \to 0} (\sqrt{G})_\rho = 1$. *(Hint: Pass to normal coordinates.)*

We shall return to the normal coordinates and their use, but for now we concentrate on the important results that can be obtained by using geodesic polar coordinates.

Our treatment of the coordinates systems on a surface was rather intuitive. It can be made more formal (and rigorous) by using the exponential map.

The first of these results is

Theorem 6.2.49 (Minding's Theorem) *Let S_1, S_2 be two regular surfaces such that $K_{S_1} \equiv K_{S_2} \equiv$ const.. Then S_1 and S_2 are locally isometric.*

In other words, any two surfaces having the same constant Gauss curvature can be bended one into the other, locally. It is easy to see that a global isometry is not always possible. (The reader is invited to provide a simple counterexample him/herself.)

Before attacking the proof of the theorem, we need some preparatory formulas. We begin from the following rather standard formula, whose demonstration is left as an exercise:

$$K = -\frac{2}{2\sqrt{EG}} \left[\left(\frac{E_v}{\sqrt{EG}} \right)_v + \left(\frac{G_u}{\sqrt{EG}} \right)_u \right]. \tag{6.2.16}$$

Exercise 6.2.50 *Prove Formula 6.2.16.*

In the case of polar geodesic coordinates, $E = 1, F = 0$, thus we obtain, passing to geodesic polar coordinates, that

$$K = -\frac{(\sqrt{G})_{rr}}{\sqrt{G}} \tag{6.2.17}$$

Exercise 6.2.51 *Verify Formula 6.2.51.*

For $K \equiv$ const. the formula above can be written in the form of a simple linear differential of order 2 with constant coefficients, namely

$$(\sqrt{G})_{rr} + K\sqrt{G} = 0, \tag{6.2.18}$$

which can be studied with standard, elementary methods (see, e.g. [325]).

We can now turn to the proof of the theorem:

Proof We begin by computing G, i.e. by solving Equation 6.2.18 for each of possible signs of K.

$K = 0$ In this case, it immediately follows from (6.2.18) that $(\sqrt{G})_{rr} = 0$, thence $(\sqrt{G})_r = g(\phi)$. Moreover, given that $\lim_{\rho \to 0}(\sqrt{G})_\rho = 1$, we have that $(\sqrt{G})_\rho \equiv 1$, thence that $\sqrt{G} = \rho + f(\phi)$. Given that, in addition, we also have that $f(\phi) = \lim_{\rho \to 0}(\sqrt{G}) = 0$, it follows that $f(\phi) = 0$ and, in consequence $G(r, \varphi) = r^2$ (and, of course, $E = 1, F = 0$).

$K > 0$ In this case, the general solution of the equation (6.2.18) is

$$\sqrt{G} = A(\phi) \cos \sqrt{K}\rho + B(\phi) \sin \sqrt{K}\rho,$$

hence, given that, $\lim_{\rho \to 0}(\sqrt{G}) = 0$, we have that $A(\phi) = 0$, hence that $(\sqrt{G})_\rho = B(\phi)\sqrt{K} \cos \sqrt{K}\rho$. Given that $\lim_{\rho \to 0}(\sqrt{G}) = 0$, it follows that $B(\phi) = \frac{1}{\sqrt{K}}$, hence that (besides having $E = 1, F = 0$) $G = \frac{1}{K} \sin^2 \sqrt{K}\rho$

$K < 0$ In this case, the general solution of the equation (6.2.18) is

$$\sqrt{G} = A(\phi) \cosh \sqrt{K}\rho + B(\varphi) \sinh \sqrt{K}\rho,$$

therefore, in a manner similar to the one used in the previous case, we obtain that $G = -\frac{1}{K} \sinh^2 \sqrt{K}\rho$ (in addition to the fact that $E = 1, F = 0$).

We can now turn to the proof of Minding's Theorem per se. To this end, we make use of the exponential map as follows: Let V_1, V_2 be normal neigborhoods of $p_1 \in S_1$ and $p_2 \in S_2$, and let $\varphi : T_{p_1}(S_1) \to T_{p_2}(S_2)$, such that $\varphi(\mathbf{e}_1) = \mathbf{f}_1, \varphi(\mathbf{e}_2) = \mathbf{f}_2$, where $\{\mathbf{e}_1, \mathbf{e}_2\}$ and $\{\mathbf{f}_1, \mathbf{f}_2\}$ are orthonormal bases of $T_{p_2}(S_2)$ and $T_{p_2}(S_2)$, respectively. Then $\psi : V_1 \to V_2$, $\psi = \exp_{p_2} \circ \varphi \circ \exp_{p_2}^{-1}$ is the required local isometry. We leave the verification of this assertion to the reader.

□

Exercise 6.2.52 *Provide the details required in the conclusion of the proof above.*

Corollary 6.2.53 *Let S be such that $K(S) = K_0$. Then, if*

1. *$K_0 = 1/a^2$, then S is locally isomorphic to a sphere of radius a;*

2. *$K_0 = 0$, then S is locally isomorphic to a plane;*

3. *$K_0 = -1/a^2$, then S is locally isomorphic to a pseudosphere.*

Remark 6.2.54 *This corollary, in conjunction with Exercise Problem 2.3.15 of Chapter 2 provides with a range of counterexamples that demonstrate that, in Minding's Theorem, "local isometry" cannot be replaced, in general, by "isometry".*

The following corollary shows that surfaces of constant curvature are homogeneous, in the following sense:

Corollary 6.2.55 *Let S be a surface of constant Gaussian curvature. Then, for any $p, q \in S$ and any $\mathbf{v} \in T_p(S)$, $\mathbf{w} \in T_q(S)$, there exists a local isometry $\varphi : S \to S$, such that $\varphi(p) = q$ and $d\varphi(\mathbf{v}) = \mathbf{w}$.*

Exercise 6.2.56 *Prove Corollary 6.2.55.*

The computations involved in the proof of Minding's Theorem are quite important, as the following particular – yet of far reaching significance in Geometry and Topology – case demonstrates:

Example 6.2.57 *If K is constant and $K < 0$, in particular in the representative case $K \equiv -1$, that is that of the Hyperbolic Plane, we obtain*

$$\sqrt{G} = A(\varphi)e^{\rho} + B(\varphi)e^{-\rho},$$

hence, for $A \equiv 1, B \equiv 1$, the first fundamental form will be

$$ds^2 = d\rho^2 + e^{2\rho}d\theta^2.$$

If we denote $x\phi, y = e^{-\rho}$, we obtain that

$$ds^2 = \frac{dx^2 + dy^2}{y^2}. \tag{6.2.19}$$

Since y cannot equal 0, and given that a plane must be simply connected, hence, in particular, connected, we must restrict ourselves either to the upper or of the lower half-plane in $\mathbb{R}^2$. We chose, mainly for convenience and by tradition, the upper-half plane $\mathbb{H}_+ \equiv \mathbb{R}_+^2 = \{(x,y) \in \mathbb{R}^2 \,|\, y > 0\}$ which we identify with the Hyperbolic Plane $\mathbb{H}^2$.[5]

Having computed the its first fundamental form, i.e. the local metric of the Hyperbolic Plane, we can now determine its geodesics. Indeed, in the new coordinates system we have $E = G = 1, F = 0$, thence $\Gamma_{11}^2 = \frac{1}{y}, \Gamma_{12}^1 = \Gamma_{22}^2 = -\frac{1}{y}$, the other of the Christoffel symbols being equal to 0. Therefore, by applying the chain rule for the derivatives, it follows that

$$y\frac{d^2y}{dx^2} + \left(\frac{dy}{dx}\right)^2 + 1 = 0.$$

which represents the differential equation of the geodesics of the Hyperbolic Plane. We can do, in fact, even better, and actually determine the geodesics: By integrating the equation above, we obtain

$$y^2\left[1 + \left(\frac{dy}{dx}\right)^2\right] = a^2,$$

and, by integrating this last equation, we get

$$(x - x_0)^2 + y^2 = a^2,$$

i.e. the equation of the circle. Given that we restrict ourselves to the upper half-plane, it follows that the geodesics are half-circles, whose centers are on $\mathbb{R} = \partial\mathbb{H}^2$, that is circles orthogonal to the real line. However, we didn't take into account a special case, namely the case $y \equiv 0$. In this instance, it follows that $ds^2 = dx^2$, that is Euclidean (half-)lines perpendicular to the real line are also geodesics (see Figure 6.16).

---

[5]This is one of the two Poincaré conformal models of the Hyperbolic Plane.

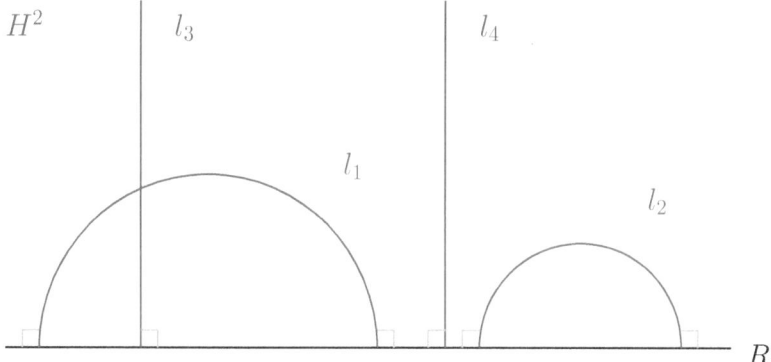

Figure 6.16 Geodesics in the Hyperbolic Plane are half-circles and rays perpendicular to $\partial \mathbb{H}^2 = \mathbb{R}$.

Exercise 6.2.58 *The second Poincaré model is the disk model of the Hyperbolic Plane: $\Delta = \{(x,y) \in \mathbb{R}^2 \,|\, x^2 + y^2 < 0\}$, endowed with the metric given by the length element*

$$ds^2 = \frac{4(dx^2 + dy^2)}{(1 - x^2 - y^2)^2}. \tag{6.2.20}$$

Determine the geodesics of the disk model.

Remark 6.2.59 *It is quite easy to construct an isometry between the two Poincaré models; this is one the two usual manner of finding the geodesics of the disk model (the other being to define them axiomatically) – see, e.g. [322], [141]. Here we request the reader to approach the problem from the differential geometric viewpoint, by emulating the computations showed in the case of the half-plane model.*

6.2.1 The Hopf-Rinow Theorem

We cannot, in good conscience, conclude the theoretical part of this chapter without connects an essential metric property of a surface with the notion of geodesics and gives a condition for their existence between any pair of points on the surface. (As we have seen, the existence of geodesics connecting two given points is not guaranteed.) We do not bring the proof of this results (or, rather, pair of results) since the proof is quite technical and rather analytic in nature. We also presume that the reader is familiar with a number of notions of Analysis and Topology.

Theorem 6.2.60 (Hopf-Rinow) *Let S be a connected surface. Then the following properties are equivalent:*

1. *Every Cauchy sequence of points on S is convergent. (In other words, S is metrically complete.)*

2. *Every geodesic can be extended indefinitely extended in both directions, or it is a closed curve. (That is to say S is geodesically complete.)*

3. *Every bounded set in S is relatively compact.*

Remark 6.2.61 *Condition (2) above can formally formulated as follows: For every $p \in S$, the exponential map $\exp_p$ is defined all of $T_p(S)$.*

The next theorem is also due to Hopf and Rinow (and in fact, it is usually the one referred to by their names or it is concatenated with the first one):

Theorem 6.2.62 (Hopf-Rinow) *Let S be a complete, connected surface, and let p, q be two points on S. Then there exists a minimal geodesic connection p and q.*

Remark 6.2.63 *The converse theorem is not true. The unit open disk in $\mathbb{R}^2$, endowed with the standard Euclidean metric is the simplest and classical counterexample.*

As a consequence of Theorem 6.2.60 we have the following

Proposition 6.2.64 *Any closed (hence any compact) surface is complete.*

Remark 6.2.65 *Here gain, the converse statement does not hold: There exist complete surfaces that are not closed.*

Exercise 6.2.66 *Construct a counterexample to Proposition 6.2.64.*

Remark 6.2.67 *Interestingly – and perhaps also surprisingly, given the rather general, abstract setting of the Hopf-Rinow Theorem – discrete versions of it exists. We find that the one of Keller and Munch [185] is especially fitted to applications in Complex Networks. While we will not present any of the technical details, not even the formal definitions, since this would take far away from the main course of the book, let us mention that the considered geodesics are relative to the path metric. Their version of the Hopf-Rinow Theorem essentially states[6] that the on a metric space there exists a (positive) weight that generates the path metric and that, with this metric, the space is complete iff it is geodesically complete.*

6.3 DISCRETIZATION OF GEODESICS

As already noted in the introduction to this chapter, for the computer scientist (and for many more practitioners in a multitude in other fields educated in the spirit of Computer Science), finding geodesics is hardly a problem: Looking for shortest paths on graphs is a standard problem, attached usually[7] using the extremely well-known *Dijkstra algorithm*.[8] However, these are shortest geodesics and moreover they are supposed to exist globally. This is far from the situation in the classical case, where straightest and frontal geodesics are not less important. Further, and perhaps more

---

[6]We drastically simplify the statement in view of our remark about.

[7]although other such algorithms do exist

[8]So well known, in fact, that most practitioners confuse the metric it computes, namely the *path metric* with the algorithm used to calculate it, and call it – quite wrongly! – "the Dijkstra metric".

importantly, such algorithms as Dijkstra are hardly relevant in Graphics, CAGD, CAD and related fields. One technical reason is that such surfaces are not graphs, evidently, and geodesics restricted (by use of graph algorithms) are far from being close to the desired ones, that should converge to the true geodesics on the smooth surface that the triangular mesh, on which such algorithms are commonly applied, is supposed to approximate. (True, one might be compromise and accept geodesics on the 1-skeleton of such meshes, with the condition of having a good estimate of the error inevitable in the departure of such a geodesic from the true one. We shall briefly return to this idea toward the end of this section.) Beyond the technical or mathematical reason, there is a practical one, namely the fact that in Graphics, etc., one is more interested to realistically model straight (or "shooting") geodesics, rather than shortest ones.

The difficulty raises not only from the fact that, as opposed to a computer algorithm, a "ball" on a surface does not "see" all the possible directions and outcomes simultaneously ab ovo, but rather needs to "guess" whether the direction of the initial "kick" will propagate it to the desired target, but also from a very basic mathematical fact, namely that geodesics are given by differential equations and that such equations have, in general, only local solutions and that even when the solutions can be extended, they might cease to be unique. In other words, a geodesic might be a short path locally, but then it can (and in generally will) bifurcate (or worse).

The *cut locus* of a surface (or, more generally, of a Riemannian manifold – see the sequel), where

Definition 6.3.1 *The cut locus of a point p on a surface S, C_p, is defined to be the closure of the set of points that can reached from p along more than one geodesic. The cut locus of a surface is the reunion $\bigcup_{p \in S} C_p$ of the cut loci of the points in S.*

How unruly the deceivingly simple graph C_p can truly be is revealed by considering antipodal points on the unit sphere. On a generic surface, it has vertices of degree 3 and higher – see Figure 6.17 for a typical example, as well as for a more detailed discussion and for an extensive bibliography. In fact, even such a drawing does not make justice to the difficulty of the problem of determining the cut locus of a surface. To wit, it is not even not known – to the best of our knowledge – for the ellipsoid, and for which a (seemingly open) conjecture goes back to 1878 [55]. (See [37] for more details. See also [118] for a deeper study.)

Exercise 6.3.2 *Let p be a point on the outside part of the equator of torus of revolution (that is in the region where $K > 0$). Sketch the cut locus of p.*

The cut locus is strongly connected to the notion of *injectivity radius*:

Definition 6.3.3 *Let M^n be a Riemannian manifold. The* injectivity radius *of M^n is defined as:* $\mathrm{InjRad}(M^n) = \inf \{x \in M^n \,|\, \mathrm{Inj(x)}\}$, *where* $\mathrm{Inj}(x) = \sup \{r \,|\, \exp_x|_{\mathbb{B}^n(x,r)}$ *is a diffeomorphism}.*

In turn, it is connected to the *convexity radius*:

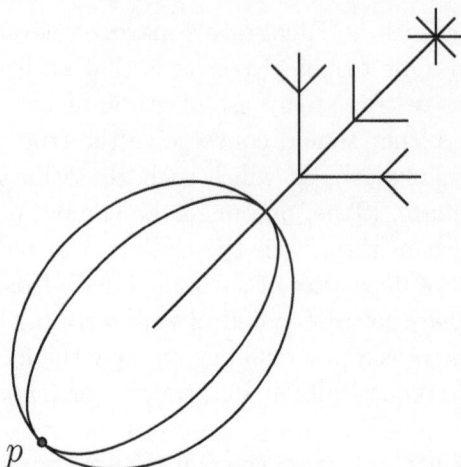

Figure 6.17 The typical cut locus on a generic surface.

Definition 6.3.4 *Let M^n be a Riemannian manifold. The convexity radius of M^n is defined as $\inf\{r > 0 \mid \beta^n(x,r)$ is convex, for all $x \in M^n\}$.*

More precisely, $\mathrm{ConvRad}(M^n) = \frac{1}{2}\mathrm{InjRad}(M^n)$ – see, e.g. [37].

Remark 6.3.5 *Note that by a classical result of Cheeger [72], an estimate for the injectivity radius is easy to obtain in terms of essential and intuitive geometric invariants. More precisely, there is a universal positive lower bound for $\mathrm{InjRad}(S)$ in terms of K_0, D and V, where K_0 is a lower bound , D is an upper bound on the diameter, and K_0 and V are lower bounds for the are on Gaussian curvature and area (in higher dimensions, volume) of S , respectively.*

It is this result (and similar ones – see also the Chapter on Ricci curvature) that make possible (and quite simple) the constructions of "nice" triangulations, that allow for good sampling an reconstruction of surfaces (and higher dimensional manifolds, even generalized ones, as well as images and more general signals), as well as their "good"[9] representation on models spaces (e.g. $\mathbb{S}^1$ (more generally, $\mathbb{S}^n$)) – see [289], [274].

The problem of determining geodesics on PL surfaces is by no means a simple one, and in fact the interest in it arouse in Pure Mathematics earlier than in its application. The pioneer in this field is Stone [312]. Unfortunately, his analysis is mathematically deeper than the space we can allocate it here might it do any justice. Therefore, regretfully, we do not elaborate on it here. Suffice to say that its influenced the first practical approach that we detail here and it is also similar, somewhat, to the second one on which we shall dwell in some detail.

The first practical truly useful approach to the problem of determining geodesic on PL surfaces is due to Polthier and Schmies [1],[10] itself an extension of previous results of the Alexandrov school (see [7], [253]), for the computation of κ_g for curves on 2-dimensional polyhedral surfaces. More precisely, let γ be a PL curve γ on a

---

[9]technically: *quasiregular*

[10]Note that the preprint behind this publication is earlier, and dates to 1998.

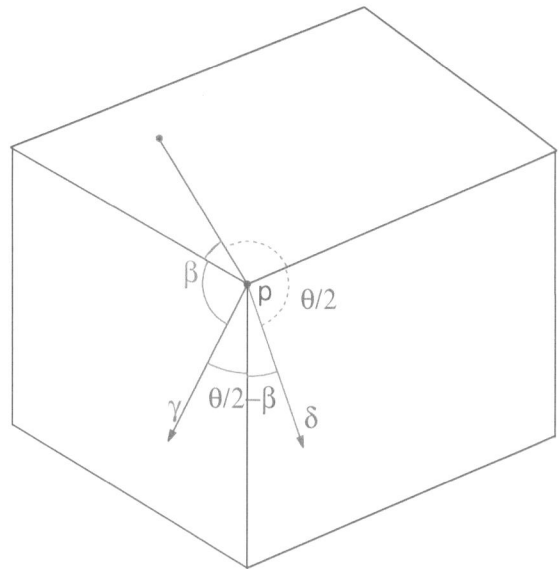

Figure 6.18 Discrete geodesic curvature at a vertex of a polyhedral surface.

polyhedral surface, and let $p \in \gamma$ be a vertex of the polyhedral surface. Then the of γ at p is given by following formula:

$$\kappa_g(\gamma, v) = \frac{2\pi}{\theta}\left(\frac{\theta}{2} - \beta\right). \tag{6.3.1}$$

where β represents a choice (i.e. the smallest) of the angle of γ at p – see Figure 6.18, and where where δ represents the *discrete straightest geodesic* (at p), defined as in Figure 6.19. In this discrete approach to geodesic curvature, $\kappa_g(e) \equiv 0$, for any edge e. To include the geodesic curvature of the edges one can apply the following formula:

$$\kappa_g(e) = \angle(\bar{N}_{v_1}, \bar{N}_{v_2}), \tag{6.3.2}$$

where $\bar{N}_{v_1}, \bar{N}_{v_2}$ represent the *computed* normals at the vertices that define e (see Figure 6.20). In [287] we have adopted Poltier's and Schmies' method by choosing the geodesic that are included in the 1-skeleton of the PL surface. Clearly, by making this simplifying assumption we allow the obtained to depart from being a true PL geodesic, but in a controlled manner, thus we are in fact computing a *quasi-geodesic*. The results of this approach are illustrated in Figure 6.21.

The practical approaches to computing geodesics (and quasi-geodesics) are far too numerous to be listed here, even more so since in many case the approach is variational or motivated by Differential Equation ideas. Perhaps the most influential among these are the (by now classical) works of Sethian [300], [189]. An author who specializes in the problem and has proposed a plethora of algorithms is He [341], [346], [69] (to enumerate just a few of his contributions). In any case, detailing even only his ideas, never mind the full gamut of suggested methods would not only take to long, but, more gravely, it would drift us away from our central goal herein, that is of trying to teach Differential Geometry via its essential interplay with discretizations motivated by concrete applications.

Thus we have chosen instead to present here, in some detail, a method motivated by ideas from *PL* Topology [15], [110], [17]. The basic idea is to use so called *normal curves*:

Definition 6.3.6 *A curve c on a triangulated orientable surface S is called* normal curve *iff all intersections of c with 2-simplices of the triangulation* $\mathcal{T}$ *are made of normal pieces, i.e. made of paths which run from one edge to a different edge of the same triangle. We assume c is smooth in the interior of the 2-cells it passes through (See Figure 6.22).*

These normal pieces should satisfy a set of *matching equations*

Consider a normal curve c and let τ, μ be two 2-cells, adjacent along an edge e. We may code each normal piece of c in τ according to the single vertex of τ it separates from the other two vertices. Since each normal piece in τ which paths through e into μ will separate either of the vertices of e, we get an equation of the form,

$$x_\tau + y_\tau = x_\sigma + y_\sigma,$$

where x_τ, x_σ are the number of normal pieces separating the vertex x in τ, μ respectively (and analogously for y) – See Figure 6.23. It turns out that every normal curve on a triangulated surface induces a set of matching equations and vice versa: Every *integral solution* all values of which are *non negative*, of a system of matching equations can be realized as a normal curve, not necessarily connected. (For a proof see [15], [110].)

The natural idea of this method is to shorten the pieces in a manner that attends the minimum length, i.e. when c becomes a *PL* geodesic. An essential role is played, as expected, by the curvature, which in this case is defined as follows:

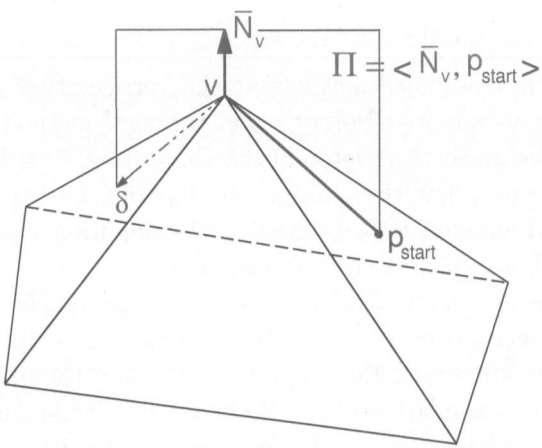

Figure 6.19 Straightest geodesic through a vertex of a polyhedral surface

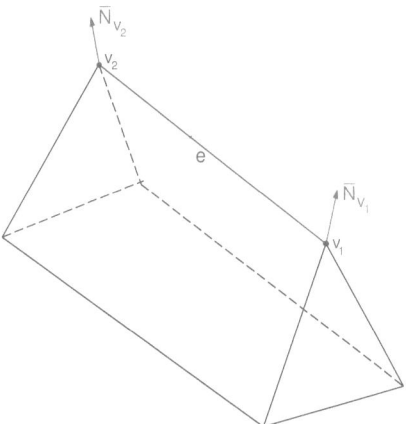

Figure 6.20 Discrete geodesic curvature along an edge of a polyhedral surface.

Definition 6.3.7 *Let c be a normal curve with respect to a triangulation $\mathcal{T}$ on a surface c. We will define the curvature of c at a point x as follows.*

1. *If x is an internal point on a segment of $c \cap \mathcal{T}^2$, we will take the curvature to be,*

$$\mathfrak{K}(x) = k_g(c, x),$$

where $k_g(c, x)$ denotes the geodesic curvature of c at x.

2. *If x is a vertex formed by the intersection of c with $\mathcal{T}^1$, and let t_1, t_2, be the two vectors tangent to c at x. The curvature at x is defined to be*

$$\mathfrak{K}(x) = \cos(\theta_1) + \cos(\theta_2), \qquad (6.3.3)$$

where θ_1, θ_2 are, respectively, the angles between t_1, t_2 and t_e, the vector tangent to the edge e at x.

The shortening algorithm is based on a specially devised curvature flow (akin to the ones we already met and to the more powerful ones that we shall introduce in the sequel):

Definition 6.3.8 *Given a normal curve c, the* semi-smooth curvature flow *is defined as*

$$\frac{\partial c}{\partial t} = \mathfrak{K}(c). \qquad (6.3.4)$$

The suggested "minimizing through straightening" algorithm can now be presented:

Let c be a rectifiable curve of finite length on triangulated surface $(S, \mathcal{T})$, where $\mathcal{T}$) denote the triangulation. Here In this section we alter the minimization process we restrict to points on the 1-skeleton of $\mathcal{T}$, i.e. on $c \cap \mathcal{T}^1$. Then

1. Normalize C with respect to $\mathcal{T}$.

2. Take a least-weight normal curve $\widehat{C}$, isotopic to C.

3. Straighten $\widehat{C}$ at all intersections with the edges of $\mathcal{T}$, by applying the semi-smooth flow

$$\frac{\partial x}{\partial t} = \cos(\theta_1) + \cos(\theta_2) , \qquad (6.3.5)$$

4. Take a subdivision of $\mathcal{T}$ to obtain a new triangulation $\mathcal{T}_1$.

5. Go to (1), while C is replaced by $\widehat{C}$.

Remark 6.3.9 *The triangulation $\mathcal{T}$ is intrinsic to the very idea of the algorithm, given that it is devised for PL surfaces. Its existence is automatically (and trivially) ensured for triangular meshes, thus it is readily applicable in Graphics applications. Smooth surfaces need first to be triangulated, and for high accuracy a "good" triangulation, i.e. one with high aspect ratio is needed (as commonly for many such*

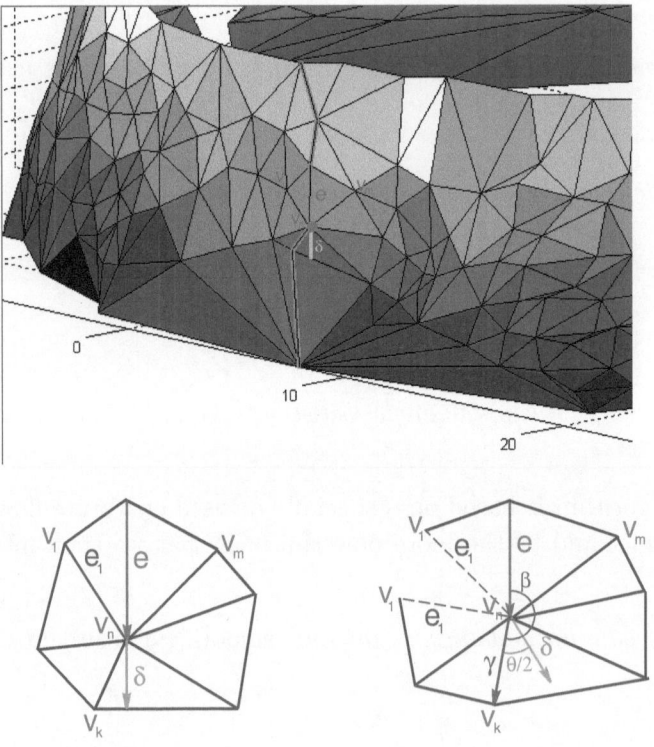

Figure 6.21 Computed quasi-geodesic on the colon surface (top). Detail: finding the straightest geodesic using Polthier and Schmies method and determining the closest quasi-geodesic (bottom) [287].

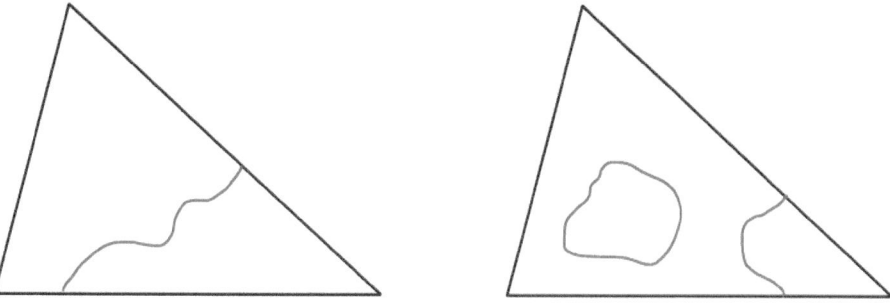

Figure 6.22 Normal (left) and not normal (right) pieces.

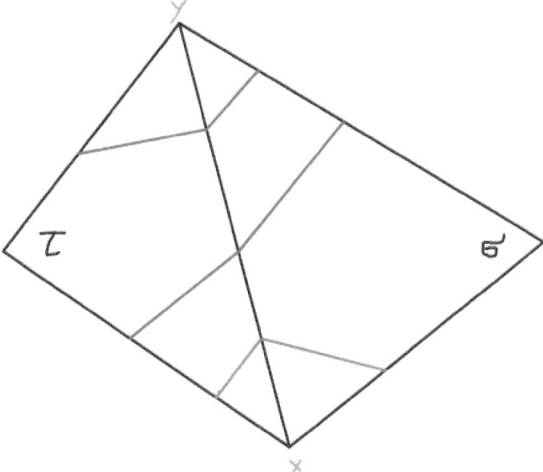

Figure 6.23 Illustration of matching equations. (In light blue a typical normal piece separating x in σ (and τ).

practical methods). One such triangulation method, that admits generalization to far more general settings[11], is the one in [289].

The efficiency of the normal curves algorithm in the computation of geodesics can be gauged in the Figures 6.24, 6.25, 6.26 and 6.27, where some results on synthetic and real surfaces[12] are presented, respectively.

After this lengthy discussion on some algorithms for the practical computation of geodesics on surfaces, with a view to Graphics and related fields, before concluding this chapter, we return briefly to the case of graphs/networks and suggest a discrete differential approach – widely divergent from the common one of Dijkstra – to the

---

[11]see [274], [275]

[12]the same colon slice as used in the illustration of the Poltier-Schmies inspired method, to be more precise

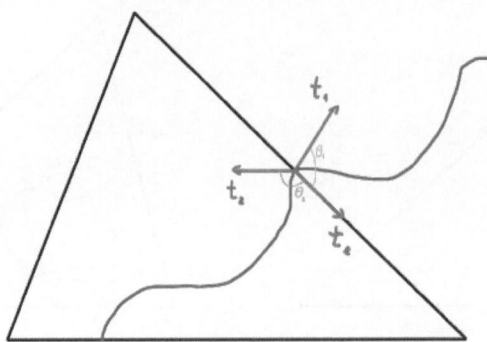

Figure 6.24 The three tangent vectors defining curvature along an edge.

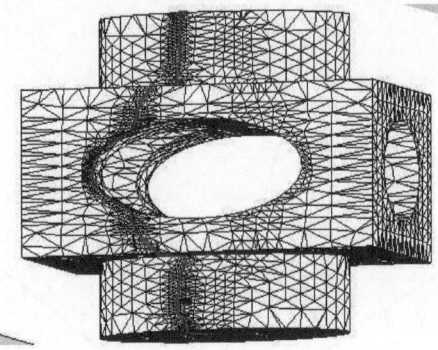

Figure 6.25 Normal approximation of a geodesic curve on a surface of high genus [17]. Results shown is obtained after 12 iterations.

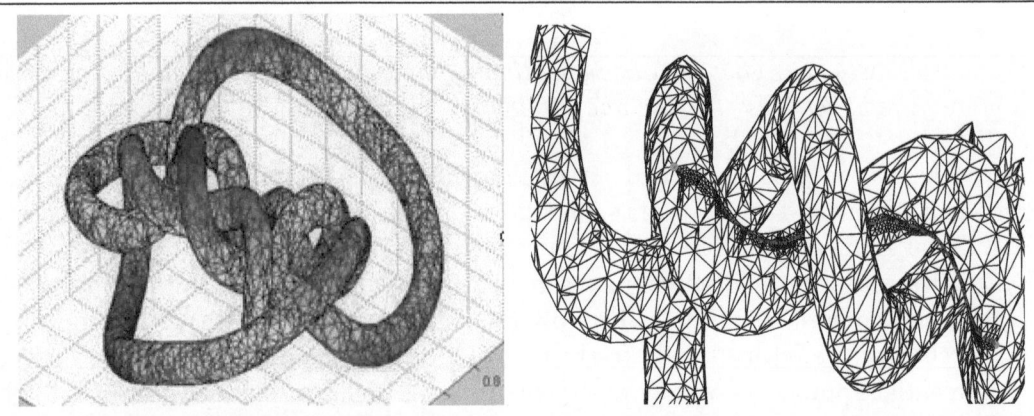

Figure 6.26 Normal geodesic curve on a knotted torus [17]. Left: The chosen surface; Right: The computed geodesic obtained after 12 iterations.

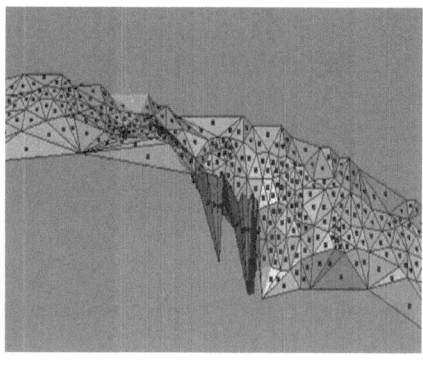 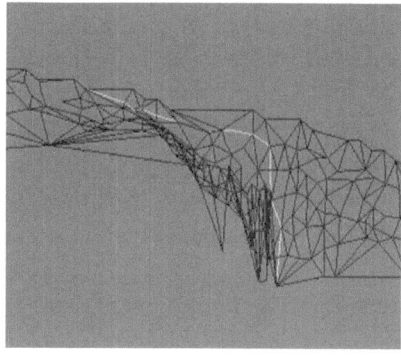

(a)   (b)

Figure 6.27 Shortest Normal curve on a Colon mesh [17]. (a) The mesh with indicating start and end points. (b) Computed shortest normal curve. Even in this fairly low resolution one can see a reasonable approximation of the shortest geodesic between the start and end points by the computed normal geodesic.

computation of geodesics on such structures. The reader might have already intuited that we propose the use of the Haantjes curvature[13] as discrete, graph version of geodesic curvature. Indeed, this is a most natural proposition, given Theorem 1.2.42 of Chapter [Curves], that shows that for curves in the Euclidean plane classical and Haantjes curvatures coincide. Since, by Lemma 6.2.10, a curve is a geodesic (that is, a straight line) iff its geodesic curvature is identically 0, it is only natural to expect the same for Haantjes geodesics in more general metric spaces. However, this is not the case in general metric spaces, and it holds only in a (rather large) class of spaces, namely in the so called *Ptolemaic spaces*, where

Definition 6.3.10 *A metric space* (X, d) *is called Ptolemaic iff*

$$d(x,y)d(z,t) \leq d(x,z)d(y,t) + d(x,t)d(y,z) \,,$$

for any quadruple $\{x, y, z, t\} \subset X$.

Essential examples of Ptolemaic spaces are the Euclidean plane $\mathbb{R}^2$ (and space $\mathbb{R}^n, n \geq 3$) and the Hyperbolic plane $\mathbb{H}^2$ (and space $\mathbb{H}^n$), and, more generally, any *inner product space* is Ptolemaic, while the spheres $\mathbb{S}^2$ ($\mathbb{S}^n, n \geq 3$) fail to be so, even locally. Clearly, combinatorial graphs are not Ptolemaic if they contain cycles of length > 3.

As already mentioned, in certain Ptolemaic spaces, the desired property holds. More precisely, we have the following

Theorem 6.3.11 (Haantjes) *Let* γ *be a curve in a Ptolemaic space of Wald (Alexandrov) curvature* $\leq K_0$, *such that* $\kappa_{H,\gamma} \equiv 0$. *Then* γ *is a geodesic.*

---

[13]Menger curvature might be a also an option, except that it is too restrictive for this application, given that it is non-trivial only in the presence of triangles and we consider only rectifiable curves, thus the two definitions coincides, wherever both are applicable.

Given our goals, it is only natural to ask which PL surfaces that are Ptolemaic. The answer is given by a Theorem of Buckle et al. ([62] Theorem A), showing that, inter alia, any finite, Ptolemaic PL simplicial complex of dimension ≥ 2 has Wald (Alexandrov) curvature ≤ 0. (The result holds, in fact for simplicial complexes composed of hyperbolic simplices, but not of spherical ones.) This is quite intuitive, at lest in the 2-dimensional case, since it is easy to see that, for instance, no steep quadrilateral pyramid (i.e. having small apex angles) can be Ptolemaic (in the intrinsic metric).

Remark 6.3.12 *The theorem above again appertains to the field of Comparison Geometry. In particular in this context, a way of seeing the connection between the Wald of the ambient space and the Haantjes curvature of a path, is to note (see [6]) that, if a curve γ (in an Alexandrov space X) satisfies $\kappa_H \leq K_0$, then γ appears, to any observer in X, to be less convex than a curve in the model space of X, having constant curvature K_0.*

In practice, we are again forced to compute, in fact, not true, but only "almost" geodesics, whic h are naturally defined as follows:

Definition 6.3.13 *A curve γ in a Ptolemaic metric space is called a κ-quasigeodesic iff $\kappa_{H,\gamma}(p) \leq \kappa$, for all points $p \in \gamma$.*

Importantly, κ-quasigeodesics represent the metric equivalent of smooth curves – see [6], Remark 6.3.

As far as applications are concerned, one might use Haantjes quasigeodesics not only on triangular meshes, but also, to give just an example, in communication networks (modeled as edge weighted graphs), for the detection of holes (whose appearance forces, e.g. the routing path from shortness (i.e. from being a geodesic), as illustrated in the toy example below in Figure 6.28.

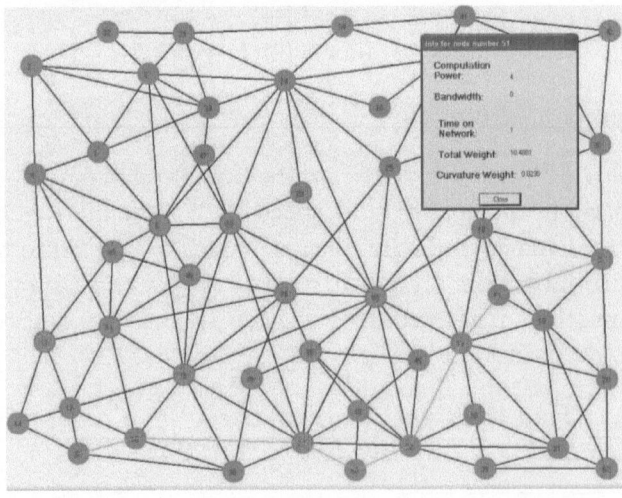

Figure 6.28 A hole avoiding Haantjes-quasigeodesic in a small model of a communication network.

CHAPTER 7

Geodesics and Curvature

As made intuitive by the exponential map, there is a strong connection between geodesics and curvature. Since, as Berger notes in [37], godesics are intrinsically difficult to find, because they are defined by means of an infimum, and not just before of difficulties involved in solving a system of differential equations. (This stumbling block of actually finding geodesics, even using numerical methods, has been made manifest in the last section of the previous chapter 2.) On the other hand, as Berger continues and notes, curvature is far easier to compute – and if not compute, then at least estimate (e.g. using upper and/or lower bounds, as we already saw). Thus one can exploit the innate connection between the two notions to establish a number of beautiful and far reaching facts.

We begin with the following rather immediate observation: If K has constant sign, then Formula 6.2.51 of the previous chapter, K expression in geodesic polar coordinates,

$$K = -\frac{(\sqrt{G})_{rr}}{\sqrt{G}},$$

has a quite intuitive geometric consequence. Indeed, let $\ell(\rho)$ denote the length of the arc on the curve $\{\rho = \text{const.}\}$ comprised, at distance ρ from a point of intersection between the geodesics γ_0, γ_1, be two geodesics determined the angles ϕ_0 and ϕ_1, respectively. (In the Euclidean case, i.e. that of curvature equal to 0, it is simple to visualize this as the length of the arc of circle between two lines which intersect, say, at the origin.) Then, of course,

$$\ell(\rho) = \int_{\phi_0}^{\phi_1} \sqrt{G(\rho, \phi)} d\phi.$$

Recall also that, by Exercise 6.2.48 of the preceding chapter,

$$\lim_{\rho \to 0}(\sqrt{G})_\rho = 1,$$

and, moreover,

$$\lim_{\rho \to 0}(\sqrt{G})_{\rho\rho} = -K\sqrt{G}.$$

DOI: 10.1201/9781003350576-7

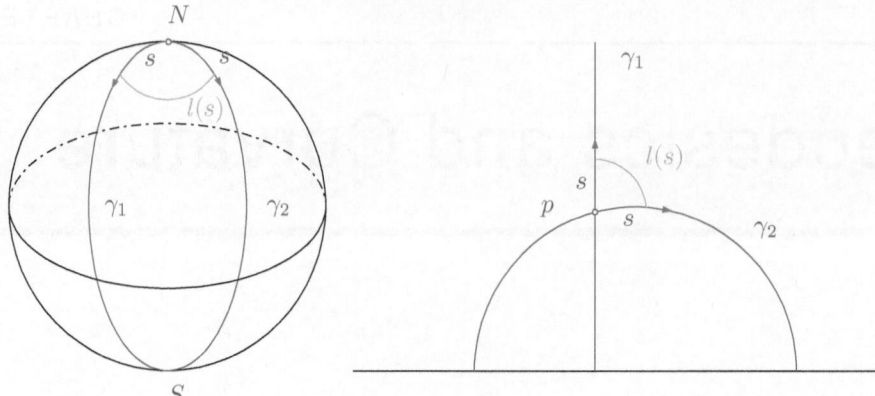

Figure 7.1 Left: Geodesics on the sphere starting from the North Pole first separate, maximum distance between them being attended on the equator, then they converge again to meet again at the South Pole. Right: Geodesics in the Hyperbolic Plane keep diverging.

Therefore, if $K < 0$, $\ell(\rho)$ increases with ρ, that is γ_0 and γ_1 grow farther and farther away from each other. On the contrary, if $K > 0$, then typically the geodesics get closer to each other, however this is not always the case, depending on K. The archetypal surface illustrating the convergent geodesics is that of the sphere, on which geodesics starting at the North Pole (i.e. meridians) keep separating till they reach the equator, only to focus again toward the South Pole, whereas the model of behavior in the case $K < 0$ is the Hyperbolic Plane (See Figure 7.1).

Exercise 7.0.1 *Provide full details for both cases.*

Remark 7.0.2 *Clearly the discussion above holds only when it is possible to apply the formula we started with, that is only if it is possible to impose geodesic polar coordinates, that is in a normal neigborhood of the intersection point of the geodesics in question.*

The first observation can be made more precise.

Theorem 7.0.3 (Bertrand-Puiseux, 1848)

$$\text{length } C(p, \varepsilon) = 2\pi\varepsilon - \frac{\pi}{3}K(p)\varepsilon^3 + o(\varepsilon^3)\,, \qquad (7.0.1)$$

This theorem clearly admits a straightforward generalization to segments of circle of measure (central angle α):

$$\text{length} C(p, \varepsilon, \alpha) = \alpha\varepsilon - \frac{\alpha}{3\sin^2\alpha}K(p)\varepsilon^3 + o(\varepsilon^3) \qquad (7.0.2)$$

Furthermore, one quite naturally[1] can prove a similar result (See Figure 7.2).

---

[1] and the fact that the two theorems were proved in the same year strengthens this assertion

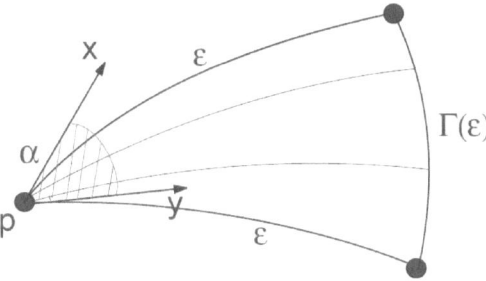

Figure 7.2 The setting of the Bertrand-Puiseux and Diguet theorems.

Theorem 7.0.4 (Diquet, 1848)

$$\text{area } B(p, \varepsilon) = \pi\varepsilon - \frac{\pi}{12}K(p)\varepsilon^4 + o(\varepsilon^4)\,. \tag{7.0.3}$$

The role of the sign of K becomes now quite manifest: In the vicinity of a point of positive curvature (hence on a surface with $K > 0$) circles are both shorter and "less massive" (or more "meager") than their Euclidean counterparts, whereas in the neighborhood of a point of negative curvature (thus, a fortiori, on a surface with $K < 0$), they are longer and "heavier" than euclidean circles (and disks) having the same radius. This fact is beautifully illustrated (in the full sense of the word) in the classical drawing of Escher below in Figure 7.3.

However, a stick points in two opposite directions, and we can express, by virtue of the Theorems of Bertrand-Puiseux and Diquet, Gaussian curvature at a point both in terms of length and aria of a geodesic circle:

$$K(p) = \lim_{\varepsilon \to 0} \frac{3}{\pi} \frac{2\pi\varepsilon - length\, C(p, \varepsilon)}{\varepsilon^3} = \lim_{\varepsilon \to 0} \frac{12}{\pi} \frac{\pi\varepsilon^2 - \text{area } B(p, \varepsilon)}{\varepsilon^4} \tag{7.0.4}$$

Thus the formulas of Bertrand-Puiseux and Diquet can be used to *define* curvature in terms of lengths and areas of "small" circles/disks. This is extremely pertinent in discrete contexts, when there are no true ε-circles exist. The typical such setting is that of graphs/networks, where $\varepsilon = 1$, i.e. one use at the 1-link of a vertex. This is far better understood (that not always realized) as the small, classical "puzzle" in Figure 7.4 below illustrates. If the networks under scrutiny are (edge and/or vertex) weighted, one can define a discrete curvature at a vertex that takes into account the weights of the nodes/edges, viewed as measures, i.e. generalized areas. In fact, the extensions to higher dimensions of these these theorems allow us to define notions of curvature in such spaces. (We shall also see, in our later chapter on the discretizations of Ricci curvature, that an approach similar in spirit was used by Ollivier [238] to define the Ricci curvature of *Markov chains*.)

Figure 7.3 "Limit Circle III" – A classical (1959) woodcut of M. S. Escher, based on a tessellation of the Poincare disk model of the Hyperblic plane. By recalling that, in this *conformal model*, all the fishes are equal in size, one can actually see the Bertrand-Puiseux and Diquet "in action" by simply counting edges and triangles in the 1-,2-,etc. neighbourhoods of any vertex, i.e. intersection of two white arcs of circle, that is hyperbolic lines. (A similar experimental verification of these theorems can be performed on the paper model of the Hyperbolic plane that we have shown in Figure 2.15 of Chapter 2.)

We now return to the proofs of Theorems 7.0.1 and 7.0.3. We begin with the following

Lemma 7.0.5 *In geodesic polar coordinates*

$$\sqrt{G} = r - \frac{r^3}{6}K(r) + R,\qquad(7.0.5)$$

where $\lim_{r\to 0}\frac{R(r,\phi)}{r^3} = 0$ *uniformly in* ϕ.

Proof Since, by Formula (6.2.18) of the previous chapter,

$$K = -\frac{(\sqrt{G})_{rr}}{\sqrt{G}},$$

How would the circle look if π = 3?

Figure 7.4 A jocose riddle regarding the number π.

it follows that

$$\frac{\partial^3(\sqrt{G})}{\partial r^3} = -K(\sqrt{G})_r - K_r(\sqrt{G}),$$

and, since as we have seen, $\lim_{r \to 0} \sqrt{G} = 0$, it follows that

$$-K(p) = \lim_{r \to 0} \frac{\partial^3(\sqrt{G})}{\partial r^3}.$$

By developing $\sqrt{G} = \sqrt{G(r, \phi)}$ into Maclaurin series and substituting into it the obtained formulas for G and its derivatives, we obtain the required formula (R being the remainder in the Maclaurin series).

□

We can now give the proofs of the theorems themselves.

Proof (*Proof of Bertrand-Puiseux Theorem*) For $r = \varepsilon$, a simple computation gives

$$\text{length } C(p, \varepsilon) = \int_0^{2\pi} \sqrt{G(r, \phi)} d\phi = 2\pi r - \frac{\pi}{3} r^3 K(p) + o(\varepsilon^3),$$

where $\lim_{\varepsilon \to 0} \frac{o(\varepsilon^3)}{\varepsilon^3} = 0$.

□

Proof (*Proof of Diquet Theorem*) A similar computation gives

$$\text{area } B(p, \varepsilon) = \int_0^{2\pi} \int_0^\varepsilon \sqrt{G(r, \phi)} dr d\phi = \int_0^{2\pi} \int_0^\varepsilon \left(r - \frac{1}{6} K(p) r^3 \right) dr d\phi$$

$$+ \int_0^{2\pi} \int_0^\varepsilon o(r^3) dr d\phi = 2\pi \left(\frac{1}{2} - \frac{K(p)}{24} r^4 \right) + o(r^4),$$

whence the desired formula immediately follows.

□

While it is more common to prove the Theorems of Bertrand-Puiseux and Diquet using geodesic polar coordinates, as just shown, it is also possible to do so by means of normal coordinates. Indeed, we can choose, in the normal coordinates, u, v to be the arc lengths of the two orthogonal parametric curves at a point p (i.e. for $r = 0$). Then, if we denote, for now, $s = r$ (just to use a more standard notation in the computations below), we have

$$\begin{cases} x = s \frac{du}{ds}(0) \\ y = s \frac{dv}{ds}(0). \end{cases} \tag{7.0.6}$$

By developing into Maclaurin series (as also done in the previous proofs) we obtain that

$$\begin{cases} u = x - \frac{1}{2}\left[\Gamma^1_{11}(0) + 2\Gamma^1_{12}(0) + \Gamma^1_{22}(0)\right]xy \\ v = y - \frac{1}{2}\left[\Gamma^2_{11}(0) + 2\Gamma^2_{12}(0) + \Gamma^2_{22}(0)\right]xy. \end{cases} \tag{7.0.7}$$

By writing r, ϕ in terms of x, y, computing the coefficients of the first fundamental form and developing them into Maclaurin series, we obtain that

$$G = r^2 + \alpha r^4 + \cdots \tag{7.0.8}$$

hence that the first fundamental form is

$$ds^2 = dx^2 + dy^2 + \alpha(ydx - xdy) + \cdots \tag{7.0.9}$$

To determine the coefficient α we must take into account that it is an intrinsic, since both the normal coordinates and x, y are. Starting from the Frobenius formula (4.1.2), and noting that $E = 1, F = 0, G = G(r\varphi)$, we obtain that

$$K = -\frac{1}{\sqrt{G}}\frac{\partial^2\sqrt{G}}{\partial r^2}. \tag{7.0.10}$$

Since, by (7.0.8), $\sqrt{G} = r\sqrt{1 + \alpha r^2 + \cdots} = r + \frac{\alpha}{2}r^3 + \cdots$, it follows that

$$\frac{\partial\sqrt{G}}{\partial r} = 1 + \frac{3}{2}\alpha r^2 + \cdots, \quad \frac{\partial^2\sqrt{G}}{\partial r^2} = 3\alpha r + \cdots, \tag{7.0.11}$$

thence

$$K = \frac{3\alpha r + \cdots}{1 + \frac{3}{2}\alpha r^2 + \cdots}, \tag{7.0.12}$$

therefore

$$K(p) = \lim_{r \to 0} K = -3\alpha, \tag{7.0.13}$$

which shows the profound geometric significance of the coefficient α (and, en passant, its intrinsic character).

Moreover, by applying the obtained value of α in the expression of $\sqrt{G}$ from (7.0.11) we can from this point on, proceed with the proofs of the Theorems of Bertrand-Puiseux and Diquet, precisely as previously.

Remark 7.0.6 *We didn't bring the full details of the computations above. The determined reader is invited to complete them.*[2]

Exercise 7.0.7 *Supply the full details of the computations leading to Formula (7.0.13).*

From here on the proofs coincide, naturally, with the ones given above.

This discussion can be extended and allow us to obtain even stronger results by making appeal to the so called *Jacobi vector fields* that are obtained by the *variation* of a geodesics. This is a notion that we do not make more precise here, since even the basic discussion would become too technical, but we rather rely on the suggestive

---

[2]More details are also given in [168].

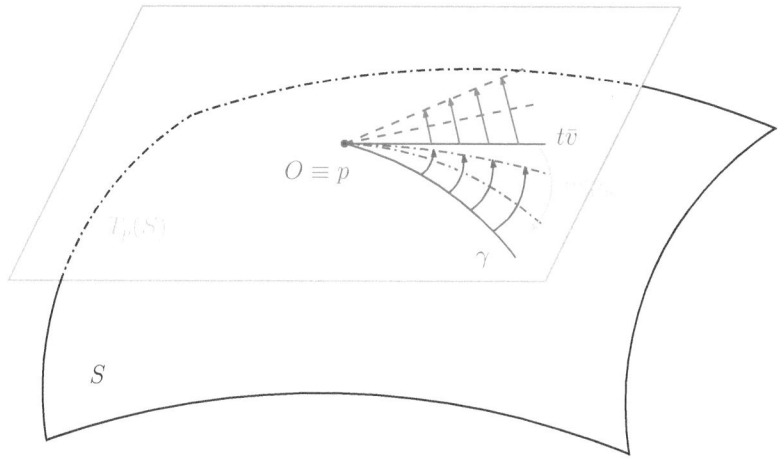

Figure 7.5 Jacobi fields are obtained through the variation of geodesics, via the exponential map.

Figure 7.5. Note that the vector field $\bar{\mathbf{y}}(t)$ is orthogonal along the geodesic γ if the initial value $\bar{\mathbf{y}}(0)$ is. Far less intuitive, but easier to operate with, certainly in the classical, smooth surfaces setting, is the characterization of Jacobi fields as solution of the *Jacobi equation*:

$$\ddot{\bar{\mathbf{y}}}(t) + K(\gamma(t))\bar{\mathbf{y}}(t) = 0 \,. \tag{7.0.14}$$

Exercise 7.0.8 *Write the Jacobi equation for the geodesics on a sphere of radius R and determine its general solution.*

While clearly not intuitively pleasing, the characterization above has the advantage of emphasizing the role of Gauss curvature in the behavior of families of geodesics. An other advantage is the fact that (7.0.14) is an ordinary differential equation[3], thus one can apply staple results on such equations to prove

Proposition 7.0.9 *Given a geodesic γ, there exists precisely one Jacobi field $\bar{\mathbf{y}}(t)$ along γ, with initial conditions $\bar{\mathbf{y}}(0) = \bar{\mathbf{y}}_0$ and $\dot{\bar{\mathbf{y}}}(0) = \dot{\bar{\mathbf{y}}}_0$. Furthermore, $\bar{\mathbf{y}}(t)$ can be realized by a one parameter family $\{\gamma_\alpha\}$ of geodesics.*

Of course, a Jacobi vector field might be *trivial*, that is $\dot{\bar{\mathbf{y}}} = \bar{0}$. These are fields that are simply parallel transported along the geodesic. By recalling that Jacobi fields are orthogonal to the geodesic, the reader can easily produce an example.

Exercise 7.0.10 *Sketch an example of a trivial Jacobi field.*

However, not all Jacobi vector fields are trivial – far from so; to edify him/herself the reader is invited to solve the following exercise:

---

[3]an *Euler equation*, more precisely

Exercise 7.0.11 *Show that the vector field* $\bar{\mathbf{y}}(s) = \sin(s)\bar{\mathbf{v}}(s)$, *where* $\bar{\mathbf{v}}(s)$ *is the parallel transport of* $\bar{\mathbf{v}}_0 = \bar{\mathbf{v}}(s)$ *along the geodesic* γ *of the unit sphere, where* γ *is a quarter of the equator, and where* s *denotes (again) the arc length. Sketch this vector field.*

On surfaces of constant curvature K Equation (7.0.14) has the special, standard form

$$\ddot{\bar{\mathbf{y}}}(t) = -K\bar{\mathbf{y}}(t),\tag{7.0.15}$$

which, for the (canonical) initial conditions $\mathbf{y}(0) = \bar{0}$ and $\|\bar{\mathbf{y}}(0)\| = 1$ has the well-known solutions

$$\bar{\mathbf{y}}(t) = \begin{cases} \frac{\sin\sqrt{K}t}{\sqrt{K}} & K > 0; \\ t\mathbf{y} & K = 0; \\ \frac{-\sinh\sqrt{-K}t}{\sqrt{-K}} & K < 0. \end{cases}\tag{7.0.16}$$

(The reader should note the similarity with some of the formulas used in the chapter on metric Gauss curvature.)

From here the expression of the length element in geodesic coordinates is easily to obtain:[4]

$$ds^2 = d\rho^2 + \begin{cases} \frac{\sin^2\sqrt{K}\rho}{K}d\sigma^2 & K > 0; \\ \rho^2 d\sigma^2 & K = 0; \\ \frac{-\sinh^2\sqrt{-K}t}{K}d\sigma^2 & K < 0. \end{cases}\tag{7.0.17}$$

where $d\sigma$ denotes the metric of the unit circle $\mathbb{S}^1 \subset T_p(S)$.

Remark 7.0.12 *Strangely enough, given that geodesics themselves have only rather late been treated with success in Graphics and its related fields, Jacobi fields have been discretized a long time ago [314], precisely in the PL setting on which the field of Graphics rests. This was done to introduce a discrete notion of Ricci curvature, and we shall elaborate more in the chapter dedicated to the discretizations of that notion. However, it should be remarked here that the notion of Jacobi field naturally extends to higher dimensions (thence their role in defining Ricci curvature).*

Remark 7.0.13 *Another classical comparison result goes back to Gauss' foundational paper [130]. Motivated by geodesy, he wished to compare the angles and areas of geodesic triangle on a surface* S *with that of its Euclidean model. More precisely, given a (small) geodesic triangle* $T \in S$, *sides* a, b, c *and angles* α, β, γ *(accordingly) consider the triangle* $T' \in \mathbb{R}^2$, *with sides equal to those of* T, *and corresponding angles* α', β', γ'. *Then the angle* α' *is given by the following formula:*

$$\alpha = \alpha' + \frac{\sigma}{12}(2K(p) + K(q) + K(r)) + o(a^4 + b^4 + c^4);\tag{7.0.18}$$

where, to preserve Gauss' notation, we set $\sigma = \text{Area}(T)$. *Clearly, one can write the similar formulas for* β' *and* γ'.

---

[4]see, e.g. [37]

Note that, for $K \equiv$ const., e.g. for $K \equiv 1/R^2$ (that is, for S being a sphere of radius R), formula (7.0.18) reduces to the Legendre formula

$$\alpha = \alpha' + \frac{\sigma}{3}K + o(a^4 + b^4 + c^4), \qquad (7.0.19)$$

which was already classical even in Gauss' time.[5]

Moreover, he obtained a similar comparison result for the areas. More precisely, the following formula holds:

$$\sigma = \sigma' \left[K(p)(s - a^2) + K(q)(s - b^2) + K(r)(s - c^2) \right] + o(a^4 + b^4 + c^4); \quad (7.0.20)$$

where $\sigma' = \text{Area}(T')$ For further results, see [106].

Clearly, Formulas (7.0.18) and (7.0.20) are both of high potential applicative value.[6]

The solution in (7.0.22) can be extended to include the case of varying curvature, which again hold in any dimension, as shown in the following pair of celebrated theorems due to Rauch [256], formulated here for the 2-dimensional case:

Theorem 7.0.14 (Rauch 1) Let S be a surface such that having Gauss curvature bounded from below, that is $K \geq K_1$, for some K_1, and let $\bar{\mathbf{y}}(t)$ be a solution of (7.0.14) satisfying the initial conditions $\mathbf{y}(0) = \bar{0}$ and $\|\bar{\mathbf{y}}(0)\| = 1$. Then

$$\|\bar{\mathbf{y}}(t)\| \leq \begin{cases} \frac{\sin t\sqrt{K_1}}{\sqrt{K_1}} & K_1 > 0; \\ t & K_1 = 0; \\ \frac{\sinh t\sqrt{-K_1}}{\sqrt{-K_1}} & K_1 < 0. \end{cases} \qquad (7.0.21)$$

which holds if $\bar{\mathbf{y}}(t) \neq \bar{0}$, for any $s \in (0, t)$.

Theorem 7.0.15 (Rauch 2) Let S be a surface such that having Gauss curvature bounded from above, that is $K \leq K_2$, for some K_1, and let $\bar{\mathbf{y}}(t)$ be a solution of (7.0.14) satisfying the initial conditions $\mathbf{y}(0) = \bar{0}$ and $\|\bar{\mathbf{y}}(0)\| = 1$. Then

$$\|\bar{\mathbf{y}}(t)\| \geq \begin{cases} \frac{\sin t\sqrt{K_2}}{\sqrt{K_2}} & K_2 > 0; \\ t & K_2 = 0; \\ \frac{\sinh t\sqrt{-K_2}}{\sqrt{-K_2}} & K_2 < 0 \end{cases} \qquad (7.0.22)$$

for any

$$0 \leq r \leq \begin{cases} \frac{\pi}{\sqrt{K_2}} & K > 0; \\ \infty & K_2 \leq 0. \end{cases}$$

The proofs of the Rauch Theorems are quite complicated and we do not bring them here.

Remark 7.0.16 The theorems above are classical examples of Comparison Geometry results, as are the Theorems of Bertrand-Puiseux and Diquet. We have encountered already this approach in the definition of Wald and Alexandrov curvatures. However, the full force of this approach will be apparent later on, when we shall employ it actually define curvature in higher dimensions and to obtain discretizations and generalizations of classical notions of curvature.

---

[5]Much more modern formulas in the same spirit can be found in [75].

[6]This is quite expected, given they were motivated, as noted above, by a practical problem.

Jacobi fields have, not surprisingly, a role in the determination of conjugate points (recall our discussion in the previous chapter). More precisely, we have the following

Proposition 7.0.17 *Let $\gamma \subset S$ be a geodesic through $p = \gamma(0)$. Then the following statements are equivalent:*

1. *$q = \gamma(t_0)$ is the conjugate of p along $\gamma|_{[0,t_0]}$.*

2. *There exists a nontrivial Jacobi field $\bar{\mathbf{y}}(t)$ along $\gamma(t), 0 \leq t \leq t_0, (t_0 \geq 0)$, such that $\bar{\mathbf{y}}(0) = \bar{\mathbf{y}}(t_0)$.*

The proof is an immediate application of the way Jacobi fields are constructed using the exponential map. Given this, we leave the proof as an exercise to the reader.

Exercise 7.0.18 *Prove Proposition 7.0.17 above.*

Exercise 7.0.19 *Applying the result obtained in Exercise 7.0.11, show that the first conjugate of any point $p \in \mathbb{S}^2$ is his antipodal point $-p$.*

A stronger result extending the proposition above is the theorem below, which we also bring without proof:

Theorem 7.0.20 *Let $p, q \in S$ and let γ a geodesic arc of ends p and q. Then γ is the shortest arc connecting p and q if and only if p admits no conjugate point $p*$ between p and q on γ.*

Exercise 7.0.21 *Verify the theorem above if $S = \mathbb{S}^2$.*

As a particular case of the previous Theorem we obtain

Theorem 7.0.22 *Let S be a surface such that $K \leq 0$. Then there are no conjugate points on any geodesic $\gamma \subset S$. Moreover, any geodesic arc between points $p, q \in S$ is the shortest between the arcs connecting p and q.*

The proof is not long, nor difficult, thus we leave it to the reader.

Exercise 7.0.23 *Prove the theorem above. (Hint: Use the Jacobi equation and apply the previous theorem.)*

We conclude the series of theorems that follow from Jacobi's equation with the following classical result, whose proof we omit:

Theorem 7.0.24 (Bonnet) *Let S be a surface such that*

$$K \geq \frac{1}{k^2} > 0;\qquad(7.0.23)$$

where $k = $ const. > 0. Then S is compact and

$$\operatorname{diam}(S) \leq \pi k.\qquad(7.0.24)$$

We delimitate the power of Bonnet's Theorem by bringing forward the following remarks:

Remark 7.0.25 *The hypothesis that $K \geq \frac{1}{k^2} > 0$ cannot be weakened, as the paraboloid $S = \{(x, y, z) \mid z = x^2 + y^2\}$ demonstrates.*

Exercise 7.0.26 *Verify the counterexample above.*

Remark 7.0.27 *The bound $\mathrm{diam}(S) \leq \pi k$ is tight, as the example of the unit sphere demonstrates.*

Exercise 7.0.28 *Validate the counterexample above.*

Remark 7.0.29 *The condition that S is complete cannot be discarded. Indeed, one has the following*

Counterexample 7.0.30 $S = \mathbb{R} \times (-\pi/2, \pi/2)$ *endowed with the metric $ds^2 = dx^2 + \cos^2 x \, dy$.*

Exercise 7.0.31 *Confirm the counterexample above.*

Exercise 7.0.32 *Show that the converse of Bonnet's Theorem does not hold. (Hint: Consider the torus of revolution; it's diameter is $\pi(r + R)$.)*

7.1   GAUSS CURVATURE AND PARALLEL TRANSPORT

Before concluding this chapter we should bring yet another connection between Gaussian curvature and geodesics, or rather with parallel transport. Consider a point p on a surface S, such that p is included in a simple region D, such that $\gamma = \partial D$ is a closed curved, contained in a coordinate patch. Suppose (for technical reasons) that the said coordinate patch is parametrized by an isothermal parametrization $f = f(u, v)$. Furthermore, suppose that γ is parametrized by arc length. Consider a vector $\vec{v}$ tangent to S at $\vec{v}(0)$ and parallel transport along γ. (See Figure 7.6). Then, on the one hand, clearly

$$\int_\gamma k(g) dl = 0 \,.$$

On the other hand, by Proposition 6.2.23 of Chapter 6,

$$\int_\gamma k(g) dl = \int_0^{l(\gamma)} \frac{1}{2\sqrt{EG}} \left(G_u \frac{dv}{ds} ds - E_v \frac{du}{ds} \right) + \int_0^{l(\gamma)} \frac{d\varphi}{ds} \,.$$

By applying well-known Green Theorem of Calculus we obtain that

$$\int_\gamma k(g) dl = - \iint_D K dA + (\varphi(l(\gamma)) - \varphi(0)) \,.$$

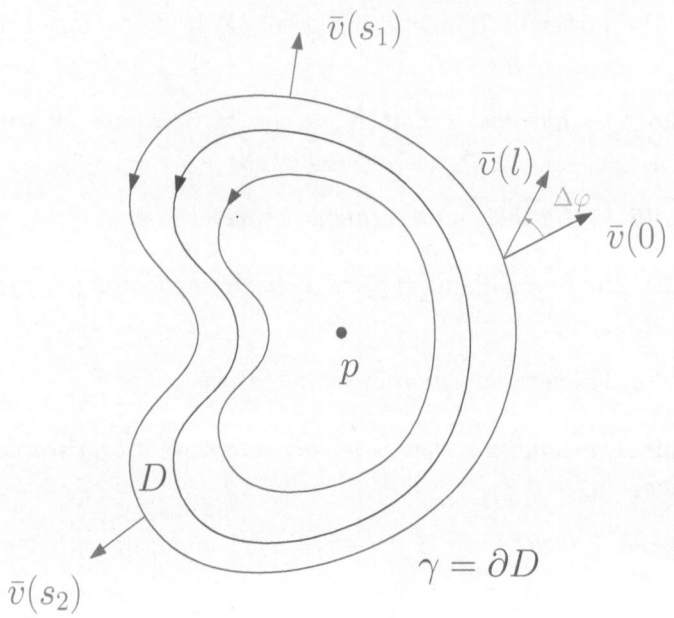

Figure 7.6 Parallel transport along a closed curve.

(Here, as before, $\varphi = \varphi(s)$ denotes a smooth determination of the angle between f_u and $\vec{v}(s)$.) To summarize, we have obtained that

$$\Delta\varphi = \varphi(l(\gamma)) - \varphi(0) = \iint_D K dA. \qquad (7.1.1)$$

Thus, the integral of Gaussian curvature over D equals precisely the discrepancy between the initial and final determination of the angle φ, as induced by the parallel transport along ∂D.[7] Given that $\Delta(\varphi)$ does not depend neither on $\vec{v}(0)$, nor on γ,[8] we can pass to the limit – using a natural approach, given original Gauss' definition of curvature – to conclude that

$$K(p) = \lim_{D \to 0} \frac{\Delta\phi}{\text{Area}(D)}; \qquad (7.1.2)$$

which represents the desired interpretation of Gaussian curvature in terms of parallel transport.

Remark 7.1.1 *This result could have been included already after Proposition 2.3 of Chapter 6, as an immediate application. However, we feared it might get lost amidst the mass of material concerning geodesics. Another natural location would have been in the Chapter dedicated to the Gauss-Bonnet Theorem. Indeed, this the context in which it appears in many textbooks. (In [105], for instance, it is "sandwiched" between the Local and the Global versions of the Gauss-Bonnet Theorem – see Chapter 9.)*

---

[7]This difference $\Delta(\varphi)$ is called the *holonomy* of the loop γ.
[8]Explain why not!

However, we felt that there it might be overlooked as well, given the number of deep, important results concentrated therein. Thus, given the fact that we dedicated a separated chapter to the connection between geodesics and curvature, we felt that this would be, from a number of reasons, the best placement for this result.

Remark 7.1.2 *This interpretation of K in terms of parallel transport might prove itself useful in defining holonomy in such discrete settings where curvature is concentrated at points, notably in the case of nodes in networks.*

CHAPTER 8

The Equations of Compatibility

We are now faced with the rather unpleasant[1] task of introducing the so called *Christoffel symbols*. However, given their essential role in formulating the equivalent for surfaces of the Serret-Frènet Equations and the Fundamental Theorem of curve theory, as well as their importance in the theory of geodesics (which will be discussed in the next chapter), it is also an undertaking which we cannot evade. However, to avoid lengthy computations and to conform to the spirit of this book, we shall keep the development of the ensuing formulas to the bare minimum.

The basic idea is to write the second order partial derivatives of a parametrization $f : U \to S$ in terms of the firsts order ones or, more geometrically put, to determine the coordinates of $f_{uu}, f_{uv}, f_{vu}, f_{vv}$ as elements of the tangent plane (at a generic point), that is to express them in terms of its basis, i.e. as linear combinations of f_u, f_v. Clearly, this is not possible as such, as acceleration depends on the way the surface curves, nontechnically put. In other words, we need to add N as a basis element to obtain the desired derivation. We thus write:

$$\begin{cases} f_{uu} = \Gamma_{11}^1 f_u + \Gamma_{11}^2 f_v + L_{11} N \\ f_{uv} = \Gamma_{12}^1 f_u + \Gamma_{12}^2 f_v + L_{12} N \\ f_{vu} = \Gamma_{21}^1 f_u + \Gamma_{21}^2 f_v + L_{21} N \\ f_{vv} = \Gamma_{22}^1 f_u + \Gamma_{22}^2 f_v + L_{22} N \end{cases} \tag{8.0.1}$$

The system (8.0.4) is called *The Gauss Formulas*, and the symbols $\Gamma_{ij}^k, i, j, k = 1, 2$ are called the *Christoffel symbols* (of S).[2] An immediate (and useful) observation is that the Γ_{21}^1 are symmetric in the lower indices, that is $\Gamma_{12}^1 = \Gamma_{21}^1, \Gamma_{12}^2 = \Gamma_{21}^2$.

In addition, we wish to express N itself in the basis $\langle f_u, f_v \rangle$, that is to write

$$\begin{cases} N_u = a_{11} f_u + a_{21} f_v \\ N_v = a_{12} f_u + a_{22} f_v \end{cases} \tag{8.0.2}$$

---

[1]made even less attractive by the knowledge that, at least in the beginning, the reader might find them quite unpalatable

[2]In case the reader wonders if we couldn't simplify the notation: This is a standard symbol, which we cannot avoid.

DOI: 10.1201/9781003350576-8

thus obtaining the so called *Weingarten formulas* (which we first encountered in Exercise 2.3.23 of Chapter 2).

The coefficients involved are quite easy to determine: First one determines the coefficients of N by taking the inner products of the Gauss Equations with N, and obtains that, pleasantly (and perhaps not unexpectedly) enough

$$L_{11} = e, \ L_{12} = L_{21} = f, \ L_{22} = g. \tag{8.0.3}$$

where e, f, g are precisely the coefficients of the second fundamental form (of S).

To find the Christoffel symbols once again takes the inner products of the Equations (8.0.4), this time with f_u and f_v, and obtain that

$$\begin{cases} \Gamma^1_{11}E + \Gamma^2_{11}F = \frac{1}{2}E_u \\ \Gamma^1_{11}F + \Gamma^2_{11}G = F_u - \frac{1}{2}E_v \\ \Gamma^1_{12}E + \Gamma^2_{12}F = \frac{1}{2}E_v \\ \Gamma^1_{12}F + \Gamma^2_{12}G = \frac{1}{2}G_u \\ \Gamma^1_{22}E + \Gamma^2_{22}F = F_v - \frac{1}{2}G_u \\ \Gamma^1_{22}F + \Gamma^2_{22}G = \frac{1}{2}G_v \end{cases} \tag{8.0.4}$$

where E, F, G are the coefficients of the first fundamental form.

As we are sure that the reader expects by now, we leave the details of the derivations above as an exercise:

Exercise 8.0.1 *Prove Formulas (8.0.2) and (8.0.4) above.*

For us, it is more important to notice that the Gauss and Weingarten equations represents the precise analog of the Serret-Frènet formulas for curves. Unfortunately, they are far more complicated to solve in order to find a parametrization f satisfying them, due to the fact that they are partial differential equations. Thus, on a practical level, they are not at par with their curve model. However, this does not mean they are less important from the theoretical viewpoint, a fact that we shall be able to make precise soon.

It is now possible to determine the Christoffel symbols from the Gauss Formulas. Indeed, note that the determinant of each of the three pairs of equations in (8.0.4) equals $EG - F^2 \neq 0$, thus the system constituting the Gauss Formulas is solvable. In consequence, the Christoffel symbols can be expresses solely in terms of the coefficients E, F, G of the first fundamental form. Since these are invariant under isometries, we have thus obtained the following

Fact 8.0.2 *All geometric notions and properties that can be expressed in terms of the Christoffel symbols are invariant under isometries.*

This is a very important observation. Not only will it be significant in the study of geodesics, but it also has an immediate – and quite important – consequence, that fully justifies the introduction of the Christoffel symbols and unpleasantness in their handling. To formulate it we need first to prove the so called *Gauss Equation*:

$$\tag{8.0.5}$$

$$\left(\Gamma^2_{12}\right)_u - \left(\Gamma^2_{11}\right)_v + \Gamma^1_{12}\Gamma^2_{11} + \Gamma^2_{12}\Gamma^2_{12} - \Gamma^2_{11}\Gamma^2_{22} - \Gamma^1_{11}\Gamma^2_{12} = -E\frac{eg - f^2}{EG - F^2} = -EK.$$

Exercise 8.0.3 *Prove the Gauss Equation 8.0.5.*

Exercise 8.0.4 *Prove the "sister" formula of the Gauss Equation, namely*

$$\left(\Gamma^1_{12}\right)_u - \left(\Gamma^1_{11}\right)_v + \Gamma^2_{12}\Gamma^1_{12} - \Gamma^2_{11}\Gamma^1_{22} = FK. \tag{8.0.6}$$

The immediate yet very important consequence of the Gauss Equation (and its companion formula) is the fact that it shows that Gauss curvature depends only on the Christoffel symbols and their derivatives[3] In other words, we just provided yet another proof of the

Theorem 8.0.5 (Theorema Egregium) *Gauss curvature is intrinsic.*

Remark 8.0.6 *Compare the route to this proof – even with the deleting of the technical stages involved – with the simple and intuitive proofs we brought in the PL (polyhedral) case.*

Sometimes, one wishes to emphasize the fact that K is a function only on E, F, G and their derivatives and further manipulates Formulas (8.0.5) and (8.0.6) to obtain the *Frobenius formula*

$$K = -\frac{1}{4(EG - F^2)^2} \begin{vmatrix} E & E_u & E_v \\ F & F_u & F_v \\ G & G_u & G_v \end{vmatrix} - \frac{1}{\sqrt{EG - F^2}} \left(\frac{\partial}{\partial v} \frac{E_v - F_u}{\sqrt{EG - F^2}} - \frac{\partial}{\partial u} \frac{F_v - G_u}{\sqrt{EG - F^2}} \right); \tag{8.0.7}$$

or the alternative (and even lengthier) *Brioschi formula*

$$K = \frac{1}{(EG - F^2)^2} \begin{vmatrix} -\frac{1}{2}G_{uu} + F_{uv} - \frac{1}{2}E_{vv} & \frac{1}{2}E_u & F_u - \frac{1}{2}E_v \\ F_v - \frac{1}{2}G_u & E & F \\ \frac{1}{2}G_v & F & G \end{vmatrix} - \begin{vmatrix} 0 & \frac{1}{2}E_v & \frac{1}{2}G_u \\ \frac{1}{2}E_v & E & F \\ \frac{1}{2}G_u & F & G \end{vmatrix}. \tag{8.0.8}$$

which we already displayed in Chapter 4.

Exercise 8.0.7 *Demonstrate the Frobenius and Brioschi formulas above.*

Remark 8.0.8 *For some strange reason, the less symmetric and more complicated formula of Brioschi has become popular amogst the engineering community who need to make appeal to classical Differential Geometry, while the highly symmetrical and elegant one of Frobenius is completely forgotten.*

Exercise 8.0.9 *Show that the converse of Theorema Egregium does not hold, namely that there exist surfaces S_1 and S_2, such that $K_1 \equiv K_2$, but S_1, S_2 are not isometric. (Hint: Consider the surfaces given by the parametrizations $\mathbf{x}(u, v) = (u\cos v, u\sin v, \log u), u > 0$ and $\tilde{\mathbf{x}}(u, v) = (u\cos v, u\sin v, v).)$*

---

[3]or, in other words, on the coefficients of the first fundamental form and its coefficients.

Before passing one, let us observe that one might like to actually formally solve the Gauss equation. However, the value of such a, endeavor would be restricted, as it is much more easy and rewarding to solve it for some concrete surfaces, like the important class of surfaces of revolution.

Exercise 8.0.10 *1. Compute the Christoffel symbols for the generic surface of revolution $f(u,v) = (\varphi(v)\cos(u), \varphi(v)\sin(u), \psi(v))$.*

2. In particular, determine the Christoffel symbols of the torus of revolution, i.e. for $\varphi(v) = R + r\cos(v), \psi(t) = r\sin(v), 0 < r < R$.

Returning (essentially) to the Gauss Formulas and solving them to determine the coefficients of N one attains the following pair of formulas known as the *Mainardi-Codazzi equations*:

$$\begin{cases} e_v - f_u = e\Gamma_{12}^1 + f\left(\Gamma_{12}^2 - \Gamma_{11}^1\right) - g\Gamma_{11}^2 \\ f_v - g_u = e\Gamma_{22}^1 + f\left(\Gamma_{22}^2 - \Gamma_{12}^1\right) - g\Gamma_{12}^2 \end{cases} \tag{8.0.9}$$

(For a nicer[4], but longer and practically unknown form of the Mainardi-Codazzi equations, see [168].)

The Gauss formula together with the Mainardi-Codazzi equations are collectively known as the *Equations of Compatibility* which give the chapter its title.

Exercise 8.0.11 *Show that if there are no umbilical points on $S = f(U)$ and if the coordinate curves are lines of curvature (i.e. if $F = f = 0$), then the Mainardi-Codazzi equation become*

$$\begin{cases} e_v = \frac{E_v}{2}\left(\frac{e}{E} + \frac{g}{G}\right) \\ g_u = \frac{G_u}{2}\left(\frac{e}{E} + \frac{g}{G}\right). \end{cases} \tag{8.0.10}$$

Since we saw in the Gauss and Weingarten formulas the analogs for surfaces of the Serret-Frènet, it is only natural to seek a surface analog also for the Fundamental Theorem. Such a result has been indeed obtained, namely

Theorem 8.0.12 (The Bonnet Theorem) *Let $E, F, G \in \mathcal{C}^2(V)$ and $e, f, g \in \mathcal{C}^1(V)$, where $V = \text{int}(V) \subset \mathbb{R}^2$, such that $E, G > 0$. Suppose that the functions E, F, G, e, f, g satisfy the Gauss and Mainardi-Codazzi equations and that $EG - F^2 > 0$. Then for every $p \in V$, there exists $p \in U \subset V$ and a diffeomorphism $\mathbf{x} : U \to \mathbf{x}(U) \subset \mathbb{R}^3$, such that the regular surface $S = \mathbf{x}(U)$ has E, F, G and e, f, g as the coefficients of its first, and second fundamental forms, respectively. Furthermore, if U is connected and if $\tilde{\mathbf{x}} : U \to \tilde{\mathbf{x}}(U) \subset \mathbb{R}^3$ is another diffeomorphism satisfying the same conditions, then there exists an orientation preserving isometry Φ of $\mathbb{R}^3$, such that $\tilde{\mathbf{x}} = \Phi \circ \mathbf{x}$.*

In other words, the equations of compatibility determine the surface, up to local isometry.

---

[4]without containing any Christofell symbol!

The proof of the Bonnet Theorem is lengthy and technical, therefore we do not bring it here. However, the interested reader can find it, for instance, in [105] or [310].

The Bonnet Theorem is often used to show that two surface are not isometric or that a surface having certain fundamental forms cannot exists, as the reader is invited to convince her/himself by solving the following exercises.

Exercise 8.0.13 *Show that the sphere $\mathbb{S}^2$ is not locally isometric to the*

1. *Plane;*

2. *Cylinder;*

3. *Saddle surface $z = x^2 - y^2$.*

Exercise 8.0.14 *Does there exists a surface $S = f(u, v)$ such that*

1. *$E = G = 1, F = 0$ and $e = 1, g = -1, f = 0$;*

2. *$E = 1, F = 0, G = \cos^2 u$ and $e = \cos^2 u, f = 0, g = 1$?*

8.1 APPLICATIONS AND DISCRETIZATIONS

Perhaps surprisingly, even such a theoretical and technical aspect of Differential Geometry has its importance in applications. To begin with, classical (smooth) theory is relevant in Imaging. Indeed, by a now standard paradigm [188], images can be viewed as surfaces in $\mathbb{R}^3$ (or, more generally as *hypersurfaces* in some $\mathbb{R}^N$). Thus understanding curvature and the way the surface, i.e. the image, is "positioned" in space is extremely important. This is an even more pertinent issue in the case of color images, given that the classical – but naive – notion of independent color channels is known to be false (see, e.g. [254]), thus rendering the computation of the geometric properties of the resulting hypersurface quite difficult.

Interestingly enough, for the discrete, applied setting, such as it arises in Graphics, Imaging, Networking, etc., a metric analog of the Compatibility Equations exists [277], that we have already mentioned in Chapter 3. Briefly recalling it, to put things better in the right perspective, given a vertex v, with metric curvature $K_W(v)$, the following system of inequalities should hold:

$$\begin{cases} \max A_0(v) \leq 2\pi; \\ \alpha_0(v; v_j, v_l) \leq \alpha_0(v; v_j, v_p) + \alpha_0(v; v_l, v_p), \quad \text{for all } v_j, v_l, v_p \sim v; \\ V_\kappa(v) \leq 2\pi; \end{cases} \quad (8.1.1)$$

Here

$$A_0 = \max_i V_0 ; \quad (8.1.2)$$

"$\sim$" denotes incidence, i.e. the existence of a connecting edge $e_i = vv_j$ and, of course, $V_\kappa(v) = \alpha_\kappa(v; v_j, v_l) + \alpha_\kappa(v; v_j, v_p) + \alpha_\kappa(v; v_l, v_p)$, where $v_j, v_l, v_p \sim v$, etc.

Interpreting the formulas (8.1.1) in the light of the Compatibility Equation, note that the first two inequalities represent the equivalent of the Mainardi-Codazzi equations, while the third one represents the equivalent of the Gauss equation.

As far as the manner in which the metric approach above comes to play a role is quite straightforward for Imaging, Graphics, CAD, etc., where the metric space is the 1-skeleton of a triangular (or, more generally, polyhedral mesh), and where angles are easily to interpret. For networks, the metric space is the nothing else network/graph itself. The importance of the Compatibility Equations in such a context arises, for instance, in *conceptual networks*, where the ambient space represents the context of a conceptual or semantic network [282].

CHAPTER 9

The Gauss-Bonnet Theorem and the Poincare Index Theorem

9.1 THE GAUSS-BONNET THEOREM

The Gauss-Bonnet Theorem, which we shall soon formulate, is, as we shall shortly see, a global result (thus one of the few such results of elementary Differential Geometry). However, the route to its proof naturally begins with a local version. However, to be able to formulate it we need to introduce (or just recall) a number of definitions.

First, we need to introduce (in more than an intuitive manner) *piecewise-regular curves*:

Definition 9.1.1 *$\gamma : [0, l] \to S$ is called a simple, closed, piecewise-regular, parameterized curve if*

1. *$\gamma(0) = \gamma(k)$;*

2. *$\gamma(t) \neq \gamma(s)$, for any $s \neq t$;*

3. *There exists a subdivision $0 = t_0 < t_1 < \cdots < t_p = l$, such that $\gamma|_{(t_i, t_{i+1})}$ is smooth and regular, for all $0 \leq i \leq p$.*

   The points $\gamma(t_i), 0 \leq i \leq p$ are called the vertices of γ and the arcs $\gamma((t_i, t_{i+1}))$ are called the regular arcs of ∂D.

As Figure 9.1 suggests (and as it should be clear from a Calculus 1 course), angles arise at vertices, and we have to take some care in they way we define them. We begin by noting that, by the regularity of γ, the limits $\lim_{t \nearrow t_i} \gamma'(t) = \gamma'_{i_-}$ and $\lim_{t \searrow t_i} \gamma'(t) = \gamma'_{i_+}$ exist and are $\neq 0$. If S is oriented (as we usually assume, and certainly in this chapter) then we can consider $0 \leq |\varphi_i| < \pi$ – the smallest angle between γ'_{i_-} and γ'_{i_+}, where the sign of φ_i is taken to be the sign of $\det(\gamma'_{i_-}, \gamma'_{i_+}, \vec{N}_i)$

DOI: 10.1201/9781003350576-9

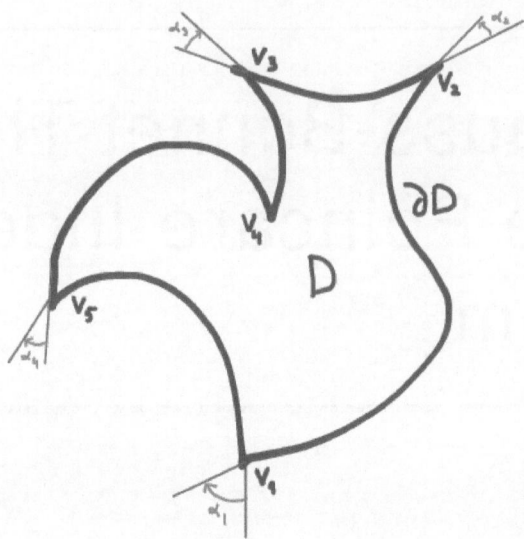

Figure 9.1 A simple region D bounded by a simple, closed, piecewise-regular curve. The vertices of ∂D and their angles are emphasized.

(where, as usual, $\vec{N}_i$ denotes the normal to S at the point $\gamma(t_i)$[1]. – See Figure 9.2. The angle φ_i is called the *exterior angle*, and $\psi_i = \pi - \varphi_i$ is called the *interior angle* at $\gamma(t_i)$. However, this determination of the angle φ_i is possible only if $\gamma(t_i)$ is not a *cusp*, i.e. when $|\varphi_i| \neq \pi$. In this exceptional case, one makes appeal to the regularity of γ that ensures that there exists $\varepsilon_0 > 0$ such that $\det(\gamma'_{i-\varepsilon}, \gamma'_{i+\varepsilon}, \vec{N}_i)$ does not change sign for any $0 < \varepsilon\varepsilon_0$, and defines the sign of φ_i as being that of $\det(\gamma'_{i-\varepsilon}, \gamma'_{i+\varepsilon}, \vec{N}_i)$. – See Figure 9.3 for the intuition behind this determination of the sign in the case of cusps. See also Figure 9.4 for a more exhaustive example.

Next, we have to make precise the notion of *region* on a surface:

Definition 9.1.2 *Given a surface S, a region $D \subset S$ is a union of connected of connected open set int$D$ and its boundary ∂D. A region is called regular if it is compact and $\partial D = \cap_1^p \gamma_i$, where $\gamma_i, 1 \leq i \leq p$ are closed, piecewise-regular curves, $\gamma_i \cap \gamma_j = \emptyset\, i \neq j$. D is called simple if it is homeomorphic with a disk and ∂D is a simple, closed, positively oriented piecewise-regular curve.*

9.1.1 The Local Gauss-Bonnet Theorem

Theorem 9.1.3 *Let U be a simply connected coordinate patch parametrized by an isothermal parametrization (that is $E = G = \lambda^2(x, y), F = 0$) compatible with the orientation of the surface S. Consider a simple region $D \subset U$, such that $\gamma = \partial D$ is positively oriented curved, parametrized by arc length, and let $v_1, \ldots, v_p$ and $\varphi_1, \ldots, \varphi_p$*

---

[1]In other words, the sign of φ_i is determined by the orientation of S, this being the reason we definitely need S to be oriented.

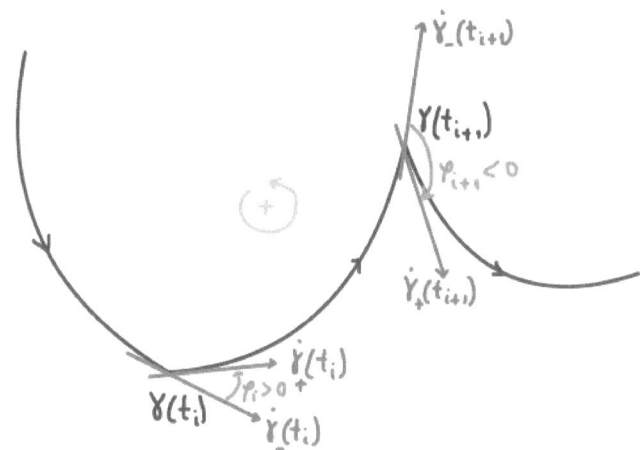

Figure 9.2 The sign of the exterior angle φ at vertex which is not a cusp.

denote the vertices and external angles of γ, respectively – see Figure 9.2. Then

$$\iint_D K dA + \sum_0^p \int_{v_i}^{v_{i+1}} k_g dl + \sum_0^p \varphi_i = 2\pi \; ; \qquad (9.1.1)$$

where k_g denotes the geodesic curvature of the regular arcs of γ.

Remark 9.1.4 *Clearly the requirement that D is parametrized by a isothermal parametrization is a technical one and we shall see later that it can be, indeed, discarded.*

The proof is not trivial, but it is not too difficult either. In fact, it is based on some technical results that we have already proved. As such, and to keep the chapter as much within the framework of this book, namely with the focus on the Geometry (and with a view to discretizations), we leave it as an exercise to the reader. (Alternatively he/she might like to avoid it and look into [105] or [168], for instance.)

Exercise 9.1.5 *Prove Theorem 9.1.3. (Hint: Use the fact that the parametrization is isothermal, as well as Proposition 6.2.23 of Chapter 6. Also, besides these technical results developed in the previous chapters, for the last term on the left side of Formula (9.1.1), one needs to make appeal to the following classical – and simple – theorem:)*

Theorem 9.1.6 (Turning Tangents Theorem[2].) *Let $\gamma \subset \mathbb{R}^2$ be a simple, piecewise-regular closed curve. Then, with the notation introduced above, we have*

$$\sum_{i=0}^p \left(\psi_i(t_{i+1}) - \psi_i(t_i) \right) + \sum_{i=0}^p \varphi_i = \pm 2\pi \; ; \qquad (9.1.2)$$

where the $\pm$ sign is determined according to the orientation of γ. (See Figures 9.3 and 9.4.)

---

[2]This represents, in fact, a generalization of the Turning Tangents, a.k.a. the *Hopf Umlaufsatz*

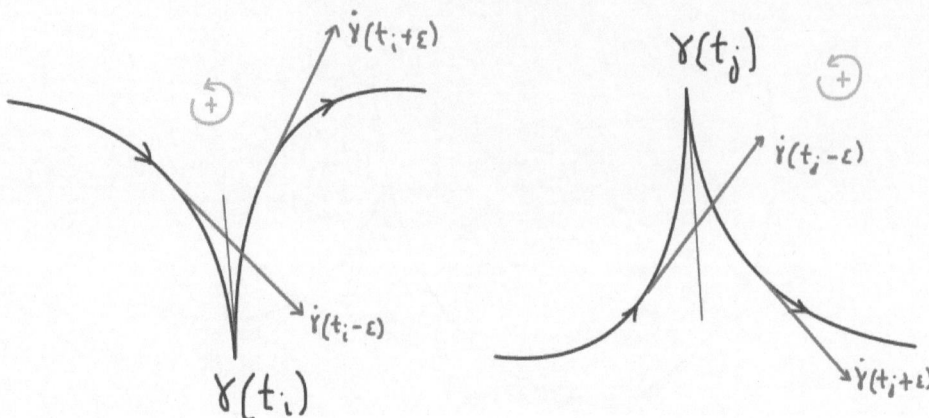

Figure 9.3 The sign of the exterior angle φ at a cusp.

The proof of this theorem is not especially hard and, moreover, it would fit, in spirit, to our book. Unfortunately, the material is growing exponentially fast and we are regretfully forced to refer the reader to [105] or [309] (for the restricted, original, theorem) and [310] (for the generalization employed here).

It is most tempting to view the Local Gauss-Bonnet Theorem as a variation of the well-known Stokes Theorem. However, this a "false friend", due to the role of the geodesic curvature, which cannot be captured by the Caculus theorem, thus it is not possible to prove it in this manner,[3] "however tempting it might be".[4]

9.1.2 The Global Gauss-Bonnet Theorem

We can now proceed to the Global Gauss-Bonnet Theorem, whose formulation is, largely, as the same suggests, a global version of the local version we just met. Nevertheless, it is also a strengthening of it, not just a duplicate of the local theorem:

Theorem 9.1.7 (Gauss-Bonnet) *Let S be an oriented surface and let $D \subset S$ be a regular region, $\partial D = \gamma_1 \cup \ldots \cup \gamma_m$, where $\gamma_1, \cdots, \gamma_m$ are simple, closed, positively oriented piecewise-regular curves.*

$$\iint_D K dA + \sum_{i=1}^{p} \int_{\gamma_i} k_g dl + \sum_{0}^{p} \varphi_i = 2\pi\chi(D);\qquad(9.1.3)$$

where $\varphi_i, i = 1, \ldots, p$ are the external angles of the curves $\gamma_1, \cdots, \gamma_m$.

The proof makes, inevitably, appeal to the local version (which might be thus viewed as a lemma for the benefit of the "big" global theorem) (See Figure 9.4). However, it also requires a far deeper, essentially topological result,[5] namely the following

---

[3]i.e. by "brute force"

[4][37]

[5]The result below is stronger, in fact, than the one needed here, but both for the sake of thoroughness as well as due to its immense importance not just in theory, but in a large variety of applications we bring it here, together with discussion and a sketch of proof.

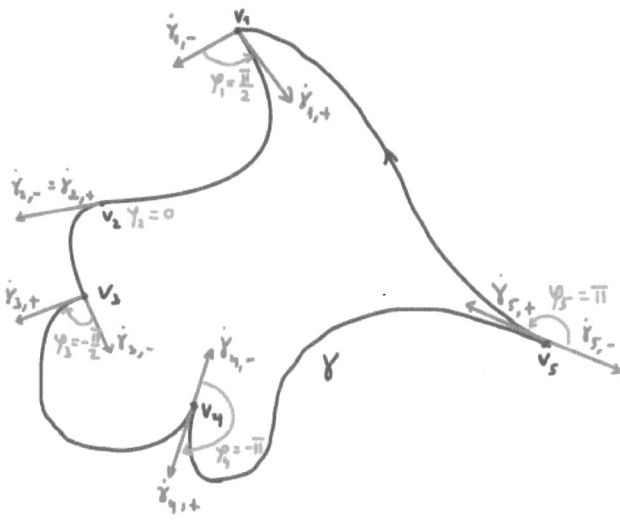

Figure 9.4 The sign of the exterior angle φ at a cusp.

Theorem 9.1.8 *Any smooth surface admits a smooth triangulation.*

This is not a trivial or immediate result, even it might seem to be so. For the basic topological aspect of the problem the interested reader is referred to [322] and the bibliography cited therein (which goes back to the foundations of Differential Topology and the works of Whitehead and Munkres). However, in practice, be it further mathematical uses, Generalized Relativity, Computer Graphics, Sampling Theory and Imaging, one would like to produce triangulations that enjoy certain "good" geometric properties (such as a high *aspect ration* or to be geodesic). Furthermore, a construction that is simple and algorithmic in nature is clearly to be preferred, certainly if aiming toward applications. As expected, the literature on the subject is truly vast, and we enumerate here – without even hoping to be exhaustive – a number of such works, especially those appertaining to the applicative vein, or with such uses: [10], [41], [89], [81], [99], [114], [205], [242], to mention just a few (since we are sure, given the importance of the problem as well as the nature of the work in the field, that we missed many good papers.)

However, the problem was solved, on a theoretical level at least, quite early[6], by Cairns [66], [67]. Moreover, while the constructions in these papers are quite technical, and they were seemingly (and sadly) largely forgotten, Cairns also suggested a very simple, geometric method for sampling smooth, closed surfaces, a method that generalizes immediately to higher dimensions. Furthermore, his construction produces precisely such "good" triangulations and, in fact, the method proposed in [10] uses

---

[6]namely in the mid 1930s

basically the same method as Cairns' original one,[7]. Moreover, the proof given in [10] is more technical[8].

While the triangulation problem is somewhat tangential to the main narrative of this book, we still feel that, given the high import of the problem in applications as well as in theory, and the fact that Cairn's method is, at core, an ingenious application of one of the first notions of Differential Geometry that we introduced in this course, it would be more than just a pity to entirely evade Cairns proof. In consequence, we bring here a sketch of it, (which can be viewed as an exercise appertaining too, or an application of, notions developed in earlier chapters).

Sketch of Proof We begin by first adapting a classical notion and introducing a couple of new definitions:

Definition 9.1.9 *Let $S \subset \mathbb{R}^3$ be a smooth, complete surface, and let be $p \in S$, and let $r > 0$. Then*

1. *A sphere $\mathbb{S}^2(p, r)$ is called* tangent *to S at p if $T_p(\mathbb{S}^2(p, r)) = T_p(S)$.*

2. *A straight line $l \subset \mathbb{R}^3$ is called l is* secant *to $X \subset S$ if $|l \cap X| \geq 2$.*

Note that the definitions above are readily extendable to higher dimensions. We can now formulate the notion of osculating spheres in a manner that allows us to define such spheres for any dimension[9]:

Definition 9.1.10 *1. $\mathbb{S}^2(p, r)$ is an* osculating sphere *at $p \in S$ iff:*

(a) *$\mathbb{S}^2(p, r)$ is tangential at x;*
and

(b) *$\mathbb{B}^2(p, r) \cap S = \emptyset$.*

2. *Let $X \subset S$. Then*

$$\omega = \omega_X = \sup\{\rho > 0 \mid \mathbb{S}^2(p, r) \text{ osculating at any } p \in X\}$$

is called the maximal osculating radius *at X.*

Remark 9.1.11 *The following facts are essential for Cairn's construction:*

1. *There exists an osculating sphere at any point of S (see [68]).*

2. *If X is compact, then $\omega_X > 0$.*

We can now proceed to main steps of triangulation method itself, which, in the compact case we consider here, rest upon the construction of a point set $A \subseteq M^n$, that is maximal with respect to the following density condition:

$$d(a_1, a_2) \geq \eta, \tag{9.1.4}$$

where

$$\eta < \omega_M. \tag{9.1.5}$$

---

[7]However, the seminal papers of Cairns are not referenced wherein.
[8]yet perhaps more algorithmic
[9]and codimension

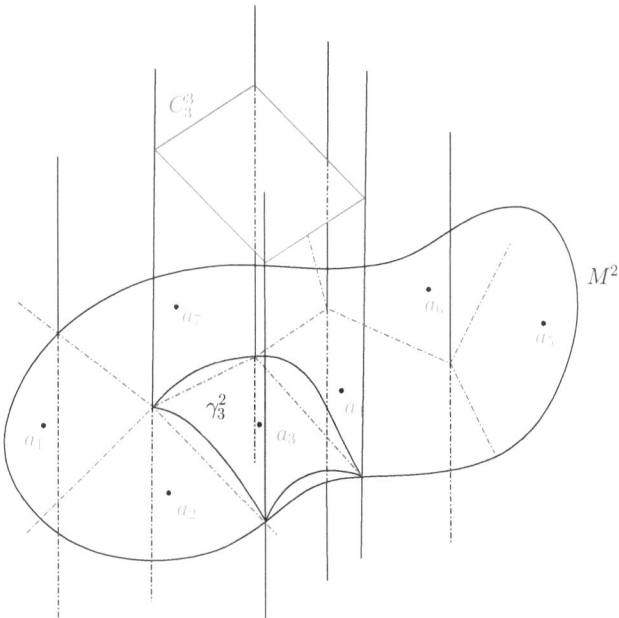

Figure 9.5 The cell complex "cut" on the surface S^2 by the Euclidean Dirichlet (Voronoi) cell complex determined by the sampling points a_i.

One makes next use of the fact that for a compact manifold M^n we have $|A| < \aleph_0$, to construct the finite cell complex "cut out of M" by the ν-dimensional Dirichlet complex, whose (closed) cells are given by:

$$\bar{c}_k = \bar{c}_k^3 = \{x \in \mathbb{R}^3 \,|\, d_{eucl}(a_k, x) \le d_{eucl}(a_i, x),\ a_i \in A,\ a_i \neq a_k\}, \qquad (9.1.6)$$

i.e. the (closed) cell complex $\{\bar{\gamma}_k^n\}$, where:

$$\{\bar{\gamma}_k^2\} = \bar{\gamma}_k = \bar{c}_k \cap S \qquad\qquad (9.1.7)$$

(See Figure 9.5.)

□

For the detailed proof, the motivated reader can consult Cairn's original paper, or he/she can, alternatively view this as a (definitely not trivial, but also quite solvable, given the determination) exercise:

Exercise 9.1.12 *Provide the full details of Cairn's construction.*

Remark 9.1.13 *As already mentioned above, Cairns method extends to higher dimensions. Furthermore, it can be refined to apply to non-compact manifolds [241], as well as manifolds with boundary [269]. Good (local) triangulations where obtained earlier in [75], for the approximation of (some of) the curvatures of higher dimensional manifolds. (We shall return to this subject in one of our following chapters.) Further triangulation results were given in [21], [54].*

Hyperbolic manifolds deserve special attention, thus specific methods were devised for them [327], [59].

Higher dimensional manifolds can be triangulated using a different approach, based on a technique of Grove and Petersen [146] (but that goes back, it would seem, at least to an idea of Colin de Verdier [100]). This method can be extended to weighted manifolds [274], with applications, for instance in Imaging – see [206] and the references therein.

We now return to the

Proof of the Gauss-Bonnet Theorem Since we wish to connect the curvature determined left side of Formula (9.1.3) to the Euler characteristic of the surface, the proof clearly should begin by applying Theorem 9.1.8 to endow S with a triangulation $\mathcal{T}$ – See Figure 9.6. Furthermore, we can refine the triangulation, by using repeated subdivisions,[10] to ensure that each triangle is contained in a isothermal coordinate patch. Moreover, by an argument already repeatedly used in Calculus, if the triangulation is positively oriented, adjacent triangles will impose opposite orientations on the common edge.

We are now ready to apply the local Gauss-Bonnet Theorem (as desired) and, if we denote the number of triangles by F, we obtain that

$$\iint_D K \, dA + \sum_i \int_{\gamma_i} k_g \, dl + \sum_{j=1}^F \sum_{m=1}^3 \varphi_{jm} = 2\pi F \,;$$

where φ_{jm}, $m = 1, 2, 3$ denote the exterior angles of the triangle T_i.

The double sum in the left s-de of the formula above can also be expressed expressed in terms of the interior angles ψ_{jm} of the triangles T_j, namely:

$$\sum_{j=1} \sum_{m=1} \varphi_{jm} = \sum_{j=1} \sum_{m=1} \pi - \sum_{j=1} \sum_{m=1} \psi_{jm} = 3\pi F - \sum_{j=1} \sum_{m=1} \psi_{jm} \,.$$

Observe that the triangulation $\mathcal{T}$ of D contains (in general[11]) both internal and external (boundary) edges and vertices, and by denote them, respectively, as E_i, E_e, V_i, V_e, we obtain. Given that the curves γ_i are closed it follows that $E_e = V_e$. Moreover, $3F = 2E_i + E_e$ (check!) thus

$$\sum_{j=1} \sum_{m=1} \varphi_{jm} = 2E_i + \pi E_e - \sum_{j=1} \sum_{m=1} \psi_{jm} \,.$$

We should next note that the vertices in V_e might be either part of the original vertices of one of the curves γ, or a new vertex induced by the triangulation $\mathcal{T}$. We denote the sets of such vertices by $V_{e,\gamma}$ and $V_{e,\mathcal{T}}$, respectively. Using this notation and observing the sum of angles around internal vertices is π, we obtain that

$$\sum_{j=1} \sum_{m=1} \varphi_{jm} = 2\pi E_i + \pi E_e - 2\pi V_i - \pi V e, \mathcal{T} - \sum_i (\pi - \varphi_i) \,.$$

---

[10] *e.g. barycentric subdivisions*
[11] What is the exceptional case?

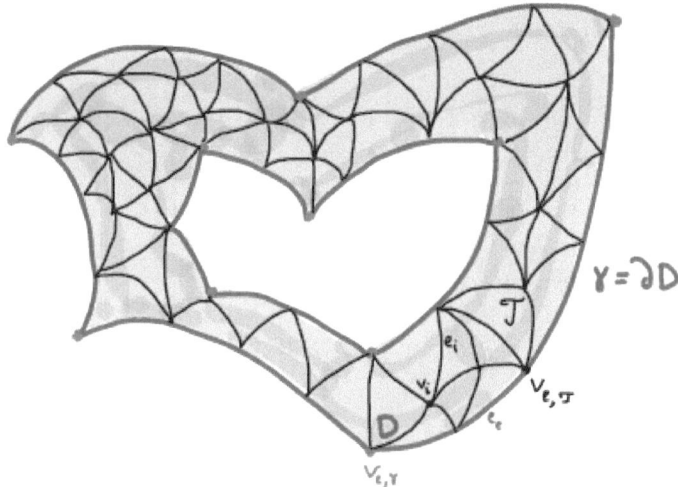

Figure 9.6 A triangulation of a domain D. Typical representatives of the various types of edges are illustrated: Internal – e_i, and external – e_e edges; internal v_i and external vertices, which might belong to γ – $v_{e,\gamma}$; or be induced by $\mathcal{T}$ – $v_{e,\mathcal{T}}$.

By some elementary manipulations (which the reader is invited to)supply them her/himself) we obtain that

$$\sum_{j=1}\sum_{m=1}\varphi_{jm} = 2\pi E - 2\pi V + \sum_i(\pi - \varphi_i)\,;$$

thus that

$$\iint_D K\,dA + \sum_{i=1}^p \int_{\gamma_i} k_g\,dl + \sum_0^p \varphi_i = 2\pi(V - E + F) = 2\pi\chi(D)\,.$$

$\square$

Remark 9.1.14 *The proof above is basically the one in [105]. However, the proof is essentially unique and there is only place for some variation regarding notations and the order of some arguments (see e.g. [168], [240].)*

In particular, if D is a simple region, we get the perfect copy of the local version in the global setting:

Corollary 9.1.15 $D \subset S$ *is a simple region. Then*

$$\iint_D K\,dA + \sum_0^p \int_{v_i}^{v_{i+1}} k_g\,dl + \sum_0^p \varphi_i = 2\pi\,. \tag{9.1.8}$$

In particular, if all the arcs $[v_i, v_{i+1}]$ are geodesic segments, then

$$\iint_D K\,dA + \sum_0^p \varphi_i = 2\pi\,. \tag{9.1.9}$$

The case $p = 3$ is important enough to deserve emphasizing as Let $T = \triangle ABC \subset S$ be a geodesic triangle. If K has a constant sign in T, then

$$\iint_D K dA + \sum_1^3 \varphi_i = 2\pi. \qquad (9.1.10)$$

Given that the exterior angles $\varphi_1, \varphi_2, \varphi_3$ are supplements of the interior angles of the triangles, that is (to use the time-honored notation) $\varphi_1 = \pi - \hat{A}$, $\varphi_2 = \pi - \hat{B}$, $\varphi_3 = \pi - \hat{C}$, the formula above may be written as

$$\iint_D K dA = (\hat{A} + \hat{B} + \hat{C}) - \pi. \qquad (9.1.11)$$

Recall that the expression on the left side of Formula (9.1.11) above is called the *defect* (or *excess*) of the triangle $\triangle ABC$, and is usually denoted by $\delta = \delta(\triangle ABC) = \delta(T)$. Thus, the sign of the defect of a geodesic triangle is the same as that of K, namely

1. $\delta > 0$, if $K > 0$;

2. $\delta = 0$, if $K = 0$;

3. $\delta < 0$, if $K < 0$.

The most significant case – and the one with the longest and deepest history – is that of $K = \text{const.}$, and most importantly for $K \equiv \pm 1$:

1. If $K \equiv 1$, i.e. if $T = \triangle ABC$ is a spherical triangle, then

   (a) $\delta(T) > 0$;
   (b) $\text{Area}(T) = \delta(T)$.

2. If $K \equiv -1$, i.e. if $T = \triangle ABC$ is a hyperbolic triangle, then

   (a) $\delta(T) < 0$;
   (b) $\text{Area}(T) = -\delta(T)$.

Note that in the Euclidean case, that is of $K \equiv 0$, area cannot be computed in terms of the defect, since it equals 0 in this case.

Remark 9.1.16 *Again with the exception of $K \equiv 0$, the defect equals the oriented area of the image of the image of T via the Gauss mapping N. This is a fact that we in fact used and exploited in our approach to the discrete proof of Theorema Egregium.*

Corollary 9.1.17 *Let $D \subset S$ be a regular region, such that ∂D has no vertex, i.e. ∂D is smooth. Then*

$$\iint_D K dA + \sum_{i=1}^p \int_{\gamma_i} k_g dl = 2\pi \chi(D); \qquad (9.1.12)$$

The next corollary is the one that is usually quoted as the Gauss-Bonnet Theorem, or the *Gauss-Bonnet Formula*:[12]

Corollary 9.1.18 *Let S be a orientable compact surface (without boundary). Then*

$$\int\int_S K dA = 2\pi\chi(S). \tag{9.1.13}$$

This is an amazing theorem! Indeed, it connects curvature (i.e. "rigid" Geometry) to Topology (that is to say "rubber Geometry"). As it is usually understood or employed, it is taken to mean that the topology of the surface determines, via the Euler characteristic, the global, or average, behavior of Gauss curvature. While this is certainly true, a stick points in two opposite directions, and we might interpret it as showing that the total curvature of a surface determines its topology. Both ways of understanding the result are, of course, correct, and various applications necessitate the use of both implications, according to the context (as we shall see in the corollaries below).

Indeed, not only is the theorem hugely important per se, it also has a large number of important consequences. In particular, we have the following

Corollary 9.1.19 *Let S be a compact surface with $K > 0$. Then S is homeomorphic to a sphere.*

In fact, one can also prove, as a corollary, a stronger result due to Hadamard ():

Theorem 9.1.20 (Hadamard) *Let S be a closed, orientable surface in $\mathbb{R}^3$, having $K > 0$. The S is convex.*[13]

The proof we are bringing below is, essentially, that of [82]. It is somewhat more difficult than most of the proofs in this text, in the sense that it makes appeal to classical results in Topology.

Proof By Corollary 9.1.19 S is homeomorphic to a sphere, thus, by the Gauss-Bonnet formula, $\int\int_S K dA = 4\pi$.

Since S is orientable, we can chose an orientation and consider the Gauss mapping $N : S \to \mathbb{S}^2$. Given that $K > 0$ it follows that $\det N \neq 0$, thus N is locally injective, therefore $N(S)$ is an open set of $\mathbb{S}^2$ by the *Invariance of Domain Theorem* (see, e.g. [229]). Given that S is compact, it follows that $N(S)$ is a compact subset of $\mathbb{S}^2$, thus it is closed (see, e.g. [230]). Thus, since it open, closed (and non-void) it follows that $N(S) = \mathbb{S}^2$.

We next show that N is (globally) injective. Indeed, if there exists $p, q \in S, \neq q$ such that $N(p) = N(q)$, then there exist an neigborhood V of q, such that $N(S\backslash V) = \mathbb{S}^2$. Given that $\int\int_{S\backslash V} K dA$ equals the area of $N(S\backslash V)$, counted with multiplicities, it follows that $\int\int_{S\backslash V} K dA \geq 0$. However, $\int\int_V K dA > 0$, therefore $\int\int_S K dA = \int\int_{S\backslash V} K dA + \int\int_V K dA > 4\pi$. We have obtained, therefore, a contradiction, hence N is injective, thus a homeomorphism.

□

---

[12]To make things even more confusing, sometimes this corollary is known as the Gauss-Bonnet Theorem, while Theorem 9.1.7 is known as the Gauss-Bonnet Formula.

[13]That is S is included in one of the half-spaces determined by any of its tangent planes.

Remark 9.1.21 *Hadamard's Theorem holds under the more relaxed condition that* $K \geq 0$, *however the proof in this case is much more difficult. (See [83].)*

The connection between curvature and geodesics is reinforced by the following two results:

Corollary 9.1.22 *Let* S *be an orientable such that* $K \leq 0$, *and let* γ_1, γ_2 *are two geodesics having a common point* p. *Then either* $\gamma_1 \cap \gamma_2 = \{p\}$, *or* $\gamma_1 \cup \gamma_2^{-1} \neq \partial R$, *where* R *is a simple region.*

Corollary 9.1.23 *Let* S *be a surface homeomorphic to a cylinder and having* $K < 0$. *Then* S *has at most one closed geodesic.*

Exercise 9.1.24 *(Hint: Use the fact that* S *is homeomorphic to* $\mathbb{R}^2 \backslash \{(0,0)\}$ *and apply the previous corollary.)*

Corollary 9.1.25 *Let* S *be a compact surface with* $K > 0$. *Then any two closed, simple geodesics* $\gamma_1, \gamma_2 \subset S$ *intersect.*

The result certainly holds for the prototype compact surface of positive curvature, i.e. the sphere, on each any two geodesics – that is great circles – intersect. The corollary shows that, due to the Gauss-Bonnet Theorem, this holds for any such surface. The proof is quite short, and we believe it is a good exercise to the reader.

Exercise 9.1.26 *Prove the corollary above. (Hint: Apply Corollary 9.1.19.)*

The last consequence of the Gauss-Bonnet Theorem that we bring here is, arguably, quite surprising, the Jacobi theorem:

Theorem 9.1.27 (Jacobi) *Let* $c : I \to \mathbb{S}^2$ *be a closed curve such that* $\kappa(s) \neq 0$, *for any* $s \in I$. *Then* c *divides* $\mathbb{S}^2$ *into two domains of equal area.*

The short theorem below is again due to Chern [82]. The reader is invited to supply the details that he/she feels are missing him/herself. (More detailed proofs can be found in [105], [310].)
Proof Define ϑ by the following system of equations:

$$\begin{cases} \kappa = \sqrt{\kappa^2 + \tau^2} \cos \vartheta \\ \tau = \sqrt{\kappa^2 + \tau^2} \sin \vartheta \end{cases} \tag{9.1.14}$$

where κ, τ denote (as usual) the curvature and torsion of c.
Then, using the standard notations, we have

$$d(-\mathbf{t} \cos \vartheta + \mathbf{b} \sin \vartheta) = (\mathbf{t} \sin \vartheta + \mathbf{b} \cos \vartheta) d\vartheta - \sqrt{\kappa^2 + \tau^2} \mathbf{N} ds \, ;$$

where $N = N(c(s))$ represents the image of c under the Gauss mapping.[14]

---

[14]also called the *normal indicatrix*

Therefore, if we denote by σ the arc length of $N = N(s)$, it follows that $d\vartheta/d\sigma$ represents the geodesic curvature of $N(s) \in \mathbb{S}^2$.

Let D be one of the domains bounded by c, and let $A(D)$ denote its area. Then, the Gauss-Bonnet Formula implies that

$$\int_{N(c)} d\vartheta + \int\int_D dA = 2\pi \, ;$$

thence that $A(D) = 2\pi$.

$\square$

Exercise 9.1.28 *In the last stage of the proof above we used the fact $\int_{N(c)} d\vartheta = 0$. Prove this assertion.*

9.2   THE POINCARÉ INDEX THEOREM

This theorem is just only slightly less well-known than the Gauss-Bonnet Theorem and it is, perhaps, even more surprising, since it connects, again, the global curvature of a surface, but this time to the vector fields on the surface. To make this statement precise we must first bring some preparatory definitions:

Definition 9.2.1 *Let* **v** *be a vector field on a (smooth) oriented surface. A point $p_0 \in S$ is called a singular point (or critical point) for* **v** *if* $\mathbf{v}(p_0) = 0$. *(Thus singular points are also called zeros of the vector field.) A singular point p_0 is called isolated if there exists a neighborhood V of p_0 that contains no other singular points.*

To every singular point of a vector field one can attach an integer number in a quite intuitive manner, as follows: Consider a small circle around the critical point, choose an arbitrary point on it and denote by $\varphi \in [0, 2\pi)$ the angle made at this point by the vector field with the positive Ox axis. Presuming the angle between the vector field and changes continuously as we describe the circle in positive trigonometric sense, we can follow the total change of φ and define the index through this change as being

$$i(\mathbf{v}, p_0) = \frac{1}{2\pi} \left[\varphi(2\pi) - \varphi(0) \right] . \tag{9.2.1}$$

See Figures 9.7 and 9.8. (We have seen a similar approach when introducing the notion of parallel transport.) The idea above can be formalized by noticing that one can write

$$i(\mathbf{v}, p_0) = \frac{1}{2\pi} \int_0^{2\pi} \frac{d\varphi}{dt} dt \, ; \tag{9.2.2}$$

(this being an integral – and idea – that those who studied Complex Analysis are familiar with.)

Remark 9.2.2 *As Berger [37] notes one can view (for infinitesimal circles only!) the index as a 1-dimensional version of the Gauss map.*

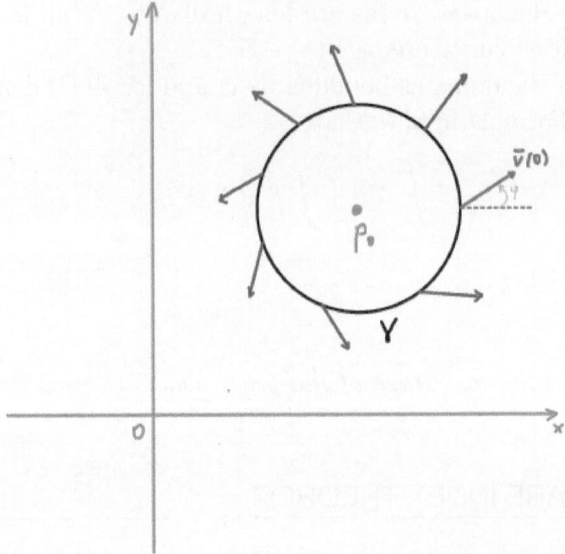

Figure 9.7 The index as the change along a small circle around the critical point of the angle between the vector field and the positive Ox axis.

However, intuitive this basic idea of the definition of the index might be, it is not easy at all to make it formally sound.

The first problem arises even if one wants to check that the index is, as expected, an integer. (Again, those who took a Complex Analysis course might recall the technicalities involved.)

The attentive reader might have noted the second problem: In the very definition of the angle φ we presumed one can compute the angle made by the vector field with the Ox axis, thus implicitly presuming that such an axis is defined. In other words, the construction, simple as it might be, can be performed only if a orthogonal coordinates system is assumed. Then one has, of course, to prove that the integer is independent on the choice of this local coordinate system. There is yet another choice that we, perhaps unconsciously made, but which also must be proven as not relevant (which one?). We summarize all this observation as an exercise for the reader.

Exercise 9.2.3　*1. Verify that $i(\mathbf{v}, p_0) \in \mathbb{Z}$.*

  2. Prove that in the definition of the index any simple closed curved γ around p_0 can be considered and that, in fact $i(\mathbf{v}, p_0)$ does not depend on the choice of γ.

  3. Demonstrate that $i(\mathbf{v}, p_0)$ is independent on the choice of the parameterization. (Hint: Use the parallel transport interpretation of Gaussian curvature to show that

$$\varphi(2\pi) - \varphi(0) = \iint_{\text{int}\gamma} K \, dA \, ; \qquad (9.2.3)$$

where γ is as above.

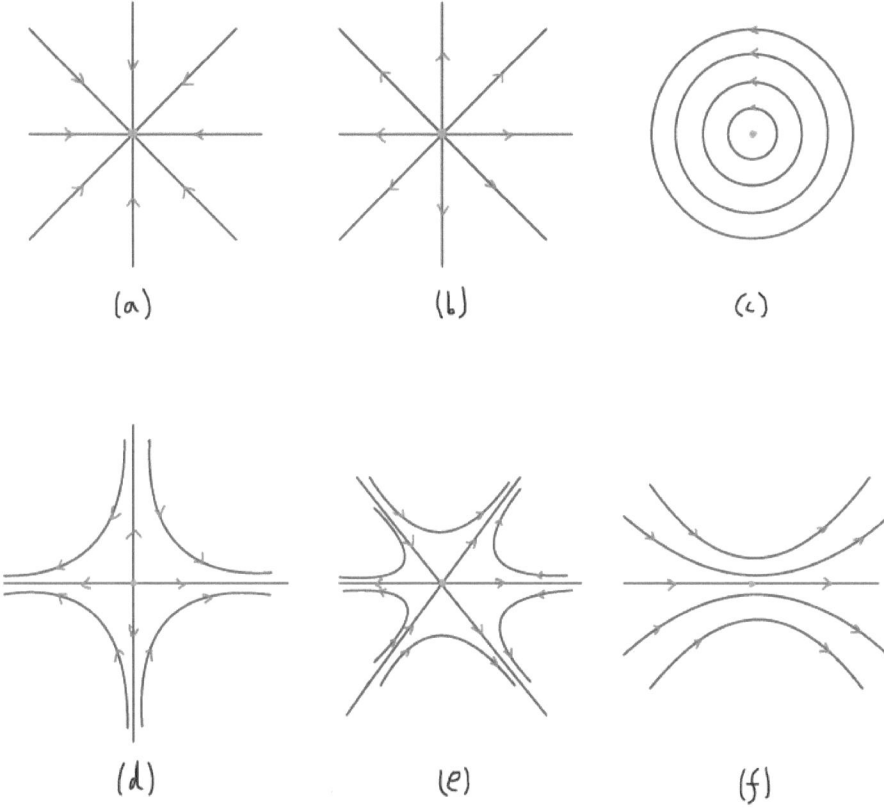

Figure 9.8 Typical critical points of a vector field, with the nomenclature inspired by gradient fields and their respective indices: (a) *sink*, $i = 1$; (b) *source* $i = 1$; (c) *center*, $i = 1$; (d) *simple saddle*, $i = -1$; (e) *monkey saddle*, $i = -2$; (f) *dipole*, $i = 2$.

Before passing on, we give in Figure 9.8 a few examples of typical critical points of vector fields and the nomenclature for each of these kinds of critical points.

As made evidently ample above, the classical, formal definition of the index of a vector point at a critical point is difficult and wrought with technicalities to ensure it is actually consistent. We therefore bring an alternative approach, du to Thurston [322]. It is not only very intuitive, it is also designed to work for cell complexes, thus ideal, as we shall shortly see, for use in the scheme of proof of the theorem. Furthermore, it is essentially a discrete approach, thus in tandem with the main drive of our book.

To the point: Given an isolated singular point p_0 of a vector field $\mathbf{v}$ on a smooth surface S, circumscribe a small polygon having p_0 in its interior and such that its edges are not tangent to the vector field (we shall return shortly to this aspect, in somewhat more detail). Next, a charge (think "electrical charge") is placed at the interior, the vertices and middles of the edges of the polygon, as follows: A "+" charge is attached to the interior of each polygon and at each of its vertices, and a

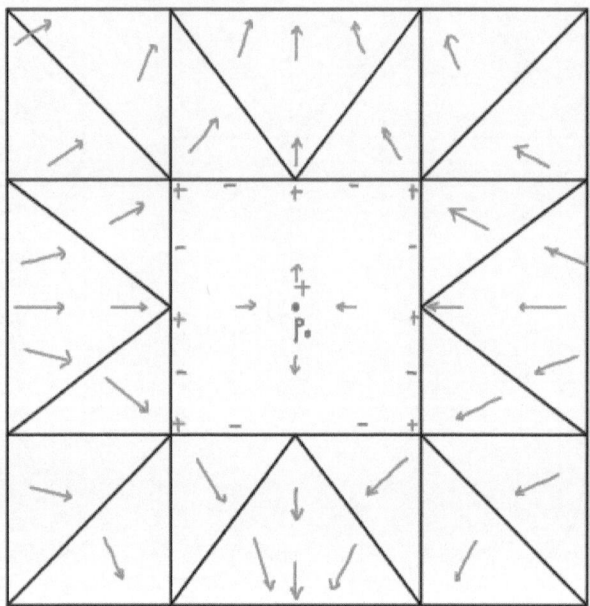

Figure 9.9 A saddle in the charges model of the index (after [322].).

"-" charge at each of its edges. The charges flow off the (boundary of) polygon, by the action of **v**, nut they will not (necessarily) cancel in the interior of the polygon (see Figure 9.9). We then define the index $i(\mathbf{v}, p)$ to be the sum of the remaining charges in the interior of the polygon. In this model, a isolated critical point is called a *sink* or a *source* if $i(\mathbf{v}, p) = +1$ (with the filed points to the/away from p_0, respectively), and a *saddle* if $i(\mathbf{v}, p) = -1$ (See Figure 9.9). As an illustration of the simple – yet powerful – intuition behind the charges-based definition of index, we suggest that the reader prove the following exercise (which can be viewed as intuition building for the proof we shall give for Theorem 9.2.5 below):

Exercise 9.2.4 *Using the charges model, prove that the Euler characteristic of a cube (hence of the sphere) is 2.*

The statement of the theorem itself is quite simple:

Theorem 9.2.5 (Poincaré Index Theorem) *Let S be a closed, orientable surface, and let **v** be a vector field on S having only isolated critical points. Then*

$$\sum_{\mathbf{v}(p)=0} i(\mathbf{v}, p) = \chi(S) \,. \tag{9.2.4}$$

Remark 9.2.6 *As expected, given its largely topological nature and the proofs seen above, the Poincaré's Theorem has an inherent extension to any dimension, and it is known, in this case, as the Hopf-Poincaré's theorem.*

Remark 9.2.7 *The requirement that S has only isolated critical points is not restrictive, since any vector field on a closed surface satisfies this property. More precisely, we have*

Lemma 9.2.8 *Given a finite triangulation[15] of a smooth surface S, there exists a smooth vector field on S, such that it triangle (2-face) has as his center a source, a sink at each vertex (0-face), and a saddle at the middle of each edge (1-face).*

Exercise 9.2.9 *Prove Lemma 9.2.8 above. (Hint: Draw a small example first.)*

The proof is hardly elementary since it too intrinsically needs the use of the Triangulation Theorem 9.1.8. In consequence, we can bring here only a sketch of the full proof, i.e. we'll be forced to omit the numerous and deep technical details behind the Differential Topology tools essential in the demonstration. (This is does not mean that the proof is faulty or inherently incomplete, rather that we make appeal to some intuitive – yet deed – topological facts.[16]) Here we follow the demonstration in [322]. A somewhat – but not essentially – different approach, making appeal to Formula (9.2.3) based definition of the index of critical points (and which might appear complete, but where all the delicate points of a topological nature are omitted) can be found in [105]. (See also [165] for a nice brief overview of the proof.)

Proof By Theorem 9.1.8 there exists a smooth triangulation of S. Then it is possible by subdivisions and *elementary moves*[17], i.e. by "jiggling" the triangulation, it is possible to ensure that the singular points (zeros) of the vector fields are interior points of some triangles and, moreover, that no triangle contains more than one such point. Next, encompass each singular point by a convex polygon P that is fully contained in the same triangle T as the given zero and that is *transverse* to the vector field, that is the field is nowhere tangent to an edge. The set $T\backslash P$ is a (topological) annulus, which can again be triangulated. The triangulation away from the polygon P can now also be made transverse to the vector field. Given our interpretation of indices in terms of electrical charges, the proof is concluded once we observe that the contribution of each polygon to the sum on the left side of (9.2.4) is precisely the index of the vector field at the singular point.

□

Besides the basic assumption that the surface admits a triangulation, we have made appeal to a number of facts, mostly quite intuitive, some of them immediate and some quite involved. Given their essential topological and combinatorial nature we do not bring here their proofs. Still, we invite the interested reader to try and prove them.

Exercise 9.2.10 *Prove that given a vector field on a closed, smooth surface, there exists a triangulation of S transverse to the vector field. (As stated, this is not a trivial challenge. Hint: Start by ensuring that the star of each vertex contains is contained in a coordinate patch of diameter $\leq \varepsilon$.)*

Exercise 9.2.11 *Let $\mathbf{v}$ be a vector field on closed, smooth surface S, and let p_0 be an isolated critical point of $\mathbf{v}$. Using the charges definition of the index, prove that $i(\mathbf{v}, p_0)$ is independent of the choice of the polygon P used to define the index.*

---

[15]or, more generally, a finite cell partition

[16]The reader has, in fact, been faced with this type of demonstration in many Calculus courses, where *Jordan's Theorem* – which is perhaps hardest to prove highly intuitive result – is used freely without proof.

[17]See [230] for the precise technical definition.

Exercise 9.2.12 *Let P_1, P_2 be star-shaped polygons, such that $P_1 \subset \text{int} P_2$. Recall that*

Definition 9.2.13 *A set D is called* star-shaped *(or* star-like*) if for any $p \in \text{int} P$ and any $q \in \partial D$, the segment $[p, q] \subset X$.*

1. *Triangulate the annulus $A = P_2 \backslash P_1$.*

2. *Let P be a polygon which is star-shaped relative to the isolated critical point p_0 of the vector field $\mathbf{v}$. Prove that ∂P can be made transverse to $\mathbf{v}$ only by small moves[18] [229]) of its vertices. Conclude that the resulting polygon will also be star-shaped.*

3. *Can the construction above be extended to the case when P_1 is not star-like?*

The best known application (in fact, quite widely popularized) of the Poincaré Index Theorem is its following

Corollary 9.2.14 *(Poincaré, 1885) Any vector field on $\mathbb{S}^2$ has at least one zero.*

This is a result with a clear (though perhaps not very practical) meteorological implication: At any time, there exists at least a point on Earth were the wind does not blow.[19] This theorem is popularized also as "The Hedgehog Theorem", given its formulation as "You can't comb a hedgehog" (there will always be a quill that pokes up).[20]

Surely the reader expects, by now, the following (quite straightforward)

Exercise 9.2.15 *Prove the corollary above.*

However, we wish to propose a more challenging exercise:

Exercise 9.2.16 *Prove Corollary 9.2.14 directly, that is without using the Poincaré Index Theorem. (Hint: Recall that the cube is a topological sphere and use the electrical charges method introduced in this section.[21]*

It would be simply sinful to conclude the overview of the Gauss-Bonnet and Poincaré Index Theorems without recapping and noticing that they connect a number of seemingly disparate filed of Mathematics: Combinatorics and Graph Theory (via definition of the Euler characteristic of a map on a surface, which, by the way, we used in the proof of the Gauss-Bonnet Theorem); Topology (if one recalls that, for oriented surfaces, $\chi = 2 - 2g$, where g is the *genus* of S, i.e. the number of "holes" of the surface); Differential Geometry (evidently!) and Differential Equations (via

---

[18] formally: *ε-moves*

[19] Given that the theorem cannot supply the location – at any time – of the quiet point, its usefulness in prognosis is quite severely limited.

[20] Given that human had is also a topological sphere, it usually follows that hair "cow licks" (i.e. hair whorls) cannot be combed, either.

[21] To the best of our knowledge this approach is due, again, to Thurston [322].

vector fields and their singularities). If anyone still needed a proof of the unity of Mathematics,[22] a better one could hardly be found!

We conclude this section with a perhaps unexpected application (yet immediate) of Theorem 9.2.5, which we bring it in the form of an exercise for the reader:

Exercise 9.2.17 *Prove, using the Poincaré Index Theorem, that if S is a smooth surface in* $\mathbb{R}^3$, *then S has at least one umblical point.*[23]

9.2.1   Discretizations of the Gauss-Bonnet Theorem

Given its great importance in so many fields, thus so many potential applications, it is little wonder that many discretizations Gauss-Bonnet Theorem have been proposed. We must restrict here, therefore, only to a selected few.

The immediate discretization that one would look for is, evidently (certainly for the readers of this book) the one for *PL* surfaces (mainly due to their ubiquitous presence in Graphics, CAD, etc.) in the guise of triangular meshes. As the reader surely must expect, given the exposition of Banchoff's elementary methods presented in Chapter 4, such an intuitive and simple adaptation does exists and it is again due to Banchoff. What still might come as a surprise is the fact that the proof is, in fact, much simpler than we one of the earlier Theorema Egregium. While this partly due to the preparatory development introduced therein, it is also a consequence of the simplicity (and power) that Topology, via the Euler characteristic infuses. More precisely, the route is through the Poincaré Index Theorem, which is only natural given Banchoff approach to curvature by means of the index (and whose proof is, as we have seen, surprisingly direct and uncomplicated). Furthermore, our preparatory material ensures that the proof above applies holds for smooth as well as for polyhedral surfaces.

Proof (*Gauss-Bonnet, smooth or polyhedral*) By Formula 4.1.11 of Chapter 4

$$K(S) = \frac{1}{2} \int_{\mathbb{S}^2} \sum_{p \in S^2} i(p, l) dA \,.$$

Thence, by the Poincaré Index Theorem

$$K(S) = \frac{1}{2} \int_{\mathbb{S}^2} \chi(S) dA = \frac{1}{2}\chi(S) \int_{\mathbb{S}^2} dA = 2\pi\chi(S) \,.$$

□

Unsurprisingly, Banchoff's method inherently extends to higher dimensions [25]. Similar results for higher dimensional *PL* manifolds can be found in [71] and the earlier (and by now classical) [75].[24] Also, a fitting Gauss-Bonnet Theorem for Alexandrov surfaces was given by Machigashira [215].

---

[22]and why it should properly be called, like in other languages "Mathematic", and not "Mathema*tics*

[23]We brought this result, without a proof, in Chapter, as Proposition 2.21.

[24]We shall encounter again this last reference in our excursus into discretizations of higher dimensional curvatures.

If the discretization is sought for applications in Complex Networks, then certainly the most basic version one would hope for would be for combinatorial graphs. Such a version indeed exists and it is due to Knill [192].[25] Knill's discretization is essentially combinatorial in nature, as it deals with *proper* (or "*simple*", as he calls them) graphs, that is undirected graphs, without multiple edges or (self-)loops. Furthermore, the considered graphs are unweighted ones (or, alternatively, all vertex and edge weights equal 1). Moreover, only finite graphs are considered (by analogy with the compactness in the smooth version of the theorem). Thus his version of the Gauss-Bonnet Theorem is a purely combinatorial one. It can also be interpreted as showing how far one can get from Differential Geometry and still retain its flavor in its most essential and intrinsic aspects.

Before formulating and proving this combinatorial version of the theorem, it is best we introduce some notations specific to Knill, as well as some definitions necessary in the formulation of the result and in its proof.

- Given a graph $G = (V, E)$, a subgraph $H < G$ is called a *clique*[26] if it is homeomorphic to some *complete graph* K_{m+1}. We denote $G_m = \{H < G \mid H$ is a clique$\}$, and we set $v_m = |G_m|$.

- We can now define the Euler characteristic as

$$\chi(G) = \sum_{m=0}^{\infty} (-1)^m v_m . \tag{9.2.5}$$

Note that the sum is, in fact, finite, by the finiteness of G.[27] Observe that this definition generalizes the definition for surfaces to any dimension and that, if $\max(v_m) = 2$, then we obtain the classical euler characteristic of a surface.

- We also denote $V_m = \#\{K_{m+1} < \mathrm{lk}(v)\}$, i.e. the number of m-cliques in the *link* of the vertex v.[28]

The numbers V_m satisfy what Knill calls the *transfer equations*:

$$\sum_{v \in V} V_{m-1} = (m+1)v_m . \tag{9.2.6}$$

Exercise 9.2.18 *Prove the transfer equations.*

- The *combinatorial curvature* of a vertex v can now be defined as

$$K(v) = \sum_{m=0}^{\infty} \frac{(-1)^m V_{m-1}}{m+1} . \tag{9.2.7}$$

(Note that we used again the infinite summation notation.)

---

[25]In fact, Knill has proposed yet another discretization of the Gauss-Bonnet Theorem [191], but the second one is too technical and departing from the main themes of this book to be included here.

[26]Note the Computer Science terminology.

[27]However, this notation allows us to economize on the number of additional notations.

[28]This topological notion can be interpreted as the 1-sphere with center v, in the combinatorial metric, and Knill denotes is as such by $S(v)$.

We can now formulate the desired version of the theorem:

Theorem 9.2.19 (Combinatorial Gauss-Bonnet Theorem)

$$\sum_{v \in V} K(v) = \chi(G) \,. \tag{9.2.8}$$

Proof From the definition of $K(v)$ we immediately have

$$\sum_{v \in V} K(v) = \sum_{v \in V} \sum_{m=0}^{\infty} \frac{(-1)^m V_{m-1}}{m+1} \,.$$

Given that the sums involved are, in fact, finite, they commute, thus

$$\sum_{v \in V} \sum_{m=0}^{\infty} \frac{(-1)^m V_{m-1}}{m+1} = \sum_{m=0}^{\infty} \sum_{v \in V} \frac{(-1)^m V_{m-1}}{m+1} \,.$$

Then, by the transfer equations, the right hand term becomes

$$\sum_{m=0}^{\infty} v_m \,,$$

i.e. equals the Euler characteristic $\chi(G)$.

□

One of the authors has also proposed a local Gauss-Bonnet for Complex Networks [344], which is not only capable of dealing with (vertex and edge) weighted networks (in contrast with Knill's approach), but it is closer to the smooth one, since it more closely mimics the classical geometric quantities. The approach is that of *PL* and metric differential geometry, which we hope it is by now quite familiar to the reader, thus we present it here in some detail. Indeed, the idea is introducing a simple, *PL* manifold like structure on networks, and extend to networks ideas and results from that more classical setting [283].

We begin from the Local Gauss-Bonnet Theorem (9.1.1). To pass to the networks case, let us first note that, in the absence of a background curvature, the very notion of angle is undefinable. Thus, for abstract (non-embedded) cells, there exists no *honest* notion of angle. Hence, the last term on the left side of above Formula (9.1.1) has no proper meaning, and thus, can be discarded. Moreover, the distances between non-adjacent vertices on the same cycle (apart from the path metric) are not defined, and thus, the second term on the left side of above Formula (9.1.1) also vanishes.

We first concentrate on the case of combinatorial (unweighted) networks. For such networks endowed with the combinatorial metric, the area of each cell is usually taken to be equal to 1. Moreover, one can naturally assume that the curvature is constant on each cell, and thus, the first term on the left side of Formula (9.1.1) reduces simply to K. In addition, given that D is a 2-cell, we have $\chi(D) = 1$. Therefore, in the absence of the definition of an angle, it is naturally to define

$$K = 2\pi - \int_{\partial D} k_g dl \,. \tag{9.2.9}$$

It is tempting to next consider ∂D as being composed of segments (on which k_g vanishes), except at the vertices, thus rendering the above expression as

$$K = 2\pi - \sum_1^n \kappa_H(v_i) \,. \tag{9.2.10}$$

Remark 9.2.20 *An alternative approach to defining the curvature of cell would be the following: Since in a Euclidean polygon, the sum of the angles equals $\pi(n-2)$, where n represents the number of vertices of the polygon, one could replace the angle sum term in (9.1.1) simply by $\pi(n-2)$.*

Moving to the general case of weighted networks, one cannot define a (non-trivial) Haantjes curvature for vertices, since, as already noted above, no proper distance between two non-adjacent vertices v_{i-1} and v_{i+1} on the same cycle can be implicitly assumed (apart from the one given by the path metric, which would produce trivial zero curvature at vertex v_i). In fact, in this general case, neither can the arc (path) $\pi = v_0, v_1, \ldots, v_n$ be truly viewed as smooth. Therefore, we have no choice but to replace the second term on the right side of above (9.1.1) by $\kappa_H(\pi)$, where it should be remembered that π represents the path $v_0, v_1, \ldots, v_n$ of chord $e = (v_0, v_n)$.

We can now define the *Haantjes-Gauss curvature of a 2-cell* $\mathfrak{c}$. Given an edge $e = (u, v)$ and a 2-cell $\mathfrak{c}$, $\partial \mathfrak{c} = (u = v_0, v_1, \ldots, v_n = v)$ *(relative to the edge $e \in \partial \mathfrak{c}$)*, we have

$$K_{H,e}(\mathfrak{c}) = 2\pi - \kappa_{H,e}(\pi) \,, \tag{9.2.11}$$

where π denotes the path $v_0, v_1, \ldots, v_n$ subtended by the chord $e = (v_0, v_n)$, and $\kappa_{H,e}(\pi)$ denotes its respective Haantjes curvature. In the sequel we shall refer to this version as the *strong* Haantjes-Gauss curvature of the cell $\mathfrak{c}$ to distinguish it from the simpler version each time such a differentiation is required.

We focus next on the general case of weighted graphs/networks. Firstly, note that it is not reasonable to attach area 1 to every 2-cell in such graphs. However, as discussed in [166, 295], it is possible to endow cells in an abstract weighted graph with weights that are both derived from the original ones and have a geometric content. For instance, in the case of unweighted social or biological networks, endowed with the combinatorial metric, one can designate to each face, instead of the canonical combinatorial weight equal to 1, a weight that *penalizes* the faces with more edges, and thus, reflecting the weaker mutual connections between the vertices of such a face. Thus, it is possible to derive a proper local Gauss-Bonnet formula for such general networks, in a manner that still retains the given data, yet captures the geometric meaning of area, volume, etc. Thus, when considering any such geometric weight $w_g(\mathfrak{c}^2)$ of a 2-cell $\mathfrak{c}^2$, the appropriate form of the first term on the left side of (9.1.1) becomes

$$Kw_g(\mathfrak{c}^2) \,,$$

and the fitting form of Formula (9.2.11)

$$K_{H,e}(\mathfrak{c}) = -\frac{1}{w_g(\mathfrak{c}^2)} \left(2\pi - \kappa_{H,e}(\pi) \right) \,. \tag{9.2.12}$$

Before passing to the problem of extending the above definition to the case of general weights, let us note that considering directed networks actually simplifies the problem, in the sense that it allows for variable curvature (and not just one with constant sign). For general edge weights, we have the problem that the *total weight* $w(\pi)$ of a path $\pi = v_0, v_1, \ldots, v_n$ is not necessarily smaller than the weight of its subtending chord $e = (v_0, v_n)$ [29], thus Haantjes' definition cannot be applied. However, we can turn this to our own advantage by reversing the roles of $w(\pi)$ and $w(v_0, v_n)$ in the definition of the Haantjes curvature and assigning a minus sign to the curvature of cycles for which this occurs. Thus, this approach actually allows us to define a variable sign Haantjes curvature of cycles (hence, a Ricci curvature as well), even if the given network is not a naturally directed one.

Note that the case when $w(v_0, v_1, \ldots, v_n) = w(v_0, v_n)$, i.e., that of zero curvature of the 2-cell $\mathfrak{c}$ with $\partial\mathfrak{c} = v_0, v_1, \ldots, v_n, v_0$ straightforwardly corresponds to the splitting case for the path metric induced by the weights $w(v_i, v_{i+1})$.

Remark 9.2.21 *Indeed, the method suggested above reduces to the use of the path metric, in most of the cases. One can always pass to the path metric and apply to it the Haantjes curvature. Beyond the complications that this might induce in certain cases, it is, in our view, less general, at least from a theoretical viewpoint, since it necessitates the passage to a metric. However, in the case of most general weights, i.e. both vertex and edge weights, one has to pass to a metric. We find the path degree metric (see e.g. [186]) especially alluring as it is both simple and has the capacity to capture, in the discrete context, essential geometric properties of Riemannian metrics. We also refer the reader to [286] for an ad hoc metric devised precisely for use on graphs in tandem with Haantjes curvature.*

9.2.2 Discretizations of the Poincaré Index Theorem

While not as known (and arguably less fundamental) as the Gauss-Bonnet Theorem, discretizations of the Poincaré Index Theorem were proposed. The most direct – and, we are convinced, expected by the reader given the course of the exposition – is the one for polyhedral surfaces due to Banchoff [26]. True, in this case, the theorem is applicable only for the height function[30] associated to a direction l. Thus it is only analogous to the Poincaré Index Theorem, not a perfect discretization. However, it is still more generic than it might appear at first view, given that the fact that the number of non-general directions is finite, thus almost all directions give rise to a height function. (Recall also Footnote 12 on page 123 in Chapter 4 regarding the generality of possible height functions.) The fitting formulation of the theorem, for the polyhedral case is

Theorem 9.2.22 (Polyhedral Index Theorem) *Let S^2 be a polyhedral surface in*

---

[29]We suggest the name *strong local metrics* for those sets of positive weights that satisfy the generalized triangle inequality $w(v_0, v_1, \ldots, v_n, v_{n+1}) < w(v_0, v_n)$, for any elementary 1-cycle $v_0, v_1, \ldots, v_n, v_0$.

[30]i.e. *Morse function*

$\mathbb{R}^3$ *and let l be a general direction for S^2. Then*

$$\sum_{p \in S^2} i(p, l) = \chi(S^2) . \tag{9.2.13}$$

Proof By eventual subdivisions we might assume that S^2 is in fact PL. But, by a simple combinatorial argument, for such a surface, $3T = 2E$.[31] Having established this elementary fact, we can now proceed to the proof itself:

$$\sum_{p \in S^2} i(p, l) = \sum_{p \in S^2} \left(1 - \frac{1}{2} (\#T \text{ s.t.} p \text{ is a middle for } l) \right)$$

$$V - \sum_{p \in S^2} (\#T \text{ s.t.} p \text{ is a middle for } l) .$$

But, since every triangle has only one vertex that is a middle for l, we can write the last expression as

$$V - \frac{1}{2}T .$$

Recalling that $3T = 2E$, i.e. $T = 2E - 2T$, we obtain that

$$\sum_{p \in S^2} i(p, l) = V - \frac{1}{2}(2E - 2T) = V - E + F = \chi(S^2) .$$

$\square$

As the reader perhaps expected, in addition to his adaptation of the Gauss-Bonnet Theorem to graphs, Knill also proposed a fitting version of the Poincaré Index Theorem.[32] Its statement is quite straightforward and largely a calque of the classical version:

Theorem 9.2.23 (Poincaré Index Theorem for Graphs) *Let $G = (V, E)$ be a graph and let $f : V \to \mathbb{R}$ be a (scalar) function. Then*

$$\sum_{v \in V} i_f(v) = \chi(G). \tag{9.2.14}$$

While we already defined the Euler characteristic of a graph, we still have to make explicit the one of index in this setting.

Definition 9.2.24 *Let $v \in V$ and let f be as above. Denote $\mathrm{lk}_f^-(v) = \{w \in \mathrm{lk}(v) \mid f(w) - f(v) < 0\}$, $\mathrm{lk}_f^+(v) = \{w \in \mathrm{lk}(v) \mid f(w) - f(v) > 0\}$. The sets $\mathrm{lk}_f^-(v)$ and $\mathrm{lk}_f^+(v)$ are called the exit set and entrance set, respectively. Then the index (of v relative to f) is defined as*

$$i_f(v) = 1 - \chi(\mathrm{lk}_f^-) . \tag{9.2.15}$$

---

[31]This is a trivial combinatorial result thus we will not give its proof. However, the incredulous reader (if one might be such) is invited to prove him/herself.

[32]Not accidentally, Knill has also suggestd a graph version for the classical Stokes Theorem.

Remark 9.2.25 *For injective functions f, $i_f(v)$ is independent of f.*

As for the Gauss-Bonnet Theorem, before proceeding to the proof proper, we need some additional notations and facts:

- We denote $V_m^- = \#\{K_{m+1} < \mathrm{lk}_f^-\}$, $V_m^+ = \#\{K_{m+1} < \mathrm{lk}_f^+\}$. In addition let W_m denote the number of k-simplices that contain vertices both from V_m^- and V_m^+. (Thus $V_m = W_m + V_m^- + V_m^+$.)

- In analogy with the transfer equations we have the *intermediate equations*

$$\sum_{v \in V} W_m(v) = m v_{m+1} \,. \tag{9.2.16}$$

Proof Denote $\chi^*(G) = \sum_{v \in V} i_f(v)$. Since passing from f to $-f$ interchanges lk^- and lk^+ while preserving the defining sum, we van prove instead prove that

$$2\chi^*(G) = 2v_0 - \sum_{v \in V} \left(\chi(\mathrm{lk}_f^-) + \chi(\mathrm{lk}_f^-) \right) = 2\chi(G) \,.$$

Indeed, by the transfer and intermediate equations we have

$$2\chi^*(G) = 2v_0 + \sum_{m=0}^{\infty} (-1)^m \sum_{v \in V} \left(\mathrm{lk}_f^-(v) + \mathrm{lk}_f^+ \right) = \sum_{m=0}^{\infty} (-1)^m \sum_{v \in V} (V_m(v) - W_m(v))$$

$$= 2v_0 + \sum_{m=0}^{\infty} (-1)^m 2v_0 + \sum_{m=0}^{\infty} (-1)^m = 2\chi(G) \,.$$

□

For more details and a more in depth look, see [193].

Remark 9.2.26 *As far as we are aware, (quite surprisingly, we might add!) there are no applications to Complex Networks of Knill's result.*[33]

---

[33]Or al least it has not truly penetrated the field.

CHAPTER 10

Higher Dimensional Curvatures

10.1 MOTIVATION AND BASICS

As we have already stated, Differential Geometry is, essentially, nothing else but the study of curvature. Indeed, we made curvature the red thread running along through this book. In fact, one might even say that if one has to chose a leitmotif for our book, then one might say that the "story" behind this text is one of the development of the notion of curvature in all its facets. Thus it is only natural, given that, this book, as any else in this world, must, after all, be finite, to concentrate in these concluding chapters on curvature in higher dimensions, in its many and various facets. We shall see that, as expected (or feared) higher dimensional generalizations are necessarily more complicated. We shall also learn however, that, most surprisingly, some of the more complicated notions are precisely the ones that lend themselves to discretization, simplification and dimensionality reduction. Given that it is not possible to give, in the confines of this book, a proper, formal derivation of curvature in higher dimension, perhaps the main reason for introducing them is precisely the fact that they allow to open a practically new world of modern, discrete Differential Geometry with a variety of novel and important applications. Thus we restrict all the technical details, that properly belong to a Riemannian Geometry book, to a modicum that allows us to proceed as fast as possible (yet still coherently) toward the geometric core intended.

Before presenting the "main cast", so to say, of our overview of higher dimensional curvatures, let us specify some notations: From here on we denote by M^n an n-dimensional $\mathcal{C}^k$-Riemannian manifold, $k \geq 2$, which we may presume, by Nash's theorem (see, e.g. [308]), to be isometrically embedded (i.e. "realized") in R^N, for some N sufficiently large. Analogously to the notation for surfaces, let $T_p(M^n)$ denote the *tangent space* at the point $p \in M^n$, and let $T_p^{\perp}(M^n)$ stand be the orthogonal complement of $T_p(M^n)$ in $T_p(\mathbb{R}^N)$, i.e. $T_p(M^n) \oplus T_p^{\perp}(M^n) = T_p(\mathbb{R}^N)$. Then M^n can be locally written as the graph of a function $f : T_p(M^n) \to T_p^{\perp}(M^n)$. For further technical details, which, however sadly, we definitely cannot include here, such as regarding *tensor*, see, for instance, [308].

DOI: 10.1201/9781003350576-10

10.1.1 The Curvature Tensor

We can now introduce the basic definition in defining curvature of Riemannian manifolds:

Definition 10.1.1 *Let M^n and f be as above, and let $p \in M^n \subset \mathbb{R}^N$. The bilinear form $II_p : T_p(M^n) \to T_p^\perp(M^n)$*

$$II_p(M^n) = (\beta_{ij})_{1 \le i,j \le n} \tag{10.1.1}$$

where $\beta_{ij} = \partial^2 f / \partial x_i \partial x_j$, $1 \le i, j \le n$; is called the second fundamental tensor *of M^n at the point p.*

This notion is a transparent generalization of the second fundamental form of surfaces. It also allows us to define the *Riemannian curvature tensor:*

Definition 10.1.2 *The* Riemannian curvature tensor *(at a point p) is defined as the tensor of 2×2-minors of $II_p(M^n)$, i.e.:*

$$R_{ijkl} = \beta_{ik}\beta_{jl} - \beta_{jk}\beta_{il}. \tag{10.1.2}$$

Remark 10.1.3 *Riemannian curvature is intrinsic. (This is a very important fact, indeed, but we shall not use it as such, only through the notion of sectional curvature that we shall define shortly.)*

For further, deeper information on the curvature tensor, see, e.g. [104], [309] or [249]. For now, however, we cannot escape the feeling that the geometric intuition behind this notion appears quite arcane (and it also appears[1] hard to manipulate in computations).

10.1.2 Sectional Curvature

As we have just seen, both the second fundamental tensor and the Riemannian curvature tensor are not very intuitive concepts. We would like, therefore, to be able to define a more geometric notion that would be essentially equivalent to that of the curvature tensor. Given that the second fundamental form is used, in the case of surfaces, to define Gauss curvature, one would like and expect a definition extrapolating the later to a concept of curvature for higher dimensional manifolds. This is, indeed the case, and the concept in question is that of *sectional curvature*, whose definition being a very direct, geometric generalization of its classical 2-dimensional counterpart:

Definition 10.1.4 *Let $p \in M^n$, and let $\Pi \subset T_p(M^n)$ be a 2-dimensional plane, and let $S = M^n \cap \left(\Pi \oplus T_p^\perp(M^n) \right)$. Then $\dim S = 2$ and we define the sectional curvature as $K(\Pi) = K_p(S)$, where $K_p(S)$ represents the Gauss curvature of S at the point p.*

---

[1]quite rightly so

Of course, if $n = 2$, K reduces to the classical Gauss curvature, thus justifying the name (and the notation).

There is a close connection between sectional curvature and the Riemannian curvature tensor (as could be expected given the fact that we wanted to substitute the first notion for the later), an interdependence revealed through the following formula:

$$K(\Pi) = \sum_{1 \leq i,j,k,l \leq n} R_{ijkl} \, \mathsf{x_i x_j x_k x_l} \, ; \qquad (10.1.3)$$

where $\{\mathsf{x}_h\}_{1 \leq h \leq n}$ is an orthonormal base of $T_p(M^n)$.

Moreover, by direct computations it is easy to show that knowledge of $K(M)$ on all tangent planes is equivalent to knowing the curvature tensor (see, e.g. [104], pp. 94-95). In particular, it implies that sectional curvature is an intrinsic notion (as expected from a straightforward generalization of Gaussian curvature). More generally, it implies that to all practical purposes, one can exchange the more technical notion of curvature tensor with the geometrically intuitive one of sectional curvature. Thus one gets an intuition of curvature similar to the one obtained for surface via the principal curvatures. In the n-dimensional case, one "cuts" the manifold using n mutually orthogonal directions.

Unfortunately, however important this relation between sectional curvature and the Riemann tensor might be, in practice it is still not easy to intuitively comprehend the curvature of a manifold, given that there are n *principal curvatures* (corresponding to the elements of $\{\mathsf{x}_h\}_{1 \leq h \leq n}$), and the symmetric polynomes (generalizing mean and Gauss curvature of surfaces) do not convey enough (and simple enough) geometric information. Moreover, even after symmetries reduce this to "only" $n^2(n^2 - 1)/12$, for $n \geq 3$ there are clearly to many numbers to deal with and a simple, geometric interpretation is highly needed. Furthermore, as we have seen in the case of surfaces, principal curvatures are not expressive enough to capture the geometry of the manifold, this is the case, a fortiori' for higher dimensional manifolds. Recall also that the equivalent of the curvature tensor is given by the curvature of sections, i.e. that produced by pairs of principal directions/curvatures. Thus we would like to have a more intuitive understanding of sectional curvature. Such a geometric interpretation is given by an analog (and slight generalization) of the Bertrand-Puiseaux formula (7.0.2) of Chapter 7, namely by:

Proposition 10.1.5 *Let M^n be as above. Denote by α the angle between geodesics starting from a common point p. Then:*

$$length \, C(p, \varepsilon, \alpha) = \alpha\varepsilon - \frac{\alpha}{3 \sin^2 \alpha} K(p)\varepsilon^3 + o(\varepsilon^3), \qquad (10.1.4)$$

where $C(p, \varepsilon, \alpha)$ denotes the arc of length α of $C(p, \varepsilon)$. In particular, for $\alpha = 2\pi$ (and $n = 2$) one gets the Bertrand-Diguet-Puiseaux formula in its classical form.

Clearly, the geometric content of the formula above can be interpreted as a measure of the rate of divergence of two geodesics that start from the same point, in directions making a given angle (see Figure 10.1.). One could codify the behavior of pairs of geodesics more precisely by using the Rauch Theorems 7.0.14 and 7.0.15, [256],

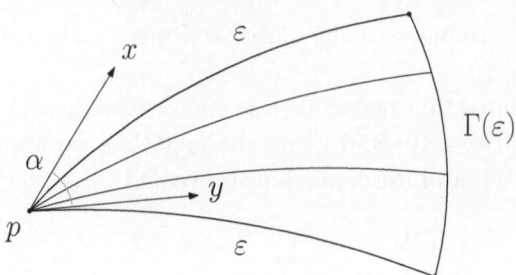

Figure 10.1 The geometric interpretation of sectional curvature ensuing from the higher dimensional version of the Bertrand-Puiseaux formula: Sectional curvature measures the divergence rate of two geodesics staring from a common point, as a function of the initial angle made by the geodesics. (After [36].)

Chapter 7. (Recall that, while formulated at that point for surfaces, they hold in any dimension, given that sectional curvature is just Gauss curvature of each 2-section and the theorems deal with pairs of geodesics only.)

Thus sectional curvature (and the curvature tensor) measure the defect of M^n from being locally Euclidean. This is done at the 2-dimensional level. More precisely, M^n is *flat* (i.e. locally Euclidean) iff $K \equiv 0$. In addition, if $K \equiv k_0$, where k_0 is a constant, then M^n is locally isometric to the simply connected space of constant sectional curvature. (Recall the special case of simply connected surfaces of constant Gauss curvature we discussed in Chapter 3.)

The divergence of geodesics property allows for natural extensions of the notion of sectional curvature to a variety of non-smooth spaces, in an approach quite different from the one adopted with Wald curvature, but still in the frame of the Comparison Geometry paradigm. The best known among these generalizations that allow to pass from a local notion to a global one is *Alexandrov curvature*, which became of one of the most influential concepts of modern Geometry, and to which we dedicated an Appendix A. An even more general approach, based on convexity properties of the distance function, is the one of Busemann [65].

Remark 10.1.6 *K behaves like a second derivative (or as a Hessian) of the metric g (see [37], p. 267, and the references therein.*

10.1.3 Ricci Curvature

Formally, the *Ricci curvature* is obtained by *contracting* the Riemannian curvature tensor:

Definition 10.1.7 *Let* $\mathbf{v} \in T_p M^n$ *be a unit vector. The* Ricci curvature *in the direction* $\mathbf{v}$ *is defined (in local coordinates) as:*

$$\mathrm{Ric}_{ij} = \sum_i R_{ijil} \tag{10.1.5}$$

It also follows that:

$$\mathrm{Ric}(\mathbf{v}) = \sum_{i=2}^{n} K(\mathbf{v}, \mathbf{x}_i) = \sum_{i=2}^{n} R(\mathbf{v}, \mathbf{x}_i, \mathbf{v}, \mathbf{x}_i), \qquad (10.1.6)$$

where $\{\mathbf{v}, \mathbf{x}_1, \ldots, \mathbf{x}_{n-1}\}$ represents an orthonormal base of $T_p M^n$.

Given the definition of Ricci curvature, the following fact is hardly surprising: Let $\langle \mathbf{v}, \mathbf{w} \rangle$ denote the plane spanned by $\mathbf{v}$ and $\mathbf{w}$. Then the following holds:

$$\mathbf{v} \cdot \mathrm{Ric}(\mathbf{v}) = \frac{n-1}{vol(\mathbb{S}^n - 2)} \int_{\mathbf{w} \in T_p(M^n),\ \mathbf{w} \perp \mathbf{v}} K(< \mathbf{v}, \mathbf{w} >); \qquad (10.1.7)$$

that is the Ricci curvature represents an average of sectional curvatures.

The geometric meaning of the Ricci curvature is encapsulated in the following version of the Bertrand-Puiseaux and Diguet formulas:

Theorem 10.1.8 ([161]) *Let M^n be as above. Denote by $d\alpha$ the n-dimensional solid angle in the direction of the vector $\mathbf{v} \in T_p(M^n)$ and by $\omega(\alpha)$ the $(n-1)$-volume generated by geodesics of length ε in $d\alpha$. Then:*

$$\mathrm{Vol}\big(\omega(\alpha)\big) = d\alpha\, \varepsilon^{n-1} \left(1 - \frac{\mathrm{Ric}(\mathbf{v})}{3} \varepsilon^2 + o(\varepsilon^2) \right). \qquad (10.1.8)$$

Given that we do not develop here the full necessary technical apparatus for the proof, we leave it to the highly motivated reader to study the original one in [161] (See Figure 10.2).

The essential role of Ricci curvature in control of volume growth is best expressed, probably, by the theorem below, which has become a classical result of Riemannian Geometry, which states that the volume of balls in a complete Riemannian manifold with a lower bound for Ricci curvature does not increase faster than the volume of balls in the model space form.

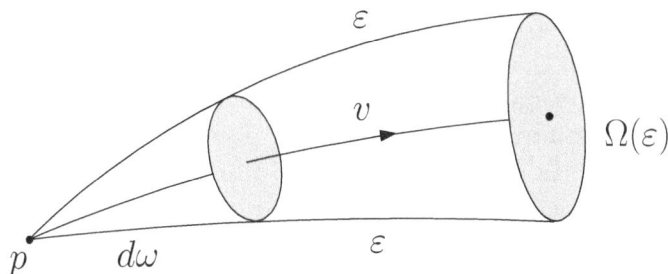

Figure 10.2 Ricci curvature as defect of the manifold from being locally Euclidean in various tangential directions (after [36]). Here $d\omega$ denotes the n-dimensional solid angle in the direction of the vector $\mathbf{v}$, $\Omega(\alpha)$ the $(n-1)$-volume generated by geodesics of length ε in $d\omega$, and $\mathrm{Ric}(\mathbf{v})$ the Ricci curvature in the direction $\mathbf{v}$.

Theorem 10.1.9 (Bishop-Gromov) *Let (M^n, g) be a complete Riemannian manifold satisfying* $\mathrm{Ric} \geq (n-1)k$. *Then, for any* $x \in M = M^n$, *the function*

$$\varphi(r) = \frac{\mathrm{Vol}B(x,r)}{\int_0^r S_K^n(t)dt}, \tag{10.1.9}$$

is nonincreasing (as function of r), where

$$S_K^n(r) = \begin{cases} \left(\sin\sqrt{\frac{K}{n-1}}r\right)^{n-1} & \text{if } K > 0 \\ r^{n-1} & \text{if } K = 0 \\ \left(\sinh\sqrt{\frac{|K|}{n-1}}r\right)^{n-1} & \text{if } K < 0 \end{cases} \tag{10.1.10}$$

The Gromov-Bishop Theorem is one of the fundamental results in Comparison Geometry. It is not only essential in the proof of deep results (see, for instance, [146] for such a result and [37] for an overview of its origins and extensions), but it also rends itself to generalizations – see [142], [328] (and the bibliography therein) – which we shall partially discuss very briefly in the sequel), which in turn allow us to obtain applicative results on triangulations of generalized manifolds [274] (see Chapter 13).

Remark 10.1.10 *The reader might very ask him/herself whether there is something not just special, but also perhaps restrictive about balls, and muse whether there might not be comparison results for volume growth of some other fundamental geometric objects. In particular, in view of the success we had in Chapter 4 using volume of tubes to compute the mean curvature of surfaces. Intuition is correct as such results do indeed exists – [137].[2] However, we would like to remind the reader that we already brought a higher-dimensional generalization of the tube formula in Theorem 4.1.23 of Chapter 4.*

Formula (10.1.8) also allows, as in the case of sectional curvature, as an interpretation of Ricci curvature as a measure of the dispersal of geodesics. However, while sectional curvatures quantifies the pairwise divergence of geodesics, Ricci curvature is concerned with their "sociology", as it measures the rate of dispersal of the geodesics staring at the same point and included in a solid angle in a given direction. (See Figure 10.3.) The intuition behind this is, essentially, the same with that for the Gaussian curvature: In positive curvature (of which the sphere is the golden standard), geodesics converge and their behavior is periodical, while in negative curvature (for which the Hyperbolic Plane is the epitome) they diverge *exponentially fast in this model case*; try and use the Bishop-Gromov Theorem to understand the behavior in the general case, using the dual geometric interpretation of Ricci curvature we just gave.

To summarize: *Ricci curvature represents a measure both of volume growth, as well as dispersal rate of geodesics.* This expressiveness is what allows Ricci curvature

---

[2]The title of this book says it all, as far the existence and importance of such results.

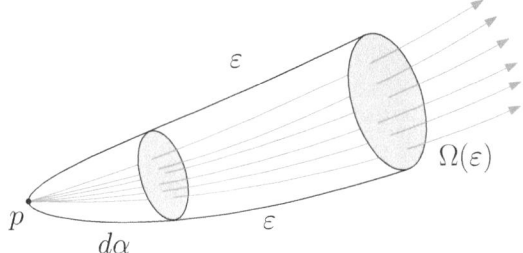

Figure 10.3 The geometric interpretation of Ricci curvature ensuing from the higher dimensional version of the Bertrand-Puiseaux formula: Ricci curvature measures the dispersal rate of (fascicles of) geodesics staring from a common point, as a function of the initial angle made by the geodesics.)

to model a large range of phenomena in a variety of setting, including discrete ones and represents the reason for its popularity and success in recent years, after being relegated for many years to the more technical recesses of Riemannian Geometry.[3]

The connection between Ricci curvature and geodesics dispersal is seen in yet another way, namely by considering the change of the element of volume as given by the Jacobian of the exponential mapping **J** along a Jacobi field (and given, precisely as in Chapter 6, by the Jacobi equation, with the curvature factor appropriately adapted – See Figure 10.4). We do not dwell deeper into this approach since its details would deviate us too far from the main route of this text and would also require more technicalities. We make thus due only with referring the reader to [178], as well as to [328], where it constitutes the basis on which a wide raging extension of the notion of Ricci curvature to much more general spaces than smooth manifolds is introduced (see also Chapter 13).

The counterpart of Remark 10.1.6 is:

Remark 10.1.11 *The Ricci curvature behaves as the Laplacian of the metric g (see [37], p. 267 and the bibliography therein).*

In fact, there is yet another deep connection between Ricci curvature and the Laplacian, which is codified through the so called *Bochner-Weizenböck Formula*

$$-\frac{1}{2}\Delta\left(||df||^2\right) = ||\mathrm{Hess}f||^2 - \langle df, \Delta df \rangle + \mathrm{Ric}(df, df). \qquad (10.1.11)$$

(Here $\mathrm{Hess}f$ denotes the Hessian of t f: $\mathrm{Hess}f = \nabla df = \nabla^2 f$ and $\langle \cdot, \cdot \rangle$ as usual, the inner product.)

Remark 10.1.12 *One unexpected consequence of the formula above is the fact, that the order two differential operator on the right side of the equation, equals an order three one, on the left side of the equation. Thus the ("proper") Laplacian captures*

---

[3]With the exception with its applications in Generalized Relativity and Cosmology via the so called *Regge Calculus*.

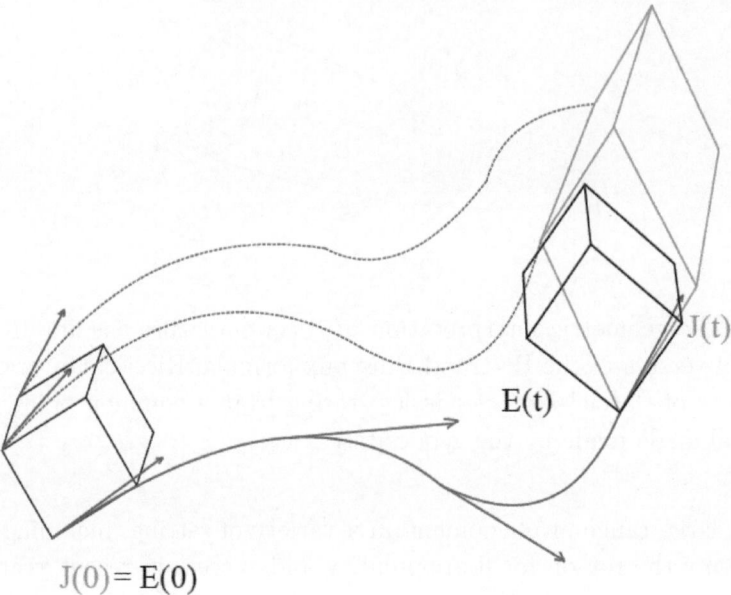

Figure 10.4 The geometric interpretation of Ricci curvature by considering Jacobi fields (after [328]): The volume of the Jacobian of the exponential mapping **J** (in gray) coincides at time $t_0 = 0$ with the elementary volume element (in black), which is preserved at time t by parallel transport, while **J** changes along a Jacobi field.

a higher order smoothness than expected! This is a surprising fact that might explain some of the very nice results that one obtains in practice (in Imaging, for instance) by using its application which we shall discuss at length in one of our future chapters.

While Formula 10.1.11 is, perhaps, the best known form of the Bochner-Weizenböck formula , there are other forms, better suited for applications (and easier to write and probably also understand). In particular, passing to differential forms proves to be beneficial, not just because it allows for a simpler expression, but also because it allows for extensions to the much larger class of weighted cell complexes (see Chapter 13).

In the language of differential forms, Formula (10.1.11) above has – for 1-forms (corresponding to functions) – the following form:

$$\frac{1}{2}\Delta\left(||\omega||^2\right) = ||D\omega||^2 - <\omega, \Delta\omega> + \mathrm{Ric}(\omega, \omega)\,. \qquad (10.1.12)$$

For forms of (higher) order p it admits a most compact and elegant expression as

$$\Box_p = dd^* + d^*d = \nabla_p^*\nabla_p + \mathrm{Curv}(R)\,, \qquad (10.1.13)$$

where $\Box_p$ is the *Riemann-Laplace operator* (or *Hodge Laplacian operator*) $\Box_p$, $\nabla_p^*\nabla_p$ is the *Bochner* (or *rough*) Laplacian and where $\mathrm{Curv}(R)$ is a complicated expression with linear coefficients in the *curvature tensor*. (Here ∇_p is, as before, the *covariant derivative* and d^* is the adjoint operator of d, thus $\Delta = dd^* + d^*d = (d+d^*)^2$.) The gist of this derivation is that, while in general the curvature term $\mathrm{Curv}(R)$ is, as we noted, quite abstruse, for $p=1$ it coincides with Ricci curvature. It is precisely this observation that is crucial in the development of a discrete Ricci curvature due to Forman [123], and which we shall explore in quite some detail in the sequel.

This is quite easily to interpret from a physical viewpoint: Imagine a metal plate heated by a candle flame at some point. The heat will propagate on the plate via the *heat kernel*, given by the Laplacian of the temperature function. However, clearly the heat will propagate on the plate with different speeds, according to the shape of the plate: It will take it more time to traverse "bumpy" or "wavy" regions than flat portions of the plate. Since 'bumpiness" is measured by curvature, it is clear that to understand heat propagation on such plate, we should "weigh" the standard (classical, Riemmanian, i.e. "flat") Laplacian by curvature. This is indeed the physical fact that Bochner-Weizenböck formula encapsulates.

However intuitive this results might be, they are hardly easy to prove[4] For a proofs, the determined reader should consults [178] (and, of course, the original works of Bochner [50] Weizenböck [339]). Note that, besides proofing the above statement, [178] emphasizes the role of Ricci curvature in the formula of the Jacobian determinant of the exponential map, thus underlining the role of Ricci curvature as a measure of growth – see also the discussion below.

Remark 10.1.13 *The connection between the heat kernel and curvature is deeper and more ramified than the one through the Bochner-Weizenböck formula. In particular, there is a deep connection between the asymptotic expansion of the heat kernel and various notions of curvature (including the introduced in this chapter). For an overview, the reader should read [37] (and, for more detail, the references cited there).*

Remark 10.1.14 *A detailed review of Ricci curvature and some of its discretizations can be found in [32]. It is more detailed than ours, as far as this specific section is concerned, and it might be used as a complement for certail aspects. However, we shall present a much more extended presentation of the main discretizations of Ricci curvature, besides presenting many more discretizations, for more classes of spaces.*

10.1.4   Scalar Curvature

Formally, *scalar curvature* is defined as the trace of the Ricci curvature:

Definition 10.1.15 *Let M^n be as above. Then the scalar curvature is defined as*

$$\mathrm{scal} = \sum_i \mathrm{Ric}_{ii}. \tag{10.1.14}$$

---

[4]This being in many instances the case: The more intuitive the result, the harder the proof, in inverse proportion.

It follows immediately from the definition that, for any orthonormal basis $\{\mathbf{x}_1, \ldots, \mathbf{x}_n\}$ of $T_p(M^n)$, the following holds:

$$\text{scal} = 2 \sum_{1 \leq i < j \leq n} K(<\mathbf{x}_i, \mathbf{x}_j>) = \frac{n(n-1)}{vol(\Pi)} \int_{\Pi \in \mathcal{P}} K(\Pi) ; \qquad (10.1.15)$$

where $\mathcal{P}$ represents the collection of 2-planes in $T_p(M^n)$.

This curvature also admits an analog of the second Bertrand-Diguet-Puiseaux formula:

Theorem 10.1.16 ([138], p. 166) *Let M^n, α, ω and ε like in the statement of Theorem 10.1.8. Then*

$$\text{Vol}\, B(p, \varepsilon) = \omega_n \varepsilon^n \left(1 - \frac{1}{6(n+2)} \text{scal}(p)\varepsilon^2 + o(\varepsilon^2)\right) ; \qquad (10.1.16)$$

where ω_n denotes the volume of the unit ball in $\mathbb{R}^n$.

That is, scalar curvature measures the defect of the manifold from being locally Euclidean at the level of volumes of small geodesic balls.

Note that, being (as it names makes quite clear) a scalar it easier to manipulate, but also it is less expressive than the other notions of curvature that precede it in this section. Indeed, it seems that between the perhaps excessive information that sectional curvature necessitates, and the lacunar one that scalar curvature can comprise, Ricci curvature strikes an ideal balance, between effectiveness and concisenesses, a fact that is reflected in the plethora of discretization it lends itself too, and which we shall explore in the following chapters.

We conclude this chapter with the following observation (which the reader probably made her/himself already):

Remark 10.1.17 *For surfaces the notions of sectional, Ricci and scalar curvature (essentially) coincide.*

CHAPTER 11

Higher Dimensional Curvatures 2

11.1 MOTIVATION

The material in this chapter might appear unpleasantly technical to the reader and certainly in contrast to the tone of most of the text so far. We could argue that, as one progresses with the subject matter, the discussion naturally becomes more technical and dry, which is of course true, as a rule. However, in this specific instance, this is not the case, at least not primarily, for the motivation behind the whole technical arsenal is one stemming from Physics, more specifically from General Relativity. Indeed, this entire direction of study stems from a – quite intuitive, by the way – paper of Tullio Regge [257] (see also [258]). His idea of producing a discrete, computationally feasible approach to Cosmology (determination of the solution of the *Einstein field equations*) via *PL* approximations has grown and become what is known as the *Regge Calculus*. The initial drive [74] of Cheeger, Müller and Schrader to solve this problem brought the to seek the mathematically general solution, which they achieved in [75]. To this end the made appeal to a generic class of curvatures, which represent the focus of our next section.

11.2 THE LIPSCHITZ-KILLING CURVATURES

We begin directly with the following technical definition:

Definition 11.2.1 *Given a Riemannian manifold M^n, the Lipschitz-Killing curvatures of M^n are defined as follows:*

$$R^j(M^n) = \frac{1}{(n-j)!2^j\pi^{j/2}(j/2)!} \sum_{\pi \in S_n} (-1)^{\epsilon(\pi)}\Omega_{\pi(1)\pi(2)} \wedge \cdots \wedge \Omega_{\pi(j-1)\pi(j)} \wedge \omega_{\pi(j+1)}\wedge$$

$$\wedge \cdots \wedge \omega_{\pi(n)}\,,$$

(11.2.1)

DOI: 10.1201/9781003350576-11

where $\Omega_{\pi(j-1)\pi(j)}$ are the curvature 2-forms and ω_{kl} denote the connection forms, and they are interrelated by the structure equations:

$$\begin{cases} d\omega_k = -\sum_i \omega_{kl} \wedge \omega_l \,, \\ d\omega_{kl} = -\sum_i \omega_{ki} \wedge \omega_{il} + \Omega_{kl} \,. \end{cases} \quad (11.2.2)$$

where $\{\omega_k\}$ is the dual basis of $\{e_k\}$. Moreover, the integral $\int_{M^n} R^j$ is also known as the integrated mean curvature (of order j).

Remark 11.2.2 *While the expression (11.2.1) above might appear daunting[1], in low dimensions Lipschitz-Killing curvatures are, in fact, quite familiar: $R^0 \equiv$ volume and $R^2 \equiv$ scalar curvature. Moreover, $R^n \equiv$ Gauss-Bonnet-Chern form, (for $n = 2k$).*

We also have the following expression of the Lipschitz-Killing curvatures:

$$R^j(M^n) = \frac{1}{\mathrm{Area}(\mathbb{S}^{n-j-1})} \int_{M^{n-1}} S_{n-j-1}(k_1(x), k_2(x), \dots, k_{n-1}(x)) d\mathcal{H}^{n-1} \,, \quad (11.2.3)$$

where $M^{n-1} = \partial M^n$, $d\mathcal{H}^{n-1}$ denotes the $(n-1)$-dimensional *Hausdorff measure*, and where the *symmetric functions* S_j are defined by:

$$S_j(k_1(x), k_2(x), \dots, k_{j-1}(x)) = \sum_{1 \leq k_{i_1} \leq k_{i_k} \leq j-1} k_{i_1}(x) \cdots k_{i_k}(x) \,, \quad (11.2.4)$$

$k_1(x), k_2(x), \dots, k_{n-1}(x)$ being the principal curvatures – see e.g. [214].

Remark 11.2.3 *Formula (11.2.3) above shows why the Lipschitz-Killing curvatures are also called the total mean curvatures (and the "S_j"-s are called the mean curvatures (of order j). It also suggests a quite direct method of obtaining a local (pointwise) version.*

In a similar manner (but technically slightly more complicated), one can define the associated boundary curvatures (or *mean curvatures*) H^j which are curvature measures on ∂M^n: Let $\{e_k\}_{1 \leq k \leq n}$ be an orthonormal frame for the tangent bundle T_{M^n} of M^n, such that, along the boundary ∂M^n, e_n coincides with the inward normal. Then, for any $2k + 1 \leq j \leq n$, we define

$$H^j = \sum_k \Omega_{j,k} \,, \quad (11.2.5)$$

where

$$\Omega_{j,k} = c_{j,k} \sum_{\pi \in S_{n-1}} (-1)^{\epsilon(\pi)} \Omega_{\pi(1)\pi(2)} \wedge \cdots \wedge \Omega_{\pi(2k-1)\pi(2k)} \wedge \omega_{\pi(2k+1),n} \wedge \cdots \wedge \omega_{\pi(j-1),n}, \omega_{\pi(j)} \wedge$$

$$(11.2.6)$$

$$\wedge \cdots \wedge \omega_{\pi(n-1)} \,,$$

---

[1]and rightly so!...

and

$$
c_{j,k} = \begin{cases} (-1)^k \left(2^j \pi^{\frac{j-1}{2}} k! \left(\frac{j-1}{2} - k \right)! (n-j)! \right)^{-1}, & j = 2p+1 \\ (-1)^k \left(2^{k+\frac{j}{2}} \pi^{\frac{j}{2}} k! (j-2k-1)! (n-j)! \right), & j = 2p. \end{cases} \tag{11.2.7}
$$

These curvatures measures are normalized by imposing the condition that:

$$
\int_{T^{n-j} \times M^j} R^j + \int_{T^{n-j} \times \partial M^j} H^j = \chi(M^j) \mathrm{Vol} T^{n-j}, \tag{11.2.8}
$$

for any flat T^{n-j}.

Remark 11.2.4 *As expected, in view of the similar elucidation of the Lipschitz-Killing curvatures, the low dimensional boundary curvatures also have quite familiar interpretations: $H^1 \equiv$ area boundary, $H^2 \equiv$ mean curvature for inward normal (as expected given the generic names for these H^j-s), etc.*

11.2.1   Curvatures' Approximation

11.2.1.1   Thick Triangulations

We now briefly present the solution of Cheeger, Müller and Schrader of the approximation of Lipschitz-Killing curvatures in PL approximation. The answer is, as we shall see, positive, with the proviso that one needs to use "good" PL approximations, that is precisely of the type that are know in Graphics as having a "high aspect ratio". Since such triangulations are essential in ascertaining the existence of a solution, but also for expressing and understanding its precise form, we bring below their technical definition, as given in [75]:

Definition 11.2.5 *Let $\tau \subset \mathbb{R}^n$; $0 \le k \le n$ be a k-dimensional simplex. The thickness φ of τ is defined as being:*

$$
\varphi = \varphi(\tau) = \inf_{\substack{\sigma \le \tau \\ \dim \sigma = j}} \frac{Vol_j(\sigma)}{\mathrm{diam}^j \sigma}. \tag{11.2.9}
$$

The infimum is taken over all the faces of τ (including τ itself), $\sigma \le \tau$, and $\mathrm{Vol}_j(\sigma)$ and $\mathrm{diam}\,\sigma$ stand for the Euclidian j-volume and the diameter of σ respectively. (If $\dim \sigma = 0$, then $\mathrm{Vol}_j(\sigma) = 1$, by convention.) A simplex τ is φ_0-thick, for some $\varphi_0 > 0$, if $\varphi(\tau) \ge \varphi_0$. A triangulation (of a submanifold of $\mathbb{R}^n$) $\mathcal{T} = \{\sigma_i\}_{i \in \mathbf{I}}$ is φ_0-thick if all its simplices are φ_0-thick. A triangulation $\mathcal{T} = \{\sigma_i\}_{i \in \mathbf{I}}$ is thick if there exists $\varphi_0 \ge 0$ such that all its simplices are φ_0-thick.

Remark 11.2.6 *The notion of thickness appears repeatedly, under many guises and in different manners (some more transparent than others) in many theoretical and applied settings, both in Mathematics, e.g. in Geometric Function Theory (Quasiregular Mappings) and Geometric Measure Theory, as well as Graphics and Imaging.*

The definition of thickness given above, as we have already mentioned, is that introduced in [75]. One reason for doing this is to preserve the "unity of style", so to say: Since we primarily present the results and techniques of [75], we find only proper to use, at least in the beginning, the same definition as that of Cheeger et al. However, they also prove in the same paper the following result that gives a more intuitive interpretation on the notion of thickness of simplices as a function of their dihedral angles in all dimensions:

Proposition 11.2.7 ([75]) *There exists a constant $c(k)$ that depends solely upon the dimension k of τ such that*

$$\frac{1}{c(k)} \cdot \varphi(\tau) \leq \min_{\sigma < \tau} \angle(\tau, \sigma) \leq c(k) \cdot \varphi(\tau), \tag{11.2.10}$$

and

$$\varphi(\tau) \leq \frac{Vol_j(\sigma)}{diam^j \sigma} \leq c(k) \cdot \varphi(\tau). \tag{11.2.11}$$

Here $\angle(\tau, \sigma)$ denotes the (*internal*) *dihedral angle*. While intuitively simple, the formal definition[2] of this notion requires some technical preliminaries:

Definition 11.2.8 *A simplicial cone $C^k \subset \mathbb{R}^k \subset \mathbb{R}^n$, is the set $C^k = \bigcap\limits_{j=1}^{k} H_j$, where H_j are open half spaces in general position, such that $0 \in H_j, j = 1, \ldots, k$. The set $L^{k-1} = C^k \bigcap \mathbb{S}^{n-1}$ is called a* spherical simplex.

Definition 11.2.9 *Consider the simplices $\sigma^k < \tau^m$, and let $p \in \sigma^k$. The normal cone is defined as $C^{\perp}(\sigma^k, \tau^m) = \{\overrightarrow{px} \mid x \in \tau^m, \overrightarrow{px} \perp \sigma^k\}$, where $\overrightarrow{px}$ denotes the ray through x and base-point p. The spherical simplex $L(\sigma^k, \tau^m)$ associated to $C^{\perp}(\sigma^k, \tau^m)$ is called the* link *of σ^k in τ^m.*

Remark 11.2.10 *$C^{\perp}(\sigma^k, \tau^m)$ does not depend upon the choice of p.*

We are now ready to formally define the notion of dihedral angle, as follows:

Definition 11.2.11 *The (internal) dihedral angle $\angle(\tau^k, \sigma^m)$ of σ in τ is the normalized volume of $L(\sigma^k, \tau^m)$, where the normalization is such that the volume of $\mathbb{S}^{n-1}$ equals 1, for any $n \geq 2$.*

Remark 11.2.12 *Note that the definition of thickness is hierarchical, in the sense that for a simplex to be thick, all its lower dimensional faces have to be thick. This is also transparent from condition 11.2.11 This is also transparent from condition 11.2.11 that expresses fatness as given by "large area/diameter" (or "volume/diameter") ratio.*

Remark 11.2.13 *Using the dihedral angle approach in assuring the "aspect ratio" of a triangular mesh is commonly used in Computational Geometry, Computer Graphics and related fields, to ensure that the constituting tetrahedra are not to "flat" or too "slim", in particular that no "slivers"' appear (see, e.g. [10], [41], [99], [114]) (see Figure 11.1). It also appears to be very promising in view of some quite recent developments (e.g. [8], [22], [103], [258]) of Regge Calculus.*

---

[2]For an alternative definition, see [305] IV. 2, IX. 15.

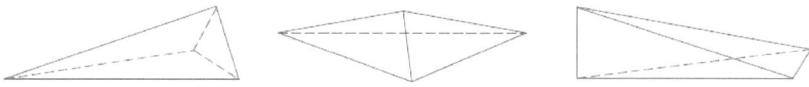

Figure 11.1 Three slivers [290].

11.2.1.2 Curvatures' Approximation Results

The approximation theory of Lipschitz-Killing curvatures (thus, by Remarks 11.2.2 and 11.2.4 of such basic quantities as volume, scalar curvature, mean (inward) curvature of the boundary) rests on the by now classical work of Cheeger et al. [75], the essential result therein being

Theorem 11.2.14 ([75]) *Let M^n be a compact Riemannian manifold, with or without boundary, and let M_i^n be a sequence of fat PL (piecewise flat) manifolds, that are secant approximations of M^n, converging to M^n in the Hausdorff metric. Denote by $\mathfrak{R}$ and $\mathfrak{R}_i$ respectively, the Lipschitz-Killing curvatures of M^n, M_i^n. Then $\mathfrak{R}_i \to \mathfrak{R}$ in the sense of measures.*

Remark 11.2.15   *1. One can thickly triangulate the smooth manifold M^n and obtain the desired approximation results for curvatures using directly and explicitly the intrinsic metric, not just PL (Euclidean) approximations [257], [75]).*

2. As noted above, Theorem 11.2.14 is given essentially in terms of the intrinsic geometry of M^n. A similar characterization of the curvature measures in terms of the extrinsic geometry (of embeddings in R^n) is given in [125].

3. In practice the condition that the triangulation necessarily becomes arbitrarily fine is, in fact, too strong if the manifold contains large flat regions. (The classical example, widely noted and exploited in Computer Graphics, is that of a round cylinder in $\mathbb{R}^3$.) Also, we have noted in [273], [290] the need and possibility of a triangulation with variable density of vertices, adapting to curvature. (Recall also that in [125] the hypothesis in the definition of fatness requiring that the mesh of the triangulation converges to zero is discarded).

Remark 11.2.16 *A similar approximation result for the so called* Einstein tensor,

$$E = \frac{\text{scal}}{2}g - \text{Ric};\qquad(11.2.12)$$

where g stands, as usual, for the Riemannian metric, was obtained by Bernig – see [2] and the references within.

The necessity of fat triangulations as a prerequisite for Theorem 11.2.14, is hardly surprising, in view of the characterization of fat triangulations as being precisely those triangulations having dihedral angles bounded from below (Proposition 11.2.7) and in view of the following expression of the Lipschitz-Killing curvatures in terms of dihedral angles (see [75] for the proof):

$$R^j = \sum_{\sigma^{n-j}} \left\{ 1 - \chi(L(\sigma^j)) + \sum_{l} \sphericalangle(\sigma^{n-j}, \sigma^{n-j+i_1}) \cdots \sphericalangle(\sigma^{n-j+i_{1-1}}, \sigma^{n-j+i_1}) \right. \quad (11.2.13)$$

$$\left. \cdot \left[1 - \chi(L(\sigma^{n-j+i_l})) \right] \right\} \mathrm{Vol}(\sigma^{n-j}),$$

(Similar, but more complicated expressions for the boundary curvatures H^j can also be written – see [75]. Note that $R^j|_{\partial M^n} = H^j$ and, in the case of PL manifolds, they represent the contribution of the $(n-j)$-dimensional simplices that belong to the boundary.)

One can extend Theorem 11.2.14 to many classes of unbounded manifolds if one can ensure the existence of thick triangulations, e.g. by using the following result:

Theorem 11.2.17 ([269]) *Let M^n be a connected, oriented n-dimensional ($n \geq 2$) submanifold of $\mathbb{R}^N$ (for some N sufficiently large), with boundary, having a finite number of compact boundary components, and such that one of the following condition holds:*

(i) M^n is of class C^r, $1 \leq r \leq \infty$, $n \geq 2$;

(ii) M^n is a PL manifold and $n \leq 4$;

(iii) M^n is a topological manifold and $n \leq 3$.

If the boundary components admit fat triangulations of fatness $\geq \varphi_0$, then there exist a global fat triangulation of M^n.

Remark 11.2.18 *In fact, the conditions on the compactness and boundedness of the boundary components in the theorem above are too strong, as indicated by the results in [272], [273], where the theorem above was shown to hold also for (hyperbolic) manifolds with infinitely many boundary components (as well as for more general spaces). The role of the conditions in question is to exclude certain "pathological" cases.*

Moreover, given the triangulation results of [229], [75] for manifolds without boundary, the following (important in the sequel) corollary follows immediately:

Corollary 11.2.19 ([269]) *Let M^n be a Riemannian manifold satisfying the conditions in the statement of Theorem 11.2.17 above.*

Then M^n admits a fat triangulation.

The immediate differential geometric consequence of Theorems 11.2.14 and 11.2.17, as well as Corollary 11.2.19 – where by "immediate" we mean here that it can be directly inferred by applying the methods of [75] – is the following result that can be considered, in a sense, as the "reverse" of the result of [75], Section 8, regarding the convergence of the boundary measures:

Theorem 11.2.20 *Let $N = N^{n-1}$ be a not necessarily connected manifold, such that $N = \partial M$, $M = M^n$, where M^n is, topologically, as in the statement of Theorem 11.2.17.*

(i) If M, N are PL manifolds, then the Lipschitz-Killing curvature measures of N can be extended to those of M. More precisely, there exist Lipschitz-Killing curvature measures $\mathfrak{R} = \{R^j\}$ on $\bar{M} = M \cup N$, such that $\mathfrak{R}|_N = \mathfrak{R}_N$ and $\mathfrak{R}|_M = \mathfrak{R}_M$, except on a regular (arbitrarily small) neighborhood of N, where $\mathfrak{R}_N$, $\mathfrak{R}_M$ denote the curvature measures of N, M respectively.

(ii) If M, N are smooth manifolds, then the same holds, but only in the sense of measures.

11.3 GENERALIZED PRINCIPAL CURVATURES

A different, quite natural, approach to the estimation of curvatures of smooth manifolds by the discrete curvature of PL (piecewise flat) approximations stems from Definition 2.2.12. Here one regards the relevant edges of a fine enough triangulation as the principal directions. Let us examine the feasibility of this method of approximating curvatures:

First, let us note that, as we already saw in Chapter 3, for PL manifolds this approach is somewhat naive, if applied directly, as it gives only very approximative results. (This is a consequence of the fine interplay between the necessity of ensuring the fatness of the triangulation, simultaneously with a good sampling of the direction in the tangential plane (or rather cone).) One can overcome this obstacle by using an extension of the notion of principal curvatures to a far larger class of geometric objects than mere smooth or even PL manifolds (see, e.g. [214]), by passing to the so called *generalized principal curvatures*. Unfortunately, in order that we may be able to present even very succinctly this tool stemming from Federer's classical work [120], we have to introduce additional of technical definitions and notations.

Let X be an arbitrary set in some $\mathbb{R}^N$ and let X_ε denote the ε-neighborhood of X. Then, for small enough $\varepsilon > 0$, ∂X_ε is a $\mathcal{C}^{1,1}$-hypersurface.[3] More precisely, ε has to be strictly smaller than the *reach* of X,

$$\text{reach}(X) = \sup\{r > 0 \,|\, \forall y \in X_r \,,\, \exists! \, x \in X \text{nearest to } y\}\,, \tag{11.3.1}$$

Moreover, the reach itself has to be strictly positive.

It follows that ∂X_ε admits principal curvatures (in the classical sense) $k_i^\varepsilon(x + \mathbf{n})$ at almost any point $p = x + \mathbf{n}$, where $\mathbf{n}$ denotes the normal unit vector (at x). Define the *generalized principal curvatures* by: $\kappa_i(\varepsilon, \mathbf{n}) = \lim_{\varepsilon \to 0} k_i^\varepsilon(x + \mathbf{n})$. Then $\kappa_i(\varepsilon, \mathbf{n})$ exist $\mathcal{H}^{N-1}$-a.a. $(x, \mathbf{n})$. (Here $\mathcal{H}$ denotes again the Hausdorff measure.)

Using this generalization of principal curvatures, one can retrieve a proper ana-

---

[3]In general, ∂X_ε is a $(N-1)$-dimensional manifold with Lipschitz outer unit normal field, for any $\varepsilon \in (0, \text{reach}(X))$, where the definition of *reach* is given in (11.3.1) – see [120].

logue of Formula (11.2.3), namely

$$C_j(X,B) = \int_{\mathrm{nor}(X)} \mathbf{1}_B \prod_{i=1}^{N-1} \frac{1}{\sqrt{1+\kappa_i(x,\mathbf{n})^2}} S_{N-1-j}(\kappa_1(x,\mathbf{n}),\cdots,\kappa_{N-1}(x,\mathbf{n})) d\mathcal{H}^{N-1}(x,\mathbf{n}),$$

(11.3.2)

where $C_j(X,B)$ denote the so called *Lipschitz-Killing curvature measures* (see [214] and the bibliography therein for details), B being a bounded Borel set in $\mathbb{R}^N$, and $\mathrm{nor}(X)$ denotes the (*unit*) *normal bundle* of X:

$$\mathrm{nor}(X) = \{(x,\mathbf{n}) \in \partial X \times \mathbb{S}^{N-1} \,|\, \mathbf{n} \in \mathrm{Nor}(X,x)\}, \tag{11.3.3}$$

where $\mathrm{Nor}(X,x) = \{\mathbf{n} \in S_{N-1}\,|\, <\mathbf{n},v> \leq 0, v \in \mathrm{Tan}(X,x))\}$ is the *normal cone* (to X at the point $x \in T$), dual to the *tangent cone* (to X at the point $x \in T$). Using the convergence properties of the generalized principal curvatures and of the Lipschitz-Killing curvature measures (again, see [214] and the bibliography therein), the desired result now follows easily.

Remark 11.3.1 *The idea of passing to a smooth (enough) surface close to the original given data set is a common one in Imaging and Graphics, even though the precise method presented above is not the standard one and appears, perhaps, to be rather technical. Moreover, this approach is similar in concept (even though based on a very different mathematical apparatus) to the one based on smoothings, approach that will be employed extensively in the sequel.*

The case of smooth manifolds follows easily by following the same scheme as in the main approach in [75].

11.4 OTHER APPROACHES

Besides the rather technical approaches we detailed above, and which are of great importance primarily (but not solely) on the theoretical side, there exist a number of other approaches, perfect easier to implement in applications and some which were devised precisely toward such ends. We briefly present below an assortment of such method.

11.4.1 Banchoff's Definition Revisited

The first, both in time and probably also in mathematical importance and relevance for applications, among the alternative approaches, is that of Banchoff, which we presented in detail and employed in Chapter 4. As we mentioned there, Banchoff has generalized his methods to higher dimensional manifolds. (In fact, his more general paper [25] precedes the 2-dimensional case one [26]). In the extended setting, scalar curvature $K(v)$ is defined at any vertex of an m-dimensional *convex cell complex* embedded in some $\mathbb{R}^n$. While, to be sure, the extended, earlier paper considers various generalizations, the essential definition is precisely the one we introduced in our chapter on the Theorema Egregium, the single difference being that the integral in the definition is taken over $\mathbb{R}^{n-1}$, instead of $\mathbb{R}^2$, thus we do not repeat it here.

It should be noted that Bloch [44], [45] extended Banchoff's work to a wider class of geometric objects. While certainly most interesting, we cannot present his ideas here since they would demand too many technical intricacies beyond the scope of this book.

While clearly the 2-dimensional version which we presented at length in Chapter 4 is implementable and, indeed, of clear importance in applications, the general case and not less Bloch's generalization are also relevant for the Persistent Homology of networks [268].

11.4.2 Stone's Sectional Curvature

Another early essay in the discretization of curvatures of higher dimensional structures is that of Stone [313], [316]. His approach is clearly influenced by ideas and techniques of PL Differential Topology, rather than Geometry (a fact that is also transparent in his other relevant work on which we shall dwell in Chapter 12). The definition he suggested is not dissimilar in spirit to the ones of [1] and [15] which we discussed at quite some length in Chapter 6) (but clearly precedes them[4]).

Stone's definition of PL sectional curvature is motivated, evidently, by the problem of geodesics in PL manifolds and more general objects. (Indeed, he first introduces the notions of *cut locus* and *open cut locus* on such structures.) We present his ideas only briefly and rather informally. The essential definition, from our point of view here, is his introduction of two curvatures, k_+ and k_- which are defined (rather informally here) as follows:

Definition 11.4.1 *Let X be a connected, locally finite simplicial complex linearly embedded in some Euclidean space $\mathbb{R}^N$. Then for and let $p \in X$ and any given tangential direction (ray) $\overrightarrow{px}$[5], one defines the following two numbers:*

$$k_+(\overrightarrow{px}) = 2\pi - \min\{\measuredangle xpy \,|\, \overrightarrow{py} \text{ tangential at } p\}\,; \tag{11.4.1}$$

and

$$k_-(\overrightarrow{px}) = 2\pi + \max\{\measuredangle xpz \,|\, \overrightarrow{pz} \text{ tangential at } p\}\,; \tag{11.4.2}$$

which are called the maximum, *respective* minimum *curvatures of X, at p, in the direction $\overrightarrow{px}$.*

Remark 11.4.2 *1. If the cut locus of p (which is defined precisely like for surfaces, since this is in fact a purely metric notion) is void, then the curvatures are defined as $k_+(\overrightarrow{px}) = k_-(\overrightarrow{px}) = 0$.*

This holds, in particular if X -s an n-manifold without boundary and when $p \in \sigma^m$, where σ^k is an m dimensional face, $m = n - 1$ or $m = n$.

---

[4]Sadly enough, his work seems to be forgotten – and certainly overlooked – by practitioners in the field if Graphics and Imaging. Part of the reason certainly is due to the fact that his work published in the 70s, was certainly avant la lettre, as it predates by a large margin the true advent of the fields in question. Another reason might be a certain technicality in his writing style, which perhaps made his papers less accessible to parts of the Computer Science community.

[5]i.e. $\overrightarrow{px}$ is included in the *tangential cone* of X at p (rather than the tangential plane, as four surfaces/manifolds)

2. $k_+(\overrightarrow{px}) \geq k_-(\overrightarrow{px})$.

These are indeed proper adaptations to simplicial complexes of the notion of sectional curvature, since for manifolds, there is a correlation between k_+ (k_-) and the maximum (minimum) sectional curvature at p in a tangent direction $\overrightarrow{px}$. This connections are proven by a number of theorems that parallel their equivalents in the classical setting (but which we shall not bring here to avoid the discussion to become overly lengthy and technical). Furthermore, if X is a 2-manifold (i.e. *PL* surface), $k_+(\overrightarrow{px}) = k_-(\overrightarrow{px})$ and they equal the classical combinatorial (defect) curvature $K(p)$.

Remark 11.4.3 *In general $k_+(\overrightarrow{px}), k_-(\overrightarrow{px})$ depend on the direction $\overrightarrow{px}$, as expected from a proper generalization of sectional curvature.*

11.4.3   Glickenstein's Sectional, Ricci and Scalar Curvatures

Motivated partly by earlier work by Cooper and Rivin [94], [261] on the *packing metric*, as well by a classical paper of Cheeger [77], Glickenstein proposed [136], the following extensions of the classical defect curvature for piecewise flat surfaces[6] that we have first encountered in Chapter 4:

Definition 11.4.4 *Let $(M, \mathcal{T}, \ell)$ be a 3-dimensional piecewise flat manifold, where M denotes the underlaying space, $\mathcal{T}$ its triangulation and ℓ the metric defined by the lengths of the edges, and let $e_{ij} = v_i, v_j$ be an edge of $\mathcal{T}$, connecting the vertices v_i and v_j. Then the edge curvature of e_{ij} is defined as*

$$K_{ij} = K(e_{ij}) = \left(2\pi - \sum_{k,l} \beta_{ij,kl} \right) \ell_{ij} ; \qquad (11.4.3)$$

where $\beta_{ij,kl}$ denotes the dihedral angle at the edge $e_{i,j}$ of the tetrahedron $T(v_i, v_j, v_k, v_l)$, $\ell_{ij} = \ell(e_{ij})$.

Given our intuition developed with geodesics, mean curvature and triangular meshes, it is easy to convince ourselves that

$$\frac{K_i}{\ell_{ij}} = 2\pi - \sum_{k,l} \beta_{ij,kl}$$

can be interpreted (cf. [257], see also [224]) as the measuring the defect in the parallel transport along (or "around") the edge e_{ij}. In consequence, one can view $\frac{K_i}{\ell_{ij}}$ as an analog, for piecewise flat 3-manfiolds, of sectional curvature or even of the Riemann curvature operator.

Remark 11.4.5 *The dihedral angles $\beta_{ij,kl}$ can be computed easily from the lengths $\ell_{ij}, \ell_{ik}, \dots$ etc.using a combination of Euclidean and spherical trigonometry.*

---

[6]known to the reader through their avatar as triangular meshes

However, remembering that edges correspond to vectors (directions) in the smooth setting, and that to directions one attaches Ricci curvature, it is natural to also give a fitting definition for Ricci curvature:

Definition 11.4.6 *Let $(M, \mathcal{T}, \ell)$ and e_{ij} as above. Then the Ricci curvature of the edge e_{ij} is defined as*

$$\mathrm{Ric}(e_{ij}) = K_{ij} \,. \tag{11.4.4}$$

Remark 11.4.7 *Interestingly enough, Glickenstein does not make this definition explicit in his papers, even though he defines Ricci flat piecewise flat manifolds by requiring that*

$$K_{ij} = 0 \,, \text{ for all edges } e_{ij} \,.$$

Thus making it clear that K_{ij} is view en lieu of $\mathrm{Ric}(e_{ij})$. (In fact, he strengthens this by accordingly defining Einstein manifolds).

The definition of scalar curvature is now immediate:

Definition 11.4.8 *Let $(M, \mathcal{T}, \ell)$ and let v_i ve a vertex of $\mathcal{T}$. Then the* scalar curvature *of the 3-dimensional piecewise flat manifold is defined as*

$$K_i = K(v_i) = \sum_j \left(2\pi - \sum_{k,l} \beta_{ij,kl} \right) \ell_{ij} = \sum_j K_{ij} \,. \tag{11.4.5}$$

Note that Glickenstein offered, in [134],[135] yet another definition of scalar curvature, again motivated by the *sphere packing metric*:

Definition 11.4.9 *Let $T_{iijkl} = T(v_i, v_j, v_k, v_l)$ be the tetrahedron with vertices v_i, v_j, v_k, v_l, and let α_{ijkl} denote the solid angle of T_{iijkl} at the vertex v_i. Then the scalar curvature at the vertex v_i by*

$$K_i = K(v_i) = 4\pi - \sum T_{i,j,k,l} \alpha_{ijkl} \,; \tag{11.4.6}$$

where the sum is taken over all the tetrahedra with v_i as a vertex.

(Recall that the solid angle at a vertex v_i equals the area of spherical triangle on the unit sphere with center at v_i, determined by the planes $\pi(v_i, v_j, v_k)$, $\pi(v_i, v_j, v_l)$ and $\pi(v_i, v_k, v_l)$.)

Remark 11.4.10 *Note the sum on the ride side of 11.4.6 is symmetric in j, k, l.*

Exercise 11.4.11 *Using elementary spherical trigonometry (see e.g. [305], [38]), compare the two notions of scalar curvature presented above.*

Glickenstein's curvatures seem ideal for applications in Graphics and Imaging (e.g. for 3-D Tomography), yet sadly we know of no such uses.

11.4.4   The Ricci Tensor of Alsing and Miller

A rather recent alternative approach, focusing on practical computations and also motivated by the Regge Calculus is that of Alsing, Miller and their collaborators [8], [9], [224]. They in their thorough program the introduced simplicial Riemann and Ricci tensors, as well as Ricci scalar and sectional and scalar curvatures. While we do not present here the intricacies of their simplicial tensors and, moreover, we shall return in the next chapter and discuss in more detail their Ricci curvature and the associated flow, we very briefly bring below their definition of the sectional, Ricci and scalar.

Definition 11.4.12 *Let $e = \{v_1, v_2\}$ an edge in a PL simplicial complex. Then the three essential types of curvature are defined as*

1. The sectional curvature of e is defined as

$$K_e = \sum_{e_{v_{1,i}}, e_{v_{2,j}} \sim e} \frac{1}{2} \left(\frac{\cos^2 \theta}{A_{e_{v_1}}} \varepsilon_{e_{v_1}} + \frac{\cos^2 \theta}{A_{e_{v_2}}} \varepsilon_{e_{v_2}} \right) ; \qquad (11.4.7)$$

where the sum is taken over all the edges that share a common vertex with e (see Figure 11.2); and where A_e denotes the dual Voronoi area [257], [9], [224] of the edge e, ε_e represents the deficit areas [257], [9], [224] of edge e, and θ_i is the angle between the edges e and e_i (again, see Figure 11.2).

2. The Ricci (scalar)[7] curvature of e is given by

$$R_e = \frac{1}{2} (R_{v_1} + R_{v_2}) ; \qquad (11.4.8)$$

where R_{v_1}, R_{v_2} denote the scalar curvatures of the vertices v_1, v_2, respectively, and where

3. The scalar curvature of a vertex v is defined as

$$R_v = \frac{1}{V_v} \sum_{e \sum v} \ell(e) \varepsilon_e ; \qquad (11.4.9)$$

where V_v is the dual volume [257], [9], [224] associated with the vertex v and $\ell(e)$ denotes, as before, the length of edge e.

11.4.5   The Metric Approach

There are at least two radical routes toward a metrization of curvatures in higher dimension. The first one represents a direct continuation of the approach we presented at the beginning of this chapter, namely the metrization of the Lipschitz-Killing curvatures.

---

[7]as it is called by Alsing and Miller

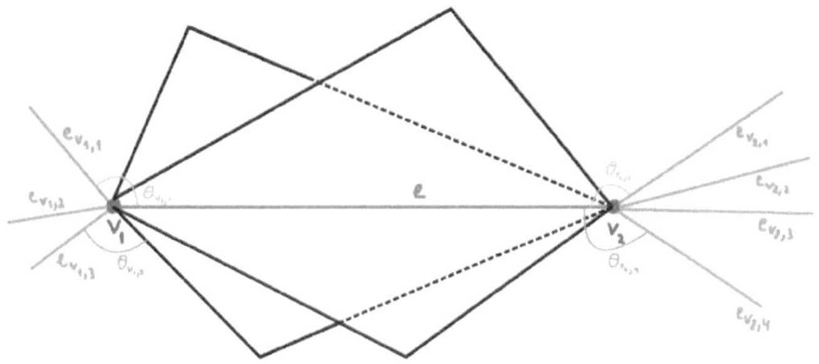

Figure 11.2 The edges and angles appearing in the definition of Alsing and Miller's simplicial curvatures.

11.4.5.1 Metrization of the Lipschitz-Killing Curvatures

Given our extensive exploration of the metric approach to curvature, it is only natural to seek for metric versions of the curvatures of higher dimensional manifolds. In fact, this quest is spurred also by the discretization of the Lipschitz-Killing which opened this chapter: Given that Regge's drive was to find a purely metric (discrete) formulation of Gravity, the presence of angles in the Lipschitz-Killing curvatures is a bit "unaesthetic". Hence one has to ask him/herself whether there is a formulation of the Cheeger, Müller and Schrader result and its consequences in purely metric terms?

One obstruction would seem to be the fact that, the definition of fatness is metric, the thickness condition (11.2.7) one actually uses in the proofs is not expressed in metric terms, but rather through angles. However, this is an obstacle easy to overcome. The solution (or, rather, solutions) represents, however, a somewhat longer aside, thus we have relegated it to a dedicated appendix. having settled this aspect, it is now possible to translate all the steps of the proofs and formulas in Subsection 11.2.1 into purely metric terms, for instance the reader might like to solve the following:

Exercise 11.4.13 *Express Formula (11.2.13) in metric terms only, (i.e. without making appeal to angles).*

One possible approach to solve this problem is, evidently, to make appeal yet once again to Formula (11.2.3) to express the curvature at a vertex by principal curvatures (which is easy – and natural – to determine for a piecewise flat surface). One can again make appeal to the methods employed in Chapter 3 to show that for piecewise flat manifolds, using metric curvatures (Menger and/or Haantjes) one can approximate well (in fact: as well as desired) smooth curvatures. However, as we have seen, these are, even in the case of surfaces, approximations with quite limited convergence properties only. The problem is that, in order to ensure convergence for the Cheeger et al. process, the triangulation has not only to converge *in mesh* to 0, it also has to remain thick. On the other hand, to ensure a "good sampling" of the directions on a surface (so to ensure good approximation of the principal curvatures), one necessarily has to produce samplings whose edges directions are arbitrarily dense in the tangent plane (cone) at a given point (vertex), in contradiction with the previous fatness constraint. (Note, however, that this problem does not exist if one is willing to settle for approximate results, correct up to a predetermined error.) Unfortunately, given the fact that, as we saw above, one cannot increase as desired the number of triangles adjacent to a vertex, there exist only a very limited number of directions to "choose" from. Thus one is quite restricted in adding new directions (thus to better "sample" the surface, so to say) without negatively affecting the thickness of the triangulation.[8]

There are a number of approaches to solve this problem, at least in practice:

- "Mix" the angles in the manner described in detail in [272], to obtain angles whose measure is close to their mean (i.e. $\pi/\deg(v)$ – where $\deg(v)$ denotes the number of triangles adjacent to the vertex v. (This "trick" works if a smoothness condition (albeit, minimal) is imposed on the manifold – it certainly holds for triangulated surfaces.)

- Add directions by considering the *PL-quasi-geodesic* and "normalize" (by projection on the normal plane) – see [287] and the references therein for details and some numerical results.

- The approximation of principal curvatures being, as already noted above, notoriously difficult – see [217] [318], [155], [99] – one seeks other, perhaps less direct strategies. Such a method does, in indeed exist, embodied by the generalized principal curvatures, as we have detailed above. Passing to smooth surfaces allows for the use of a wide scale of well developed and finely honed methods of Graphics and related fields. Furthermore, it clearly compensates for its departure from the given (data) set of discrete/geometric *PL* object by its generality whence its applicative potential in a very general setting.

Remark 11.4.14 *As we have seen above, we can compute the principal curvatures of S via those of the smoother surface S_ε, at least up to some infinitesimal distortion. However, to determine the full curvature tensor, in the case of higher dimensional manifolds, suffices to determine the Gauss curvature of all the 2-sections (see observation above). So, for the full reconstruction of the curvature tensor it is not necessary to determine the principal curvatures, suffices to find the sectional curvatures. Thus the question arises whether it is possible to determine other important curvatures (e.g. Ricci and scalar) as well as the Lipschitz-Killing curvatures, via the sectional curvatures, without appeal to principal curvatures. One such line of attack of this problem is clearly outlined in Chapter 3. Furthermore, we have tackled this problem in [147] and gave a partial, possible solution and some of its consequences, which we shall present in the next chapter.*

Given the lengthy and somewhat "brute force" (largely imposed upon us both by the basic mathematical method and by computational constraints), it is natural to

---

[8]In practice (Graphics, etc.) even angles $\pi/12$ are already problematically small!...

ask the question whether it is possible to compute the Lipschitz-Killing curvatures, starting from the defining Formula (11.2.1), that is if one can compute the necessary curvature 2-forms and connection 1-forms. The answer seems to be positive, even though, till recently this was only a mainly theoretical possibility (see [307]). However, after the appearance of the computational exterior Differential Calculus, introduced by Gu [148] and Gu and Yau [149], [150] and embraced and developed since then by many others, this approach appears quite feasible, at least in dimensions 2 and 3.

11.4.5.2   A Metric Gauss-Bonnet Theorem and PL Curvatures

We briefly sketch below a different strategy of defining a notion of Ricci curvature for PL and more general polyhedral manifolds that we introduced in [283]. It is based on Haantjes curvature in its role of geodesic curvature and a new, non-metric ingredient, namely the discretization of the classical Local Gauss-Bonnet Theorem we discussed in Chapter 9.

To begin with, less us notice that, in the case of piecewise flat manifolds, the simplest and most direct approach is to view, for each triangle T adjacent to an edge e, its *Menger curvature* $\kappa_M(T)$ as the sectional curvature of the section (plane) T. Since the Menger curvature represents a measure of the "flatness" of a triangle, $\kappa_M(T)$ gives a measure for the spreading of geodesics in the plane T, thus it provides, indeed, a discrete sectional curvature. Equipped with this metric version of sectional curvature, and using the same notations as in [147], it is immediate to obtain the fitting versions of Ricci and scalar curvatures:

$$\mathrm{Ric}_M(e) = \sum_{T_e \sim e} \kappa_M(T_e), \tag{11.4.10}$$

where $T_e \sim e$ denote the triangles adjacent to the edge e; and

$$\mathrm{scal}_M(v) = \sum_{e_k \sim v} \mathrm{Ric}_M(e_k) = \sum_{T \sim v} \kappa_M(T); \tag{11.4.11}$$

where, $e_k \sim v, T \sim v$ stand for all the edges adjacent to the vertex v and all the triangles T having v as a vertex, respectively. (See Figure 11.3) Note that $\mathrm{Ric}_M(e)$ captures, in accordance to the intuition behind $\kappa_M(T)$ the geodesic dispersion rate aspect of Ricci curvature.

While this manner of extending a simple metric curvature to the context of piecewise flat manifolds (and, in fact, of PL, in general), is quite direct and intuitive, it has also two limitations that hinder its usefulness in the study of such manifolds. The first such impediment is the fact that, being a discretization of the curvature notion for planar curves, Menger curvature of triangles is, intrinsically, always positive. Therefore, it can represent only a discretization of the absolute value of sectional curvature (hence of Ricci and scalar curvatures as well). The second obstruction in extending this approach to more general types of meshes/manifolds resides in the fact that, Menger curvature, by its very definition, is restricted solely to triangles, thus is not applicable to more general cells.

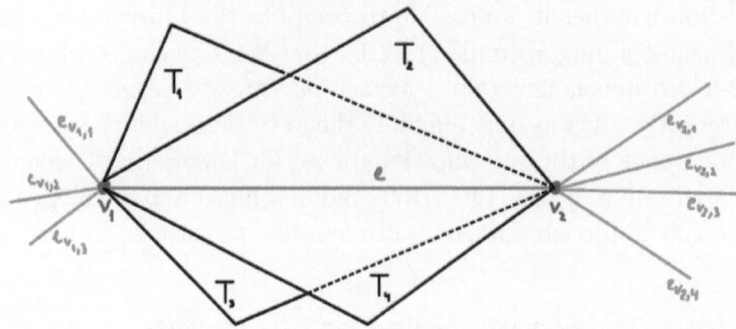

Figure 11.3 The geometric setting of the Menger-Ricci curvature of an edge.

However, making instead appeal to Haantjes curvature provides us with a solution for both problems. The first one is solved trivially, by the very definition of Haantjes curvature. The second problem that we singled out above also gets is resolution, as we have seen in Chapter 9. Thus one can define the metric equivalents of sectional, Ricci and scalar curvatures, by making the simple observation (first made by Stone) that, in PL manifolds – which should be noted that are not necessarily piecewise flat – 2-cells play the role of two sections, thus Gauss curvature of 2-cells is identified with the sectional curvature of the cell (PL section). Therefore, equipped with the definition and notations of the Chapter 9, we can now easily adapt Formulas (11.4.10) and (11.4.11) above and define (using what we called in Chapter 9 the strong Haantjes-Gauss curvature)

$$\mathrm{Ric}_H = \sum_{\mathfrak{c} \sim e} \kappa_{H,e}(\mathfrak{c}), \tag{11.4.12}$$

where the sum is taken over all the 2-cells $\mathfrak{c}$ adjacent to e; and

$$\mathrm{scal}_H(v) = \sum_{e_k \sim v} \mathrm{Ric}_H(e_k) = \sum_{\mathfrak{c} \sim v} \kappa_H(\mathfrak{c}); \tag{11.4.13}$$

that is by summing the *Haantjes-Ricci curvatures* of all the edges or, alternatively, the *Haantjes-scalar curvatures* of all the two cells adjacent to v.

We can provide yet a further generalization, by noting that in the discretization of Haantjes curvature one is not restricted to use a single edge as a discretization of the chord, and that, we can consider, in fact, any two vertices u, v that can be connected by a path π. Among the simple paths $\pi_1, \ldots, \pi_m$ connecting the vertices, the shortest one, i.e., the one for which $l(\pi_{i_0}) = \min\{l(\pi_1), \ldots, l(\pi_m)\}$ is attended represents the *metric segment* of ends u and v. Therefore, given any two such vertices, we can define the Haantjes-Ricci curvature in the direction $\overline{uv}$ to be

$$\mathrm{Ric}_H(\overline{uv}) = \sum_{1}^{m} K_{H,\overline{uv}}^i = \sum_{1}^{m} \kappa_{H,\overline{uv}}(\pi_i) \tag{11.4.14}$$

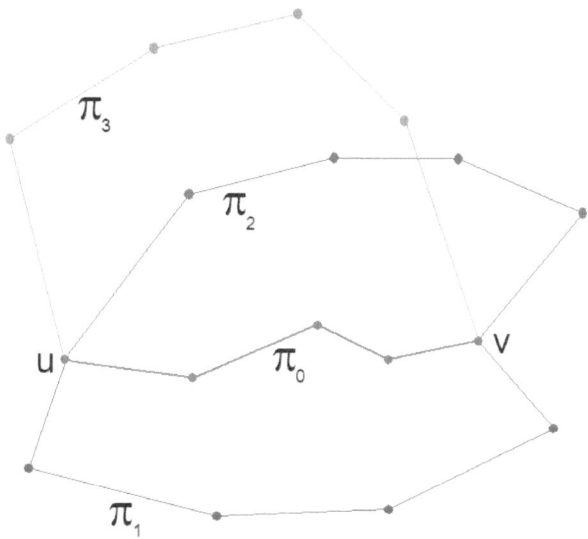

Figure 11.4 The geometric setting of the Haantjes-Ricci curvature of a path.

where $K^i_{H,\overline{uv}}$ denotes the Haantjes-Ricci curvature of the cell $\mathfrak{c}_i$, where $\partial \mathfrak{c}_i = \pi_i \pi_0^{-1}$, relative to the direction $\overline{uv}$, and where

$$\kappa^2_H(\pi_i) = \frac{l(\pi_i) - l(\pi_0)}{l(\pi_0)^3} \ , \tag{11.4.15}$$

and where the paths $\pi_1, \ldots, \pi_m$ satisfy the condition that $\pi_i \pi_0^{-1}$ is an *elementary cycle*. This represents a locality condition in the network setting. (See Figure 11.4.)

Remark 11.4.15 *We shall return to these ideas in the next chapter, when we will explore their extensions to weighted networks.*

We conclude this chapter with the following remark: While the basic motivation for proposing various types of discrete curvatures for PL manifolds stems from the Regge Calculus, that is to say from Theoretical Physics, there also arouse a need for these types of curvatures into more "mundane", applicative domains. Indeed, in Computer Graphics, Computer Aided Geometric Design, Imaging, etc. it has become customary lately to compute so called *volumetric curvatures* (see, e.g. [236], [150] and the bibliography therein). In most instances, this amounts, in fact, to the computation of the curvatures (Gauss and mean) of surfaces evolving in time. We take this opportunity to note that, while surely this approach has its merit and applications, a proper "volumetric", i.e. 3-dimensional curvature (, would entail the computation of a namely *sectional*, *scalar* and *Ricci* curvatures.[9] Clearly, the approaches to the

---

[9]In all fairness, we should add though that, for Computer Graphics, where usually the data is already embedded in $\mathbb{R}^3$, thus endowed with the Euclidean (flat) geometry of the ambient space, such computations are, therefore, rather meaningless.

computation of PL (and polyhedral) manifolds considered above are more relevant in practice for the applicative fields considered above. This is particularly true for 3-dimensional manifolds, with further emphasis on the types of cube grids (red "voxels") traditionally employed in CT and MRI Medical Imaging. We shall thus further explore meaningful notions of curvatures for such volumetric images in the next chapter.

CHAPTER 12

Discrete Ricci Curvature and Flow

Until quite recently, Ricci curvature has been quite a specious notion in the field of Differential Geometry. An exception, quite important but hardly "visible" from the viewpoint of Computer Science, is that, already mentioned, of its role in Cosmology and General Relativity, in the garb of Einstein manifolds and Regge Calculus. However, this has changed with the emergence of G. Perelman's far reaching work on Hamilton's Ricci flow and the Geometrization Conjecture [247], [248]. While started far earlier [314], the search for the discretization of this notion has recently gained fast momentum, spurred by Perelman's groundbreaking results. Some of these discretization have already become mainstream tools in Computer Science, Imaging and Networking. We overview a number of them below (some will be presented in the following chapter). The order of our exposition is "from geometric to discrete", so to say: We start with the case closest to the classical setting, namely that of n-dimensional PL manifolds, then pass to the 2-dimensional, combinatorial version of the Ricci curvature and flow, proceed to the reduction to the 1-dimensional case of networks, beginning with a metric version and concluding with the broadened definition due to Ollivier [238], [239].

12.1 PL MANIFOLDS – FROM COMBINATORIAL TO METRIC RICCI CURVATURE

Given not only their function as basic geometric structures in Graphics and, no less, their role as the very foundation of a whole – and important – branch of Topology, namely the homonymous PL (and Combinatorial) Topology, which represents one of the mathematical roots from which Discrete Differential Geometry has sprung, PL manifolds represent the very first and most natural generalization of smooth manifolds on which one would seek to define a notion of discrete Ricci curvature. Given this ontogeny and his work similar in spirit that we already cited in Chapters 7 and 11, it is not truly surprising that the first discretization of Ricci curvature is also due to Stone [314], [315]. His discretization was a purely combinatorial one. Inspired by Stone's pioneering work the authors also proposed in [147] a concept of metric Ricci curvature, thus extending Stone's original one and bridging the gap between the classical, smooth setting and the restricted, combinatorial one. However, our definition is not based on Stone's one, as such, but rather rests on his adaptation of the notion of Jacobi fields (see Chapter 7). In consequence, we do not first detail Stone's version of Ricci curvature, but rather proceed directly to the metric one and present his, for comparison, in due time. Before proceeding on, let us apprise the reader that we presents the results in some detail, as we see them also as a recap of many of the ideas and techniques of Metric Geometry we encountered throughout the book.

12.1.1 Definition and Convergence

To begin with, and before trying to prove any (expected/desired) of its properties, we have to be able to properly define Ricci curvature for PL manifolds. As already mentioned above, This is indeed possible, not just for PL manifolds but also for more general polyhedral ones – and in a quite natural manner – combining ideas of Stone [314], [315] and metric curvatures. For this one makes appeal to the definition of Ricci curvature as the mean of sectional curvatures:

$$\mathrm{Ric}(e_1) = \mathrm{Ric}(e_1, e_1) = \sum_{i=2}^{n} K(e_1, e_i), \qquad (12.1.1)$$

for any orthonormal basis $\{e_1, \ldots, e_n\}$, and where $K(e_1, e_j)$ denotes the sectional curvature of the 2-sections containing the directions e_1.

First, one has, of course, to be able to define (*variational*) *Jacobi fields* (see below). This is where we rely upon Stone's work. However, we do not need the whole force of this technical apparatus, only to determine the relevant 2-sections and, of course, to decide what a direction at a vertex of a PL manifold is.

In fact, in Stone's work, combinatorial Ricci curvature is defined both for the given simplicial complex $\mathcal{T}$, and also for its *dual complex* $\mathcal{T}^*$. In the later case, cells – playing here the role of the planes in the classical setting of which sectional curvatures are to be averaged – are considered. However, his approach for the given complex, where one computes the Ricci curvature $\mathrm{Ric}(\sigma, \tau_1 - \tau_2)$ of an n-simplex σ in the direction of two adjacent $(n-1)$-faces, τ_1, τ_2, is not natural in a geometric context (even if useful in his purely combinatorial one), except for the 2-dimensional case, where it coincides with the notion of Ricci curvature in a direction (i.e., in this case, an edge – see also Remark 12.1.6 below). Passing to the dual complex will not restrict us, since $(\mathcal{T}^*)^* = \mathcal{T}$ and, moreover – and more importantly – considering *thick* triangulations enables us to compute the more natural metric curvature for the dual complex and use the fact that the dual of a thick triangulation is thick, as we shall detail below. Working only with thick triangulations does not restrict us, however, at least in dimension ≤ 4, since any triangulation admits a "thickening" – see Chapter 11.[1]

---

[1]This holds, as already mentioned, for any PL manifold of dimension ≤ 4, and in all dimensions for smoothable PL manifolds, as well for any manifold of class $\geq \mathcal{C}^1$. Since the proof of the main result regarding our metric Ricci curvature, regarding manifolds of dimension higher than 3, holds

To be able to define and estimate the Ricci curvature of $\mathcal{T}$ and $\mathcal{T}^*$ and the connection between them, we have to make appeal in an essential manner to the fatness of the given complex. We begin by noting – using formula (B.0.9) of Appendix B – that, since the length of the edge l_{ij}^*, dual to the edge l_{ij} common to the faces f_i, f_j equals $r_i + r_j$, the first barycentric subdivision[2] of a thick triangulation is thick. Note that, for planar triangulations, and also for higher dimensional complexes embedded in some $\mathbb{R}^N$, one can realize the dual complex (also in $\mathbb{R}^N$) by constructing the dual edges l_{ij}^* orthogonal to the middle of the respective l_{ij}-s. To show the thickness of the dual simplices, one has also to make appeal to the characterization of thickness in terms of dihedral angles (Conditions (11.2.10) and (11.2.11) of Chapter 11). Importantly for us, the notion of thickness also makes sense for general cells:

Definition 12.1.1 *Let* $\mathfrak{c} = \mathfrak{c}^k$ *be a k-dimensional cell. The thickness (or fatness) of* $\mathfrak{c}$ *is defined as:*

$$\varphi(\mathfrak{c}) = \min_{\mathfrak{b}} \frac{\mathrm{Vol}(\mathfrak{b})}{\mathrm{diam}^l(\mathfrak{b})}, \qquad (12.1.2)$$

where the minimum is taken over all the l-dimensional faces of $\mathfrak{c}$, $0 \leq k$. *(If* $\dim \mathfrak{b} = 0$, *then* $\mathrm{Vol}(\mathfrak{b}) = 1$, *by convention.)*

Therefore, we can summarize the discussion above as

Lemma 12.1.2 *The dual complex* $\mathcal{T}^*$ *of a thick (simplicial) complex* $\mathcal{T}$ *is thick.*

Remark 12.1.3 $\mathcal{T}$ *and* $\mathcal{T}^*$ *converge together in the Gromov-Hausdorff metric, more precisely*

$$\lim_{\delta(\mathcal{T}) \to 0} (\mathcal{T}) = \lim_{\delta(\mathcal{T}^*) \to 0} (\mathcal{T}^*); \qquad (12.1.3)$$

where $\delta(\mathcal{T}), \delta(\mathcal{T}^*)$ *denote the mesh of* $\mathcal{T}, \mathcal{T}^*$, *respectively.*

Remark 12.1.4 *We should again stress here the crucial role of the thickness of the triangulation, as far as geometry is concerned: Thickness ensures, by its definition, the fact that no degeneracy of the simplices occurs, hence no collapse and degeneracy of the metric can take place. Conversely, in its absence no uniform estimates for the edge lengths can be made, hence convergence of (dual) meshes and, as we shall see shortly, of their metric Ricci curvatures, cannot be guaranteed.*

Returning to the definition of Ricci curvature for simplicial complexes: Given a vertex v_0, in the dual of a n dimensional simplicial complex, a *direction* at v_0 is just an oriented edge $e_1 = v_0 v_1$. Since, there exist precisely n 2-cells, $\mathfrak{c}_1, \ldots, \mathfrak{c}_n$, having e_1 as an edge and, moreover, these cells form part of n relevant variational (Jacobi) fields (see [314]), the Ricci curvature at the vertex v, in the direction e_1 is simply

$$\mathrm{Ric}(v) = \sum_{i=1}^{n} K(\mathfrak{c}_i). \qquad (12.1.4)$$

---

only for manifolds admitting smoothings, restricting ourselves only to such manifolds does not represent any further hindrance.

[2] needed in the construction of the dual complex – see e.g. [169]

Observe that the index "i" in the definition (12.1.4) above runs from 1, and not from 2, as expected judging from the classical (smooth) setting. This is due to the fact that we defined Ricci curvature by passing to the dual complex, with its simple but demanding (so to say) combinatorics. (For the implications of this fact, see Theorem 12.1.29 and Remark 12.1.30 below.)

Remark 12.1.5 *Note that we followed [314] only in determining the variational fields, but not in his definition of Ricci curvature. Indeed, Stone considers a direction at a vertex v_0 to be the union of two edges e_1, e_2 in the dual complex, where $e_1 = (v_0, v_1), e_2 = (v_1, v_2)$, and where the direction is determined by the lexicographical order. Then (according to [314], pp. 16-17) the relevant variational fields are given by the $2n$ distinct 2-cells $\mathfrak{c}_1, \ldots, \mathfrak{c}_{2n}$, containing the edges e_1 and e_2, but only $2n - 1$ relevant ones, since one of the cells is enumerated twice. Hence, the Ricci curvature at v in the direction $e_1 e_2$ is to be taken as the total defect of these $2n - 1$ cells, as follows:*

$$\mathrm{Ric}^*(v_0, e_1 - e_2) = 8n - \sum_{j=1}^{2n-1} \left\{ |\partial \mathfrak{c}_j| \,\middle|\, e_1 < \mathfrak{c}_j \text{ or } e_2 < \mathfrak{c}_j \right\}. \tag{12.1.5}$$

(See Figure 12.1.) This approach is necessary in the combinatorial case. However, it is more difficult than our approach and it would produce unnecessary complications in determining the relevant analogs of the number $n - 1$ of 2-sections of the classical, smooth case. Moreover, it is quite possible that, in any practical implementation, the advantages obtained by considering larger variational fields would be undermined by "noise" added by considering such order 2 (or larger) neigborhoods of the given vertex. However, computing Ricci curvature according to this scheme is still possible, using our metric approach (but see also the following Remark 12.1.6).

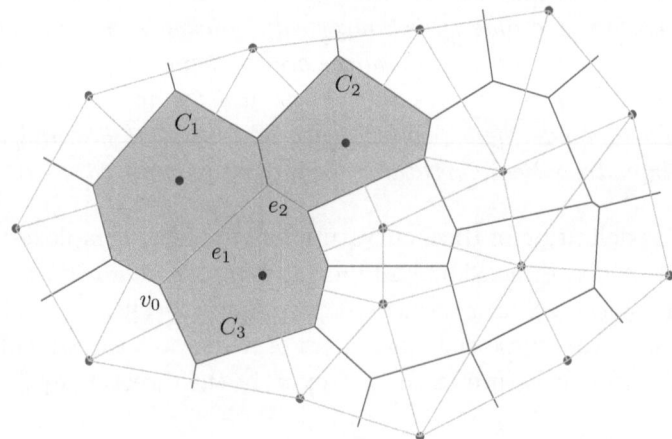

Figure 12.1 Part of the simplicial complex $\mathcal{T}$ and its dual cell complex $\mathcal{T}^*$. The (variational) Jacobi field given by $2n - 1$ cells ($n = 2$), in the direction $e_1 - e_2$, at the vertex v_0 is emphasized.

Remark 12.1.6 *It is still possible (by dualization) to compute Ricci curvature according, more-or-less, to Stone's ideas, at least for the 2-dimensional case. Indeed, according to [315],*

$$\mathrm{Ric}(\sigma, \tau_1 - \tau_2) = 8n - \sum_{j=1}^{2n-1} \left| \{\beta_j \mid \beta_j < \tau_1 \text{ or } \beta_j < \tau_2; \dim \beta_j = n-2\} \right|. \quad (12.1.6)$$

This definition of Ricci curvature is a combinatorial defect one[3]. This is evident from its expression, but made more transparent by the 2-dimensional case: Indeed, in this case, the simplices β_j are 0-dimensional, i.e. vertices, and $N(\beta_j)$ is just the number of 2-simplices having β_j as a common vertex, hence $\mathrm{Ric}(\sigma, \tau_1 - \tau_2)$ represents nothing but the total combinatorial defect at these $2n - 1$ vertices. (See also [314], p. 17. for the similar interpretation of Ric^*.)

In consequence, using the approach of the original proof of Hilbert and Cohn-Vossen [163], (and following methods well established in Graphics, etc.), we can consider, instead of the combinatorial defect, the angular defect of the cell $\mathfrak{c}_j$ dual to the vertex β_j. This, of course, applies both for our way – as well as Stone's – of determining a direction.

However, this approach to the definition of PL Ricci curvature is far less intuitive (and apparently has lesser geometric content, so to speak) in dimension ≥ 3. This is the reason why, for our present study, we have made use of the dual complex.

To determine – using solely metric considerations – the sectional curvatures $K(\mathfrak{c}_i)$ of the cells $\mathfrak{c}_i$, we shall employ the called (modified) Wald curvature K_W (see Chapter 3). To extend it from quadruples to more general cells, we first should note that the role of the abstract open sets U in the definition of Wald curvature (Definition 3.2.10 of Chapter 3) is naturally played by the cells $\mathfrak{c}_i$. We can state this as a formal definition, for the record:

Definition 12.1.7 *Let $\mathfrak{c}$ be a cell with vertex set $V_{\mathfrak{c}} = \{v_1, \ldots, v_p\}$. The embedding curvature $K(\mathfrak{c})$ of $\mathfrak{c}$ is defined as:*

$$K(\mathfrak{c}) = \min_{\{i,j,k,l\} \subseteq \{1,\ldots,p\}} \kappa(v_i, v_j, v_k, v_l). \quad (12.1.7)$$

Remark 12.1.8 *(1) Evidently, the definition above presumes that cells in the dual complex have at least 4 vertices. However, except for some utterly degenerate (planar) cases, this condition always holds. (Moreover, it can be easily corrected by truncation of the problematic vertices.)*

(2) Obviously, one can use the same method as above to compute the Ricci curvature (of $\mathcal{T}^*$), according to Stone's original approach for determining a directions in cell complexes.

Remark 12.1.9 *On a more abstract note, we should remark here that, given its*

---

[3]presumably inspired by the classical definition of Gauss curvature as the angular defect at a vertex — see, e.g. [HC-V].

(metric) intrinsic nature, K_W "behaves well", so to speak, under Gromov-Hausdorff convergence, thus it allows us (cf. Appendix A) to view the whole problem of defining and computing Ricci for PL (polyhedral) manifolds, (and in particular its applications in Graphics, Regge Calculus, etc.) in the modern context of Alexandrov spaces.

To return to the main problem of this section: From the definitions and results above we obtain – first discretely, at finite scale bounded away from zero, then passing to the limit – the following result connecting between the Ricci curvatures of a simplicial (polyhedral) complex and its dual:

Theorem 12.1.10 *Let $\mathcal{T}, \mathcal{T}^*$ be as in Lemma 2.1.12 above. Then*

$$\lim_{\mathrm{mesh}(\mathcal{T})\to 0} \mathrm{Ric}(\sigma) = \lim_{\mathrm{mesh}(\mathcal{T}^*)\to 0} C \cdot \mathrm{Ric}^*(\sigma^*), \qquad (12.1.8)$$

where $\sigma \in \mathcal{T}$ and where $\sigma^* \in \mathcal{T}^*$ is (as suggested by the notation) the dual of σ.

Note that, equipped with general triangulations, one can only formulate the somewhat weak result above. Indeed, the precise constant C is hard to determine. The thickness condition, that ensures a metric "quasi-regularity" of the triangulation, supplies us only with weak estimates. To obtain stronger ones, one should be able to control the regularity of the combinatoric structure, as well. (This is evident, but it will become even clearer in the sequel.) It should be noted in this context that, at least in Graphics, mesh improvement techniques allow us to consider such "combinatorial almost regular" triangulations. Moreover, while the desired constant C_1 is, of course, $C_1 = 1$, and some first experimental results hint that, at least for certain "nice" triangulations, this is indeed the case, we can't guarantee a better result – see the remark following the preceding theorem.

We now easily obtain the following theorem:

Theorem 12.1.11 *Let M^n be a (smooth) Riemannian manifold and let $\mathcal{T}$ be a thick triangulation of M^n. Then*

$$\mathrm{Ric}_{\mathcal{T}} \to C_1 \cdot \mathrm{Ric}_{M^n}, \text{ as } \mathrm{mesh}(\mathcal{T}) \to 0, \qquad (12.1.9)$$

where the convergence is the weak convergence (of measures).

We have encountered similar, results for other types of curvature, in Chapter 11. (See also [42], for the fitting results so called *Einstein measures*.)

Proof The theorem follows easily from Lemma 2.1.12, the fact that $\mathrm{Ric}(v)$ is defined in a purely metric, intrinsic manner, from the fact that intrinsic properties are preserved under Gromov-Hausdorff limits (see [142]), and also from Theorem 12.1.10 above.

□

12.1.2   The Bonnet-Myers Theorem

Having introduced a metric Ricci curvature for PL manifolds, one naturally wishes to verify that this represents, indeed, a proper notion of Ricci curvature, and not just an approximation of the classical notion. According to the synthetic approach to Differential Geometry (see, e.g. [142], [328]), a proper notion of Ricci curvature should satisfy adapted versions of the main, essential theorems that hold for the classical notions. Among such theorems the first and foremost is the Bonnet-Myers Theorem[4] (see, e.g., [37], [73]).

Theorem 12.1.12 (Bonnet-Myers) *Let M^n be a complete Riemannian manifold. If*

1. $K_M \geq K_0 > 0$,

   or

2. $\mathrm{Ric} \geq (n-1)K_0$;

then every geodesic of length $\geq \pi/\sqrt{K_0}$ has conjugate points, hence $\mathrm{diam}(M^n) \leq \pi/\sqrt{K_0}$. Therefore, it follows that M^n is compact.

Remark 12.1.13 *The first part of the theorem is known as the Bonnet Theorem, while the second one is called the Myers Theorem.*

Indeed, fitting versions for combinatorial cell complexes and weighted cell complexes were proven, respectively, by Stone [314], [315], and Forman [123] (whose work we shall discuss in its own dedicated section and in another section regarding its Complex Networks applications). Moreover, the Bonnet part of the Bonnet-Myers theorem, that is the one appertaining to the sectional curvature, was also proven for PL manifolds, again by Stone – see [316], [313].

12.1.2.1   The 2-Dimensional Case

For the special – yet of main importance in applications – case of 2-dimensional manifolds, such a result is easy to prove, given the fact that Ricci and sectional curvature essentially coincide. More precisely, we can formulate the following theorem:

Theorem 12.1.14 (Bonnet-Myers for PL 2-manifolds – Combinatorial) *Let M^2_{PL} be a complete, connected 2-dimensional PL manifold such that*
    (i) There exists $d_0 > 0$, such that $\mathrm{mesh}(M^2_{PL}) \leq d_0$, (where $\mathrm{mesh}(M^2_{PL})$ denotes the mesh of the 1-skeleton of M^2_{PL}, i.e. the supremum of the edge lengths).
    (ii) $K_{Comb}(M^2_{PL}) \geq K_0 > 0$.
    Then M^2_{PL} is compact and, moreover

$$\mathrm{diam}(M^2_{PL}) \leq \begin{cases} 2\pi d_0, & k_0 \geq (2 - \sqrt{2})\pi\,; \\ 4\pi^3 d_0/[(2\pi - d_0)(4\pi k_0 - k_0^2)^{1/2}], & \text{else}\,; \end{cases} \qquad (12.1.10)$$

---

[4]also called, sometimes, the Bonnet-Schoenberg-Myers Theorem

where K_{Comb} denotes the combinatorial Gauss curvature of M_{PL}^2,

$$K_{Comb}(v_i) = 2\pi - \sum_{p=1}^{m_i} \alpha_p(v_i) \tag{12.1.11}$$

where $\alpha_1, \ldots, \alpha_{m_i}$ are the (interior) face angles adjacent to the vertex v_i.

Remark 12.1.15 *Condition (i), that ensures that the set of vertices of the PL manifold is "fairly dense"[5] is nothing but the necessary and quite common density condition for good approximation both of distances and of curvature measures. The mere existence of such a d_0 is evident for a compact manifold, however it can't be presumed a priori for a general manifold, hence has to be postulated. Recall, moreover, that to ensure a good approximation of curvature, this density factor has to be properly chosen , thus tighter estimates for the mesh of the triangulation can be obtained from (12.1.10) along with better curvature approximation. No less importantly, an adequate choice of the vertices of the triangulation, also ensures, via the thickness property, the non-degeneracy of the manifold (and of its curvature measures).*

A first proof of Theorem 12.1.14 is easy to give, if one is ready to make appeal to Stone's previous work [316]:

Proof We make appeal to Theorem 3 of [316]. Indeed, in the two dimensional case, the so called *maximum* and *minimum curvatures*, k_+, respective k_- (Definition 11.4.1, Chapter 11) at the vertices of M_{PL}^2 coincide with the combinatorial Gauss curvature. Moreover, due to the fact that here we are concerned solely with 2-dimensional simplicial complexes (*PL* manifolds), conditions (1) and (2) of Theorem 3 of [316] are equivalent, respectively, to our conditions (ii) and (i) above. Therefore, the conditions in the statement of Theorem 3, [316] are satisfied and, by condition (ii), the theorem follows immediately.

□

Remark 12.1.16 *It is easy to see that the theorem above extends to more general polyhedral surfaces, since such surfaces admit simplicial subdivisions. However, during this subdivision, k_+ and k_- do not change, since the only relevant contributions to these quantities occur at the vertices, and depend only on the angles at these vertices, more precisely on the normal geometry (see [316], p. 12), that suffer no change during the subdivision process.*

The proof above suffers from the disadvantage of making use of Stone's maximum and minimum curvatures, even though, in this context making appeal to them is rather natural. In particular, the bound (12.1.10) is rather weak, as compared to the one for the classical case, but it is the only one supplied by Stone's result we made appeal to, namely Theorem 3 of [316]. We can, however, provide a different proof,

---

[5]in Stone's formulation ([313])

independent of Stone's work, but at the price of using some heavy (albeit classical) machinery, that, moreover, takes us away, so to say, from the discrete methods.

Proof The basic idea is again to consider a *smoothing* M^2 of M_{PL}^2. Since, by [229], Theorem 4.8, smoothings approximate arbitrarily well both distances and angles[6] on M_{PL}^2, defects are also arbitrarily well approximated. Given that the combinatorial curvature of M_{PL}^2 is bounded from below, it follows that so will be the sectional (i.e. Gauss) curvature of M^2.

Unfortunately, the Gaussian curvature of M^2 is positive only on isolated points (the set of vertices of M_{PL}^2), so we cannot apply the classical Bonnet theorem yet. However, we can ensure that M^2 is arbitrarily close to a smooth surface M_+^2, having curvature Gaussian curvature $K(M_+^2) > 0$.[7] Therefore, the classical Bonnet Theorem can be applied for M_+^2, hence M_{PL}^2 is compact and its diameter has the same upper bound (again using the same arguments as before[8]) as that of M_+^2 (and M^2), namely

$$\text{diam}(M_{PL}^2) \leq \frac{\pi}{\sqrt{K_0}} . \tag{12.1.12}$$

$\square$

Remark 12.1.17 *Apparently, the bound for diameter given by the proof above, is tighter than the one obtained by Stone in [316]. Nevertheless, we should keep in mind that, in practice, one is more likely to encounter PL surfaces as approximations of smooth ones.[9] However, the larger the mesh of the approximating surface (i.e. the "rougher" the approximation), the larger the deviation of the approximating triangles from the tangent planes (at the vertices), hence the more likely is to obtain large combinatorial curvature. Hence, there is a correlation between size of the simplices and curvature, even though not a straightforward one.*

Since the leitmotif of the previous section was metric (Wald) curvature, it is natural to ask whether a fitting version of the Bonnet-Myers Theorem exists for this type of curvature? The answer is – at least in dimension 2 – positive: We can, indeed state an analog of Myers' Theorem, in terms of the Wald curvature:

Theorem 12.1.18 (Bonnet-Myers for PL 2-manifolds – Metric) *Let M_{PL}^2 be a complete, connected 2-dimensional PL manifold such that*
    (i') There exists $d_0 > 0$, such that $\text{mesh}(M_{PL}^2) \leq d_0$;
    (ii') $K_W(M_{PL}^2) \geq K_0 > 0$.
Then M_{PL}^2 is compact and, moreover

$$\text{diam}(M_{PL}^2) \leq \frac{\pi}{\sqrt{K_0}} . \tag{12.1.13}$$

---

[6]More precisely, they are δ-approximation and, for δ small enough, also ε-approximations of M_{PL}^2 – for details see [229]

[7]This is easily seen by adding spherical "roofs" (of low curvature) over the faces, and then slightly modifying the construction, to ensure that the curvature will be positive also on the "sutures" of the said roofs, corresponding to the edges of the original PL manifold.

[8]i.e. δ- and ε-approximations (see [229] for definitions).

[9]and, obviously, PL surfaces are PL approximations of their own smoothings

Proof 12.1 *Since distances (and angles) are arbitrarily well approximated by smoothings, it follows that so are metric quadruples (including their angles), hence so is Wald curvature. Since, by Theorem 3.2.19 of Chapter 3 the Wald curvature at any point of non-trivial geometry M^2, namely at a vertex v, $K_W(v)$ equals the classical (Gauss) curvature $K(v)$ (and, of course, this is also true a fortiori at all the other points, where both the smooth and the PL manifold are flat). Therefore, the Gauss curvature of M^2 approximates arbitrarily well the Wald curvature of M_{PL}^2, hence we can apply the same argument as in Proof 2 above to show that M_{PL}^2 is, indeed, compact and, furthermore, satisfies the upper bound (12.1.12).*

Remark 12.1.19 *Like the previous theorem, the result above can be extended to polyhedral manifolds, and even in a more direct fashion, since Wald curvature does not take into account the number of sides of the faces incident to a vertex, but only their lengths.*

This result, as well as its generalization to higher dimensions (see Theorem 12.1.21 below) is hardly surprising, given the fact that, by [64], Theorem 3.6, Myers' theorem holds for general Alexandrov spaces of curvature $\geq K_0 > 0$, and since Wald-Berestovskii curvature is essentially equivalent to the Rinow curvature, hence to the Alexandrov curvature (see Appendix A). Rather, we give, in the special case of PL surfaces (manifolds) a simpler, more intuitive proof of the Burago-Gromov-Perelman extension of Myers' Theorem. However, in higher dimension, none of the arguments applied in both proofs of Theorem 12.1.14 are applicable, at least not without imposing further conditions:

1. Regarding the first proof:

   - In dimensions higher than 2, k_+ and k_- do not, necessarily equal each other (see [316], Example 4, p. 14) and, a fortiori, they fail to equal the combinatorial Gauss curvature. They do, however, according to Stone [316], resemble in their behavior the minimum, respective maximum sectional curvature at a point common to two 2-planes, that contain a given (fixed) tangent vector at the point in question.

     An important proviso should be added, however: While for the general PL simplicial complexes, the equality between k_+ and k_- fails to hold, it is true for the most relevant – at least as far as our analysis is concerned – case, namely that of PL manifolds without boundary (see [316], Example 3, p. 13). Therefore, in the light of facts above, it follows that, while for a fairly general and important setting the connection is straightforward, it is not clear how to compare, in the general case, our proposed metric discretization of Ricci curvature, with the maximal and minimal curvatures of Stone (hence to combinatorial curvature, whenever they equal it – and each other).[10]

---

[10]A natural attempt would be to use straightforward extensions of k_+ and k_- – let us denote them, for convenience, Ric_{min} and Ric_{max}. However, it is less evident (at least at this point in time) how expressive these definitions would prove to be.

Remark 12.1.20 *It is true that the lower bound on k_+, as considered in Theorem 3 of [316] has a simple expression, in any dimension, via a topological condition (cf. Lemma 5.1 of [316]), namely that the intersection of any (PL) geodesic segment of ends p and q with the 2-skeleton of M^2_{PL} is precisely the set $\{p, q\}$ (with the exception, of course, of the case when the segment is contained in a simplex. However, since the metric information contained in this new condition is void (or rather thoroughly encrypted, so to say) it has no apparent advantage for application in conjunction with metric curvature.*

- For an application of the Stone's methods in combination with the metric curvature approach to any dimension, one would have to make appeal to Jacobi fields, as defined in [314]. However, as discussed in the previous section, this would probably led to numerical instability.

2. As far as the second proof is concerned:

  - No smoothing of a *PL* manifold necessarily exists in dimension higher than $n > 4$ and, even if it exists, it is not necessarily unique, for $n \geq 4$ – see [322].

    However, if such a smoothing exists, then the second proof of Theorem 12.1.14 (and of Theorem 12.1.18) extends to any dimension, and we obtain the following *PL* (metric) versions of the classical results:

Keeping this facts in mind, we can now formulate the metric adaptation of the classical Bonnet Theorem:

Theorem 12.1.21 (PL Bonnet – metric) *Let M^n_{PL} be a complete, n-dimensional PL, smoothable manifold without boundary, such that*
  (i") There exists $d_0 > 0$, such that $\mathrm{mesh}(M^n_{PL}) \leq d_0$;
  (ii") $K_W(M^n_{PL}) \geq K_0 > 0$,
  where $K_W(M^n_{PL})$ denotes the sectional curvature of the "combinatorial sections", i.e. the cells c_i (see Section 1 above).
  Then M^n_{PL} is compact and, moreover

$$\mathrm{diam}(M^2_{PL}) \leq \frac{\pi}{\sqrt{K_0}} . \tag{12.1.14}$$

Proof We should note in the beginning that the "rounding" argument of Proof 2 of Theorem 12.1.14 is not easy to extend directly – if at all – to higher dimension. Instead, a more subtle argument has to be devised. To this end we make appeal again to Stone's paper [316], and we build the spherical simplicial complex $M^n_{Sph,\rho}$ associated to the given *PL* (or rather piecewise-flat) complex M^n_{PL}. This is constructed as follows: Consider the sphere of radius $R = R(\sigma)$ and radius $O = O(\sigma)$, circumscribed to a given simplex σ, and its image $\sigma^* = \sigma^*(R^*)$ on a sphere of radius $R^* = R^*(\sigma), R^* \geq R$, via the central projection from O. We denote by $M^n_{Sph,\rho}$ the simplicial complex obtained by remetrization of M^n_{PL} by the replacement of each σ by its spherical counterpart σ^*. Then, by Lemma 5.5 of [316], for large enough $R^* > R$, the following holds for any pair of points $p, q \in M_{PL}^n$: $\mathrm{dist}_{M_{PL}^n}(p, q) \leq C\mathrm{dist}_{M_{Sph,\rho}^n}(p^*, q^*)$, for a certain constant C, where p^*, q^* denote the spherical images of p, q. Since the curvature at each vertex of the spherical simplex obtained by central projection of the simplices of M_{PL}^n onto their circumscribed spheres is smaller than the corresponding one (at the same vertex) in the PL (piecewise flat manifold), this holds a fortiori for $M_{Sph,\rho}^n$. It follows from the classical Bonnet theorem (after applying the necessary smoothing) that $\mathrm{diam}(M_{PL}^n) < \mathrm{diam}(M_{Sph,\rho}^n)$.

$\square$

Remark 12.1.22 *An approach similar to the one used in the proof above was also employed by Cheeger [77] in a rather similar context. (We should stress here that, as a byproduct of the results in this paper, we also address – using our own methods – a problem posed by Cheeger in [77], Remark 3.5.)*

We should underline the fact that, if we approach the problem of PL Ricci curvature from the viewpoint of the first part of the paper, that is of PL (secant) approximations of smooth manifolds, then the situation changes dramatically. Indeed, even when such a smoothing M^n ($n \geq 3$) exists, it is not probable that its sections provided by M_{PL}^n, in the manner indicated above, suffice to approximate well enough – let alone reconstruct – the Ricci curvature of M^n. In simple words, "there are not enough directions" in M_{PL}^n to allow us to infer from the metric curvatures of a PL approximation, those of a given smooth manifold M^n (in fact, not not even a good approximation), hence we are faced again with a problem that we already mentioned in conjunction with the first proof, namely that of insufficient "sampling of directions" in PL approximations, while, on the other hand, increasing of the number of directions, i.e. of 2-dimensional sections (simplices) generates a decrease of the precision of the approximation, due to the triangulation's loss of thickness.

To summarize: All the considerations above show us that, unfortunately, in higher dimensional, no general analog of Myers' Theorem for PL manifolds can be obtained by applying solely smoothing arguments). It is true that *a* Ricci curvature of the smooth manifold M^n is obtained in terms of that of M_{PL}^n, however, it is not clear, in view of the paucity of sectional directions (i.e. possible 2-sections), how precisely is this connected to its discrete counterpart. Therefore, we can obtain, at best, an approximation result (with limits imposed by the thickness constraint – see discussion above).

In conclusion, from the discussion above is transparent that, unfortunately, at this point in time, we can offer no proof for the general case, that is for non-smoothable PL manifolds of dimension $n \geq 4$. To obtain such a proof for Bonnet's Theorem, one should adapt Stone's methods, as developed in [316], while for a comprehensive generalization of Myers' theorem, one has the apparently more difficult task of accordingly modify, for the metric case, the purely combinatorial methods of [314]. A quite different approach, but one that would allow us to extend the metric approach to quite general weighted CW complexes, would be to adjust Forman's methods developed in [123] to our case. The essential step in this direction would be to find relevant geometric content (e.g. lengths, area, volume) for Forman's "standard weights" associated to each cell.

12.1.2.2 Wald Curvature and Alexandrov Spaces

In this section we bring a proof of a more general case in a more common language in the most of the contemporary literature, albeit at the price of using some extraneous and powerful techniques and results. For this, we first need to refer the reader to Applendix A.2, where we point out that Wald's curvature is essentially equivalent with the much more modern notion of Alexandrov curvature, at least for spaces in which there exists "sufficiently many" minimal geodesics. The reason we prefer working with the Wald curvature, is that it is computable and, moreover, that it has even simpler, more practical approximations – see Chapter 3.

It is, however, important to notice that one has take into account the "discrete" nature of the types of spaces considered, hence to compute solely the Wald curvature of the 1-star neighborhood of a vertex, as already stressed above, and not to consider (ever) smaller neighborhoods, as perhaps natural in other contexts. This, however, agrees with the method of computing discrete curvature as angular defect. A positive consequence of this fact is that any such neighborhood becomes a region having the same Alexandrov curvature bounded from below as the computed Wald one. Moreover, by the Alexandrov-Topogonov Theorem (see, e.g. [252], Theorem 43 and its proof, pp. 837-840), the whole surface becomes a space of curvature (Wald or Alexandrov) bounded from below. However, taking into account only "discrete" neighborhoods is very important when equating the Wald and Alexandrov curvature, since it allows to avoid the blow-up of Alexandrov curvature at the vertices during smoothing. However, if one still wishes to consider smaller-and-smaller neighborhood of the vertices (motivated, perhaps, by other applications then Imaging and Graphics, such as those in Regge Calculus), one can resort to the basic approach of Brehm and Kühnel [56], that is "rounding" the edges by cylinders of radius ε (without any change in curvature) and replacing the polyhedral cones at the vertices by smooth "caps", up to a predetermined admissible error of, say, ε_1. Note that such a "filtration" of K_W by Gaussian curvature (of the approximating smooth surfaces) is in concordance with common practices in Imaging, Vision and, indeed, in many applicative fields. In addition, considering only this "discrete" neighborhoods is very important when equating the Wald and Alexandrov curvature, since it also allows us to avoid the blow-up of Alexandrov curvature at the vertices during smoothing.

If one is willing to make appeal to the theory of Alexandrov spaces, then, by using this equivalence of Wald and Alexandrov curvatures with the above mentioned provisos, a result of the desired type follows immediately:

Theorem 12.1.23 (Bonnet-Myers – Alexandrov Spaces) Let M_{PL}^n be a complete, connected PL manifold, such that $K_W(M_{PL}^n) \geq K_0 > 0$.

Then M_{PL}^n is compact and, moreover

$$\operatorname{diam}(M_{PL}^2) \leq \frac{\pi}{\sqrt{K_0}} \, . \tag{12.1.15}$$

Proof The theorem follows from [252], Corollary 47, p. 840 and from the fact that M_{PL}^n is locally compact.

$\square$

Thick Cell Complexes Determining weather a general PL complex has Wald curvature bounded from below can be, in practice, quite difficult. However, in the special case of thick complexes (see Definition 12.1.1) one can determine a simple criterion as follows.

Lemma 12.1.24 *Let $M = M_{PL}^n$ be a complete, connected PL manifold thickly embedded in some $\mathbb{R}^N$, such that $K_W(M^2) \geq K_0 > 0$, where M^2 denotes the 2-skeleton of M. Then there exists $K_1 > 0$ such that $K_W(M_{PL}^n) \geq K_1 > 0$.*

Sketch of Proof We indicate a proof the only for the case $n = 3$; the general case follows by a simple inductive argument. Consider an edge e belonging to the 1-skeleton of M_{PL}^n. We have to show that $K_W(Q) \geq K_0 > 0$, for any quadruple incident to e. If Q is one of the quadruples determined by the original cells of M^2 (such as Q_1 in Figure 12.2) the condition is fulfilled trivially since $K_W(M^2) \geq K_0 > 0$. Otherwise the edges of Q are either edges of the original cells (see Figure 12.2), or diagonals of such cells (e.g. d in Figure 12.2), or they connect vertices belonging to two different cells of the given complex (such as $\tilde{e}$ in Figure 12.2 connecting between vertices of the cells $\mathfrak{c}_2$ and $\mathfrak{c}_3$). But, it is quite standard to show that, by the fatness of the cells $\mathfrak{c}_i$, there exists a constant c_1 such that $\frac{1}{c_1}e \leq d \leq c_1 e$. In a similar manner, using the boundedness from below of the angles as an equivalent definition of thickness, one can show that the fatness of the embedding implies that there exists a c_2 such that $\frac{1}{c_2}e \leq \tilde{e} \leq c_2 e$.

The desired conclusion follows from the two double inequalities above and from the continuity of the determinant function that defines the Wald curvature.

$\square$

The fitting version of Bonnet-Myers now follows as a direct corollary:

Theorem 12.1.25 (Bonnet-Myers – Thick Complexes) *Let $M = M_{PL}^n$ be a complete, connected PL manifold thickly embedded in some $\mathbb{R}^N$, such that $K_W(M^2) \geq K_0 > 0$, where M^2 denotes the 2-skeleton of M. Then M_{PL}^n is compact and, moreover*

$$\mathrm{diam}(M_{PL}^2) \leq \frac{\pi}{\sqrt{K_0}} \,. \tag{12.1.16}$$

Remark 12.1.26 *We should keep in mind that, while the theorem usually is applied to piecewise flat manifolds since, as we have seen throughout the book, this is the case of most interest, both for theoretical ends as well as application oriented ones, the proof extends – mutatis mutandis – to the case of spaces whose simplices are modeled after spherical or hyperbolic spaces.*

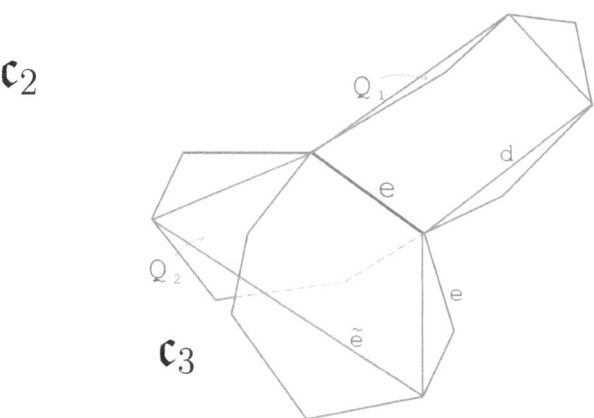

Figure 12.2 Thickness of metric quadruples adjacent to an edge in an embedded piece-wise flat 3-manifold.

We conclude this section with an open research problem for the advanced and motivated reader:

Problem 12.1.27 *Develop a fitting notion of "Einstein metric" associated to the metric Ricci curvature introduced in this paper. (Hint: Such a metric will probably be related to the stationary points of the Regge functional –see, e.g. [136]).*

12.1.3 A Comparison Theorem

Up to this point, we have not yet defined the sectional curvature $K(\mathfrak{c})$ of a cell $\mathfrak{c}$. In light of our preceding discussion and results, the following definition is quite natural:

Definition 12.1.28 *Let $M = M_{PL}^n$ be an n-dimensional PL manifold (without boundary). The scalar metric curvature scal_W of M is defined as*

$$\mathrm{scal}_W(v) = \sum K_W(\mathfrak{c}), \qquad (12.1.17)$$

the sum being taken over all the cells of M^* incident to the vertex v of M^*.

Using this definition and the results of Section 2, we immediately[11] obtain, the following generalization of the classical curvature bounds comparison in Riemannian geometry (compare also with [2], Theorem 1):

Theorem 12.1.29 (Comparison theorem) *Let $M = M_{PL}^n$ be an n-dimensional*

---

[11]and, in fact, quite trivially, since the result holds, regardless of the specific definition for the curvature of a cell

PL manifold (without boundary), such that $K_W(M) \geq K_0 > 0$, i.e. $K(\mathfrak{c}) \geq K_0$, for any 2-cell of the dual manifold (cell complex) M^*. Then

$$K_W \underset{\geq}{\lessgtr} K_0 \Rightarrow \mathrm{Ric}_W \underset{\geq}{\lessgtr} nK_0 . \qquad (12.1.18)$$

Moreover

$$K_W \underset{\geq}{\lessgtr} K_0 \Rightarrow \mathrm{scal}_W \underset{\geq}{\lessgtr} n(n+1)K_0 . \qquad (12.1.19)$$

Remark 12.1.30   *1. Inequality (12.1.19) can be formulated in the seemingly weaker form:*

$$\mathrm{Ric}_W \underset{\geq}{\lessgtr} nK_0 \Rightarrow \mathrm{scal}_W \underset{\geq}{\lessgtr} n(n+1)K_0 , \qquad (12.1.20)$$

2. *Note that in all the inequalities above, the dimension n appears, rather then $n-1$ as in the smooth, Riemannian case (hence, for instance one has in (12.1.19), $n(n+1)K_0$, instead of $n(n-1)K_0$[12] as in the classical case). This is due to our definition (12.1.4) of Ricci (and scalar) curvature, via the dual complex of the given triangulation, hence imposing standard and simple combinatorics, at the price of allowing for only for such weaker bounds.[13]*

Remark 12.1.31 *Before concluding this section, let us note that a more recent purely combinatorial version of the Bonnet-Myers Theorem has been put forward by Trout [326].*

12.2   RICCI CURVATURE AND FLOW FOR 2-DIMENSIONAL *PL* SURFACES

The increasing popularity and remarkable success of Ricci curvature in a variety of applicative fields, stretching far beyond the predictable Regge Calculus, is largely due to its almost chameleonic capability to adapt itself to an amazing variety of mathematical contexts. In particular, why its represents, in the classical setting, a notion devised for manifolds of dimension 3 and higher, it lends itself to a meaningful restriction and adaptation to lower dimensions. While the 1-dimensional versions of Ricci curvature are somewhat newer (but growing in acceptance), the 2-dimensional restriction has become by now classical, as it was introduced, in a pure – and classical – mathematical setting by Hamilton [158], and adapted to *PL* manifolds by Chow and Luo [88]. It is precisely this discrete version that is by far the most successful and, in consequence, the best know. It is, therefore, the one to whom we dedicate the next section.

12.2.1   Combinatorial Surface Ricci Flow

Surface conformal mappings preserve infinitesimal circles as shown in Figure 12.3. The left frame shows a Riemann mapping from a male facial surface onto the planar

---

[12]but, on the other hand, this holds even if $n = 3!$...

[13]without affecting the analog of the Bonnet-Myers Theorem – see Section 2 above.

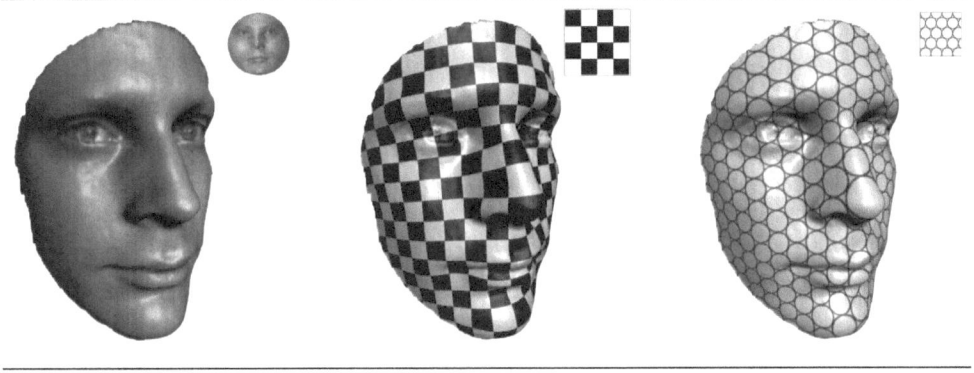

Figure 12.3 Conformal mapping maps infinitesimal circles to infinitesimal circles.

unit disk; the middle frame displays the checker-board texture mapping, where all the right corner angles are well preserved; the right frame demonstrates the mapping transforms the small circles on the texture image onto circles on the surface. This intuition inspires the combinatorial conformal mapping based on circle packing, where the infinitesimal circles are replaced by circles with finite radii as shown in Figure 12.4. The left frame illustrates a circle packing: given a planar domain $\Omega \subset \mathbb{C}$, we compute a triangulation $\mathcal{T}$, associate each vertex v_i with a circle C_i. The circle C_i is centered at v_i and with radius γ_i. For each edge $[v_i, v_j]$ in the triangle, the two circles at the end vertices C_i and C_j are tangential to each other. Then we modify the circle radii, keep

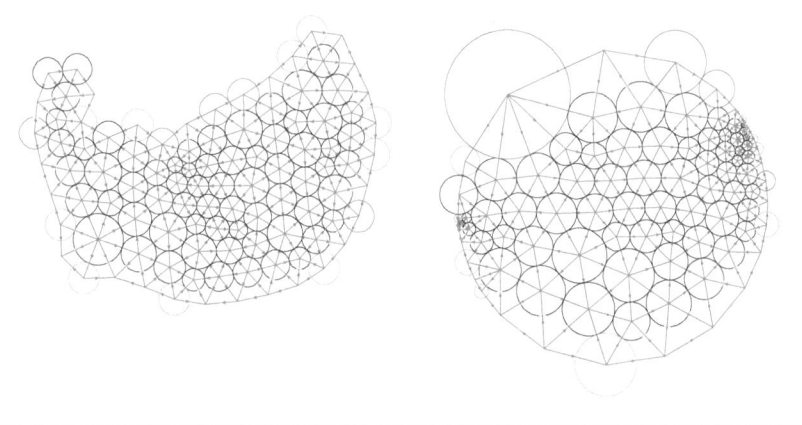

Figure 12.4 Combinatorial Riemann mapping based on tangential circle packing.

the tangential relation to deform the triangulation to another shape. As shown in the right frame, for each boundary vertex v_i, the curvature K_i is defined as angle deficit (π minus the adjacent corner angles), K_i is proportional to the radius γ_i, $K_i/\gamma_i = c$, then the triangulation covers a convex domain Ω^*. By a scaling, Ω^* approximates the unit disk. This circle packing induces piecewise linear mapping $\varphi : \Omega \to \Omega^*$, which preserves the combinatorial structure of the triangulation, maps each vertex v_i of $\mathcal{T}$ to a vertex $\varphi(v_i)$ in $\varphi(\mathcal{T})$. For each interior point $p \in \Omega$, we find a triangle covering it $p \in [v_i, v_j, v_k]$, the bary-centric coordinates of p is given by

$$p = \lambda_i v_i + \lambda_j v_j + \lambda_k v_k, \quad \lambda_i + \lambda_j + \lambda_k = 1, \quad \lambda_i, \lambda_j, \lambda_k \geq 0,$$

then

$$\varphi(p) = \lambda_i \varphi(v_i) + \lambda_j \varphi(v_j) + \lambda_k \varphi(v_k).$$

If p is shared by two triangles, then we can define the $\varphi(p)$ using any one of them, the results are consistent. Thurston conjectured that when the triangulations $\{\mathcal{T}_n\}$ get refiner and refiner, the piecewise linear maps $\{\varphi_n\}$ will converge to the real Riemann mapping from Ω to the unit disk $\mathbb{D}$. Rodin and Sullivan proved Thurston's conjecture. This result is generalized to the combinatorial surface Ricci flow theory.

Suppose S is a topological surface, $V \subset S$ is a finite set of distinct points on S, $\mathcal{T}$ is a triangulation of the point set V. For each edge of $\mathcal{T}$, we associate it with a positive number, $l : E(\mathcal{T}) \to \mathbb{R}_{\geq 0}$, such that the triangle inequality holds on every face $[v_i, v_j, v_k]$,

$$l_{ij} + l_{jk} > l_{ki}, \quad l_{jk} + l_{ki} > l_{ij}, \quad l_{ki} + l_{ij} > l_{jk}, \tag{12.2.1}$$

then we call l an *edge length function*. We call $(S, V, \mathcal{T}, l)$ a *discrete metric surface*. If each triangle face is a Euclidean triangle, namely the discrete metric surface is obtained by isometrically glue a set of Euclidean triangles along their common edges, then we call the discrete surface is with the *Euclidean $\mathbb{E}^2$ background geometry*. Similalary, if each face is a spherical $\mathbb{S}^2$ or hyperbolic $\mathbb{H}^2$ triangle, then we say the surface is with the spherical or hyperbolic background geometry.

Given a discrete metric surface $(S, V, \mathcal{T}, l)$ with Euclidean (spherical or hyperbolic) background metric, we define the discrete Gaussian curvature as angle deficit – See Figure 12.5:

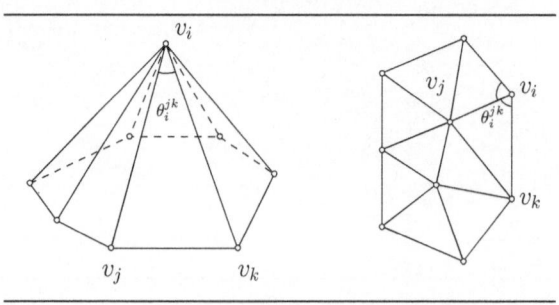

Figure 12.5 Discrete curvature.

Definition 12.2.1 (Discrete Curvature) *Suppose* $(S, V, \mathcal{T}, l)$ *is a discrete metric surface with* $\mathbb{E}^2$ *(*$\mathbb{S}^2$ *or* $\mathbb{H}^2$*)background metric, the discrete curvature* $K : V \to \mathbb{R}$ *is defined as*

$$K(v_i) = \begin{cases} 2\pi - \sum_{jk} \theta_i^{jk} & v_i \notin \partial S \\ \pi - \sum_{jk} \theta_i^{jk} & v_i \in \partial S \end{cases} \qquad (12.2.2)$$

where θ_i^{jk}'*s are the corner angles adjacent to* v_i.

Similar to the smooth surfaces, the Gauss-Bonnet Theorem holds for discrete metric surfaces as well.

Theorem 12.2.2 (Discrete Gauss-Bonnet Theorem) *Given a discrete metric surface* $(S, V, \mathcal{T}, l)$*, the total curvature satisfies:*

$$\sum_{v \in \partial S} K(v) + \sum_{v \notin \partial S} K(v) + \lambda A(S) = 2\pi \chi(S), \qquad (12.2.3)$$

where $A(S)$ *is the area of the surface,* λ *equals* $+1$, 0 *or* -1 *if the surface is with* $\mathbb{S}^2$, $\mathbb{E}^2$ *or* $\mathbb{H}^2$ *background geometry respectively.*

Proof Here we give a brief proof for discrete metric surface with the $\mathbb{E}^2$ background geometry only. Assume the discrete surface S is closed. Its total discrete curvature equals to

$$\sum_{v \in S} K(v) = \sum_{v_i \in S} \left(2\pi - \sum_{jk} \theta_i^{jk} \right) = 2\pi |V| - \pi |F| = 2\pi \left(|V| - \frac{1}{2} |F| \right).$$

On the other hand, each triangle has 3 edges, each edge is shared by two faces, therefore $3|F| = 2|E|$,

$$\chi(S) = |V| + |F| - |E| = |V| - \frac{1}{2} |F|.$$

Thus $\sum_{v \in S} K(v) = 2\pi \chi(S)$.

Suppose the discrete surface S has boundaries, we double the surface to get a closed symmetric surface $\bar{S}$, then $\chi(\bar{S}) = 2\chi(S)$,

$$\sum_{\bar{v} \in \bar{S}} K(\bar{v}) = 2 \sum_{v \in S} K(v),$$

therefore Eqn. (12.2.3) holds.

□

For smooth metric surfaces, the Gaussian curvature is determined by the Riemannian metric. For discrete metric surfaces, the discrete curvature is determined by the edge lengths via cosine laws. As shown in Figure 12.6, triangles with different background geometries have different cosine laws:

$$\begin{array}{rcll} \cos l_i & = & \frac{\cos \theta_i + \cos \theta_j \cos \theta_k}{\sin \theta_j \sin \theta_k} & \mathbb{S}^2 \\[6pt] \cosh l_i & = & \frac{\cosh \theta_i + \cosh \theta_j \cosh \theta_k}{\sinh \theta_j \sinh \theta_k} & \mathbb{H}^2 \\[6pt] 1 & = & \frac{\cos \theta_i + \cos \theta_j \cos \theta_k}{\sin \theta_j \sin \theta_k} & \mathbb{E}^2 \end{array} \qquad (12.2.4)$$

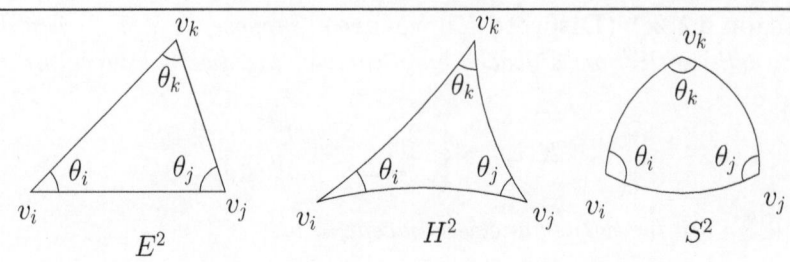

Figure 12.6 Different cosine laws for Euclidean, hyperbolic and spherical triangles.

Definition 12.2.3 (Thurston's Circle Packing Metric) *Given a discrete surface $(S, V, \mathcal{T})$, we associate each vertex v_i with a circle with radius γ_i. On edge e_{ij}, the two circles intersect at the angle of $\phi_{ij} \in [0, \frac{\pi}{2})$. The edge lengths are*

$$l_{ij}^2 = \gamma_i^2 + \gamma_j^2 + 2\gamma_i \gamma_j \cos \varphi_{ij}.$$

The radii and the intersection angle (Γ, Φ) defines the circle packing metric on $(S, V, \mathcal{T})$, where

$$\Gamma = \{\gamma_i | \forall v_i\}, \Phi = \{\varphi_{ij} | \forall e_{ij}\}.$$

If all the intersection angles φ_{ij} are zeros, the Thurston's circle packing is called the tangential circle packing.

Figure 12.7 shows the tangential circle packing (left) and the Thurston's circle packing (right).

Given a discrete surface with a circle packing metric (Γ, Φ), we can deform the circle radii Γ but keep the intersection angles unchanged Φ, this is an analogy to conformal metric deformation. The deformation will change the discrete curvature accordingly. This leads to the discrete version of the surface Ricci flow.

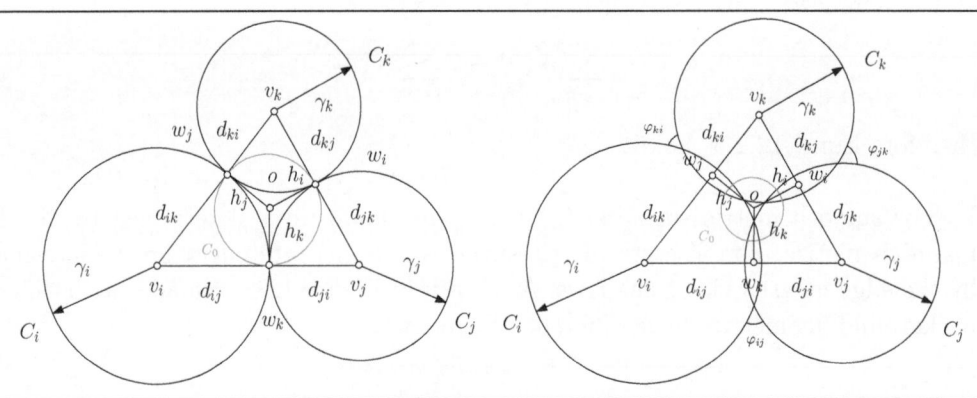

Figure 12.7 Tangential circle packing (left) and Thurston's circle packing (right).

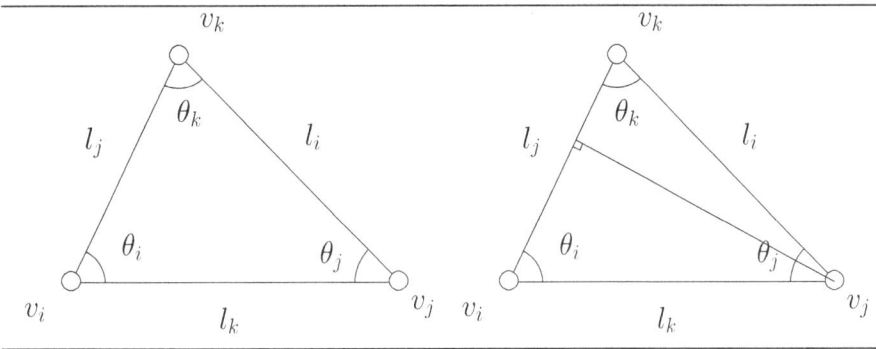

Figure 12.8 Derivative cosine law.

Discrete Ricci Energy on Triangle We first define discrete Ricci energy on each triangular face.

Definition 12.2.4 (Discrete Conformal Factor) *Given a discrete surface $(S, V, \mathcal{T})$ with a Turston's circle packing metric (Γ, Φ), the discrete conformal factor is defined as a function $\mathbf{u} : V \to \mathbb{R}$,*

$$
u_i = \begin{cases}
\log \gamma_i & \mathbb{R}^2 \\
\log \tanh \frac{\gamma_i}{2} & \mathbb{H}^2 \\
\log \tan \frac{\gamma_i}{2} & \mathbb{S}^2
\end{cases}
$$

Theorem 12.2.5 (Derivative Cosine Law) *Given a Euclidean triangle with edge lengths (l_i, l_j, l_k) and the corner angles $(\theta_i, \theta_j, \theta_k)$, then*

$$
\frac{d\theta_i}{dl_i} = \frac{l_i}{A}, \quad \frac{d\theta_i}{dl_j} = -\cos\theta_k \frac{l_i}{A}, \tag{12.2.5}
$$

where the area $A = l_j l_k \sin\theta_i$.

Proof We treat the edge lengths as variables and the angles are functions of edge lengths,

$$
\theta_i(l_i, l_j, l_k) : \mathbb{R}^3_{>0} \to (0, \pi).
$$

By Euclidean cosine law,

$$
\begin{aligned}
\frac{\partial}{\partial l_i}\left(2l_j l_k \cos\theta_i\right) &= \frac{\partial}{\partial l_i}\left(l_j^2 + l_k^2 - l_i^2\right) \\
-2l_j l_k \sin\theta_i \frac{d\theta_i}{dl_i} &= -2l_i \\
\frac{d\theta_i}{dl_i} &= \frac{l_i}{A}
\end{aligned}
$$

Next, as shown in the right frame of Figure 12.8, the edge

$$l_j = l_i \cos \theta_k + l_k \cos \theta_i$$

therefore,

$$
\begin{aligned}
\frac{\partial}{\partial l_j} \left(2 l_j l_k \cos \theta_i \right) &= \frac{\partial}{\partial l_j} \left(l_j^2 + l_k^2 - l_i^2 \right) \\
2 l_j &= 2 l_k \cos \theta_i - 2 l_j l_k \sin \theta_i \frac{d\theta_i}{dl_j} \\
\frac{d\theta_i}{dl_j} &= \frac{l_k \cos \theta_i - l_j}{A} = -\frac{l_i \cos \theta_k}{A} \\
&= -\frac{d\theta_i}{dl_i} \cos \theta_k
\end{aligned}
$$

□

Now, we use derivative cosine law to deduce the derivative relation between the angle and the circle radii. By the definition of circle packing metric,

$$l_i^2 = \gamma_j^2 + \gamma_k^2 + 2 \cos \varphi_{jk} \gamma_j \gamma_k$$

, taking the derivative with respect to γ_j on both sides,

$$
\begin{aligned}
\frac{\partial}{\partial \gamma_j} l_i^2 &= \frac{\partial}{\partial \gamma_j} \left(\gamma_j^2 + \gamma_k^2 + 2 \gamma_j \gamma_k \cos \varphi_{jk} \right) \\
2 l_i \frac{dl_i}{d\gamma_j} &= 2 \gamma_j + 2 \gamma_k \cos \varphi_{jk} \\
\frac{dl_i}{d\gamma_j} &= \frac{2 \gamma_j^2 + 2 \gamma_j \gamma_k \cos \varphi_{jk}}{2 l_i \gamma_j} \\
&= \frac{\gamma_j^2 + \gamma_k^2 + 2 \gamma_j \gamma_k \cos \varphi_{jk} + \gamma_j^2 - \gamma_k^2}{2 l_i \gamma_j} \\
&= \frac{l_i^2 + \gamma_j^2 - \gamma_k^2}{2 l_i \gamma_j}
\end{aligned}
$$

Let $u_i = \log \gamma_i$, then by chain rule $\frac{d\theta}{du} = \frac{d\theta}{dl} \frac{dl}{d\gamma} \frac{d\gamma}{du}$, then we obtain the derivative relation between $d\theta$ and $d\mathbf{u}$:

$$
\begin{pmatrix} d\theta_1 \\ d\theta_2 \\ d\theta_3 \end{pmatrix} = \frac{-1}{A}
\begin{pmatrix} l_1 & 0 & 0 \\ 0 & l_2 & 0 \\ 0 & 0 & l_3 \end{pmatrix}
\begin{pmatrix} -1 & \cos \theta_3 & \cos \theta_2 \\ \cos \theta_3 & -1 & \cos \theta_1 \\ \cos \theta_2 & \cos \theta_1 & -1 \end{pmatrix}
$$

$$
\begin{pmatrix} 0 & \frac{l_1^2 + r_2^2 - r_3^2}{2 l_1 r_2} & \frac{l_1^2 + r_3^2 - r_2^2}{2 l_1 r_3} \\ \frac{l_2^2 + r_1^2 - r_3^2}{2 l_2 r_1} & 0 & \frac{l_2^2 + r_3^2 - r_1^2}{2 l_2 r_3} \\ \frac{l_3^2 + r_1^2 - r_2^2}{2 l_3 r_1} & \frac{l_3^2 + r_2^2 - r_1^2}{2 l_3 r_2} & 0 \end{pmatrix}
\begin{pmatrix} r_1 & 0 & 0 \\ 0 & r_2 & 0 \\ 0 & 0 & r_3 \end{pmatrix}
\begin{pmatrix} du_1 \\ du_2 \\ du_3 \end{pmatrix}
$$

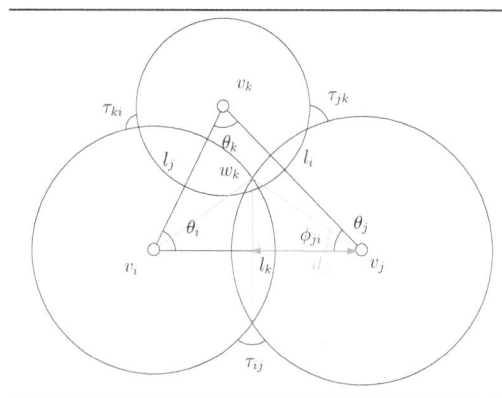

Figure 12.9 Derivative relation between the angle and the conformal factor.

Lemma 12.2.6 *As shown in Figure 12.9, the discrete surface is with a Thurston's circle packing metric, then*

$$\frac{dl_k}{du_j} = d_{ji}. \tag{12.2.6}$$

Proof As shown in Figure 12.9, from the defintion of circle packing metric $l_k^2 = \gamma_i^2 + \gamma_j^2 + 2\cos\tau_{ij}\gamma_i\gamma_j$, we obtain

$$2l_k \frac{dl_k}{d\gamma_j} = 2\gamma_j + 2\gamma_i \cos\tau_{ij}$$

$$r_j \frac{dl_k}{d\gamma_j} = \frac{2\gamma_j^2 + 2\gamma_i\gamma_j \cos\tau_{ij}}{2l_k}$$

$$= \frac{\gamma_j^2 + \gamma_i^2 + 2\gamma_i\gamma_j \cos\tau_{ij} + \gamma_j^2 - \gamma_i^2}{2l_k}$$

$$= \frac{l_k^2 + \gamma_j^2 - \gamma_i^2}{2l_k}$$

In the triangle $[v_i, v_j, w_k]$,

$$\frac{dl_k}{du_j} = 2\frac{l_k\gamma_j \cos\phi_{ji}}{2l_k} = \gamma_j \cos\phi_{ji} = d_{ji}$$

□

As shown in Figure 12.10, there is a unique circle orthogonal to three circles (v_i, γ_i), which is called the *power circle*. The *power center* is the power circle center o, the distance from o to edge $[v_i, v_j]$ is called *the height h_k*.

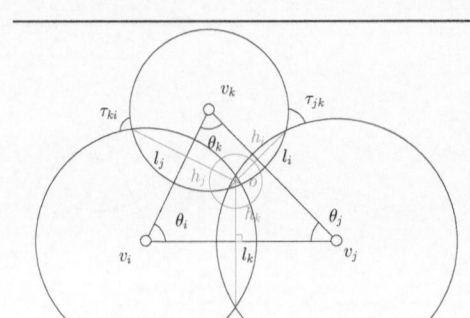

Figure 12.10 The derivative relation between the angles and the conformal factors under the Thurston's circle packing metric.

Theorem 12.2.7 (Symmetry) *Given a discrete surface* $(S, V, \mathcal{T})$ *with a Thurston's circle packing metric* (Γ, Φ), *then on each triangle*

$$\frac{d\theta_i}{du_j} = \frac{d\theta_j}{du_i} = \frac{h_k}{l_k}$$

$$\frac{d\theta_j}{du_k} = \frac{d\theta_k}{du_j} = \frac{h_i}{l_i} \qquad (12.2.7)$$

$$\frac{d\theta_k}{du_i} = \frac{d\theta_i}{du_k} = \frac{h_j}{l_j}$$

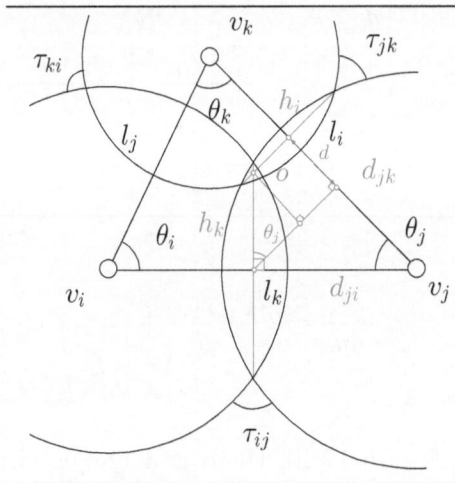

Figure 12.11 The derivative relation between the angles and the conformal factors under the Thurston's circle packing metric.

Proof From Figure 12.11, we directly obtain

$$
\frac{\partial \theta_i}{\partial u_j} = \frac{\partial \theta_i}{\partial l_i}\frac{\partial l_i}{\partial u_j} + \frac{\partial \theta_i}{\partial l_k}\frac{\partial l_k}{\partial u_j} = \frac{\partial \theta_i}{\partial l_i}\left(\frac{\partial l_i}{\partial u_j} - \frac{\partial l_k}{\partial u_j}\cos\theta_j\right)
$$

$$
= \frac{l_i}{A}(d_{jk} - d_{ji}\cos\theta_j) = \frac{dl_i}{l_i l_k \sin\theta_j} = \frac{h_k \sin\theta_j}{l_k \sin\theta_j} = \frac{h_k}{l_k}
$$

□

Lemma 12.2.8 *For any three non-obtuse angles* $\varphi_{ij}, \varphi_{jk}, \varphi_{ki} \in [0, \frac{\pi}{2})$ *and any three positive numbers* γ_i, γ_j *and* γ_k, *there is a configuration of 3 circles in Euclidean geometry, unique upto isometry, having radii* γ_i*'s and meeting in angles* φ_{ij}*'s.*

Proof By the definition of circle packing metric, we have

$$
max\{\gamma_i^2, \gamma_j^2\} < \gamma_i^2 + \gamma_j^2 + 2\gamma_i\gamma_j\cos\varphi_{ij} \le (\gamma_i + \gamma_j)^2
$$

hence $max\{\gamma_i, \gamma_j\} < l_k \le \gamma_i + \gamma_j$, so $l_k \le r_i + r_j < l_i + l_j$.

□

For any three non-obtuse angles $\varphi_{ij}, \varphi_{jk}, \varphi_{ki} \in [0, \frac{\pi}{2})$, any $(u_i, u_j, u_k) \in \mathbb{R}^3$, there is a configuration of 3 circles in Euclidean geometry, with the radii $(e^{u_i}, e^{u_j}, e^{u_k})$ and the meeting angles $(\varphi_{ij}, \varphi_{jk}, \varphi_{ki})$. Suppose the corner angles are $(\theta_i, \theta_j, \theta_k)$, then they are the functions of the conformal factors (u_i, u_j, u_k) and the intersection angles. We always fix the intersection angles, and obtain the following lemma.

Lemma 12.2.9 *The differential 1-form*

$$
\omega = \theta_i du_i + \theta_j du_j + \theta_k du_j
$$

is well defined and closed in $\Omega := \{(u_1, u_2, u_3) \in \mathbb{R}^3\}$.

Proof By symmetry relation Eqn. (12.2.7), we have $\frac{\partial \theta_i}{\partial u_j} = \frac{\partial \theta_j}{\partial u_i}$, so

$$
d\omega = \left(\frac{\partial \theta_i}{\partial u_j} - \frac{\partial \theta_j}{\partial u_i}\right)du_j \wedge du_i + \left(\frac{\partial \theta_j}{\partial u_k} - \frac{\partial \theta_k}{\partial u_j}\right)du_k \wedge du_j + \left(\frac{\partial \theta_k}{\partial u_i} - \frac{\partial \theta_i}{\partial u_k}\right)du_i \wedge du_k,
$$

therefore $d\omega = 0$.

□

Lemma 12.2.10 *The Ricci energy on the triangle*

$$
E(u_1, u_2, u_3) = \int_{(0,0,0)}^{(u_1, u_2, u_3)} \theta_1 du_1 + \theta_2 du_2 + \theta_3 du_3 \tag{12.2.8}
$$

is well defined in $\mathbb{R}^3$, *and strictly concave on the subspace* $u_1 + u_2 + u_3 = 0$.

Proof By lemma 12.2.8 and lemma 12.2.9, the differential 1-form $\omega = \sum \theta_i du_i$ is closed in $\mathbb{R}^3$, $\mathbb{R}^3$ is simply connected, therefore ω is exact, the integration is independent of the choice of the path, hence the Ricci energy $E(u_1, u_2, u_3)$ is well defined. because of $\theta_1 + \theta_2 + \theta_3 = \pi$,

$$\frac{\partial \theta_i}{\partial u_i} = -\frac{\partial \theta_i}{\partial u_j} - \frac{\partial \theta_i}{\partial u_k} = -\frac{\partial \theta_j}{\partial u_i} - \frac{\partial \theta_k}{\partial u_i}$$

The gradient $\nabla E = (\theta_1, \theta_2, \theta_3)$, the Hessian matrix is

$$H = \begin{pmatrix} \frac{\partial \theta_1}{\partial u_1} & \frac{\partial \theta_1}{\partial u_2} & \frac{\partial \theta_1}{\partial u_3} \\ \frac{\partial \theta_2}{\partial u_1} & \frac{\partial \theta_2}{\partial u_2} & \frac{\partial \theta_2}{\partial u_3} \\ \frac{\partial \theta_3}{\partial u_1} & \frac{\partial \theta_3}{\partial u_2} & \frac{\partial \theta_3}{\partial u_3} \end{pmatrix} = - \begin{pmatrix} \frac{h_3}{l_3} + \frac{h_2}{l_2} & -\frac{h_3}{l_3} & -\frac{h_2}{l_2} \\ -\frac{h_3}{l_3} & \frac{h_3}{l_3} + \frac{h_1}{l_1} & -\frac{h_1}{l_1} \\ -\frac{h_2}{l_2} & -\frac{h_1}{l_1} & \frac{h_2}{l_2} + \frac{h_1}{l_1} \end{pmatrix}$$

$-H$ is diagonal dominant with the null space $(1, 1, 1)$. Therefore, on the subspace $u_1 + u_2 + u_3 = 0$, the Hessian matrix is strictly negative definite and the discrete Ricci energy $E(u_1, u_2, u_3)$ is strictly concave.

$\square$

Discrete Ricci Energy on Surface Now, we consider the discrete Ricci energy defined on the whole discrete surface $(S, V, \mathcal{T})$ with the Thurston's circle packing metric (Γ, Φ).

Our goal is the discrete conformal factor $\mathbf{u}$, such that the corresponding circle packing metric induces the desired target curvature $\bar{\mathbf{K}}$. This can be solved using the combinatorial surface Ricci flow.

Definition 12.2.11 (Normalized Combinatorial Surface Ricci Flow) *Suppose the target curvature is given, $\bar{K} : V \to (-\infty, 2\pi)$, satisfying the Gauss-Bonnet condition: $\sum_{v_i \in V} \bar{K}_i = 2\pi\chi(S)$, the normalized combinatorial surface Ricci flow is defined as $\forall v_i \in V$,*

$$\frac{du_i}{dt} = \bar{K}_i - K_i. \tag{12.2.9}$$

We expect that when t goes to ∞, the discrete conformal factor $\mathbf{u}(t) \to \bar{\mathbf{u}}$, and $\mathbf{K}(t) \to \bar{\mathbf{K}}$. Therefore, we need to consider the existence, the uniqueness of the solution to the flow, and the convergence of the flow.

Definition 12.2.12 (Circle Packing Laplace-Beltrami Operator) *Given a discrete surface $(S, V, \mathcal{T})$ with a circle packing metric (Γ, Φ), for each edge $[v_i, v_j]$ shared by two faces $[v_i, v_j, v_k]$ and $[v_j, v_i, v_l]$, the power centers are o_k and o_l on each face respectively, the height from o_k to $[v_i, v_j]$ is h_k and the height from o_l to $[v_j, v_i]$ is h_l, then the edge weight is given by*

$$w_{ij} := \frac{h_k}{l_{ij}} + \frac{h_l}{l_{ij}}. \tag{12.2.10}$$

The Laplace-Beltrami operator of the circle packing metric (Γ, Φ) is defined as: given $f : V \to \mathbb{R}$,

$$\Delta f(v_i) = \sum_j w_{ij}(f(v_j) - f(v_i)). \tag{12.2.11}$$

Lemma 12.2.13 *Given a discrete surface* $(S, V, \mathcal{T})$ *with a Turston's circle packing metric* (Γ, Φ)*, the following relations hold:*

1. *Symmetry*

$$\frac{\partial K_i}{\partial u_j} = \frac{\partial K_j}{\partial u_i} = -w_{ij} \tag{12.2.12}$$

where the edge weight w_{ij} *is defined in Eqn. (12.2.10);*

2. *Discrete Poisson Equation*

$$d\mathbf{K} = \Delta d\mathbf{u}, \tag{12.2.13}$$

Δ *is the discrete Laplace-Beltrami operator of the circle packing metric,* $\mathbf{u} = (u_1, u_2, \dots, u_n)$ *and* $\mathbf{K} = (K_1, K_2, \dots, K_n)$ *are the vector representations of the discrete conformal factor and the discrete curvature.*

Proof Suppose an $[v_i, v_j]$ is shared by two faces $[v_i, v_j, v_k]$ and $[v_j, v_i, v_l]$, then

$$\frac{\partial K_i}{\partial u_j} = -\frac{\partial \theta_i^{jk}}{\partial u_j} - \frac{\partial \theta_i^{jl}}{\partial u_j} = -w_{ij} = -\frac{\partial \theta_j^{ik}}{\partial u_i} - \frac{\partial \theta_j^{il}}{\partial u_i} = \frac{\partial K_j}{\partial u_i},$$

where the edge weight w_{ij} is defined in Eqn. (12.2.10). This proves the symmetry Eqn. (12.2.12).

By discrete Gauss-Bonnet theorem 12.2.2, the total curvature is constant, hence $\sum_j \frac{K_j}{\partial u_i} = 0$, hence

$$\frac{\partial K_i}{\partial u_i} = -\sum_{j \neq i} \frac{\partial K_j}{\partial u_i} = -\sum_{j \neq i} \frac{\partial K_i}{\partial u_j} = \sum_{j \neq i} w_{ij}.$$

This shows the Poisson equation in Eqn. (12.2.13).

$\square$

Definition 12.2.14 (Discrete Ricci Energy) *Given a discrete surface* $(S, V, \mathcal{T})$ *with a Thurston's circle packing metric* (Γ, Φ)*, the discrete surface Ricci energy is defined as*

$$E(\mathbf{u}) := \int_0^{\mathbf{u}} \sum_{v_i \in V} K_i du_i. \tag{12.2.14}$$

Lemma 12.2.15 *Given a discrete surface* $(S, V, \mathcal{T})$ *with a Thurston's circle packing metric* (Γ, Φ)*, the discrete surface Ricci energy is strictly convex on the hyperplane* $\sum_{v_i \in V} u_i = 0$.

Proof Define differential 1-form $\omega = \sum_{v_i \in V} K_i du_i$, then by lemma 12.2.8, ω is defined on $\mathbb{R}^n$, where n is the number of vertices. By the symmetry Eqn. (12.2.12), ω is closed $d\omega = 0$. Because $\mathbb{R}^n$ is simply connected, ω is exact, therefore the discrete surface Ricci energy Eqn. (12.2.14) is well defined.

The gradient of the energy is the discrete curvature $\nabla E(\mathbf{u}) = \mathbf{K}$, hence the Hessian is $\left(\frac{\partial K_i}{\partial u_j} \right)$. By Eqn. (12.2.13), the Hessian matrix has the null space $(1, 1, \dots, 1)$,

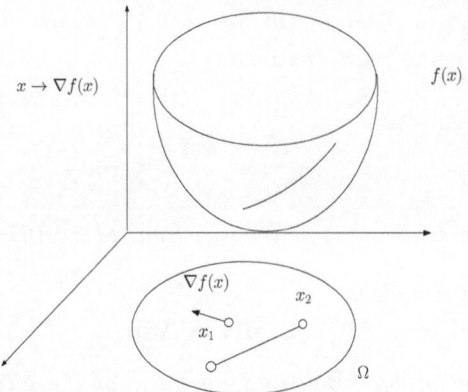

Figure 12.12 Gradient map of a convex function.

and is strictly positive definite on the complmentary space $\sum_{v_i \in V} u_i = 0$, hence the discrete Ricci energy is strictly convex on the hyper-plane.

$\square$

Lemma 12.2.16 *Suppose $\Omega \subset \mathbb{R}^n$ is a convex domain, $f : \Omega \to \mathbb{R}$ is a C^1 strictly convex function, then the gradient map*

$$\nabla f : \Omega \to \nabla f(\Omega), \quad x \mapsto \nabla f(x), \quad x \in \Omega$$

is injective.

Proof As shown in Figure 12.12, suppose $x_1, x_2 \in \Omega$, $x_1 \neq x_2$ but $\nabla f(x_1) = \nabla f(x_2)$. Because Ω is convex, the line segment $(1-t)x_1 + tx_2$, $t \in [0,1]$ is contained in Ω. Construct a convex function $g(t) = f((1-t)x_1 + tx_2)$,

$$g''(t) = (x_2 - x_1)^T Hess((1-t)x_1 + tx_2)(x_2 - x_1) > 0,$$

hence $g'(t)$ is monotonously increasing. But

$$g'(0) = \langle \nabla f(x_1), x_2 - x_1 \rangle = \langle \nabla f(x_2), x_2 - x_1 \rangle = g'(1),$$

contradiction.

$\square$

This leads to the global rigidity theorem as follows:

Theorem 12.2.17 (Global Rigidity) *Suppose $(S, V, \mathcal{T})$ is a discrete surface with a Thurston' circle packing metric (Γ, Φ), all intersection angles are non-obtuse, $\varphi_{ij} \in [0, \frac{\pi}{2})$. Given the target curvature $(\bar{K}_1, \bar{K}_2, \cdots, \bar{K}_n)$, satisfying the Gauss-Bonnet condition $\sum_i \bar{K}_i = 2\pi\chi(M)$. If there is a solution to the combinatorial surface Ricci flow Eqn. 12.2.9, $(\bar{u}_1, \bar{u}_2, \cdots, \bar{u}_n) \in \Omega(M), \sum_i \bar{u}_i = 0$ exists, then it is unique.*

Proof The discrete Ricci energy E on $\Omega \cap \{\sum_{v_i \in V} u_i = 0\}$ is convex,

$$\nabla E(u_1, u_2, \cdots, u_n) = (K_1, K_2, \cdots K_n).$$

Use lemma 12.2.16, the gradient mapping

$$\nabla E : \Omega \cap \left\{ \sum_{v_i \in V} u_i = 0 \right\} \to \left\{ K : \sum_{v_i \in V} K_i = 2\pi\chi(S) \right\}$$

is injective, hence the solution is unique.

□

The existence is proven by Thurston, a target curvature is realizable by the circle packing metric if and only if it satisfies the condition in the following Thurston's theorem.

Theorem 12.2.18 (Thurston) *Suppose $(S, V, \mathcal{T})$ is a closed, connected discrete surface with a circle packing metric (Γ, Φ), I is a proper subset of vertices of V, here the intersection angle is a map $\Phi : E \to [0, \frac{\pi}{2})$. Then for any Thurston's circle packing metric, we have*

$$\sum_{i \in I} K_i(u) > - \sum_{(e,v) \in Lk(I)} (\pi - \Phi(e)) + 2\pi\chi(F_I),$$

where F_I is the CW-subcomplex of cells whose vertices are in I and

$$Lk(I) = \{(e, v) | v \in I, e \cap I = \emptyset, (e, v) \ form \ a \ triangle\}$$

Definition 12.2.19 (Convergent Solution) *A solution to combinatorial surface Ricci flow Eqn. (12.2.9) is called convergent if*

1. $\lim_{t \to \infty} K_i(t) = K_i(\infty) \in (-\infty, 2\pi)$ *exists for all $v_i \in V$, and*

2. $\lim_{t \to \infty} \gamma_i(t) = \gamma_i(\infty) \in \mathbb{R}_{>0}$ *exists for all $v_i \in V$.*

Definition 12.2.20 (Convergent exponentially fast) *A convergent solution to combinatorial surface Ricci flow Eqn. (12.2.9) is called convergent exponentially fast if there are positive constants c_1, c_2 so that for all time $t \geq 0$,*

$$|K_i(t) - K_i(\infty)| \leq c_1 e^{-c_2 t},$$

and

$$|\gamma_i(t) - \gamma_i(\infty)| \leq c_1 e^{-c_2 t}.$$

Chow and Luo proved the convergence of the combinatorial surface Ricci flow.

Theorem 12.2.21 (Chow and Luo) *Suppose $(S, V, \mathcal{T})$ is a closed, connected discrete surface with an initial circle packing metric (Γ, Φ), the solution to the normalized combinatorial surface Ricci flow Eqn. (12.2.9) in the Euclidean geometry with the given initial value exists for all time and converges if and only if for any proper subset $I \subset V$,*

$$\sum_{i \in I} \bar{K}_i > - \sum_{(e,v) \in Lk(I)} (\pi - \Phi(e)) + 2\pi\chi(F_I).$$

Furthermore, if the solution converges, then it converges exponentially fast to the metric of target curvature $\bar{K}$.

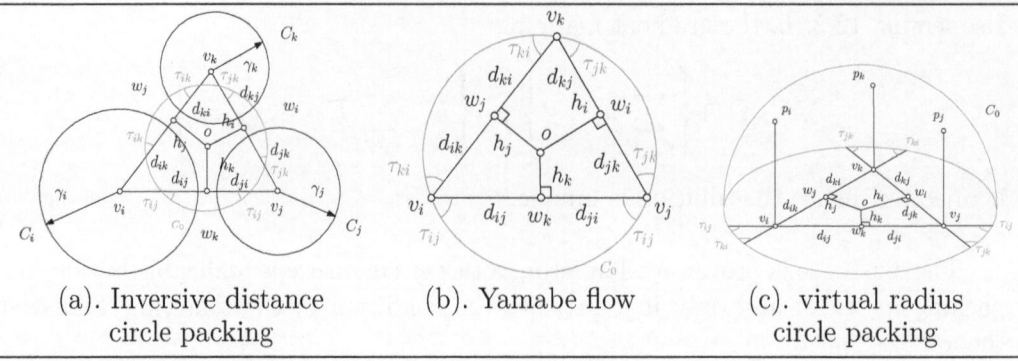

(a). Inversive distance
circle packing

(b). Yamabe flow

(c). virtual radius
circle packing

Figure 12.13 Generalized circle packing metric.

In fact, the normalized combinatorial surface Ricci flow Eqn. (12.2.9) is the gradient flow of the concave energy,

$$\tilde{E}(\mathbf{u}) := \sum_{v_i \in V} \bar{K}_i u_i - \int^{\mathbf{u}} \sum_{v_i \in V} K_i du_i,$$

which can be optimized directly using Newton's method.

Generalized Circle Packing Metric   Thurston's circle packing can be generalized directly. Figure 12.13 shows the inversive distance circle packing (left), the combinatorial Yamabe flow (middle) and the virtual radius circle packing (right).

Given a discrete surface $(S, V, \mathcal{T})$, we associate each vertex v_i with a circle C_i with radius γ_i. On every edge $[v_i, v_j]$, two circles are disjoint. We define the inversive distance function defined on the edge set $\eta : E \to (0, \infty)$, which is the generalization of the $\cos \varphi_{ij}$ in the Thurston's circle packing. The edge length of $[v_i, v_j]$ is defined as

$$l_{ij}^2 = \gamma_i^2 + \gamma_j^2 + 2\eta_{ij}\gamma_i\gamma_j \qquad (12.2.15)$$

In the left frame of Figure 12.13, the power circle is orthogonal to three vertex circles, the power center and the heights are defined similarly. The definition of the circle packing edge weight Eqn. (12.2.10), the Laplace-Beltrami operator Eqn. (12.2.11) are in the exact same formulae, the symmetry relation Eqn. (12.2.12) and the Poisson equation Eqn. (12.2.13) still hold.

The combinatorial Yamabe flow is defined in a simpler form,

$$l_{ij}^2 = 2\eta_{ij}\gamma_i\gamma_j = 2e^{u_i}\eta_{ij}e^{u_j}. \qquad (12.2.16)$$

The power circle is the circum-cirle of the triangle, the power center is the circle center, the heights are the line segments through the power center orthogonal to the edges. The edge weights Eqn. (12.2.10) become the conventional cotangent edge weights

$$w_{ij} = \cot \theta_k^{ij} + \cot \theta_l^{ji}.$$

Table 12.1 Parameters for different schemes of circle packing.

Scheme	ε_i	ε_j	η_{ij}
Tangential Circle Packing	$+1$	$+1$	$+1$
Thurston's Circle Packing	$+1$	$+1$	$[0,1]$
Inversive Distance Circle Packing	$+1$	$+1$	$(0,\infty)$
Yamabe Flow	0	0	$(0,\infty)$
Virtual Distance Circle Packing	-1	-1	$(0,\infty)$
Mixed Type	$\{-1,0,+1\}$	$\{-1,0,+1\}$	$(0,\infty)$

The Laplace-Beltrami operator Eqn. (12.2.11) is the conventional discrete Laplace-Beltrami operator. The symmetry relation Eqn. (12.2.12) and the Poisson equation Eqn. (12.2.13) still hold.

For virtual radius circle packing, we associate each vertex v_i with a circle C_i with a imaginary radius $\sqrt{-1}\gamma_i$. For each edge, we associate the inversive distance $\eta_{ij} \in (0,\infty)$. The edge length of $[v_i, v_j]$ is defined as

$$l_{ij}^2 = -\gamma_i^2 - \gamma_j^2 + 2\eta_{ij}\gamma_i\gamma_j \qquad (12.2.17)$$

In the right frame of Figure 12.13, we vertically lift v_i by γ_i to p_i, and lift v_j, v_k to p_j, p_k similarly. The power circle is the equator of the hemisphere through p_i, p_j, p_k. The heights are the line segments through the power center and orthogonal to the edges. The definition of the circle packing edge weight Eqn. (12.2.10), the Laplace-Beltrami operator Eqn. (12.2.11) are in the exact same formulae, the symmetry relation Eqn. (12.2.12) and the Poisson equation Eqn. (12.2.13) still hold.

Unified Combinatorial Surface Ricci Flow   All the above circle packing schemes can be unified into a consistent framework [348, 78]. Furthermore, they can be mixed together to define the mixed circle packing scheme.

The *discrete conformal factor* is defined as $u : V \to \mathbb{R}$,

$$u_i = \begin{cases} \log \gamma_i & \mathbb{E}^2 \\ \log \tanh \frac{\gamma_i}{2} & \mathbb{H}^2 \\ \log \tan \frac{\gamma_i}{2} & \mathbb{S}^2 \end{cases}$$

The *edge lengths* are given by

$$u_i = \begin{cases} l_{ij}^2 = 2\eta_{ij}e^{u_i+u_j} + \varepsilon_i e^{2u_i} + \varepsilon_j e^{2u_j} & \mathbb{E}^2 \\ \cosh l_{ij} = \frac{4\eta_{ij}e^{u_i+u_j}+(1+\varepsilon_i e^{2u_i})(1+\varepsilon_j e^{2u_j})}{(1-\varepsilon_i e^{2u_i})(1-\varepsilon_j e^{2u_j})} & \mathbb{H}^2 \\ \cos l_{ij} = \frac{-4\eta_{ij}e^{u_i+u_j}+(1-\varepsilon_i e^{2u_i})(1-\varepsilon_j e^{2u_j})}{(1+\varepsilon_i e^{2u_i})(1+\varepsilon_j e^{2u_j})} & \mathbb{S}^2 \end{cases}$$

For each triangle $[v_i, v_j, v_k]$, its area equals

$$A = \frac{1}{2}\sin\theta_i s(l_j)s(l_k), \qquad (12.2.18)$$

where the auxiliary function

$$s(x) = \begin{cases} x & \mathbb{E}^2 \\ \sinh x & \mathbb{H}^2 \\ \sin x & \mathbb{S}^2 \end{cases}$$

Another auxiliary function is defined as

$$\tau(i,j,k) = \begin{cases} \frac{1}{2}(l_i^2 + \epsilon_j \gamma_j^2 - \epsilon_k \gamma_k^2) & \mathbb{E}^2 \\ \cosh l_i \cosh^{\epsilon_j} \gamma_j - \cosh^{\epsilon_k} \gamma_k & \mathbb{H}^2 \\ \cos l_i \cos^{\epsilon_j} \gamma_j - \cos^{\epsilon_k} \gamma_k & \mathbb{S}^2 \end{cases}$$

Lemma 12.2.22 *The corner angles* $(\theta_i, \theta_j, \theta_k)$ *are the functions of* (u_i, u_j, u_k), *the Jacobi matrix is given by*

$$\frac{\partial(\theta_i, \theta_j, \theta_k)}{\partial(u_i, u_j, u_k)} = -\frac{1}{2A} L\Theta L^{-1} D, \tag{12.2.19}$$

where the matrix L *is*

$$L = \begin{pmatrix} s(l_i) & 0 & 0 \\ 0 & s(l_j) & 0 \\ 0 & 0 & s(l_k) \end{pmatrix} \tag{12.2.20}$$

Θ

$$\Theta = \begin{pmatrix} -1 & \cos\theta_k & \cos\theta_j \\ \cos\theta_k & -1 & \cos\theta_i \\ \cos\theta_j & \cos\theta_i & -1 \end{pmatrix} \tag{12.2.21}$$

matrix D *is*

$$D = \begin{pmatrix} 0 & \tau(i,j,k) & \tau(i,k,j) \\ \tau(j,i,k) & 0 & \tau(j,k,i) \\ \tau(k,i,j) & \tau(k,j,i) & 0 \end{pmatrix} \tag{12.2.22}$$

The lemma can be proved by direct computation. It can be verified that the Jacobi matrix is symmetric. The Jacobi matrix has the geometric interpretation.

Lemma 12.2.23 *For the triangle with* $\mathbb{E}^2$, $\mathbb{H}^2$ *and* $\mathbb{S}^2$ *background geometry, there is a power circle, orthogonal to three vertex circles. The distances from the power center to the edges are* h_i, h_j *and* h_k *respectively, then*

$$\frac{\partial\theta_1}{\partial u_2} = \frac{\partial\theta_2}{\partial u_1} = \frac{h_3}{l_3}$$

$$\frac{\partial\theta_1}{\partial u_2} = \frac{\partial\theta_2}{\partial u_1} = \frac{\tanh h_3}{\sinh^2 l_3}\sqrt{2\cosh^{\epsilon_1} r_1 \cosh^{\epsilon_2} r_2 \cosh l_3 - \cosh^{2\epsilon_1} r_1 - \cosh^{2\epsilon_2} r_2}$$

$$\frac{\partial\theta_1}{\partial u_2} = \frac{\partial\theta_2}{\partial u_1} = \frac{\tan h_3}{\sin^2 l_3}\sqrt{-2\cos^{\epsilon_1} r_1 \cos^{\epsilon_2} r_2 \cos l_3 + \cos^{2\epsilon_1} r_1 + \cos^{2\epsilon_2} r_2}$$

It can be proved by direct computation. By the symmetry, we can define the Ricci energy on each face,

Definition 12.2.24 *Given a discrete surface $(S, V, \mathcal{T})$ with circle packing metric $(\Gamma, \eta, \varepsilon)$, for each triangle $f = [v_i, v_j, v_k]$ with inner angles $(\theta_i, \theta_j, \theta_k)$, the entropy energy for the face is given by*

$$E_f(u_i, u_j, u_k) := \int^{(u_i, u_j, u_k)} \theta_i du_i + \theta_j du_j + \theta_k du_k. \tag{12.2.23}$$

By lemma 12.2.23, the differential 1-form $\omega = \theta_i du_i + \theta_j du_j + \theta_k du_k$ is closed and exact on a simply connected domain, hence the energy is well defined.

Definition 12.2.25 (Unified Discrete Surface Ricci Energy) *Given a discrete surface $(S, V, \mathcal{T})$ with circle packing metric $(\Gamma, \eta, \varepsilon)$, the Ricci entropy energy is given by*

$$E(\mathbf{u}) := \int^{\mathbf{u}} \sum_{v_i \in V} (\bar{K}_i - K_i) du_i. \tag{12.2.24}$$

The surface Ricci energy can be represented as the summation of face energies,

$$E(\mathbf{u}) = \sum_{v_i \in V} (\bar{K}_i - 2\pi) u_i + \sum_{f \in F} E_f(\mathbf{u}).$$

Definition 12.2.26 (Unified Normalized Discrete Surface Ricci Flow) *Given a discrete surface $(S, V, \mathcal{T})$ with circle packing metric $(\Gamma, \eta, \varepsilon)$, given the target curvature $\bar{K} : V \to (-\infty, 2\pi)$, satisfying the Gauss-Bonnet condition $\sum_{v_i \in V} \bar{K}_i = 2\pi \chi(S)$, the unified normalized discrete surface Ricci flow is defined as $\forall v_i \in V$*

$$\frac{du_i}{dt} = \bar{K}_i - K_i. \tag{12.2.25}$$

For general schemes, the existence of the solution is the discrete surface Ricci flow is a challenging problem. Because it is very likely that during the flow, some triangles become degenerated. A natural idea to avoid the degeneracy is to update the triangulation during the flow, such that the triangulation is always Delaunay. It turns out that this method successfully prevents the flow from degeneracy and leads to the celebrated discrete uniformization theorem [151, 152]. Details can be found in Chapter 20.

12.2.2 The Metric Ricci Flow for Surfaces

Clearly, after seeing the force of the combinatorial Ricci flow for surfaces, and having also in mind the metric Ricci curvature introduced in Section 12.1, one cannot but consider a similar flow for polyhedral manifolds. Furthermore, in doing so one would make the connection between the combinatorial curvature and flow and the dynamic research in the geometry of Alexandrov spaces [259]. There exists yet one more reason, though, for attempting a metric Ricci flow for PL surfaces, namely to gain a better understanding of some of the properties of the combinatorial flow [279]. (As we shall see, our approach toward the metric flow will be through transforming it into a smooth one. Therefore, connecting it to the combinatorial one, would let us infer properties of the later from those of the classical flow.)

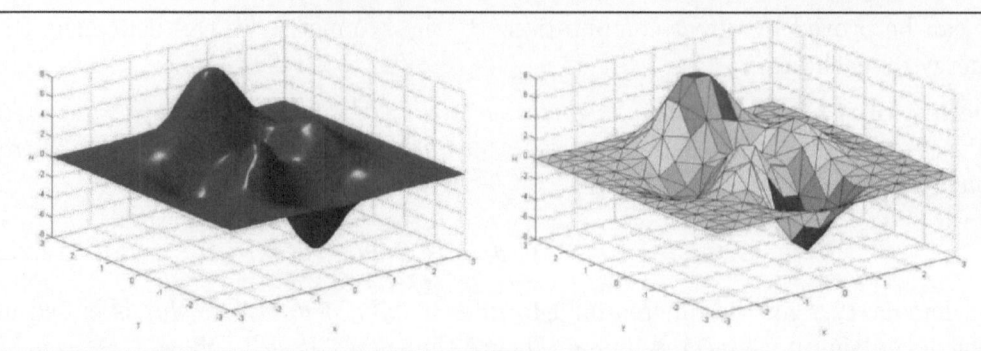

Figure 12.14   A smooth surface (left) as the smoothing of one of its PL approximations (right) [283].

12.2.2.1   Smoothings and Metric Curvatures

We begin by noting that, combining our previous observations on smoothings, δ-approximations and the work of Hühnels, we actually obtain a positive answer to the question – not posed so far, to the best of our knowledge – whether the metric curvature version of Brehm and Hühnels's basic result also holds, namely we have proved (see Figure 12.14):

Proposition 12.2.27 ([279]) *Let S^2_{Pol} be a compact polyhedral surface without boundary. Then there exists a sequence $\{S^2_m\}_{m\in\mathbb{N}}$ of smooth surfaces (homeomorphic to S^2_{Pol}), such that*

1. *(a) $S^2_m = S^2_{Pol}$ outside the $\frac{1}{m}$-neighborhood of the 1-skeleton of S^2_{Pol},*

   (b) The sequence $\{S^2_m\}_{m\in\mathbb{N}}$ converges to S^2_{Pol} in the Hausdorff metric;

2. *$K(S^2_{Pol}) \to K_W(S^2_{Pol})$, where the convergence is in the weak sense.*

The proposition above shows that, in fact, the metric approach to curvature is essentially equivalent to the combinatorial (angle-based) one, as far as polyhedral surfaces (in $\mathbb{R}^3$) are concerned.[14] In particular, as far as approximations of smooth surfaces in $\mathbb{R}^3$ are concerned, both approaches render, in the limit, the classical Gauss curvature.) However, it should be stressed that the metric approach is more general, given that it is applicable to a very large class of abstract metric spaces.

Remark 12.2.28 *The converse implication – namely that Gaussian curvature $K(\Sigma)$ of a smooth surface Σ may be approximated arbitrarily well by the Wald curvatures $K_W(\Sigma_{Pol,m})$ of a sequence of approximating polyhedral surfaces $\Sigma_{Pol,m}$ – is, as we have already mentioned above, quite classical. (For other approaches to curvatures convergence, see, among the extensive literature dedicated to the subject, [75] and [52], [91], for the theoretical and applicative viewpoints, respectively.)*

---

[14]This holds, of course, up to the specific type of convergence for the metric and combinatorial curvature, namely pointwise and in measure, respectively.

Remark 12.2.29 *The metric approach adopted here renders, in fact, a somewhat stronger result than that given in [56], since no embedding in $\mathbb{R}^3$ is assumed, just in some $\mathbb{R}^N$ (even though as we have already noted above, this represents only a slight improvement). More importantly, no change in the geometry of the 1-skeleton is made, not even in the neighborhoods of the vertices.*

12.2.2.2 A Metric Ricci Flow

From the metric and curvature approximations results above, it follows that one can study the properties of the metric Ricci flow via those of its smooth counterpart, by passing to a smoothing of the polyhedral surface. The heavier machinery of metric curvature considered above pay off, in the sense that, by using it, the flow is purely metric and, moreover, the curvature at each stage (that is, for every "t") is given – as in the classical context – in an intrinsic manner, i.e. solely in terms of the metric.

The Flow   In direct analogy with the classical flow

$$\frac{dg_{ij}(t)}{dt} = -2K(t)g_{ij}(t)\,. \tag{12.2.26}$$

we define the *metric* Ricci flow as

$$\frac{dl_{ij}}{dt} = -2K_i l_{ij}\,, \tag{12.2.27}$$

where $l_{ij} = l_{ij}(t)$ denote the edges (1-simplices) of the triangulation (PL or piecewise flat surface) incident to the vertex $v_i = v_i(t)$, and $K_i = K_i(t)$ denotes the curvature at the same vertex. We shall discuss below in detail what proper notion of curvature should be chosen to render this type of metric flow meaningful.

We also consider the normalized flow

$$\frac{dg_{ij}(t)}{dt} = (K - K(t))g_{ij}(t)\,, \tag{12.2.28}$$

and its metric counterpart

$$\frac{dl_{ij}}{dt} = (\bar{K} - K_i)l_{ij}\,, \tag{12.2.29}$$

where $K, \bar{K}$ denote the average classical, respectively metric, sectional (Gauss) curvature of the initial surface: S_0, $K = \int_{S_0} K(t)dA / \int_{S_0} dA$, and $\bar{K} = \frac{1}{|V|}\sum_{i=1}^{|V|} K_i$, respectively. (Here $|V|$ denotes, as usually, the number of the vertex set of S_{Pol}.)

Our approach to this type of metric flow should be evident in light of the previous discussion: Using smoothings, we approximate the metric Ricci curvature by its classical counterpart, thus replacing the metric flow with its smooth model.

Remark 12.2.30 *Since for PL surfaces (hence for their smoothings) $K_W(p) = 0$ for all points apart from vertices, the ensuing Ricci flow is stationary, except at vertices where the change rate is quite drastic. In this aspect, the metric Ricci flow introduced here resembles the combinatorial, rather than the smooth (classical) one. This should* *not be too surprising, since, as already stated, the initial motivation of considering the metric flow (and, a fortiori, smoothings) was to gain a better understanding of some of the properties of the combinatorial flow.*

An advantage of the method introduced here is that it automatically solves, by passing to infinitesimal distances, the asymmetry in equation 12.2.27, that is caused by the fact that the curvature on two different vertices acts, so to say, on the same edge. We shall address this issue again shortly.

The main theoretical disadvantage of the approach embraced here is that it is not easy to extend it to higher dimensional manifolds. Indeed, it is not clear how to correctly define a flow for general high-dimensional manifolds (both for theoretical reasons and because of their applications, such as Medical Imaging, Video, etc.), not least because 3-dimensional analogs of all the relevant results on the Ricci flow for smooth surfaces have yet to be obtained. Moreover, even defining a metric Ricci curvature in dimension 3 and higher is a non-trivial endeavor. See, however, the discussion and results in the following sub-section.

Therefore, developing a purely metric Ricci flow – that is one that does not make appeal to smoothings – is highly desirable. One basic observation that has to be made is that one hast to take care of the lack of symmetry that we mentioned when we fist introduced the metric flow. From symmetry reasons, a natural way of defining the flow while addressing this is issue is (using the same notation was before):

$$\frac{dl_{ij}}{dt} = -\frac{K_i + K_j}{2} l_{ij} , \qquad (12.2.30)$$

where in this case, K_i, K_j denote, of course, the Wald curvature at the vertices v_i and v_j, respectively. (Note that, in fact, this expression appears also in the practical method of computing the combinatorial curvature, where it is derived via the use of a conformal factor [150].)

An important benefit of such a purely metric flow would be that, as in the case of combinatorial Ricci flow of Chow and Luo [88], equation (12.2.26) becomes – due to the fact that K_i depends only the lengths of the edges l_{ij}, and not on their derivatives – an ODE, instead of a PDE, thence they are easier to study and enjoy better properties. In particular (and of importance in applications) the metric flow will have the backward existence property.

For further observations regarding the metric flow, as well as for some other possible improvements, see [278].

Existence and Uniqueness For the classical Ricci flow $dg_{ij}(t)/dt = 2K(t)g_{ij}(t))$ the (local) existence and uniqueness hold, on some *maximal* time interval $[0, T]; 0 < T \leq \infty$ (see, e.g. [87], as well as [158], for the original, different proof.[15]) Moreover, the backward uniqueness of a solution (if existing) has been proven by Kotschwar [194] (see also [323] for a sketch of the proof). Beyond the theoretical importance, the

---

[15]See also [323], Theorems 5.2.1 and 5.2.2 and the discussion following them for short exposition of the main steps of the proof.

existence and uniqueness of the backward flow would allow us to find surfaces in the conformal class of a given circle packing (Euclidean or Hyperbolic).

More importantly, the use of purely metric approach, based on the Wald curvature (or any of other equivalent metric curvatures, for that matter), rather than the combinatorial (and metric) approach of [88], allows us to give a first, tentative, purely theoretical at this point, answer to Question 2, p. 123, of [88], namely whether there exists a Ricci flow defined on the space of all piecewise constant curvature metrics (obtained via the assignment of lengths to a given triangulation of 2-manifold). Since, by Hamilton's results [175] (and those of Chow [77], for the case of the sphere), the Ricci flow exists for all compact surfaces, it follows from the arguments above that the fitting metric flow exits for surfaces of piecewise constant curvature. In consequence, given a surface of piecewise constant curvature (usually a mesh with edge lengths satisfying the triangle inequality for each triangle), one can evolve it by the Ricci flow, either forward, as in the papers mentioned above, to obtain, after the suitable area normalization, the polyhedral surface of constant curvature conformally equivalent to it; or backward – if possible – to find the "primitive" family of surfaces, including the "original" surface conformally equivalent to the given one, (where, by "original", we mean the surface obtained via the backward Ricci flow, at time T).

Remark 12.2.31 *Note that is not necessarily true that all the surfaces obtained via the backward flow are embedded (or, indeed, embeddable) in $\mathbb{R}^3$ – a see the discussion below.*

Proposition 12.2.32 *Let (S_{Pol}^2, g_{Pol}) be a compact polyhedral 2-manifold without boundary, having bounded metric curvature. Then there exists $T > 0$ and a smooth family of polyhedral metrics[16] $g(t), t \in [0, T]$, such that*

$$\begin{cases} \frac{\partial g}{\partial t} = -2K(t)g(t) & t \in [0, T]; \\ g(0) = g_{Pol}. \end{cases} \tag{12.2.31}$$

(Here $K(t)$ denotes the Wald curvature induced by the metric $g(t)$.)

Moreover, both the forward and the backward (when existing) Ricci flows have the uniqueness of solutions property, that is, if $g_1(t), g_2(t)$ are two Ricci flows on $(S_{Pol}^2,$ such that there exists $t_0 \in [0, T]$ such that $g_1(t_0) = g_2(t_0)$, then $g_1(t) = g_2(t)$, for all $t_0 \in [0, T]$.

In fact, by combining our method with a result of Shi [302], we can extend the proposition above to complete polyhedral surfaces, as follows:

Proposition 12.2.33 *Let (S_{Pol}^2, g_{Pol}) be a complete polyhedral surface, such that $0 < K_W \leq K_0$. Then there exists a (small) T as above, such that there exists a unique solution of (12.2.26) for any $t \in [0, T]$.*

---

[16]see [169], [279]

Convergence Rate A further type of result, highly important both from the theoretical viewpoint and for computer-driven applications, is that of the convergence rate. For the Ricci flow mostly applied in such s setting, namely for the combinatorial Ricci flow, it was proven in [88] that, in the case of background Euclidean (Theorem 1.1) or Hyperbolic (Theorem 1.2) metric, the solution – if it exists – converges, without singularities, exponentially fast to a metric of constant curvature. Using the classical results of [158] and [86], since we already know that the solution exists and it is unique (see the subsection below for the nonformation of singularities), we are able to control the convergence rate of the curvature:

Theorem 12.2.34 *Let* (S^2_{Pol}, g_{Pol}) *be a compact polyhedral 2-manifold without boundary. Then the normalized metric Ricci flow converges to a surface of constant metric curvature. Moreover, the convergence rate is*

1. *exponential, if* $K < 0$; $\chi(S^2_{Pol}) < 0$;

2. *uniform; if* $K = 0$;

3. *exponential, if* $K > 0$.

Recall that convergence rate of solutions is defined as follows:

Definition 12.2.35 *A solution of (12.2.29) is said to be convergent iff*

1. $\lim_{t \to \infty} K_i(t) = K_i(\infty)$, *for all* $1 \le u \le |V|$, *where* $K_i(\infty) \in (0, 2\pi)$;

2. $\lim_{t \to \infty} l_{ij}(t) = l_{ij}(\infty)$, $l_{ij}(\infty) > 0$.

A convergent solution is said to converge exponentially fast iff there exists constants c_1, c_2, *such that, fora any* $t \ge 0$, *the following inequalities hold:*

1. $|K_i(t) - \bar{K}_i| \le c_1 e^{-c_2 t}$;

2. $|l_{ij}(t) - \bar{l}_{ij}| \le c_1 e^{-c_2 t}$

(The fitting definition for the flow (12.2.27) is immediate.)

Remark 12.2.36 *A more realistic model for (gray-scale as well as color) images should be based on surfaces with boundary. Similar results can be obtained for this type of surfaces (see [57], [58]), however we defer for further study the detailed analysis, in this model, of the metric Ricci flow of images.*

Singularities Formation. In the classical (smooth) case, by [158], Theorems 1.1 and 5.1, the Ricci flow evolves without singularities formation, even for surfaces of low genus. Also, by [88], Theorem 5.1, the combinatorial Ricci flow evolves without singularities on compact surfaces of genus ≥ 2. However, on surfaces of low genus, that are extremely important in Graphics and Imaging, singularities do form [152].

Combining our smoothing technique with Hamilton's classical results mentioned above, as well was with a more recent result of Topping [351], we obtain the following results:

Proposition 12.2.37 *Let (S^2_{Pol}, g_{Pol}) be a complete polyhedral 2-manifold, with at most a finite number of hyperbolic cusps (punctures), having bounded metric curvature and satisfying the noncollapsing condition below.*

There exists $r_0 > 0$, such that, for all $x \in M$ the following holds:

$$\mathrm{Vol}_g\left(B_g(x, r_0)\right) \geq \varepsilon > 0. \tag{12.2.32}$$

(Here, as usual, $B_g(x, r_0)$ denotes the open ball, in the metric g, of center x and radius r_0.)

Then there exists a unique Ricci flow contracting the cusps. Furthermore, the curvature remains bounded at all times during the flow.

Embeddability in $\mathbb{R}^3$ In Graphics, where one of the main problems solved via the combinatorial Ricci flow is that of registration, by producing, via the flow, a conformal mapping from the given surface to one of the model surfaces (see, e.g. [150], [352]), the embeddabilty – not necessarily isometric – is both trivial and not of real interests. However, this aspect is highly significant in Image Processing (see [353]), and, in fact, the results below were motivated precisely by these applicative aspects of the Ricci flow.

Here we mainly consider a problem regarding smooth surfaces, since, by now, the connection with the version for polyhedral surfaces is, we hope, quite clear. We should note that, by [229], Theorem 8.8, any δ-approximation of an embedding is also an embedding, for small enough δ. Since, as we have already mentioned, smoothing represent δ-approximations, the possibility of using results regarding smooth surfaces to deduce facts regarding polyhedral embeddings is proven. (The reverse implication – namely from smooth to PL and polyhedral manifolds – follows from the fact that the *secant approximation* is a δ-approximation if the simplices of the PL approximation satisfy a certain nondegeneracy condition – see [229], Lemma 9.3.)

In the following S^2_0 denotes a smooth surface of positive Gauss curvature, and let S^2_t denote the surface obtained at time t from S^2_0 via the Ricci flow.

Proposition 12.2.38 *Let S^2_0 be the unit sphere $\mathbb{S}^2$, equipped with a smooth metric g, such that $K(g) > 0$. Then the surfaces S^2_t are (uniquely, up to a congruence) isometrically embeddable in $\mathbb{R}^3$, for any $t \geq 0$.*

In fact, the result above can be slightly strengthened as

Corollary 12.2.39 *Let S^2_0 be a compact smooth surface. If $\chi(S^2_0) > 0$, then there exists some $t_0 \geq 0$, such that the surfaces S^2_t are isometrically embeddable in $\mathbb{R}^3$, for any $t \geq t_0$.*

In contrast, for (complete) surfaces uniformized by the hyperbolic plane we have only a negative result:

Proposition 12.2.40 *Let (S^2_0, g_0) be a complete smooth surface, and consider the normalized Ricci flow on it. If $\chi(S^2) < 0$, then there exists some $t_0 \geq 0$, such that the surfaces S^2_t are not isometrically embeddable in $\mathbb{R}^3$, for any $t \geq t_0$.*

For full proofs, further related facts and comments, as well as some tentative experimental confirmation of these results, see [279].

We conclude this section with an open research problem, motivated both by the results in this section and by the previous one on Ricci curvature for polyhedral manifolds:

Problem 12.2.41 *Formulate and prove similar results for the metric Ricci flow for polyhedral manifolds of dimension ≥ 3. (Hint: This is a non-trivial problem, given the intricacies of the Ricci flow for smooth manifolds, even in dimension 3. Thus one should perhaps (a) Restrict to 3-dimensional manifolds; (b) Consider first only (the simpler) PL manifolds.)*

12.2.3 Combinatorial Yamabe Flow

Given the theoretical importance and success in applications of the combinatorial Ricci flow, it is only natural to muse about a combinatorial *Yamabe flow* (or *scalar flow*), which should be even simpler to implement, given that it considers only the scalar curvature, thus not a tensorial quantity, hence that on triangular meshes (and networks) is concentrated at vertices. Recall that, in the classical setting of smooth Riemannian manifolds, the Yamabe flow is defined as

$$\frac{\partial g_{ij}}{\partial t} = -\text{scal}g_{ij}\,. \tag{12.2.33}$$

(Here g_{ij} denotes, as before, the Riemannian curvature.)

Luo's Combinatorial Yamabe Flow on Surfaces The first combinatorial Yamabe flow is due to Luo [210], in a natural continuation of his study [88] of his combinatorial Ricci flow. He first defines it as follows:

Definition 12.2.42 *The combinatorial Yamabe flow on a triangulated surface $(S, \mathcal{T})$, $\mathcal{T} = (V, E)$ is defined as the following system of ODE's*

$$\begin{cases} \frac{du_i}{dt} = -u_i K_i \\ u_i(0) = 1 \end{cases} \tag{12.2.34}$$

where $u : V \to \mathbb{R}_+^*$ is a conformal factor, V denoting the vertex set of the triangulation, and were we denoted $u_i = u(v_i)$, $K_i = K_i(v_i)$, both being considered at time t. (Thus K_i denotes the curvature evolved by the conformal factor.)

Note that, in many instances, is more convenient instead to consider (as we have already seen in the case of the combinatorial Ricci flow) the *normalized flow*

$$\begin{cases} \frac{du_i}{dt} = -u_i(K_i - \bar{K}) \\ u_i(0) = 1\,; \end{cases} \tag{12.2.35}$$

where, as before, $\bar{K}$ denotes the average curvature.

Clearly, it easier and more intuitive to deal with a flow of the curvature only. A result to this end represents the main result of [210]. More precisely, it is shown that, for a fixed metric of a piecewise flat, the curvature evolves according to the following equation:

$$\frac{dK_i}{dt} = \sum_{j=1}^{|V|} c_{ij} K_j ; \qquad (12.2.36)$$

where $(c_{ij}) 1 \leq i, j \leq |V|$ is a symmetric. positively semi-negative definite matrix. (In consequence $\sum_{i=1}^{|V|} K_i^2$ is decreasing function of the time t.) In other words, the curvature flow is a combinatorial heat flow.

Further results, regarding the local rigidity[17] of the metric under the flow, and singularities formation are also obtained. We do not dwell upon them here, and refer the interested reader to the original article [210].

Glickenstein's Combinatorial Yamabe Flow on Surfaces   The extension of the combinatorial Yamabe flow to piecewise flat 3-manifolds, obtained via sphere packings[18] (hence also extension of the combinatorial Ricci flow) was devised by [134]. (See also [136].) In his definition, the Yamabe flow is given by

$$\frac{dr_i}{dt} = -K_i r_i ; \qquad (12.2.37)$$

where r_i denotes the radius of the sphere centered at v_i. (Thus, as in the case of circle packings, the distance l_{ij} between the vertices v_i and v_j is given by $l_{ij} = r_i + r_j$.)

As Glickenstein notes, it is relatively simple to implement this form of the Yamabe flow, at least for triangulations with few vertices. Moreover, his first experiments have shown, that, as expected, the metric evolves toward one having constant one. However, some paradoxical results were obtained, at least in the case of *vertex homogeneous triangulations* (i.e where all vertices have the same degree). To wit, in all the experiments the considered piecewise flat 3-torus evolved toward one endowed with a positively constant curvature (thus in contrast with expected zero curvature).

Dynamic Combinatorial Surface Yamabe Flow   The conventional combinatorial surface Yamabe flow preserves the triangulation during the flow, therefore some triangles may become degenerated, hence some target curvature may not be realizable. It is highly non-trivial to determine the existence of the solution to the flow for a prescribed target curvature.

Fundamentally, the triangulations for a polyhedral surface can be treated as local parameters for a smooth surface, which should be adapted to the metric structure. Isothermal parameters are preferred for many theoretic deductions, analogically, Delaunay triangulations are required for real applications. We can modify conventional combinatorial Yamabe flow by dynamically updating the triangulations

---

[17] In contrast, global rigidity is not achieved, at least to the best of our knowledge.

[18] This are not necessarily actual sphere packings.

to be Delaunay during the flow. This simple modification guarantees the existence of the solution to the dynamic combinatorial surface Yamabe flow for any target curvature $\bar{K} : V \to (-\infty, 2\pi)$, as long as $\bar{K}$ satisfies the Gauss-Bonnet condition $\sum_{v_i \in V} \bar{K}_i = 2\pi\chi(S)$. This leads to the discrete uniformization theorems for polyhedral surfaces with both Euclidean and hyperbolic background geometries [151], [152]. The algorithmic details can be found in Chapter 20.

12.3 RICCI CURVATURE AND FLOW FOR NETWORKS

The success of the Ricci curvature and flow discretizations (mainly of the combinatorial one) in dimension 2, encourages one to even more drastically reduce the dimension, to the lowest possible, namely to the setting of graphs/networks. Clearly, such an adaptation of a notion defined originally for manifolds of dimension 3 and higher necessitates quit different tools. Indeed, the approach of the most expressive such discretizations is a *synthetic* one, namely by capturing, in the 1-dimensional case, certain essential properties of the classical Ricci curvature.

12.3.1 Metric Ricci Curvature of Networks

It is most natural to consider, in view of the metric approach we embraced in Sections 12.1 and 12.1 above, to also consider the possibility of defining a metric version (or even versions) of the Ricci curvature and define their fitting flows. This is, indeed, possible, and we have in fact prepared the ground for the definition of network Ricci curvature in Chapters 6 and 11. More precisely, we should recall Formulas (11.4.10) and (11.4.12) of Chapter 11, defining Menger- and Haantjes Ricci curvatures, respectively:

$$\mathrm{Ric}_M(e) = \sum_{T_e \sim e} \kappa_M(T_e) \,,$$

where $T_e \sim e$ denote the triangles adjacent to the edge e; and

$$\mathrm{Ric}_H(e) = \sum_{\mathfrak{c} \sim e} \kappa_{H,e}(\mathfrak{c}) \,,$$

where the sum is taken over all the 2-cells $\mathfrak{c}$ adjacent to e.

In fact, one could even consider a *semi-local* variant of the Haantjes-Ricci curvature by using the extended variant of the Ricci curvature, taking into account longer paths, not just edges, that we defined in Formula (12.3.1), Chapter 11:

$$\mathrm{Ric}_H(\overline{uv}) = \sum_1^m K_{H,\overline{uv}}^i = \sum_1^m \kappa_{H,\overline{uv}}(\pi_i)$$

where $K_{H,\overline{uv}}^i$ denotes the Haantjes-Ricci curvature of the cell $\mathfrak{c}_i$, where $\partial\mathfrak{c}_i = \pi_i\pi_0^{-1}$, relative to the direction $\overline{uv}$, and where

$$\kappa_H^2(\pi_i) = \frac{l(\pi_i) - l(\pi_0)}{l(\pi_0)^3} \,,$$

Figure 12.15 The Menger-Ricci curvature based segmentation (right) of a classical test image (left).

and where the paths $\pi_1, \ldots, \pi_m$ satisfy the condition that $\pi_i \pi_0^{-1}$ is an elementary cycle.

These metric curvatures, introduced in [292] (a considerable extension of the earlier conference paper [111]) already have proven their efficiency in addressing a number of fundamental, as well as application-specific problems in the intelligence of networks. In the first category we mention the fundamental problem of network sampling and *backbone* extraction and *core* detection [28], while to the second one belongs [264], where these types of curvature (as well as the Ollivier-Ricci and Forman-Ricci curvature[19] are shown to be efficient in detecting financial markets crushes). The applications above make appeal to the straightforward, local definitions of Menger- and Haantjes-Ricci curvatures. The extend definition, which rakes into account longer paths, not just edges, proved to be useful in the understanding of semantic networks [69]. Furthermore, since (black and white) images can be viewed as square grids (the duals to the pixels underlying the image) one can use these types of curvature for sampling of images. Furthermore, since edges contours in images are associated to high curvature, one can use the graph curvatures above for image segmentation (see Figure 12.15).

To each of these curvatures one can associate a *normalized* flow, thus ensuring that the metric does not contract (i.e. the weights do not converge to zero) and the network does not collapse to a node in the limit. This flow is defined, in analogy with

---

[19]see our next chapter

that for surfaces (see, e.g. [149]), as

$$\frac{\partial \omega(e)}{\partial t} = - \left(\text{Ric} \left(\omega(e) \right) - \overline{\text{Ric}} \right) \cdot \omega(e) \, ; \tag{12.3.1}$$

where $\overline{\text{Ric}}$ denotes the mean Ricci curvature, and Ric stands for the (graph) Menger- or Haantjes-Ricci curvature, according to the chosen curvature. Also, since in the setting of networks time is also assumed to evolve in discrete steps and each "clock" (that is, time step) has a length of 1, the Ricci flow takes the following form:

$$\tilde{\omega}(e) - \omega(e) = - \left(\text{Ric} \left(\omega(e) \right) - \overline{\text{Ric}} \right) \cdot \omega(e) \, ; \tag{12.3.2}$$

where $\tilde{\omega}(e)$ denotes the new (updated) value of $\omega(e)$, and $\omega(e)$ is the original i.e. given - one.

These flows have proven their efficiency in speeding-up the improving the sampling and backbone detection capabilities of a network, as well as indicating its limit geometry (see also our next chapter).

Here again we propose a research problem that arises naturally in light of the definitions and results we just presented:

Problem 12.3.1 *Formulate and prove results similar to those in Section 12.2.2.2 for the Menger-Ricci and Haantjes-Ricci flows.*[20]

12.3.2 Ollivier Ricci Curvature

We conclude this chapter with what os, undoubtedly, the best known and the most successful, so far, curvature notion for networks, namely the so called *Ollivier-Ricci curvature*. Indeed, it and its related flow, have been applied with no small measure of success in various fields, for instance in the deeper understanding of Cancerous Networks [266] and the intelligence of Finacial markets Networks; and for a number of tasks, such as understanding the network topology and backbone extraction [243], network alignment [233], community detection [235], [303].

The idea that resides behind it is deeply geometric, but stemming from a quite different vein than the metric curvatures that we systematically developed, based on the very basic concepts of Differential Geometry, from the very beginning of this book. It rather is reminiscent in spirit with the Bertrand-Diguet-Puiseaux Theorem, in that it compares spheres and balls with those in a model spaces. However, it does so not by volume comparison but in a novel way that, as many great ideas, leaves one astounded and asking him/herself how come no one has come up with this simple thought before? More precisely, it is based on the observation that "in positive curvature, balls are closer than their centers are" [238], and that they are farther away in negative curvature – see (Figure 12.16). This fact is based upon results in classical Riemannanian geometry that can be formalized in a technical formula reminiscent (as expected) of the one due to Bertrand, Diguet and Puiseaux. More precisely, consider

---

[20]Evidently, developing the theoretical foundations for the Haantjes-Ricci flow would also solve, as essentially a particular case, the problem for the Menger-Ricci flow.

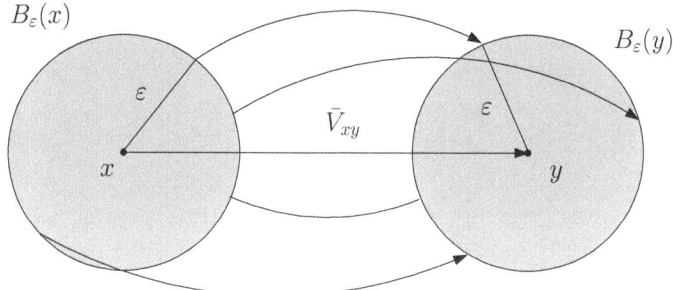

Figure 12.16 Ollivier's coarse curvature interpretation (after [238], [239]: Given two close points x and y in a Riemannian manifold of dimension n, defining a tangent vector, one can consider the parallel transport in the direction. Then points on an infinitesimal ball (sphere) $B(x)$ centered at x, are transported to points on the corresponding ball (sphere) $B(y)$. Moreover, they are transported, on the average, by a distance equal to $d(x, y)$. It follows that in spaces of positive curvature balls are closer than their centers are, whereas in spaces of negative curvature they are more distant.

the ball $B(x) = B(x, \varepsilon)$ and its parallel transport to $B(y) = B(y, \varepsilon)$ (see Figure 12.17 and its caption.). Then the average distance between (the points of) the balls $B(x), B(y)$ is

$$\delta \left[1 - \frac{\varepsilon^2}{2(n+2)} \mathrm{Ric}(\bar{\mathbf{v}}_{xy}) + O(\varepsilon^3 + \varepsilon^2 \delta) \right] ; \qquad (12.3.3)$$

where $\delta = d(x, y)$. (For a proof see [238].)

However, its precise formulation is less important in the discrete context, since Ollivier's idea was to generalize the notion of "closeness" from balls – or rather volumes of balls – to general measures supported on balls. (One should also keep in mind that Ollivier's notion of curvature was devised for *Markov chains*.)

Thus one is conducted to the problem of computing the distance between measures. While there are a number of possible choices (see [132]), in this context is natural[21] to consider the Wasserstein (a.k.a. the "earthmover" or the *optimal transportation* (ODT)) distance W_1, with *cost function* $c(x, y) = |x-y|$. (For further details see, e.g. [328].) We concentrate here only on this *synthetic* definition of curvature for networks:

Definition 12.3.2 *Let $(\mathcal{N}, d)$ be a network, endowed with some metric d. Let m_x, m_y be measures of centered at the nodes x and y, respectively. Then the coarse Ricci curvature (or Ollivier-Ricci curvature) of the pair of vertices x and y is defined as*

$$\kappa(x, y) = 1 - \frac{W_1(m_x, m_y)}{d(x, y)} . \qquad (12.3.4)$$

---

[21]even more so given the fact that the quest for discrete Ricci curvature was largely motivated by Optimal Transportation

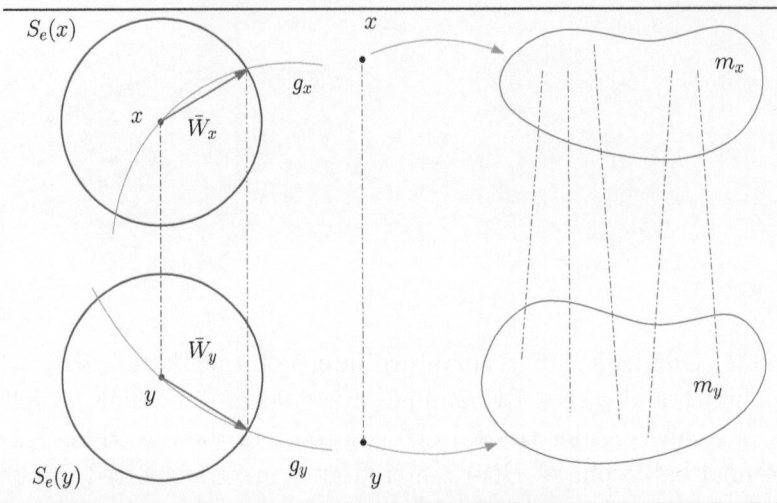

Figure 12.17 In positive (negative) curvature, balls are closer (farther) than their centers are. Right: To generalize this idea to metric measure spaces, one has to replace the spheres/balls, by measures m_x, m_y. On the average, points will be transported by a distance equal to $(1 - \kappa)d(x, y)$, where $\kappa = \kappa(x, y)$ represents the coarse (Ollivier) curvature along the geodesic segment xy. (After [238].)

that is the Ollivier-Ricci curvature represents a measure of the defect between the underlying distance and the Wasserstein transportation distance.

Remark 12.3.3 *Clearly, by its very definition, Ollivier's Ricci curvature captures in the discrete context, the volume growth aspect off Ricci curvature.*

Recall that the *Wasserstein distance* of order p W_p is defined as follows:

Definition 12.3.4 *Given μ, ν be two probability measures on X where X is a Polish space,[22], the Wasserstein p-distance is defined as*

$$W_p(\mu, \nu) = \left(\inf \int_X d((x, y)^p d\pi(x, y) \right)^{\frac{1}{p}}, \qquad (12.3.5)$$

where the infimum is taken over all the transference plans between μ and ν;

and where transference plans are defined as follows:

Definition 12.3.5 *A transference plan (or coupling) of the measures μ and ν is a measure on $X \times X$, with marginals μ and ν, i.e. such that, for any measurable sets $A, B \subset X$, the following properties hold:*

$$\pi[A \times X] = \mu[A], \pi[X \times B] = \nu[B];$$

[22]i.e. a separable, completely metrizable topological space [230]

i.e.

$$\sum_{y \in X} \pi(x, y) = \mu(y), \sum_{x \in X} \pi(x, y) = \nu(y).$$

For networks, which are modeled by (locally finite) graphs, it is only natural to consider the W_1 distance (since no higher order smoothness can be presumed), which in this case takes the following simple form:

$$W_1(\mu, \nu) = \min_{\mathcal{N}} \sum d(x, y) \pi(x, y); \qquad (12.3.6)$$

where, again, the minimum is taken over all the transference plans between μ and ν.

It is easier to understand, the behavior of a network discretization of a classical notion by contemplating first the simplest case, that is of (undirected) *combinatorial graphs*, i.e. with all (vertex and edge) weights equal to 1. While such a reduction greatly affects the resulting geometry of the network, it also reduces it to its bare "skeleton" and thus allows us to grasp the essential intuition behind the notion, that might be well hidden between the intricacies of actual computations with various, real-life weights. In this case, it is also natural to take $\varepsilon = 1$, thus the balls in the classical definition become the balls of radius 1, in the discrete metric with which $\mathcal{N}$ is endowed, i.e. where the distance between any two adjacent vertices equals 1, i.e, $d(x, y) = 1$, for any $x \sim y$. The measure of each such ball, given by the Brownian motion is the natural one, namely

$$m_x = \frac{1}{\deg(x)}. \qquad (12.3.7)$$

Equipped with this discrete measure, it is easy to see that triangles and quadruples decrease transportation cost between neighboring vertices x and y, thus increase the Ollivier-Ricci curvature of the edge $e = \overline{xy}$. On the other hand in the presence of hexagons (i.e. if an edge appertains to a hexagon), the Ollivier-Ricci curvature is not definable.

Exercise 12.3.6 *Verify the assertions above regarding the role of triangles, and quadruples and hexagons. What is the effect of pentagons?*

This property is usually seen as an advantage of Ollivier-Ricci curvature, in the sense that, while is a *graph curvature*, i.e. it is defined solely in terms of the vertices and edges of a given network, it "senses" cycles, at least those of order 3,4, and 5, thus transcending is purely graph definition. (This being in contrast with the "pure" graph nature of the Menger-Ricci and Haantjes-Ricci curvatures, as well as with the (reduced) Forman-Ricci curvature that we shall introduce and discus at length in our next chapter.) However, this can also be construed as a deficiency: For one, as we have seen above, this capability stops to pentagons and this represents a serious drawback since, in contrast to the common wisdom till not long ago, even small networks contain higher order cycles – see, e.g. [334].[23] In fact, one might argue that, from a purely

---

[23]In fact, in gene networks even cycles of length i the order of thousands are observable!

graph sense, this might be an undesired property, since one might to restrict the be-havior of an edge-defined notion only to the edges, and not to higher order structures. Even if such a "puristic" might be discarded by most, there is another problem, much more relevant in practice that the detection of triangles (for instance) engenders: By its definition, Ollivier-Ricci curvature is restricted to 1-dimensional cycles, thus it cannot distinguish between a "full" or an "empty" triangle, that is between a 2-face and its 1-skeleton (of length 3). However, such a distinction is extremely important in modeling various phenomena, most importantly in Biological Networks (e.g. for gene and protein interaction networks), as well as in Social Networks. This importance is easy to exemplify in a social network where, three people that collaborate, say, to write a paper should be distinguished from three persons that only work in pairs. (The first case is modeled by a "full" triangle, while the second by an "empty" one). Such *multiplex networks*, as they are sometimes called, are more-and-more prevalent yet, as we have seen, the Ollivier-Ricci curvature is descriptive enough to handle them.

It is also very important to note that, the special case of the combinatorial graphs discussed above is deceivingly simple, and in fact it is quite hard to compute "by hand" the coarse curvature, even for relatively simple graphs. Therefore, one needs to pass to computer-driven methods, and certainly for real-life networks, even medium sized. In applications, one would like to generalize the measure m_x as

$$m_x = m_x^\alpha(y) = \begin{cases} \alpha & , y = x \,; \\ (1-\alpha)/\deg(x) & , y \sim x \,; \\ 0 & , \text{else} \,; \end{cases} \qquad (12.3.8)$$

where $\alpha \in [0,1]$. For this measure, the Wasserstein distance W_1 becomes

$$W_1(m_x^\alpha, m_y^\alpha) = \min_{\mathcal{N}} \sum_{x_i, y_j \in V_\mathcal{N}} M(x_i, y_j) \,; \qquad (12.3.9)$$

where $M(x_i, y_j)$ represents the amount of mass transported from vertex x_i to y_j, along the shortest path (geodesic) that minimizes the transport distance. To do this, one must make appeal to by *Linear Programming* (LP). More precisely, we assume that the transportation plan of $W_1(m_x^\alpha, m_y^\alpha)$ is given by variables represented by an $m \times n$ matrix (M_{ij}), such that $M_{ij} \geq 0$. This transportation plan needs to preserve total mass and we are looking for the optimal plan that minimizes the total transportation distance. The LP formulation is as follows.

$$\text{Min:} \quad \sum_j \sum_i d(x_i, y_j) \rho_{ij} m_x^\alpha(x_i)$$

$$\text{S.t.:} \quad \sum_j \rho_{ij} = 1 \quad \forall i, \quad 0 \leq \rho_{ij} \leq 1 \quad \forall i,j \qquad (12.3.10)$$

$$\sum_i \rho_{ij} m_x^\alpha(x_i) = m_x^\alpha(x_j) \text{ and } \sum_i \rho_{ij} m_x^\alpha(x_i) = m_y^\alpha(y_j) \,; \; \forall j \,.$$

The main obstruction in implementing the Wasserstein distance, hence in the compu-tation of the Ollivier-Ricci curvature, resides in its heavy computational complexity.

To overcome this obstacle, we follow [233] and consider instead of W_1, the *average transportation distance* (ATD) $A(m_x, m_y)$, in which an equal amount of mass is transported from each neighbor x_i of x to each neighbor y_j of y, and the mass of x is transfered to y. Thus, by removing the LP step in the computation, the computational complexity of the Ollivier-Ricci curvature, or rather of the *modified Ollivier-Ricci* (or *ATD-Ollivier-Ricci*) curvature.

$$\kappa_A(x, y) = 1 - \frac{A(m_x, m_y)}{d(x, y)}, \tag{12.3.11}$$

is substantially improved.

Remark 12.3.7 *As we have seen above, computing the W_1, while quite natural in the weighted graphs context, is also difficult, thus we preferred to replace it with that of the total-variation distance. Seemingly paradoxical, it is far easier to compute the W_2 distance between histograms, once they have been replaced by an equivalent normal distribution, using an approach going back to Fréchet [124] (see also [213]). More precisely, we have the following formula:*

$$W_2(X_1, X_2)^2 = ||\mu_1 - \mu_2||^2 + \text{tr}\left(\Sigma_1 + \Sigma_2 - 2(\Sigma_1^{1/2}\Sigma_2\Sigma_1^{1/2})^{1/2}\right); \tag{12.3.12}$$

for any two Gaussian distributions X_1, X_2, with respective means μ_1, μ_2 and covariance matrices Σ_1, Σ_2.

To normalize the histograms we made appeal to the so called Box-Cox transform [354]:

$$y = y^\beta = \frac{x^\beta - 1}{\beta}; \tag{12.3.13}$$

where the best coefficient β in our setting was experimentally to be determined to be $\beta = 0.5$.

The computation of W_2 allows network applications [218], as well as Imaging ones [156], [318], [237].

Remark 12.3.8 *The Ollivier-Ricci curvature does not lend itself immediately to generalization to higher dimensions. In this direction one must mention the paper of Asoodeh, Gao and Evans [20]. Unfortunately, if the classical, 1-dimensional computation of the Wasserstein distance is hard, its extension to higher dimensions is proportionally much harder and, in fact, we do not know if their approach lends to concretization in any practical setting. We have offered, en passant, in [295] a more pragmatic approach to this problem, based on the computation of the Wasserstein distance on the dual graph, an idea which is quite natural from a topological and geometric viewpoint. We should definitely also mention here the recent work of Gu and Lei.*

Remark 12.3.9 *As common with the discrete Ricci flow, and especially with graph-based ones, singularities tend to form quite rapidly. A possible way of overcoming this problem, at least as far as the Ollivier-Ricci flow is concerned was very recently suggested in [196].*

For further insights into the Ollivier-Ricci curvature and its relations with other type of graph curvatures, as well as for references to some of the vast – and ever growing – theoretical literature regarding the combinatorial Ollivier-Ricci curvature, see [32], [293].

CHAPTER 13

Weighted Manifolds and Ricci Curvature Revisited

The title (hence also the content) of this chapter might appear outlandish to the unenlightened/unaware reader. However, weighted manifolds in fact commonly encountered for instance, in Imaging, and none the less than in modeling the commonest, simplest and most natural example of image, namely *grayscale images*. Unfortunately, the surface model which we discussed at some length in the previous chapters, and its discretization, the stick model (which we also encountered already and shall further use later in this very chapter), are not capable of dealing directly, in an intrinsic and geometrically meaningful manner with data that is not simply distributed along a geometric structure (usually a Riemannian surface/manifold), but it also incorporates essential measures, probability measures being clearly the archetypal ones, but more general one naturally arise in Imaging and Pattern Recognition. However, besides the surface model which we discussed at some length in the previous chapters, and its discretization, the stick model (which we also encountered already and shall further use later in this very chapter), such images can be also modeled as weighted manifolds (surfaces) (a.k.a. *manifolds with densities*). In point of fact, such an image can be viewed as nothing more then the graph of a distribution (the grayscale) over a very basic type manifold (a rectangle) – see Figure 13.1.[1] Moreover, in the context of Medical Imaging, densities appear at even a more basic, intrinsic level: Indeed, the density of many types of MRI images equals the very proton density. Therefore, modeling such images by manifolds with density represents, as far as the physical acquiring process is concerned, the proper approach and, as such, one can expect more accurate results. Furthermore, such weights (or densities) may appear as uncertainties intrinsic to the acquiring of the image (for instance in Ultrasonography). They also arise in Graphics, e.g. such as luminosity over a surface/mesh. Another, perhaps less known example, appears in the context of texture classification and mapping. In addition, such measures represent useful tools in many concrete settings, as ad hoc tools employed at various stages of the implementation of a variety of tasks, such as

---

[1]In fact, some of our reader might have a revelation similar to Molière's character from "Le Bourgeois gentilhomme" who discovers, to his amazement that he has been speaking prose.

DOI: 10.1201/9781003350576-13

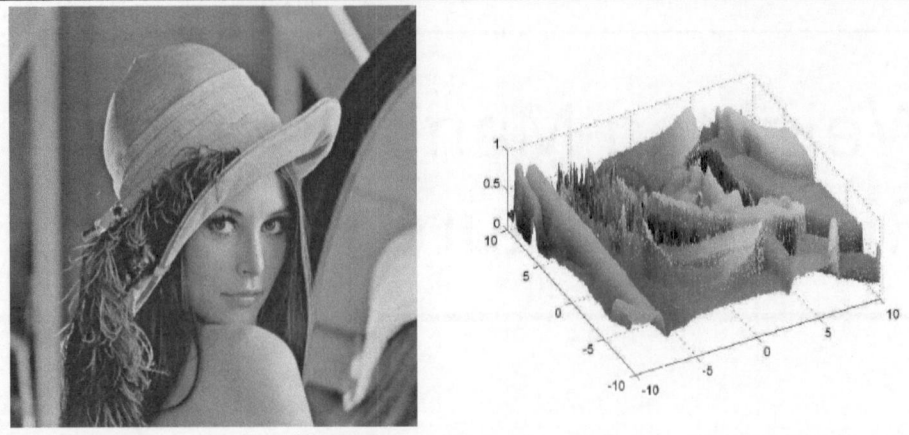

Figure 13.1 A black-and white image (left) can be viewed as a distribution – given by the image intensity – over the plane/grid of the pixels (right).

smoothing, (elastic) registration, warping, segmentation, etc. (see, e.g. [14]). Recently, the importance of the weighted manifolds in Deep Learning, and their connection to Optimal Transport was explored and emphasized in [199], [200].

13.1   WEIGHTED MANIFOLDS

The importance of *manifolds with densities* (or, more generally, *metric measure spaces*) and their Ricci curvature was first emphasized by Gromov [142], [143], partly motivated by the connection between Ricci curvature and volume growth and collapse under the Ricci flow.

The simplest generalization of Ricci curvature to such a setting is due to Bakry, Émery and Ledoux, and it can be easily expressed and computed. More precisely, they consider *smooth metric measure spaces* (or *weighted manifolds*):

Definition 13.1.1 *A smooth metric measure spaces is a pair* (M, Ψ), *where* M *is a Riemannian manifold and* Ψ *denotes a smooth, positive density function* $\Psi = \Psi(x)$, *that induces weighted* n- *and* $(n-1)$-*volumes, e.g. in the classical cases* $n = 2$ *and* $n = 3$, *volume, area and length. More precisely, the volume, area and length elements* dV, dA, ds *of the weighted manifold* (M^n, Ψ) *are given by:*

$$dV = \Psi dV_0, dA = \Psi dA_0, ds = \Psi ds_0 \,, \tag{13.1.1}$$

where dV_0 *represents the natural (Riemannian) volume element of* M, *etc.*

Usually (following Gromov [142]), density functions of the type $\Psi(x) = e^{-\varphi(x)}$ are considered. (However, more general density functions have also been studied – see [228].)

The Bakry, Émery and Ledoux generalization [23], [24] to manifolds with density (M, Ψ) of Ricci curvature is (quite simply, but with a deep geometric/analytic motivation – see e.g. [228]) defined as:

$$\text{Ric}_\varphi = \text{Ric} + \text{Hess}\varphi, \tag{13.1.2}$$

(where Hess denotes the Hessian matrix). In particular, for 2-dimensional manifolds (i.e. surfaces) Ricci curvature reduces, essentially, to Gaussian curvature K, more precisely $K = \frac{1}{2}\text{Ric}$.

Remark 13.1.2 *Note that here we adopted, in concordance with Forman's work on which we shall elaborate in detail in the sequel, the "+" convention for the sign of the Hessian and Laplacian commonly, since this is more intuitive, at least in the context of Imaging and where weights, that is grayscale values are always positive.*

Surprisingly enough, albeit its natural definition in Imaging and its simple, and readily usable formula, there are very such few implementations. See [206] for a first application of this approach to image sampling and reconstruction.

13.1.1 The Curvature-Dimension Condition of Lott-Villani and Sturm

The most influential and far reaching generalization (or, rather, generalizations, as there exist extensions to even more abstract spaces) of the notion of Ricci curvature is that developed, independently, by Lott and Villani [209] and Sturm [314, 315]. We do not bring it here in its most general form, but rather restrict ourself only to the manifold case, which is closer to the scope of this book. We begin by specifying the type of object on which our generalized curvature will be defined.

Definition 13.1.3 *A smooth metric measure space (or weighted manifold) is a triples (M, d, ν), where $M = M^n$ is a complete, connected n-dimensional Riemannian manifold, d is the geodesic (intrinsic) distance on M, and ν is a measure of the form: $\nu(dx) = e^{-V(x)}\text{Vol}(dx)$, where $V : M^n \to \mathbb{R}$ is a smooth function.*

The *generalized Ricci tensor* is then defined as follows:

$$\text{Ric}_{N,\nu} = \begin{cases} \text{Ric} + \text{Hess}(\psi), & N = \infty, \\ \text{Ric} + \text{Hess}(\psi) - \frac{\nabla\psi \otimes \nabla\psi}{N-n}, & n < N < \infty, \\ \text{Ric} + \text{Hess}(\psi) - \infty \, (\nabla\psi \otimes \nabla\psi), & N = n, \\ -\infty, & N < n \end{cases} \tag{13.1.3}$$

where, by convention, $\infty \cdot 0 = 0$, and where $\nabla V \otimes \nabla V$ is a quadratic form on TM^n (hence it operates on tangent vectors – as does the classical Ricci curvature Ric), and $\nabla^2 V$ is the Hessian matrix Hess, defined as:

$$(\nabla V \otimes \nabla V)_x(v) = (\nabla V(x) \cdot v)^2.$$

Therefore

$$\text{Ric}_{N,\nu}(\dot{\gamma}) = (\text{Ric} + \nabla^2 V)(\dot{\gamma}) - \frac{(\nabla V \cdot \dot{\gamma})^2}{N - n}.$$

Also, here the number N is the so called *effective dimension* and is to be inputed.

Lott's words Intuitively, "M has dimension n but pretends to have dimension N. (Identity theft)". Admittedly, the notion of effective dimension is, at least at first view, somewhat mystifying, however the need for such a parametric dimension stems, in particular, from the desire to further extend the notion of Ricci curvature to length spaces, for which no innate notion of dimension exists.

Remark 13.1.4 *1. If $N < n$ then $\mathrm{Ric}_{N,\nu} = -\infty$*

2. If $N = n$ then, by convention, $0 \times \infty = 0$, therefore (13.1.3) is still defined even if $\nabla V = 0$, in particular $\mathrm{Ric}_{n,\mathrm{Vol}} = \mathrm{Ric}$ (since, in this case $V \equiv 0$).

3. If $N = \infty$ then $\mathrm{Ric}_{\infty,\nu} = \mathrm{Ric} + \nabla^2 V$.

We can now bring our next (and main, of this section) definition, that represents the natural extension of the Ricci curvature boundedness condition for the classical case:[2]

Definition 13.1.5 *(M, d, ν) satisfies the curvature-dimension estimate $\mathrm{CD}(K, N)$ iff there exist $K \in \mathbb{R}$ and $N \in [1, \infty]$, such that $\mathrm{Ric}_{N,\nu} \geq K$ and $n \leq N$. (If $\nu = d\mathrm{Vol}$, then the first condition reduces to the classical $\mathrm{Ric} \geq K$.)*

This is, indeed a "true" extension of Ricci curvature (bounds) a fact proven by the fact that it satisfies a proper generalization of the important Bishop-Gromov inequality which we introduced in Chapter 10:

Theorem 13.1.6 (Generalized Bishop-Gromov Inequality) *Let M be a Riemannian manifold equipped with a reference measure $\nu = e^{-V}\mathrm{Vol}$ and satisfying a curvature-dimension condition $\mathrm{CD}(K, N)$, $K \in \mathbb{R}, 1 < N < \infty$.*
Then, for any $x \in M$, the function

$$\varphi(r) = \frac{\nu\left[B(x, r)\right]}{\int_0^r S_K^N(t)dt}, \tag{13.1.4}$$

is non-increasing (as function of r), where

$$S_K^N(t) = \begin{cases} \left(\sin\sqrt{\frac{K}{N-1}}t\right)^{N-1} & \text{if } K > 0 \\ t^{N-1} & \text{if } K = 0 \\ \left(\sinh\sqrt{\frac{|K|}{N-1}}t\right)^{N-1} & \text{if } K < 0 \end{cases} \tag{13.1.5}$$

For a proof see, e.g. [328], p. 499-500. In the same see the comprehensive yet pleasantly readable monograph can also be found extensions to the weighted manifold case of other classical theorems, as well as further developments and generalizations, in addition to a deep and thorough background.

---

[2]that is, it constitutes the geometric approach to the generalization of Ricci curvature, as opposed to the more technical one of Formula (13.1.3).

Remark 13.1.7 *An extension of the $CD(K, N)$ to graphs has been developed in [51]. Unfortunately, the necessary technical apparatus makes it too cumbersome to be used in networking applications.*

Remark 13.1.8 *The fact that this notion of curvature (and its further extension to metric measure spaces) represents a "correct" generalization of the classical Ricci curvature is proven also by the fact that it plays the role of the later in an essential task in many applications, namely that of sampling [274].*

We conclude this section by underlying the fact that we did not include it solely for the importance it has in the mathematical development of the notion of generalized Ricci curvature. Indeed, while its defining formula is quite possibly frightening as it makes use to the somewhat unclear notion of effective dimension, it is largely precisely this very notion that makes, in our view, the generalized Ricci curvature important in Machine Learning and Big Data applications, since it does not need to arbitrarily presume a dimension of the data.[3] Instead, it allows us to explore the data in many dimensions, and such a filtration allows us to really understand its structure.

13.1.2   Corwin et al.

Instead of extending and generalizing the ideas of Bakry-, Émery and Ledoux as done in the previous section, one could adopt an opposite stance and rather than generalizing, try and better understand the fundamental idea of theirs by exploring its implementation in the basic, yet motivational case of curves and surfaces. This route was followed by Corwin et al. [171] (see also [228]). Clearly the first task was to explore the

13.1.2.1   Curvature of Curves in Weighted Surfaces

Clearly, the adaption of formula (...) to this 1-dimensional case is given by

Definition 13.1.9 *Let $S_\varphi = (S, \psi), \psi = e^\varphi$ be a weighted surface, and let c be a curve in S. Then the* weighted curvature *of c is defined as*

$$\kappa_\varphi = \kappa - \frac{d\varphi}{d\bar{\mathbf{n}}}, \tag{13.1.6}$$

where $\bar{\mathbf{n}}$ denotes, as always, the unit normal to the curve.

Once provided with this definition, we can proceed and define geodesics on weighted surfaces. However, due to the role of the density ψ, we have first to introduce an intermediate notion:

---

[3]Such instances where one one cannot determine the intrinsic ("real") dimension of a set of sampled data points, that is without a priori assumed knowledge (e.g. curve, surface, volumetric data, etc.), arise, for instance in Medical Imaging (and Medicine and Biology in General), Psychology, Astronomy.

Definition 13.1.10 *Given a number $A_0 \geq 0$, an* isoperimetric curve *for A_0 is a curve γ such that*

$$\ell(\gamma) = \min\{\ell(c) \mid \text{Area}(\text{int}(c) = A_0\} \,.$$

We have the following characterization of isoperimetric curves:

Proposition 13.1.11 *Let γ be an isoperimetric curve. Then $\kappa_\varphi(\gamma) \equiv \text{const.}$.*

We can now proceed and define the geodesics, in a manner inspired by Lemma 6.2.10 (1), Chapter 6:

Definition 13.1.12 *An isoperimetric curve is called a* geodesic $\kappa_\varphi(c) \equiv 0$.

Unfortunately, we cannot truly justify this definition, nor prove the preceding Proposition 13.1.11, since we didn't develop the necessary notions.[4]

Furthermore, while a precise formula for its computation is rather easy to obtain, we do not bring it here, but rather leave it as an exercise for the reader.

Exercise 13.1.13 *Develop a formula for the computation of κ_φ. (Hint: Use polar coordinates.)*

The influence of the density is highly nontrivial and produces some phenomena that are quite alien to our intuition stapled by our experience with "standard" surfaces:

Facts 13.1.14 *Consider the plane $\mathbb{R}^2$ endowed with the density $\psi = e^\phi = e^{-ar^2+c}$ (i.e., we consider a slight generalization of the Gauss plane $\mathcal{G} = (\mathbb{R}^2, \varphi_0)$, where $\varphi_0 = \frac{1}{2\pi}e^{\frac{-r^2}{2}}$.) Then*

Straight lines are not geodesics, except those passing through the origin.

2. *There exists a unique geodesic circle centered at the origin, of radius r_0.*

3. *All other closed geodesics, except those determined in (1) and (2) above, are closed curves with dihedral symmetry (with the symmetry center at the origin).*

Exercise 13.1.15 *Using the formula obtained in the previous exercise, prove assertions (1) and (2) above.*

13.1.2.2 The Mean Curvature of Weighted Surfaces

The same intuition that brought us to Definition 13.1.6 motivates us to give the following two definitions, generalizing the classical ones:

Definition 13.1.16 *Let $S_\varphi = (S, \psi), \psi = e^\varphi$ be a weighted surface. Then the weighted mean curvature of S_φ is defined as*

$$H_\varphi = H - \frac{1}{n-1}\frac{d\varphi}{d\bar{N}} \,; \tag{13.1.7}$$

where $\bar{N}$ denotes, as usually, the unit normal to S.

---

[4]More precisely we didn't consider *variations*. (This is also one of the reasons we did not discuss minimal surfaces.)

Definition 13.1.17 *Let* $S_\varphi = (S, \psi), \psi = e^\varphi$ *be a weighted surface. Then the weighted principle curvature of* S_φ *(at a point* $p \in S$*) are defined as*

$$\kappa_{\varphi,i} = \kappa_i - \frac{d\varphi}{d\bar{N}}, i = 1, 2;$$ (13.1.8)

where κ_i*,* $i = 1, 2$ *are the principal curvatures of* S*.*

We have the following results regarding the mean curvature:

Proposition 13.1.18 *The plane* $\mathbb{R}^2$ *endowed with the density* $\psi = e^\phi = e^{-ar^2+c}$ *has constant mean curvature.*

Proposition 13.1.19 *Consider the sphere* $S_R = \mathbb{S}^2(0, R)$ *in the space* $\mathbb{R}^3$ *endowed with the density* $\psi = e^\phi = e^{-ar^2+c}$*. Then*

$$H_{S_R} = aR - \frac{1}{R}.$$ (13.1.9)

thus combining, quite elegantly the mean curvature of $\mathbb{S}^2(0, R)$ *and the contribution of the weight.*

Exercise 13.1.20 *Prove Propositions 13.1.18 and 13.1.19 above.*

Remark 13.1.21 *Definitions 13.1.16 and 13.1.17 readily generalize to hypersurfaces of any dimension, and so do Propositions 13.1.18 and 13.1.19.*

13.1.2.3 *Gauss Curvature of Weighted Surfaces*

Quite unsurprisingly[5] we also have the following generalization of Gauss curvature:

Definition 13.1.22 *Let* $S_\varphi = (S, \psi), \psi = e^\varphi$ *be a weighted surface. Then the weighted Gauss curvature of* S_φ *is defined as*

$$K_\varphi = K - \Delta\varphi;$$ (13.1.10)

where $\Delta\varphi$ *denotes, of course, the Laplacian of* φ*.*

The fact that K_φ and κ_φ represent the proper generalizations of Gauss and geodesic curvature, respectively, is proven by the fact that they satisfy the fitting generalizations of the Gauss-Bonnet theorems:

Theorem 13.1.23 (Weighted Local Gauss-Bonnet Theorem) *With the same notations as in the classical Local Gauss-Bonnet Theorem we have*

$$\iint_D K_\varphi dA + \sum_0^p \int_{v_i}^{v_{i+1}} k_\varphi dl + \sum_0^p \varphi_i = 2\pi.$$ (13.1.11)

---

[5]This might have been arrived immediately from the formula for the weighted of Ricci curvature of Bakry and Émery,

and

Theorem 13.1.24 *Under the same assumptions and with the same notations as in the classical Gauss-Bonnet Theorem we have*

$$\iint_D K_\varphi dA = 2\pi\chi.$$ (13.1.12)

In other words, K_φ and κ_φ represent "good" extensions of the classical notions to the setting of weighted surfaces, since they do not give birth to a "weighted topology", which remains, as it should, the one of the underlaying surface.

Exercise 13.1.25 *Prove the theorems above. (Hint: To prove Theorem 13.1.11 combine the classical Local Gauss-Bonnet Theorem and Stokes Theorem.)*

Corollary 13.1.26 *The plane $\mathbb{R}^2$ endowed with the density $\psi = e^\phi = e^{-ar^2+c}$ has constant $K_\varphi \equiv 4a$.*

Exercise 13.1.27 *Prove the corollary above.*

Remark 13.1.28 *Further results regarding to the areas and perimeters of simple, closed curves in the weighted plane above are also proven in [171].*

Exercise 13.1.29 *Find the formula for the area of a simple, closed curve of constant κ_φ.*

13.2 FORMAN-RICCI CURVATURE

As we have noted above, the weighted manifold approach to generalized Ricci curvature does not lend itself adaptations to discrete structures, that are, on the one hand, desirably natural and easily applicable, and on the other hand rigorous and of mathematical depth. Fortunately, another perspective, due to Forman [123], enjoys both these capabilities. Rather than dealing with smooth manifolds and their possible approximations, Forman's setting is that (positively) *weighted quasiconvex CW complexes*, a type of structure that includes (and generalizes) graphs/networks, triangular meshes and square grids, i.e. all the essential underlying structures considered in Complex Networks, Graphics and Imaging, as well as more general simplicial and polyhedral complexes. Not to encumber the reader with further technicalities, we do not formally define such structures here, but rather refer him/her to, e.g. [160].

13.2.1 The General Case

The essential geometric idea residing behind Forman's definition of curvature for *CW* complexes is to exploit the notion of *parallelism* of faces, which is defined as follows:

Definition 13.2.1 *Let $\alpha_1 = \alpha_1^p$ and $\alpha_2 = \alpha_2^p$ be two p-cells. α_1 and α_2 are said to be* parallel *($\alpha_1 \parallel \alpha_2$) iff either: (i) there exists $\beta = \beta^{p+1}$, such that $\alpha_1, \alpha_2 < \beta$; or (ii) there exists $\gamma = \beta^{p-1}$, such that $\alpha_1, \alpha_2 > \gamma$ holds, but not both simultaneously. (For example, in Figure 13.2, e_1, e_2, e_3, e_4 are all the edges parallel to e_0.)*

Note that parallelism isolates precisely those faces whose contribution is intrinsic, as desired from, e.g. a proper discretization of Ricci curvature. – See Figure 13.2.

In the simples case, namely the combinatorial one, this prompts the following (rather natural) definition:

Definition 13.2.2 *Let $\alpha = \alpha^p$ be a p-dimensional cell in a CW complex. The combinatorial p-th curvature function $\mathcal{F}_p(\alpha)$ of α is defined as*

$$\mathcal{F}_p(\alpha) = \#\{(p+1)-\text{cells } \beta > \alpha\}+\#\{(p-1)-\text{cells } \gamma < \alpha\}-\#\{\text{parallel neighbors of } \alpha\} \tag{13.2.1}$$

Since in the classical setting, Ricci curvature operates on vectors, and in *CW* complexes their role is take by the edges, we are conducted to define, in the special case $p = 1$ the (discrete) *Forman-Ricci curvature* as follows:

Definition 13.2.3 *Let $\alpha = \alpha^1$ be a 1-cell (i.e. an edge). Then the* Forman Ricci curvature *of α is defined as:*

$$\text{Ric}_F(\alpha) = \mathcal{F}_1(\alpha). \tag{13.2.2}$$

I the combinatorial case considered above this becomes simply

$$\mathcal{F}_p(\alpha) = \#\{2-\text{cells } \beta > \alpha\} + \#\{0-\text{cells } \gamma < \alpha\} - \#\{\text{parallel edges of } \alpha\} \tag{13.2.3}$$

where *alpha* is a 1-cell (i.e. edge).

To generalize the formula above to general weights, one has to make recourse to the Bochner-Weizenböck Formula (10.1.13), Chapter 10 and to properly discretize it in the setting of *CW* complexes. Before proceeding further, let us note the similarity between Formula (13.1.2) above and the Bochner-Weizenböck Formula (10.1.13). Thus, the approach of Bakry-Émery and Forman's one are not as dissimilar as their expressions (and uses) might suggest. In this context, it is proper to mention already, in view of one of our main applications, that the coupling with a Laplacian, that opens further directions for possible applications, represents an advantage, besides its computational simplicity (which is manifest in the combinatorial case), of Forman's curvature for networks over Ollivier's one.

Let $\alpha = \alpha^p$ be a p-dimensional cell (in a *CW* complex) and let $w(\alpha)$ denote its weight. While general weights are possible, making the combinatorial Ricci curvature extremely versatile, it is suffices (cf. [123]), Theorem 2.5 and Theorem 3.9) to restrict oneself only to so called *standard weights*:

Definition 13.2.4 *The set of weights $\{w_\alpha\}$ is called a* standard set of weights *iff there exist $w_1, w_2 > 0$ such that given a p-cell α^p, the following holds:*

$$w(\alpha^p) = w_1 \cdot w_2^p.$$

(Note that the combinatorial weights $w_\alpha \equiv 1$ represent a set of standard weights, with $w_1 = w_2 = 1$.)

The formula for the curvature functions is (the less intuitive) following one

$$\mathcal{F}(\alpha^p) = \omega(\alpha^p) \left[\left(\sum_{\beta^{p+1} > \alpha^p} \frac{\omega(\alpha^p)}{\omega(\beta^{p+1})} + \sum_{\gamma^{p-1} < \alpha^p} \frac{\omega(\gamma^{p-1})}{\omega(\alpha^p)} \right) \right.$$

$$\left. - \sum_{\substack{\alpha_1^p \| \alpha^p, \alpha_1^p \neq \alpha^p}} \left| \sum_{\substack{\beta^{p+1} > \alpha_1^p \\ \beta^{p+1} > \alpha^p}} \frac{\sqrt{\omega(\alpha^p)\omega(\alpha_1^p)}}{\omega(\beta^{p+1})} - \sum_{\substack{\gamma^{p-1} < \alpha_1^p \\ \gamma^{p-1} < \alpha^p}} \frac{\omega(\gamma^{p-1})}{\sqrt{\omega(\alpha^p)\omega(\alpha_1^p)}} \right| \right]; \qquad (13.2.4)$$

where $\alpha < \beta$ denotes α being a face of β and $\alpha_1 \| \alpha_2$ parallel faces α_1 and α_2. Needles to say, the discretization of the Bochner-Weizenböck Formula van not be achieved without a fitting discretization of the p-Laplacians and, in fact, Forman's route is to first find discretization for the Laplace operators $\Box_p$, and derive the curvature functions based on. Unfortunately, Forman's derivation is lengthy and convoluted, so we are not able to bring it here. Nevertheless, the interested and determined reader is encouraged to read Forman's original paper [123]. Neither do we bring here the formulas of the Bochner and Riemann Laplacians, even though they have a number of useful applications in Graphics and Complex Networks, as to not depart from the .main goal of this chapter (and of the book in general).

Remark 13.2.5 *It would have been tempting to simply extend the definition curvature functions from the case of weights 1, as given by Formula (13.2.1), to the general weight one, by replacing, e.g. its first term by the total weight of the $(p+1)$-faces having α as a face, etc. While such an approach is common in Combinatorics, it is not the proper route to adopt, and not only because of the desire to obtain parallel formulas for the Laplacians. Indeed, experiments on simple grids as emerging in Imaging (see below) show that this approach produces quite weak results, as drastically diverges from the smooth case.*

13.2.2 Two-Dimensional Complexes

For weighted 2-dimensional complexes the formula of the Forman-Ricci curvature (i.e. the case $p = 1$ of (13.2.4)) is:

$$\mathrm{Ric}_F(e) = w(e) \left[\left(\sum_{e \sim f} \frac{w(e)}{w(f)} + \sum_{v \sim e} \frac{w(v)}{w(e)} \right) \right. \qquad (13.2.5)$$

$$\left. - \sum_{\hat{e} \| e} \left| \sum_{\hat{e}, e \sim f} \frac{\sqrt{w(e) \cdot w(\hat{e})}}{w(f)} - \sum_{v \sim e, v \sim \hat{e}} \frac{w(v)}{\sqrt{w(e) \cdot w(\hat{e})}} \right| \right].$$

(See Figure 13.2). Let us emphasize again here that motivation residing behind the importance of the parallel edges stems from the fact that Ricci curvature is an intrinsic quantity, thus edges like e_1 and e_2 in Figure 13.2 should appear in its definition, whereas e_0 who represents an extrinsic quantity (being the minimal distance in the grid realization of the underlying graph) is extrinsic. – See also Figure 13.3.

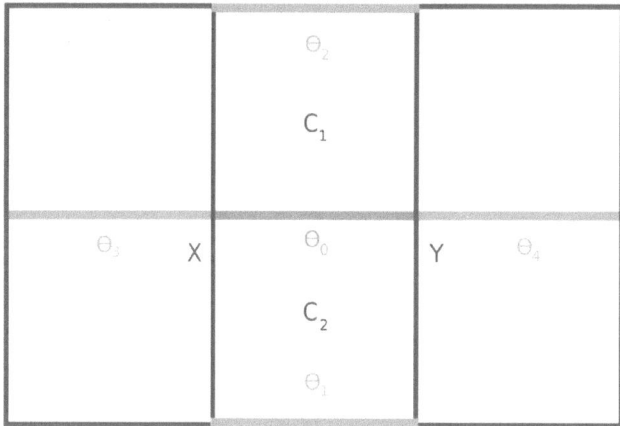

Figure 13.2 A square grid, such as naturally arising in Imaging. The edges parallel to e_0 are those having with it a common "child" (e_1 and e_2), or a common "parent" (e_3 and e_4).

There are two special cases extremely important in applications. The first case is that of square grids, as they arise most naturally in Imaging. Due to the specific (and evident) parallelism relationship, the resulting formula attains the following simple form (see also Figure 13.2):

$$\mathrm{Ric}_F(e_0) = w(e_0)\left[\left(\frac{w(e_0)}{w(c_1)} + \frac{w(e_0)}{w(c_2)}\right) - \left(\frac{\sqrt{w(e_0)w(e_1)}}{w(c_1)} + \frac{\sqrt{w(e_0)w(e_2)}}{w(c_2)}\right)\right]. \quad (13.2.6)$$

The second special case is that of PL manifolds and, more generally, of graphs and networks where the only 2-cycles are triangles. This type of structure is, again, relevant in Graphics and related fields. Moreover, it is precisely the type of graph (network) that we developed and discussed above. The fitting formula is

$$\mathrm{Ric}_F(e) = w(e)\left[\sum_{t>e}\frac{w(e)}{\omega(t)} + \left(\frac{w(v_1)}{w(e)} + \frac{w(v_2)}{w(e)}\right)\right. \quad (13.2.7)$$

$$\left. -\sum_{\tilde{e}\sim e}\left|\sum_{t>e,\,t>\tilde{e}}\frac{\sqrt{w(e)w(\tilde{e})}}{w(t)} - \sum_{v<e,\,v<\tilde{e}}\frac{w(v)}{\sqrt{w(e)w(\tilde{e})}}\right|\right].$$

where "$t > e$" denotes that the edge e is a face of the triangle (2-cycle) t.

To differentiate between the 1- and 2-dimensional versions of Forman-Ricci curvature, we shall refer to the former as the *graph* or *reduced Forman-Ricci curvature*, and to the later as the *full Forman-Ricci curvature* (or simply the *Forman-Ricci curvature*).

Note that, for combinatorial weights, the addition to the complex X of each triangle $\mathbf{t}$ adjacent to an edge e increases the Forman Ricci curvature by 3, that is

$$\mathrm{Ric}_F(e|X + t) = \mathrm{Ric}_F(e|X) + 3. \quad (13.2.8)$$

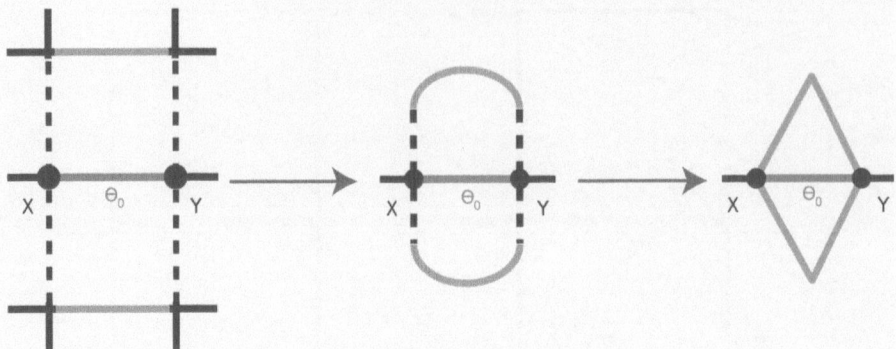

Figure 13.3 The role of parallel edges in Forman's curvature: Ricci curvature is an intrinsic quantity, thus only those depicted in as curved arc in the middle picture count. Their identification with edges in a square grid appear on the left, while the meaning of parallel edges in the 1-dimensional case, obtained by the degeneration of square grid, is shown on the right.

(For details, see [335].)

As for other discretization of Ricci curvature, we can define the fitting scalar curvature, in a similar manner:

$$\kappa_{\mathrm{Ric}_F}(v) = \mathrm{Ric}_F(v) = \sum_{e_k \sim v} \mathrm{Ric}_F(e_k); \tag{13.2.9}$$

where, again, $e_k \sim v$ stands for all the edges e_k adjacent to the vertex v.

Remark 13.2.6 *Note that, while we concentrated above on Forman's Ricci curvature, one can substitute [294] (see also the very recent important theoretical contribution [180]) instead Ollivier's discretization, which as we saw in the previous chapter is widely employed in Network Theory. We preferred Forman's curvature for a number of reasons (one of which we mentioned in the previous chapter), perhaps not the least of them being its clear computational advantages, as well as for the easy manner in which it extends to higher dimension (see also the discussion in the preceding chapter). One of these additional advantages we already mentioned above, namely that by its defining formula it comes coupled with a Laplacian, even though we do not explore this direction here.*

The most immediate application of the Forman-Ricci curvature is to Imaging, where the square grid of the pixels immediately suggests itself as a testing ground for the applicative capabilities of this type of discrete curvature, even more so given its simple expression (13.2.6) in this case.[6] Evidently, the immediate application of the

---

[6]Indeed, square – and more general, cubical complexes – occupy, due to the easy expression of parallelism in their case, a special place in Forman's original paper [123] as well. For the importance and role of cube complexes in "Pure" Mathematics, see, for instance [278], and the references therein. (Note that the open problem on cube complexes attacked there is solved using Alexandrov and Wald curvatures, rather than the Forman-Ricci one.)

Figure 13.4 Forman-Ricci curvature as an edge detector, illustrated on the classical "Cameraman" test image.

Forman-Ricci curvature as an edge detector, mainly in such instances where the data is "very discrete", most notably in Medical Imaging [288], [291]. As we have shown in the mentioned papers, as an edge detector the Forman-Ricci curvature outperforms the Gauss curvature computed standardly using the classical differential operators as implemented by **MATLAB**. The reader can gage its efficiency from the simple example of the standard test image in Figure 13.4 Beyond this basic, yet effective use, stronger results for denoising of images have been obtained by combining the Forman-Ricci curvature (and the fitting Bochner Laplacian) with some other methods, such as *constrained local means* and *non-local means* [355]. Another Imaging applications concerns the classification of the various types of textures, with special insight on the so called stochastic textures. In this task, the Forman-Ricci curvature clearly outperforms the Ollivier curvature [27]. Another successful application of the Forman-Ricci (again, in tandem with the Forman Laplacians) is in change detection in Medical Imaging (with clear medical implications) as well in aerial images, were it also proved to be quite potent in distinguishing man-made objects in satellite and aerial images.

Clearly, it is impossible to consider Ricci curvature without also appraising the benefits of the Ricci flow. Indeed, given its geometric interpretation (see Figure 13.5) its application to image smoothing and denoising is almost mandatory. In fact, one can also use the *backward Ricci flow*, given that, while the classical flow is given by a PDE, thence not reversible, its discretization is expressed as an ODE, thus reversible. This allows for a *forward-and-backward* (*FAB*) Ricci flow [112], which permits smoothing of images without blurring the edges. In particular, such an approach (especially if used in conjunction with the Bocher-Laplacian curvature) is extremely useful in change detection in aerial images [112], [306]. A further, unrelated application is for the *high dynamic range* (*HDR*) *compression* [112], [306].

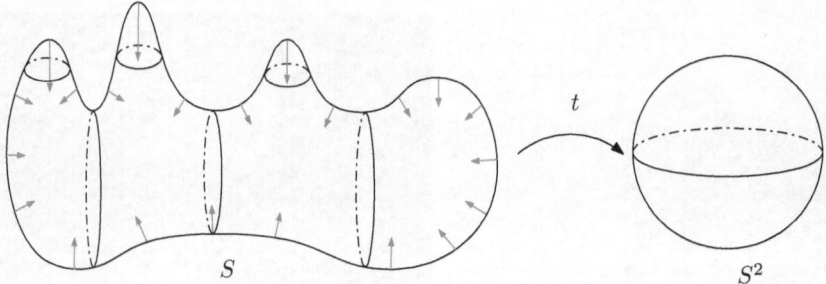

Figure 13.5   The Ricci flow of a smooth surface evolves the higher-curvature points faster, thus obtaining in the limit a round sphere.

Remark 13.2.7 *The smoothing properties and its efficiency in imaging of Ricci flow can be inferred from the fact that it represents the Beltrami-Laplace flow on the metric of the underlying space. However, for images, which need to evolve over a fixed grid, reconstruction of the image from its metric is necessary, which for the short time flow is indeed possible, using, e.g. the so called Poisson solver [18], [306] which are based on the image's gradient. Unfortunately, recovering the said gradients is a quite challenging task, and the success of the method is thus rather limited.*

Clearly, the triangular meshes for which the Forman-Ricci curvature has the form (13.2.7), are the natural ground, so to say, of Graphics applications. Unfortunately, we did only very few experiment in this setting, except in the related case of images (where a natural subdivision into triangles is applied).

Of course, it would be most natural to consider the case of 3-dimensional cubical grids, i.e. of voxels, which is the essential in MRI imaging and CT tomography (see Figure 13.6). Incipient experimental results show high potentiality for diagnosis of cancer. Here again, one can consider the easier to handle tetrahedral meshes.

Exercise 13.2.8 *Write the formula for the Forman-Ricci curvature of tetrahedral meshes – See Figure 13.7.*

Remark 13.2.9 *One could make use of the higher dimensional Forman curvature functions, in classification of analysis endowed with further features (color, texture, luminosity), interpreted as additional dimensions.*

There exists yet another natural ground for the application of the Forman-Ricci curvature and flow, which it is even "more discrete" that of images, thus making this type of curvature even more relevant (and even demanded). We refer to the case of multiplex networks and hypernetworks which we presented in some detail earlier. The imperative use of the Forman-Ricci curvature in this setting is quite imperative, since such objects lack any manifold structure (see Figure 13.8).

We have explored the use of the Forman-Ricci curvature and flow for such networks, from a number of theoretical viewpoints [335], [265], [295] and for a variety

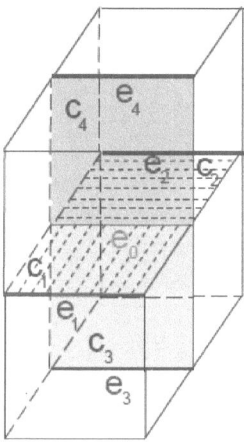

Figure 13.6 Parallel cells (and edges) in a cubical grid, as encountered in volumetric imaging.

of tasks e.g. backbone extraction, sampling [295], [28], and embedding [337], and applications, such as modeling cellular dynamics in differentiation and cancer evolution [183].

Studying such generalized networks by means of the Forman-Ricci curvature flow has yet another important, yet perhaps unexpected advantage. it stems from what, originally, was a lacuna of the Forman-Ricci curvature: Given that, in dimension 2, Ricci and Gauss classical curvatures coincide, one would expect Forman-Ricci curvature to satisfy a version of the essential Gauss-Bonnet Theorem. However, this is not the case, even though other essential properties of the classical notion (e.g. Meyers Theorem) are reciprocated by the considered discrete version [123]. To remedy this lacuna, Bloch [46] proposed a *modified Forman-Ricci curvature* that does indeed satisfy this important theorem. To this end he introduced a number of *auxiliary functions*,[7] to help define the Forman curvature functions. In the case of cell complexes which represent our object of interest, only the 0- and 2-dimensional auxiliary functions are relevant, as we shall see shortly, the curvature functions which he denotes by R_0 and R_2 are

$$R_0(v) = 1 + \frac{3}{2}A_0(v) - A_0^2(v)\,, \ R_2(t) = 1 + 6B_2(t) - B_2^2(t)\,; \qquad (13.2.10)$$

where A_0, B_2 are the aforementioned auxiliary functions, which are defined in the following simple and combinatorially intuitive manner:

$$A_0(x) = \#\{y \in F_1, x < y\}\,, \ B_2(x) = \#\{z \in F_1, z < x\}\,. \qquad (13.2.11)$$

---

[7]albeit not unique

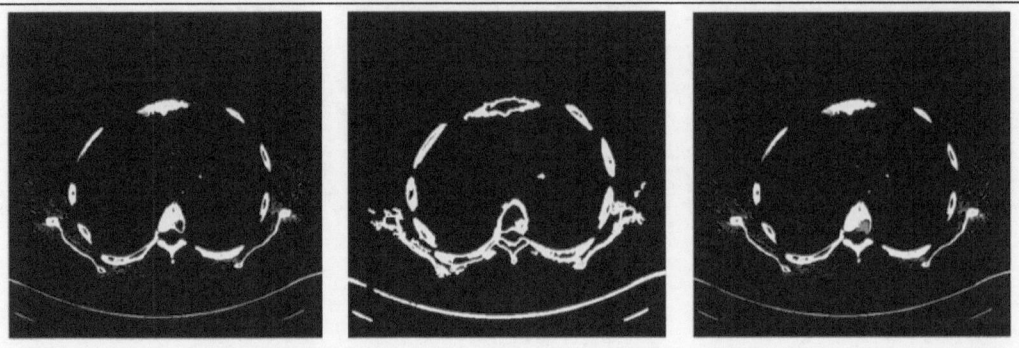

Figure 13.7 Ricci flow as a change detector in a CT medical image. Left: Original image; Middle: Forman-Ricci curvature; Right: The detection result. (Courtesy of E. Appleboim.)

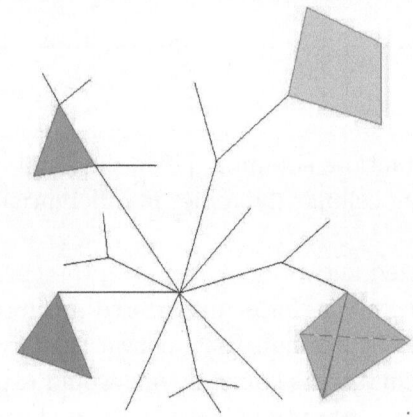

Figure 13.8 Higher order correlations.

Using the functions above, Bloch obtains the following discrete version of the Gauss-Bonnet Theorem:

$$\sum_{v \in F_0} R_0(v) - \sum_{e \in F_1} \mathrm{Ric}_F(e) + \sum_{f \in F_2} R_2(t) = \chi(X) \,. \qquad (13.2.12)$$

After this theoretical detour we can return to the advantage alluded to above: Using the Euler characteristic we can define the *prototype* (or *gauge*) network associated to a given one, namely:

Definition 13.2.10 *(Prototype networks) Let X be a 2-dimensional polyhedral complex with Euler characteristic χ as given by the Bloch-Gauss-Bonnet formula (13.2.12). Then we define X to be*

1. *Spherical, if $\chi > 0$;*

2. *Euclidean, if $\chi = 0$;*

3. *Hyperbolic, if $\chi < 0$.*

This definition prompts us to propose a consistent notion of background geometry for networks (viewed as polyhedral complexes, or rather as their 2-skeleta) that allows for the study of long term network evolution. This notion enables us to study the topological complexity of a networks, as well as intertwined geometric properties, such as dispersion of geodesics, recurrence, volume growth, etc., specific to each geometric type. For more details and some first suggestive examples see [335], and for more examples on larger real life networks, [28].

Yet another application (which, unfortunately, given its potential, in particular in *Persistent Homology*, has not been sufficiently explored) issues from the connection between Forman-Ricci curvature and the algebraic topological properties of the underlying complex [123]. We do bring here more details and refer the reader to the short paper [297].

We conclude this section by noting that there is yet another possible route to the application of Forman-Ricci curvature to Persistent Homology, this time via the connection between it and the Banchoff polyhedral Morse Theory which we essentially presented in Chapter 9 – see [268].

13.2.3   The Forman-Ricci Curvature of Networks

While not considered by Forman, it is also natural to ponder the limiting case of 1-dimensional complexes, i.e. graphs or networks. In this case, given the absence of higher dimensional (i.e., in this case, 2-dimensional) faces, the expression of Forman-Ricci curvature reduces to the following elementary formula:

$$\mathbf{F}(e) = w(e) \left(\frac{w(v_1)}{w(e)} + \frac{w(v_2)}{w(e)} - \sum_{e(v_1) \sim e, \, e(v_2) \sim e} \left[\frac{w(v_1)}{\sqrt{w(e)w(e(v_1))}} + \frac{w(v_2)}{\sqrt{w(e)w(e(v_2))}} \right] \right) ;$$

(13.2.13)

where $\mathbf{F}$ denotes the Forman-Ricci curvature for networks[8], and where, again,

- e denotes the edge under consideration between two nodes v_1 and v_2;

- $w(e)$ denotes the weight of the edge e under consideration;

- $w(v_1), w(v_2)$ denote the weights associated with the nodes v_1 and v_2, respectively;

- $e(v_1) \sim e$ and $e(v_2) \sim e$ denote the edges incident to the nodes v_1 and v_2, respectively, after *excluding* the edge e under consideration which connects the two nodes v_1 and v_2, i.e. $e(v_1), e(v_2) \neq e$.

See Figure 13.9 for a simple illustration, as well as Figure 13.10 for a concrete social network example.

---

[8]notice the change of notation for clarity reasons

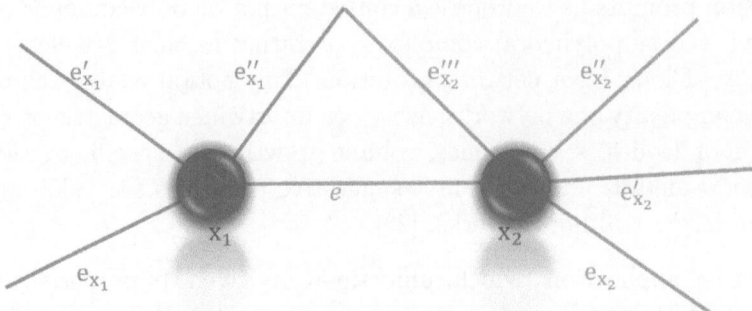

Figure 13.9 The edges and vertices contributing to computation of the 1-dimensional Forman Ricci curvature of an edge.

The geometric meaning of the formula above is easy to grasp if one first reduces it to the combinatorial case, that is that of weights equal to 1. In this case it reduces to the following very simple formula:

$$\mathbf{F}(e) = 4 - \deg(v_1) - \deg(v_2); \tag{13.2.14}$$

where v_1, v_2 are the end nodes of the edge e. Therefore, it is evident that, at least in this case, $\mathbf{F}(e)$ measures the spreading or dispersion at the vertices of the edge e, thus representing a measure of the flow through the edge e. This interpretation extends to the general weights case, even though the understanding of the various cases is more involved – see [154].

While, as discussed above, we are mainly interested in edge-centric measures, and more specifically in Ricci curvature, one can also define the Forman-scalar curvature, in manner similar to the PL manifolds case which we already encountered earlier:

$$\kappa_{\mathbf{F}(v)} = \mathbf{F}(v) = \sum_{e_k \sim v} \mathrm{Ric}_F(e_k); \tag{13.2.15}$$

where $e_k \sim v$ denotes the edges e_k adjacent to the vertex v.

The natural applications of the reduced Ricci-Forman curvature are, of course, in Complex Networks, in all their multiple facets. They proved to be useful in various applications in Chemistry [293], Genetics [356], and other aspects of Biology [293], Brain Networks analysis and its application to the understanding of Autism [116] and predicting intelligence [208], Financial Networks [264], Deep Learning [357], etc. Furthermore, it has been successfully applied to a number of networks' specific task, such as sampling and backbone detection [28], as well as denoising and change detection [336].

We also applied the 1-dimensional Forman-Ricci curvature in Medical Imaging [201]. More precisely, we used in its natural role of edge detector. Unexpectedly, but happily enough, it is precisely it discrete nature that renders this curvature as powerful tool in the task at hand. This is due, in part, to the fact that the ossicles, as being the smallest bones in the human body, occupy only approximately 200 voxels, thus rendering more classical approaches based on approximations of classical operators as ineffectual (and even producing artifacts).

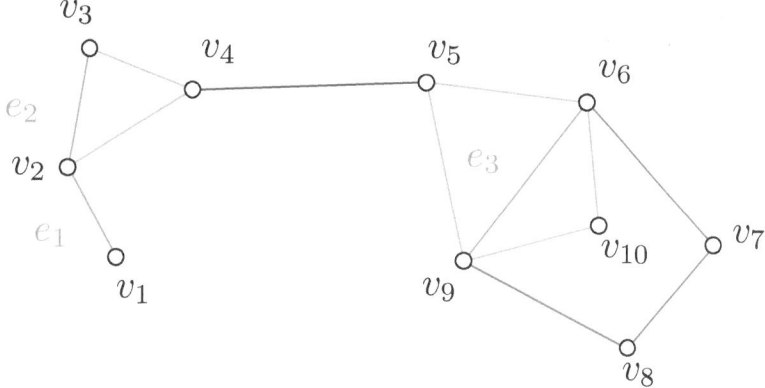

Figure 13.10 A small combinatorial network (i.e. with all vertex and edge weights equal to 1) based on the well-known "Dolphins" social network [195], emphasizing the differences between the three types of Ricci curvature considered. $\mathrm{Ric}_H(e_1)$ is zero, since there are no cycles (faces) adjacent to e_1 and $\mathbf{F}(e_1)$ also equals 0 (as it is really seen from the degrees of its end vertices) and, moreover, so does $\mathrm{Ric}_F(e_1)$, since in this case it coincides, due to the absence of adjacent faces, with $\mathbf{F}(e_1)$. $\mathrm{Ric}_H(e_2) = \sqrt{2}$, given that there is only one face – a triangle – adjacent to e_2. Degree counting easily renders $\mathbf{F}(e_2) = -1$, and $\mathrm{Ric}_F(e_2) = 2$. In the case of e_3, degree counting renders $\mathbf{F}(e_3) = -4$. Since the edge e_3 is adjacent to 2 triangles and a quadrangle, $\mathrm{Ric}_H(e_3) = 2\sqrt{2} + \sqrt{3}$, while given the fact that there is only one edge parallel to e_3, namely $(v_7 v_8)$, we obtain that $\mathbf{F}(e_2) = -4$.

Remark 13.2.11 *In truth, we could have presented the Ollivier-Ricci curvature in this chapter, as well as in the preceding one. Indeed, by its very definition, one considers measures (attached to the vertices) than are transported along the edges, according to their lengths, i.e. to the distances between these vertices, thus Ollivier-Ricci curvature is defined by essentially viewing networks as metric measure spaces.*

We should also mention that there is yet another manner to regard networks as metric measure spaces – see [281]. (The approach suggested therein is rather technical and along a different vein, so we do not bring it here. However, let us note, that it facilitates the analysis of networks at many scales.)

These observations, mainly those regarding the Ollivier- and Forman-Ricci curvatures, which are the most commonly used in networks' intelligence, suggest that, while most researchers still view graphs/networks as metric spaces, the model of metric measure spaces for weighted networks, is the most natural and expressive way of describing the properties of such geometric objects.[9]

---

[9]To this end, a natural idea is to incorporate the node weights and edge weights into one expressive metric, thus rendering any weighted network into a "honest to God" metric space, whose geometric properties (curvature, geodesics, embeddings, etc.) can than be investigated with (more-or-less) classical tools. An example of such a comprehensive metric is the so called *degree path metric* – see, e.g. [186]. Another well-known such metric is the *resistance metric* (see, e.g. [102]). (Yet another, somewhat different approach is the one suggested in [281].)

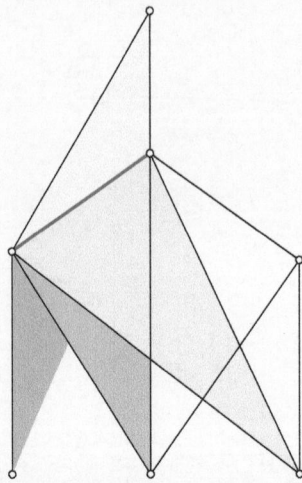

Figure 13.11 Computation of the combinatorial Forman-Ricci curvature of an edge in the poset-based model: The red edge is adjacent to the depicted four triangles. Thus, by Formula (13.2.3), the combinatorial Forman-Ricci curvature of this edge is $4 - 1 + 2 = 5$.

13.2.3.1 *From Networks to Simplicial Complexes*

The simple Graph Forman-Ricci curvature introduced above, while alluring in his clarity and directness, lacks the full power of the "Full" Forman-Ricci curvature, since it only models a 1-dimensional nerve of the underlying considered complex. It is therefore desirable to be able to return to the standard setting and, moreover, to do this in a unified manner, common too all type of networks, directed or undirected, generalized or not. As a matter of fact, such an approach is possible, that canonically transforms any type of network, into a 2-dimensional simplicial complex, i.e. the most classical object of Combinatorial and Algebraic Topology, whose (combinatorial) Forman-Ricci curvature can easily be computed [344]. This method is based again on the work of Bloch [46] and the construction is quite simple: One adds a triangle (2-simplex) for each path of length 2, a tetrahedron (3-simplex) for each path of length 3, etc. (See Figure 13.11 for the computation of Forman-Ricci curvature of the resulting 2-complex.) The technical detail can be found, of course, in Bloch's paper, while experiments on real life hypernetwork are brought in [344].

II

Differential Geometry, Computational Aspects

CHAPTER 14

Algebraic Topology

This chapter focuses on the fundamental concepts and computational algorithms in surface algebraic topology. Algebraic topology studies the topology of a space using algebraic methods. In general, special groups are associated with the space, algebraic structures of these groups convey the topological information [160].

14.1 INTRODUCTION

Topology plays a fundamental role in many engineering and medical fields. As shown in Figure 14.1, the vase surface is of genus 5, namely it has 5 "handles". There are two loops on each handle, the dark (red) and the light (green) ones. The surface separates the whole ambient space $\mathbb{R}^3$ into two connected components, one is inside the surface with finite volume, denoted as $\mathcal{I}$ and the other is outside the surface with infinite volume, denoted as $\mathcal{O}$. Each dark (red) curve can shrink to a point in $\mathcal{O}$, but not in $\mathcal{I}$; conversely, each light (green) curve can shrink to a point in $\mathcal{O}$, but not in $\mathcal{O}$. The dark (red) curves are called the *tunnel loops*, the light (green) curves the *handle loops*. Each handle has a pair of handle and tunnel loops. In practice, handles on a surface can be detected by finding the handle loops and the tunnel loops [101].

In medical imaging field, the human organ surfaces can be extracted from volumetric CT or MRI images. First, the images are segmented into foreground and background; second, the boundary surface of the foreground volume is extracted. Due to the image noises and the computational inaccuracy, the reconstructed surfaces always have fake handles, which are called *topological noises*. It is crucial to detect and remove these spurious topological features. This process is called *topological denoise*. As shown in Figure 14.2, in virtual colonoscopy, the human colon surface is reconstructed from abdominal CT scans. There are hundreds of fake handles, which can be detected by computing the handle loops and the tunnel loops. Furthermore, all the fake handles can be removed by the following surgery: cut each fake handle along its handle loop to produce two holes (boundary components), then fill each hole by a topological disk. Topological denoise is essential for many medical imaging applications.

The basic philosophy for algebraic topology is to associate algebraic groups with topological spaces, and maps between these algebraic objects are associated to continuous maps between spaces. In this way, we translate topology to algebra, and

DOI: 10.1201/9781003350576-14

Figure 14.1 The handle loops and the tunnel loops on high genus surfaces.

study the topology by analyzing the group structures and the group homomorphisms. Namely, algebraic topological method is a functor $F : \mathfrak{C}_1 \to \mathfrak{C}_2$ between the topology category $\mathfrak{C}_1$ and the algebra category $\mathfrak{C}_2$, where

$$\begin{aligned} \mathfrak{C}_1 &= \{Topological\ Spaces, Continuous\ Mappings\} \\ \mathfrak{C}_2 &= \{Groups, Homomorphisms\} \end{aligned}$$

F preserves the identity morphisms and composition of morphisms,

$$F(id_X) = id_{F(X)} \quad \forall X \in \mathfrak{C}_1,$$
$$F(f \circ g) = F(f) \circ F(g) \quad \forall f : X \to Y, g : Y \to Z \in \mathfrak{C}_1$$

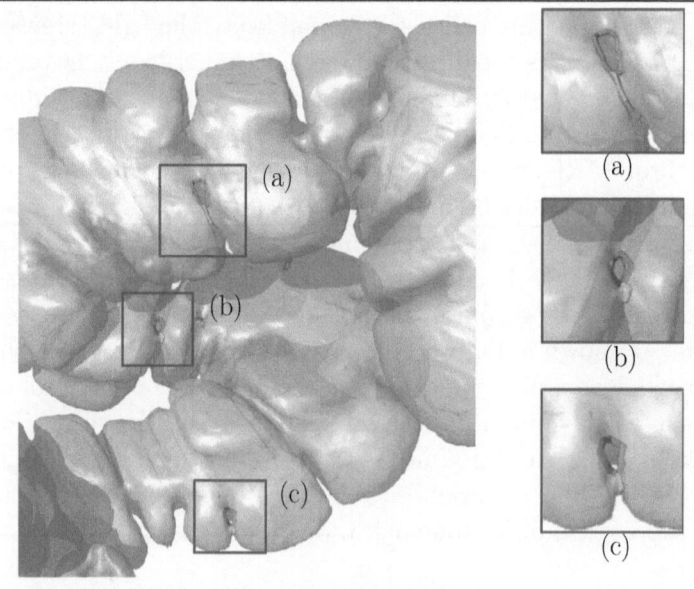

Figure 14.2 Topological denoise in virtual colonoscopy.

14.2 SURFACE TOPOLOGY

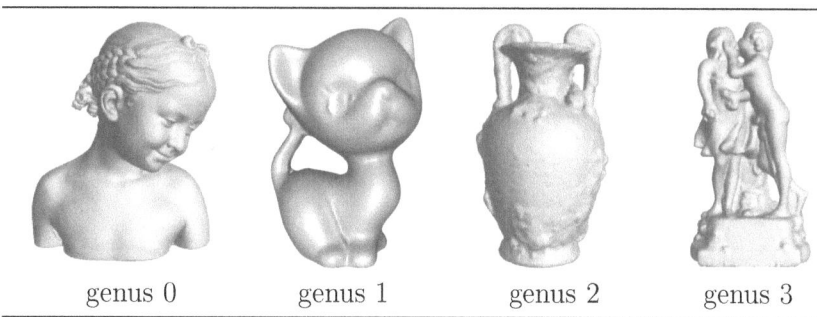

genus 0	genus 1	genus 2	genus 3

Figure 14.3 Topological classification for close surfaces .

In real physical world, the surfaces are everywhere, since the boundaries of all the solid volumes are surfaces. Surfaces can be classified according to their topologies.

Definition 14.2.1 (Topological Equivalence) *Given two manifolds M_1 and M_2, a continuous map $f : M_1 \to M_2$ is called a* homeomorphism, *if it is invertable, and its inverse $f^{-1} : M_2 \to M_1$ is also continuous. Two manifolds are* homeomorphic *to each other, or* topologically equivalent *to each other, if there is a homeomorphism between them.*

Orientable Surfaces   Figure 14.3 shows several closed surfaces embedded in $\mathbb{R}^3$. Each of them separates the whole space $\mathbb{R}^3$ into two connected components, the finite component is called the *interior volume*, the infinite component is called the *exterior volume*. At each point of the surface, the normal vector points from the interior to the exterior. Therefore, each surface has two sides, the outside and the inside. We can paint the outside without touching the inside. In this situation, we say the surface is *orientable*. Furthermore, each surface has several handles, the number of handles is called the *genus* of the surface. Two orientable closed surfaces are topologically equivalent, if and only if they have the same genus.

Figure 14.4 shows surfaces with boundaries. Each surface still have two distinct sides, one cannot go from one side to the other without crossing the boundaries. Therefore, all the surfaces in the figure are still orientable. The left surface has no handle, so it is of genus zero. Its boundary has one connected component. The middle surface is more complicated, its boundary has 3 connected components. In fact, it has one handle. For two orientable surfaces with boundaries, they are topologically equivalent, if and only if they have the same genus and the same number of boundary connected components.

Non-orientable Surfaces   A surface with only one side is called *non-orientable*. The Möbius band is one of the most common non-orientable surfaces. Fix a point p on the Möbius band, one can travel on the surface without crossing the boundary and

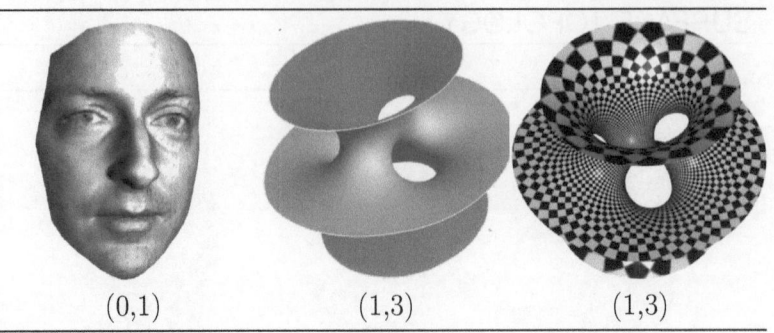

(0,1)   (1,3)   (1,3)

Figure 14.4 Topological classification for surfaces with boundaries, represented as (genus, boundary).

reach the same point p again but on the opposite side. Now he has finished one loop γ on the surface. If he continues marching along the same path on surface, he can return to p again, this time on the same side. Now, he has finished the same loop twice. The loop γ is the generator of the fundamental group of the Möbius band.

The simplest closed non-orientable surface is the projective plane $\mathbb{RP}^2$, which can be abstractly defined as follows:

Definition 14.2.2 (Projective Plane) *All straight lines through the origin in $\mathbb{R}^3$ form a two dimensional manifold, which is called the projective plane $\mathbb{RP}^2$.*

Figure 14.5 visualizes the construction of a projective plane. Each line through the origin in $\mathbb{R}^3$ intersects the interior of the lower hemisphere at a single point and the equator at two antipodal points. Therefore, the projective plane can be obtained from the (closed) lower hemisphere by identifying two antipodal points on the equator. Then we cut off a spherical cap from the hemisphere to obtain a topological annulus and a topological disk (the spherical cap) as shown in the top row. We call the cutting curve as the spherical cutting locus. The outer circle of the annulus is the equator, the inner circle is the spherical cutting locus, denoted as ρ. In the second row, the annulus is divided into two topological rectangles A and B. We call the two cutting curves as the annular cutting locus. Each rectangle has two horizontal sides, one is the equator, the other is the spherical cutting locus. Each rectangle also has two vertical sides, corresponding to the annular cutting locus. The rectangle A is flipped to align the antipodal points on the equator with those on B. The two rectangles are glued along the original equator, such that the corresponding antipodal points are attached to each other. Now, one bigger rectangle is obtained, whose both horizontal sides are the spherical cutting locus and vertical sides are the annular cutting locus. The boarder between A and B in the bigger rectangle is the semi-equator, denoted as γ. Then the vertical sides of the bigger rectangle are glued together following the original annular cutting locus to obtain a Möbius band with a single boundary curve, which is the original spherical cutting locus ρ. In the Möbius band, γ becomes a loop which generates its fundamental group. We can directly glue back the spherical cap along the original spherical cutting locus (ρ). The final resulting surface is a projective

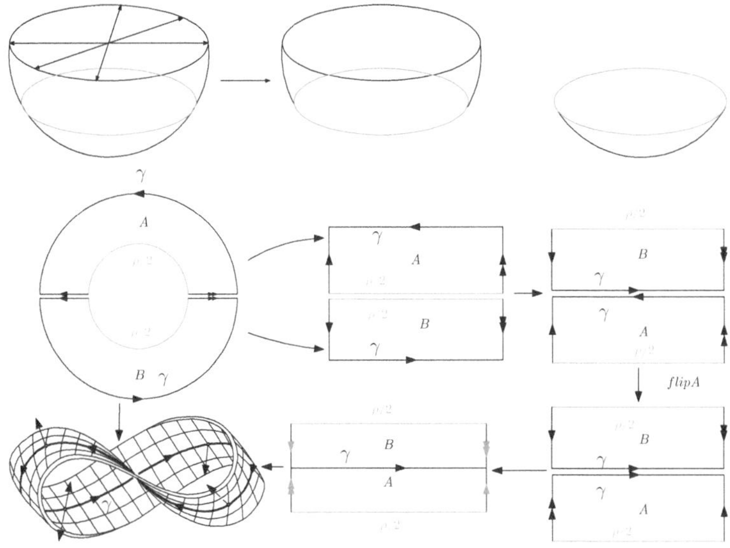

Figure 14.5 The construction of a projective plane.

plane $\mathbb{RP}^2$. The projective plane has no boundary, therefore it is a closed surface, but with only one side, therefore non-orientable. Now we compute its fundamental group, see section 14.3 for the concept and theorems of fundamental group.

Lemma 14.2.3 $(\pi_1(\mathbb{RP}^2))$ *The fundamental group of the real projective plane is*

$$\pi_1(\mathbb{RP}^2, q) = \langle \gamma | \gamma^2 \rangle$$

Proof From above construction, as shown in Figure 14.5, $\mathbb{RP}^2 = M \cup C$, M is a Möbius band, C is the spherical cap. $M \cap C$ is the spherical cutting locus ρ. $\pi_1(M) = \langle \gamma \rangle$, $\pi_1(C) = \langle e \rangle$, $\pi_1(M \cap C) = \langle \rho \rangle$. If we cut the Möbuis along γ, we obtain the original annulus, the inner boundary ρ is homotopic to the outer boundary γ^2. By Serfeit-Van Kampan theorem 14.5.7, we obtain the generator of $\pi_1(\mathbb{RP}^2)$ is γ, the relator of $\pi_1(\mathbb{RP}^2)$ is $\rho = \gamma^2$.

□

A projective plane with a hole is called a *crosscap*, namely a Möbius band.

Surface Topology   Surfaces with complicated topologies can be constructed by the connected sum of simple building blocks.

Definition 14.2.4 (connected Sum) *The connected sum (Figure 14.6) $S_1 \oplus S_2$ is formed by deleting the interior of disks D_i and attaching the resulting punctured surfaces $S_i - D_i$ to each other by a homeomorphism $h : \partial D_1 \to \partial D_2$, so*

$$S_1 \oplus S_2 = (S_1 - D_2) \cup_h (S_2 - D_2).$$

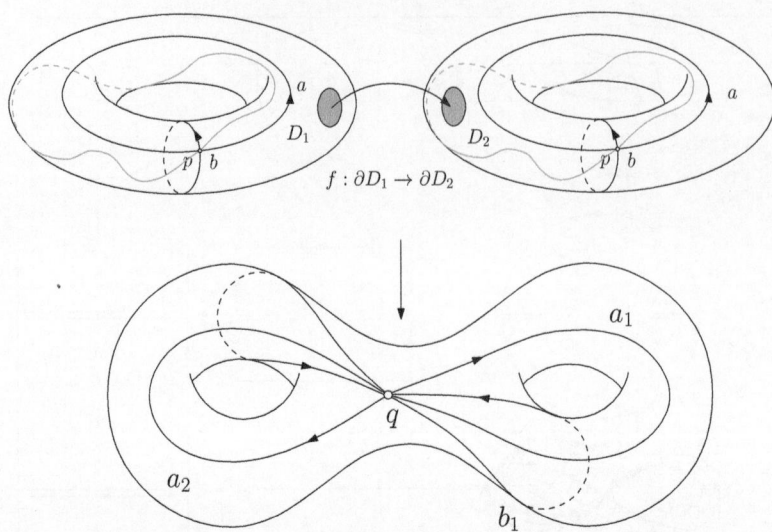

Figure 14.6 Connected sum.

There are only two types of building block, one is the torus, the other is the projective plane. Oriented surfaces are constructed by tori; non-oriented surfaces are built from projective planes. Figure 14.7 shows a genus 8 closed surface, which is the connected sum of 8 tori.

Theorem 14.2.5 (surface Topology) *Any closed connected surface is homeomorphic to exactly one of the following surfaces: a sphere, a finite connected sum of tori, or a finite sum of projective planes, namely a sphere with a finite number of disjoint discs removed and with cross caps glued in their places. The sphere and connected sums of tori are orientable surfaces, whereas surfaces with crosscaps are non-orientable.*

Figure 14.7 A Genus eight Surface, constructed by connected sum of tori.

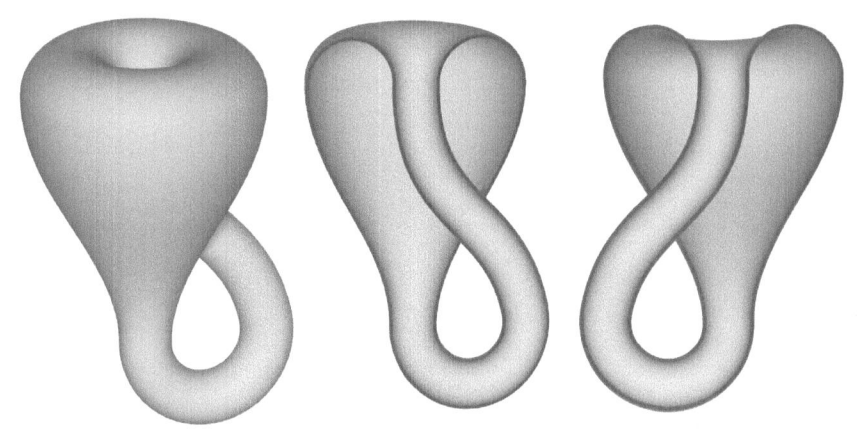

Figure 14.8 A Klein bottle is the connected sum of two projective planes.

Equivalently, any closed surface is the connected sum

$$S = S_1 \oplus S_2 \oplus \cdots \oplus S_g,$$

if S is orientable, then each S_i is a torus, otherwise if S is non-orientable, then each S_i is a projective plane. Figure 14.8 shows a symmetric Klein bottle, if we divide the bottle by the symmetry plane into two halves, each half is a Möbius band. This shows the Klein bottle is the connected sum of two projective planes.

14.3 FUNDAMENTAL GROUP

Suppose S is a topological space, $p \in S$ is the base point, all the oriented closed curves (loops) on S through p can be classified by homotopy. All the homotopy classes form the so-called *fundamental group* of S, or *the first homotopy group*, denoted as $\pi_1(S, p)$. The group structure of $\pi_1(S, p)$ fully determines the topology of S, if S is a topological surface.

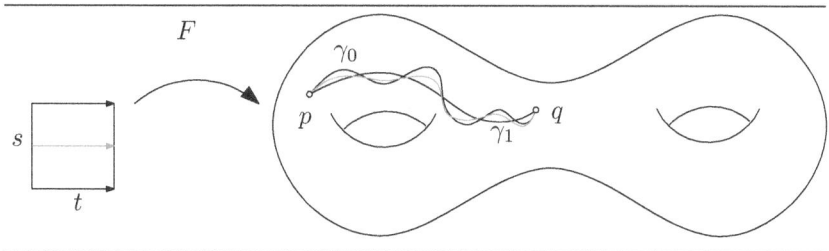

Figure 14.9 Path homotopy.

Definition 14.3.1 (Curve) *Let S be a topological space with a base point $p \in S$, a curve is a continuous mapping $\gamma : [0,1] \to S$. A closed curve through p is a curve, such that $\gamma(0) = \gamma(1) = p$. A closed curve is also called a loop.*

Among all the loops, there is a special one that stays at the base point p for any time $t \in [0,1]$, denoted as $e(t) \equiv p$.

Figure 14.9 explains the concept of homotopy.

Definition 14.3.2 (Homotopy) *Let $\gamma_1, \gamma_2 : [0,1] \to S$ be two curves. A homotopy connecting γ_1 and γ_2 is a continuous mapping $F : [0,1] \times [0,1] \to S$, such that*

$$F(0,t) = \gamma_1(t), F(1,t) = \gamma_2(t).$$

We say γ_1 is homotopic to γ_2 if there exists a homotopy between them.

Lemma 14.3.3 *Homotopy relation is an equivalence relation.*

Proof First, we show any loop is homotopic to itself, $\gamma \sim \gamma$. We construct the homotopy as $F(s,t) = \gamma(t)$. Second, we show if $\gamma_1 \sim \gamma_2$, then $\gamma_2 \sim \gamma_1$. Suppose $F(s,t)$ is the homotopy from γ_1 to γ_2, then $F(1-s,t)$ is the homotopy from γ_2 to γ_1.

□

Corollary 14.3.4 *All the loops through the base point can be classified by the homotopy relation. The homotopy class of a loops γ is denoted as $[\gamma]$.*

Given two loops through the base point, we can concatenate them to form a bigger loop, which is the product of the two loops.

Definition 14.3.5 (Loop product) *Suppose γ_1, γ_2 are two loops through the base point p, the product of the two loops is defined as*

$$\gamma_1 \cdot \gamma_2(t) = \begin{cases} \gamma_1(2t) & 0 \le t \le \frac{1}{2} \\ \gamma_2(2t-1) & \frac{1}{2} \le t \le t \end{cases},$$

as shown in Figure 14.10.

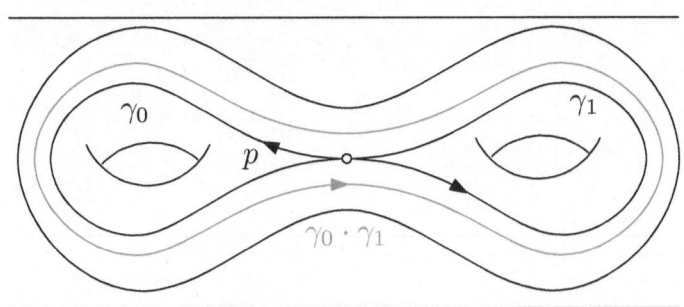

Figure 14.10 Loop product.

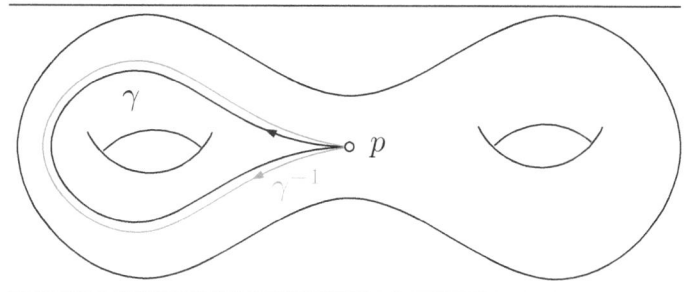

Figure 14.11 Loop inversion.

Definition 14.3.6 (Loop inverse) *Suppose γ is a loop on S, its inverse is obtained by reversing the orientation, $\gamma^{-1}(t) = \gamma(1-t)$, as shown in Figure 14.11.*

Definition 14.3.7 (Group) *A group $(\mathcal{G}, \cdot)$ consists of a set $\mathcal{G}$ and a product operator $\cdot : \mathcal{G} \times \mathcal{G} \to \mathcal{G}$, satisfy the following conditions:*

1. *Closure: $\forall g_1, g_2 \in \mathcal{G}$, the product $g_1 \cdot g_2 \in \mathcal{G}$;*

2. *Identity element: $\exists e \in \mathcal{G}$, such that $\forall g \in \mathcal{G}$, $g \cdot e = g$ and $e \cdot g = g$;*

3. *Inverse element: $\forall g \in \mathcal{G}$, $\exists g^{-1} \in \mathcal{G}$, such that $g \cdot g^{-1} = g^{-1} \cdot g = e$;*

4. *Associativity: $\forall g_1, g_2, g_3 \in \mathcal{G}$,*

$$(g_1 \cdot g_2) \cdot g_3 = g_1 \cdot (g_2 \cdot g_3).$$

We may omit the product operator and use $g_1 g_2$ to represent $g_1 \cdot g_2$. In general, the group is not commutative, namely the product depends on the order. There may be two elements $g_1, g_2 \in \mathcal{G}$, such that $g_1 \cdot g_2 \neq g_2 \cdot g_1$.

Definition 14.3.8 (Abelian Group) *A group $(\mathcal{G}, \cdot)$ is Abelian if $\forall g_1, g_2 \in \mathcal{G}$,*

$$g_1 \cdot g_2 = g_2 \cdot g_1.$$

All the homotopy classes of the loops on the topological space S through the base point $p \in S$ form the set $\mathcal{G}$, and the concatenation of the loops is the product operator $\cdot$, then we can define the concept of the fundamental group of S. The topological information of S is conveyed by the structure of the fundamental group.

Definition 14.3.9 (Fundamental Group) *Given a topological space S, fix a base point $p \in S$, the set of all the loops through p is Γ, the set of all the homotopy classes is $\Gamma/\sim$. The product is defined as:*

$$[\gamma_1] \cdot [\gamma_2] := [\gamma_1 \cdot \gamma_2],$$

the unit element is defined as [e], *the inverse element is defined as*

$$[\gamma]^{-1} := [\gamma^{-1}],$$

then the set $\Gamma/\sim$ *with the product operator* · *forms a group, the fundamental group of* S, *and is denoted as* $\pi_1(S, p)$.

We can easily verify that the fundamental group is well defined. The key observation is that if $\gamma \sim \gamma'$ and $\tau \sim \tau'$, then $\gamma \cdot \tau \sim \gamma' \cdot \tau'$, therefore the product of two homotopy classes is independent of the choices of the representatives of each class.

Suppose S is a two dimensional manifold, namely a surface, then the fundamental group $\pi_1(S, p)$ fully determines the topology of S. Namely, two surfaces are homeomorphic if and only if their fundamental groups are isomorphic.

14.4 WORD GROUP REPRESENTATION

In practice, the fundamental group of a topological space can be represented symbolically, so that a computer can manipulate the group by symbolic computation. First, we define a special word group formed by some symbols with special relations, then we show for each fundamental group, we can find an isomorphic word group. Thus, we can use the word group as the representation of the fundamental group.

Let $G = \{g_1, g_2, \cdots, g_n\}$ be n symbols, a word generated by G is a sequence

$$w = g_{i_1}^{e_1} g_{i_2}^{e_2} \cdots g_{i_k}^{e_k}, \quad g_{i_\alpha} \in G, e_\alpha \in \mathbb{Z},$$

where for positive integer power e_k

$$g_{i_k}^{e_k} := \underbrace{g_{i_k} \cdot g_{i_k} \cdots g_{i_k}}_{e_k}.$$

We say G is a set of *generators*. Given two words $w_1 = \alpha_1 \cdots \alpha_{n_1}$ and $w_2 = \beta_1 \cdots \beta_{n_2}$, the *product* operator · is defined as the concatenation:

$$w_1 \cdot w_2 = \alpha_1 \cdots \alpha_{n_1} \beta_1 \cdots \beta_{n_2}.$$

The empty word $\emptyset$ is also treated as the *identity* (unit) element. The *inverse* of a word is defined as

$$\left(g_{i_1}^{e_1} g_{i_2}^{e_2} \cdots g_{i_k}^{e_k}\right)^{-1} = g_{i_k}^{-e_k} g_{i_{k-1}}^{-e_{k-1}} \cdots g_{i_1}^{-e_1}.$$

We can easily verify that all the words with the product operator form a group, freely generated by G, denoted as

$$\langle g_1, g_2, \cdots, g_n \rangle.$$

More complicated relations can be introduced by relator. The set of relators is a set of special words, $R = \{R_1, R_2, \cdots, R_m\}$, such that each relator R_k can be replaced by the empty word $\emptyset$.

Definition 14.4.1 (word equivalence relation) *Two words are equivalent if we can transform one to the other by finite many steps of the following two types of elementary transformations:*

1. *Insert a relator anywhere,*

$$\alpha_1 \cdots \alpha_i \alpha_{i+1} \cdots \alpha_l \mapsto \alpha_1 \cdots \alpha_i R_k \alpha_{i+1} \cdots \alpha_l.$$

2. *If a subword is a relator, remove it from the word,*

$$\alpha_1 \cdots \alpha_i R_k \alpha_{i+1} \cdots \alpha_l \mapsto \alpha_1 \cdots \alpha_i \alpha_{i+1} \cdots \alpha_l.$$

All the words are classified by the word equivalence relation, all the equivalence classes with the concatenation operator form a group.

Definition 14.4.2 (Word Group) *Given a set of generators G and a set of relators R, all the equivalence classes of the words generated by G form a group under the concatenation, denoted as*

$$\langle g_1, g_2, \cdots, g_n | R_1, R_2, \cdots, R_m \rangle.$$

The word group is well defined. As before, if there is no relator, then the word group is a *free group*. In the following, we will show that the fundamental group of a surface can be represented as a word group, and explain the topological meanings of generators and relators.

14.5   FUNDAMENTAL GROUP CANONICAL REPRESENTATION

Suppose S is a C^2 surface embedded in $\mathbb{R}^3$, two C^1 curves γ_1 and γ_2 on the surface intersect at a point $q \in S$, as shown in Figure 14.12, we can define the intersection index at q as follows:

Definition 14.5.1 (Intersection Index) *Suppose $\gamma_1(t), \gamma_2(\tau) \subset S$ intersect at $q \in S$, the tangent vectors satisfy*

$$\frac{d\gamma_1(t)}{dt} \times \frac{d\gamma_2(\tau)}{d\tau} \cdot \mathbf{n}(q) > 0,$$

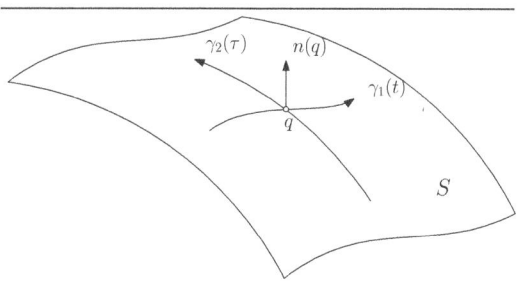

Figure 14.12   Intersection index.

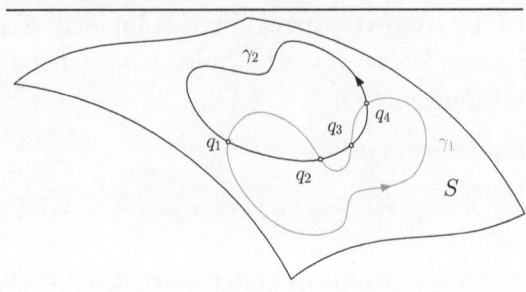

Figure 14.13 Algebraic intersection number between γ_1 and γ_2.

then the index of the intersection point q of γ_1 and γ_2 is $+1$, denoted as $Ind(\gamma_1, \gamma_2, q) = +1$. If the mixed product is zero or negative, then the index is 0 or -1 respectively.

Two curves may have multiple intersection points, as shown in Figure 14.13, the total sum of all the intersection indices is called the *algebraic intersection number* between the two curves.

Definition 14.5.2 (Algebraic Intersection Number) *The algebraic intersection number of $\gamma_1(t), \gamma_2(\tau) \subset S$ is defined as*

$$\gamma_1 \odot \gamma_2 := \sum_{q_i \in \gamma_1 \cap \gamma_2} Ind(\gamma_1, \gamma_2, q_i).$$

As shown in Figure 14.14, the algebraic intersection number of two loops is invariant under free homotopy deformation, namely homotopy deformation without fixing the based point.

Lemma 14.5.3 *Suppose $\gamma_1, \gamma_2, \gamma_3 \subset S$ are loops on S, γ_1 is free homotopic to γ_2, then*

$$\gamma_1 \odot \gamma_3 = \gamma_2 \odot \gamma_3.$$

Proof Suppose F is the homotopy between γ_1 and γ_2, $F(0, \tau) = \gamma_1(\tau)$ and $F(1, \tau) = \gamma_2(\tau)$, then $F(t, \cdot) \odot \gamma_3$ changes continuously with respect to t, but $F(t, \cdot) \odot \gamma_3$ is

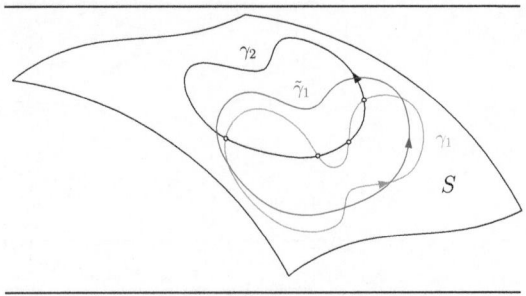

Figure 14.14 Algebraic intersection number is homotopically invariant.

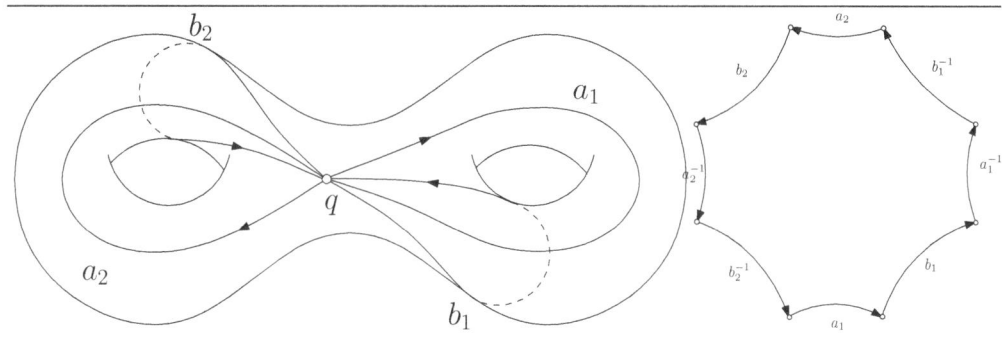

Figure 14.15 Canonical fundamental group representation and the canonical polygonal schema.

always an integer, therefore it is constant,

$$\gamma_1 \odot \gamma_3 = F(0, \cdot) \odot \gamma_3 = F(1, \cdot) \odot \gamma_3 = \gamma_2 \odot \gamma_3.$$

$\square$

Figure 14.15 shows a genus 2 oriented compact surface S, through the base point $q \in S$, on each handle, there are two loops a_i and b_i, such that the algebraic intersection number of a_i and b_i is $+1$, that of a_i and a_j is 0, that of b_i and b_j is 0, that of a_i and b_j on different handles is also 0. By deforming a_i's and b_j's, we can slice the surface to get a topological octagon, its boundary is

$$a_1 b_1 a_1^{-1} b_1^{-1} a_2 b_2 a_2^{-1} b_2^{-1}.$$

The set of loops $\{a_1, b_1, a_2, b_g\}$ is called a canonical basis of $\pi_1(S, q)$. The number of handles of a surface is called the *genus* of the surface. The above observation can be directly generalized to high genus surfaces.

Definition 14.5.4 (Canonical Basis) *Suppose S is a compact, oriented surface, there exists a set of generators of the fundamental group $\pi_1(S, p)$,*

$$G = \{[a_1], [b_1], [a_2], [b_2], \cdots, [a_g], [b_g]\}$$

such that

$$a_i \odot b_j = \delta_i^j, \quad a_i \odot a_j = 0, \quad b_i \odot b_j = 0,$$

where $a_i \odot b_j$ represents the algebraic intersection number of loops a_i and b_j, δ_{ij} is the Kronecker symbol, then G is called a set of canonical basis of $\pi_1(S, p)$.

The canonical basis is non-unique. Figure 14.16 shows another set of canonical basis (after free homotopic deformations). In fact, there are infinite many sets of canonical basis on the same surface. Therefore, surface fundamental group has canonical representations.

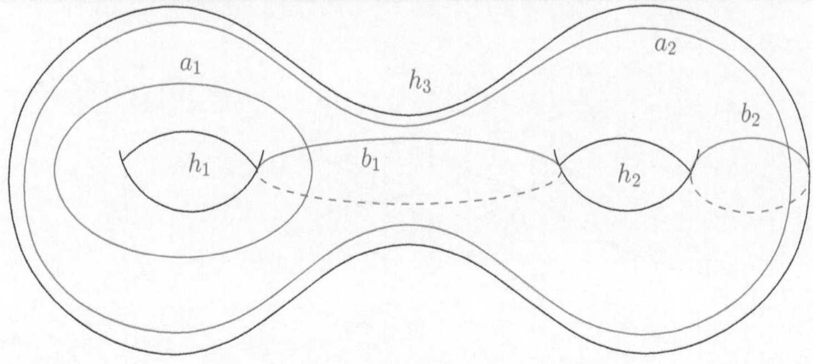

Figure 14.16 Canonical representation of $\pi_1(S)$.

Theorem 14.5.5 (Surface Fundamental Group Canonical Representation)
Suppose S is a compact, oriented surface with genus g, $q \in S$ is the base point, the fundamental group has a canonical representation,

$$\pi_1(S, p) = \langle a_1, b_1, a_2, b_2, \cdots, a_g, b_g | \Pi_{i=1}^g [a_i, b_i] \rangle,$$

where a_i's and b_j's form a set of canonical basis through q, furthermore

$$[a_i, b_i] = a_i b_i a_i^{-1} b_i^{-1}.$$

Using the canonical representation of surface fundamental groups, we can show the following theorem:

Theorem 14.5.6 *Suppose $\pi_1(S_1, p_1)$ is isomorphic to $\pi_2(S_2, p_2)$, then S_1 is homeomorphic to S_2, and vice versa.*

Proof For each surface, find a set of canonical basis, slice the surface along the basis to get a $4g$ polygonal schema as shown in Figure 14.15, then construct a homeomorphism between the polygonal schema with consistent boundary condition. The inverse is given by theorem 14.5.5 directly.

□

In order to prove the canonical representation theorem 14.5.5, we need the following Seifert-van Kampen theorem, which allows us to solve topological problems in the divide-and-conquer approach.

Theorem 14.5.7 (Seifert-Van Kampen) *Topological space M is decomposed into the union of U and V, $M = U \cup V$, the intersection between U and V is W, $W = U \cap V$, where U, V and W are path connected. Pick a base point $p \in W$, the fundamental groups of U, V and W are*

$$\pi_1(U, p) = \langle u_1, \cdots, u_k | \alpha_1, \cdots, \alpha_l \rangle$$
$$\pi_1(V, p) = \langle v_1, \cdots, v_m | \beta_1, \cdots, \beta_n \rangle$$
$$\pi_1(W, p) = \langle w_1, \cdots, w_p | \gamma_1, \cdots, \gamma_q \rangle$$

then the $\pi_1(M,p)$ is given by

$$\pi_1(M,p) = \langle u_1, \ldots, u_k, v_1, \ldots, v_m | \alpha_1 \ldots \alpha_l, \beta_1, \ldots, \beta_n,$$
$$i(w_1)j(w_1)^{-1}, \ldots, i(w_p)j(w_p)^{-1} \rangle$$

where $i : W \to U$, $j : W \to V$ are the inclusion maps.

This means the generators of the union is the union of generators, the relators of the union is the union of relators and the generators of the intersection. The proof is elementary, the students can prove it as an exercise.

In the following, we consider the oriented surfaces only, and represent it as the connected sum of tori to prove the canonical representation of the fundamental group. We prove the fundamental group of the torus first.

Lemma 14.5.8 *The fundamental group of a closed torus is*

$$\pi_1(T^2, p) = \langle a, b | aba^{-1}b^{-1} \rangle.$$

Proof As shown in Figure 14.17, we find a canonical set of basis of T^2, the latitude a and the longitude b, slice T^2 along a and b to obtain a canonical polygonal schema, which is a topological rectangle with boundary $aba^{-1}b^{-1}$. Because the boundary of the rectangle can shrink to a single point on the rectangle, this means $aba^{-1}b^{-1}$ is homotopic to e. This shows the relator of $\pi_1(T^2, p)$ is $aba^{-1}b^{-1}$.

In the following, we show $\{a, b\}$ are the generators of $\pi_1(T^2, p)$. Given a loop $\gamma \subset S$, it may intersect a and b at multiple points. Suppose $0 \le t_1 < t_2 \cdots t_n < 1$, and $\gamma(t_k)$'s are the intersection points between γ and a or between γ and b. We homotopically deform γ such that all the intersection points are moved to the base point p, namely after the deformation $\gamma(t_k) = p$, $k = 1, 2, \ldots, n$. Then the intersection points divide γ into curve segments $\{\gamma_1, \gamma_2, \ldots, \gamma_n\}$, each γ_k is the restriction of $\gamma(t)$ on $[t_k, t_{k+1}]$. By construction, γ_k is a loop and has no interior intersection points with a or b, namely, the end points of γ_k are at the corners of the rectangle, and the interior of γ_k is contained in the interior of the rectangle. Therefore γ_k is homotopic to a boundary segment of the rectangle, namely γ_k can be represented as the product

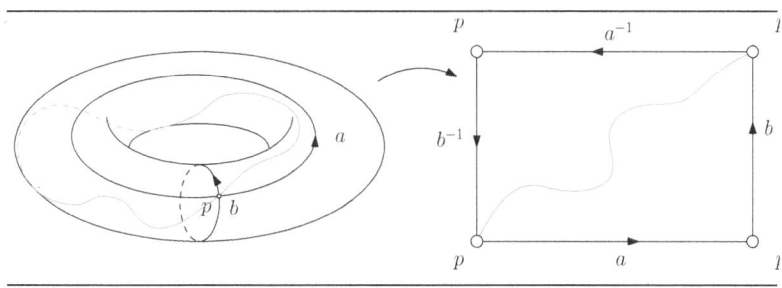

Figure 14.17 $\pi_1(T^2, p) = \langle a, b | aba^{-1}b^{-1} \rangle.$

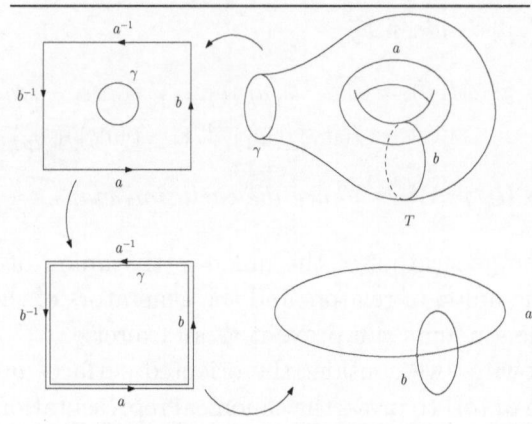

Figure 14.18 Punctured torus, fundamental group $\pi_1(T \setminus \{q\}, p) = \langle a, b \rangle$.

of a, b and their inverses. As shown in the right frame of the Figure 14.17, the red curve represents a γ_k, which is homotopic to the boundary segment ab. The whole loop γ is the product of $\gamma_1 \gamma_2 \ldots \gamma_n$, hence the whole loop can be represented as the product of a, b and their inverses. This means a and b are the generators of $\pi_1(T^2, p)$.

$\square$

Next, we study the fundamental group of a punctured torus. As shown in Figure 14.18, we punch a hole on the torus surface, the puncture is a single point $\{q\}$. Then we enlarge the puncture, the boundary of the hole is γ. we further enlarge the hole, until exhaust the whole surface area, and only leave the 1-dimensional skeleton, $a \vee b$, where $a \vee b$ is obtained by glue the loop a and the loop b at a single point. The generators of $a \vee b$ are a and b, and no non-trivial loops can shrink to a point, hence there is no relators. Therefore $\pi_1(T^2 \setminus \{q\}, p) = \langle a, b \rangle$.

Now we can prove the main theorem of surface topology, the canonical fundamental group representation theorem 14.5.5.

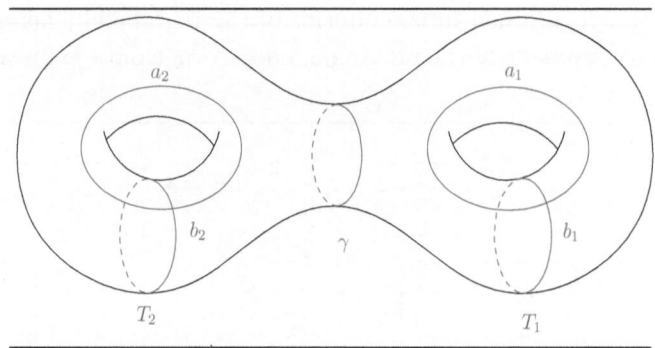

Figure 14.19 Divide conquer method to compute the fundamental group.

Proof We want to show that for a surface $S = \oplus_{i=1}^{g} T^2$,

$$\pi_1(S, p) = \langle a_1, b_1, \cdots, a_g, b_g | \Pi_{i=1}^{g}[a_i, b_i] \rangle \qquad (14.5.1)$$

We use mathematical induction to prove it.

When $g = 0$, the fundamental group of S is $\pi_1(S, p) = \langle e \rangle$.

When $g = 1$, lemma 14.5.8 shows Equation (14.5.1) holds.

When $g = 2$, as shown in Figure 14.19, the surface is the connected sum of two tori, $S = T_1^2 \oplus T_2^2$. Let $T_1 = T_1^2 \setminus D_1$, $T_2 = T_2^2 \setminus D_2$ be the punctured tori, where D_1 and D_2 are disks, then $S = T_1 \cup T_2$, the intersection is the loop γ, $\gamma = T_1 \cap T_2$. Choose a base point $p \in \gamma$, then we have the fundamental groups of T_1 and T_2 are

$$\begin{aligned} \pi_1(T_1, p) &= \langle a_1, b_1 \rangle \\ \pi_1(T_2, p) &= \langle a_2, b_2 \rangle \end{aligned}$$

and the fundamental group of $T_1 \cap T_2$ is

$$\pi_1(T_1 \cap T_2, p) = \langle \gamma \rangle$$

Furthermore, in $\pi_1(T_1, p)$, $[\gamma] = a_1 b_1 a_1^{-1} b_1^{-1}$; in $\pi_1(T_2, p)$, $[\gamma] = (a_2 b_2 a_2^{-1} b_2^{-1})^{-1}$. so by Seifert-Van Kampen theorem 14.5.7, we have

$$\pi_1(T_1 \cup T_2) = \langle a_1, b_1, a_2, b_2 | [a_1, b_1][a_2, b_2] \rangle.$$

where $[a_k, b_k] = a_k b_k a_k^{-1} b_k^{-1}$.

Suppose Equation (14.5.1) holds for genus less than or equal to $g - 1$ cases, then for genus equals to g case,

$$S = (T_1^2 \oplus T_2^2 \ldots T_{g-1}^2) \oplus T_g^2.$$

Let T_k be the k-th punctured torus, $T_k = T_k^2 \setminus D_k$, where $D_k \subset T_k^2$ is a disk, the surface S is the union

$$S = (T_1 \cup T_2 \ldots T_{g-1}) \cup T_g.$$

The intersection is

$$\gamma = (T_1 \cup T_2 \ldots T_{g-1}) \cap T_g.$$

We choose the base point $p \in \gamma$, then we have

$$\begin{aligned} \pi_1(T_1 \cup T_2 \ldots T_{g-1}, p) &= \langle a_1, b_1, \cdots a_{g-1}, b_{g-1} \rangle \\ \pi_1(T_g, p) &= \langle a_g, b_g \rangle \end{aligned}$$

The fundamental group of the intersection is

$$\pi_1((T_1 \cup T_2 \ldots T_{g-1}) \cap T_g, p) = \langle \gamma \rangle.$$

In $\pi_1(T_1 \cup T_2 \ldots T_{g-1}, p)$, $[\gamma] = \pi_{k=1}^{g-1}[a_k, b_k]$; in $\pi_1(T_g, p)$, $[\gamma] = [a_g, b_g]^{-1}$. so by Seifert-Van Kampen theorem 14.5.7, we have

$$\pi_1((T_1 \cup T_2 \ldots T_{g-1}) \cup T_g, p) = \langle a_1, b_1, a_2, b_2, \ldots, a_{g-1}, b_{g-1}, a_g, b_g | \Pi_{i=1}^{g}[a_i, b_i] \rangle.$$

By induction, Equation (14.5.1) holds for any genus $g \geq 0$.

$\square$

14.6 COVERING SPACE

The continuous mappings between two topological spaces induce homomorphisms between their fundamental groups.

Definition 14.6.1 (Subgroup) *Suppose* $(\mathcal{G}, \cdot)$ *is a group,* $H \subset \mathcal{G}$ *is a subset of group elements that satisfies the four group requirements, then* H *is called a subgroup of* $\mathcal{G}$.

Definition 14.6.2 (Coset) *For a subgroup* H *of a group* $\mathcal{G}$ *and an element* g *of* $\mathcal{G}$, *define* gH *to be the left coset of* H,

$$gH := \{gh : h \in H\}.$$

Similarly Hg *to be the right coset*

$$Hg := \{hg : h \in H\}.$$

Definition 14.6.3 (Normal Subgroup) *Suppose* $(\mathcal{G}, \cdot)$ *is a group,* N *is a subgroup of* $\mathcal{G}$, *such that*

$$gNg^{-1} = N, \quad \forall g \in \mathcal{G},$$

then N *is called a normal subgroup.*

If N is a normal subgroup of $\mathcal{G}$, then its left coset equals to its right coset, $gN = Ng$.

Definition 14.6.4 (Quotient Group) *For a group* $(\mathcal{G}, \cdot)$ *and a normal subgroup* N *of* $\mathcal{G}$, *the product of two cosets of* N *in* $\mathcal{G}$ *is given by*

$$(Ng_1) \cdot (Ng_2) := N(g_1 \cdot g_2).$$

The set of cosets of N *in* $\mathcal{G}$ *with the product operator form the quotient group of* N *in* $\mathcal{G}$, *denoted as* $\mathcal{G}/N$.

It can be easily verified that the quotient group is well-defined, the product of cosets is independent of the choice of the representatives.

Definition 14.6.5 (Covering Space) *Suppose* $\tilde{S}$ *and* S *are topological spaces, a continuous map* $p : \tilde{S} \to S$ *is surjective, such that for each point* $q \in S$, *there is a neighborhood* U *of* q, *its preimage* $p^{-1}(U) = \cup_i \tilde{U}_i$ *is a disjoint union of open sets, and the restriction of the map* p *on each open set* $\tilde{U}_i$ *is a local homeomorphism from* $\tilde{U}_i$ *to* U, *then* $(\tilde{S}, p)$ *is called a* covering space *of* S, p *is called the* projection map.

Definition 14.6.6 (Deck Transformation) *Suppose* $(\tilde{S}, p)$ *is a covering space of* S, *an automorphism of* $\tilde{S}$, $\tau : \tilde{S} \to \tilde{S}$, *is called a* deck transformation, *if it satisfies* $p \circ \tau = p$, *where* $p : \tilde{S} \to S$ *is the . All the deck transformations form a group, the* covering group, *or the* deck transformation group, *and denoted as* $Deck(\tilde{S})$.

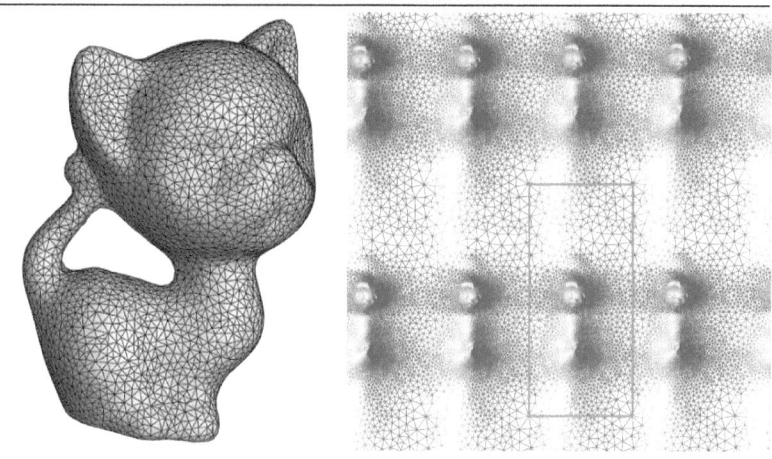

Figure 14.20 The universal Covering Space of a genus one surface, the kitten model.

Suppose $\tilde{q} \in \tilde{S}$, $p(\tilde{q}) = q$. The projection map $p : \tilde{S} \to S$ induces a homomorphism between their fundamental groups, $p_\# : \pi_1(\tilde{S}, \tilde{q}) \to \pi_1(S, q)$, if $p_\#\pi_1(\tilde{S}, \tilde{q})$ is a normal subgroup of $\pi_1(S, q)$ then we have the following theorem:

Theorem 14.6.7 (Covering Group Structure) *The quotient group of* $\frac{\pi_1(S)}{p_\#\pi_1(\tilde{S}, \tilde{q})}$ *is isomorphic to the deck transformation group of* $\tilde{S}$.

$$\frac{\pi_1(S, q)}{p_\#\pi_1(\tilde{S}, \tilde{q})} \cong Deck(\tilde{S}).$$

Definition 14.6.8 (Universal Covering Space) *If a covering space* $\tilde{S}$ *is simply connected (i.e.* $\pi_1(\tilde{S}, \tilde{q}) = \{e\}$), *then* $\tilde{S}$ *is called a* universal covering space *of* S.

For the universal covering space

$$\pi_1(S, q) \cong Deck(\tilde{S}, \tilde{q}).$$

Namely, the fundamental group of the base space S is isomorphic to the deck transformation group of the universal covering space $\tilde{S}$.

Definition 14.6.9 (Fundamental Domain) *Suppose* $(\tilde{S}, p)$ *is a covering space of* S, $q \in S$ *is a point on* S, $p^{-1}(q) = \{\tilde{q}_1, \tilde{q}_2, \ldots, \tilde{q}_n, \ldots\}$ *is the* orbit *of* q *in the covering space* $\tilde{S}$. *Let* $\tilde{D} \subset \tilde{S}$ *is a simply connected domain in* $\tilde{S}$, *such that* D *intersects each orbit at exactly one point, then* D *is called a* fundamental domain *of* S *in* $\tilde{S}$.

Figure 14.21 shows the universal covering space of a genus two closed surface. Different fundamental domains are rendered using different colors. One set of canonical fundamental group generators is shown in the left frame, which correspond to the boundary of each fundamental domain, a canonical polygonal schema.

Any path connected topological manifold has a universal covering space, which is formed by all the path homotopy classes.

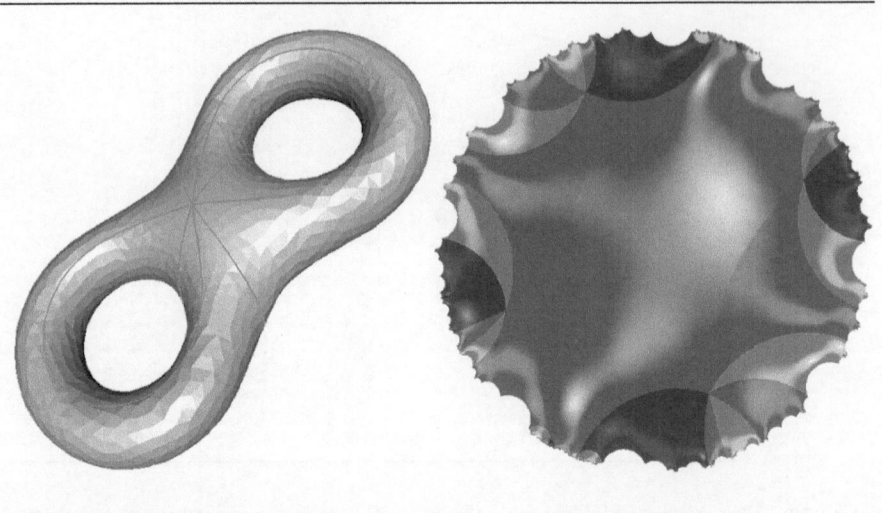

Figure 14.21 The universal Covering Space of a genus two surface.

Theorem 14.6.10 *Suppose a topological manifold S is path connected, then there is a universal covering space $(\tilde{S}, p)$ of S.*

Proof Fix a base point $q \in S$, consider the space Γ of all the paths starting from q, and quotient Γ by the homotopy equivalence relation to obtain $\tilde{\Gamma}$,

$$\Gamma := \{\gamma : [0,1] \to S | \gamma(0) = q\}, \quad \tilde{\Gamma} := \Gamma / \sim,$$

namely $\tilde{\Gamma}$ is the space of all the homotopy classes of paths starting from the base point q. The projection $p : \tilde{\Gamma} \to S$ maps a path class of γ to its target point,

$$p([\gamma]) = \gamma(1),$$

the target point is independent of the choice of γ, therefore p is well defined. Now, we

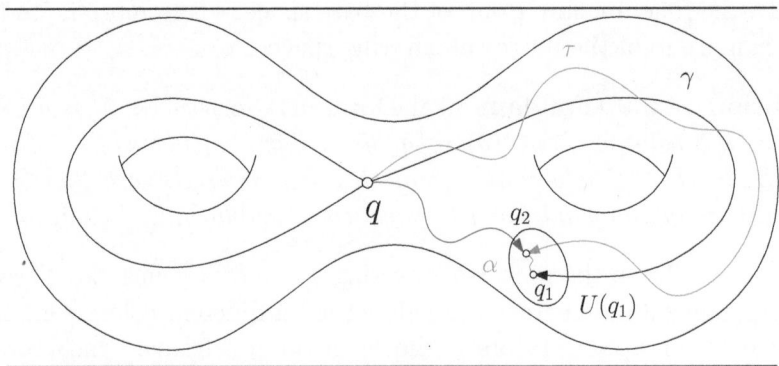

Figure 14.22 Universal Covering Space Construction.

define the topology of $\tilde{\Gamma}$. As shown in Figure 14.22, given a path $\gamma \in \Gamma$, it target point is $\gamma(1) = q_1$. Let $U(q_1) \subset S$ be an open set of q_1. For each point $q_2 \in U(q_1)$, there is a path $\alpha \subset U(q_1)$ from q_1 to q_2. A path $\tau \in \Gamma$ is homotopic to $\gamma \cdot \alpha$, then we say $[\tau] \in \tilde{\Gamma}$ is in the neighborhood of $[\gamma]$. Namely, we define a neighborhood $\tilde{U}([\gamma]) \subset \tilde{\Gamma}$ of $[\gamma]$ as

$$\tilde{U}([\gamma]) := \{[\tau]|\tau \sim \gamma \cdot \alpha, \alpha(0) = \gamma(1), \alpha \subset U(\gamma(1))\}.$$

All such kind of neighborhoods $\tilde{U}([\gamma])$ define the topological basis of $\tilde{\Gamma}$, therefore $\tilde{\Gamma}$ is a topological space. By definition, $(\tilde{\Gamma}, p)$ is a covering space of S. The base point of $\tilde{\Gamma}$ is denoted as $\tilde{q} = [e]$, where e is the trivial loop $e(t) \equiv q$, $p(\tilde{q}) = q$.

In the figure, there is another path ρ from q to q_2, which is not homotopic to τ, therefore $[\rho]$ and $[\tau]$ are different points in $\tilde{\Gamma}$. Consider a loop $\tilde{\gamma}$ in the covering space $\tilde{\Gamma}$, starting from and ending at $\tilde{q}$. The projection of $\tilde{\gamma}$ is a loop γ on S, by our construction, $\tilde{\gamma}(1) = [\gamma]$. Furthermore, $\tilde{\gamma}(0) = \tilde{q}$ implies $\tilde{\gamma}(0) \sim \tilde{\gamma}(1)$, namely $\gamma \sim e$. We can homotopically deform γ to e and homotopically deform $\tilde{\gamma}$ accordingly. The homotopy of γ is denoted as a family of loops $\gamma_s : [0, 1] \to S$, $s \in [0, 1]$. We use γ_τ^t to represent the restriction of γ_s on the time interval $[0, t]$, then we define the loop in $\tilde{\Gamma}_s$ as

$$\tilde{\gamma}_s(t) = [\gamma_s^t].$$

Then $\tilde{\gamma}_s$ gives the homotopy between $\tilde{\gamma}$ and $\tilde{q}$. This shows $\tilde{\Gamma}$ is simply connected, therefore it is the universal covering space of S.

□

As shown in Figure 14.23, suppose $p : \tilde{S} \to S$ is a covering space of S, $\gamma \subset S$ is a loop on the base space S, $q \in S$ is the base point. We cover the loop by a sequence of open sets $U_0, U_1, \ldots, U_{n-1}$, such that the consecutive sets overlap, $U_k \cap U_{k+1} \neq \emptyset$. The orbit of q is $p^{-1}(q) = \{\tilde{q}_k\}$. Because the projection map p is local homeomorphic, we can find a neighborhood $\tilde{U}_0$ of $\tilde{q}_0 \in \tilde{S}$, such that $p(\tilde{U}_0) = U_0$, $\tilde{q}_0 \in \tilde{U}_0$, the restriction of p on $\tilde{U}_0$ is bijective. Then we can map $\gamma \cap U_0$ to $\tilde{U}_0$ by p^{-1}, namely lift γ to $\tilde{U}_0$. Similarly, we project U_1 back to $\tilde{U}_1$, and continue to lift $\gamma \cap U_1$ to $\tilde{U}_1$. We repeat this procedure, project U_k back to $\tilde{U}_k$, and lift $\gamma \cap U_k$ to $\tilde{U}_k$, $k = 0, 1, \ldots, n-1$, until the whole loop γ is lifted to a unique path $\tilde{\gamma} \subset \tilde{\Gamma}$ starting from $\tilde{q}_0$. The lifting process

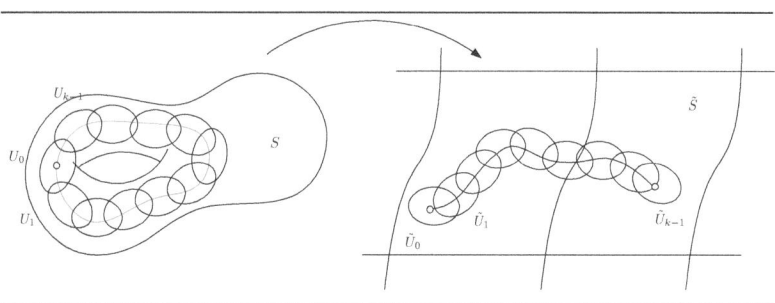

Figure 14.23 Lifting a loop to the Universal Covering Space.

makes the following diagram commutes:

$$
\begin{CD}
[0,1] @>\tilde{\gamma}>> \tilde{S} \\
@V id VV @VV p V \\
[0,1] @>\gamma>> S
\end{CD}
$$

Lemma 14.6.11 *Let $(\tilde{S}, p)$ be the universal covering space of S, $q \in S$ be the base point. The orbit of base is $p^{-1}(q) = \{\tilde{q}_k\}$. γ_1 and γ_2 are two loops through the base point, their lifts are $\tilde{\gamma}_1$ and $\tilde{\gamma}_2$, both starting from $\tilde{q}_k$. $\gamma_1 \sim \gamma_2$ if and only if the end points of $\tilde{\gamma}_1$ and $\tilde{\gamma}_2$ coincide.*

Proof If the end points of $\tilde{\gamma}_1$ and $\tilde{\gamma}_2$ coincide, because $\tilde{S}$ is simply connected, $\tilde{\gamma}_1$ is homotopic to $\tilde{\gamma}_2$, the projection map is continuous, then their images γ_1 and γ_2 is homotopic.

Reversely, suppose γ_1 is homotopic to γ_2, we can lift γ_1 to $\tilde{\gamma}_1$. Since the projection map is locally homeomorphic, it can lift the homotpic deformations to the covering space, therefore $\tilde{\gamma}_1$ is homotopic to $\tilde{\gamma}_2$, they share the same starting and ending points. □

As shown in Figure 14.24, two non-homotopic loops γ_1, γ_2 are on the genus one surface, both of them are lifted to two paths $\tilde{\gamma}_1$ and $\tilde{\gamma}_2$ on the universal covering space. After the alignment, the two paths start from the same fundamental domain, but end at different fundamental domains.

14.7 COMPUTATIONAL ALGORITHMS

Triangle Mesh   In practice, surfaces are represented as *triangular meshes*. As shown in Figure 14.25, the King David's sculpture surface is densely triangulated, then each triangle is represented by a Euclidean triangle.

Definition 14.7.1 (Triangular mesh) *Suppose S is a topological surface, V is a set of distinct points on S, then (S, V) is called a* marked surface; *$\mathcal{T}$ is a triangulation*

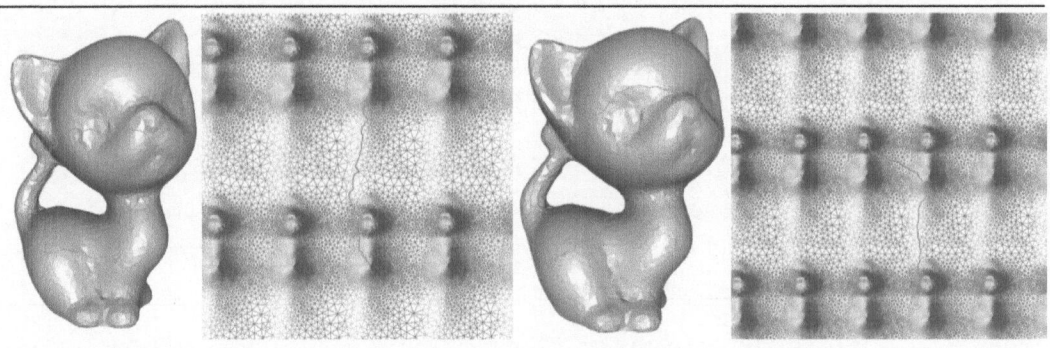

Figure 14.24 Loop lifting to the universal Covering Space.

Figure 14.25 Triangular mesh.

of the point set V, d is a Riemannian metric such that each face of $\mathcal{T}$ is a Euclidean triangle. Then $(S, V, \mathcal{T}, d)$ is called a discrete metric surface, or a triangular mesh.

We use v_i to denote a vertex, and $[v_i, v_j]$ an oriented edge from the source vertex v_i to the target vertex v_j, $[v_j, v_i]$ is the oriented edge with the opposite orientation. Similarary, each face is represented by its vertices $[v_i, v_j, v_k]$, the permutation of three vertices gives the orientation of the face. Two permutations, say $[v_i, v_j, v_k]$ and $[v_j, v_k, v_i]$, represent the same orientation if they differ by an even number of swaps; they represent the opposite orientations, say $[v_i, v_j, v_k]$ and $[v_j, v_i, v_k]$ if they differ by an odd number of swaps. We call an oriented edge as a *half-edge*, and an non-oriented edge as *edge*. Each oriented face $[v_i, v_j, v_k]$ has three half-edges $[v_i, v_j]$, $[v_j, v_k]$ and $[v_k, v_i$.

Definition 14.7.2 (Orientation) *Suppose $(S, V, \mathcal{T}, d)$ is a discrete metric surface, if one can choose an orientation for each face, such that for any edge shared by two faces, the two adjacent half-edges are with opposite orientations, then we say the triangular mesh is orientable.*

If the closed triangular mesh is embedded in $\mathbb{R}^3$, we can define the normal to each face as the unit vector orthogonal to the support plane through the face and points from the inside to the outside. This naturally defines an orientation of the triangular mesh, such that:

1. Each edge has two opposite oriented edges (half-edges).

2. Each face is counter clockwisely oriented with respect to the normal of the surface.

Most surface topological computations are based on triangle meshes.

Each triangular mesh $M = (S, V, \mathcal{T}, d)$ has a natural *Poincaré dual mesh* $\bar{M}$, which is another cell decomposition of the metric surface (S, V, d). The dual mesh is constructed as follows:

1. For each simplex σ of M (vertex, edge, face), compute its barycenter $\tilde{\sigma}$ (mean of the vertices);

2. Given a k-dimensional simplex σ, its dual $\bar{\sigma}$ is a $(2 - k)$-dimensional simplex, for each face $f \in M$,

$$\bar{\sigma} \cap f = \text{Conv}\{\tilde{\tau} | \sigma \subset \tau, \tau \subset f\}.$$

where Conv is the convex hull of a point set.

Intuitively, each face $\sigma \in M$ is dual to a vertex $\bar{\sigma} \in \bar{M}$, which is the barycenter of the vertices of σ; each edge $e \in M$ shared by two faces σ_1 and σ_2 is dual to an edge $\bar{e} \in \bar{M}$, $\bar{e} \cap \sigma_k$ is the line segment connecting the barycenter of the edge e and the barycenter of the face σ_k, $k = 1, 2$; each vertex $v \in$ is dual to a cell $\bar{v}$, for each face $f \in M$, $\bar{v} \cap f$ is the convex hull of the barycenter of the face f and those the edges of f containing v.

CW-Cell Decomposition   In the above discussion, we always consider the loops through the base point. This requirement can be relaxed in the following way: given a loop γ, we find a path τ from the base point q to the starting point of γ, then $\tau \cdot \gamma \cdot \tau^{-1}$ is a loop through the base point. Therefore, we can omit the base point and use free homotopy class to construct the fundamental group.

Most concepts in surface fundamental group theory can be explicitly computed. The surface is represented as a triangle mesh. In order to improve the computational efficiency, we first introduce the concept of *CW-cell decomposition*, which decompose an n-dimensional manifold S_n into the union of $(n-1)$-dimensonal skeleton S_{n-1} with n dimensional disks D_n's, where we call each n dimensional disk as a n-cell; then the $(n-1)$ skeleton is further decomposed into the union of an $(n-2)$-dimensional skeleton S_{n-2} and $(n-1)$-dimensional disks. We recursively decompose each k-dimensional skeleton as the union of $(k-1)$-dimensional skeleton and the k dimensional disks, until k equals to zero. The zero dimensional skeleton S_0 is a set of disjoint points.

Definition 14.7.3 (CW-cell Decomposition) *Suppose M is a n-dimensional manifold, its CW-cell decomposition is recursively defined as follows:*

1. *0-skeleton S_0 is the union of a set of 0-cells.*

2. *k-skeleton S_k*

$$S_k = S_{k-1} \cup D_k^1 \cup D_k^2 \cdots \cup D_k^{n_k},$$

   such that

$$\partial D_k^i \subset S_{k-1}.$$

   The k-skeleton is constructed by gluing k-cells to the $k - 1$ skeleton, all the boundaries of the k-cells are in the $k - 1$ skeleton.

3. *$S_n = M$.*

Suppose we have an n-dimensional manifold M with a CW-cell decomposition, then its fundamental group equals to that of the 2-dimensional skeleton.

Theorem 14.7.4 (CW-cell decomposition) *Suppose* M *is an* n-*dimensional manifold with a CW-cell decomposition,*

$$\pi_1(S_2) = \pi_1(S_3) \cdots \pi_1(S_n) = \pi_1(M).$$

Proof $M = S_n$, therefore $\pi_1(M) = \pi_1(S_n)$. By CW-cell decomposition, $S_k = S_{k-1} \cup D_k^1 \cup D_k^2 \cdots \cup D_k^{n_k}$. $S_{k-1} \cup D_k^2 \cup D_k^2 \cdots \cup D_k^{n_k}$ intersect D_k^1 at the boundary of D_k^1, ∂D_k^1. When $k \geq 3$, $\pi_1(\partial D_k^1) = \langle e \rangle$, $\pi(D_k^1) = \langle e \rangle$, therefore

$$\pi_1(S_k) = \pi_1(S_{k-1} \cup D_k^1 \cup D_k^2 \cdots \cup D_k^{n_k}) = \pi_1(S_{k-1} \cup D_k^2 \cdots \cup D_k^{n_k})$$

Repeat this argument n_k times, each time remove one k-dimensional disk, eventually only the $(k-1)$-dimensional skeleton is left, we obtain $\pi_1(S_k) = \pi_1(S_{k-1})$. Similarly, we have

$$\pi_1(S_n) = \pi_1(S_{n-1}) = \cdots = \pi_1(S_2).$$

□

The fundamental group of the 2-skeleton can be further computed as follows:

Lemma 14.7.5 *Suppose* $S_2 = S_1 \cup D_2^1 \cup D_2^2 \cdots \cup D_2^{n_2}$, *and* $\pi_1(S_1) = \langle g_1, g_2, \ldots, g_k \rangle$, *then*

$$\pi_1(S_2) = \left\langle g_1, g_2, \ldots, g_k | \ [\partial D_2^1], [\partial D_2^2], \ldots, [\partial D_2^{n_2}] \right\rangle. \tag{14.7.1}$$

Proof Since the 1-skeleton S_1 is a graph, $g_1, g_2, \ldots, g_k$ are the independent loops on the graph, they form the generators of $\pi_1(S_1)$. Since any non-trivial loop on S_1 cannot shrink to a point on S_1, $\pi_1(S_1)$ has no relators, therefore $\pi_1(S_1) = \langle g_1, g_2, \ldots, g_k \rangle$. For each 2-dimensional disk D_2^i, $\pi_1(D_2^i) = \langle e \rangle$. The intersection between S_1 and D_2^i is a circle ∂D_2^i, it is included in S_1, assume its representation in $\pi_1(S_1)$ is $[\partial D_2^i]$. By using the Seifert-Van Kampen theorem 14.5.7, we obtain formula in Eqn. (14.7.1).

□

Given a two dimensional manifold (surface) S, we can compute a special CW-cell decomposition with a single disk, $S = S_1 \cup D_2$, in this case the fundamental group has the form

$$\pi_1(S) = \langle g_1, g_2, \ldots, g_k | \ [\partial D_2] \rangle.$$

The 1-skeleton S_1 can be computed as a *cut graph* of the surface.

Definition 14.7.6 (Cut Graph) Γ *is a graph on a surface* S, *such that* $S \setminus \Gamma$ *is a topological disk, then* Γ *is called a* cut graph *of* S.

Cut Graph   Suppose M is a triangular mesh. The *cut graph* of the mesh can be computed by a linear algorithm. We compute the dual cell decomposition $\bar{M}$. We treat $\bar{M}$ as a graph and compute a spanning tree $\bar{T}$, which is a tree connecting all the vertices of $\bar{M}$. The dual of $\bar{T}$ is the set of all the faces of M connected by some edges. Since $\bar{T}$ can shrink to a point, its dual is a topological disk. All the primal edges whose dual edges are not on the spanning tree $\bar{T}$ form a graph Γ, which is the desired cut graph of M.

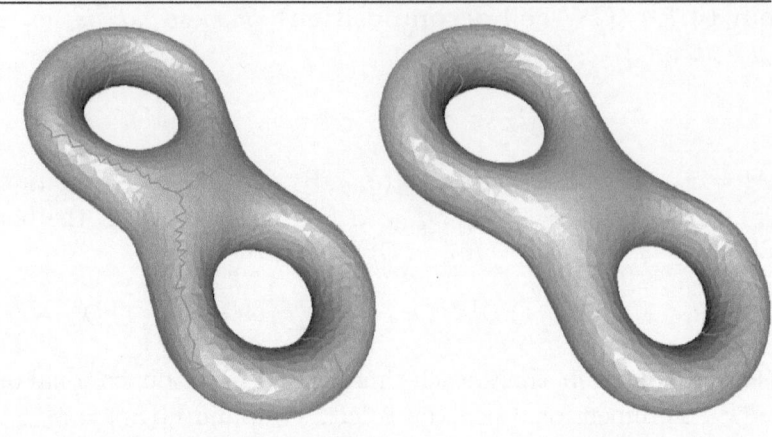

Figure 14.26 A cut graph of a genus two surface.

Each vertex v_i in Γ is adjacent to some edges in Γ, the number of adjacent edges is called the *topological degree* of v_i. We call a vertex v_i as a *node* of the graph, if its degree is not equal to 2. The nodes divide the graph Γ into *segments*. A segment is called a *dangling segment* if one of its nodes is with topological degree one. We repeatly remove all the dangling segments, until all the nodes are with degree greater than 2. The algorithmic details can be found in Alg. 3.

Since the choice of the spanning tree $\bar{T}$ is non-unique, the algorithm may produce different cut graphs. The cut graph Γ may have many dangling edges, which can be easily pruned. Figure 14.26 shows one cut graph of a genus two surface obtained by this method.

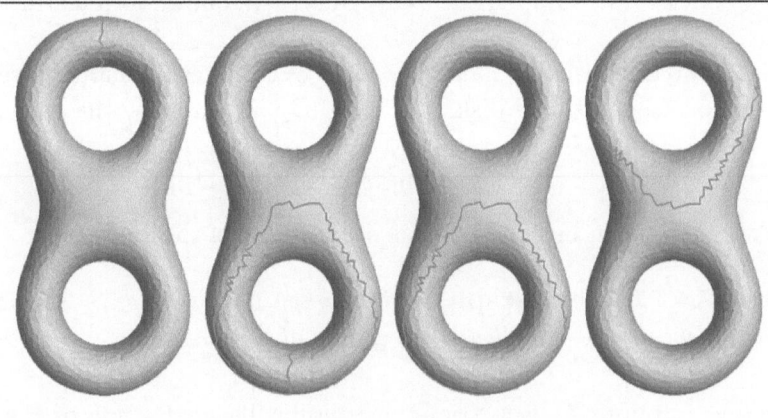

Figure 14.27 Fundamental group generators of a genus two surface.

Algorithm 3 Cut Graph

Require: A closed triangle mesh M

    Compute the dual mesh $\bar{M}$ of the input mesh M;

    Compute a spanning tree $\bar{T}$ of $\bar{M}$;

    The cut graph is given by $\Gamma := \{e \in M | \bar{e} \notin \bar{T}\}$.

    for each $v \in \Gamma$ **do**

        Compute the topological degree of v;

        if the degree of v isnot equal to 2 **then**

            label v as a node;

        end if

    end for

    Divide Γ by the nodes into segments;

    repeat

        for each node v_i of Γ **do**

            if the degree of v_i equals to 1 **then**

                Find the segment s_i adjacent to v_i;

                Find the other end node v_j of s_i;

                $\Gamma \leftarrow \Gamma \setminus s_i$;

                The degree of v_j minus 1

            end if

        end for

    until the degrees of all nodes are greater than 2

Ensure: Γ is a cut graph of M

Fundamental Group The cut graph algorithm 3 finds a CW-cell decomposition $M = S_1 \cup D_2$, where the 1-skeleton S_1 is the cut graph Γ, the disk D_2 is $M \setminus \Gamma$. The generators of $\pi_1(M)$ equal to those of $\pi_1(\Gamma)$. The computation of the generators of Γ is based on the following lemma.

Lemma 14.7.7 *Suppose Γ is a graph, $T \subset \Gamma$ is a spanning tree of G,*

$$\Gamma \setminus T = \{e_1, e_2, \ldots, e_k\}.$$

Each oriented edge $e_i \cup T$ includes a unique loop g_i, then

$$\pi_1(\Gamma) = \langle g_1, g_2, \ldots, g_k \rangle.$$

Proof Each g_i includes e_i, which is not on any g_j, $j \neq i$, therefore g_i cannot be represented as the product of other g_j's.

Given any loop $\gamma \subset \Gamma$, represented as a consecutive sequence of oriented edges. Suppose γ doesn't go through any oriented edge in $\Gamma \setminus T$, the γ is a trivial loop. Otherwise, we list the consecutive oriented edges in γ, and only show those in $\Gamma \setminus T$ as

$$\ldots e_{i_1} \ldots e_{i_2} \ldots e_{i_3} \ldots e_{i_k} \ldots,$$

then γ is homotopic to $g_{i_1} \cdot g_{i_2} \ldots g_{i_k}$. Namely, g_i's form a set of generators of $\pi_1(\Gamma)$.

$\square$

Given a closed triangle mesh M, we first compute its cut graph Γ using the Algorithm 3, then compute a spanning tree of T of the cut graph. For each edge e_i not in the tree, the union of T and e_i has a unique loop g_i, all such kind of γ_i's form the generators of $\pi_1(\Gamma)$. We then compute the representation of the boundary of the fundamental domain D in $\pi_1(\Gamma)$, which gives the relator for the fundamental group $\pi_1(M)$. The algorithmic details can be found in Alg. 4. Figure 14.27 shows the generators of the cut graph in Figure 14.26 of the genus two surface, which are also the generators of the fundamental group of the surface.

Algorithm 4 Fundamental Groups of Closed Oriented Surfaces

Require: A closed triangle mesh M

   Compute the cut graph Γ of M using the Algorithm 14.26;

   Compute a spanning tree T of Γ;

   Compute $\{e_1, e_2, \ldots, e_k\} \leftarrow \Gamma \setminus T$;

   for each $e_i = [v_i^-, v_i^+] \in \Gamma \setminus T$ **do**

      Find the leaves of the starting/ending vertex v_i^-/v_i^+ of e_i in T;

      Trace the path $\gamma_i^- \subset T$ from v_i^- to the root of the tree;

      Trace the path $\gamma_i^+ \subset T$ from v_i^+ to the root of the tree;

      Construct the loop $g_i \leftarrow (\gamma_i^-)^{-1} \cup e_i \cup \gamma_i^+$;

      $G \leftarrow G \cup g_i$;

   end for

   Compute the fundamental domain $D \leftarrow M \setminus \Gamma$;

   Set the relator to be empty $R \leftarrow \emptyset$;

   for each $e \in \partial D$ **do**

      if e equals to $e_i \in \Gamma \setminus T$ **then**

         $R \leftarrow R \cup g_i$;

      else if e equals to e_i^{-1}, $e_i \in \Gamma \setminus T$ **then**

         $R \leftarrow R \cup g_i^{-1}$;

      end if

   end for

Ensure: The fundamental group $\pi_1(M) = \langle G | R \rangle$.

Universal Covering Space   The universal covering space is useful for many computational topological tasks, such as homotopy detection, canonical fundamental group generators and so on. Suppose we have computed the CW-cell decomposition of a closed high genus surface $S = S_1 \cup D_2$, we can construct a finite portion of the universal covering space of the surface by gluing the copies of D_2 along the corresponding boundary segments in a specific way.

   By Algorithm 3, we can compute the cut graph Γ of a triangle mesh M. Suppose the segments of Γ are $\{s_1, s_2, \ldots, s_k\}$. We assign an orientation to each segment. Then we cut the surface along Γ to obtain the topological disk D, the fundamental domain, its boundary ∂D is in Γ, therefore ∂D is a product of s_i's and their inverses. It is obvious that each s_i and s_i^{-1} are in ∂D once. Then we initialize the universal covering space $\tilde{M}$ as one copy of D. At each iteration, we glue one copy of D to the current covering space $\tilde{M}$ along just one segment $s_i \subset \partial\tilde{M}$ and $s_i^{-1} \subset \partial D$. Then we update the boundary of the updated covering space $\tilde{M}$, if we find two consecutive segments on $\partial\tilde{M}$ are inverse to each other, say s_j and s_j^{-1}, then we sew the covering space up along these two segments. This will ensure the covering space $\tilde{M}$ is always simply connected. We can repeat this procedure several steps, until the universal covering space is large enough for our purposes. The algorithmic details can be found in Alg. 5.

Algorithm 5 Universal Covering Space

Require: A closed triangle mesh M

  Compute the cut graph Γ of M using the Algorithm 3;

  Slice M along Γ to obtain a fundamental domain D;

  Initialize the universal covering space $\tilde{M} \leftarrow D$;

  repeat

    Choose an oriented segment $s_i \subset \partial\tilde{M}$;

    Glue a copy of D with $\tilde{M}$ along s_i, $\tilde{M} \leftarrow \tilde{M} \cup_{\partial\tilde{M} \supset s_i \sim s_i^{-1} \subset \partial D} D$;

    Trace the boundary of $\tilde{M}$ as a sequence of segments $\{s_{i_1}, s_{i_2}, \ldots, s_{i_n}\}$

    for each segment s_{i_k} of $\partial\tilde{M}$ **do**

      if s_{i_k} and $s_{i_{k+1}}$ are inverse to each other **then**

        Sew $\tilde{M}$ up along s_{i_k} and $s_{i_{k+1}}$;

        Remove s_{i_k} and $s_{i_{k+1}}$ from $\partial\tilde{M}$;

      end if

    end for

  until The universal covering space $\tilde{M}$ is large enough;

Ensure: A finite portion of the universal covering space $\tilde{M}$ of the surface M.

Homotopy Detection Given two loops γ_1 and γ_2 on a surface, we can first find a shortest path τ between them, then construct a loop $\gamma = \gamma_1 \tau \gamma_2^{-1} \tau^{-1}$, γ_1 is homotopic to γ_2 if and only if γ is trivial. Then we can lift γ to the universal covering space to $\tilde{\gamma}$. If $\tilde{\gamma}$ is a loop in $\tilde{S}$, then γ is a trivial loop in S, $\gamma_1 \sim \gamma_2$. This algorithm is straight forward, the major difficulty is to determine the size of the finite portion of the universal covering space at the beginning. One way to improve it is to enlarge the universal covering space dynamically during the lifting process. Another way to improve the efficiency is to homotopicaly shorten the loop γ to reduce the size of the required finite porting of the universal covering space. Algorithmic details can be found in Alg. 6.

    A completely different approach is based on surface Ricci flow. First we deform the Riemannian metric of the surface to be hyperbolic using discrete surface Ricci flow [151, 152], then homotopically deform γ_1 and γ_2 to geodesics under the hyperbolic metric. Because in each homotopy class, there is a unique geodesic under the hyperbolic metric, γ_1 is homotopic to γ_2 if and only if the two geodesics coincide. Details for discrete surface Ricci flow algorithms can be found in chapter 20.

Algorithm 6 Homotopy Detection

Require: A closed triangle mesh M, two loops γ_1 and γ_2
Ensure: Verify whether γ_1 is homotopic to γ_2.

    Compute the shortest path τ between γ_1 and γ_2;

    Construct the loop $\gamma \leftarrow \gamma_1 \tau \gamma_2^{-1} \tau^{-1}$, represented as a list of vertices $v_0, v_1, \ldots, v_{n-1}$;

    Construct a finite portion of the universal covering space $\tilde{M}$ of M using Algorithm 5;

    Find a vertex $\tilde{v}_0 \in \tilde{M}$, the projection of $\tilde{v}_0$ is v_0;

    for $k \leftarrow 1$ to $n-1$ **do**

        Find $\tilde{v}_k \in \tilde{M}$ adjacent to $\tilde{v}_{k-1}$ such that its projection is v_k;

    end for

    Find $\tilde{v}_n \in \tilde{M}$ adjacent to $\tilde{v}_{n-1}$ such that its projection is v_0;

    return whether $\tilde{v}_n$ coincides with $\tilde{v}_0$

CHAPTER 15

Homology and Cohomology Group

This chapter first introduces the fundamental concepts, theorems in simplicial homology/cohomology, simplicial mapping, and their applications in fixed point theorem, Poincaré-Hopf index theorem, then focuses on the computational algorithms in both algebraic and combinatorial approaches.

15.1   SIMPLICIAL HOMOLOGY

The method of simplicial homology is to triangualte the topological space, then study different dimensional cycles and boundaries. The boundaries must be cycles, but the cycles may not be bounaries. The differences between the each dimensional cycles and bounaries convey the topological information of the original space. The homology groups are Abelian, therefore they are easier to compute using linear algebra. On the other hand, comparing to homotopy theory, homology gives less topological information.

A triangulation of a topological space is formulated as a simplicial complex.

Definition 15.1.1 (Simplex) *Suppose $k + 1$ points are in the general positions in $\mathbb{R}^n$, $v_0, v_1, \cdots, v_k$, the standard simplex $[v_0, v_1, \cdots, v_k]$ is the minimal convex set including all of them,*

$$\sigma = [v_0, v_1, \cdots, v_k] = \left\{ x \in \mathbb{R}^n \Big| x = \sum_{i=0}^{k} \lambda_i v_i, \sum_{i=0}^{k} \lambda_i = 1, \lambda_i \geq 0 \right\},$$

we call $v_0, v_1, \cdots, v_k$ as the vertices of the simplex σ. Suppose $\tau \subset \sigma$ is also a simplex, then we say τ is a facet of σ.

DOI: 10.1201/9781003350576-15

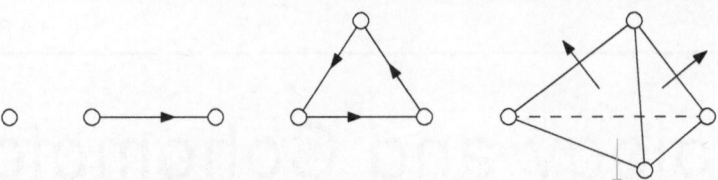

Figure 15.1 Different dimensional simplices.

Definition 15.1.2 (Simplicial complex) *A simplicial complex* Σ *is a union of simplices, satisfies the following conditions:*

1. *If a simplex* σ *belongs to* Σ, *then all its facets also belong to* Σ.

2. *If* $\sigma_1, \sigma_2 \subset \Sigma$, $\sigma_1 \cap \sigma_2 \neq \emptyset$, *then their intersection is also a common facet.*

Given an abstract simplicial complex K, we can construct is topological realization $|K|$ by taking the union of the interiors of its simplices in the standard form, such that the interiors are disjoint.

Definition 15.1.3 (Topological Realization) *The topological realization* $|K|$ *of an abstract simplicial complex* K *is the space obtained by the following procedure:*

1. *For each* $\sigma \in K$, *take a copy of the standard n-simplex, where* $n + 1$ *is the number of elements of* σ. *Denote this simplex by* $\Delta\sigma$. *Label its vertices with the elements of* σ.

2. *Whenever* $\sigma \subset \tau \in K$, *identify* $\Delta\sigma$ *with a subset of* $\Delta\tau$, *via the face inclusion which sends the elements of* σ *to the corresponding elements of* τ.

Figure 15.2 illustrates the concept of simplicial complex: the left frame shows a simplicial complex, where the intersections of simplices are their facets; the union of simplices in the right frame is not a simplicial complex, where the intersection between 2-simplices is not a facet of any of them. Given a manifold, we can study different dimensional sub-manifolds in it. The manifold is triangualted and approximated by a simplicial complex. A k-dimensional sub-manifold is approximated by a k-dimensional chain, which is an integeral linear combination of k-simplices. The integer coefficient of each simplex represent the multiplicity and the orientation of the simplex in the

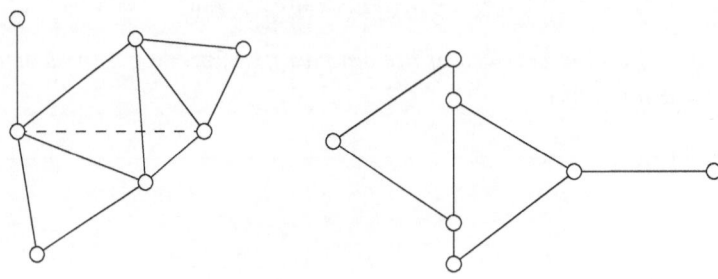

Figure 15.2 Left: a simplicial complex; right: not a simplicial complex.

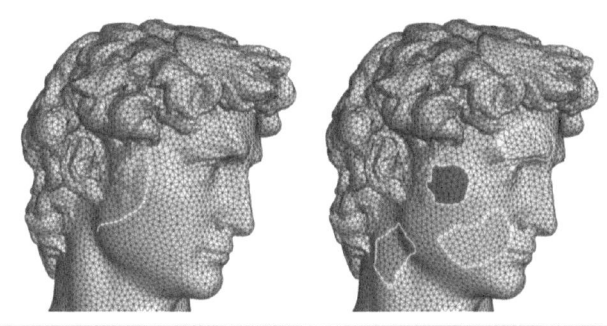

Figure 15.3 1-chains and 2-chains on a simplicial complex.

chain. Naturally, the linear combination of two chains is still a chain, therefore all the k-dimensional chains form a linear space, the so-called k-chain space. The dimension of the k-chain space equals to the number of k-simplicies in the complex. As shown in Figure 15.3, the Michelangelo's King David sculpture is approximated by a two dimensional simplicial complex. In the left frame, an oriented curve on the surface is represented as a sequence of oriented edges, namely a 1-chain. If we change the curve to the opposite direction, all the coefficients will be negated accordingly. In the right frame, several two dimensional patches are represented as a collection of oriented faces, namely a 2-chain. Different patches are in different colors, each patch itself is a 2-chain, their summation is also a 2-chain. The normals to all the faces point outward. If we negate the coefficients of all the faces, then their normal point inward.

Definition 15.1.4 (Chain Space) *A k-chain is a linear combination of all k-simplicies in Σ, $\sigma = \sum_i \lambda_i \sigma_i, \lambda_i \in \mathbb{Z}$. The k dimensional chain space is the linear space formed by all k-chains, denoted as $C_k(\Sigma, \mathbb{Z})$.*

As shown in Figure 15.4, the boundary operator extracts the boundary of a chain. Given an oriented curve, its boundary consists its target point and its source point.

Figure 15.4 Boundary operator.

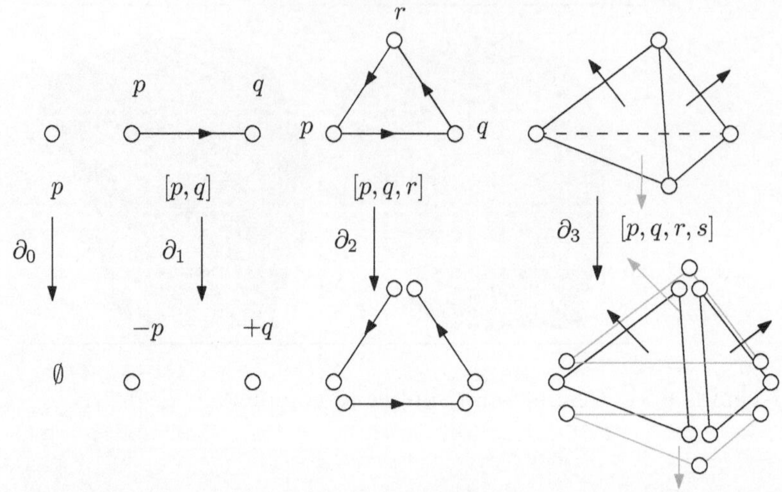

Figure 15.5 Boundary operator.

Given an oriented surface patch, its boundary is a closed loop, the orientation of the loop and the normal to the patch satisfy the right hand rule. The boundary operator can be directly generalized to higher dimensional sub-manifolds. In the setting of simplicial complex, the boundary operator can be defined directly on the chains, and treated a linear maps between chain spaces.

Figure 15.5 shows the boundaries of different dimensional simplices. The orientations between the simplex and its boundary facets have special consistency. The boundary of a chain is the linear combination of the boundaries of the simplces with the same coefficients.

Definition 15.1.5 (Boundary Operator) *The n-th dimensional boundary operator $\partial_n : C_n(\Sigma) \to C_{n-1}(\Sigma)$ is a linear operator, such that on a n-simplex*

$$\partial_n[v_0, v_1, v_2, \cdots, v_n] = \sum_i (-1)^i [v_0, v_1, \cdots, v_{i-1}, v_{i+1}, \cdots, v_n].$$

The boundary operator on a n-chain, $\sigma = \sum_i \lambda_i \sigma_i$ where σ_i is an n-simplex, is defined as the linear combination of the boundaries of all the simplices:

$$\partial_n \sum_i \lambda_i \sigma_i := \sum_i \lambda_i \partial_n \sigma_i.$$

The boundary of an n-chain is an $(n-1)$-chain. Figure 15.6 shows the difference between closed 1-chains and the open 1-chains. The boundary of a closed 1-chain is empty, namely zero 0-chain. In general, closed k-chains have zero boundaries, therefore they are in the kernel of the boundary operator ∂_k.

Definition 15.1.6 (closed chain) *A k-chain $\sigma \in C_k(\Sigma)$ is called a closed k-chain, if $\partial_k \sigma = 0$, namely $\sigma \in \ker \partial_k$.*

Since $\partial_k : C_k(\Sigma, \mathbb{Z}) \to C_{k-1}(\Sigma, \mathbb{Z})$ is a linear operator, its kernel is a linear subspace of $C_k(\Sigma)$, which is the set of all the closed k-chains, denoted as $Z_k(\Sigma)$.

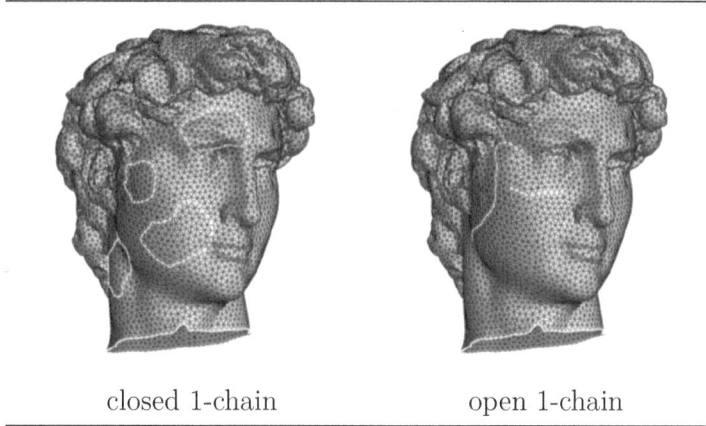

| closed 1-chain | open 1-chain |

Figure 15.6 Closed 1-chain and an open 1-chain.

In the left frame of Figure 15.6, each closed 1-chain bounds a 2-chain, namely these closed 1-chains are the boundaries of some 2-chains. So we also call them exact 1-chains or boundary 1-chains, which are in the image of the boundary operator ∂_2. This can be generalized to higher dimensional chains.

Definition 15.1.7 (Exact Chain) *A k-chain $\sigma \in C_k(\Sigma)$ is called an exact k-chain or boundary k-chain, if there exists a $(k+1)$-chain $\tau \in C_{k+1}(\Sigma)$, such that $\partial_{k+1}\tau = \gamma$, namely $\sigma \in Img\ \partial_{k+1}$.*

Because $\partial_{k+1} : C_{k+1}(\Sigma, \mathbb{Z}) \to C_k(\Sigma, \mathbb{Z})$ is a linear operator, its image is also a linear subspace of $C_k(\Sigma)$, which is the set of all the exact (boundary) k-chains, denoted as $B_k(\Sigma)$.

Intuitively, the boundary of a 2-dimensional patch is a closed loop, the boundary of a 3-dimensional volume is a 2 dimensional closed surface. Therefore, it is natural that the boundary of boundary is empty.

Theorem 15.1.8 (Boundary of Boundary) *The composition of boundary operators $\partial_k \circ \partial_{k+1} : C_{k+1}(\Sigma, \mathbb{Z}) \to C_{k-1}(\Sigma, \mathbb{Z})$ is zero,*

$$\partial_k \circ \partial_{k+1} = 0.$$

Namely, the boundary of a boundary is empty.

Proof It is sufficient for us to prove it on a $(k + 1)$-simplex, $\sigma = [v_0, v_1, \ldots, v_{k+1}]$. We use $\sigma_i, 0 \leq i \leq k + 1$, to denote the k-simplex,

$$\sigma_i := [v_0, \ldots, v_{i-1}, v_{i+1}, \ldots, v_{k+1}],$$

and $\sigma_{i,j}, 0 \leq i < j \leq k + 1$, to denote the $(k - 1)$-simplex

$$\sigma_{i,j} = [v_0, \ldots, v_{i-1}, v_{i+1}, \ldots, v_{j-1}, v_{j+1}, \ldots, v_{k+1}].$$

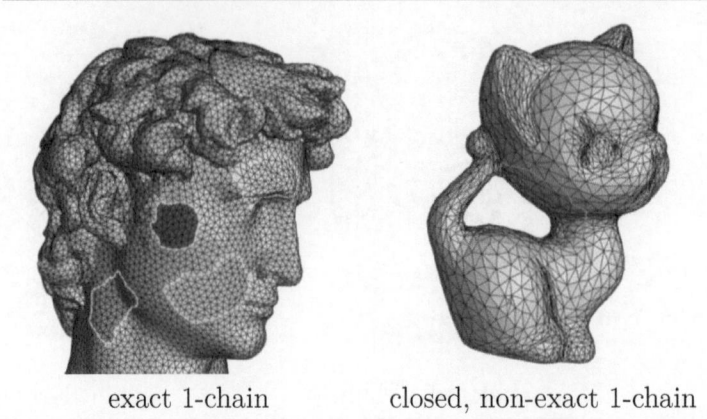

exact 1-chain   closed, non-exact 1-chain

Figure 15.7 Left frame: exact 1-chains; right frame: closed, non-exact 1-chain.

Then $\partial_{k+1}\partial_k\sigma$ is a linear combination of $\sigma_{i,j}$'s. We need to show for each $\sigma_{i,j}$, its corresponding coefficient is 0. By direct computation, in the first step, the boundary operator ∂_{k+1} acts on σ, we obtain terms $(-1)^i\sigma_i$. In the second step, the boundary operator ∂_k acts on σ_i's and obtain $\sigma_{i,j}$'s. We focus on two special terms $(-1)^j\sigma_j$ and $(-1)^i\sigma_i$ after the first step. Because $i < j$, v_i is in the i-th position in σ_j, hence $\partial_k(-1)^j\sigma_j$ has a term $(-1)^{i+j}\sigma_{i,j}$. But v_j is in the $(j-1)$-th position in σ_i, thus $\partial_k(-1)^i\sigma_i$ has a term $(-1)^{i+j-1}\sigma_{i,j}$. Therefore, the two terms of $\sigma_{i,j}$ cancel each other, the coefficient of $\sigma_{i,j}$ in $\partial_{k+1}\partial_k\sigma$ is 0. Since $\sigma_{i,j}$ is arbitrarily chosen, $\partial_{k+1}\partial_k\sigma$ is 0.

$\square$

This theorem shows exact chains are closed, $B_k \subset Z_k$. But the reverse may not be true, Z_k may be larger than B_k. The left frame in Figure 15.7 shows the exact 1-chains on David's head model, the right frame shows a closed, but non-exact 1-chain on a genus one kitten model. In fact, we can verify that any loop on the David's face surface must bound a surface patch, namely any closed 1-chain must be exact; but on the kitten surface, we can find loops which doesn't bound any surface patch, namely closed 1-chain may not be exact. This shows that the difference between the closed and exact chains conveys the topological information of the underlying space, which is the key idea of homology theory.

Two closed k-chains σ_1 and σ_2 are *homologous*, if they differ by an exact k-chain, namely there is a $(k+1)$-chain τ, such that

$$\sigma_1 - \sigma_2 = \partial_{k+1}\tau.$$

All the homologous classes under the addition form an Abelian group, the so-called homology group. Closed k-chains form the kernel space of the boundary operator ∂_k, exact k-chains form the image space of ∂_{k+1}.

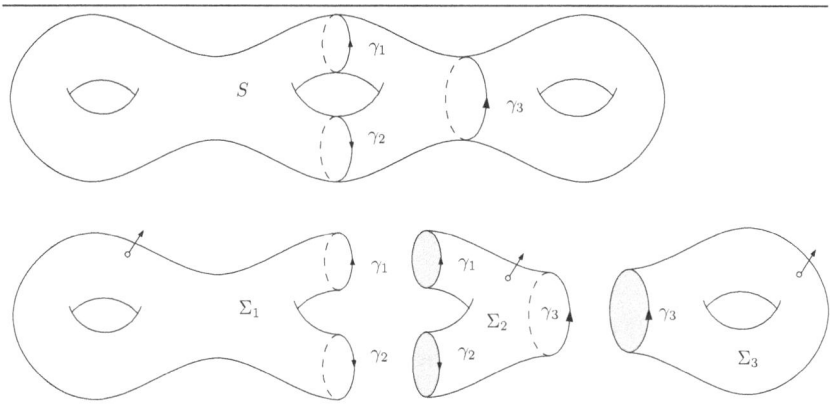

Figure 15.8 Homology vs. homotopy.

Definition 15.1.9 (Homology Group) *The k dimensional homology group $H_k(\Sigma, \mathbb{Z})$ is the quotient space of ker ∂_k by the image space of img ∂_{k+1}.*

$$H_k(\Sigma, \mathbb{Z}) = \frac{ker\ \partial_k}{img\ \partial_{k+1}}.$$

Namely, the quotient group $H_k(\Sigma) = Z_k(\Sigma)/B_k(\Sigma)$.

15.2   HOMOLOGY VS. HOMOTOPY

Given a topological space S, we have defined the fundamental group $\pi_1(S, p)$, the first homotopy group, and the first homology group $H_1(S, \mathbb{Z})$. The fundamental group is non-Abelian, but H_1 is Abelian. In general, the fundamental group has more complicated structure and conveys more topological information.

Figure 15.8 shows the fact that the homotopy classification is refiner than homologous classification. From the figure, we can deduce the following

$$\partial \Sigma_1 = \gamma_1 - \gamma_2, \quad \partial \Sigma_2 = \gamma_3 - \gamma_1 + \gamma_2, \quad \partial \Sigma_3 = -\gamma_3.$$

We obtain that γ_1 and γ_2 are not homotopic to each other but homologous; γ_3 is not homotopic but homologous to $\gamma_1 - \gamma_2$; γ_3 is not homotopic to e, but homologous to 0. This means $[\gamma_3]$ is non-trival in $\pi_1(S, p)$, but trivial in $H_1(S, \mathbb{Z})$. Therefore, the homology group ignores γ_3, but the fundamental group can detect γ_3. In fact, by ignoring loops like γ_3, the commutators, the non-Abelian group $\pi_1(S, p)$ becomes the Abelian group $H_1(S, \mathbb{Z})$.

Definition 15.2.1 (Commutator) *Suppose $(\mathcal{G}, \cdot)$ is a group, any two elements $g_1, g_2 \in \mathcal{G}$, the commutator of g_1 and g_2 is*

$$[g_1, g_2] = g_1 g_2 g_1^{-1} g_2^{-1}.$$

The product $g_1 g_2 = g_2 g_1$, if and only if $[g_1, g_2]$ is the identity element. The distance between a group $\mathcal{G}$ to an Abelian group is measured by the commutator subgroup.

Definition 15.2.2 (Commutator Subgroup) *The commutator subgroup $[\mathcal{G}, \mathcal{G}]$ is the subgroup generated by all the commutators,*

$$[\mathcal{G}, \mathcal{G}] = \langle [g_1, g_2] | g_1, g_2 \in \mathcal{G} \rangle.$$

The group $(\mathcal{G}, \cdot)$ is Abelian, if and only if its commutator subgroup is trivial, $[\mathcal{G}, \mathcal{G}] = \{e\}$. It is easy to verify that the commutator subgroup is a normal subgroup, the quotient group is Abelian. This process is called Abelianization.

Lemma 15.2.3 (Abelianization) *Suppose $(\mathcal{G}, \cdot)$ is a group, then its commutator subgroup $[\mathcal{G}, \mathcal{G}]$ is a normal subgroup, the quotient group*

$$\mathcal{G}' := \frac{\mathcal{G}}{[\mathcal{G}, \mathcal{G}]}$$

is an Abelian group.

The fundamental group $\pi_1(S, q)$ in general is non-Abelian, the first homology group $H_1(S, \mathbb{Z})$ is Abelian. They differ by the commutator subgroup of $\pi_1(S, q)$. As shown in Figure 15.9, the non-trivial loop γ cannot shrink to a point,

$$\gamma \sim a_1 b_1 a_1^{-1} b_1^{-1} \sim (a_2 b_2 a_2^{-1} b_2^{-1})^{-1}$$

it is homopotic to $[a_1, b_1]$ and $[a_2, b_2]^{-1}$. So $[\gamma]$ is in the commutator subgroup $[\pi_1(S, q), \pi_1(S, q)]$. If we cut the surface along γ, then we can see γ is the boundary of the right (also the left) connected component. Hence in $H_1(S, \mathbb{Z})$, $[\gamma]$ is 0. This shows the topological meaning of commutators in $\pi_1(S, q)$. The following lemma is easy to prove.

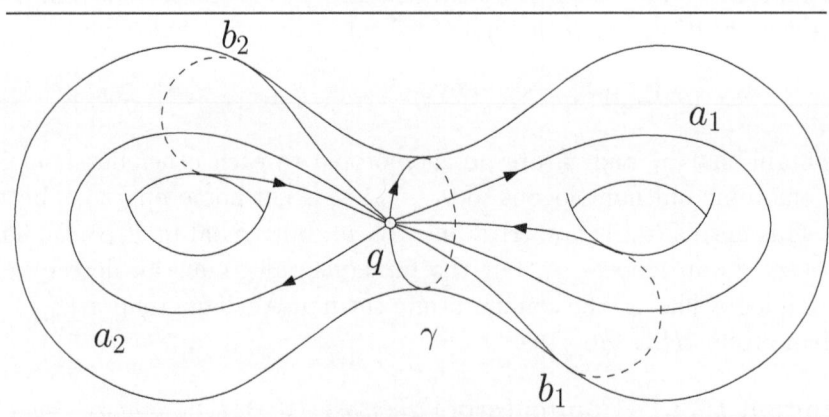

Figure 15.9 The topological meaning of the commutators of the fundamental group.

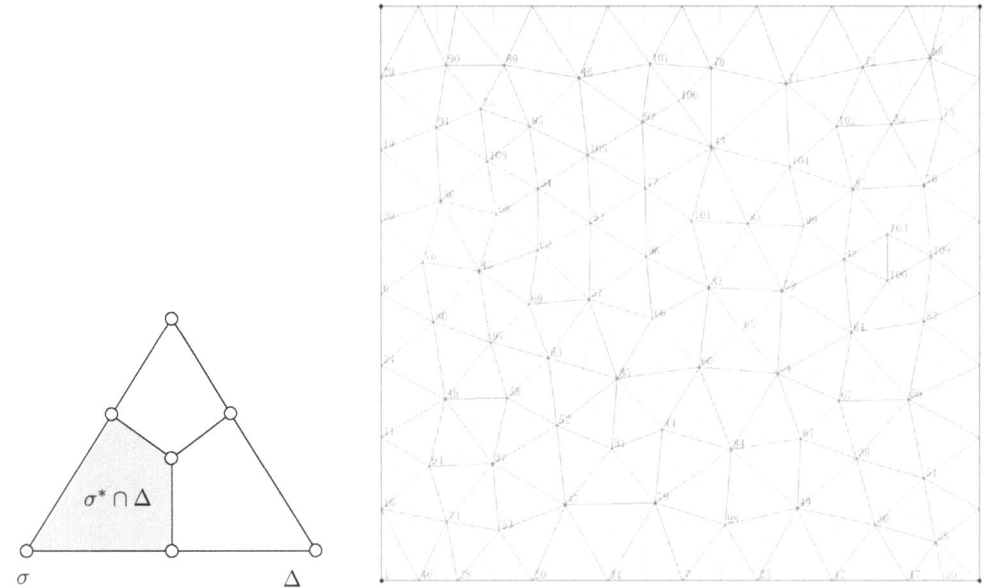

Figure 15.10 Poincaré Duality.

Lemma 15.2.4 *Suppose S is an n-manifold, the first homology group $H_1(S,\mathbb{Z})$ is the Abelianization of the fundamental group,*

$$H_1(S,\mathbb{Z}) = \frac{\pi_1(S,q)}{[\pi_1(S,q),\pi_1(S,q)]}.$$

where $[\pi_1(S,p),\pi_1(S,p)]$ is the commutator subgroup of $\pi_1(S,p)$.

In general, the fundamental group encodes more information than the first homology group, but is more difficult to compute.

Poincaré Duality Suppose Σ is a triangluation of an n dimensional manifold, σ a simplex of Σ. Let Δ be a top-dimensional simplex of Σ containing σ, so we can think of σ as a subset of the vertices of Δ. Define the dual cell σ^* corresponding to σ so that $\Delta \cap \sigma^*$ is the convex hull in Δ of the barycentres of all subsets of the vertices of Δ that contain σ. All the dual cells form a cell decomposition of the manifold, denoted as Σ^*. Each k-simplex σ in Σ corresponds to a unique $(n-k)$-cell σ^* in Σ^*. Figure 15.10 shows the Poincaré's duality, the manifold is a planar square. The Delaunay triangulation is Σ, the Voronoi diagram is the dual cell decomposition Σ^*. Each k-simplex σ corresponds to a $(2-k)$ cell σ^*, that is also the unique $(2-k)$ cell intersecting σ. By the Poincaré duality, we can easily prove the following theorem.

Theorem 15.2.5 *Suppose M is an n dimensional closed manifold, then $H_k(M,\mathbb{Z}) \cong H_{n-k}(M,\mathbb{Z})$.*

Proof The intersection map $C_k(\Sigma) \times C_{n-K}(\Sigma^*) \to \mathbb{Z}$ gives an isomorphism $C_k(\Sigma) \to C^{n-k}(\Sigma^*)$. Furthermore, $C^{n-k}(\Sigma^*)$ is the dual space of $C_{n-k}(\Sigma^*)$, Σ and Σ^* are different cell decompositions of the same manifold, $C_{n-k}(\Sigma^*)$ equals to $C_{n-k}(\Sigma)$. □

We can prove the following theorem on homology groups of a closed surface.

Theorem 15.2.6 *Suppose M is a genus g closed surface, then $H_0(M, \mathbb{Z}) \cong \mathbb{Z}$, $H_1(M, \mathbb{Z}) \cong \mathbb{Z}^{2g}$, $H_2(M, \mathbb{Z}) \cong \mathbb{Z}$.*

Proof Suppose M is an n dimensional manifold with k connected components, then any two points in the same connected components can be connected by a path, then they are homologous. Therefore, each connected component corresponds to one homologous class in $H_0(M, \mathbb{Z})$. Namely, if $H_0(M, \mathbb{Z}) = \mathbb{Z}^k$, then M has k connected components. By Poincaré duality, $H_n(M, \mathbb{Z})$ also equals to $\mathbb{Z}^k$. If M is a genus g closed surface, by the fact that $H_1(M, \mathbb{Z})$ is the Abelianization of $\pi_1(M)$, we obtain $H_1(M, \mathbb{Z})$ is $\mathbb{Z}^{2g}$. □

15.3 SIMPLICIAL COHOMOLOGY

In linear algebra, we know that given a linear space $\mathcal{L}$ all the linear functionals of $\mathcal{L}$ form the dual linear space $\mathcal{L}^*$. The k-th homology group $H_k(\Sigma, \mathbb{Z})$ of a simplicial complex Σ is a linear space, its dual space is called the k-th cohomology group of Σ, and denoted as $H^k(\Sigma, \mathbb{Z})$. Cohomology groups are more abstract than homology groups, but more convenient for the purposes of analysis.

Definition 15.3.1 (Cochain Space) *Suppose Σ is a simplicial complex, a k-cochain on Σ is a linear function $\omega : C_k(\Sigma, \mathbb{Z}) \to \mathbb{Z}$. A k-cochain is also called*

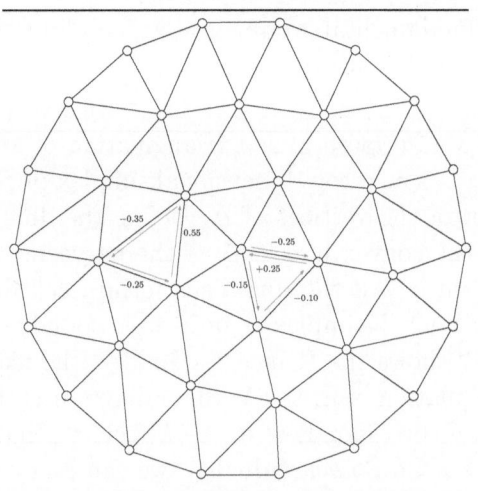

Figure 15.11 A 1-cochain on a simplicial complex.

a k-form. The k-cochain space $C^k(\Sigma, \mathbb{Z})$ is a linear space formed by all the linear functions defined on $C_k(\Sigma, \mathbb{Z})$,

$$C^k(\Sigma, \mathbb{Z}) := \{\omega : C_k(\Sigma, \mathbb{Z}) \to \mathbb{Z} |\ \omega\ is\ linear\}.$$

Figure 15.11 illustrates the concept of 1-chain. The simplicial complex Σ is a triangulation of a planar disk. ω is a 1-chain defined on Σ. For each 1-simplex σ, an oriented edge, we label $\omega(\sigma)$ on σ.

The boundary operator is the linear operator between chain spaces. The coboundary operator is dual to the boundary operator, which is a linear operator between cochain spaces.

Definition 15.3.2 (Coboundary Operator) *The coboundary operator $d^k : C^k(\Sigma, \mathbb{Z}) \to C^{k+1}(\Sigma, \mathbb{Z})$ is a linear operator, such that $\forall \sigma \in C_{k+1}(\Sigma, \mathbb{Z})$ and $\omega \in C^k(\Sigma, \mathbb{Z})$,*

$$(d^k \omega)(\sigma) := \omega \circ \partial_{k+1}(\sigma).$$

For an example, in Figure 15.11, ω is a 1-form defined on Σ, then $d^1\omega$ is a 2-form, such that

$$d^1\omega([v_0, v_1, v_2]) = \omega \circ \partial_2([v_0, v_1, v_2]) = \omega([v_0, v_1]) + \omega([v_1, v_2]) + \omega([v_2, v_0]).$$

Intuitively, the coboundary operators are discrete analogies to differential operators: d^0 corresponds to the gradient operator, d^1 to the curl operator, d^2 to the divergence operator.

Suppose $\omega \in C^k(\Sigma)$ is a k-form, $\sigma \in C_k(\Sigma)$ is a k-chain, we denote the pair

$$\langle \omega, \sigma \rangle := \omega(\sigma) \quad or \quad \int_\sigma \omega := \langle \omega, \sigma \rangle.$$

By the definition of the coboundary operator, the classical Stokes theorem can be directly generalized to the simplicial setting as follows:

Theorem 15.3.3 (Stokes) *Suppose Σ is a simplicial complex, given $(k-1)$-form $\omega \in C^{k-1}(\Sigma, \mathbb{Z})$ and k-chain $\sigma \in C_k(\Sigma, \mathbb{Z})$, we have*

$$\int_\sigma d^{k-1}\omega = \int_{\partial_k \sigma} \omega.$$

Namely, $\langle d^{k-1}\omega, \sigma \rangle = \langle \omega, \partial_k \sigma \rangle$.

Simplicial 1-forms are discrete analogy to smooth vector fields. From classical field theory, we know gradient vector fields and curl-free vector fields play important roles. A gradient field must be a curl-free field, the inverse may not be true. The simplicial analogies to curl-free fields and gradient fields are closed forms and exact forms respectively.

Definition 15.3.4 (Closed Forms) *Suppose Σ is a simplicial complex, a k-form $\omega \in C^k(\Sigma, \mathbb{Z})$ is closed, if $d_k \omega = 0$. Namely, closed k-form $\omega \in ker\ d_k$.*

All the closed k-forms form the kernel of d^k. Since $d^k : C^k(\Sigma, \mathbb{Z}) \to C^{k+1}(\Sigma, \mathbb{Z})$ is a linear map, its kernel ker d^k is a sub-linear space, denoted as $Z^k(\Sigma, \mathbb{Z})$. Similarly,

Definition 15.3.5 (Exact Forms) *Suppose Σ is a simplicial complex, a k-form $\omega \in C^k(\Sigma, \mathbb{Z})$ is exact, if there exists a $k - 1$ form $\rho \in C^{k-1}(\Sigma, \mathbb{Z})$, such that*

$$\omega = d^{k-1}\rho.$$

Namely, exact k-form $\omega \in img\ d^{k-1}$.

All the exact k-forms form the image of d^{k-1}. Since $d^{k-1} : C^{k-1}(\Sigma, \mathbb{Z}) \to C^k(\Sigma, \mathbb{Z})$ is a linear map, its image img d^{k-1} is a sub-linear space, denoted as $B^{k-1}(\Sigma, \mathbb{Z})$. Similar to homology theory, $\partial_k \circ \partial_{k+1} = 0$, for cohomology, we have $d^{k+1} \circ d^k = 0$.

Theorem 15.3.6 *Suppose Σ is a simplicial complex, the coboundary operators $d^{k-1} : C^{k-1}(\Sigma, \mathbb{Z}) \to C^k(\Sigma, \mathbb{Z})$ and $d^k : C^k(\Sigma, \mathbb{Z}) \to C^{k+1}(\Sigma, \mathbb{Z})$,*

$$d^k \circ d^{k-1} = 0.$$

This implies that all exact forms are closed. For example, the gradient fields are curl-free.
Proof For any $\omega \in C^{k-1}(\Sigma, \mathbb{Z})$, $\sigma \in C_{k+1}(\Sigma, \mathbb{Z})$, then

$$\langle d^k \circ d^{k-1}\omega, \sigma \rangle = \langle d^{k-1}\omega, \partial_{k+1}\sigma \rangle = \langle \omega, \partial_k \circ \partial_{k+1}\sigma \rangle,$$

since $\partial_k \circ \partial_{k+1}$ is 0, then $d^k \circ d^{k-1}$ equals to 0.

□

One the complex plane $\mathbb{C}$, given any closed form ω, we can construct a function $f(p) = \int_0^p \omega$, where the integration path can be arbitrarily chosen, then ω equals to the gradient of f. This means on $\mathbb{C}$, closed 1-forms must be exact, $B^1(\mathbb{C}) = Z^1(\mathbb{C})$. Now, we consider the punctured plane $\mathbb{C} \setminus \{0\}$, and a special 1-form

$$\omega = \frac{xdy - ydx}{x^2 + y^2},$$

the curl of ω is

$$d\omega = \frac{x^2 + y^2 - 2x^2}{(x^2 + y^2)^2} dx \wedge dy - \frac{x^2 + y^2 - 2y^2}{(x^2 + y^2)^2} dy \wedge dx = 0,$$

where d is the exterior differential operator. This shows ω is a closed 1-form. Now, let's integrate ω along the unit circle $\gamma(\theta) = (\cos\theta, \sin\theta)$,

$$\oint_\gamma \omega = \int_0^{2\pi} \frac{\cos\theta d\sin\theta - \sin\theta d\cos\theta}{\cos^2\theta + \sin^2\theta} = \int_0^{2\pi} d\theta = 2\pi.$$

This shows ω is not the gradient (differential) of any function, namely ω is not exact. This shows $B^1(\mathbb{C} \setminus \{0\}) \neq Z^1(\mathbb{C} \setminus \{0\})$. This example reveals the key idea of cohomology: the difference between exact forms and closed forms indicates the topology of the underlying space. Two 1-forms $\omega_1, \omega_2 \in C^1(\Sigma, \mathbb{Z})$ are cohomologous, if they differ by a gradient of a 0-form f,

$$\omega_1 - \omega_2 = d_0 f.$$

All the cohomologous classes under the addition form a group.

Definition 15.3.7 (Cohomology Group) *Suppose Σ is a simplicial complex, the k-dimensional cohomology group of Σ is defined as*

$$H^n(\Sigma, \mathbb{Z}) = \frac{ker \; d^k}{img \; d^{k-1}}.$$

Duality between Homology and Cohomology   The homology group and the cohomology group, $H_k(\Sigma, \mathbb{Z})$ and $H^k(\Sigma, \mathbb{Z})$, are dual to each other. Suppose $\omega \in Z^k(\Sigma, \mathbb{Z})$ is a closed k-form, $\sigma \in Z_k(\Sigma, \mathbb{Z})$ is a closed k-chain, then the pair $\langle \omega, \sigma \rangle$ is a bilinear operator, $\langle, \rangle : H^k(\Sigma, \mathbb{Z}) \times H_k(\Sigma, \mathbb{Z}) \to \mathbb{Z}$,

$$\langle [\omega], [\sigma] \rangle = \omega(\sigma).$$

It is easy to verify that the bilinear operator $\langle, \rangle$ is well defined. By direct computations, let ρ be a $(k-1)$-form, τ a $(k+1)$-chain, then

$$\langle \omega + d^{k-1}\rho, \sigma + \partial_{k+1}\tau \rangle = \langle \omega, \sigma \rangle + \langle \omega, \partial_{k+1}\tau \rangle + \langle d^{k-1}\rho, \sigma \rangle + \langle d^{k-1}\rho, \partial_{k+1}\tau \rangle$$

$$= \langle \omega, \sigma \rangle + \langle d^k \omega, \tau \rangle + \langle \rho, \partial_k \sigma \rangle + \langle \rho, \partial_k \partial_{k+1}\tau \rangle = \langle \omega, \sigma \rangle,$$

where we use the closedness of ω and σ, and $\partial^2 = 0$. This leads to the following duality theorem.

Theorem 15.3.8 (Duality between homology and cohomology) *Suppose Σ is a simplicial complex, then the k-th homology group is isomorphic to the k-th cohomology group,*

$$H_k(\Sigma, \mathbb{Z}) \cong H^k(\Sigma, \mathbb{Z}).$$

Proof Using the bilinear operator, each element $[\omega] \in H^k(\Sigma, \mathbb{Z})$ is treated as a linear function, $f_\omega : H_k(\Sigma, \mathbb{Z}) \to \mathbb{Z}$,

$$f_\omega([\sigma]) := \langle [\omega], [\sigma] \rangle.$$

If we treat $H_k(\Sigma, \mathbb{Z})$ as a linear space, then $H^k(\Sigma, \mathbb{Z})$ is the dual space of $H_k(\Sigma, \mathbb{Z})$, namely $H^k(\Sigma) = (H_k(\Sigma))^*$. By linear algebra, a finite dimensional linear space is isomorphic to its dual space.

$\square$

For each basis of the homology group, we can find the dual basis for the corresponding basis of the cohomology group.

Definition 15.3.9 (Dual Cohomology Basis) *Suppose a homology basis of H_k $(\Sigma, \mathbb{Z})$ is $\{\sigma_1, \sigma_2, \cdots, \sigma_n\}$, the dual cohomology basis of $H^k(\Sigma, \mathbb{Z})$ is $\{\omega_1, \omega_2, \cdots, \omega_n\}$, such that*

$$\langle \omega_i, \gamma_j \rangle = \delta_i^j, \quad \forall 1 \le i, j \le n,$$

where δ_i^j is the Kronecker symbol.

Cohomology was introduced by H. Whitney in order to represent stiefel whitney characteristic class [226]. In the above discussion, the coefficient group is integer group $\mathbb{Z}$, which can be replaced by other Abelian group, such as the real number group $\mathbb{R}$ and so on.

15.4   SIMPLICIAL MAPPING

In practice, all the surfaces are approximated by simplical complexes (polyhedral surfaces). All the continuous mappings are approximated by simplicial mappings (piecewise linear maps). A simplicial map between two simplicial complexes maps a simplex to a simplex (may with different dimension). Simplicial mapping is the foundation for animations in computer graphics. It is also the foundation for finite element analysis in computational mechanics and so on.

Definition 15.4.1 (simplicial mapping) *Suppose M and N are simplicial complexes, $f : M \to N$ is a continuous map, if $\forall \sigma \in M$, σ is a simplex, $f(\sigma)$ is a simplex, then f is called a* simplicial mapping.

The restriction of a simplicial map on a simplex is an affine map, which is convenient to represent and easy to analyze. In order to improve the approximation precision, sometimes we need to subdivide the source and the target complexes. Then we can show that any continuous mapping between surfaces can be approximated by simplicial mappings to arbitrary precision by subdivisions.

For each simplex, we can add its gravity center (barycenter), and subdivide the simplex to multiple ones. The resulting complex is called the .

Theorem 15.4.2 (Simplicial Approximation) *Suppose M and N are simplicial complexes embedded in $\mathbb{R}^n$, $f : M \to N$ is a continuous mapping. Then for any $\epsilon > 0$, there exist some gravity subdivisions $\tilde{M}$ and $\tilde{N}$, and a simplicial mapping $\tilde{f} : \tilde{M} \to \tilde{N}$, such that*

$$\forall p \in |M|, |f(p) - \tilde{f}(p)| < \epsilon.$$

The proof is elementary, the students can prove it as an excise. Figure 15.12 shows a continuous map from the unit disk to itself, which is approximated by a simplicial mapping between two simplical complexes (triangular meshes). The mapping restricted on each triangular face is linear. If the triangulation is highly refined, the mapping looks very smooth.

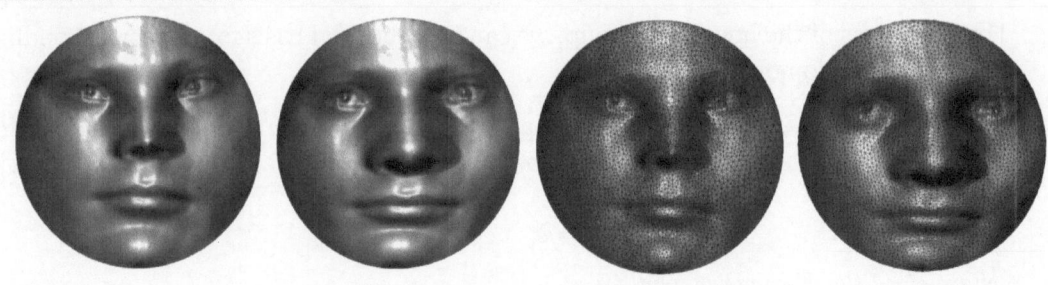

Figure 15.12   A simplicial mapping from the planar unit disk to itself.

Definition 15.4.3 (chain map) *Suppose $f : K \to L$ is a simplicial map, it induces the n-chain map $f_n : C_n(K) \to C_n(L)$ defined by its values on the simplices, as follows:*

$$f_n([v_0, v_1, \ldots, v_n]) := [f(v_0), f(v_1), \ldots, f(v_n)].$$

The following lemma is easy to prove.

Lemma 15.4.4 *The chain map induced by the composition $g \circ f : K \to M$ of two simplicial maps $f : K \to L$ and $g : L \to M$ is the composition of the chain maps induced by these simplicial maps:*

$$(g \circ f)_k = g_k \circ f_k.$$

Furthermore, we can show that the simplicial map and the boundary operator commute.

Lemma 15.4.5 *Suppose $f : K \to L$ is a simplicial map, along with the chain maps of f, the boundary operators can be put in the diagram:*

$$
\begin{array}{ccc}
C_k(K) & \xrightarrow{\ f_k\ } & C_k(L) \\
\downarrow{\scriptstyle \partial_k^K} & & \downarrow{\scriptstyle \partial_k^L} \\
C_{k-1}(K) & \xrightarrow{\ f_{k-1}\ } & C_{k-1}(L)
\end{array}
$$

The diagram commutes, namely

$$\partial_k^L f_k = f_{k-1}\partial_k^K, \quad k = 1, 2, \ldots, n$$

or more compactly $\partial f = f \partial$.

This lemma can be generalized such that the following diagram commutes, namely the computational results are path independent.

$$
\begin{array}{ccccccc}
\cdots \longrightarrow & C_{k+1}(K) & \xrightarrow{\partial_{k+1}} & C_k(K) & \xrightarrow{\partial_k} & C_{k-1}(K) & \longrightarrow \cdots \\
& \downarrow{\scriptstyle f_{k+1}} & & \downarrow{\scriptstyle f_k} & & \downarrow{\scriptstyle f_{k-1}} & \\
\cdots \longrightarrow & C_{k+1}(L) & \xrightarrow{\partial_{k+1}} & C_k(L) & \xrightarrow{\partial_k} & C_{k-1}(L) & \longrightarrow \cdots
\end{array}
$$

By $\partial f = f \partial$, we can show the chain maps take cycles to cycles $f_k(Z_k(K)) \subset Z_k(L)$, and boundaries to boundaries $f_k(B_k(K)) \subset B_k(L)$. Thus we are able to define the homology map $f_* : H_*(K) \to H_*(L)$ induced by the simplicial map $f : K \to L$.

Definition 15.4.6 (Homology Map) *Given a simplicial map $f : K \to L$, it induces chain maps $f_k : C_k(K) \to C_k(L)$, $k = 0, 1, 2, \ldots$, which can be restricted to cycles $f_k : Z_k(K) \to Z_k(L)$ and boundaries $f_k : B_k(K) \to B_k(L)$. The homology map $f_{k,*} : H_k(K) \to H_k(L)$ is defined as*

$$f_{k,*}([\sigma]) := [f_k(\sigma)].$$

The homology map $f_* : H_*(K) \to H_*(L)$ induced by f is defined as

$$f_* := (f_{0,*}, f_{1,*}, \ldots, f_{n,*}).$$

The cohomology map can be defined accordingly.

Definition 15.4.7 (Cohomology Map) *Given a simplical map* $f : K \to L$*, it induces the homology group* $f_* : H_*(K) \to H_*(L)$*. The cohomology map* $f^{k,*} : H^k(L) \to H^k(K)$ *is defined as follows: suppose* $[\sigma] \in H_k(K)$*,* $[\omega] \in H^k(L)$*,*

$$f^{k,*}[\omega]([\sigma]) = \omega(f_{k,*}[\sigma]) = \omega(f_k(\sigma)).$$

Similarly the cohomomogy map $f^* : H^*(L) \to H^*(K)$ *is defined as*

$$f^* := (f^{0,*}, f^{1,*}, \ldots, f^{n,*}).$$

Mapping Degree   Suppose K and L are two closed surfaces. $H_2(K, \mathbb{Z}) = \mathbb{Z}$, $H_2(L, \mathbb{Z}) = \mathbb{Z}$, suppose $[K]$ is the generator of $H_2(K, \mathbb{Z})$, which is the union of all the faces. Similarly, $[L]$ is the generator of $H_2(L, \mathbb{Z})$. $f : K \to L$ is a continuous map. Then f induces the homology map $f_* : H_*(K, \mathbb{Z}) \to H_*(L, \mathbb{Z})$,

$$f_{2,*} : H_2(K, \mathbb{Z}) \to H_2(L, \mathbb{Z}), \quad f_{2,*} : \mathbb{Z} \to \mathbb{Z},$$

must has the form $f_{2,*}(z) = cz, c \in \mathbb{Z}$.

Definition 15.4.8 (Mapping Degree) *In the above setting, the homology map* $f_*([K]) = c[L]$*, then the integer* c *is called the* degree *of the map.*

The degree of the map $f : K \to L$ is the total (algebraic) number of the pre-images $f^{-1}(q)$ for an arbitrary point $q \in L$. Furthermore, it is independent of the choice of q. In the following, we use the concept of the mapping degree to prove the Gauss-Bonnet theorem, which claims the total Gaussian curvature of a surface equals to 2π times the Euler characteristic number of the surface.

Suppose S is a C^2 genus g closed surface embedded in $\mathbb{R}^3$, $G : S \to \mathbb{S}^2$ is the Gauss map, which maps each point $p \in S$ to its normal $\mathbf{n}(p)$. The degree of the Gauss map is an integer, which is invariant under continuous deformations of the source surface S. We deform the surface continuously to the shape as shown in the left frame in Figure 15.13, and choose a point q on the Gauss sphere $\mathbb{S}^2$. From the figure, there are $g + 1$ preimages of q, $G^{-1}(q) = \{p_1, p_2, \cdots, p_g, g_{g+1}\}$, where p_{g+1} is in a convex region of the surface with positive Gaussian curvature; each p_k, $k = 1, 2, \ldots, g$, is in

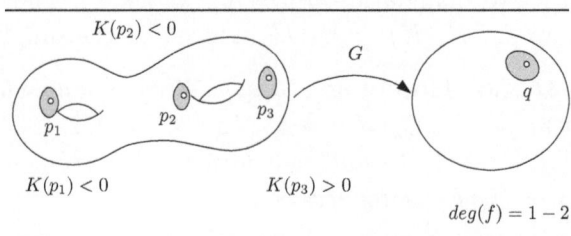

Figure 15.13   Map degree of the Gauss map.

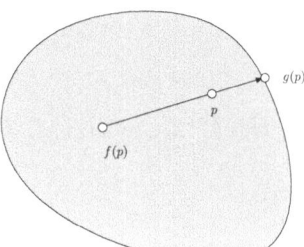

Figure 15.14 Brouwer fixed point.

the saddle region of the k-th handle with negative Gaussian curvature. The Gaussian curvature equals to the Jacobian of the Gauss map. The image of a surface region with positive Gaussian curvature wraps the Gauss sphere with positive (consistent) orientation; the image of a surface region with negative Gaussian curvature wraps the Gauss sphere with negative (reverse) orientation. Therefore, the total algebraic number of pre-images equal to $1 - g$, the degree of the Gauss map is $1 - g$. The Gauss map wraps the surface S around $\mathbb{S}^2$ with $1 - g$ layers, so the total area of the image $G(S)$ equals to the sphere area 4π multiplies the total number of layers $1 - g$, namely $4\pi(1 - g) = 2\pi\chi(S)$, which is the total Gaussian curvature of S. This proves the classical surface Gauss-Bonnet theorem.

15.5 FIXED POINT

In this section, we apply homology/cohomology theory to prove several fixed point theorems, which play fundamental roles in mathematics.

Theorem 15.5.1 (Brouwer Fixed Point) *Suppose $\Omega \subset \mathbb{R}^n$ is a compact convex set, $f : \Omega \to \Omega$ is a continous map, then there exists a fxied point $p \in \Omega$, such that $f(p) = p$.*

Proof Assume $f : \Omega \to \Omega$ has no fixed point, namely $\forall p \in \Omega$, $f(p) \neq p$. We construct $g : \Omega \to \partial\Omega$, a ray starting from $f(p)$ through p and intersect $\partial\Omega$ at $g(p)$, $g|_{\partial\Omega} = id$. i is the inclusion map, $(g \circ i) : \partial\Omega \to \partial\Omega$ is the identity,

$$\partial\Omega \xrightarrow{\ i\ } \Omega \xrightarrow{\ g\ } \partial\Omega$$

$(g \circ i)_* : H_{n-1}(\partial\Omega, \mathbb{Z}) \to H_{n-1}(\partial\Omega, \mathbb{Z})$ is $z \mapsto z$. But $H_{n-1}(\Omega, \mathbb{Z}) = 0$, then $(g \circ i)_* = 0$. Contradiction.

$\square$

Definition 15.5.2 (Index of Fixed Point) *Suppose M is an n-dimensional topological space, p is a fixed point of $f : M \to M$. Choose a neighborhood $p \in U \subset M$, $f_* : H_{n-1}(\partial U, \mathbb{Z}) \to H_{n-1}(\partial U, \mathbb{Z})$,*

$$f_* : \mathbb{Z} \to \mathbb{Z}, z \mapsto \lambda z,$$

where λ is an integer, the algebraic index of p, $Ind(f, p) =: \lambda$.

Given a compact topological space M, and a continuous automorphism $f : M \to M$, it induces homomorphisms

$$f_{k,*} : H_k(M, \mathbb{Z}) \to H_k(M, \mathbb{Z}),$$

each $f_{k,*}$ is represented as a matrix.

Definition 15.5.3 (Lefschetz Number) *The Lefschetz number of the automorphism $f : M \to M$ is given by*

$$\Lambda(f) := \sum_k (-1)^k Tr(f_{k,*} | H_k(M, \mathbb{Z})).$$

The Lefschetz Fixed Point theory connects the total algebraic indices of fixed points with Lefschetz number. In the following, we only prove a simpler version of the theorem for the existence of fixed points, the Lefschetz fixed point theorem:

Theorem 15.5.4 (Lefschetz Fixed Point) *Given a continuous automorphism of a compact topological space $f : M \to M$, if its Lefschetz number is non-zero, then there is a fixed point $p \in M$, $f(p) = p$.*

Proof We triangulate M and use a simplicial map to approximate f, then f induces a series of chain maps $f_k : C_k(M, \mathbb{Z}) \to C_k(M, \mathbb{Z})$. If

$$\sum_k (-1)^k Tr(f_k | C_k) \neq 0,$$

then there must be a k-simplex $\sigma \in C_k(M, \mathbb{Z})$, such that $f_k(\sigma) = \lambda \sigma$, where λ is a non-zero integer, by Brouwer fixed point theorem 15.5.1, f must have a fixed point in σ.

By definition $C_k = C_k/Z_k \oplus Z_k$, Z_k is the closed chain space; $Z_k = B_k \oplus H_k$, B_k is the exact chain space, H_k is the homology group. The restriction of ∂_k on C_k/Z_k to B_{k-1}, $\partial_k : C_k/Z_k \to B_{k-1}$, is isomorphic. The following diagram commutes,

$$
\begin{array}{ccc}
C_k/Z_k & \xrightarrow{f_k} & C_k/Z_k \\
\partial_k \downarrow & & \downarrow \partial_k \\
B_{k-1} & \xrightarrow{f_{k-1}} & B_{k-1}
\end{array}
$$

Therefore, $\partial_k \circ f_k \circ \partial_k^{-1} = f_{k-1}$ restricted on B_{k-1}. Take the trace of both sides,

$$Tr(\partial_k \circ f_k \circ \partial_k^{-1} | B_{k-1}) = Tr(f_k | C_k/Z_k) = Tr(f_{k-1} | B_{k-1}),$$

By $C_k = C_k/Z_k \oplus B_k \oplus H_k$, we have

$$
\begin{aligned}
Tr(f_k | C_k) &= Tr(f_k | C_k/Z_k) + Tr(f_k | Z_k) \\
&= Tr(f_k | C_k/Z_k) + Tr(f_k | B_k) + Tr(f_k | H_k) \\
&= Tr(f_{k-1} | B_{k-1}) + Tr(f_k | B_k) + Tr(f_k | H_k)
\end{aligned}
$$

This directly leads to

$$\sum_k (-1)^k Tr(f_k|C_k) = \sum_k (-1)^k Tr(f_k|H_k) = \Lambda(f). \qquad (15.5.1)$$

If $\Lambda(f) \neq 0$, $\sum_k (-1)^k Tr(f_k|C_k) \neq 0$, there exists a fixed point. The key point is that, the left hand side depends on the triangulation, but the Lefschetz number doesn't depend on the triangulation, therefore it is more general.

□

From the above discussion, it is obvious that the Lefschetz number solely depends on the homotopy class of the mapping f. Therefore, we can choose the simplest mapping in the homotopy class to reduce the computational cost.

Lemma 15.5.5 *Suppose M is a compact oriented surface, $f : M \to M$ is a continuous automorphism of M, f is homotopic to the identity map of M, then the Lefschetz number of f equals to the Euler characteristic number of M,*

$$\Lambda(f) = \chi(S).$$

Proof We construct a triangulation of M and use a simplicial map to approximate the automorphism. Then

$$\Lambda(f) = \Lambda(Id) = |V| + |F| - |E| = \chi(S).$$

□

We continue to use cohomology to study the zero points of a vector field on a surface, especially the Poincaré-Hopf theorem.

Definition 15.5.6 (Isolated Zero Point) *Given a smooth tangent vector field $\mathbf{v}$: $S \to TS$ on a smooth surface S, $p \in S$ is called a zero point, if $\mathbf{v}(p) = \mathbf{0}$. If there is a neighborhood $U(p)$, such that p is the unique zero in $U(p)$, then p is called an isolated zero point.*

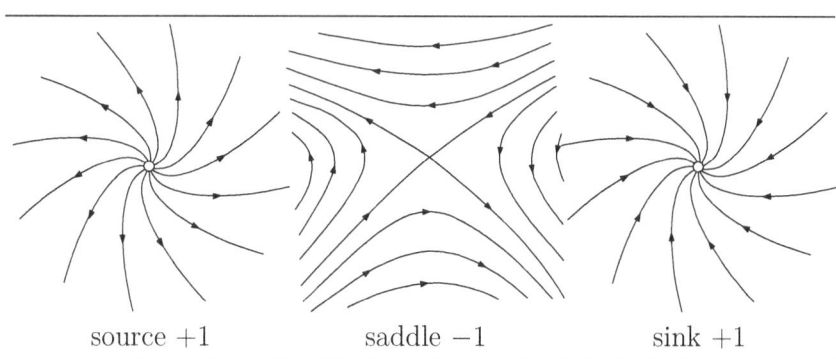

source +1   saddle −1   sink +1

Figure 15.15 Indices of zero points.

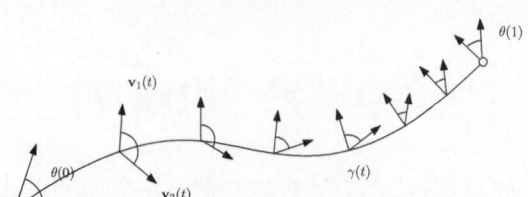

Figure 15.16 The angle function $\theta(t)$ between two vector fields along a curve.

Definition 15.5.7 (Index of Zero Point) *Given an isolated zero point $p \in S$, choose a small disk $B(p, \varepsilon)$, define a map $\varphi : \partial B(p, \varepsilon) \to \mathbb{S}^1$, $q \mapsto \mathbf{v}(q)/|\mathbf{v}(q)|$. This map induces a homomorphism $\varphi_\# : \pi_1(\partial B) \to \pi_1(\mathbb{S}^1)$, $\varphi_\#(z) = kz$, where the integer k is called the index of the zero point p, denoted as $Ind(p, \mathbf{v})$.*

Figure 15.15 shows the typical indices of zero points, the source and sink points have $+1$ indices, the saddle point has -1 index.

Theorem 15.5.8 (Poincaré-Hopf Index Theorem) *Assume S is a compact, oriented smooth surface, $\mathbf{v}$ is a smooth tangent vector field with isolated zero points. If S has boundaries, then on the boundary $\mathbf{v}$ is along the exterior normal direction, then we have*

$$\sum_{p \in Z(\mathbf{v})} Ind(p, \mathbf{v}) = \chi(S),$$

where $Z(\mathbf{v})$ is the set of all zeros, $\chi(S)$ is the Euler characteristic number of S.

Proof Given two vector fields $\mathbf{v}_1$ and $\mathbf{v}_2$ with different isolated zeros. We construct a triangulation T, such that each face contains at most one zero. Define two 2-forms, Ω_1 and Ω_2, on each triangle face Δ,

$$\Omega_k(\Delta) = Ind(p, \mathbf{v}_k), \quad p \in \Delta \cap Z(\mathbf{v}_k), \quad k = 1, 2.$$

As shown in Figure 15.16, along a path $\gamma(t)$, the function $\theta(t)$ is the angle from $\mathbf{v}_1 \circ \gamma(t)$ to $\mathbf{v}_2 \circ \gamma(t)$. Define a one form ω,

$$\omega(\gamma) := \frac{1}{2\pi} \int_\gamma \dot{\theta}(\tau) d\tau.$$

Given a triangle Δ, the relative rotation of $\mathbf{v}_2$ with respect to $\mathbf{v}_1$ is given by

$$\omega(\partial \Delta) = d\omega(\Delta)$$

then we get

$$\Omega_2 - \Omega_1 = d\omega.$$

Therefore, Ω_1 and Ω_2 are cohomologous. This implies $\int_S \Omega_1 = \int_S \Omega_2$. The total index of zeros of a vector field

$$\sum_{p \in Z(\mathbf{v}_k)} Ind(p, \mathbf{v}_k) = \int_S \Omega_k, \quad k = 1, 2.$$

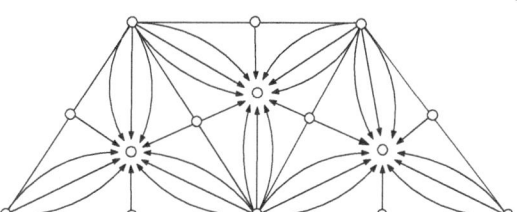

Figure 15.17 A special vector field.

Therefore the total index of $\mathbf{v}_1$ equals to that of $\mathbf{v}_2$. Namely, the total index is independent of the choice of the vector field. Hence we can construct a special vector field, and compute its total index.

As shown in Figure 15.17, on the special vector field, each vertex becomes a source, each face center is a sink, each edge center is a saddle point. Therefore, the total index of all the zero points is

$$\sum_{p \in Z(v)} \mathrm{Ind}(p, \mathbf{v}) = |V| + |F| - |E| = \chi(S).$$

□

There is a natural relation between the fixed points and the zero points of a vector field, therefore the Lefschetz fixed point theorem and the Poincaré-Hopf index theorem. Given a smooth tangent vector field $\mathbf{v}$, we can define a one parameter family of automorphisms, $\varphi_t : S \to S$, such that

$$\frac{\partial \varphi_t(p)}{\partial t} = \mathbf{v} \circ \varphi_t(p) \quad \varphi_0(p) = p.$$

Then φ_t is an automorophism homotopic to the identity. According to lemma 15.5.5, the total index of the fixed points of φ_t is $\chi(S)$. Each fixed point of φ_t corresponds to a zero point of $\mathbf{v}$ at the same location and with the same index. This gives the intrinsic relation between the Lefschetz fixed point theorem and the Poincaré-Hopf index theorem.

15.6 COMPUTATIONAL ALGORITHMS

Homology Group   Given a closed surface of genus g, we compute a triangulation, namely a simplicial complex Σ. In order to compute the first homology group $H_1(\Sigma, \mathbb{Z})$, we can use the combinatorial method in Algorithm 4 to compute a set of fundamental group basis $\{\gamma_1, \ldots, \gamma_{2g}\}$, which is also the basis of $H_1(\Sigma, \mathbb{Z})$. We can also compute $H_1(\Sigma, \mathbb{Z})$ using the following algebraic method.

Each boundary operator: $\partial_k : C_k(\Sigma, \mathbb{Z}) \to C_{k-1}(\Sigma, \mathbb{Z})$ is a linear map between linear spaces $C_k(\Sigma, \mathbb{Z})$ and $C_{k-1}(\Sigma, \mathbb{Z})$, therefore it can be represented as an integer matrix. Suppose there are n_k k-simplices in Σ, denoted as $\{\sigma_1^k, \sigma_2^k, \ldots, \sigma_{n_k}^k\}$.

$$C_k(\Sigma, \mathbb{Z}) := \left\{ \sum_{i=1}^{n_k} \lambda_i \sigma_i^k \mid \lambda_i \in \mathbb{Z} \right\}.$$

The boundary operator ∂_k is represented as a $n_{k-1} \times n_k$ matrix, $\partial_k = ([\sigma_i^{k-1}, \sigma_j^k])$, where

$$[\sigma_i^{k-1}, \sigma_j^k] = \begin{cases} +1 & +\sigma_i^{k-1} \in \partial_k \sigma_j^k \\ -1 & -\sigma_i^{k-1} \in \partial_k \sigma_j^k \\ 0 & \sigma_i^{k-1} \notin \partial_k \sigma_j^k \end{cases}$$

Definition 15.6.1 (Cominatorial Laplace Operator) *Suppose Σ is a simplicial complex, the boundary operator is $\partial_k : C_k(\Sigma, \mathbb{Z}) \to C_{k-1}(\Sigma, \mathbb{Z})$, the linear operator $\Delta_k : C_k(\Sigma, \mathbb{Z}) \to C_k(\Sigma, \mathbb{Z})$ is defined as:*

$$\Delta_k := \partial_k^T \partial_k + \partial_{k+1} \partial_{k+1}^T,$$

Δ_k is called the combinatorial Laplace operator.

Lemma 15.6.2 (Combinatorial Laplace Operator) *Suppose Σ is an n dimensional simplicial complex, $\Delta_k : C_k(\Sigma, \mathbb{Z}) \to C_k(\Sigma, \mathbb{Z})$ is the combinatorial Lapalce operator, $0 \le k \le n$, then any eigen vector of Δ_k corresponding to the zero eigen value is in $H_k(\Sigma, \mathbb{Z})$.*

Proof Suppose $\xi \in C_k(\Sigma, \mathbb{Z})$, such that $\Delta_k \xi = 0$, then

$$\xi^T \Delta_k \xi = \xi^T \partial_k^T \partial_k \xi + \xi^T \partial_{k+1} \partial_{k+1}^T \xi = \|\partial_k \xi\|^2 + \|\partial_{k+1}^T \xi\|^2,$$

hence $\partial_k \xi = 0$, $\xi \in \ker \partial_k$; $\partial_{k+1}^T \xi = 0$, $\xi \notin \operatorname{img} \partial_{k+1}$. Therefore, $\xi \in H_k(\Sigma, \mathbb{Z})$.
□

The eigen decomposition of Δ_k can be carried out using the Smith normal form.

Definition 15.6.3 (Smith Normal Form) *Let A be a nonzero $m \times n$ integer matrix. There exist invertible $m \times m$ and $n \times n$-matrices S, T, such that the product SAT is*

$$\begin{pmatrix} \alpha_1 & 0 & 0 & \cdots & & 0 \\ 0 & \alpha_2 & 0 & \cdots & & 0 \\ 0 & 0 & \ddots & & & 0 \\ \vdots & & & \alpha_r & & \vdots \\ & & & & 0 & \\ & & & & & \ddots \\ 0 & & \cdots & & & 0 \end{pmatrix}$$

and the diagonal elements α_i satisfy $\alpha_i \mid \alpha_{i+1}$, $\forall 1 \le i < r$. This is the Smith normal form of the matrix A.

The elements α_i are unique up to multiplication by a unit and are called the elementary divisors. They can be computed (up to multiplication by a unit) as

$$\alpha_i = \frac{d_i(A)}{d_{i-1}(A)},$$

where $d_i(A)$ (called i-th determinant divisor) equals to the greatest common divisor of all $i \times i$ minors of the matrix A and $d_0(A) := 1$. The Smith normal form method is general to arbitrary dimensional homology group, but it is expensive to compute. For the first homology group $H_1(\Sigma, \mathbb{Z})$, the combinatorial algorithm 4 based on CW-cell decomposition is more efficient.

Cohomology Group   Given a closed surface of genus g, we triangulate it to a simplicial complex, denoted as Σ. We then compute a set of basis of $H_1(\Sigma, \mathbb{Z})$, denoted as

$$\{\gamma_1, \gamma_2, \cdots, \gamma_{2g}\}.$$

For each γ_i, we slice Σ along γ_i to obtain a simplicial complex Σ_i with two boundary components $\partial \Sigma_i = \gamma_i^+ - \gamma_i^-$. We then construct a function $f_i : M_i \to \mathbb{R}$,

$$f_i(v) = \begin{cases} 1 & v \in \gamma^+ \\ 0 & v \in \gamma^- \\ \text{rand} & v \notin \gamma^+ \cup \gamma^- \end{cases}$$

Then we construct a 1-form $\omega_i := df_i$. For each edge e on the boundary $e \in \gamma_i^+ \cup \gamma_i^-$, $\omega_i(e) = 0$. Therefore, ω_i can be treated as a 1-form defined on the original simplicial complex Σ.

Lemma 15.6.4 *In the above setting, the closed 1-forms*

$$\{\omega_1, \omega_2, \cdots, \omega_{2g}\}$$

is a set of basis of $H^1(\Sigma, \mathbb{R})$.

Proof By the construction, for any loop γ, the algebraic intersection number $\gamma_i \odot \gamma$ equals to the integration of ω_i along γ,

$$\gamma_i \odot \gamma = \int_\gamma \omega_i.$$

Suppose $\tau = \sum_i \lambda_i \gamma_i$, then the corresponding 1-form is $\sum_i \lambda_i \omega_i$:

$$\tau \odot \gamma = \sum_i \lambda_i \gamma_i \odot \gamma = \sum_i \lambda_i \int_\gamma \omega_i = \int_\gamma \sum_i \lambda_i \omega_i.$$

Suppose γ_i's are canonical fundamental group generators, $\{a_1, b_1, \ldots, a_g, b_g\}$, and the produced 1-forms are $\{\alpha_1, \beta_1, \ldots, \alpha_g, \beta_g\}$. Given a loop $\gamma = \sum_{i=1}^g \lambda_i a_i + \mu_i b_i$, then

$$\int_\gamma \alpha_i = \mu_i, \quad \int_\gamma \beta_i = -\lambda_i.$$

It is obvious that α_i, β_i are linearly independent, and form a basis of $H^1(\Sigma, \mathbb{R})$. In general cases, two sets of basis of $H_1(\Sigma, \mathbb{Z})$ differ by a non-degenerated linear transformation:

$$\gamma_i = \sum_{j=1}^g \lambda_{ij} a_i + \mu_{ij} b_i,$$

in turn, we have

$$\omega_i = \sum_{j=1}^{g} \lambda_{ij} \alpha_i + \mu_{ij} \beta_i,$$

hence ω_i's are linearly independent and form a basis of $H^1(\Sigma, \mathbb{R})$.

□

The details of the computation of the surface first cohomology group basis is in Algorithm 7.

Algorithm 7 Cohomology Group

Require: A closed genus g triangle mesh Σ

  Compute a set of basis of $H_1(\Sigma, \mathbb{Z})$ using Algorithm 4, γ_i, $i = 1, \ldots, 2g$;

  for each homology basis γ_i **do**

    slice Σ along γ_i to obtain Σ_i, $\partial \Sigma_i = \gamma_i^+ - \gamma_i^-$;

    set a 0-form $\tau_i \in C^0(\Sigma_i, \mathbb{R})$, such that $\tau_i(v) \leftarrow 1$, $\forall v \in \gamma_i^+$ and $\tau_i(w) \leftarrow 0$, $\forall w \in \gamma_i^-$;

    $\omega_i \leftarrow d\tau_i$;

  end for

Ensure: All $\{\omega_1, \omega_2, \cdots, \omega_{2g}\}$ form a basis of $H^1(M, \mathbb{R})$

CHAPTER 16

Exterior Calculus and Hodge Decomposition

The simplicial homology of a simplicial complex is the difference between the closed chains and the exact chains. The simplicial cohomology is the difference between the closed forms and the exact forms. In this chapter, we generalize the concept of cohomology to the de Rham cohomology using exterior calculus on smooth manifolds, then introduce the important Hodge decomposition theorem and the cohomology group of harmonic differential forms. Finally, we explain the discrete Hodge theory, discrete harmonic forms and the computational algorithms for discrete Hodge decomposition.

16.1 EXTERIOR DIFFERENTIALS

Definition 16.1.1 (Manifold) *A manifold is a topological space M covered by a set of open sets $\{U_\alpha\}$, $M \subset \bigcup_\alpha U_\alpha$. A homeomorphism $\phi_\alpha : U_\alpha \to \mathbb{R}^n$ maps U_α to the Euclidean space $\mathbb{R}^n$. (U_α, ϕ_α) is called a* coordinate chart *of M. The set of all charts $\{(U_\alpha, \phi_\alpha)\}$ form the* atlas *of M. Suppose $U_\alpha \cap U_\beta \neq \emptyset$, then*

$$\phi_{\alpha\beta} = \phi_\beta \circ \phi_\alpha^{-1} : \phi_\alpha(U_\alpha \cap U_\beta) \to \phi_\beta(U_\alpha \cap U_\beta)$$

is a transition map.

If all transition maps are smooth, $\phi_{\alpha\beta} \in C^\infty(\mathbb{R}^n)$, then the manifold is a *differential manifold* or a *smooth manifold*.

Definition 16.1.2 (Tangent Vector) *A tangent vector ξ at the point p is an association to every coordinate chart $(x^1, x^2, \cdots, x^n)$ at p an n-tuple $(\xi^1, \xi^2, \cdots, \xi^n)$ of real numbers, such that if $(\tilde{\xi}^1, \tilde{\xi}^2, \cdots, \tilde{\xi}^n)$ is associated with another coordinate system $(\tilde{x}^1, \tilde{x}^2, \cdots, \tilde{x}^n)$, then it satisfies the transition rule*

$$\tilde{\xi}^i = \sum_{j=1}^n \frac{\partial \tilde{x}^i}{\partial x^j}(p)\xi^j.$$

A smooth vector field ξ assigns a tangent vector for each point of M, it has local

DOI: 10.1201/9781003350576-16

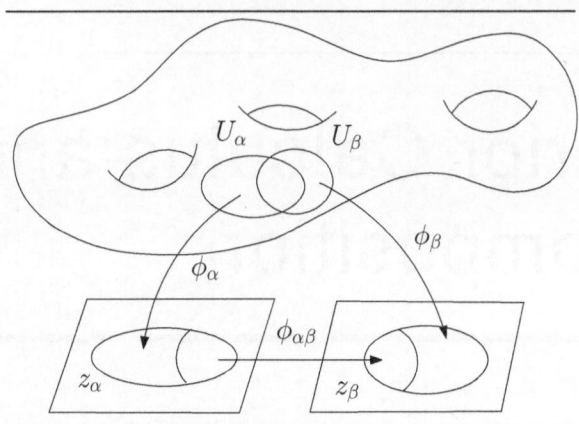

Figure 16.1 A manifold.

representation

$$\xi(x^1, x^2, \cdots, x^n) = \sum_{i=1}^{n} \xi_i(x^1, x^2, \cdots, x^n)\frac{\partial}{\partial x_i}.$$

$\{\frac{\partial}{\partial x_i}\}$ represents the vector fields of the velocities of iso-parametric curves on M. They form a basis of all vector fields.

Definition 16.1.3 (Push-forward) *Suppose $\phi : M \to N$ is a differential map from M to N, $\gamma : (-\epsilon, \epsilon) \to M$ is a curve, $\gamma(0) = p$, $\gamma'(0) = \mathbf{v} \in T_pM$, then $\phi \circ \gamma$ is a curve on N, $\phi \circ \gamma(0) = \phi(p)$, we define the tangent vector*

$$\phi_*(\mathbf{v}) = (\phi \circ \gamma)'(0) \in T_{\phi(p)}N,$$

as the push-forward tangent vector of $\mathbf{v}$ induced by ϕ.

Definition 16.1.4 (Differential 1-form) *The tangent space T_pM is an n-dimensional vector space, its dual space T_p^*M is called the cotangent space of M at p. Suppose $\omega \in T_p^*M$, then $\omega : T_pM \to \mathbb{R}$ is a linear function defined on T_pM, ω is called a* differential 1-form *at p.*

A differential 1-form field has the local representation

$$\omega(x^1, x^2, \cdots, x^n) = \sum_{i=1}^{n} \omega_i(x^1, x^2, \cdots, x^n)dx_i,$$

where $\{dx_i\}$ are the differential forms dual to $\{\frac{\partial}{\partial x_j}\}$, such that

$$\langle dx_i, \frac{\partial}{\partial x_j}\rangle = dx_i\left(\frac{\partial}{\partial x_j}\right) = \delta_{ij},$$

here δ_{ij} is the Kronecker symbol. High order exterior differential forms can be defined based on tangent vectors and differential 1-forms.

Definition 16.1.5 (Tensor) *A tensor Θ of type (m, n) on a manifold M is a correspondence that associates to each point $p \in M$ a multi-linear map*

$$\Theta_p : \underbrace{T_pM \times T_pM \cdots \times T_pM}_{m} \times \underbrace{T_p^*M \times T_p^*M \cdots \times T_p^*M}_{n} \to \mathbb{R},$$

where the tangent space T_pM appears m times and the cotangent space T_p^*M appears n times.

Definition 16.1.6 (exterior m-form) *An exterior m-form is a tensor ω of type $(m, 0)$, which is skew symmetric in its arguments, namely*

$$\omega_p(\xi_{\sigma(1)}, \xi_{\sigma(2)}, \cdots, \xi_{\sigma(m)}) = (-1)^{|\sigma|}\omega_p(\xi_1, \xi_2, \cdots, \xi_m)$$

for any tangent vectors $\xi_1, \xi_2, \cdots, \xi_m \in T_pM$ and any permutation $\sigma \in S_m$, where S_m is the permutation group, $|\sigma|$ is 0 if σ is an even permutation, and 1 if σ is an odd permuation.

The linear space of all m-forms on a manifold Σ is denoted as $\Omega^m(\Sigma)$. The local representation of a differential form ω in $(x^1, x^2, \cdots, x^m)$ is

$$\omega = \sum_{1 \leq i_1 < i_2 < \cdots < i_m \leq n} \omega_{i_1 i_2 \cdots i_m} dx^{i_1} \wedge dx^{i_2} \wedge \cdots \wedge dx^{i_m} = \omega_I dx^I,$$

ω_I is a function of the reference point p, ω is said to be differentiable, if each ω_I is differentiable.
The high order differential forms can be obtained by the wedge product of differential 1-forms.

Definition 16.1.7 (Wedge product) *A coordinate free representation of wedge product of m_1-form ω_1 and m_2-form ω_2 is defined as $(\omega_1 \wedge \omega_2)(\xi_1, \xi_2, \cdots, \xi_{m_1+m_2})$ equals*

$$\sum_{\sigma \in S_{m_1+m_2}} \frac{(-1)^{|\sigma|}}{m_1! m_2!} \omega_1\left(\xi_{\sigma(1)}, \cdots, \xi_{\sigma(m_1)}\right) \omega_2\left(\xi_{\sigma(m_1+1)}, \cdots, \xi_{\sigma(m_1+m_2)}\right)$$

Given k differential 1-forms, their exterior wedge product is given by:

$$\omega_1 \wedge \omega_2 \cdots \wedge \omega_k(v_1, v_2, \cdots, v_k) = \begin{vmatrix} \omega_1(v_1) & \omega_1(v_2) & \cdots & \omega_1(v_k) \\ \omega_2(v_1) & \omega_2(v_2) & \cdots & \omega_2(v_k) \\ \vdots & \vdots & \ddots & \vdots \\ \omega_k(v_1) & \omega_k(v_2) & \cdots & \omega_k(v_k) \end{vmatrix}$$

Exterior differential form is anti-symmetric, suppose $\sigma \in S_k$ is a permutation, then

$$\omega_{\sigma(1)} \wedge \omega_{\sigma(2)} \wedge \cdots \wedge \omega_{\sigma(k)} = (-1)^{|\sigma|}\omega_1 \wedge \omega_2 \wedge \cdots \wedge \omega_k.$$

A smooth map between manifolds can pull back a differential form defined on the target to a differential form on the source.

Definition 16.1.8 (Pull back) *Suppose $\phi : M \to N$ is a differentiable map from M to N, ω is an m-form on N, then the pull-back $\phi^*\omega$ is an m-form on M defined by*

$$(\phi^*\omega)_p(\xi_1, \cdots, \xi_m) = \omega_{\phi(p)}(\phi_*\xi_1, \cdots, \phi_*\xi_m), \quad p \in M$$

for $\xi_1, \xi_2, \cdots, \xi_m \in T_pM$, where $\phi_*\xi_j \in T_{\phi(p)}N$ is the push forward of $\xi_j \in T_pM$.

Differential forms can be integrated on manifolds. First, we define how to integrate a differential form in the Euclidean space. Suppose that $U \subset \mathbb{R}^n$ is an open set,

$$\omega = f(x)dx^1 \wedge dx^2 \wedge \cdots \wedge dx^n,$$

then

$$\int_U \omega := \int_U f(x)dx^1dx^2 \cdots dx^n.$$

The we generalize the integration of a differential form on an open set on a manifold. Suppose $U \subset M$ is an open set of a manifold M, a chart $\phi : U \to \Omega \subset \mathbb{R}^n$, then

$$\int_U \omega := \int_\Omega (\phi^{-1})^*\omega.$$

Integration is independent of the choice of the charts. Let $\psi : U \to \psi(U)$ be another chart, with local coordinates $(u_1, u_2, \cdots, u_n)$, then

$$\int_{\phi(U)} f(x)dx^1dx^2 \cdots dx^n = \int_{\psi(U)} f(x(u))det\left(\frac{\partial x^i}{\partial u^j}\right) du^1du^2 \cdots du^n.$$

Now we can define the integration of a differential form on the whole manifold by means of partition of unity. Consider a covering of M by coordinate charts $\{(U_\alpha, \phi_\alpha)\}$ and choose a partition of unity $\{f_i\}$, $i \in I$, such that $f_i(p) \geq 0$,

$$\sum_i f_i(p) \equiv 1, \forall p \in M.$$

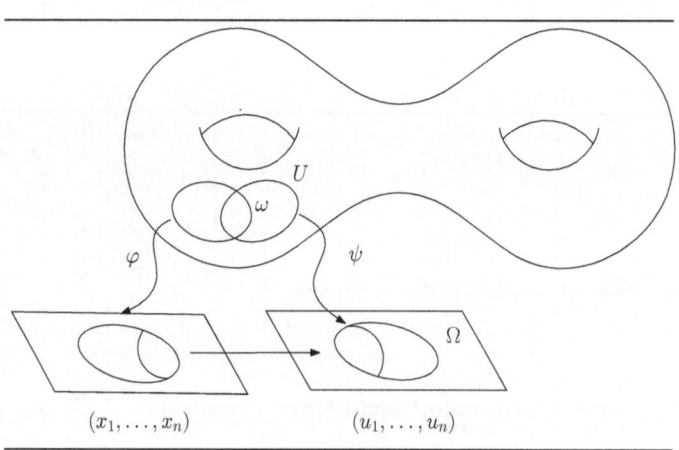

Figure 16.2 Integration of a differential form is independent of the choice of chart on a manifold.

Then $\omega_i = f_i \omega$ is an n-form on M with compact support in some U_α, we can set the integration as

$$\int_M \omega = \sum_i \int_M \omega_i.$$

16.2 DE RHAM COHOMOLOGY

Now we can define the exterior derivative of differential forms.

Definition 16.2.1 (Exterior Derivative) *Suppose $f : M \to \mathbb{R}$ is a differentiable function, then the exterior derivative of f is a 1-form,*

$$df = \sum_i \frac{\partial f}{\partial x_i} dx^i.$$

The exterior derivative of an m-form on M is an $(m+1)$-form on M defined in local coordinates by

$$d\omega = d(\omega_I dx^I) = (d\omega_I) \wedge dx^I,$$

where $d\omega_I$ is the differential of the function ω_I.

In a local chart, the exterior derivative of a differential 1-form is given by:

$$d\left(\sum \omega_i dx_i\right) = \sum_{i,j} \left(\frac{\partial \omega_j}{\partial x_i} - \frac{\partial \omega_i}{\partial x_j}\right) dx_i \wedge dx_j.$$

The exterior derivative of a differential k-form can be deduced by exterior derivatives of 1-forms:

$$d(\omega_1 \wedge \omega_2 \cdots \wedge \omega_k) = \sum (-1)^{i-1} \omega_1 \wedge \cdots \wedge \omega_{i-1} \wedge d\omega_i \wedge \omega_{i+1} \wedge \cdots \wedge \omega_k.$$

In general cases, for $\alpha \in \Omega^k(M)$ and $\beta \in \Omega^l(M)$,

$$d(\alpha \wedge \beta) = d\alpha \wedge \beta + (-1)^k \alpha \wedge d\beta, \tag{16.2.1}$$

This is called the *Leibniz Rule*. The classical *Stokes theorem* plays a fundamental role in geometry and topology.

Theorem 16.2.2 (Stokes) *Let M be an n-manifold with the boundary ∂M and ω be a differentialble $(n-1)$-form with compact support on M, then*

$$\int_{\partial M} \omega = \int_M d\omega.$$

The proof can be reduced to the classical Stokes theorem in Euclidean spaces.

Corollary 16.2.3 *Suppose Σ is a differential manifold, then the composition of the exterior differential operators, $d^{k-1} : \Omega^{k-1}(\Sigma) \to \Omega^k(\Sigma)$ and $d^k : \Omega^k(\Sigma) \to \Omega^{k+1}(\Sigma)$, is zero,*

$$d^k \circ d^{k-1} = 0.$$

Proof Assume ω is a $(k-1)$ differential form, σ is a $(k+1)$ chain, from Stokes theorem, we have

$$\int_\sigma d^k \circ d^{k-1}\omega = \int_{\partial_{k+1}\sigma} d^{k-1}\omega = \int_{\partial_k \circ \partial_{k+1}} \omega = 0,$$

since $\partial_k \circ \partial_{k+1} = 0$.

$\square$

Let $\Omega^k(\Sigma)$ be the space of all differential k-forms, $d^k : \Omega^k(\Sigma) \to \Omega^{k+1}(\Sigma)$ be exterior differential operator.

Definition 16.2.4 (Closed form) *A k-form $\omega \in \Omega^k(\Sigma)$ is called a closed form, if $d^k\omega = 0$, namely $\omega \in ker\, d^k$.*

Definition 16.2.5 (Exact Form) *A k-differential form $\omega \in \Omega^k(\Sigma)$ is called an exact form, if there is a $(k-1)$-form $\tau \in \Omega^{k-1}(\Sigma)$, such that $\omega = d^{k-1}\tau$, namely $\omega \in img\, d^{k-1}$.*

Since $d^k \circ d^{k-1} = 0$, exact forms are closed, $img\, d^{k-1} \subset ker\, d^k$.

Definition 16.2.6 (de Rham Cohomology) *Assume Σ is a differential manifold, then the de Rham complex is*

$$\Omega^0(\Sigma, \mathbb{R}) \xrightarrow{d^0} \Omega^1(\Sigma, \mathbb{R}) \xrightarrow{d^1} \Omega^2(\Sigma, \mathbb{R}) \xrightarrow{d^2} \Omega^3(\Sigma, \mathbb{R}) \xrightarrow{d^3} \cdots$$

The k-dimensional de Rham cohomology of Σ is defined as

$$H_{dR}^k(\Sigma, \mathbb{R}) := \frac{ker\, d^k}{img\, d^{k-1}}.$$

The de Rham cohomology group of a smooth manifold is isomorphic to its simplicial cohomology group.

Theorem 16.2.7 *The de Rham cohomology group $H_{dR}^m(\Sigma)$ is isomorphic to the simplicial cohomology group $H^m(\Sigma, \mathbb{R})$*

$$H_{dR}^m(\Sigma) \cong H^m(\Sigma, \mathbb{R}).$$

In practice, we use simplical cohomology groups to approximate corresponding de Rham cohomology groups. If the triangulations satisfy special conditions, then the simplicial approximations converge to the smooth counter parts.

16.3 HODGE STAR OPERATOR

Suppose M is an n-dimensional Riemannian manifold with a Riemannian metric **g**, we can locally find oriented orthonormal basis of vector fields, and choose parameterization, such that

$$\left\{ \frac{\partial}{\partial x_1}, \frac{\partial}{\partial x_2}, \cdots, \frac{\partial}{\partial x_n} \right\}$$

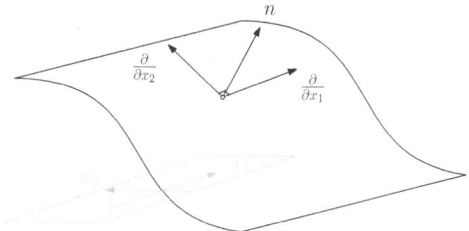

Figure 16.3 Orthonormal frame field.

form an oriented orthonormal basis, $\langle \partial/\partial x_i, \partial/\partial x_j \rangle_{\mathbf{g}} = \delta_{ij}$. Let

$$\{dx_1, dx_2, \cdots, dx_n\}$$

be the dual 1-form basis, $\langle dx_i, \partial/\partial x_j \rangle = \delta_{ij}$.

Definition 16.3.1 (Hodge Star Operator) *The Hodge star operator* $\star : \Omega^k(M) \to \Omega^{n-k}(M)$ *is a linear operator*

$$\star(dx_1 \wedge dx_2 \wedge \cdots \wedge dx_k) = dx_{k+1} \wedge dx_{k+2} \wedge \cdots \wedge dx_n. \qquad (16.3.1)$$

Let $\sigma = (i_1, i_2, \cdots, i_n)$ be a permutation $\sigma \in S_n$, then the Hodge star operator

$$\star(dx_{i_1} \wedge dx_{i_2} \wedge \cdots \wedge dx_{i_k}) = (-1)^{|\sigma|} dx_{i_{k+1}} \wedge dx_{i_{k+2}} \wedge \cdots \wedge dx_{i_n},$$

where $(-1)^{|\sigma|}$ is $+1$ if σ is an even permutation; $(-1)^{|\sigma|}$ is -1 if σ is an odd permutation.

Definition 16.3.2 (L^2-norm of k-forms) *Let* $\eta, \zeta \in \Omega^k(M)$ *are two k-forms on M, then the norm is defined as*

$$(\eta, \zeta) = \int_M \eta \wedge \star \zeta.$$

$\Omega^k(M)$ with the L^2-norm is a Hilbert space. Equivalently, we can define the Hodge star operator under local natural coordinates. Suppose the Riemannian metric $\mathbf{g}$ has a matrix presentation (g_{ij}), which gives the inner product in the tangent space T_pM,

$$g_{ij} = \langle \partial/\partial x_i, \partial/\partial x_j \rangle_{\mathbf{g}}.$$

Its inverse matrix is (g^{ij}), satisfies

$$\sum_{j=1}^{n} g_{ij} g^{jk} = \delta_i^k.$$

Definition 16.3.3 (Dual Inner Product) *Given an n-dimensional Riemannian manifold $(M, \mathbf{g})$, the dual inner product $\langle,\rangle_{\mathbf{g}} : T_p^*(M) \times T_p^*(M) \to \mathbb{R}$ is defined as follows: $\forall \omega, \eta \in T_p^*(M)$, $\omega = \sum_{i=1}^n \omega_i dx^i$, $\eta = \sum_{i=1}^n \eta_i dx^i$, then*

$$\langle \omega, \eta \rangle_{\mathbf{g}} = \sum_{i,j=1}^n g^{ij} \omega_i \eta_j. \tag{16.3.2}$$

Because (g^{ij}) is symmetric, the dual inner product is commutative:

$$\langle \omega, \eta \rangle_{\mathbf{g}} = \langle \eta, \omega \rangle_{\mathbf{g}}. \tag{16.3.3}$$

Let $\{\theta_1, \theta_2, \cdots, \theta_n\}$ is a set of orthonormal basis

$$\langle \theta_i, \theta_j \rangle_{\mathbf{g}} = \delta_i^j.$$

We use $\{\theta_i\}$ to construct the basis of $\Omega^k(M)$,

$$\Omega^k(M) := \mathrm{Span}\{\theta_{i_1} \wedge \theta_{i_2} \wedge \cdots \wedge \theta_{i_k} | i_1 < i_2 < \cdots < i_k\}.$$

We generalize the dual inner product to k-differential forms $\langle,\rangle_{\mathbf{g}} : \Omega^k(M) \times \Omega^k(M)$ as follows:

$$\langle \theta_{i_1} \wedge \cdots \wedge \theta_{i_k}, \theta_{j_1} \wedge \cdots \wedge \theta_{j_k} \rangle_{\mathbf{g}} = \delta_{i_1 \cdots i_k}^{j_1 \cdots j_k},$$

where the multi-index Kronecker delta is defined as

$$\delta_{i_1 \cdots i_k}^{j_1 \cdots j_k} = \sum_{\sigma \in S_k} (-1)^{|\sigma|} \delta_{i_{\sigma(1)}}^{j_1} \cdots \delta_{i_{\sigma(k)}}^{j_k} = \sum_{\sigma \in S_k} (-1)^{|\sigma|} \delta_{i_1}^{j_{\sigma(1)}} \cdots \delta_{i_k}^{j_{\sigma(k)}},$$

namely, if $j_1 \cdots j_k$ is an even (odd) permutation of $i_1 \cdot i_k$, then the multi-index Kronecker delta equals to $+1$ (-1); otherwise, the multi-index Kronecker delta equals to 0. Let $G = \det(g_{ij})$, then in the local coordinates, the Riemannian volume element is defined as

$$\omega_{\mathbf{g}} = \sqrt{G} dx^1 \wedge dx^2 \wedge \cdots \wedge dx^n. \tag{16.3.4}$$

By using the dual inner product operator and the Riemannian volume element, the Hodge Star Operator $,\star : \Omega^k(M) \to \Omega^{n-k}(M)$, can be equivalently defined as

$$\omega \wedge \star \eta = \langle \omega, \tau \rangle_{\mathbf{g}} \omega_{\mathbf{g}}. \tag{16.3.5}$$

Therefore,

$$\star(1) = \omega_{\mathbf{g}}, \quad \star \omega_{\mathbf{g}} = 1.$$

Suppose ζ and η are differential k-forms, $0 \le k \le n$, their L^2-norm can be written as

$$(\zeta, \eta) := \int_M \zeta \wedge \star \eta = \int_M \langle \zeta, \eta \rangle_{\mathbf{g}} \omega_{\mathbf{g}}. \tag{16.3.6}$$

Because the dual inner product is commutative Eqn. (16.3.3), the L^2 norm of differential forms is symmetric,

$$(\zeta, \eta) = \int_M \langle \zeta, \eta \rangle_{\mathbf{g}} \omega_{\mathbf{g}} = \int_M \langle \eta, \zeta \rangle_{\mathbf{g}} \omega_{\mathbf{g}} = (\eta, \zeta). \tag{16.3.7}$$

The Hodge star operator has *linearity* property, for $\lambda_1, \lambda_2 \in \mathbb{R}$ and $\omega_1, \omega_2 \in \Omega^k(M)$,

$$\star(\lambda_1\omega_1 + \lambda_2\omega_2) = \lambda_1 \star \omega_1 + \lambda_2 \star \omega_2. \tag{16.3.8}$$

Hodge start operator is commutative with 0-forms, $f \in \Omega^0(M)$ and $\omega \in \Omega^k(M)$,

$$\star(f\omega) = f(\star\omega). \tag{16.3.9}$$

The double Hodge start operator satisfies the property: for any $\omega \in \Omega^k(M)$,

$$\star(\star\omega) = (-1)^{k(n-k)}\omega. \tag{16.3.10}$$

This induces the definition of the inverse of Hodge star operator on a Riemannian manifold,

$$\star^{-1} = (-1)^{k(n-k)} \star. \tag{16.3.11}$$

In the following, we examine the Hodge star operator on a metric surface using the above two definitions.

Definition 16.3.4 (Isothermal Coordinates) *Suppose $(S, \mathbf{g})$ is a surface with a Riemannian metric, local coordinates (u, v) are called* isothermal, *if the metric can be represented as*

$$\mathbf{g} = e^{2\lambda(u,v)}(du^2 + dv^2),$$

where $\lambda : S \to \mathbb{R}$ is a function, and called the conformal factor.

Theorem 16.3.5 (Isothermal Coordinates) *Suppose $(S, \mathbf{g})$ is a surface with a Riemannian metric, for each point $p \in S$, there is a neighborhood $U(p)$ with an isothermal coordiantes on it.*

S.Chern gave an elementary proof for the existence of isothermal coordinates on surfaces [84]. Under the isothermal coordiantes, the Riemannian volume element is given by

$$\omega_{\mathbf{g}} = e^{2\lambda}du \wedge dv.$$

We construct an orthonormal basis of vector fields

$$\frac{\partial}{\partial x_1} = e^{-\lambda}\frac{\partial}{\partial u}, \quad \frac{\partial}{\partial x_2} = e^{-\lambda}\frac{\partial}{\partial v},$$

and the dual differential 1-forms are

$$dx_1 = e^{\lambda}du, \quad dx_2 = e^{\lambda}dv.$$

By the definition of Hodge Star Eqn. (16.3.1),

$$\star dx_1 = dx_2, \quad \star du = dv$$
$$\star dx_2 = -dx_1, \quad \star dv = -du$$

$$\star(1) = dx_1 \wedge dx_2 = e^{2\lambda}du \wedge dv, \quad \star(dx_1 \wedge dx_2) = 1.$$

Given a 1-forms $\omega = \omega_1 du + \omega_2 dv$ and $\tau = \tau_1 du + \tau_2 dv$, its wedge product is

$$\omega \wedge \tau = (\omega_1 \tau_2 - \omega_2 \tau_1) du \wedge dv.$$

The dual inner product is

$$\langle \omega, \tau \rangle_{\mathbf{g}} = \sum_{i,j=1}^{2} g^{ij} \omega_i \tau_j = e^{-2\lambda(u,v)}(\omega_1 \tau_1 + \omega_2 \tau_2).$$

By the second definition of Hodge star Eqn. (16.3.5), we have

$$(\omega_1 du + \omega_2 dv) \wedge \star du = \langle \omega, du \rangle_{\mathbf{g}} \omega_{\mathbf{g}} = (e^{-2\lambda}\omega_1)(e^{2\lambda} du \wedge dv) = \omega_1 du \wedge dv$$
$$(\omega_1 du + \omega_2 dv) \wedge \star dv = \langle \omega, dv \rangle_{\mathbf{g}} \omega_{\mathbf{g}} = (e^{-2\lambda}\omega_2)(e^{2\lambda} du \wedge dv) = \omega_2 du \wedge dv$$

This shows $\star du = dv$, similarly $\star dv = -du$. Therefore, in general cases,

$$\star(\omega_1 du + \omega_2 dv) = \omega_1 dv - \omega_2 du. \tag{16.3.12}$$

By direct computation, we obtain

$$\star \star \omega = -\omega. \tag{16.3.13}$$

Figure 16.4 visualizes the Hodge star operator. We have two electric charge on a square, the level sets of potential are show as blue curves, the electric field lines (integration curves of gradients of potentials) are red curves. These two families of curves are orthogonal to each other at every point. In fact, at each point, by rotating the tangent vector of the red curve by an angle $\frac{\pi}{2}$ we can obtain that of the blue curve.

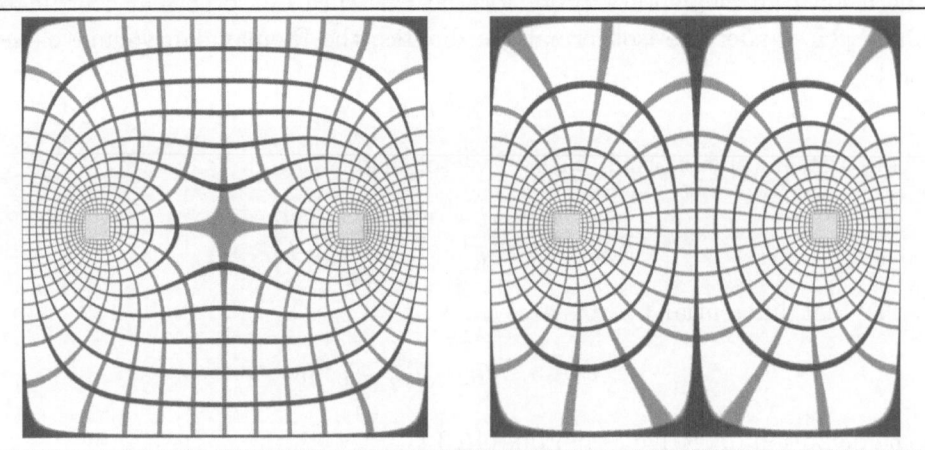

Figure 16.4 Hodge star operator.

16.4 HODGE DECOMPOSITION

Definition 16.4.1 *Suppose $(M, \mathbf{g})$ is an n-dimensional Riemannian manifold, the codifferential operator $\delta : \Omega^k(M) \to \Omega^{k-1}(M)$ is defined as*

$$\delta = (-1)^k \star^{-1} d\star, \qquad (16.4.1)$$

where $d : \Omega^{k-1}(M) \to \Omega^k(M)$ is the exterior derivative, $\star : \Omega^i(M) \to \Omega^{n-i}(M)$ the Hodge star operator, $\star^{-1}$ is given by Eqn. (16.3.11).

We can further expand the definition of δ in Eqn. (16.4.1), by Eqn. (16.3.11),

$$\star^{-1}d\star = (-1)^{(n-k+1)(n-(n-k+1))} \star d\star$$
$$= (-1)^{nk+n+k+1} \star d\star$$

where we use the fact $k^2 \equiv k \mod 2$, therefore we obtain another formula of the codifferential operator

$$\delta = (-1)^{kn+n+1} \star d \star . \qquad (16.4.2)$$

The co-differential operator has the niloponent property,

$$\delta \circ \delta = 0. \qquad (16.4.3)$$

This can be deduced directly from $d \circ d = 0$. The codifferential opeartor δ and the exterior differential operator d are adjoint to each other on closed manifolds.

Lemma 16.4.2 *Suppose $\partial M = \emptyset$, the codifferential operator δ is the adjoint of the exterior derivative operator d, in that*

$$(\delta\omega, \eta) = (\omega, d\eta). \qquad (16.4.4)$$

Proof We denote the grade of ω by $|\omega|$, and set $|\eta| = |\omega| - 1$. We also assume $\partial M = \emptyset$. Using

$$d(\eta \wedge \star\omega) = d\eta \wedge \star\omega + (-1)^{|\eta|}\eta \wedge d \star \omega,$$

we have

$$(d\eta, \omega) = \int_M d\eta \wedge \star\omega = \int_M d(\eta \wedge \star\omega) - (-1)^{|\omega|-1} \int_M \eta \wedge d \star \omega$$
$$= \int_{\partial M} \eta \wedge \star\omega - (-1)^{|\omega|-1} \int_M \eta \wedge d \star \omega$$
$$= (-1)^{|\omega|} \int_M \eta \wedge d \star \omega = (-1)^{|\omega|} \int_M \eta \wedge \star \star^{-1} d \star \omega$$
$$= (-1)^{|\omega|}(\eta, \star^{-1}d \star \omega) = (\eta, (-1)^{|\omega|} \star^{-1} d \star \omega)$$
$$= (\eta, \delta\omega).$$

Because $(d\eta, \omega) = (\omega, d\eta)$, and $(\eta, \delta\omega) = (\delta\omega, \eta)$, we obtain $(\delta\omega, \eta) = (\omega, d\eta)$.

$\square$

Definition 16.4.3 (Laplace-Beltrami Operator) *The Laplace-Beltrami operator* $\Delta : \Omega^k(M) \to \Omega^k(M)$ *is defined as:*

$$\Delta := d\delta + \delta d.$$

The *Hodge-Dirac* operator is defined as $d + \delta$. We can see

$$(d + \delta)^2 = d^2 + d\delta + \delta d + \delta^2 = d\delta + \delta d = \Delta.$$

Lemma 16.4.4 *The Laplace-Beltrami operator is symmetric*

$$(\Delta\zeta, \eta) = (\zeta, \Delta\eta) \tag{16.4.5}$$

and non-negative

$$(\Delta\eta, \eta) \geq 0. \tag{16.4.6}$$

Proof Because d and δ are adjoint to each other, for $\omega, \eta \in \Omega^k(M)$,

$$\begin{aligned}
(\Delta\omega, \eta) &= (d\delta\omega, \eta) + (\delta d\omega, \eta) \\
&= (\delta\omega, \delta\eta) + (d\omega, d\eta) \\
&= (\omega, \Delta\eta).
\end{aligned}$$

Furthermore

$$(\Delta\eta, \eta) = (\delta\eta, \delta\eta) + (d\eta, d\eta).$$

□

The kernel of the Laplace-Beltrami operator consists of harmonic differential forms.

Definition 16.4.5 (Harmonic forms) *Suppose $\omega \in \Omega^k(M)$, then ω is called a k-harmonic form, if*

$$\Delta\omega = 0.$$

Intuitively, a harmonic differential form is dual to a harmonic vector field, which is both curl-free and divergence-free.

Lemma 16.4.6 *ω is a harmonic form, if and only if*

$$d\omega = 0, \quad \delta\omega = 0. \tag{16.4.7}$$

Proof Suppose $\omega \in \Omega^k(M)$ is harmonic, then

$$0 = (\Delta\omega, \omega) = (d\omega, d\omega) + (\delta\omega, \delta\omega).$$

This shows both $d\omega = 0$ and $\delta\omega = 0$.

□

This shows the space of all harmonic forms is the intersection of the kernel of $d : \Omega^k(M) \to \Omega^{k+1}(M)$ and $\delta : \Omega : \Omega^k(M) \to \Omega^{k-1}(M)$, therefore they form a linear space.

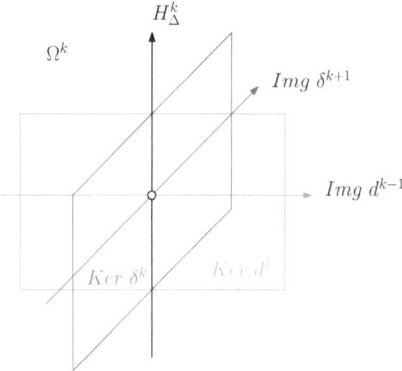

Figure 16.5 Hodge decomposition.

Definition 16.4.7 (Harmonic form group) *All harmonic k-forms form a group, the so-called harmonic form group, denoted as $H_\Delta^k(M)$.*

The Hodge decomposition theorem plays a fundamental role in differential geometry and topology.

Theorem 16.4.8 (Hodge Decomposition) *Suppose M is a n-dimensional compact orientable Riemannian manifold, then we have the following decomposition*

$$\Omega_k(M) = img\ d^{k-1} \oplus img\ \delta^{k+1} \oplus H_\Delta^k(M). \qquad (16.4.8)$$

for any $0 \le k \le n$.

Namely, if $\partial M = \emptyset$, then $\forall \omega \in \Omega^k(M)$, $\exists \alpha \in \Omega^{k-1}(M)$, $\beta \in \Omega^{k+1}(M)$, $\gamma \in \Omega^k(M)$, $d\gamma = 0$, $\delta\gamma = 0$, such that

$$\omega = d\alpha + \delta\beta + \gamma. \qquad (16.4.9)$$

In order to prove the Hodge decomposition theorem, we need some theoretical tools from the classical PDE literature, more details can be found in [333].

Lemma 16.4.9 (Compactness of Laplace-Beltrami Operator) *Let $\{\omega_k\}$ be a sequence in $\Omega_p(M)$ such that $|\omega_k| < C$ and $|\Delta\omega_k| < C$ hold for all k and for some $C > 0$. Then there exists a Cauchy subsequence $\{\omega_{k_n}\}$ of $\{\omega_k\}$ in $\Omega_p(M)$.*

Definition 16.4.10 (Weak Solution to Laplacian Equation) *Suppose $l : \Omega_k(M) \to \Omega_k(M)$ is a bounded linear functional on $\Omega_k(M)$, which satisfies*

$$l(\Delta\gamma) = \langle \alpha, \gamma \rangle$$

for any $\gamma \in \Omega_k(M)$, then we call it a weak solution *to $\Delta\omega = \alpha$.*

The following regularity theorem of Laplace operator claims that if there is a weak solution to the Laplacian equation $\Delta\omega = \alpha$, then there must be a classical solution.

Theorem 16.4.11 (Regularity of Laplace Operator) *Let* $\alpha \in \Omega_k(M)$ *and a bounded linear functional* $l : \Omega_k(M) \to \Omega_k(M)$ *be a weak solution to* $\Delta\omega = \alpha$, *then there exists a* $\omega \in \Omega_k(M)$ *such that* $l(\beta) = \langle \omega, \beta \rangle$ *for any* $\beta \in \Omega_k(M)$. *Consequently,* $\Delta\omega = \alpha$.

Lemma 16.4.12 *The group of harmonic forms* $ker(\Delta)$ *is finite dimensional.*

Proof Suppose $\dim(\ker(\Delta)) = \infty$, then we can find orthonormal basis $\{\alpha_k\}$ of $\ker(\Delta)$. By Lemma of compactness of Laplace-Beltrami operator 16.4.9, there exists a Cauchy subsequence $\{\alpha_{k_n}\}$ of $\{\alpha_k\}$. But this is impossible, since all the mutual distance between each pair of α_k's is $\sqrt{2}$.

$\square$

Lemma 16.4.13 *The following inequality holds on* $ker(\Delta)^\perp$: $\exists C > 0$, *such that*

$$|\gamma| < C|\Delta\gamma|, \quad \forall \gamma \in ker(\Delta)^\perp. \tag{16.4.10}$$

Proof If not, then there exists a sequence $\{\gamma_j\}$ in $\ker(\Delta)^\perp$ with $|\gamma_j| = 1$ and $|\Delta\gamma_j| < \frac{1}{j}$. Therefore by Lemma 16.4.9, a subsequence of γ_j is Cauchy. Denote $\bar{\gamma}$ as its limit. Because $\ker(\Delta)$ is closed, its orthogonal complement $\ker(\Delta)^\perp$ is also closed in Hilbert space. Since $\ker(\Delta)^\perp$ is closed, we have $|\bar{\gamma}| = 1$ and $\bar{\gamma} \in \ker(\Delta)^\perp$. Define a functional f on $\Omega_k(M)$ by $f(\omega) = \langle \bar{\gamma}, \omega \rangle$. Then f is a bounded linear functional. And by passing to this subsequence, we have

$$f(\Delta\omega) = \langle \bar{\gamma}, \Delta\omega \rangle = \lim_{j \to \infty} \langle \gamma_j, \Delta\omega \rangle = \lim_{j \to \infty} \langle \Delta\gamma_j, \omega \rangle = 0.$$

Therefore, f is a weak solution to $\Delta\omega = 0$, which implies $\Delta\bar{\gamma} = 0$ by theorem 16.4.11. But $|\bar{\gamma}| = 1$, $\bar{\gamma} \in \ker(\Delta)$ contradicts with $\ker(\Delta)^\perp \cap \ker(\Delta) = 0$.

$\square$

Lemma 16.4.14 $ker(\Delta)^\perp \subset img(\Delta)$.

Proof For any $\alpha \in \ker(\Delta)^\perp$, we define a linear functional l_α on $\text{img}(\Delta)$ by

$$l_\alpha(\gamma) := \langle \alpha, \beta \rangle,$$

where $\gamma = \Delta\beta$, then l_α is a well-defied linear functional on $\text{img}(\Delta)$.

Now we show the boundedness of l_α on $\text{img}(\Delta)$. For any $\beta \in \Omega_k(M)$, we can decompose β as $\beta = \beta_1 + \beta_2$, $\beta_1 \in \ker(\Delta)$, $\beta_2 \in \ker(\Delta)^\perp$, (Eqn. 16.4.11 ensures the existence of such a decomposition) then

$$|l_\alpha(\Delta\beta)| = |\langle \alpha, \beta \rangle| = |\langle \alpha, \beta_1 \rangle + \langle \alpha, \beta_2 \rangle| = |\langle \alpha, \beta_2 \rangle|$$
$$\leq |\alpha||\beta_2| \leq C|\alpha||\Delta\beta_2|$$
$$= C|\alpha||\Delta\beta|$$

where the first equality is the definition, the second equality comes from $\alpha \in \ker(\Delta)^{\perp}$, $\beta_1 \in \ker(\Delta)$, the first inequality follows from Cauchy-Schwarz inequality and the second inequality from lemma 16.4.13. Therefore, $|l_{\alpha}(\gamma)| \leq C|\alpha| \cdot |\gamma|$, $\gamma = \Delta\beta$, l_{α} is bounded on $\text{img}(\Delta)$.

By Hahn-Banach theorem, l_{α} on $\text{img}(\Delta)$ can be extended to a bounded linear functional on $\Omega_k(M)$, satisfying $\tilde{l}_{\alpha}(\Delta\beta) = \langle \alpha, \beta \rangle$ for all $\beta \in \Omega_k(M)$. Because $\tilde{l}_{\alpha}$ is bounded, it is a weak solution to $\Delta\omega = \alpha$, by theorem 16.4.11, there is a classical solution, therefore $\alpha \in \text{img}(\Delta)$, hence $\ker(\Delta)^{\perp} \subset \text{img}(\Delta)$.

□

Now we prove the Hodge decomposition theorem.

Proof The proof is divided into 4 major steps:

1. By lemma 16.4.12, $H_{\Delta}^k(M) = \ker(\Delta)$ is finite dimensional, which implies $H_{\Delta}^k(\Delta)$ is closed in the Hilbert space $\Omega_k(M)$. Therefore, we have the orthogonal decomposition

$$\Omega_k(M) = \ker(\Delta) \oplus \ker(\Delta)^{\perp}. \qquad (16.4.11)$$

2. We will prove that

$$\ker(\Delta)^{\perp} = \text{img}(\Delta).$$

The inclusion $\text{img}(\Delta) \subset \ker(\Delta)^{\perp}$ is obvious. To see that, we note that if $\Delta\omega = \beta$ and $\alpha \in \ker(\Delta)$ then

$$\langle \beta, \alpha \rangle = \langle \Delta\omega, \alpha \rangle = \langle \omega, \Delta\alpha \rangle = 0,$$

so $\beta \in \ker(\Delta)^{\perp}$. Lemma 16.4.14 proves the other inclusion $\text{img}(\Delta) \subset \ker(\Delta)^{\perp}$.

3. we will show

$$\text{img}(d^{k-1}) \perp \text{img}(\delta^{k+1}).$$

Given $\alpha \in \Omega_{k-1}(M)$ and $\beta \in \Omega_{k+1}(M)$, $(d\alpha, \delta\beta) = (d^2\alpha, \beta) = (\alpha, \delta^2\beta) = 0$.

4. We will show

$$\text{img}(d^{k-1}) \oplus \text{img}(\delta^{k+1}) = \ker(\Delta)^{\perp}.$$

Since $\Delta = d\delta + \delta d$, we have

$$\text{img}(\Delta) \subset \text{img}(d^{k-1}) \oplus \text{img}(\delta^{k+1}).$$

Suppose $\beta \in \ker(\Delta)$ is harmonic, by Eqn. (16.4.7), $\delta\beta = 0$ and $d\beta = 0$. For any $\alpha \in \Omega_{k-1}(M)$, $(d\alpha, \beta) = (\alpha, \delta\beta) = 0$, hence $\text{img}(d^{k-1}) \perp \ker(\Delta)$. Similarly, For any $\gamma \in \Omega_{k+1}(M)$, $(\delta\gamma, \beta) = (\gamma, d\beta) = 0$, hence $\text{img}(\delta^{k+1}) \perp \ker(\Delta)$. Therefore,

$$\text{img}(d^{k-1}) \oplus \text{img}(\delta^{k+1}) \subset \ker(\Delta)^{\perp}$$

This completes the proof of Hodge decomposition Eqn.(16.4.8).

□

Hodge decomposition implies the isomorphisms between the de Rham cohomology group and the group of harmonic differentials.

Theorem 16.4.15 *Given an n-dimensional compact orientable Riemannian manifold M, suppose $\omega \in \Omega^k(M)$ is a closed k-form, its harmonic component is $h(\omega)$, then the map:*

$$h : H_{dR}^k(M, \mathbb{R}) \to H_{\Delta}^k(M).$$

is isomorphic.

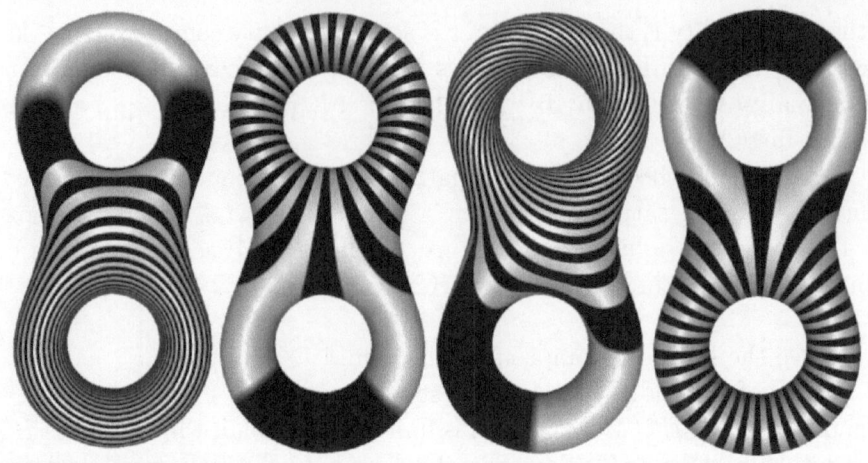

Figure 16.6 Harmonic 1-form group basis on a genus two surface.

Proof By Hodge decomposition

$$\ker d^k = \operatorname{img} d^{k-1} \oplus H_\Delta^k.$$

Suppose $\omega_i \in \ker d^k$, $i = 1, 2$, then $\omega_i = \alpha_i + \beta_i$, $\alpha_i \in \operatorname{img} d^{k-1}$, $\beta_i \in H_\Delta^k(M)$, $h(\omega_i) = \beta_i$. Furthermore, assume ω_1 and ω_2 are cohomologous, namely $\omega_2 - \omega_1 = d^{k-1}\eta$, for some $\eta \in \Omega^{k-1}(M)$, then

$$\omega_2 = \alpha_2 + \beta_2 = (\alpha_1 + d^{k-1}\eta) + \beta_1,$$

due to the uniqueness of the Hodge decomposition, we have $\alpha_2 = \alpha_1 + d^{k-1}\eta$ and $\beta_2 = \beta_1$. Therefore, $h(\omega_1) = h(\omega_2)$. Namely each cohomologous class has a unique harmonic form. Therefore, the mapping $h : H_{dR}^k \to H_\Delta^k$ is isomorphic.

$\square$

Figure 16.6 shows a basis of the harmonic 1-form group $H_\Delta^1(M)$ of a genus two closed surface. Any harmonic 1-form is a linear combination of these four base 1-forms. Intuitively, among all the vector fields within the same cohomologous class, the harmonic 1-form is the smoothest one.

16.5 DISCRETE HODGE THEORY

In this section, we discuss how to compute harmonic 1-forms on discrete polyhedral surfaces, namely triangle meshes.

Dual Mesh Figure 16.7 shows one example of primal mesh and dual mesh, which is produced by the conventional Delaunay triangulation and the Voronoi diagram on the plane.

Definition 16.5.1 (Delaunay Triangulation) *Suppose a distinct planar point set is given $P = \{p_1, p_2, \ldots, p_n, \ldots\}$, a triangulation T of P is called the* Delaunay

Figure 16.7 Mesh and dual mesh, Delaunay triangulation and Voronoi diagram.

triangulation *of P, if it satisfies the empty circle condition: for each triangle of T, the interior of its circum-circle doesn't contain any point in P.*

Definition 16.5.2 (Voronoi Diagram) *Suppose a distinct planar point set is given* $P = \{p_1, p_2, \ldots, p_n, \ldots\}$, *the cell decomposition D is called the* Voronoi diagram *of P,*

$$\mathbb{R}^2 = \bigcup_{p_i \in P} W(p_i),$$

where each $W(p_i)$ *is the Voronoi cell of* p_i,

$$W(p_i) = \{q \in \mathbb{R}^2 \mid |q - p_i| \leq |q - p_j|, \quad \forall p_j \in P\}.$$

The Delaunay triangulation T and the Voronoi diagram D are Poincaré dual to each other. Each face $f \in T$ is dual to a vertex $\bar{f} \in D$, which is the circum-center of the face; each vertex $v \in T$ is dual to a cell $\bar{v} \in D$, which is Voronoi cell of v; each edge $e \in T$ is dual to an edge $\bar{e} \in D$, which bisects e and connecting the circum-centers of the faces adjacent to e. The Voronoi diagram and the Delaunay triangulation can be generalized to the Riemannian surfaces, where the Euclidean distance is replaced by geodesic distance, the straight line segments by geodesic segments. A polyhedral

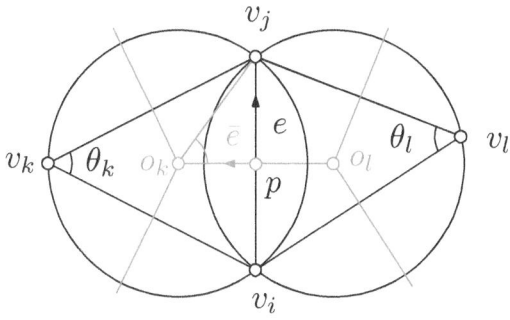

Figure 16.8 Cotangent edge weight.

surface has a flat Riemannian metric with cone singularities (vertices). The Voronoi diagram and the Delaunay triangulation of the vertices can be directly defined on the surface. Then we can assume the polyhedral surface is represented as a triangle mesh with the Delaunay triangulation, its Poincaré dual cell decomposition is the dual Voronoi diagram. Each Voronoi vertex is the circum-center of the dual triangle face.

Given a k-dimensional cell σ in the primal or dual mesh, we use $|\sigma|$ to represent the Hausdorff measure of σ, such as the edge length and the face area and so on. Especially, we define the measure of a 0-cell to be 1.

Lemma 16.5.3 *Given a Delaunay triangulation T and the Voronoi diagram D on a polyhedral surface, the ratio between the Voronoi edge $\bar{e}$ and the Delaunay edge e is*

$$\frac{|\bar{e}|}{|e|} = \frac{1}{2}(\cot\theta_k + \cot\theta_l) = w_e, \tag{16.5.1}$$

where w_e is called the cotangent edge weight *of edge e.*

Proof As shown in Figure 16.8, suppose a prime edge $e = [v_i, v_j]$ is shared with two faces $[v_i, v_j, v_k]$ and $[v_j, v_i, v_l]$, their circum-centers are o_k and o_l respectively. The dual edge is $\bar{e} = [o_l, o_k]$. The orientation is crucial, the orientation of the primal edge, the orientation of the dual edge and the normal of the surface satisfy the right-hand rule. The primal edge and the dual edge intersect at p. The inner angle of the left triangle at v_k is θ_k, the inner angle of the right triangle at v_l is θ_l. The angle $\angle v_j o_k p$ equals to θ_k, $\angle v_j o_l p$ equals to θ_l, therefore

$$\frac{|po_k|}{|pv_j|} = \cot\theta_k, \quad \frac{|po_l|}{|pv_i|} = \cot\theta_l.$$

Furthermore, $[o_l, o_k]$ is the bisector of $[v_i, v_j]$, therefore $|pv_i| = |pv_j|$, therefore

$$\frac{|\bar{e}|}{|e|} = \frac{|po_k| + |po_l|}{|pv_i| + |pv_j|} = \frac{1}{2}(\cot\theta_k + \cot\theta_l).$$

$\square$

Discrete Hodge Star Operator   By utilizing the Poincaré dual cell complex, we can generalize the Hodge star operator to the discrete setting.

Definition 16.5.4 (Discrete Hodge Star Operator) *Suppose M is an closed polyhedral surface with a triangulation (triangle mesh), $\bar{M}$ is the Poincaré dual cell complex (dual mesh), given a simplicial k-form $\omega \in C^k(M, \mathbb{R})$, its Hodge star is a simplicial $(2-k)$-form $\star\omega \in C^{2-k}(\bar{M}, \mathbb{R})$, such that for every k-simplex $\sigma \in C_k(M, \mathbb{Z})$ and its dual cell $\bar{\sigma} \in C_{n-k}(\bar{M}, \mathbb{Z})$,*

$$\frac{\omega(\sigma)}{|\sigma|} = \frac{\star\omega(\bar{\sigma})}{|\bar{\sigma}|}. \tag{16.5.2}$$

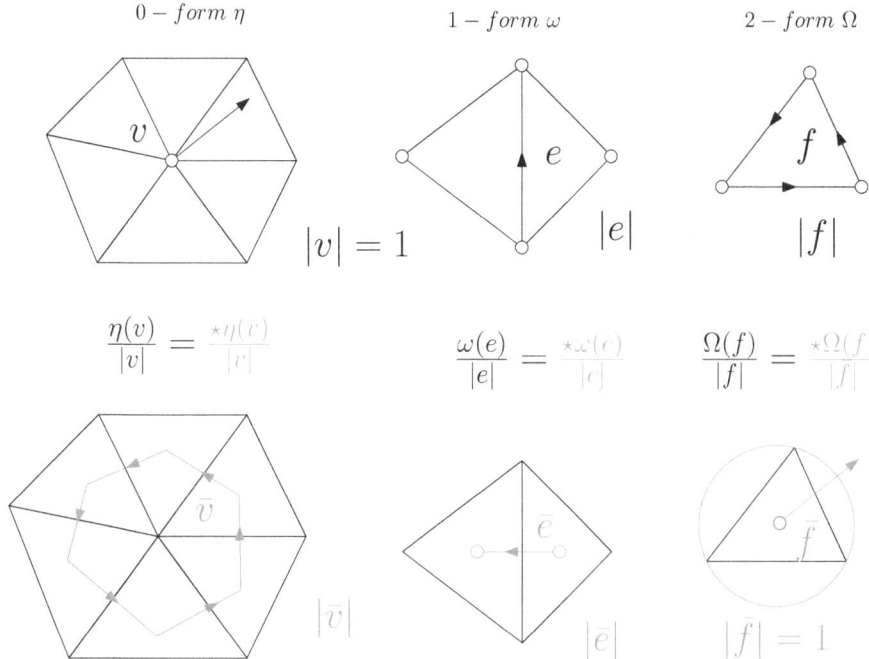

Figure 16.9 Discrete Hodge star operator.

Figure 16.9 shows the discrete Hodge star operator defined on the primal mesh and its dual mesh. Suppose $e = [v_i, v_j]$, then

$$\star\omega(\bar{e}) = \frac{1}{2}(\cot\theta_k + \cot\theta_l)\omega(e) = w_{ij}\omega(e),$$

where w_{ij} is the cotangent edge weight of the edge $[v_i, v_j]$. Suppose $\eta \in C^0(M, \mathbb{R})$ is a simplicial 0-form, then $\star\eta \in C^2(\bar{M}, \mathbb{R})$ is a simplicial 2-form on the dual mesh, for each vertex $v \in C_0(M, \mathbb{Z})$, its dual cell is $\bar{v} \in C_2(\bar{M}, \mathbb{Z})$, then $\star\eta(\bar{v}) = |\bar{v}|\eta(v)$. Similarly, suppose $\Omega \in C^2(M, \mathbb{R})$ then $\star\Omega \in C^0(\bar{M}, \mathbb{R})$, for each face $f \in C_2(\mathbb{Z})$, its dual vertex $\bar{f} \in C_0(\bar{M}, \mathbb{Z})$, then we have $\star\Omega(\bar{f}) = \Omega(f)/|f|$.

In the discrete setting, the properties of the Hodge star operator still holds:

$$\star^{-1}\star = 1, \quad \star^{-1} = (-1)^{k(n-k)}\star, \quad \star\star = (-1)^{k(n-k)} \tag{16.5.3}$$

Lemma 16.5.5 *Suppose* $\eta \in C^k(\bar{M}, \mathbb{R})$, *then* $\star\eta \in C^{n-k}(M, \mathbb{R})$, *then for each* $\sigma \in C_{n-1}(M, \mathbb{Z})$, $\bar{\sigma} \in C_k(\bar{M}, \mathbb{Z})$, *then*

$$\frac{\star\eta(\sigma)}{|\sigma|} = (-1)^{k(n-k)}\frac{\eta(\bar{\sigma})}{|\bar{\sigma}|}. \tag{16.5.4}$$

Proof Let $\omega = \star\eta$, then $\omega \in C^{n-k}(M, \mathbb{R})$, then

$$\star\omega = \star(\star\eta) = (-1)^{k(n-k)}\eta,$$

therefore

$$(-1)^{k(n-k)}\frac{\eta(\bar{\sigma})}{|\bar{\sigma}|} = \frac{\star\omega(\bar{\sigma})}{|\bar{\sigma}|} = \frac{\omega(\sigma)}{|\sigma|} = \frac{\star\eta(\sigma)}{|\sigma|}.$$

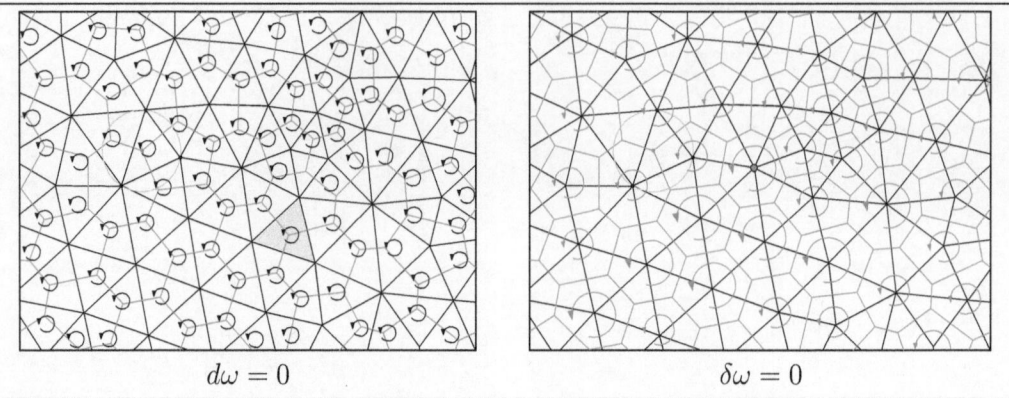

$$dw = 0 \qquad\qquad \delta w = 0$$

Figure 16.10 Discrete harmonic 1-form.

□

Furthermore, we can generalize the codifferential operator δ : $C^k(M, \mathbb{R}) \to C^{k-1}(M, \mathbb{R})$ to the discrete cases as

$$\delta = (-1)^k \star^{-1} d\star = (-1)^{kn+n+1} \star d \star. \qquad (16.5.5)$$

Discrete Hodge Decomposition   We prove the Hodge decomposition theorem in the discrete surface setting.

As shown in Figure 16.11, suppose $\Omega \in C^2(M, \mathbb{R})$ is a 2-form, then $\delta\Omega \in C^1(M, \mathbb{R})$ is a 1-form, $e = [v_i, v_j]$, $\bar{e} = [\bar{f}_k, \bar{f}_\Delta]$, sincer n and k are both equal to 2,

$$\delta\Omega(e) = (-1)^{nk+n+1}(\star d\star)\Omega(e) = (-1)(\star d\star)\Omega(e),$$

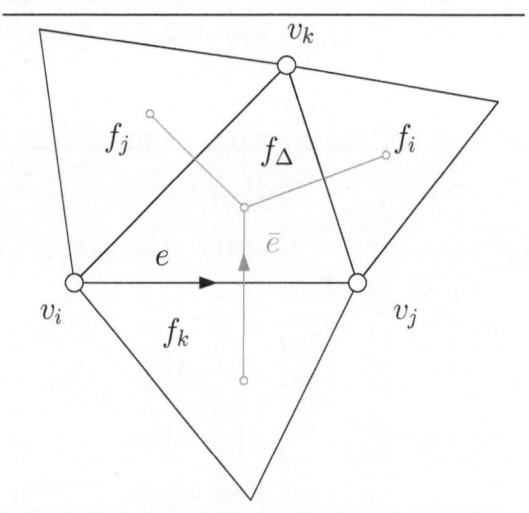

Figure 16.11 Discrete co-differential operator, δ : $C^2(M, \mathbb{R}) \to C^1(M, \mathbb{R})$.

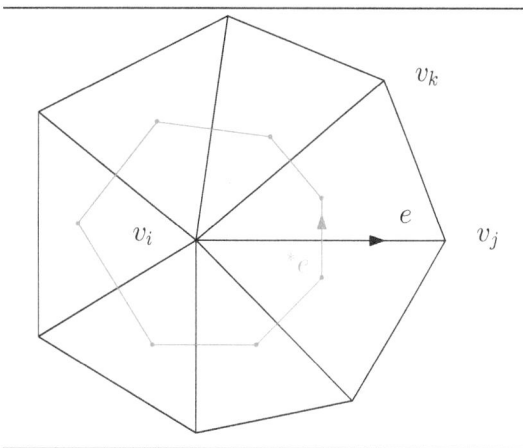

Figure 16.12 Discrete co-differential operator, $\delta : C^1(M, \mathbb{R}) \rightarrow C^0(M, \mathbb{R})$.

Since $(d \star \Omega)$ is a 1-form on the dual mesh, by Eqn. (16.5.4)

$$\star(d \star \Omega)(e) = (-1)\frac{|e|}{|\bar{e}|}(d \star \Omega)(\bar{e}) = \frac{-1}{w_{ij}}(d \star \Omega)(\bar{e})$$

Combine the above two equations, we obtain

$$\delta\Omega(e) = (-1)(\star d\star)\Omega(e) = \frac{1}{w_{ij}}(d \star \Omega)(\bar{e})$$

$$= \frac{1}{w_{ij}}(d \star \Omega)([\bar{f}_k, \bar{f}_\Delta]) = \frac{1}{w_{ij}}(\star\Omega)(\partial[\bar{f}_k, \bar{f}_\Delta])$$

$$= \frac{1}{w_{ij}}\left\{\star\Omega(\bar{f}_\Delta) - \star\Omega(\bar{f}_k)\right\} = \frac{1}{w_{ij}}\left\{\frac{\Omega(f_\Delta)}{|f_\Delta|} - \frac{\Omega(f_k)}{|f_k|}\right\}.$$

This gives the formula for the codifferential operator $\delta : C^2(M, \mathbb{R}) \rightarrow C^1(M, \mathbb{R})$

$$\delta\Omega([v_i, v_j]) = \frac{1}{w_{ij}}\left\{\frac{\Omega(f_\Delta)}{|f_\Delta|} - \frac{\Omega(f_k)}{|f_k|}\right\}. \tag{16.5.6}$$

As shown in Figure 16.12, suppose $\omega \in C^1(M, \mathbb{R})$ is a 1-form, then $\delta\omega \in C^0(M, \mathbb{R})$ is a 0-form. Sincer n and k equal to 2 and 1 respectively,

$$\delta\omega(v_i) = (-1)^{nk+n+1}(\star d\star)\omega(v_i) = (-1)(\star d\star)\omega(v_i),$$

Since $(d \star \omega)$ is a 2-form on the dual mesh, by Eqn. (16.5.4)

$$\star(d \star \Omega)(v_i) = \frac{|v_i|}{|\bar{v}_i|}(d \star \omega)(\bar{v}_i) = \frac{1}{|\bar{v}_i|}(d \star \omega)(\bar{v}_i)$$

Combine the above two equations, we obtain

$$\delta\omega(v_i) = (-1)(\star d\star)\omega(v_i) = \frac{-1}{|\bar{v}_i|}(d\star\omega)(\bar{v}_i)$$

$$= \frac{-1}{|\bar{v}_i|}(\star\omega)(\partial\bar{v}_i) = \frac{-1}{|\bar{v}_i|}\sum_{v_i\sim v_j}(\star\omega)(\bar{e}_{ij})$$

$$= \frac{-1}{|\bar{v}_i|}\sum_{v_i\sim v_j}\frac{|\bar{e}_{ij}|}{|e_{ij}|}\omega(e_{ij}) = \frac{-1}{|\bar{v}_i|}\sum_{v_i\sim v_j}w_{ij}\omega(e_{ij}).$$

This gives the formula for the codifferential operator $\delta : C^1(M,\mathbb{R}) \to C^0(M,\mathbb{R})$,

$$\delta\omega(v_i) = -\frac{1}{|\bar{v}_i|}\sum_{v_i\sim v_j}w_{ij}\omega([v_i, v_j]). \tag{16.5.7}$$

On smooth manifolds, a differential form ω is harmonic, if and only if $d\omega = 0$ and $\delta\omega = 0$, equivalently $d\star\omega = 0$. We generalize harmonic form to the discrete cases,

Definition 16.5.6 (Discrete Harmonic 1-form) *Suppose M is a closed triangle mesh, its dual mesh is $\bar{M}$. A simplicial 1-form $\omega \in C^1(M,\mathbb{R})$ is harmonic, if and only if*

$$d\omega = 0, \quad \delta\omega = 0.$$

As shown in Figure 16.10, suppose 1-form ω is harmonic, then for every face $f = [v_i, v_j, v_k]$, $d\omega(f) = 0$, namely

$$\omega([v_i, v_j]) + \omega([v_j, v_k]) + \omega([v_k, v_i]) = 0; \tag{16.5.8}$$

and for each vertex v_i, $\delta\omega(v_i) = 0$, equivalently

$$-\frac{1}{|\bar{v}_i|}\sum_{v_i\sim v_j}w_{ij}\omega([v_i, v_j]) = 0, \tag{16.5.9}$$

where w_{ij} is the cotangent edge weight. We can now prove the discrete Hodge decomposition theorem on polyhedral surfaces.

Theorem 16.5.7 (Discrete Hodge Decomposition) *Suppose M is a closed polyhedral surface with a Delaunay triangulation, $\omega \in C^1(M,\mathbb{R})$ is a simplicial 1-form, then there is a unique decomposition:*

$$\omega = d\eta + \delta\Omega + h$$

where η is a 0-form, Ω a 2-form and h a discrete harmonic 1-form.

Proof Assume such kind of decomposition exists, then

$$d\omega = d^2\eta + d\delta\Omega + dh = d\delta\Omega,$$

this leads to the equation

$$d\delta\Omega = d\omega.$$

As shown in Figure 16.11, and using the formula Eqn. (16.5.6), on each face f_Δ, we have an equation:

$$d\delta\Omega(f_\Delta) = \frac{F_i - F_\Delta}{w_{jk}} + \frac{F_j - F_\Delta}{w_{ki}} + \frac{F_k - F_\Delta}{w_{ij}} = d\omega(f_\Delta) = \omega(\partial f_\Delta), \qquad (16.5.10)$$

where $F_i = -\Omega(f_i)/|f_i|$, w_{ij} is the cotangent edge weight. Because the triangulation is Delaunay, in generic cases, all the cotangent edge weights are positive, therefore the coefficient matrix of the linear system formed by the Equations (16.5.10) is diagonal dominant, and is positive-definite in the complement space of $\sum_i F_i = 0$, the solution of Ω exists and is unique up to a constant, the co-exact 1-form $\delta\Omega$ is unique. Similarly,

$$\delta\omega = \delta d\eta + \delta^2\Omega + \delta h = \delta d\eta,$$

therefore on each vertex v_i, using Equation (16.5.7), we have

$$\delta\omega(v_i) = \delta d\eta(v_i),$$

expend it

$$-\frac{1}{|\bar{v}_i|} \sum_{v_i \sim v_j} w_{ij}\omega([v_i, v_j]) = -\frac{1}{|\bar{v}_i|} \sum_{v_i \sim v_j} w_{ij}(\eta_j - \eta_i), \qquad (16.5.11)$$

where $\eta_j = \eta(v_j)$. Since the triangulation is Delaunay, the coefficient matrix of the above linear system is diagonal dominant, and positive definite on the compliment $\sum_i \eta_i = 0$, the solution of η exists and is unique up to a constant, so the exact 1-form $d\eta$ is unique.

Now let $h = \omega - \delta\Omega - d\eta$, then

$$dh = d\omega - d\delta\Omega - d^2\eta = 0, \quad \delta h = \delta\omega - \delta^2\Omega - \delta d\eta = 0,$$

hence h is harmonic. Since $d\eta$ and $\delta\Omega$ are unique, hence the decomposition is unique.

$\square$

The proof gives us a computational algorithm for discrete Hodge decomposition.

Algorithm 8 Discrete Hodge Decomposition

Require: A closed triangle mesh M, a 1-form $\omega \in C^1(M, \mathbb{R})$

  Compute the cotangent edge weight on each edge using Eqn (16.5.1);

  Solve the coexact component $\delta\Omega$ using Eqn. (16.5.10) with the constraint $\sum_i \Omega(f_i) = 0$;

  Solve the exact component $d\eta$ using Eqn. (16.5.11) with the constraint $\sum_i \eta(v_i) = 0$;

  Set the harmonic component $h \leftarrow \omega - \delta\Omega - d\eta$.

Ensure: The Hodge decomposition of $\omega = \delta\Omega + d\eta + h$, where Ω is a 2-form, η a 0-form and h the desired harmonic form.

The discrete Hodge decomposition algorithm can be applied to compute a set of basis of the harmonic cohomology group $H^1_\Delta(M, \mathbb{R})$ using the following algorithm 9:

Algorithm 9 Harmonic Cohomology Group Basis

Require: A closed genus g triangle mesh M

Compute a set of basis of $H^1(M, \mathbb{R})$ using Algorithm 7, η_i, $i = 1, \ldots, 2g$;

    for each cohomology basis η_k **do**

        Compute the Hodge decomposition of η_k using Alg. 8;

        Set 1-form ω_k as the harmonic component of η_k;

    end for

Ensure: All $\{\omega_1, \omega_2, \cdots, \omega_{2g}\}$ form a basis of $H^1_\Delta(M, \mathbb{R})$

Discrete Wedge Product   In this section, we use the finite element method (FEM) to compute the Wedge product and the Hodge star operator on polyhedral surfaces.

Lemma 16.5.8 *As shown in Figure 16.13, given a Euclidean triangle* $\Delta = [\mathbf{v}_i, \mathbf{v}_j, \mathbf{v}_k]$, *the edge vector is* $\mathbf{e}_i = \mathbf{v}_k - \mathbf{v}_j$, *suppose* ω_1 *and* ω_2 *are constant 1-forms on* Δ, *then*

$$\int_\Delta \omega_1 \wedge \omega_2 = \frac{1}{6} \begin{vmatrix} \omega_1(\mathbf{e}_i) & \omega_1(\mathbf{e}_j) & \omega_1(\mathbf{e}_k) \\ \omega_2(\mathbf{e}_i) & \omega_2(\mathbf{e}_j) & \omega_2(\mathbf{e}_k) \\ 1 & 1 & 1 \end{vmatrix} \tag{16.5.12}$$

Proof By the definition of wedge product,

$$\int_\Delta \omega_1 \wedge \omega_2 = \frac{1}{2} \omega_1 \wedge \omega_2(\mathbf{e}_i, \mathbf{e}_j)$$

$$= \frac{1}{6} [\omega_1 \wedge \omega_2(\mathbf{e}_i, \mathbf{e}_j) + \omega_1 \wedge \omega_2(\mathbf{e}_j, \mathbf{e}_k) + \omega_1 \wedge \omega_2(\mathbf{e}_k, \mathbf{e}_i)]$$

$$= \frac{1}{6} \left\{ \begin{vmatrix} \omega_1(\mathbf{e}_i) & \omega_1(\mathbf{e}_j) \\ \omega_2(\mathbf{e}_i) & \omega_2(\mathbf{e}_j) \end{vmatrix} + \begin{vmatrix} \omega_1(\mathbf{e}_j) & \omega_1(\mathbf{e}_k) \\ \omega_2(\mathbf{e}_j) & \omega_2(\mathbf{e}_k) \end{vmatrix} + \begin{vmatrix} \omega_1(\mathbf{e}_k) & \omega_1(\mathbf{e}_i) \\ \omega_2(\mathbf{e}_k) & \omega_2(\mathbf{e}_i) \end{vmatrix} \right\}$$

$$= \frac{1}{6} \begin{vmatrix} \omega_1(\mathbf{e}_i) & \omega_1(\mathbf{e}_j) & \omega_1(\mathbf{e}_k) \\ \omega_2(\mathbf{e}_i) & \omega_2(\mathbf{e}_j) & \omega_2(\mathbf{e}_k) \\ 1 & 1 & 1 \end{vmatrix}$$

$\square$

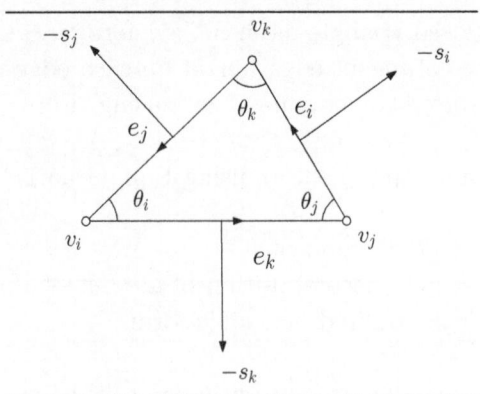

Figure 16.13 Discrete wedge product.

Lemma 16.5.9 *As shown in Figure 16.13, given a Euclidean triangle* $\Delta = [\mathbf{v}_i, \mathbf{v}_j, \mathbf{v}_k]$, *the edge vector is* $\mathbf{e}_i = \mathbf{v}_k - \mathbf{v}_j$, *suppose* ω_1 *and* ω_2 *are constant 1-forms on* Δ, *then*

$$\int_\Delta \omega_1 \wedge \star \omega_2 = \frac{1}{2} \sum_i \cot \theta_i \omega_1(\mathbf{e}_i) \omega_2(\mathbf{e}_i). \tag{16.5.13}$$

Proof We use the vector representation of the closed 1-form: assume $\omega = a\,dx + b\,dy$, then the corresponding vector is $\mathbf{w} = a\partial_x + b\partial_y$, then $\omega(\mathbf{e}_i)$ equals to the inner product $\langle \mathbf{w}, \mathbf{e}_i \rangle$. We use ω_i to represent $\omega(\mathbf{e}_i)$ and claim the vector representation of ω is

$$\mathbf{w} = \frac{1}{2A}(\omega_k \mathbf{s}_j - \omega_j \mathbf{s}_k) = \frac{1}{2A}(\omega_j \mathbf{s}_i - \omega_i \mathbf{s}_j) = \frac{1}{2A}(\omega_i \mathbf{s}_k - \omega_k \mathbf{s}_i),$$

where A is the area of the triangle. More symmetrically,

$$\mathbf{w} = -\frac{1}{6A} \begin{vmatrix} \omega_i & \omega_j & \omega_k \\ \mathbf{s}_i & \mathbf{s}_j & \mathbf{s}_k \\ 1 & 1 & 1 \end{vmatrix}.$$

Since

$$\langle \mathbf{w}, \mathbf{e}_k \rangle = \frac{1}{2A}\omega_k \langle \mathbf{s}_j, \mathbf{e}_k \rangle = \omega_k = \omega(\mathbf{e}_k)$$

similarly $\langle \mathbf{w}, \mathbf{e}_j \rangle = \omega_j$ and $\langle \mathbf{w}, \mathbf{e}_i \rangle = \omega_i$.

Given constant 1-forms ω_1 and ω_2, the corresponding vectors are $\mathbf{w}_1$ and $\mathbf{w}_2$,

$$\mathbf{w}_1 = \frac{1}{2A}(\omega_k^1 \mathbf{s}_j - \omega_j^1 \mathbf{s}_k) \quad \mathbf{w}_2 = \frac{1}{2A}(\omega_k^2 \mathbf{s}_j - \omega_j^2 \mathbf{s}_k).$$

By direct computation

$$\int_\Delta \omega_1 \wedge \star \omega_2 = A\langle \mathbf{w}_1, \mathbf{w}_2 \rangle$$

$$= \frac{1}{4A}\left\{ \omega_k^1 \omega_k^2 \langle \mathbf{s}_j, \mathbf{s}_j \rangle + \omega_j^1 \omega_j^2 \langle \mathbf{s}_k, \mathbf{s}_k \rangle - (\omega_k^1 \omega_j^2 + \omega_j^1 \omega_k^2)\langle \mathbf{s}_j, \mathbf{s}_k \rangle \right\}$$

$$= \frac{1}{4A}\left\{ -\omega_k^1 \omega_k^2 \langle \mathbf{s}_j, \mathbf{s}_i + \mathbf{s}_k \rangle - \omega_j^1 \omega_j^2 \langle \mathbf{s}_k, \mathbf{s}_i + \mathbf{s}_j \rangle - (\omega_k^1 \omega_j^2 + \omega_j^1 \omega_k^2)\langle \mathbf{s}_j, \mathbf{s}_k \rangle \right\}$$

$$= -\omega_k^1 \omega_k^2 \frac{\langle \mathbf{s}_j, \mathbf{s}_i \rangle}{4A} - \omega_j^1 \omega_j^2 \frac{\langle \mathbf{s}_k, \mathbf{s}_i \rangle}{4A} - \frac{\langle \mathbf{s}_k, \mathbf{s}_j \rangle}{4A}(\omega_k^1 \omega_k^2 + \omega_j^1 \omega_j^2 + \omega_k^1 \omega_j^2 + \omega_j^1 \omega_k^2)$$

$$= -\omega_k^1 \omega_k^2 \frac{\langle \mathbf{s}_j, \mathbf{s}_i \rangle}{4A} - \omega_j^1 \omega_j^2 \frac{\langle \mathbf{s}_k, \mathbf{s}_i \rangle}{4A} - \frac{\langle \mathbf{s}_k, \mathbf{s}_j \rangle}{4A}(\omega_k^1 + \omega_j^1)(\omega_k^2 + \omega_j^2)$$

$$= -\omega_k^1 \omega_k^2 \frac{\langle \mathbf{s}_j, \mathbf{s}_i \rangle}{4A} - \omega_j^1 \omega_j^2 \frac{\langle \mathbf{s}_k, \mathbf{s}_i \rangle}{4A} - \omega_i^1 \omega_i^2 \frac{\langle \mathbf{s}_k, \mathbf{s}_j \rangle}{4A}$$

$$= \frac{1}{2}\left(\omega_i^1 \omega_i^2 \cot \theta_i + \omega_j^1 \omega_j^2 \cot \theta_j + \omega_k^1 \omega_k^2 \cot \theta_k \right)$$

$\square$

Discrete Conjugate Harmonic 1-form   Suppose S is a genus g smooth surface, a set of harmonic 1-form basis is $\{\omega_1, \omega_2, \ldots, \omega_{2g}\}$. The conjugate 1-form $\star\omega_i$ is also harmonic, therefore can be represented as a linear combination of the basis,

$$\star\omega_i = \lambda_{i,1}\omega_1 + \lambda_{i,2}\omega_2 + \cdots + \lambda_{i,2g}\omega_{2g},$$

where the coefficients $\lambda_{i,1}, \ldots, \lambda_{i,2g} \in \mathbb{R}$. This leads to the linear system,

$$
\begin{pmatrix} \omega_1 \wedge {}^*\omega_i \\ \omega_2 \wedge {}^*\omega_i \\ \vdots \\ \omega_{2g} \wedge {}^*\omega_i \end{pmatrix} = \begin{pmatrix} \omega_1 \wedge \omega_1 & \omega_1 \wedge \omega_2 & \cdots & \omega_1 \wedge \omega_{2g} \\ \omega_2 \wedge \omega_1 & \omega_2 \wedge \omega_2 & \cdots & \omega_2 \wedge \omega_{2g} \\ \vdots & \vdots & \vdots & \vdots \\ \omega_{2g} \wedge \omega_1 & \omega_{2g} \wedge \omega_2 & \cdots & \omega_{2g} \wedge \omega_{2g} \end{pmatrix} \begin{pmatrix} \lambda_{i,1} \\ \lambda_{i,2} \\ \vdots \\ \lambda_{i,2g} \end{pmatrix} \quad (16.5.14)
$$

We take the integration of each element on both the left and the right sides, and solve the following linear system to obtain the coefficients.

$$
\begin{pmatrix} \int_M \omega_1 \wedge {}^*\omega_i \\ \int_M \omega_2 \wedge {}^*\omega_i \\ \vdots \\ \int_M \omega_{2g} \wedge {}^*\omega_i \end{pmatrix} = \begin{pmatrix} \int_M \omega_1 \wedge \omega_1 & \cdots & \int_M \omega_1 \wedge \omega_{2g} \\ \int_M \omega_2 \wedge \omega_1 & \cdots & \int_M \omega_2 \wedge \omega_{2g} \\ \vdots & & \vdots \\ \int_M \omega_{2g} \wedge \omega_1 & \cdots & \int_M \omega_{2g} \wedge \omega_{2g} \end{pmatrix} \begin{pmatrix} \lambda_{i,1} \\ \lambda_{i,2} \\ \vdots \\ \lambda_{i,2g} \end{pmatrix} \quad (16.5.15)
$$

The details are summarized in Algorithm 10.

Algorithm 10 Conjugate Harmonic 1-forms

Require: A closed genus g triangle mesh M, a set of harmonic 1-form basis $\{\omega_1, \omega_2, \ldots, \omega_{2g}\}$;
    for each base harmonic 1-form ω_i **do**
        for each base harmonic 1-form ω_j **do**
            Compute the integration $\int_M \omega_i \wedge \omega_j$ using Eqn. (16.5.12);
            Compute the integration $\int_M \omega_i \wedge \star\omega_j$ using Eqn. (16.5.13);
        end for
    end for
    Solve the linear system Eqn. (16.5.15) to obtain $\lambda_{i,j}$'s;
    for each base harmonic 1-form ω_i **do**
        Set $\star\omega_i \leftarrow \lambda_{i,1}\omega_1 + \cdots + \lambda_{i,2g}\omega_{2g}$;
    end for
Ensure: The conjugate harmonic 1-forms $\{\star\omega_1, \star\omega_2, \ldots, \star\omega_{2g}\}$.

Codifferential Operator on Boundary   For surfaces with boundaries, the above computational methods need to be modified in the neighborhoods of the boundaries. As shown in Figure 16.14 left frame, suppose one edge $[v_i, v_j]$ is on the boundary, and adjacent to a single triangle face $[v_i, v_j, v_k]$, then the cotangent edge weight is modified as

$$w_{ij} = \frac{1}{2}\cot\theta_k, \quad (16.5.16)$$

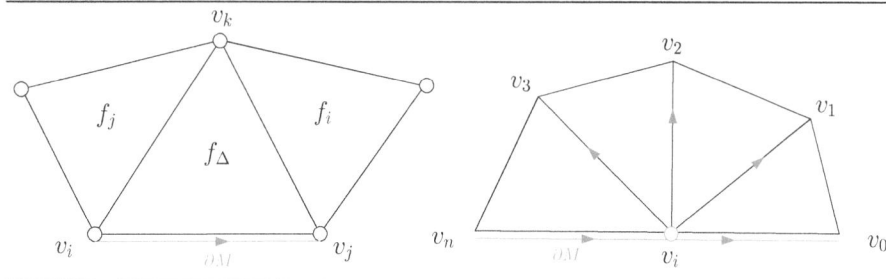

Figure 16.14   Boundary situation.

where θ_k is the corner angle at the vertex v_k in the face $[v_i, v_j, v_k]$. Given a 2-form $\Omega \in C^2(M, \mathbb{R})$, the codifferential operator $\delta : C^2(M, \mathbb{R}) \to C^1(M, \mathbb{R})$ on the boundary edge $[v_i, v_j] \subset \partial M$ is given by

$$\delta\Omega([v_i, v_j]) = \frac{1}{w_{ij}}\left\{\frac{\Omega(f_\Delta)}{|f_\Delta|} - 0\right\} \tag{16.5.17}$$

Let $\omega \in C^1(M, \mathbb{R})$, $\delta : C^1(M, \mathbb{R}) \to C^0(M, \mathbb{R})$, as shown in the right frame, given a vertex on the boundary $v_i \in \partial M$,

$$\delta\omega(v_i) = (-1)\frac{1}{|\bar{v}_i|}\sum_{v_i \sim v_j} w_{ij}\omega([v_i, v_j]) = (-1)\frac{1}{|\bar{v}_i|}\sum_{v_i \sim v_j} w_{ij}\omega([v_i, v_j]), \tag{16.5.18}$$

where $[v_i, v_0]$ and $[v_i, v_n]$ are on the boundary.

CHAPTER 17

Harmonic Map

This chapter focuses on the computational algorithms for harmonic maps, the fundamental concepts and theorems for harmonic maps between Riemannian manifolds can be found in [298]. The theoretical treatments of surface harmonic maps can be found in [179].

Because surface harmonic maps are diffeomorphic under mild conditions, they are very suitable for surface parameterization or registration. Figure 17.1 shows a harmonic map form a simply connected human facial surface onto the planar unit disk, the mapping is a homeomorphism without any folding. The figure also shows that the harmonic map preserves local shapes, minimizes the geometric distortions. The infinitesimal ellipses on the surface are mapped to infinitesimal circles on the plane. The eccentricities of the ellipses are very small, this shows the geometric distortion is very small. Furthermore, the computation of harmonic map is stable, the results continuously depend on the boundary conditions. Therefore, it has been applied broadly in engineering and medical fields.

Figure 17.2 shows a harmonic map from a genus zero closed surface to the unit sphere. The map is diffeomorphic and preserves local shapes. In fact it is a conformal map. Figure 17.3 shows a harmonic map from the kitten surface to a metric graph, which induces a measured foliation on the surface. Different kitten surfaces embed in $\mathbb{R}^3$ differently, namely each pair of them differ by an isometric deformation. However, the foliations are consistent on the surfaces. This shows the harmonic map is intrinsic, solely depends on the Riemannian metric of the surface and is independent of the embedding.

17.1 PLANAR HARMONIC MAPS

Harmonic Function   Given a planar domain $\Omega \subset \mathbb{R}^2$, consider the electric potential $u : \Omega \to \mathbb{R}$. The gradient of the potential induces electric currents, and produces heat. The heat power is represented as the *harmonic energy* of the potential function u:

$$E(u) := \int_\Omega \langle \nabla u, \nabla u \rangle dx dy. \tag{17.1.1}$$

DOI: 10.1201/9781003350576-17

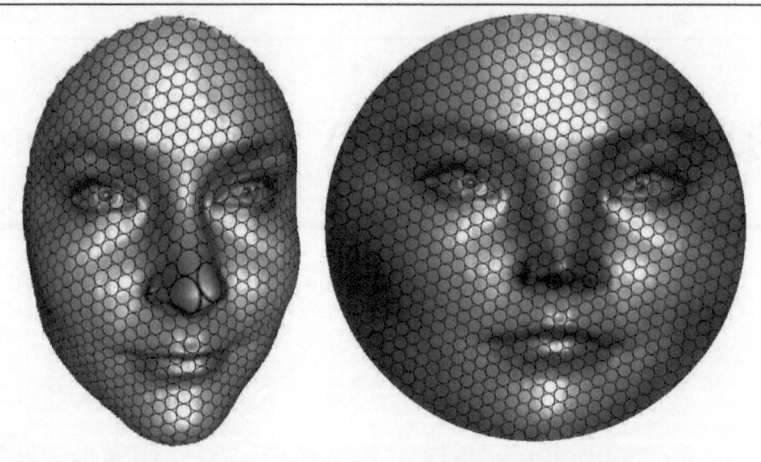

Figure 17.1 Harmonic map between topological disks.

In nature, the distribution of u minimizes the heat power, which is represented as a harmonic function.

Definition 17.1.1 (Harmonic Function) *Given a planar domain $\Omega \subset \mathbb{R}^2$, a C^1 function $f : \Omega \to \mathbb{R}$ is called harmonic, if it optimizes the harmonic energy Eqn. (17.1.1).*

We use the variational principle to deduce the Euerl-Larange equation for a harmonic function $u : \Omega \to \mathbb{R}$, whose restriction on the boundary $\partial\Omega$ equals to a prescribed function $g : \partial\Omega \to \mathbb{R}$. Assume a test function $h \in C_0^\infty(\Omega)$ has zero boundary value,

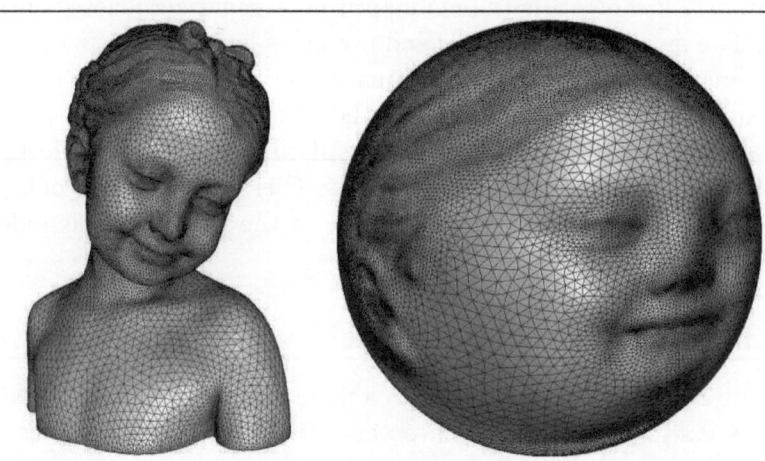

Figure 17.2 Harmonic map between topological spheres.

Figure 17.3 Harmonic map induced foliations.

then $E(u + \varepsilon h) \geq E(u)$ for small ε,

$$\frac{d}{d\varepsilon} \int_\Omega \langle \nabla u + \varepsilon \nabla h, \nabla u + \varepsilon \nabla h \rangle dxdy \Big|_{\varepsilon=0} = 2 \int_\Omega \langle \nabla u, \nabla h \rangle dxdy = 0.$$

By equality

$$\nabla \cdot (h\nabla u) = \langle \nabla h, \nabla u \rangle + h \nabla \cdot \nabla u,$$

we obtain

$$\int_\Omega \langle \nabla u, \nabla h \rangle = \int_\Omega h\Delta u dxdy - \int_\Omega \nabla \cdot (h\nabla u) dxdy = \int_\omega h\Delta u dxdy,$$

Let $h = \Delta u$, then the L^2 norm of Δu equals to 0. We obtain the Laplace equation with a Dirichlet boundary condition,

$$\begin{cases} \Delta u & \equiv & 0 \\ u|_{\partial\Omega} & = & g \end{cases}$$

Many physical phenomena, such as steady temperature field, static electric field, elastic deformation and so on, are governed by this Laplace equation.

Harmonic functions have many good properties.

Theorem 17.1.2 (Mean Value) *Assume $\Omega \subset \mathbb{R}^2$ is a planar open set, $u : \Omega \to \mathbb{R}$ is a harmonic function, then for any $p \in \Omega$,*

$$u(p) = \frac{1}{2\pi\varepsilon} \oint_\gamma u(q) ds, \tag{17.1.2}$$

where γ is a circle centered at p, with radius ε.

Proof u is harmonic, du is a harmonic 1-form, its Hodge star $^*du$ is also harmonic.

Define the conjugate function v, $dv = {}^*du$, then $\varphi(z) := u + \sqrt{-1}v$ is holomorphic. By Cauchy integration formula,

$$\varphi(z) = \frac{1}{2\pi i} \oint_\gamma \frac{\varphi(\zeta)}{\zeta - z} dz \qquad (17.1.3)$$

Hence, by taking the real part, we obtain the mean value property of harmonic function.

□

Corollary 17.1.3 (Maximal value principle) *Assume $\Omega \subset \mathbb{R}^2$ is a planar domain, and $u : \overline{\Omega} \to \mathbb{R}$ is a non-constant harmonic function, then u can't reach extremal values in the interior of Ω.*

Proof Assume p is an interior point in Ω, p is a maximal point of u, $u(p) = C$. By mean value property, we obtain that for any point q on the circle $B(p, \varepsilon)$, $u(q) = C$, where ε is arbitrary, therefore u is constant in a neighborhood of p. Therefore, $u^{-1}(C)$ is open. On the other hand, u is continuous, $u^{-1}(C)$ is closed, hence $u^{-1}(C) = \Omega$. Contradiction.

□

Corollary 17.1.4 *Suppose $\Omega \subset \mathbb{R}^2$ is a planar domain, $u_1, u_2 : \Omega \to \mathbb{R}$ are harmonic functions with the same boundary value, $u_1|_{\partial\Omega} = u_2|_{\partial\Omega}$, then $u_1 = u_2$ on Ω.*

Proof $u_1 - u_2$ is also harmonic, with 0 boundary value, the maximal and minimal values of $u_1 - u_2$ must be on the boundary, therefore both of them are 0, hence u_1 and u_2 equal to each other in Ω.

□

Theorem 17.1.5 (Rado) *Suppose a harmonic map $\varphi : (S, \mathbf{g}) \to (\Omega, du^2 + dv^2)$ satisfies:*

1. *the planar domain Ω is convex*

2. *the restriction of $\varphi : \partial S \to \partial\Omega$ on the boundary is homoemorphic,*

then φ is diffeomorphic in the interior of S.

Proof By regularity theory of harmonic maps, we get the smoothness of the harmonic map. Assume $\varphi : (x, y) \to (u, v)$ is not homeomorphic, then there is an interior point $p \in \Omega$, the Jacobian matrix of φ is degenerated at p, there are two constants $a, b \in \mathbb{R}$, not being zeros simultaneously, such that

$$a\nabla u(p) + b\nabla v(p) = 0.$$

By $\Delta u = 0, \Delta v = 0$, the auxiliary function $f(q) = au(q) + bv(q)$ is also harmonic. By $\nabla f(p) = 0$, p is an saddle point of f. Consider the level set of f near p

$$\Gamma = \{q \in \Omega | f(q) = f(p) - \varepsilon\}$$

Γ has two connected components, intersecting ∂S at 4 points.

But Ω is a planar convex domain, $\partial\Omega$ and the line $au + bv = const$ have two intersection points. By assumption, the mapping φ restricted on the boundary $\varphi : \partial S \to \partial\Omega$ is homeomorphic. Contradiction.

□

17.2 SURFACE HARMONIC MAPS

Definition 17.2.1 (Harmonic Energy) *Let (Σ_1, z) and (Σ_2, u) be two Riemann surfaces, with Riemannian metrics $\sigma(z)dzd\bar{z}$ and $\rho(u)dud\bar{u}$. Given a C^1 map $u : \Sigma_1 \to \Sigma_2$, then the harmonic energy of u is defined as*

$$E(z, \rho, u) := \int_{\Sigma_1} \rho^2(u)(u_z \bar{u}_{\bar{z}} + \bar{u}_z u_{\bar{z}}) \frac{i}{2} dz \wedge d\bar{z}$$

where $u_z := \frac{1}{2}(u_x - iu_y)$, $u_{\bar{z}} := \frac{1}{2}(u_x + iu_y)$ and $dz \wedge d\bar{z} = -2idx \wedge dy$.

Definition 17.2.2 (Harmonic Map) *If the C^1 map $u : \Sigma_1 \to \Sigma_2$ minimizes the harmonic energy, then u is called a harmonic map.*

Theorem 17.2.3 (Euerl-Larange Equation for Harmonic Maps) *Suppose $u : \Sigma_1 \to \Sigma_2$ is a C^2 harmonic map between Riemannian surfaces, then*

$$u_{z\bar{z}} + \frac{2\rho_u}{\rho} u_z u_{\bar{z}} = 0. \tag{17.2.1}$$

Geodesics are special harmonic maps, harmonic maps are generalized geodesics. The geodesic equation has the similar form:

$$\ddot{\gamma} + \frac{2\rho_\gamma}{\rho} \dot{\gamma}^2 \equiv 0.$$

Proof Suppose u is harmonic, u_t is a variation in a local coordinates system,

$$u + t\varphi, \quad \varphi \in C^0 \cap W_0^{1,2}(\Sigma_1, \Sigma_2)$$

we obtain

$$\frac{d}{dt} E(u + t\varphi)\Big|_{t=0} = 0,$$

$$0 = \frac{d}{dt} \Big\{ \int \rho^2(u + t\varphi)((u + t\varphi)_z (\bar{u} + t\bar{\varphi})_{\bar{z}}$$

$$+ (\bar{u} + t\bar{\varphi})_z (u + t\varphi)_{\bar{z}}) idzd\bar{z} \Big\}\Big|_{t=0}$$

$$= \int \Big\{ \rho^2(u)(u_z \bar{\varphi}_{\bar{z}} + \bar{u}_{\bar{z}} \varphi_z + \bar{u}_z \varphi_{\bar{z}} + u_{\bar{z}} \bar{\varphi}_z)$$

$$+ 2\rho(\rho_u \varphi + \rho_{\bar{u}} \bar{\varphi})(u_z \bar{u}_{\bar{z}} + \bar{u}_z u_{\bar{z}}) \Big\} idzd\bar{z}.$$

We set $\varphi = \frac{\psi}{\rho^2(u)}$,

$$\rho^2 \varphi_z = \psi_z - \frac{2\psi}{\rho}(\rho_u u_z + \rho_{\bar{u}} \bar{u}_z)$$

$$\rho^2 \varphi_{\bar{z}} = \psi_{\bar{z}} - \frac{2\psi}{\rho}(\rho_u u_{\bar{z}} + \rho_{\bar{u}} \bar{u}_{\bar{z}})$$

$$\rho^2 \bar{\varphi}_z = \bar{\psi}_z - \frac{2\bar{\psi}}{\rho}(\rho_u u_z + \rho_{\bar{u}} \bar{u}_z)$$

$$\rho^2 \bar{\varphi}_{\bar{z}} = \bar{\psi}_{\bar{z}} - \frac{2\bar{\psi}}{\rho}(\rho_u u_{\bar{z}} + \rho_{\bar{u}} \bar{u}_{\bar{z}})$$

$$\bar{u}_{\bar{z}}\rho^2\varphi_z = \psi_z\bar{u}_{\bar{z}} - \frac{2\psi}{\rho}(\rho_u u_z\bar{u}_{\bar{z}} + \rho_{\bar{u}}\bar{u}_z\bar{u}_{\bar{z}})$$

$$\bar{u}_z\rho^2\varphi_{\bar{z}} = \psi_{\bar{z}}\bar{u}_z - \frac{2\psi}{\rho}(\rho_u u_{\bar{z}}\bar{u}_z + \rho_{\bar{u}}\bar{u}_{\bar{z}}\bar{u}_z)$$

$$u_{\bar{z}}\rho^2\bar{\varphi}_z = \bar{\psi}_z u_{\bar{z}} - \frac{2\bar{\psi}}{\rho}(\rho_{\bar{u}}\bar{u}_z u_{\bar{z}} + \rho_u u_z u_{\bar{z}})$$

$$u_z\rho^2\bar{\varphi}_{\bar{z}} = \bar{\psi}_{\bar{z}} u_z - \frac{2\bar{\psi}}{\rho}(\rho_{\bar{u}}\bar{u}_{\bar{z}} u_z + \rho_u u_{\bar{z}} u_z)$$

$$\frac{2}{\rho}(\rho_u\psi + \rho_{\bar{u}}\bar{\psi})(u_z\bar{u}_{\bar{z}} + \bar{u}_z u_{\bar{z}})$$

$$= \frac{2\psi}{\rho}\rho_u(u_z\bar{u}_{\bar{z}} + \bar{u}_z u_{\bar{z}}) + \frac{2\bar{\psi}}{\rho}\rho_{\bar{u}}(\bar{u}_z u_{\bar{z}} + u_z\bar{u}_{\bar{z}})$$

Take summation,

$$\bar{u}_{\bar{z}}\rho^2\varphi_z + u_z\rho^2\bar{\varphi}_{\bar{z}} = \left(\psi_z\bar{u}_{\bar{z}} - \frac{2\psi}{\rho}\rho_{\bar{u}}\bar{u}_z\bar{u}_{\bar{z}}\right) + \left(\bar{\psi}_{\bar{z}} u_z - \frac{2\bar{\psi}}{\rho}\rho_u u_{\bar{z}} u_z\right)$$

$$\bar{u}_z\rho^2\varphi_{\bar{z}} + u_{\bar{z}}\rho^2\bar{\varphi}_z = \left(\psi_{\bar{z}}\bar{u}_z - \frac{2\psi}{\rho}\rho_{\bar{u}}\bar{u}_{\bar{z}}\bar{u}_z\right) + \left(\bar{\psi}_z u_{\bar{z}} - \frac{2\bar{\psi}}{\rho}\rho_u u_z u_{\bar{z}}\right)$$

The above equation becomes

$$0 = 2\Re\int\left(\bar{\psi}_{\bar{z}} u_z - \frac{2\bar{\psi}}{\rho}\rho_u u_{\bar{z}} u_z\right) i dz d\bar{z}$$

$$+ 2\Re\int\left(\psi_{\bar{z}}\bar{u}_z - \frac{2\psi}{\rho}\rho_{\bar{u}}\bar{u}_{\bar{z}}\bar{u}_z\right) i dz d\bar{z}$$

If $u \in C^2$, we can integrate by parts, $(u_z\bar{\psi})_{\bar{z}} = u_{z\bar{z}}\bar{\psi} + u_z\bar{\psi}_{\bar{z}}$,

$$0 = 2\Re\int\left(u_{z\bar{z}} + \frac{2\rho_u}{\rho}u_{\bar{z}} u_z\right)\bar{\psi} i dz d\bar{z}$$

$$+ 2\Re\int\left(\bar{u}_{z\bar{z}} + \frac{2\rho_{\bar{u}}}{\rho}\bar{u}_{\bar{z}}\bar{u}_z\right)\psi i dz d\bar{z}$$

Therefore,

$$0 = 2\Re\int\left(u_{z\bar{z}} + \frac{2\rho_u}{\rho}u_{\bar{z}} u_z\right)\bar{\psi} i dz d\bar{z}$$

□

Theorem 17.2.4 (Hopf Differential of Harmonic Maps) *Let* $u : (\Sigma_1, \lambda^2(z)dzd\bar{z})$
$\to (\Sigma_2, \rho^2(u)dud\bar{u})$ *is harmonic, then the Hopf differential of the map*

$$\Phi(u) := \rho^2 u_z\bar{u}_z dz^2$$

is holomorphic quadratic differential on Σ_1. *Furthermore* $\Phi(u) \equiv 0$, *if and only if* u
is holomorphic or anti-holomorphic.

Proof If u is harmonic, then

$$\frac{\partial}{\partial \bar{z}}(\rho^2 u_z \bar{u}_z) = \rho^2 u_{z\bar{z}} \bar{u}_z + \rho^2 u_z \bar{u}_{z\bar{z}} + 2\rho\rho_u u_{\bar{z}} u_z \bar{u}_z + 2\rho\rho_{\bar{u}} \bar{u}_{\bar{z}} u_z \bar{u}_z$$
$$= (\rho^2 u_{z\bar{z}} + 2\rho\rho_u u_{\bar{z}} u_z)\bar{u}_z + (\rho^2 \bar{u}_{z\bar{z}} + 2\rho\rho_{\bar{u}} \bar{u}_{\bar{z}} \bar{u}_z) u_z = 0.$$

Therefore, $\Phi(u)$ is holomorphic. If $\Phi(u) = \rho^2 u_z \bar{u}_z \equiv 0$, then either $u_z = 0$ or $\bar{u}_z = 0$. Since the Jacobian determinant equals to

$$|u_z|^2 - |u_{\bar{z}}|^2 > 0,$$

therefore $\bar{u}_z = 0$, namely $u_{\bar{z}} = 0$, u is holomorphic or anti-holomorphic.

We need to further prove that u cannot be holomorphic on a subset Ω of Σ_1 and anti-holomorphic on the complement of Ω. Define auxiliary functions

$$H := \frac{\rho^2(u(z))}{\lambda^2(z)} u_z \bar{u}_{\bar{z}}, \quad L := \frac{\rho^2(u(z))}{\lambda^2(z)} \bar{u}_z u_{\bar{z}}.$$

u is holomorphic, equivalent to $L \equiv 0$; u is anti-holomorphic, equivalent to $H \equiv 0$. Furthermore, it is easy to show:

$$\Delta \log H = 2K_1 - 2K_2(H - L), \quad \Delta \log L = 2K_1 + 2K_2(H - L),$$

so that H and L have isolated zeros, unless they are zero everywhere. Hence u is entirely holomorphic or anti-holomorphic.

$\square$

Lemma 17.2.5 *A holomorphic quadratic differential ω is on the unit sphere, then ω is zero.*

Proof Choose two charts z and $w = \frac{1}{z}$. Let $\omega = \varphi(z)dz^2$, then

$$\varphi(z)dz^2 = \varphi\left(\frac{1}{w}\right)\left(\frac{dz}{dw}\right)^2 dw^2 = \varphi\left(\frac{1}{w}\right)\frac{1}{w^4}dw^2.$$

since ω is globally holomorphic, when $w \to 0$,

$$\varphi\left(\frac{1}{w}\right)\frac{1}{w^4} < \infty,$$

hence $z \to \infty$, $\varphi(z) \to 0$. By Liouville theorem, $\varphi \equiv 0$.

$\square$

Theorem 17.2.6 (Spherical Harmonic Maps) *Harmonic maps between genus zero closed metric surfaces must be conformal.*

Proof Suppose $u : \Sigma_1 \to \Sigma_2$ is a harmonic map, then $\Phi(u)$ must be a holomorphic quadratic differential. Since Σ_1 is of genus zero, therefore $\Phi(u) \equiv 0$. Hence u is holomorphic.

$\square$

Definition 17.2.7 (Möbius Transformation) *A Möbius transformation* $\varphi : \hat{\mathbb{C}} \to \hat{\mathbb{C}}$ *has the form*

$$z \mapsto \frac{az + b}{cz + d}, \quad a, b, c, d \in \mathbb{C}, \quad ad - bc = 1.$$

Given $\{z_0, z_1, z_2\}$, there is a unique Möbius transformation, that maps them to $\{0, 1, \infty\}$,

$$z \mapsto \frac{z - z_0}{z - z_2} \frac{z_1 - z_2}{z_1 - z_0}.$$

Theorem 17.2.8 (Uniquess of Spherical Conformal Automorphisms)
Suppose $f : \mathbb{S}^2 \to \mathbb{S}^2$ *is a biholomorphic automorphism, then* f *must be a Möbius transformation.*

Proof By stereo-graphic projection, we map the sphere to the extened complex plane $\hat{\mathbb{C}} = \mathbb{C} \cup \{\infty\}$. First, the number of poles of f must be finite. Suppose there are infinite poles of f, because $\mathbb{S}^2$ is compact, there must be accumulation points, then f must be a constant value function.

Let $z_1, z_2, \ldots, z_n$ be the finite poles of f, with degrees $e_1, e_2, \ldots, e_n$. Let $g = \Pi_i (z - z_i)^{e_i}$, then fg is a holomorphic function on $\mathbb{C}$, therefore fg is entire, namely, fg is a polynomial. Therefore,

$$f = \frac{\sum_{i=1}^m a_i z^i}{\sum_{j=1}^n b_j z^j},$$

if $m > 1$ then f has multiple zeros, contradict to the condition that f is an automorphism. Therefore $m = 1$. Similarly $n = 1$.

□

17.3 DISCRETE HARMONIC MAP

In practice, the surfaces are approximated by polyhedral surfaces, the maps are approximated by simplical maps (piecewise lienar maps). We can use Finite Element Method to compute discrete harmonic maps, which converge to the smooth harmonic map when the triangulations are subdivided infinitely.

Definition 17.3.1 (Barycentric Coordinates) *Suppose* Δ *is a planar Euclidean triangle with vertices* $\mathbf{v}_i, \mathbf{v}_j$ *and* $\mathbf{v}_k$, $\mathbf{p} \in \Delta$ *is a point in the triangle, then the barycentric coordinates of* $\mathbf{p}$ *with respect to* Δ *are* $(\lambda_i, \lambda_j, \lambda_k)$, *such that*

$$\mathbf{p} = \lambda_i \mathbf{v}_i + \lambda_j \mathbf{v}_j + \lambda_k \mathbf{v}_k, \quad \lambda_i + \lambda_j + \lambda_k = 1,$$

where λ_i *is the ratio between the signed area of the triangle* $[\mathbf{p}, \mathbf{v}_j, \mathbf{v}_k]$ *and that of triangle* $[\mathbf{v}_i, \mathbf{v}_j, \mathbf{v}_k]$.

If $\mathbf{p}$ is inside the triangle, then all three components of its barycentric coordinates are non-negative.

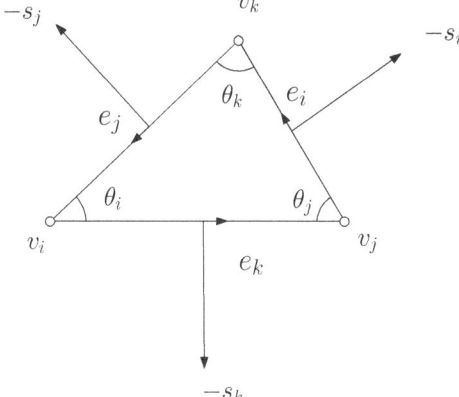

Figure 17.4 Discrete harmonic energy.

Lemma 17.3.2 *Given a Euclidean triangle $\Delta = [\mathbf{v}_i, \mathbf{v}_j, \mathbf{v}_k]$, the linear function is defined as*

$$f(\mathbf{p}) = \lambda_i f(\mathbf{v}_i) + \lambda_j f(\mathbf{v}_j) + \lambda_k f(\mathbf{v}_k),$$

where $(\lambda_i, \lambda_j, \lambda_k)$ are the barycentric coordinates of $\mathbf{p}$ with respect to Δ. Then the gradient of f is given by

$$\nabla f(\mathbf{p}) = \frac{1}{2A}(\mathbf{s}_i f(\mathbf{v}_i) + \mathbf{s}_j f(\mathbf{v}_j) + \mathbf{s}_k f(\mathbf{v}_k)), \qquad (17.3.1)$$

where

$$\mathbf{s}_i = \mathbf{n} \times (\mathbf{v}_k - \mathbf{v}_j), \quad \mathbf{s}_j = \mathbf{n} \times (\mathbf{v}_i - \mathbf{v}_k), \quad \mathbf{s}_k = \mathbf{n} \times (\mathbf{v}_j - \mathbf{v}_i).$$

Its harmonic energy on the triangle is

$$\int_\Delta \langle \nabla f, \nabla f \rangle dA = \frac{\cot \theta_i}{2}(f_j - f_k)^2 + \frac{\cot \theta_j}{2}(f_k - f_i)^2 + \frac{\cot \theta_j}{2}(f_k - f_i)^2. \quad (17.3.2)$$

Proof Select an arbitrary point $\mathbf{p} \in \Delta$, $\mathbf{p} = \lambda_i \mathbf{v}_i + \lambda_j \mathbf{v}_j + \lambda_k \mathbf{v}_k$. By the definition of barycentric coordinates,

$$\lambda_i = \frac{1}{2A}\langle \mathbf{p} - \mathbf{v}_i, \mathbf{s}_i \rangle, \quad \lambda_j = \frac{1}{2A}\langle \mathbf{p} - \mathbf{v}_j, \mathbf{s}_j \rangle, \quad \lambda_k = \frac{1}{2A}\langle \mathbf{p} - \mathbf{v}_k, \mathbf{s}_k \rangle,$$

where A is the area of the triangle. Therefore, the linear function is

$$\begin{aligned}
f(p) &= \lambda_i f_i + \lambda_j f_j + \lambda_k f_k \\
&= \frac{1}{2A}\langle \mathbf{p} - \mathbf{v}_i, f_i \mathbf{s}_i \rangle + \frac{1}{2A}\langle \mathbf{p} - \mathbf{v}_j, f_j \mathbf{s}_j \rangle + \frac{1}{2A}\langle \mathbf{p} - \mathbf{v}_k, f_k \mathbf{s}_k \rangle \\
&= \frac{1}{2A}\langle \mathbf{p}, f_i \mathbf{s}_i + f_j \mathbf{s}_j + f_k \mathbf{s}_k \rangle - \frac{1}{2A}(\langle \mathbf{v}_i, f_i \mathbf{s}_i \rangle + \langle \mathbf{v}_j, f_j \mathbf{s}_j \rangle + \langle \mathbf{v}_k, f_k \mathbf{s}_k \rangle).
\end{aligned}$$

Hence, the gradient of the linear function is

$$\nabla f = \frac{1}{2A}(f_i \mathbf{s}_i + f_j \mathbf{s}_j + f_k \mathbf{s}_k).$$

Compute the harmonic energy,

$$\int_\Delta \langle \nabla f, \nabla f \rangle dA = \frac{1}{4A} \langle f_i \mathbf{s}_i + f_j \mathbf{s}_j + f_k \mathbf{s}_k, f_i \mathbf{s}_i + f_j \mathbf{s}_j + f_k \mathbf{s}_k \rangle$$

$$= \frac{1}{4A} \left(\sum_i \langle \mathbf{s}_i, \mathbf{s}_i \rangle f_i^2 + 2 \sum_{i<j} \langle \mathbf{s}_i, \mathbf{s}_j \rangle f_i f_j \right)$$

$$= \frac{1}{4A} \left(-\sum_i \langle \mathbf{s}_i, \mathbf{s}_j + \mathbf{s}_k \rangle f_i^2 + 2 \sum_{i<j} \langle \mathbf{s}_i, \mathbf{s}_j \rangle f_i f_j \right)$$

$$= -\frac{1}{4A} (\langle \mathbf{s}_i, \mathbf{s}_j \rangle (f_i - f_j)^2 + \langle \mathbf{s}_j, \mathbf{s}_k \rangle (f_j - f_k)^2 + \langle \mathbf{s}_k, \mathbf{s}_i \rangle (f_k - f_i)^2)$$

Because

$$\frac{\langle \mathbf{s}_i, \mathbf{s}_j \rangle}{2A} = -\cot \theta_k, \quad \frac{\langle \mathbf{s}_j, \mathbf{s}_k \rangle}{2A} = -\cot \theta_i, \quad \frac{\langle \mathbf{s}_k, \mathbf{s}_i \rangle}{2A} = -\cot \theta_j,$$

thus the harmonic energy is

$$\int_\Delta \langle \nabla f, \nabla f \rangle dA = \frac{\cot \theta_k}{2} (f_i - f_j)^2 + \frac{\cot \theta_i}{2} (f_j - f_k)^2 + \frac{\cot \theta_j}{2} (f_k - f_i)^2.$$

□

Lemma 17.3.3 *Suppose M is a polyhedral surface with a triangulation (a triangle mesh), $f : M \to \mathbb{R}$ is a piecewise linear function, the harmonic energy is*

$$\int_M \langle \nabla f, \nabla f \rangle dA = \sum_{v_i \sim v_j} w_{ij} (f(v_i) - f(v_j))^2, \tag{17.3.3}$$

where w_{ij} is the cotangent edge weight for $[v_i, v_j]$.

Proof The harmonic energy of f on the whole surface is

$$\int_M \langle \nabla f, \nabla f \rangle dA = \sum_{\Delta_i} \int_{\Delta_i} \langle \nabla f, \nabla f \rangle dA$$

$$= \sum_{[v_i, v_j, v_k]} \frac{\cot \theta_i^{jk}}{2} (f_j - f_k)^2 + \frac{\cot \theta_j^{ki}}{2} (f_k - f_i)^2 + \frac{\cot \theta_k^{ij}}{2} (f_i - f_j)^2$$

$$= \sum_{[v_i, v_j]} \frac{1}{2} (\cot \theta_k^{ij} + \cot \theta_l^{ji}) (f_i - f_j)^2.$$

□

Definition 17.3.4 (Discrete Harmonic Function) *Suppose M is a polyhedral surface with a triangulation (a triangle mesh), $f : M \to \mathbb{R}$ is a piecewise linear function. If f minimizes the discrete harmonic energy Eqn. (17.3.4), then f is called a discrete harmonic function.*

Lemma 17.3.5 *Suppose M is a polyhedral surface with a triangulation (a triangle mesh), $f : M \to \mathbb{R}$ is a discrete harmonic function, then it satisfies the discrete Laplace equation*

$$\sum_{v_i \sim v_j} w_{ij}(f(v_j) - f(v_i)) = 0, \quad \forall v_i \in M \tag{17.3.4}$$

where w_{ij} is the cotangent edge weight for $[v_i, v_j]$.

Proof Since f is discrete harmonic, the partial derivative of the discrete harmonic energy with respect to the function value at each vertex is 0. Take the derivative of the discrete harmonic energy

$$E(f) = \sum_{v_i \sim v_j} w_{ij}(f_j - f_i)^2$$

with respect to f_i, we obtain

$$\frac{\partial E(f)}{\partial f_i} = \sum_{v_i \sim v_j} 2w_{ij}(f_i - f_j) = 0.$$

□

Definition 17.3.6 (Discrete Laplace-Beltrami Operator) *Suppose M is a polyhedral surface with a triangulation (a triangle mesh), $f : M \to \mathbb{R}$ is a piecewise linear function, the discrete Laplace-Beltrami operator acting on f is defined as $\Delta f : M \to \mathbb{R}$,*

$$\Delta f(v_i) := \sum_{v_i \sim v_j} w_{ij}(f(v_j) - f(v_i)). \tag{17.3.5}$$

Therefore, the function f is harmonic if and only it satisfies the Lapalce-Beltrami equation,

$$\Delta f = 0, \tag{17.3.6}$$

with appropriate boundary conditions.

Equivalently, 0-form $f \in C^0(M, \mathbb{R})$ is harmonic, then its exteiror differential df is harmonic, therefore $\delta df = 0$, for each vertex $v_i \in M$, according Eqn. (16.5.7),

$$\delta df(v_i) = \frac{-1}{|\bar{v}_i|} \sum_{v_i \sim v_j} \omega_{ij} df([v_i, v_j]) = \frac{1}{|\bar{v}_i|} \sum_{v_i \sim v_j} \omega_{ij}(f(v_i) - f(v_j)).$$

Therefore, we obtain $\Delta f = 0$ again.

17.4 COMPUTATIONAL ALGORITHM

Disk Harmonic Map   Finding a harmonic map is reduced to solving a Laplace equation with some appropriate boundary conditions. As shown in the Figure 17.1, the human facial surface S is embedded in $\mathbb{R}^3$ and with the induced Euclidean metric **g**. The harmonic map $u : S \to \mathbb{D}^2$ maps the surface onto the planar unit disk. First we map the boundary of the surface onto the unit circle. We denote the boundary ∂S

as a curve in $\mathbb{R}^3$, $\gamma(s)$, where s is the arc length parameter. The boundary map is $g : \partial S \to \mathbb{S}^1$, such that

$$g(\gamma(s)) = (\cos\theta(s), \sin\theta(s)),$$

we can choose $\theta(s)$ to be a linear function of s. The Euler-Lagrange equation for the harmonic map Eqn. (17.2.1) is

$$u_{z\bar{z}} + \frac{\rho_u}{\rho} u_z u_{\bar{z}} = u_{z\bar{z}} = 0,$$

where $\rho(u)$ is the constant 1, since the target surface is the planar disk with the Euclidean metric $du d\bar{u}$. Therefore, we obtain the conventional linear Laplace equation $\Delta u = 0$ with Dirichlet boundary condition $u|_{\partial S} = g$.

In practice, the surfaces are approximated by triangulated polyhedral surfaces, as shown in Figure 17.5. The normal cycle theory gives the sufficient conditions for the convergence of discrete surfaces to the smooth surface. Given a C^2 surface S embedded in $\mathbb{R}^3$ with the induced Euclidean metric, we first densely sample the surface and compute the geodesic Delaunay triangulation, then we replace each geodesic triangle by a Euclidean triangle to obtain a discrete polyhedral surface. If the geodesic circumcircle radii uniformly converge to zero, also all the inner angles are bounded away from zero, then the discrete surfaces converge to the smooth surface. The convergence is with respect to the Riemannian metric and the Laplace-Beltrami operator.

By using the discrete Laplace-Beltrami operator, the partial differential equation is converted to a sparse linear equation system. Suppose the triangle mesh is M, its boundary ∂M is a cyclic sequence of vertices $\{v_0, v_1, \ldots, v_{n-1}\}$. First, we construct the boundary map $g : \partial M \to \mathbb{S}^1$. Let the total length of the boundary is $s = \sum_{i=0}^{n-1} |e_i|$, where $e_i = [v_i, v_{i+1}]$. The arc length parameter for v_k is $s_k = \sum_{i=0}^{k-1} |e_i|$. For each ver-

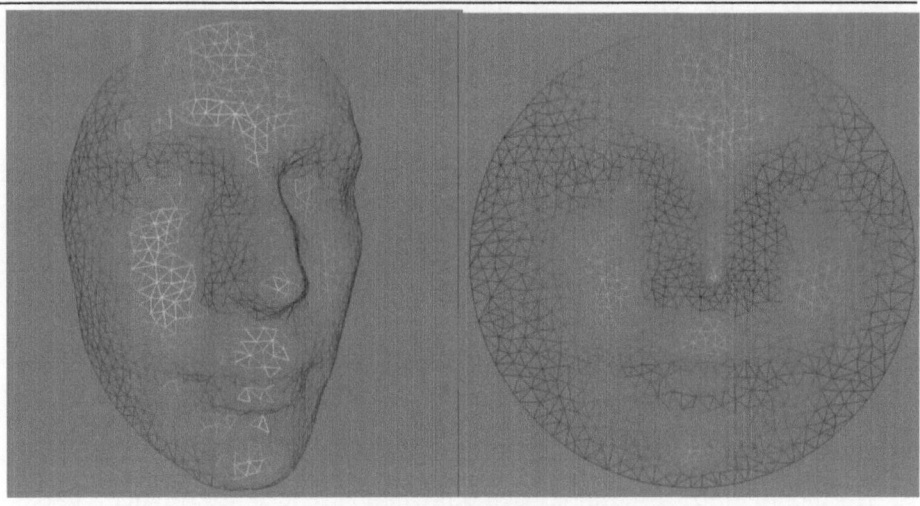

Figure 17.5 A harmonic map from the human facial surface to the planar disk.

tex $v_k \in \partial M$, we define

$$u(v_k) = (\cos \theta_k, \sin \theta_k), \quad \theta_k = 2\pi \frac{s_k}{s}. \tag{17.4.1}$$

Second, we compute the images of each interior vertex by solving the equation $\Delta u(v_i) = 0$, namely

$$\Delta u(v_i) = \sum_{v_i \sim v_j} w_{ij} (u(v_j) - u(v_i)) = 0, \tag{17.4.2}$$

where w_{ij} is the cotangent-edge weight of $[v_i, v_j]$. The details can be found in the Algorithm 11.

Algorithm 11 Algorithm for Topological Disk Harmonic Map

Require: A triangle mesh M with the disk topology;
    Extract the boundary of M, $\partial M = \{v_0, \ldots, v_{n-1}\}$;
    Compute the total length of ∂M, $s = \sum_{i=0}^{n-1} |e_i|$, where $e_i = [v_i, v_{i+1}]$;
    for each vertex $v_k \in \partial M$ **do**
        Compute the arc-length parameter of v_k, $s_k = \sum_{i=0}^{k-1} |e_i|$;
        Set the boundary condition, using Eqn. (17.4.1);
    end for
    for each edge $[v_i, v_j]$ **do**
        Compute the cotangent edge weight using Eqn. (16.5.1);
    end for
    for each interior vertex $v_i \in M \setminus \partial M$ **do**
        Construct the linear Equation (17.4.2);
    end for
    Solve the linear system to obtain $u(v_i)$'s;
Ensure: The harmonic map $u : M \to \mathbb{D}^2$.

Spherical Harmonic Map   The spherical harmonic map is more complicated. Because the target surface is a curved Riemannian manifold, the partial differential equation becomes non-linear. Furthermore, the spherical harmonic mapping is not unquie, they differ by a 6 dimensional Möbius transformation group. In order to tackle these problems, we use the so-called *non-linear heat diffusion* method.

Suppose S is a closed genus zero surface with a Riemannian metric $\mathbf{g}$. The mapping $u : S \to \mathbb{S}^2$ is treated as a vector-valued function $u : S \to \mathbb{R}^3$ and restricted on the unit sphere $\mathbb{S}^2$. The map is initialized by the conventional Gauss map, which maps each point $p \in S$ to each normal $\mathbf{n}(p) \in \mathbb{S}^2$. The Laplacian of the vector-valued function is given by $\Delta_{\mathbf{g}} u : S \to \mathbb{R}^3$, where $\Delta_{\mathbf{g}}$ is the Laplace-Beltrami operator defined on the surface. Then we use heat diffusion to reduce the harmonic energy,

$$\frac{du(p, t)}{dt} = -P_{u(p,t)}[\Delta_{\mathbf{g}} u(p, t)], \tag{17.4.3}$$

where $P_q : \mathbb{R}^3 \to T_q\mathbb{S}^2$ is the projection operator, which projects a vector v in $\mathbb{R}^3$ to the tangent plane of the unit sphere at the point $q \in \mathbb{S}^2$, equivalently

$$P_q(v) = v - \langle v, q \rangle q, \quad \forall q \in \mathbb{S}^2, \quad \forall v \in \mathbb{R}^3. \tag{17.4.4}$$

Intuitively, we use the Gauss map as the initial map, then update it along the negative Laplacian direction to reduce the harmonic energy. But we need to ensure the image points are on the target manifold, we can only move the image of each point inside the tangential space of the target manifold, therefore we project the Laplacian to the tangent plane, then update the map along the projected Laplacian. Because of this projection operation, the whole process is non-linear.

The resulting mappings differ by a Möbius transformation. In order to enusre the uniqueness of the solution, we need to normalize the map by adding the following constraint:

$$\int_S u(p,t) dA_{\mathbf{g}}(p) = 0.$$

Namely, we require the mass center of the image is at the center of the sphere. This will remove 3 freedoms of the Möbius transformation group, the left 3 freedoms come from the rotation group $SO(3)$. Since the computation doesn't induce any rotation, the solution is guaranteed to be unique.

In discrete cases, the input genus zero closed surface is represented by a triangle mesh M. Its Poincaré dual mesh is $\bar{M}$. We compute the normal to each face, then define the normal to the vertex as the linear combination of the normals to the adjacent faces:

$$\mathbf{d}(v_i) := \sum_{f_j \sim v_i} \mathbf{n}(f_j)|f_j|, \quad \mathbf{n}(v_i) := \frac{\mathbf{d}(v_i)}{|\mathbf{d}(v_i)|}. \tag{17.4.5}$$

The mapping $\mathbf{u} : M \to \mathbb{S}^2$ is initialized as the Gauss map $\mathbf{u}(v_i) = \mathbf{n}(v_i)$. In each iteration, we compute the Laplacian of the map,

$$\Delta\mathbf{u}(v_i) = \sum_{v_i \sim v_j} w_{ij}(\mathbf{u}(v_j) - \mathbf{u}(v_i)). \tag{17.4.6}$$

We update the image of each vertex along the negative direction of the projected Lapalcian, $-P_{\Delta\mathbf{u}(v_i)}[\Delta\mathbf{u}(v_i)]$, where the projection is given in Eqn. (17.4.4). The center of the mass of the images can be computed as

$$c(\mathbf{u}) = \sum_{v_i} \mathbf{u}(v_i)|\bar{v}_i|, \tag{17.4.7}$$

where $|\bar{v}_i|$ is the area of the dual cell of the vertex v_i. Then we adjust the image of vertex as $\mathbf{u}(v_i) - c(\mathbf{u})$ and normalize it to be the unit norm. By repeating this procedure, the harmonic energy of the map decreases monotonously, the maps converge to the harmonic map. The algorithm details can be found in Algorithm 12.

Algorithm 12 Algorithm for Spherical Harmonic Map

Require: A genus zero closed triangle mesh M;
        The step length λ, the threshold ε;
Ensure: A harmonic map $\mathbf{u} : M \to \mathbb{S}^2$.
  for each edge $[v_i, v_j]$ **do**
      Compute the cotangent edge weight using Eqn. (16.5.1);
  end for
  for each vertex $v_i \in M$ **do**
      Compute the vertex normal $\mathbf{n}(v_i)$, using Eqn. (17.4.5);
      Initialize the map as the Gauss map $\mathbf{u}(v_i) \leftarrow \mathbf{n}(v_i)$
  end for
  repeat
      for each vertex $v_i \in M$ **do**
          Compute the Laplacian of the map $\Delta\mathbf{u}(v_i)$ using Eqn. (17.4.6);
          Project the Laplacian to the tangent plane of $\mathbb{S}^2$, using Eqn. (17.4.4);
      end for
      for each vertex $v_i \in M$ **do**
          Update the image of v_i, $\mathbf{u}(v_i) \leftarrow \mathbf{u}(v_i) - \lambda P_{\Delta\mathbf{u}(v_i)}[\Delta\mathbf{u}(v_i)]$
      end for
      Compute the mass center using Eqn. (17.4.7);
      for each vertex $v_i \in M$ **do**
          Shift the center $\mathbf{u}(v_i) \leftarrow \mathbf{u}(v_i) - c(\mathbf{u})$;
          Normalize the image $\mathbf{u}(v_i) \leftarrow \mathbf{u}(v_i)/|\mathbf{u}(v_i)|$;
      end for
  until $\max_{v_i} \| P_{\mathbf{u}(v_i)}[\Delta\mathbf{u}(v_i)] \| < \varepsilon$

CHAPTER 18

Riemann Surface

This chapter introduces the basic concepts and theorems in Riemann surface theory. More thorough treatments can be found in [107].

18.1 RIEMANN SURFACE

Definition 18.1.1 (Holomorphic Function) *Suppose a complex function* $w :$ $\mathbb{C} \to \mathbb{C}$ *maps* $z = x + iy$ *to* $w = u + iv$, *if the function satisfies the Cauchy-Riemann equation:*

$$\frac{\partial u}{\partial x} = \frac{\partial v}{\partial y}, \quad \frac{\partial u}{\partial y} = -\frac{\partial v}{\partial x} \tag{18.1.1}$$

the it is called a holomorphic function. *If* w *is invertible, and* w^{-1} *is also holomorphic, then the function is called* bi-holomorphic.

The complex differential operators are defined as follows:

$$\frac{\partial}{\partial z} = \frac{1}{2} \left(\frac{\partial}{\partial x} - i \frac{\partial}{\partial y} \right), \quad \frac{\partial}{\partial \bar{z}} = \frac{1}{2} \left(\frac{\partial}{\partial x} + i \frac{\partial}{\partial y} \right). \tag{18.1.2}$$

Also, we can simply write the above operators as ∂_z and $\partial_{\bar{z}}$ for convenience. The Cauchy-Riemann equation can be simply written as

$$\partial_{\bar{z}} w = 0. \tag{18.1.3}$$

Suppose the function $w(z)$ is bi-holomorphic, then we have

$$dw = w_z dz + w_{\bar{z}} d\bar{z} = w_z dz.$$

The source z-plane has the Euclidean metric $dz d\bar{z}$, similarly the target w-plane has the metric $dw d\bar{w}$, then the pull-back metric induced by the mapping $w(z)$ is

$$dw d\bar{w} = |w_z|^2 dz d\bar{z},$$

which differs from the original metric $dz d\bar{z}$ by a scalar function $|w_z|^2$, therefore the mapping $w(z)$ is conformal, namely angle-preserving.

DOI: 10.1201/9781003350576-18

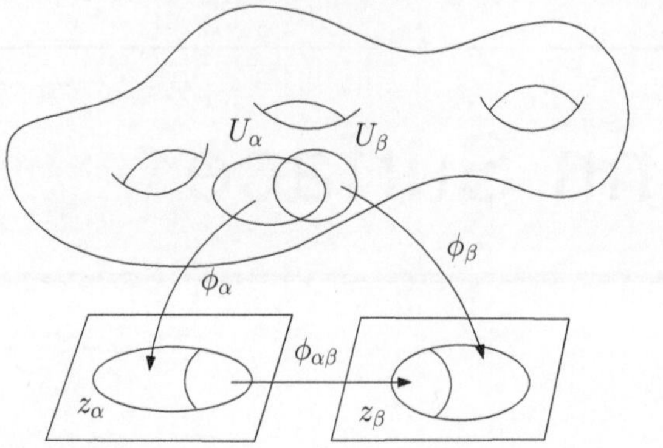

Figure 18.1 A Riemann surface.

Definition 18.1.2 (Meromorphic Function) *Suppose $f : \mathbb{C} \to \mathbb{C}$ is a complex function, $f(z) = p(z)/q(z)$, where $p(z)$ and $q(z)$ are holomorphic functions, then $f(z)$ is called a* meromorphic function.

Definition 18.1.3 (Laurent Series) *The* Laurent series *of a meromorphic function about a point z_0 is given by*

$$f(z) = \sum_{n=k}^{\infty} a_n (z - z_0)^n,$$

the series

$$\sum_{n \geq 0}^{\infty} a_n (z - z_0)^n$$

is called the analytic part *of the Laurent series; the series*

$$\sum_{n < 0} a_n (z - z_0)^n$$

is called the principal part *of the Laurent series. a_{-1} is called the* residue *of f.*

Definition 18.1.4 (Zeros and Poles) *Given a meromorphic function $f(z)$, if its Laurent series has the form*

$$f(z) = \sum_{n=k}^{\infty} a_n (z - z_0)^n,$$

if $k > 0$, then z_0 is called a zero point *of order k; if $k < 0$, then z_0 is called a* pole *of order k; if $k = 0$, then z_0 is called a* regular point. *We denote $\nu_p(f) = k$.*

Definition 18.1.5 (Conformal Atlas) *Suppose S is an oriented surface, equipped with an atals $\mathcal{A} = \{(U_\alpha, \varphi_\alpha)\}$, where each chart has complex coordinates $\varphi_\alpha : U_\alpha \to \mathbb{C}$, denoted as z_α, furthermore all the coodinate transitions are bi-holomorphic,*

$$\varphi_{\alpha\beta} : \varphi_\alpha(U_\alpha \cap U_\beta) \to \varphi_\beta(U_\alpha \cap U_\beta), \quad z_\alpha \mapsto z_\beta,$$

then the atlas is called a conformal atlas.

Definition 18.1.6 (Conformal Structure) *Suppose S is an oriented surface, $\mathcal{A}_1, \mathcal{A}_2$ are two conformal atlases of S, if their union $\mathcal{A}_1 \cup \mathcal{A}_2$ is also a conformal atlas, then we say $\mathcal{A}_1$ is conformal equivalent to $\mathcal{A}_2$. The conformal equivalence relation classifies all the conformal atalses of S into equivalence classes, each conformal equivalence class $[\mathcal{A}]$is called a* conformal structure *of the surface S.*

Because the local coordinate transisitions of a Riemann surface is bi-holomorphic, the angles are well defined on the whole surface.

Definition 18.1.7 (Riemann Surface) *An oritented topological surface with a conformal structure is called a Riemann surface (Figure 18.1).*

A surface with a Riemannian metric is called a *Riemannian surface*. A Riemannian metric structure is stronger than a conformal structure, because the metric can measure both the angles and the areas, but the conformal structure can only measure the angles. Furthermore, a Riemannian metric can induce a compatible conformal structure.

Theorem 18.1.8 *Suppose $(S, \mathbf{g})$ is an oriented Riemannian surface, then S is a Riemann surface.*

Proof For any point $p \in S$, there is a neighborhood U_α, $p \in U_\alpha$, such that we can find an isothermal coordinates (x_α, y_α) on U_α, the metric has the local representation,

$$\mathbf{g} = e^{2\lambda(x_\alpha, y_\alpha)}(dx_\alpha^2 + dy_\alpha^2),$$

we use the complex coordinate $z_\alpha = x_\alpha + iy_\alpha$, then all such kind of isothermal coordinate charts $\{(U_\alpha, z_\alpha)\}$ form a conformal atlas.

□

We can define conformal mapping between Riemann surfaces.

Definition 18.1.9 (Holomorphic Map) *Suppose $f : S \to T$ is a map between two Riemann surfaces, $(U_\alpha, \varphi_\alpha)$ is a chart of S and (V_β, ψ_β) is a chart of T,*

$$
\begin{array}{ccc}
S & \xrightarrow{\ f\ } & T \\
{\scriptstyle \varphi_\alpha}\downarrow & & \downarrow{\scriptstyle \psi_\beta} \\
\varphi_\alpha(U_\alpha) & \xrightarrow{\psi_\beta \circ f \circ \varphi_\alpha^{-1}} & \psi_\beta(V_\beta)
\end{array}
$$

then the local representation of the map is

$$\varphi_\beta \circ f \circ \varphi_\alpha^{-1} : \varphi_\alpha(U_\alpha) \to \psi_\beta(V_\beta).$$

If all the local representations are bi-holomorphic, then f is called a biholomorphic map between the two Riemann surfaces, namely a conformal map.

The concepts of holomorphic and meromorphic functions can be generalized to Riemann surfaces.

Definition 18.1.10 (holomorphic/Meromorphic Function) *Suppose a Riemann surface* $(S, \{(U_\alpha, \varphi_\alpha)\})$ *is given. A complex function is defined on the surface* $f :$ $S \to \mathbb{C} \cup \{\infty\}$. *If on each local chart* $(U_\alpha, \varphi_\alpha)$, *the local representation of the functions* $f \circ \varphi_\alpha^{-1} : \mathbb{C} \to \mathbb{C}(\cup\{\infty\})$ *is holomorphic (meromorphic), then* f *is called a holomorphic (meromorphic) function defined on* S.

A memromorphic function can be treated as a holomorphic map from the Riemann surface to the unit sphere.

Proposition 18.1.11 *Suppose* $(S, \{z_\alpha\})$ *is a compact Riemann surface, if* $f : S \to \mathbb{C}$ *is holomorphic, then* f *is constant.*

Proof Because S is compact, $|f|$ reaches its maximum at some point $p \in S$. By the max norm theorem, if the maximal norm is obtained at an interior point, then holomorphic function is constant.

$\square$

18.2 MEROMORPHIC DIFFERENTIAL

Definition 18.2.1 (Meromorphic Differential) *Given a Riemann surface* $(S,$ $\{z_\alpha\})$, ω *is a meromorphic differential of order* n, *if it has local representation,*

$$\omega = f_\alpha(z_\alpha)(dz_\alpha)^n,$$

where $f_\alpha(z_\alpha)$ *is a meromorphic function,* n *is an integer; if* $f_\alpha(z_\alpha)$ *is a holomorphic function, then* ω *is called a* holomorphic differential *of order* n.

A holomorphic differential of order 1 is called a *holomorphic 1-form*; A holomorphic differential of order 2 is called a *holomorphic quadratic differential*; A meromorphic differential of order 4 is called a *meromorphic quartic differential*.

Theorem 18.2.2 (Holomorphic q-differential) *Assume* S *is a compact Riemann surface with genus* g, q *is an integer, then the dimension of the complex linear space* Ω^q *of holomorphic q-differentials is as follows:*

- *when* $g = 0$,

$$dim \ \Omega^q = \begin{cases} 0, & q \geq 1 \\ 1 - 2q, & q \leq 0 \end{cases}$$

- *when* $g = 1$, $dim \ \Omega^q = 1, \forall q \in \mathbb{Z}$

- *when* $g > 1$,

$$dim \ \Omega^q = \begin{cases} 0, & q < 0 \\ 1, & q = 0 \\ g, & q = 1 \\ (2q - 1)(g - 1) & q > 1 \end{cases}$$

Here we only prove the dimension of the linear space of holomorphic 1-forms.

Proposition 18.2.3 *Suppose S is a compact Riemann surface, the real component and the imaginary component of a holomorphic 1-form are two conjugate real harmonic 1-forms,*

$$\omega + i \star \omega. \tag{18.2.1}$$

Proof Suppose the holomorphic 1-form has the local represenation $f(z)dz$, where f is a holomorphic function $f(x, y) = u(x, y) + iv(x, y)$, then f satisfies the Cauchy-Riemann equation,

$$u_x - v_y = 0, \quad u_y + v_x = 0.$$

The complex differential form

$$f(z)dz = (u + iv)(dx + idy) = (udx - vdy) + i(vdx + udy),$$

by the definition of Hodge star operator,

$$\star(udx - vdy) = udy + vdx.$$

Therefore,

$$f(z)dz = \omega + i \star \omega.$$

where $\omega = udx - vdy$. Next, we show ω is harmonic,

$$d\omega = u_y dy \wedge dx - v_x dx \wedge dy = -(u_y + v_x)dx \wedge dy = 0.$$

and

$$\delta\omega = -\star d \star (udx - vdy) = -\star d(udy + vdx) = -\star (u_x - v_y)dx \wedge dy = 0.$$

The conjugate of a harmonic 1-form is also harmonic, so $\star\omega$ is harmonic. □

This implies the group of all the holomorphic 1-forms Ω^1 is $2g$ real dimensional, where g is the genus of the Riemann surface.

Theorem 18.2.4 *Suppose S is a genus g closed Riemann surface, all the holomorphic 1-forms on S form a $2g$ real dimensional group $\Omega^1(S)$.*

Proof According to Hodge theorem 16.4.15, the harmonic 1-form cohomology group $H_\Delta^1(S, \mathbb{R})$ is isomorphic to the de Rham cohomology group $H_{dR}^1(S, \mathbb{R})$, and according to the proposition 18.2.3, $\Omega^1(S)$ is isomorphic to the harmonic 1-form cohomology group $H_\Delta^1(S, \mathbb{R})$. Therefore, $\Omega^1(S)$ is isomorphic to $H_{dR}^1(S, \mathbb{R})$, which is $2g$ real dimensional.

□

Figure 18.2 Holomorphic 1-form group basis of a genus two surface.

The above proof gives us a computational algorithm. Given a closed surface S of genus g, we first compute a basis of harmonic 1-form group, then compute the conjugate of each base harmonic 1-form. Each pair of the base harmonic form and its conjugate form a holomorphic 1-form. The details can be found in Algorithm 13. Figure 18.2 shows a set of holomorphic 1-form group basis.

Algorithm 13 Algorithm for Holomorphic 1-form group basis

Require: A closed genus g triangle mesh M.
Ensure: A set of holomorphic 1-form basis $\{\omega_1 + i \star \omega_1, \ldots, \omega_{2g} + i \star \omega_{2g}\}$;
 Compute a set of basis of harmonic 1-form basis, $\{\omega_1, \ldots, \omega_{2g}\}$, using Algorithm 9;
 Compute the conjugate harmonic 1-form, $\{\star\omega_1, \ldots, \star\omega_{2g}\}$, using Algorithm 10;
 for each base harmonic 1-form ω_i **do**
  form a holomorphic 1-form, $\omega_i + i \star \omega_i$.
 end for

Definition 18.2.5 (Trajectories of Holomorphic 1-forms) *Suppose S is a Riemann surface, ω is a holomorphic 1-form, for a point $p \in S$ and a tangent direction $\mathbf{v} \in T_pS$, if $\omega(\mathbf{v})$ is real (imaginary), the $\mathbf{v}$ is called a* horizontal (vertical) direction *of ω at p. A curve γ is called a* horizontal (vertical) trajectory *of ω, if its tangent vectors are horizontal (vertical) everywhere.*

Definition 18.2.6 (Natural Coordinates of Holomorphic 1-forms) *Suppose S is a Riemann surface, ω is a holomorphic 1-form, U is a neighborhood on S, fix a point $q \in U$, then for any point $p \in U$, define the* natural coordinates *of ω at p as*

$$\zeta(p) := \int_q^p \omega.$$

Since ω is holomorphic, the integration is path independent in the neighborhood,

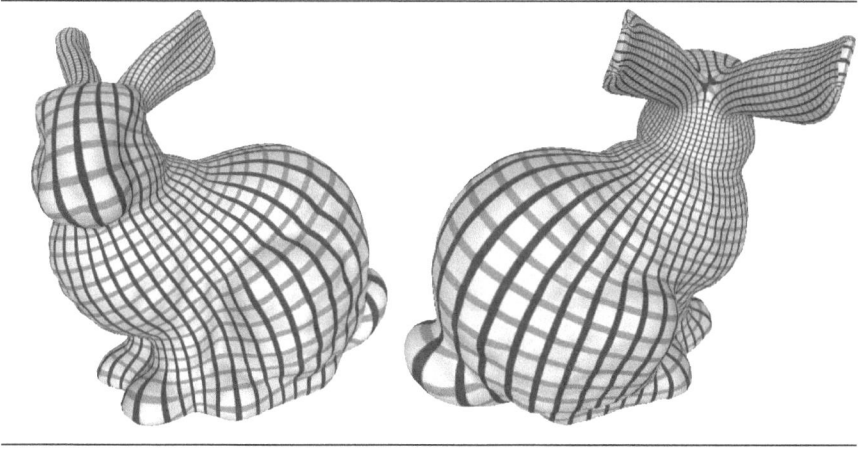

Figure 18.3 Horizontal and vertical trajectories of a holomorphic 1-form.

therefore the natural coordinates are well defined. The horizontal (vertical) trajectories of ω are the horizontal (vertical) lines of the natural coordinates of ω. Figure 18.3 shows the horizontal and vertical trajectories of a holomorphic 1-form on the Stanford bunny model.

Definition 18.2.7 (Zeros and Poles of Meromorphic Differentials) *Given a Riemann surface $(S, \{z_\alpha\})$, ω is a meromorphic differential with local representation,*

$$\omega = f_\alpha(z_\alpha)(dz_\alpha)^n.$$

If z_α is a pole (or a zero) of f_α with order k, then z_α is called a pole (or a zero) *of the meromorphic differential ω of order k.*

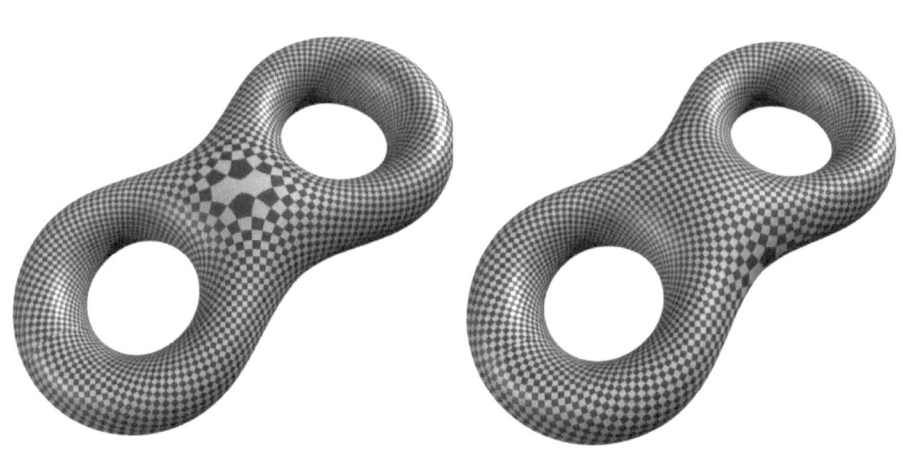

Figure 18.4 The zeros on different holomorphic 1-forms on a genus two surface.

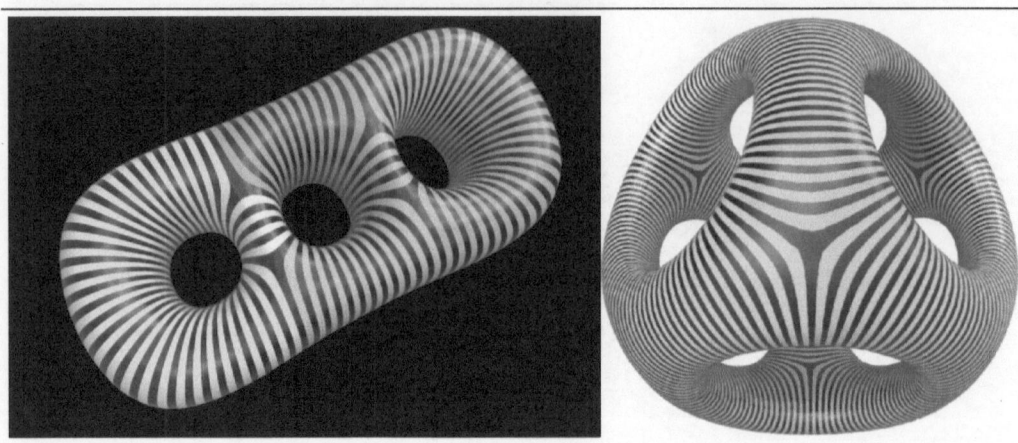

Figure 18.5 Horizontal trajectories of holomorphic quadratic differentials.

We use $Sing_\omega$ to denote the singularity set of ω. Locally near a regular point p, the differential $\omega = f(z)dz^n$ can be represented as the n-th power of a 1-form $h(z)dz$ where $h^n(z) = f(z)$ and thus $h(z) = \sqrt[n]{f(z)}$ coincides with one of n possible branches of the n-th root. We call this n-valued 1-form the n-th *roots* of ω, which is a globally well-defined multi-valued meromorphic 1-form on S.

Definition 18.2.8 (Trajectories of meromorphic differentials) *Given a meromorphic n-differential ω on S we define n distinct line fields on $S \setminus Sing_\omega$ as follows: at each non-singular point z there are exactly n distinguished directions dz at which $\omega = f(z)(dz)^n$ attains real values. Integral curves of these line fields are called trajectories of ω.*

Suppose ω is a meromorphic quadratic differential, dz is a *horizontal (vertical)* *direction* if $f(z)(dz)^2 > 0$ $(f(z)(dz)^2 < 0)$. Integral curves of horizontal direction are called *horizontal (vertical) trajectories*.

Definition 18.2.9 (Strebel Differential) *A meromorphic quadratic differential is called a Strebel differential, if all of its horizontal trajectories are finite.*

Figure 18.5 shows the horizontal trajectories of Strebel differentials. Note that the vertical trajectories of a Strebel differential may not necessarily be finite.

18.3 RIEMANN-ROCH THEOREM

Definition 18.3.1 (Divisor) *The Abelian group freely generated by points on a Riemann surface is called the* divisor group, *every element is called a* divisor, *which has the form*

$$D = \sum_p n_p p.$$

The degree *of a divisor is defined as* $\deg(D) = \sum_p n_p$. *Suppose* $D_1 = \sum_p n_p p$, $D_2 = \sum_p m_p p$, *then* $D_1 \pm D_2 = \sum_p(n_p \pm m_p)p$; $D_1 \leq D_2$ *if and only if for all* p, $n_p \leq m_p$.

Definition 18.3.2 (Meromorphic Function Divisor) *Given a meromorphic funciton f defined on a Riemann surface S, its divisor is defined as*

$$(f) = \sum_p ord_p(f)p,$$

where $ord_p(f)$ is defined as k if p is the k-order zero point of f; $-k$ if p is the k order pole of f.

The divisor of a meromorphic differential ω is defined in the similar way.

Definition 18.3.3 (Meromorphic Differential Divisor) *Suppose ω is a meromorphic differential on a Riemann surface S, suppose $p \in S$ is a point on S, we define the order of ω at p as*

$$ord_p(\omega) = ord_p(f_p),$$

where f_p is the local representation of ω in a neighborhood of p, $\omega = f_p(dz)^n$. Its divisor is defined as

$$(\omega) = \sum_p ord_p(\omega)p.$$

Let $\mathcal{M}_0(S)$ be the vector space of all meromorphic functions on S; $\mathcal{M}_1(S)$ be the vector space of all meromorphic 1-forms on S. Given a divisor D, we define a linear space over $\mathbb{C}$ of meromorphic functions as

$$L(D) := \{f \in \mathcal{M}_0(S) | (f) \geq -D\},$$

its dimension is denoted as $l(D)$; another linear space over $\mathbb{C}$ of meromorphic differentials as

$$I(D) := \{\omega \in \mathcal{M}_1(S) | (\omega) \geq D\},$$

its dimension is $i(D)$.

Theorem 18.3.4 (Riemann-Roch) *Suppose D is divisor on a compact Riemann surface with genus g, then*

$$l(D) - i(D) = deg(D) + 1 - g. \tag{18.3.1}$$

On a genus g compact Riemann surface S, suppose f is a meromorphic function, then (f) is called a *principle divisor*. The divisor of a meromorphic 1-form is called a *canonical divisor*. For any meromorphic 1-form ω, the sum of all the residues of ω vanishes. df/f is a meromorphic differential, hence every principle divisor (f) has degree zero, $deg((f)) = 0$. Suppose ω is a holomorphic 1-form, then $deg((\omega)) = 2g - 2$, since its real component is a harmonic 1-form, which is dual to a vector field. According the Poincare-Hopf index theorem 15.5.8, the total index of zeros is $2 - 2g$. Suppose ω is a meromorphic 1-form, then $deg((\omega)) = 2g - 2$, since it differs from the divisor of a holomorphic 1-form by a principle divisor.

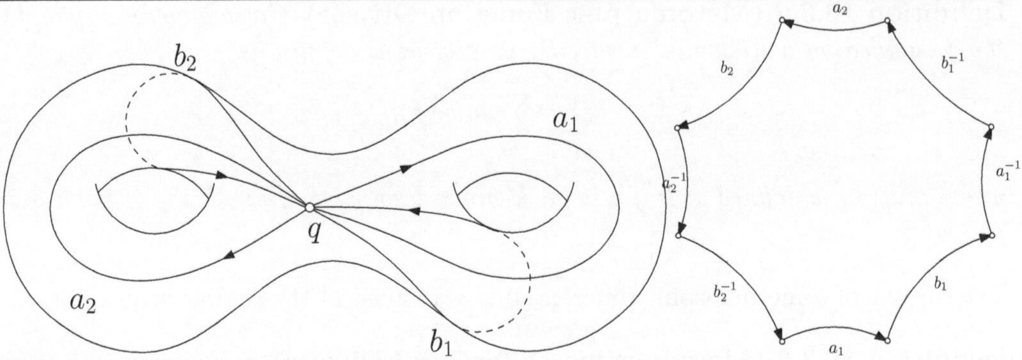

Figure 18.6 Canonical fundamental group basis.

18.4 ABEL-JACOBIAN THEOREM

Suppose $\{a_1, b_1, \ldots, a_g, b_g\}$ is a set of canonical basis for the homology group $H_1(S, \mathbb{Z})$ as shown in Figure 18.6. Each a_i and b_i represent the curves around the inner and outer circumferences of the ith handle.

Let $\{\omega_1, \omega_2, \ldots, \omega_g\}$ be a normalized basis of Ω^1, the linear space of all holomorphic 1-forms over $\mathbb{C}$. The choice of basis is dependent on the homology basis chosen above; the normalization signifies that

$$\int_{a_i} \omega_j = \delta_{ij}, \quad i, j = 1, 2, \ldots, g.$$

For each curve γ in the homology group, we can associate a vector λ_γ in $\mathbb{C}^g$ by integrating each of the g 1-forms over γ,

$$\lambda_\gamma = \left(\int_\gamma \omega_1, \int_\gamma \omega_2, \ldots, \int_\gamma \omega_g \right)$$

We define a $2g$-real-dimensional lattice Λ in $\mathbb{C}^g$,

$$\Lambda = \left\{ \sum_{i=1}^g s_i \, \lambda_{a_i} + \sum_{j=1}^g t_j \, \lambda_{b_j}, \quad s_i, t_j \in \mathbb{Z} \right\} \tag{18.4.1}$$

Definition 18.4.1 (Jacobian) *The Jacobian of the Riemann surface S, denoted as $J(S)$, is the compact quotient $\mathbb{C}^g / \Lambda$.*

The Abel-Jacobi map embeds the Riemann surface S in the Jacobian $J(S)$.

Definition 18.4.2 (Abel-Jacobi Map) *Fix a base point $p_0 \in S$. The Abel-Jacobi map is a map $\mu : S \to J(S)$. For every point $p \in S$, choose a curve c from p_0 to p; the Abel-Jacobi map μ is defined as follows:*

$$\mu(p) = \left(\int_{p_0}^q \omega_1, \int_{p_0}^q \omega_2, \ldots, \int_{p_0}^q \omega_g \right) \quad \mod \Lambda, \tag{18.4.2}$$

where the integrals are all along c.

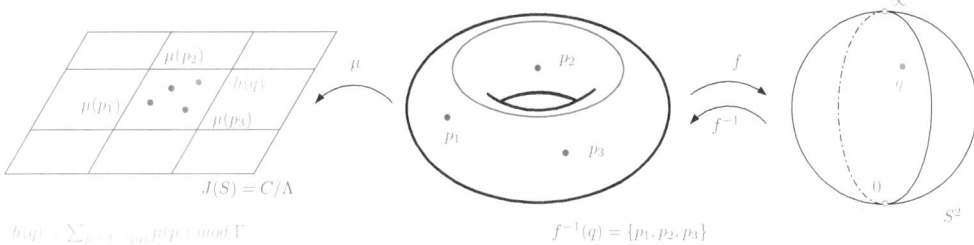

Figure 18.7 Proof of Abel theorem.

It can be shown $\mu(p)$ is well-defined, that the choice of curve c doesn't not affect the value of $\mu(p)$.

The following Abel theorem characterizes the principle divisor.

Theorem 18.4.3 (Abel) *Let D be an divisor of degree 0 on S, then D is the divisor of a meromorphic function f if and only if $\mu(D) = 0$ in the Jacobian $J(S)$.*

Proof As shown in Figure 18.7, suppose S is a torus, its Jacobian $J(S)$ is a flat torus $\mathbb{C}/\Lambda$, $\mu : S \to \mathbb{C}/\Lambda$ is the Abel-Jacobi map. Assume $f : S \to \mathbb{C} \cup \{\infty\}$ is a meromorphic function, thus f can be treated as a holomorphic mapping from the flat torus to the unit sphere, $f : S \to \mathbb{S}^2$. We construct a mapping from the sphere to the flat torus $h : \mathbb{S}^2 \to \mathbb{C}/\Lambda$: for any $q \in \mathbb{S}^2$,

$$h(q) = \sum_{p_i \in f^{-1}(q)} \mu(p_i) \qquad \mod \Lambda.$$

The mapping h is holomorphic, it can be lifted to their universal covering spaces, such that the following diagram commutes:

$$
\begin{array}{ccc}
\mathbb{S}^2 & \xrightarrow{\tilde{h}} & \mathbb{C} \\
\downarrow{\scriptstyle id} & & \downarrow{\scriptstyle \pi} \\
\mathbb{S}^2 & \xrightarrow{h} & \mathbb{C}/\Lambda
\end{array}
$$

The map $\tilde{h} : \mathbb{S}^2 \to \mathbb{C}$ is treated as a holomorphic function. Because $\mathbb{S}^2$ is compact, $\tilde{h}$ is bounded, according to Louville theorem $\tilde{h}$ is constant, hence $h = \pi \circ \tilde{h}$ is also constant. Therefore,

$$\mu((f)) = h(0) - h(\infty) = 0.$$

Now, suppose S is of high genus g, we can similarly construct the holomorphic map $h = \sum_{p_i \in f^{-1}(q)} \mu(p_i) \mod \Gamma$ from $\mathbb{S}^2$ to $J(S)$, then lift it to the universal covering space $\tilde{h} : \mathbb{S}^2 \to \mathbb{C}^g$. Because $\mathbb{S}^2$ is compact, the holomorphic function $\tilde{h}$ is bounded, thus constant in $\mathbb{C}^g$, therefore h is constant. $\mu((f)) = h(0) - h(\infty) = 0$.

$\square$

CHAPTER 19

Conformal Mapping

This chapter focuses on computational algorithms for conformal mappings of surfaces with different topologies based on holomorphic differentials, including topological quadrilateral, topological annulus, topological disk, topological poly-annulus and torus. These algorithms have been widely used for a broad range of applications in engineering and medical imaging fields.

19.1   TOPOLOGICAL QUADRILATERAL

Definition 19.1.1 (Topological Quadrilateral) *Suppose S is an oriented genus zero surface with a single boundary, which is piecewise smooth. There are four marked boundary points $\{v_0, v_1, v_2, v_3\}$ sorted counter clockwisely. The surface with the marked boundary points is called a* topological quarilateral, *and denoted as* $Q(v_0, v_1, v_2, v_3)$.

Theorem 19.1.2 (Topological Qudrilateral Conformal Module) *Suppose $Q(v_0, v_1, v_2, v_3)$ is a topological quadrilateral with a Riemannian metric* **g**, *then there exists a unique conformal map $\varphi : S \to \mathbb{C}$, such that φ maps Q to a rectangle, $\varphi(v_0) = 0$, $\varphi(v_1) = 1$. The height of the image rectangle is called the* conformal modulus *of the surface.*

The inverse of the conformal modulus of a topological quadrilateral Q is called the *extremal length* of Q. Figure 19.1 left frame shows a human facial surface with four marked boundary points, which is a topological quadrilateral. The middle frame shows the conformal mapping which maps the surface onto a planar rectangle. The width of the rectangle is 1, the height is the conformal modulus of the topological quadrilateral. The mapping pulls back a checkerboard texture image on the rectangle to the facial surface, all the right corner angles of the checkers are well preserved on the surface, this demonstrates the conformality of the mapping.

We want to compute the conformal map $\varphi : Q \to R$, where R is the planar rectangle with the unit width and the unkown height. Assume the boundary of Q consists of four consecutive segments $\partial Q = \gamma_0 + \gamma_1 + \gamma_2 + \gamma_3$, such that $\partial \gamma_k = v_{k+1} - v_k$, where $k + 1$ means modulo 4. We compute two harmonic functions $f_0, f_1 : Q \to \mathbb{R}$

DOI: 10.1201/9781003350576-19

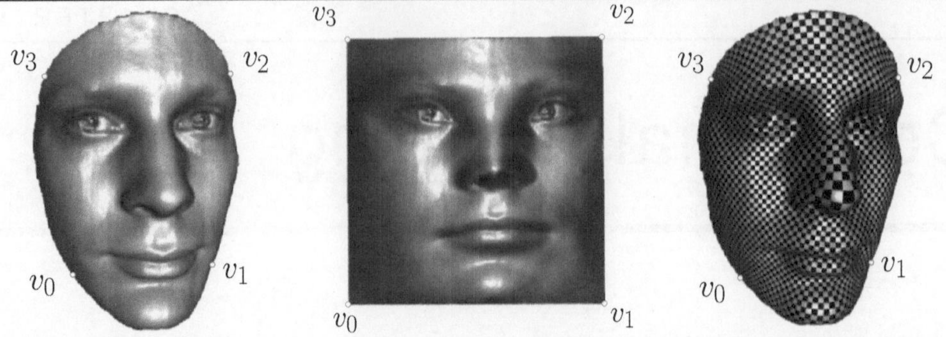

Figure 19.1 The conformal modulus of a topological quadrilateral.

with Dirichlet and Neumann boundary conditions,

$$\begin{cases} \Delta f_0 & = 0 \\ f_0|_{\gamma_1} & = 1 \\ f_0|_{\gamma_3} & = 0 \\ \frac{\partial f_0}{\partial \mathbf{n}}|_{\gamma_0 \cup \gamma_2} & = 0 \end{cases} \quad \begin{cases} \Delta f_1 & = 0 \\ f_1|_{\gamma_0} & = 0 \\ f_1|_{\gamma_2} & = 1 \\ \frac{\partial f_1}{\partial \mathbf{n}}|_{\gamma_1 \cup \gamma_3} & = 0 \end{cases} \qquad (19.1.1)$$

Then the differentials df_0 and df_1 are two exact harmonic 1-forms and orthogonal to each other. Furthermore, we can find a scalar $\lambda \in \mathbb{R}$, such that the Hodge star of df_0 and df_1 satisfy the following equations,

$$\star df_0 = \lambda df_1, \quad \star df_1 = -\frac{1}{\lambda} df_0.$$

We can construct holomorphic 1-forms,

$$\omega_0 = df_0 + i \star df_0, \quad \omega_1 = df_1 + i \star df_1.$$

Then we define two conformal mappings,

$$\varphi_0(p) = \int_{v_0}^{p} \omega_0, \quad \varphi_1(p) = \int_{v_0}^{p} \omega_1,$$

where the integration is path independent. Then the image $R_0 = \varphi_0(Q)$ is a rectangle with width 1 and height λ, $R_1 = \varphi_1(Q)$ is another rectangle with width $1/\lambda$ and height 1. By direct computation,

$$\int_Q df_k \wedge \star df_k = \int_Q \left((\partial_x f_k)^2 + (\partial_y f_k)^2 \right) d\Omega_{\mathbf{g}} = E(f_k), \quad k = 0, 1, \qquad (19.1.2)$$

the left hand side equals to the harmonic energy. Harmonic energy is invariant under conformal transformation. f_k is the horizontal coordinate function on R_k, its harmonic energy equals to the area of R_k, hence we have

$$E(f_0) = \lambda, \quad E(f_1) = \frac{1}{\lambda}.$$

Therefore, $\lambda = \sqrt{E(f_0)/E(f_1)}$.

In discrete case, the Laplace equation (19.1.1) with Dirichlet and Neumann boundary conditions can be solved using the discrete Laplace-Beltrami operator Eqn. (17.3.5). The discrete Laplace-Beltrami operator Eqn. (17.3.5) on the boundary vertices automatically satisfies the Neumann boundary condition. The harmonic energy Eqn. (19.1.2) is approximated by the discrete harmonic energy Eqn. (17.3.4). The algorithmic pipeline can be found in Alg. 15.

Algorithm 14 Integration

Require: A topological disk M, a base vertex v_0, a holomorphic 1-form ω.
Ensure: A conformal map $\varphi : M \to \mathbb{C}$, $\varphi(v_i) = \int_{v_0}^{v_i} \omega$;
  for each vertex $v_i \in M$ **do**
    Label v_i as unprocessed;
  end for
  Label v_0 as processed;
  Enqueue $Q \leftarrow v_0$;
  while Q is unempty **do**
    $v \leftarrow Q.pop()$;
    for each vertex w adjacent to v **do**
      if w is unprocessed **then**
        $\varphi(w) \leftarrow \varphi(v) + \omega([v, w])$;
        Label w as processed;
        Enqueue $Q \leftarrow w$;
      end if
    end for
  end while

Figure 19.2 shows another example of conformal modulus of a topological quadrilateral using this algorithm.

The extremal length theory can be generalized to the combinatorial setting. Given a topological quadrilateral, represented as a triangle mesh, we can associate each

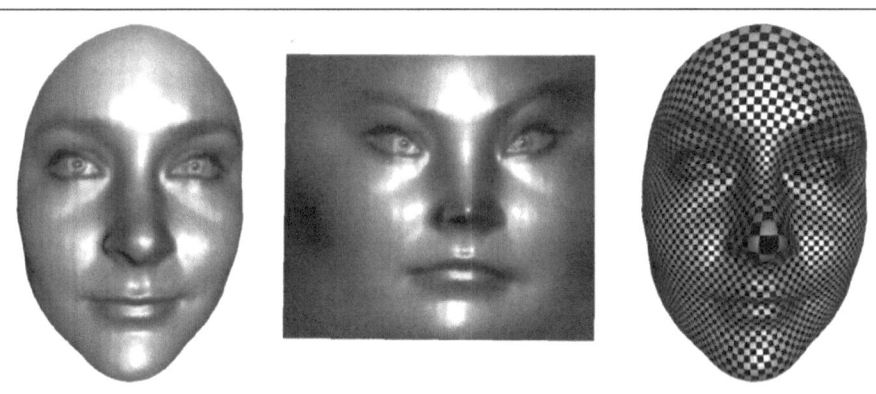

Figure 19.2   The conformal modulus of a topological quadrilateral.

Algorithm 15 Extremal Length

Require: A topological quadrilateral $Q(v_0, v_1, v_2, v_3)$.

Ensure: A conformal map $\varphi : Q \to R$, where R is a rectangle with the unit width;

Compute the harmonic functions equation (19.1.1) using the discrete Laplace-Beltrami operator Eqn. (17.3.5);

Compute the harmonic energies in Eqn. (19.1.2) using the discrete harmonic energy Eqn. (17.3.4);

Compute the conformal modulus $\lambda = \sqrt{E(f_0)/E(f_1)}$;

Construct holomorphic 1-form $\omega = df_0 + i\lambda df_1$;

Compute the conformal map $\varphi(v_i) = \int_{v_0}^{v_i} \omega$ using Alg. 14.

---

vertex with a circle, such that for every edge the two circles of the end vertices are tangential to each other. This type of configuration is called a circle packing. By adjusting the circle radii, we can find a special circle packing, which isometrically embeds the triangle mesh onto a planar rectangle, as shown in the left frame of Figure 19.3. This is the circle packing version of the combinatorial extremal length. Similarly, we can replace the circles with squares to realize square tiling version of combinatorial extremal length, as shown in the right frame of the same figure. The circle packing can be carried out by the discrete surface Ricci flow 20, the square tiling can be obtained by quadratic programming.

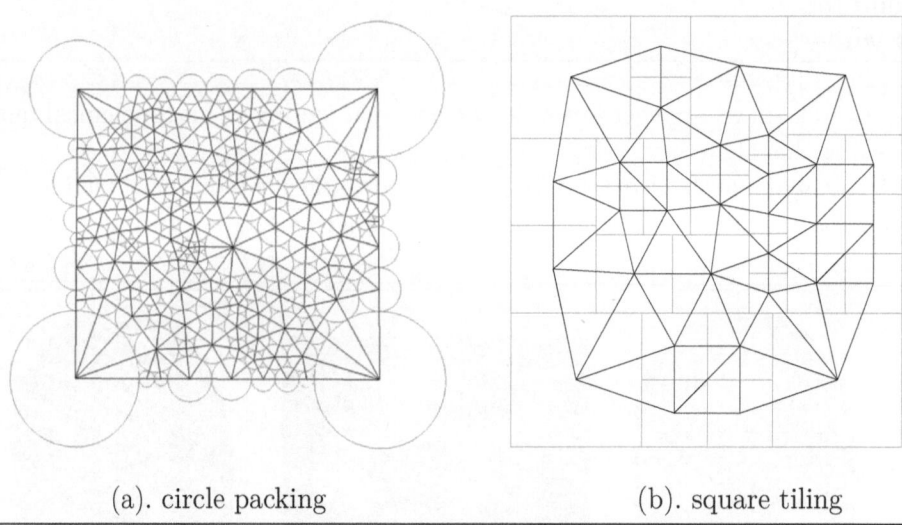

(a). circle packing   (b). square tiling

Figure 19.3 Combinatorial extremal lengths.

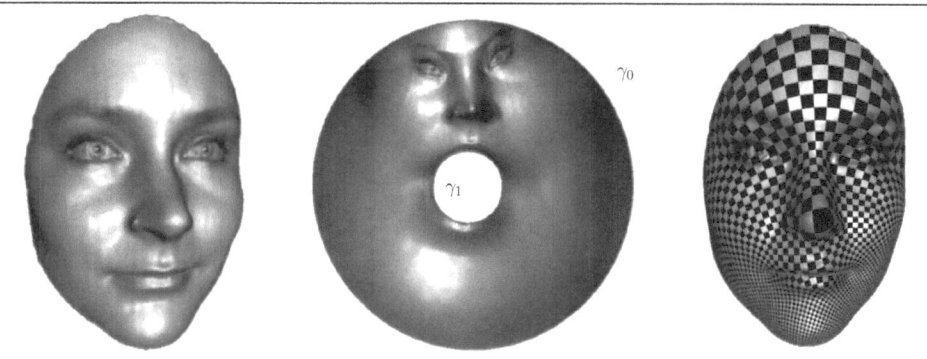

Figure 19.4 Conformal modulus of a topological annulus.

19.2 TOPOLOGICAL ANNULUS

Definition 19.2.1 (Topological Annulus) *Suppose S is a genus zero surface with two boundary components, then S is called a* topological annulus.

In practice, we assume the topological annulus is C^2 smooth with a Riemannian metric **g**, the boundary is piecewise smooth. As shown in Figure 19.4, a toplogical annulus can be conformally mapped onto a canonical planar annulus with the unit outer radius. The mapping is unique up to a rotation. The inner circle radius is the conformal modulus of the surface. The checkerboard texture on the plane is pulled back onto the surface, the right corner angles demonstrates the conformality of the map.

Theorem 19.2.2 (Topological Annulus Conformal Modulus) *Suppose S is a topological annulus with a Riemannian metric **g**, the boundary of S consists of two loops $\partial S = \gamma_0 - \gamma_1$, then there exists a conformal map $\varphi : (S, \mathbf{g}) \to \mathbb{A}$, where $\mathbb{A} \subset \mathbb{C}$ is a canonical annulus, $\varphi(\gamma_0)$ is the unit circle, $\varphi(\gamma_1)$ is another concentric circle with radius r. The mapping φ is unique up to a planar rotation. The scalar $-\log r$ is called the* conformal modulus *of the topological annulus.*

As shown in Figure 19.5, the topological annulus S has one exterior boundary loop γ_0 and an interior boundary loop γ_1. The first homology group of S is 1-dimensional, $H_1(S, \mathbb{Z}) = \mathbb{Z}$, $[\gamma_0]$ is the generator. We find the shortest path τ from γ_1 to γ_0. We slice S along τ to obtain a topological quadrilateral $\bar{S}$, the four sides are τ^-, γ_0, τ^+ and γ_1 as shown in the right frame.

Figure 19.6 shows the holomorphic 1-form basis on the topological annulus S. The left frame displays the exact harmonic 1-form, the middle frame illustrates the non-exact harmonic 1-form, the right frame is the holomorphic 1-form. We compute a harmonic function $f : S \to \mathbb{R}$ with the Dirichlet boundary condition,

$$
\begin{cases}
\Delta f &= 0 \\
f|_{\gamma_0} &= 0 \\
f|_{\gamma_1} &= 1
\end{cases}
\tag{19.2.1}
$$

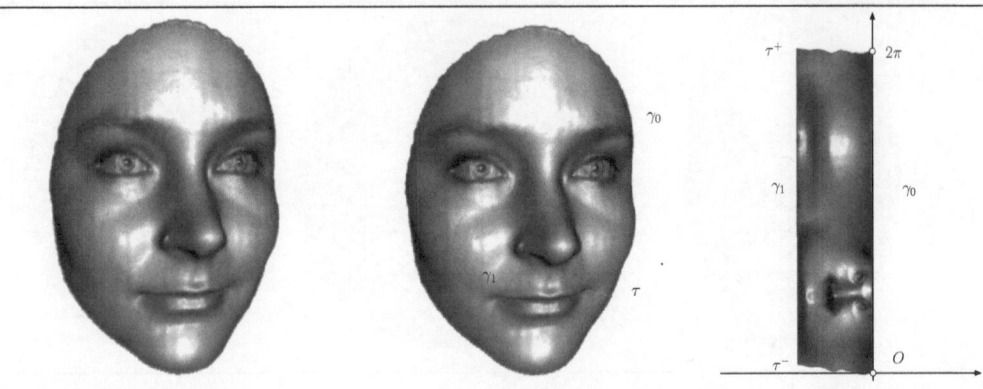

Figure 19.5 The boundary curves γ_0, γ_1 and the cutting curve τ on the topological annulus.

then df is an exact harmonic 1 form, whose integration along τ is 1. The exact harmonic 1-form df is visualized by the level set of the harmonic function f.

The non-exact harmonic 1-form is the basis of $H^1_\Delta(S, \mathbb{R})$, we can compute it using the method similar to the algorithm 9. We set a function $g : \bar{S} \to \mathbb{R}$, such that

$$g(p) = \begin{cases} 1 & p \in \tau^+ \\ 0 & p \in \tau^- \\ \text{rand} & \text{otherwise} \end{cases} \qquad (19.2.2)$$

Since g are constant on τ^+ and τ^-, therefore dg is a differential 1-form well defined on the original surface S, denoted as η. By construction, η is a closed 1-form, and its integration along γ_0 equals to 1, hence it is the basis of the first cohomology group $H^1(S, \mathbb{R})$. According to Hodge theorem 16.4.15, we can find a function $h : S \to \mathbb{R}$,

Figure 19.6 The holomorphic 1-form on the topological annulus.

such that $\zeta := \eta + dh$ is harmonic, hence $\delta\zeta = 0$,

$$\delta\eta = -\delta dh. \tag{19.2.3}$$

The harmonic 1-form satisfies the Neumann boundary condition. Suppose $p \in \gamma_0$ or $p \in \gamma_1$, a tangent vector $\mathbf{v} \in T_p S$ is orthogonal to the boundary, then $\zeta(\mathbf{v}) = 0$.

We construct two holomorphic 1-forms,

$$\omega_0 := df + i \star df, \quad \omega_1 := \zeta + i \star \zeta,$$

their integration on $\bar{S}$ give two conformal mappings φ_0 and φ_1 from $\bar{S}$ to a planar rectangle R_0 and R_1,

$$\varphi_k(p) = \int_q^p \omega_k, \quad k = 0, 1, \tag{19.2.4}$$

where the base point q is the intersection point of γ_0 and τ. Figure 19.5 right frame shows the image of φ_1. Figure 19.6 right frame shows the checkerboard texture mapping induced by φ_1, the right corner angles demonstrates the conformality of φ_1. By translation, rotation and scaling, we can transform R_1 to the position as shown in the right frame of Figure 19.5, $\varphi_1(\gamma_0)$ is along the imaginary axis, the height is 2π, $\varphi_1(\gamma_1)$ is mapped to the vertical line with real coordinate equals to $-2\pi\lambda$. This mapping is denoted as φ_2. Then we use the complex exponential map $z \mapsto \exp(z)$ to map the rectangle to the canonical annulus $\mathcal{A}$. The composed map

$$\exp \circ \varphi_2 \circ \varphi_1 : S \to \mathcal{A}$$

is called the circular conformal map for the topological annulus. The circular conformal map transforms the surface $(S, \mathbf{g})$ to the canonical annulus $\mathcal{A}$, as shown in the middle frame of Figure 19.4. The right frame shows the checkerboard texture mapping induced by this conformal map, the right corner angles shows the conformality of the map.

The conjugate harmonic 1-form of ζ and the exact harmonic 1-form df satisfy the relation

$$\star\zeta = \lambda df, \tag{19.2.5}$$

where $\lambda \in \mathbb{R}$ is a scalar. Then R_1 is a rectangle with the unit width, and the height is λ; R_0 is a rectangle with the unit height, and the width is $\frac{1}{\lambda}$. The harmonic energy of f is

$$E(f) = \int_S df \wedge \star df, \tag{19.2.6}$$

which equals to the area of R_0, namely $\frac{1}{\lambda}$. The L^2 norm (harmonic energy) of ζ is

$$E(\zeta) = \int_S \zeta \wedge \star\zeta$$

equals to the area of R_1, namely λ. Therefore, we obtain

$$\lambda = \sqrt{E(\zeta)/E(f)}. \tag{19.2.7}$$

Algorithm 16 Topological Annulus Conformal Mapping

Require: A topological annulus $(S, \mathbf{g})$;

Ensure: A circular conformal map $\varphi : S \to \mathcal{A}$, where $\mathcal{A}$ is an annulus with the unit outer radius;

1. Trace the boundary loops of S, $\partial S = \gamma_0 - \gamma_1$;
2. Compute the shortest path τ from γ_1 to γ_0;
3. Slice S along τ to obtain a topological quadrilateral $\bar{S}$;
4. Compute the harmonic function f by solving the Laplace equation (19.2.1) using the discrete Laplace-Beltrami operator Eqn. (17.3.5);
5. Compute the harmonic energy of f in Eqn. (19.2.6) using Eqn. (17.3.4);
6. Compute a random function g on $\bar{S}$ using Eqn. (19.2.2), set $\eta \leftarrow dg$;
7. Hodge decompose η to obtain the harmonic form ζ by solving Eqn. (19.2.3) using the discrete operator Eqn. (16.5.11);
8. Compute the L^2 norm (harmonic energy) of ζ using Eqn. (16.5.13);
9. Compute the scalar λ using Eqn. (19.2.7);
10. Construct the holomorphic 1-form $\omega_1 = \zeta + i\lambda df$;
11. Compute the conformal map $\varphi_1 : \bar{S} \to R_1$, $\varphi_1(p) = \int_q^p \omega_1$ using Alg. 14;
12. Find the translation, rotation and scaling map φ_2, translate R_1 to the canonical position;
13. Compose the maps φ_1, φ_2 and exp to obtain the resulting conformal map $\varphi \leftarrow \exp \circ \varphi_2 \circ \varphi_1$;

The algorithmic details for the conformal mapping of a topological annulus can be found in algorithm 16.

Figure 19.7 shows another example of the topological annulus conformal mapping, the inner circle radius of the image annulus gives the conformal modulus of the input surface.

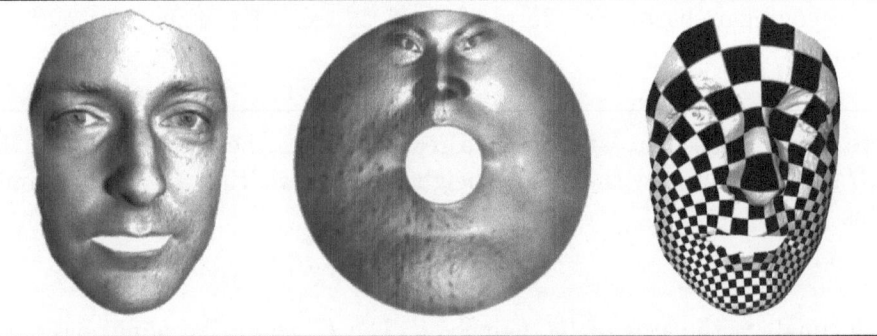

Figure 19.7 Conformal modulus of topological annulus.

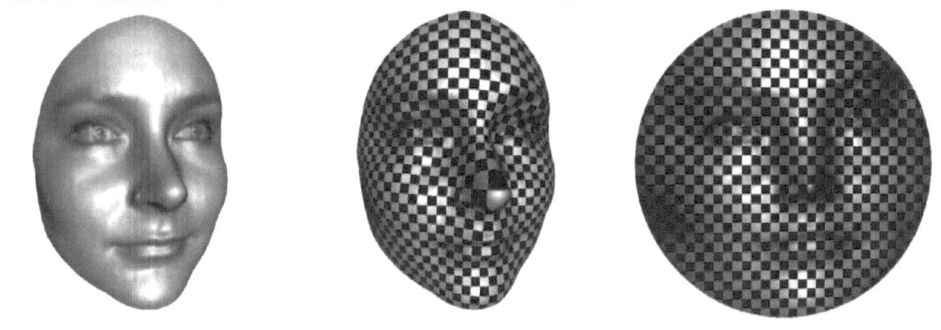

Figure 19.8 A Riemann mapping of a topological disk.

19.3   RIEMANN MAPPING FOR TOPOLOGICAL DISK

Definition 19.3.1 (Topological Disk) *Suppose S is an oriented, genus zero surface with a single boundary component, then S is called a topological disk.*

We can further assume S is C^2 smooth with a Riemannian metric **g**, the boundary is piecewise smooth.

Definition 19.3.2 (Möbius Transformation) *Suppose $\varphi : \mathbb{D}^2 \to \mathbb{D}^2$ maps the unit disk on the complex plane $\mathbb{C}$ to itself with the form*

$$\varphi(z) = e^{i\theta} \frac{z - z_0}{1 - \bar{z}_0 z}, \quad |z_0| < 1,$$

then φ is called a Möbius transformation.

The famous Riemann mapping theorem claims that any topological disk with a Riemannian metric can be conformally mapped onto the unit planar disk, the mapping is unique up to a Möbius transformation. Figure 19.8 shows one Riemann mapping example. The left frame shows a human facial surface captured by a 3D scanner. The right frame displays the image of the Riemann mapping. The middle frame demonstrates the conformality using the checkerboard texture mapping.

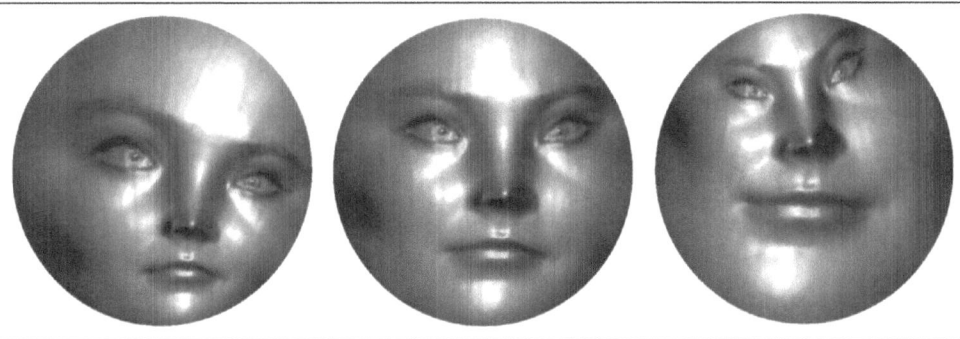

Figure 19.9 Möbius transformations.

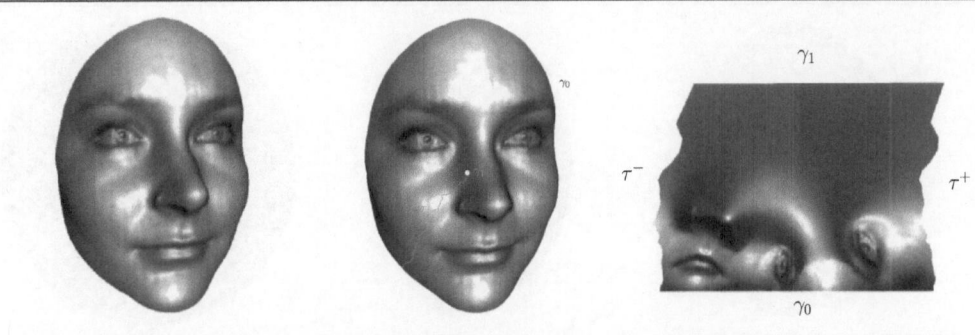

Figure 19.10 The topological disk is punctured to become a topological annulus and conformally mapped to a flat cylinder.

Theorem 19.3.3 (Riemann Mapping) *Suppose S is a topological disk with a Riemannian metric $\mathbf{g}$, then there exists a conformal mapping $\varphi : (S, \mathbf{g}) \to \mathbb{D}^2$, where $\mathbb{D}^2 \subset \mathbb{C}$ is the canonical planar unit disk. The mapping is unique up to a Möbius transformation.*

Figure 19.9 shows the Möbius transformations of the Riemann mapping image. Each transformation maps different point z_0 to the center, and is composed with a planar rotation. In practice, we can fix an interior point $p \in S$ and a boundary point $q \in \partial S$, and normalize the Riemann mapping by requiring $\varphi(p) = 0$ and $\varphi(q) = 1$.

The computational algorithm is as follows: first we puncture a small hole on the surface S, and convert the surface into a topological annulus $\bar{S}$. Second, we compute the conformal mapping from the punctured surface $\bar{S}$ onto a canonical planar annulus $\mathcal{A}$. Third, we fill the center circular hole of the annulus $\mathcal{A}$. We shrink the size of the puncture, such that the diameter is less than $1/n$, this produces a topological annulus $\bar{S}_n$. For each $\bar{S}_n$, we compute the conformal mapping $\varphi_n : \bar{S}_n \to \mathcal{A}_n$ with the consistent normalization condition $\varphi_n(q) = 1$, where $q \in \partial \bar{S}_n$. The limit of the mapping sequence $\{\varphi_n\}$ is the desired Riemann mapping, which maps the center of the puncture to the center of the planar disk and $q \in \partial S$ to 1.

Figure 19.10 shows the pucture process. The left frame is the original topological

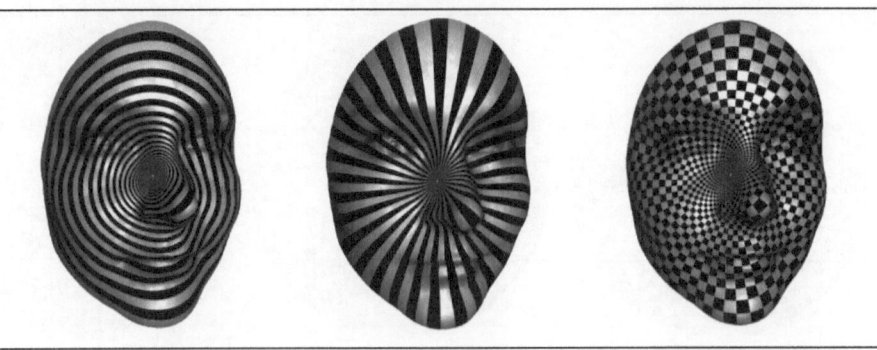

Figure 19.11 Conformal modulus of topological annulus.

disk S, we puncture a hole to get the topological annulus $\bar{S}$ in the middle frame. The exterior boundary of $\bar{S}$ is γ_0, the interior boundary is γ_1. Then we compute the shortest path connecting γ_1 to γ_0. We slice the punctured surface $\bar{S}$ along τ to get a topological disk $\hat{S}$. As shown in Figure 19.11, we compute the exact harmonic 1-form df as shown in the left frame by solving the Laplace equation with Dirichlet boundary condition Eqn. (19.2.1), and the harmonic 1-form group basis ζ by solving the equations (19.2.2) and (19.2.3). We next compute the scalar λ using Eqn.(19.2.7) such that $\star\zeta = \lambda df$, and construct the holomorhic 1-form $\omega = \zeta + i\lambda df$. The integration of ω on $\hat{S}$ induces a conformal map from the topological quadrilateral $\bar{S}$ to a flat rectangle $\mathcal{R}$,

$$\varphi_1 : \hat{S} \to \mathcal{R}, \quad p \mapsto \int_q^p \omega,$$

where p is the intersection point between γ_0 and τ. The right frame of Figure 19.10 shows the flat rectangle $\mathcal{R}$. We use a planar rotation and translation φ_2 to map the lower left corner of $\mathcal{R}$ to the origin, and the lower right corner to $2\pi i$, then use an exponential map exp to map the planar rectangle $\varphi_2(\mathcal{R})$ to the canonical planar annulus. The composition map $\exp \circ \varphi_2 \circ \varphi_1$ maps the punctured surface $\bar{S}$ to the planar circular annulus, then we fill the center hole to get the Riemann mapping as shown in Figure 19.8. By changing the puncture positions, we obtain different Riemann mappings, which differ by Möbius transformations as shown in Figure 19.9. The algorithmic details can be found in Alg. 17.

Algorithm 17 Riemann Mapping

Require: A topological disk $(S, \mathbf{g})$;

Ensure: A Riemann mapping $\varphi : S \to \mathbb{D}^2$, where $\mathbb{D}^2$ is the unit planar disk;

   1. Find the vertex v_0 with the furthest distance to the boundary;

   2. Puncture the surface at v_0, remove its one-ring neighborhood, to obtain the topological annulus $\bar{S}$;

   3. Compute the conformal map φ from $\bar{S}$ to an planar annulus $\mathcal{A}$ using Alg. 16;

   4. Set $\varphi(v_0) \leftarrow 0$, fill the inner hole of $\mathcal{A}$;

    In practice, we choose the one-ring neighborhood of a vertex as the puncture, the vertex is the furthest one from the boundary γ_0. Algorithm 18 uses the width-first searching method to find the furthest vertex from the boundary.

19.4 TOPOLOGICAL POLY-ANNULUS SLIT MAP

Definition 19.4.1 (Topological Poly-Annulus) *Suppose S is an oriented genus zero surface with multiple boundadry components, then S is called a* topological poly-annulus.

In practice, we assume S is C^2 smooth, with a Riemannian metric $\mathbf{g}$, the boundary components are piecewise smooth. As shown in Figure 19.12 left frame, a human facial surface with the eyes and the mouth regions removed is a topological poly-annulus S. The middle frame shows a conformal mapping $\varphi : (S, \mathbf{g}) \to \mathcal{A}$ which

---

Algorithm 18 Furthest Vertex From the Boundary

Require: A triangle mesh S with boundary $\partial S \neq \emptyset$;
Ensure: The vertex with the maximal combinatorial distance from ∂S;
  for each $v \in S$ **do**
    set $v.distance \leftarrow \infty$ and $v.processed \leftarrow false$;
  end for
  for each $v \in \partial S$ **do**
    set $v.distance \leftarrow 0$ and $v.processed \leftarrow true$;
    Enqueue v to the priority queue $Q \leftarrow v$
  end for
  while Q is not empty **do**
    $v \leftarrow Q.pop()$;
    for each $w \sim v$ **do**
      if $!w.processed$ **then**
        Set $w.distance \leftarrow v.distance + 1$ and $w.processed \leftarrow true$;
        Enqueue w to the priority queue $Q \leftarrow w$;
      end if
    end for
  end while
  The last vertex in Q is the furthest one.

---

maps S onto a planar annulus $\mathcal{A}$, with concentric circular slits. $\varphi(\gamma_0)$ is the exterior circular boundary of $\mathcal{A}$, $\varphi(\gamma_1)$ is the interior circular boundary of $\mathcal{A}$, γ_2 and γ_3 are mapped to concentric circular slits. The right frame demonstrates the conformality of φ by a checkerboard texture mapping.

Theorem 19.4.2 (Poly-Annulus Circular Slit Map) *Suppose $(S, \mathbf{g})$ is a poly-annulus, $\partial S = \gamma_0 - \gamma_1 \cdots - \gamma_n$, then there exists a conformal mapping $\varphi : (S, \mathbf{g}) \to \mathcal{A}$, which maps S to the canonical planar annulus with concentric circular slits. $\varphi(\gamma_0)$ is the exterior circular boundary of $\mathcal{A}$, $\varphi(\gamma_1)$ is the interior circular boundary of $\mathcal{A}$, $\varphi(\gamma_k)$'s are concentric circular slits, $k = 2, \ldots, n$. All such kind of conformal mappings differ by a planar rotation.*

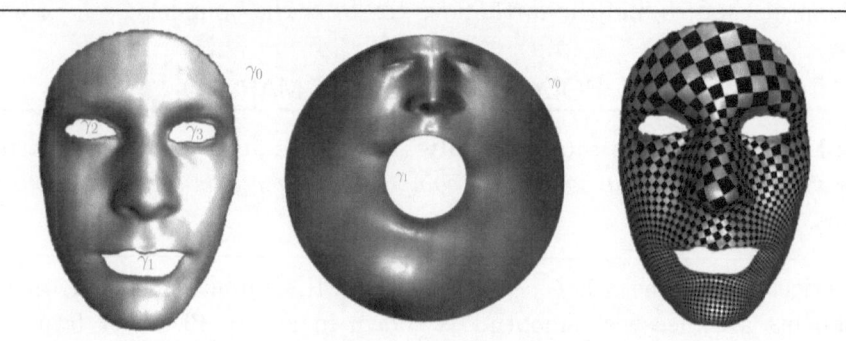

Figure 19.12   Circular slit map of a topological poly-annulus.

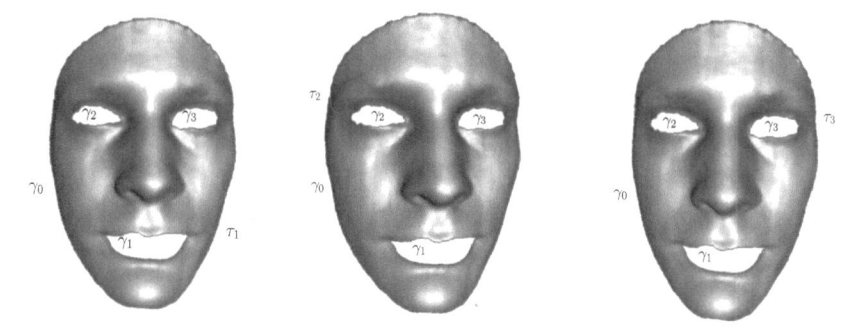

Figure 19.13 The shortest paths from inner boundary components γ_1, γ_2 and γ_3 to the outer boundary component γ_0.

In the following, we explain the procedure to compute the circular slit map of the poly-annulus $(S, \mathbf{g})$. Suppose the boundary of the surface is

$$\partial S = \gamma_0 - \gamma_1 \cdots - \gamma_n,$$

where γ_0 is the exterior boundary component of S, γ_k's are interior boundary components, $k = 1, 2, \ldots, n$. As shown in Figure 19.13, we compute the shortest path τ_k from γ_k to γ_0. Then we slice the surface along τ_k's to obtain a topological disk, denoted as $\bar{S}$.

We compute the basis of the exact harmonic 1-form group of $(S, \mathbf{g})$. For each $k \in \{1, 2, \ldots, n\}$, we compute a harmonic function $f_k : (S, \mathbf{g}) \to \mathbb{R}$ by solving the Laplace equation with the Dirichlet boundary condition:

$$\begin{cases} \Delta f_k &= 0 \\ f|_{\gamma_k} &= 1 \\ f|_{\gamma_i} &= 0, \quad \forall i \neq k \end{cases} \qquad (19.4.1)$$

This can be solved using the discrete Laplace-Beltrami operator Eqn. (17.3.5). Figure 19.14 displays the exact harmonic 1-form group basis df_k's by illustrating the level sets.

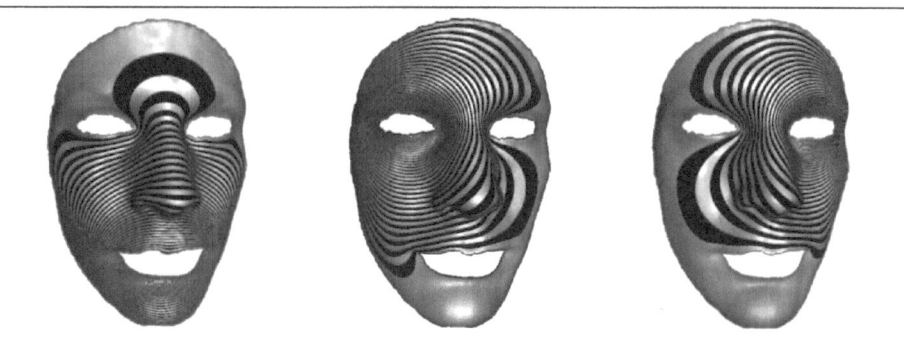

Figure 19.14 Exact harmonic form group basis.

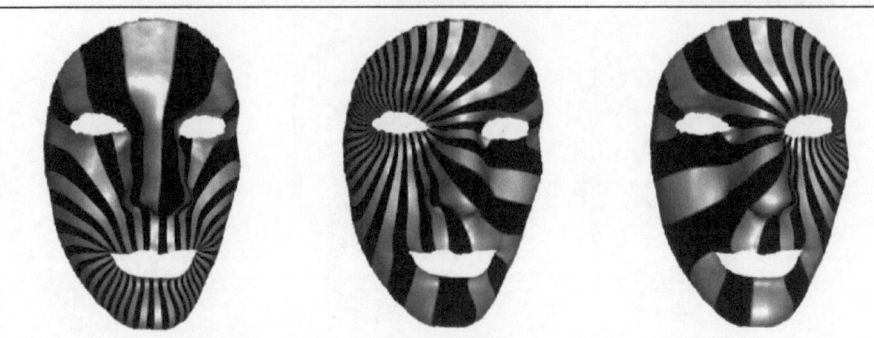

Figure 19.15 Non-exact harmonic form group basis.

The basis of the harmonic 1-form cohomology group $H^1_\Delta(S, \mathbb{R})$ can be obtained as follows. For each $k \in \{1, 2, \ldots, n\}$, we slice the original surface S along τ_k to get $\bar{S}_k$, construct a random function $g_k : \bar{S}_k \to \mathbb{R}$,

$$g_k(p) = \begin{cases} 1 & p \in \tau_k^+ \\ 0 & p \in \tau_k^- \\ \text{rand} & \text{otherwise} \end{cases} \tag{19.4.2}$$

Since g_k is constant on τ_k^+ and τ_k^-, therefore dg_k is a differential 1-form well defined on the original surface S, denoted as η_k. By construction, η_k is a closed 1-form,

$$\int_{\gamma_0} \eta_k = 1, \quad \int_{\gamma_k} \eta_k = -1, \quad \int_{\gamma_i} \eta_k = 0, \quad \forall i \neq 0, k.$$

Hence $\{\eta_1, \eta_2, \ldots, \eta_n\}$ form the basis of the first cohomology group $H^1(S, \mathbb{R})$. According to Hodge theorem 16.4.15, we can find a function $h_k : S \to \mathbb{R}$, such that $\zeta_k := \eta_k + dh_k$ is harmonic, hence $\delta\zeta_k = 0$,

$$\delta\eta_k = -\delta dh_k. \tag{19.4.3}$$

This can be solved using the discrete co-differential operator Eqn. (16.5.11). The harmonic 1-form ζ_k satisfies the Neumann boundary condition. Suppose $p \in \gamma_k$, a tangent vector $\mathbf{v} \in T_pS$ is orthogonal to the boundary, then $\zeta_k(\mathbf{v}) = 0$. Figure 19.15 shows the basis of the non-exact harmonic 1-form group $H^1_\Delta(S, \mathbb{R})$.

Figure 19.16 Holomorphic 1-forms $df_k + i \star df_k$, $k = 1, 2, \ldots, n$.

In the next step, we compute the conjugates of the exact harmonic 1-forms. We examine an exact harmonic 1-form df_k, the boundaries are level sets of f_k by observing Figure 19.14. Its conjugate harmonic form $\star df_k$ should satisfy the Neumann boundary condition, therefore

$$\star df_k = \lambda_{k,1}\zeta_1 + \lambda_{k,2}\zeta_2 + \cdots + \lambda_{k,n}\zeta_n.$$

We construct a linear system,

$$
\begin{pmatrix} \int_S \zeta_1 \wedge {}^*df_k \\ \int_S \zeta_2 \wedge {}^*df_k \\ \vdots \\ \int_S \zeta_n \wedge {}^*df_k \end{pmatrix}
=
\begin{pmatrix} \int_S \zeta_1 \wedge \zeta_1 & \cdots & \int_S \zeta_1 \wedge \zeta_n \\ \int_S \zeta_2 \wedge \zeta_1 & \cdots & \int_S \zeta_2 \wedge \zeta_n \\ \vdots & & \vdots \\ \int_S \zeta_n \wedge \zeta_1 & \cdots & \int_S \zeta_n \wedge \zeta_n \end{pmatrix}
\begin{pmatrix} \lambda_{k,1} \\ \lambda_{k,2} \\ \vdots \\ \lambda_{k,n} \end{pmatrix}
\tag{19.4.4}
$$

the integration $\int_S \zeta_i \wedge \star df_k$ can be computed using Eqn. (16.5.13), the integration $\int_S \zeta_i \wedge \zeta_j$ using Eqn. (16.5.12). We solve the linear system to obtain the coefficients, then construct the holomorphic 1-forms

$$\eta_k := df_k + i \star df_k.$$

The integration of η_k on $\bar{S}_k$ gives a conformal map, $\psi_k : \bar{S}_k \to \mathbb{C}$,

$$\psi_k(p) = \int_{q_k}^{p} \eta_k, \quad k = 1, 2, \ldots, n,$$

where q_k is the intersection point between γ_0 and τ_k. Figure 19.16 shows the holomorphic 1-forms η_k.

Similarly, we can compute the conjugates of the non-exact harmonic 1-form basis. The 1-form ζ_k satisfies the Neumann boundary condition, then its conjugate $\star\zeta_k$ should satisfies the Dirichlet boundary condition, therefore

$$\star\zeta_k = \mu_{k,1}df_1 + \mu_{k,2}df_2 + \cdots + \mu_{k,n}df_n.$$

We construct a linear system,

$$
\begin{pmatrix} \int_S df_1 \wedge {}^*\zeta_k \\ \int_S df_2 \wedge {}^*\zeta_k \\ \vdots \\ \int_S df_n \wedge {}^*\zeta_k \end{pmatrix}
=
\begin{pmatrix} \int_S df_1 \wedge df_1 & \cdots & \int_S df_1 \wedge df_n \\ \int_S df_2 \wedge df_1 & \cdots & \int_S df_2 \wedge df_n \\ \vdots & & \vdots \\ \int_S df_n \wedge df_1 & \cdots & \int_S df_n \wedge df_n \end{pmatrix}
\begin{pmatrix} \mu_{k,1} \\ \mu_{k,2} \\ \vdots \\ \mu_{k,n} \end{pmatrix}
\tag{19.4.5}
$$

the integration $\int_S df_i \wedge \star\zeta_k$ can be computed using Eqn. (16.5.13), the integration $\int_S df_i \wedge df_j$ using Eqn. (16.5.12). By solving the linear system, we can obtain the linear combination coefficients $\mu_{k,i}$'s, then construct the holomorphic 1-form

$$\omega_k := \zeta_k + i \star \zeta_k.$$

The integration of ω_k on $\bar{S}$ gives a conformal map, $\varphi_k : \bar{S} \to \mathbb{C}$,

$$\varphi_k(p) = \int_{q_k}^{p} \omega_k, \quad k = 1, 2, \ldots, n,$$

where q_k is the intersection point between γ_0 and τ_k. Figure 19.17 shows the holomorphic 1-forms ω_k, the conformality of φ_k is demonstrated by the checkerboard texture mapping.

Once we have obtained the holomorphic 1-form group basis $\{\omega_1, \omega_2, \ldots, \omega_n\}$ of $\Omega^1(S)$. We choose two boundary loops, say γ_0 and γ_1, we find a holomorphic 1-form ω, $\omega = \sum_{k=1}^{n} \lambda_k \omega_k$, $\lambda_k \in \mathbb{R}$, satisfying the conditions

$$\mathrm{Img} \int_{\gamma_0} \omega = 2\pi, \quad \mathrm{Img} \int_{\gamma_1} \omega = -2\pi, \quad \mathrm{Img} \int_{\gamma_k} \omega = 0, \quad 2 \le k \le n. \qquad (19.4.6)$$

This induces a linear system,

$$\begin{pmatrix} 2\pi \\ -2\pi \\ 0 \\ \vdots \\ 0 \end{pmatrix} = \begin{pmatrix} \int_{\gamma_0} \star\zeta_1 & \int_{\gamma_0} \star\zeta_2 & \cdots & \int_{\gamma_0} \star\zeta_n \\ \int_{\gamma_1} \star\zeta_1 & \int_{\gamma_1} \star\zeta_2 & \cdots & \int_{\gamma_1} \star\zeta_n \\ \int_{\gamma_2} \star\zeta_1 & \int_{\gamma_2} \star\zeta_2 & \cdots & \int_{\gamma_2} \star\zeta_n \\ \vdots & \vdots & & \vdots \\ \int_{\gamma_n} \star\zeta_1 & \int_{\gamma_n} \star\zeta_2 & \cdots & \int_{\gamma_n} \star\zeta_n \end{pmatrix} \begin{pmatrix} \lambda_1 \\ \lambda_2 \\ \vdots \\ \lambda_n \end{pmatrix} \qquad (19.4.7)$$

note that the $n + 1$ linear equations are linearly dependent on each other, the rank of the linear system is n. Once we obtain the λ_k's, we can construct ω, then define a conformal mapping $\varphi : \bar{S} \to \mathcal{A}$ as

$$\varphi_0^1(p) = \exp\left(\int_q^p \omega \right), \qquad (19.4.8)$$

where the base point q is the intersection point of γ_0 and τ_1. φ_0^1 maps τ_k^+ to τ_k^-, therefore in fact φ_0^1 is defined on the original surface S. φ_0^1 maps γ_0 to the exterior boundary of the annulus $\mathcal{A}$, γ_1 to the interior boundary, and each γ_k, $2 \le k \le n$, to a concentric circular slit. The left column in Figure 19.18 shows φ_0^1, the checkerboard texture mapping demonstrates the conformality of the slit map. In fact, we can choose arbitrarily two boundary loops γ_i and γ_j, use the same algorithm to construct the circular slit map φ_i^j. Figures 19.18 and 19.19 show various φ_i^j's.

Figure 19.17 The holomorphic 1-forms $\omega_k = \zeta_k + i \star \zeta_k$, $k = 1, 2, \ldots, n$.

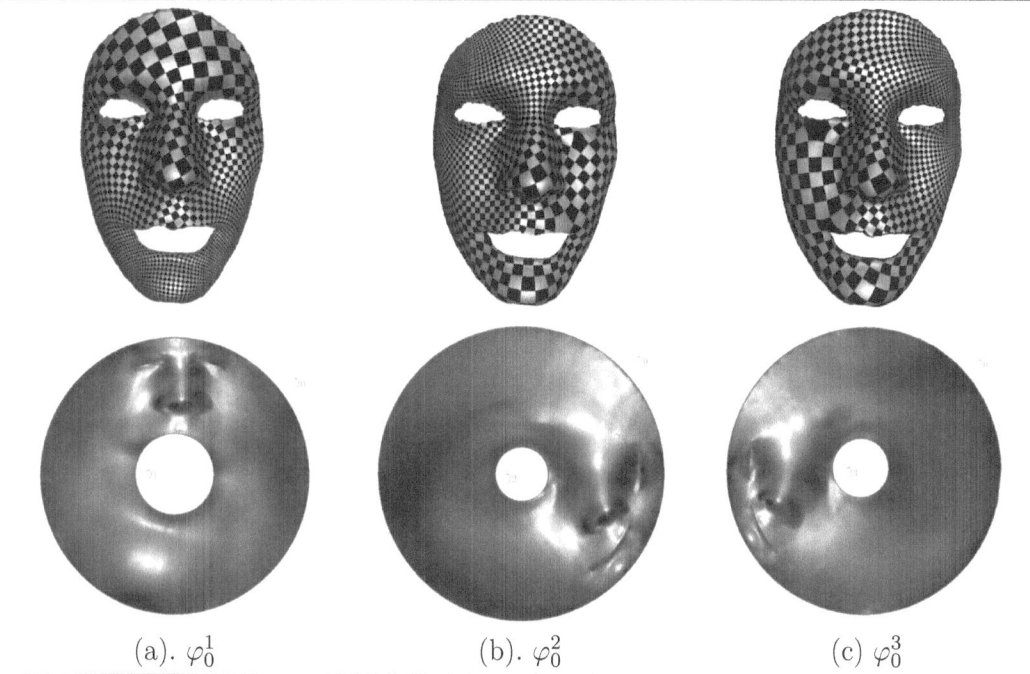

(a). φ_0^1 (b). φ_0^2 (c) φ_0^3

Figure 19.18   The circular slit maps of a poly-annulus I.

19.5   KOEBE'S ITERATION FOR POLY ANNULUS

Definition 19.5.1 (Circle Domain) *A circle domain $\mathcal{A}$ is the unit planar disk $\mathbb{D}^2$ with a finite number of circular holes, namely*

$$\mathcal{A} = \mathbb{D}^2 \setminus \bigcup_{k=1}^{n} B(c_i, r_i),$$

where the disks $B(c_i, r_i)$'s are disjoint and contained in $\mathbb{D}^2$.

Theorem 19.5.2 (Poly-annulus Circular Map) *Suppose $(S, \mathbf{g})$ is a poly annulus, $\partial S = \gamma_0 - \gamma_1 \cdots - \gamma_n$, then there exists a conformal mapping $\varphi : (S, \mathbf{g}) \to \mathcal{A}$, which maps S to a circular domain, each boundary loop γ_k is mapped onto a planar circle. Furthermore, such kind of mappings differe by Möbius transformations.*

The inner circle centers and the radii denoted as

$$\{(c_1, r_1), (c_2, r_2), \ldots, (c_n, r_n)\}$$

form another representation of the conformal modulus of the poly-annulus. We can normalize the circular map by a Möbius transformation, such that

1. The center c_1 is the center of the unit disk;

2. The center c_2 is on the real axis.

Algorithm 19 Circular Split Maps for Topological Poly-Annulus

---

Require: A topological poly-annulus $(S, \mathbf{g})$;

Ensure: A circular slit map $\varphi_0^1 : S \to \mathcal{A}$, where $\mathcal{A}$ is an annulus with concentric circular slits;

Trace the boundary loops of S, $\partial S = \gamma_0 - \gamma_1 \cdots - \gamma_n$;

Compute the shortest paths τ_k from γ_k to γ_0, $k = 1, 2, \ldots, n$;

Slice S along τ_k's to obtain a topological disk $\bar{S}$;

for each $k \in \{1, 2, \ldots, n\}$ **do**

Compute the harmonic function f_k by solving Eqn. (19.4.1) using the discrete Laplace-Beltrami operator Eqn. (17.3.5);

end for

for each $k \in \{1, 2, \ldots, n\}$ **do**

Slice S along τ_k to obtain $\bar{S}_k$;

Compute a random function g_k on $\bar{S}_k$ using Eqn. (19.4.2), set $\eta_k \leftarrow dg_k$;

Hodge decompose η_k to obtain the harmonic form ζ_k by solving Eqn. (19.4.3) using the discrete co-differential operator Eqn. (16.5.11);

end for

for each $k \in \{1, 2, \ldots, n\}$ **do**

Compute the conjugate form $\star\zeta_k$ by solving Eqn. (19.4.5);

Set the holomorphic 1-form $\omega_k \leftarrow \zeta_k + i \star \zeta_k$;

end for

Construct a holomorphic 1-form ω by solving Eqn. (19.4.7);

Construct the circular slit map φ by integrating ω on $\bar{S}$ Eqn. (19.4.8) using Alg. 14;

Therefore, we see the conformal modulus requires $3n - 3$ parameters, this is the dimensional of the Teichmüller space of topological poly-annulus with n inner holes.

Figure 19.20 shows a human facial surface with 15 inner holes, which is conformally mapped onto a planar circle domain, each boundary component is mapped to a circle. The checkerboard texture mapping induced by the map shows the conformality. Such kind of conformal mapping is not unique, all of them differ by Möbius transformations.

The conformal circular map from a poly-annulus to a circle domain can be obtained by Koebe's iteration. As shown in Figure 19.19, the input surface S is a poly-annulus, its boundary is

$$\partial S = \gamma_0 - \gamma_1 \cdots - \gamma_n,$$

where the exterior boundary component is γ_0 and the interior boundary components are $\gamma_1, \gamma_2, \ldots, \gamma_n$.

Filling inner holes First, we set S_0 as S, $S_0 \leftarrow S$, with boundary components

$$\partial S_0 = \gamma_0 - \gamma_1 \cdots - \gamma_n,$$

and compute the circular slit map $\varphi_1 : S_0 \to \mathbb{D}^2$ using Alg. 16, which maps γ_0 and γ_1 to the outer and the inner circles, the other γ_k's to concentric circular arcs, $1 < k \leq n$.

Second, we fill the inner hole $\varphi_1(\gamma_1)$ by a planar disk D_1,

$$S_1 := \varphi_1(S_0) \cup D_1,$$

then S_1 has $(n-1)$ boundaries,

$$\partial S_1 = \varphi_1(\gamma_0) - \varphi_1(\gamma_2) \cdots - \varphi_1(\gamma_n).$$

Then we compute a circular slit map $\varphi_2 : S_1 \to \mathbb{D}^2$, which maps $\varphi_1(\gamma_0)$ and $\varphi_1(\gamma_2)$ to the outer and the inner circles, and fill $\varphi_2 \circ \varphi_1(\gamma_2)$ by a disk D_2 to obtain S_2,

$$S_2 := \varphi_2(S_1) \cup D_2,$$

with boundary components

$$\partial S_2 = \varphi_2 \circ \varphi_1(\gamma_0) - \varphi_2 \circ \varphi_1(\gamma_3) \cdots - \varphi_2 \circ \varphi_1(\gamma_n).$$

We repeat this procedure, each time we fill a hole. At the k-th step, we compute the circular slit map $\varphi_k : S_{k-1} \to \mathbb{D}^2$, which maps $\varphi_{k-1} \circ \varphi_{k-2} \cdots \varphi_1(\gamma_0)$ to the outer unit circle, and $\varphi_{k-1} \circ \varphi_{k-2} \cdots \varphi_1(\gamma_k)$ to the inner circle, then we fill the inner circle by a disk D_k to obtain

$$S_k := \varphi_k(S_{k-1}) \cup D_k,$$

with boundary components,

$$\partial S_k = \varphi_k \circ \varphi_{k-1} \cdots \varphi_1(\gamma_0) - \varphi_k \circ \varphi_{k-1} \cdots \varphi_1(\gamma_{k+1}) \cdots - \varphi_k \circ \varphi_{k-1} \cdots \varphi_1(\gamma_n).$$

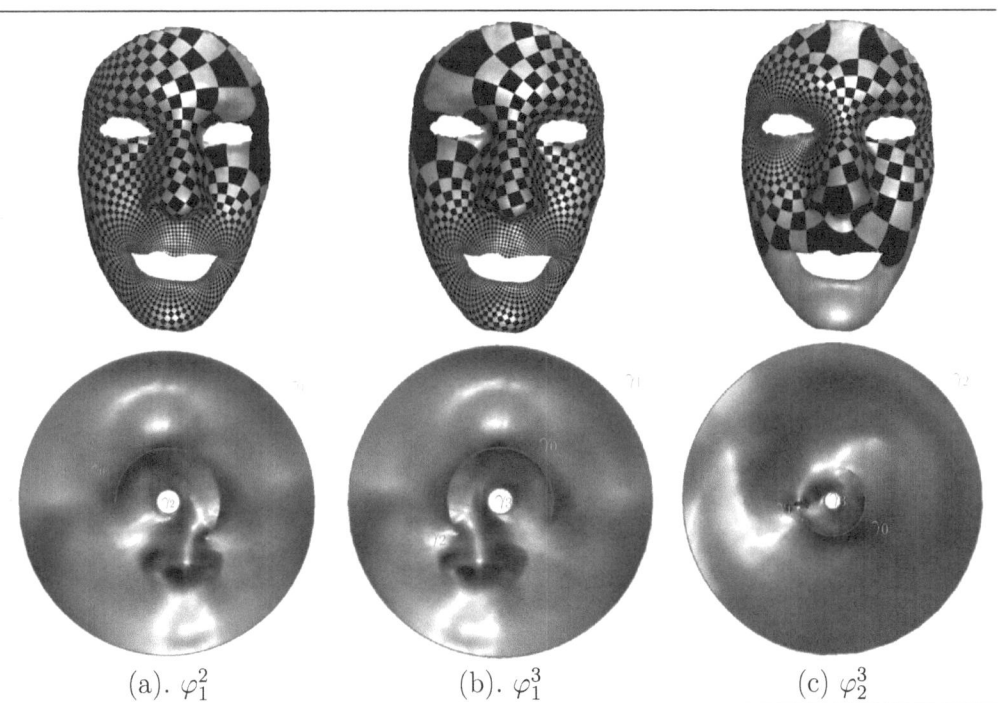

(a). φ_1^2 (b). φ_1^3 (c) φ_2^3

Figure 19.19 The circular slit maps of a poly-annulus II.

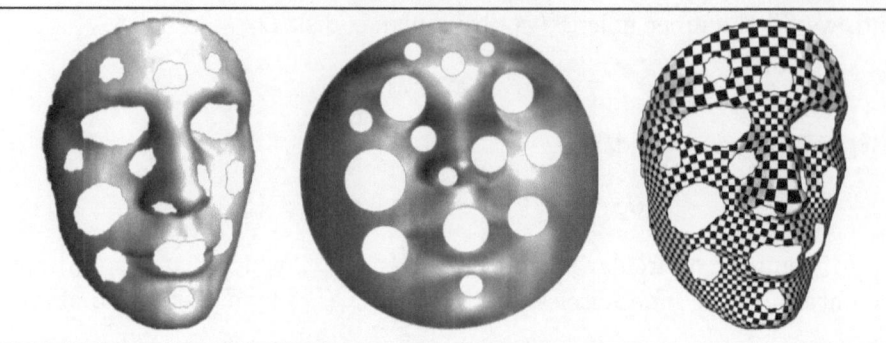

Figure 19.20 A poly-annulus with 15 inner holes is conformally mapped onto a circle domain.

We repeat this procedure n times, each step fills one hole, then S_n becomes the unit planar disk. Figure 19.21 shows the first step of the hole filling algorithm for a poly-annulus with 3 inner holes; Figures 19.22 and 19.23 show the second and the third steps. The algorithmic details can be found in Alg. 20.

Koebe's Iteration   The input poly annulus S has n inner holes, after the hole filling procedure, it becomes a topological disk S_n. The original internal boundary γ_k is mapped onto a closed loop

$$\tau_k := \varphi_n \circ \varphi_{n-1} \cdots \circ \varphi_1(\gamma_k), \quad k = 1, 2, \ldots, n,$$

bounding a simply connected domain, which is denoted as D_{k-1}, namely $\partial D_{k-1} = \tau_k$. We set S^0 as S_n, $S^0 \leftarrow S_n$, with n domains,

$$D_k^0 = D_k, \quad k = 0, 1, \ldots, n - 1.$$

We remove D_0^0 and map the surface $S^0 \setminus D_0^0$ to the canonical annulus using Alg. 16,

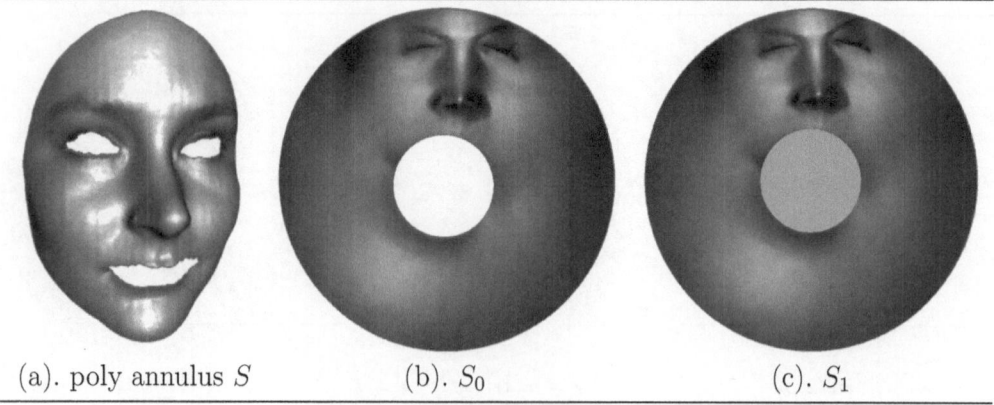

(a). poly annulus S (b). S_0 (c). S_1

Figure 19.21 Hole-filling using slit maps, first iteration.

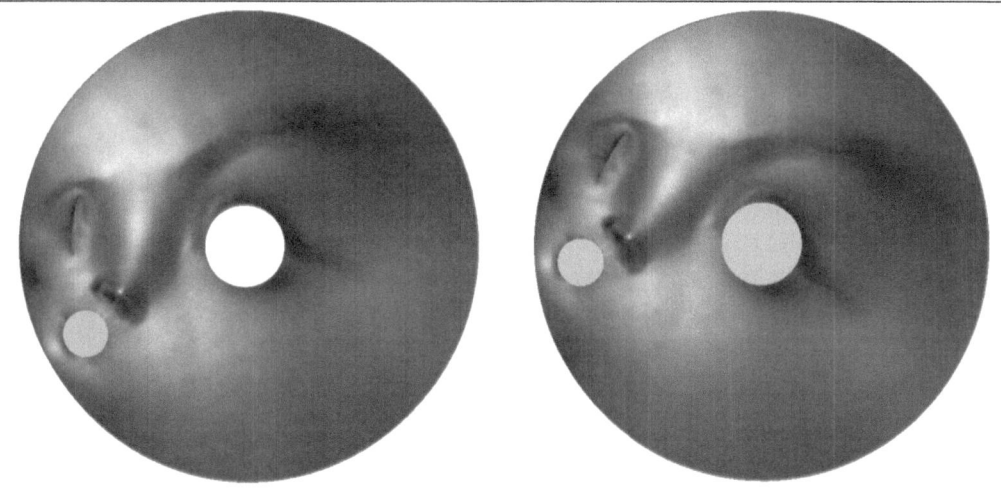

Figure 19.22 Hole filling using slit maps, second iteration.

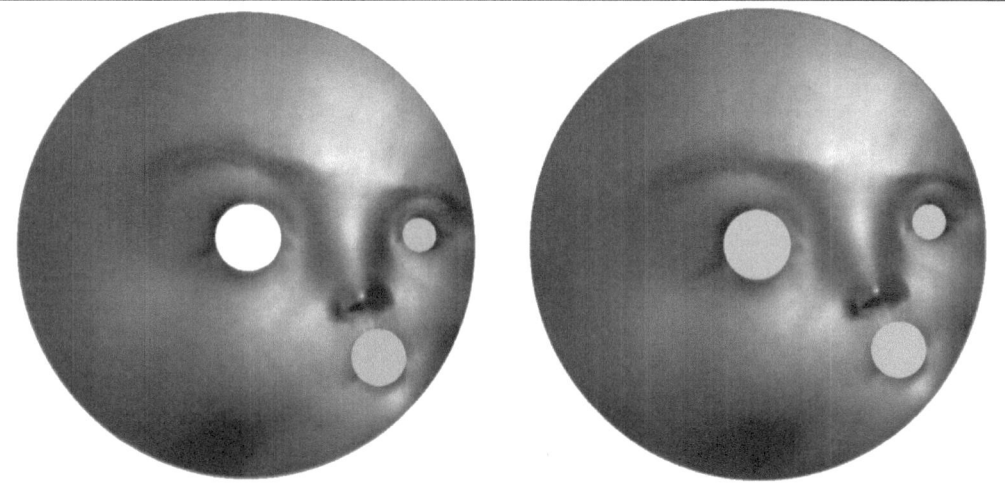

Figure 19.23 Hole filling using slit maps, third iteration.

$\varphi^0 : S^0 \setminus D_1^0 \to \mathbb{D}^2$, then fill a disk D^0 to the annulus image $\varphi^0(S^0 \setminus D_1^0)$ to obtain S^1,

$$S^1 := \varphi^0(S^0 \setminus D_0^0) \cup D^0.$$

The domains on S^1 corresponding to the original inner inner holes on S^1 are denoted $D_0^1, D_1^1, \ldots, D_{n-1}^1$,

$$D_i^1 = \varphi^0(D_i^0), \quad i \neq 0, \quad D_0^1 = D^0.$$

Then we compute $\varphi^1 : S^1 \setminus D_1^1 \to \mathbb{D}^2$, and fill a disk D^1 to the inner circular hole of the annulus $\varphi^1(S^1 \setminus D_1^1)$ and obtain

$$S^2 := \varphi^1(S^1 \setminus D_1^1) \cup D^1,$$

Algorithm 20 Topological Poly-Annulus Hole Filling Using Circular Slit Maps

---

Require: A topological poly-annulus $(S, \mathbf{g})$;
Ensure: A topological disk with all the inner holes are filled;

Trace the boundary loops of S, $\partial S = \gamma_0 - \gamma_1 \cdots - \gamma_n$;
$S_0 \leftarrow S$, $\varphi_0 \leftarrow$ id;
for each $k \in \{1, 2, \ldots, n\}$ **do**

Compute the circular slit map $\varphi_k : S_{k-1} \to \mathbb{D}_2$ using Alg. 19, such that $\varphi_{k-1} \circ \varphi_{k-2} \cdots \varphi_0(\gamma_0)$ is mapped to the outer circle; $\varphi_{k-1} \circ \varphi_{k-2} \cdots \varphi_0(\gamma_k)$ to the inner circle;

Fill the center hole $S_k \leftarrow \varphi_k(S_{k-1}) \cup D_k$
end for
return S_n;

---

with the interior domains

$$D_i^2 = \varphi^1(D_i^1), \quad i \neq 1, \quad D_1^2 = D^1.$$

We repeat this procedure, at the k-th step, the input surface is S^k, with interior domains $D_0^k, D_1^k, \ldots, D_{n-1}^k$, we puncture a hole D_r^k, $r = k \mod n$ and map it to the annulus by $\varphi^k : S^k \setminus D_r^k \to \mathbb{D}^2$, then we use a disk D^r to fill the inner circular hole of the annulus, to obtain

$$S^{k+1} := \varphi^k(S^k \setminus D_r^k) \cup D^r, \quad r \equiv k \mod n,$$

with the interior domains

$$D_i^{k+1} = \varphi^k(D_i^k), \quad i \neq r, D_r^{k+1} = D^r.$$

We repeat this procedure, the inner domains get rounder and rounder, and converge to conical circular disks eventually. Figure 19.24 shows the first two iterations of the Koebe's iteration algorithm on the facial surface in the left frame of Figures 19.21 and 19.25 show the 3rd to 7th iterations. Figures 19.26, 19.27 and 19.28 show another example of circular domain conformal mapping using Koebe's iteration. The algorithmic details can be found in Alg. 21.

Convergence Analysis Now we estimate the convergence rate of Koebe's iteration. As shown in Figure 19.29, the middle frame in the top row shows the initial planar domain C_0, $\infty \in C_0$. Its complement consists of $D_{0,1}, D_{0,2}, \cdots, D_{0,n}$, the boundary components are $\partial D_{0,i} = \Gamma_{0,i}$, $i = 1, 2, \cdots, n$. The left frame in the top row shows the circle domain $\mathcal{C}$, the complement of $\mathcal{C}$ consists of $D_1, D_2, \cdots, D_n$, where D_i's are disks, $\partial D_i = \Gamma_i$ is a canonical circle. There is a biholomorphic function $f : C_0 \to \mathcal{C}$. In the neighborhood of ∞, $f(z) = z + O(z^{-1})$.

By Riemann mapping theorem, as shown in the right frame in the top row, there is a Riemann mapping

$$h_1 : \hat{\mathbb{C}} \setminus D_{0,1} \to \hat{\mathbb{C}} \setminus \mathbb{D}^2,$$

Algorithm 21 Koebe's Iteration

Require: A topological poly-annulus $(S, \mathbf{g})$;

Ensure: A conformal mapping $\varphi : S \to \mathbb{D}^2$, φ maps S to a circle domain;

    Fill all the inner holes using Alg. 20 to obtain S^0, with n filled domains bounded by the original boundaries, $D_0^0, D_1^0, \ldots, D_{n-1}^0$;

    for each $k \in \{0, 1, 2, \ldots, n, \ldots\}$ **do**

        Remove a interior domain D_r^k, $k \equiv r \mod n$;

        Compute a conformal map $\varphi^k : S^k \setminus D_r^k \to \mathbb{D}^2$ using Alg. 16;

        Fill the center hole $S^{k+1} \leftarrow \varphi^k(S^k \setminus D_r^k) \cup D^r$;

        if all the domains D_r^{k+1}, $r = 0, \ldots, n-1$ are circular enough **then**

            Break;

        end if

    end for

    return φ_n and S_n;

where $\mathbb{D}^2$ is the unit disk, h_1 maps $\gamma_{0,1}$ to $\Gamma_{1,1}$, C_0 to C_1 satisfying the normalization condition $h_1(\infty) = \infty$, $h_1'(\infty) = 1$, and

$$D_{1,k} = h_1(D_{0,k}), \quad k = 2, \cdots, n.$$

Repeat this procedure, at $k \leq n$ step, construct a Riemann mapping,

$$h_k : \hat{\mathbb{C}} \setminus D_{k-1,k} \to \hat{\mathbb{C}} \setminus \mathbb{D},$$

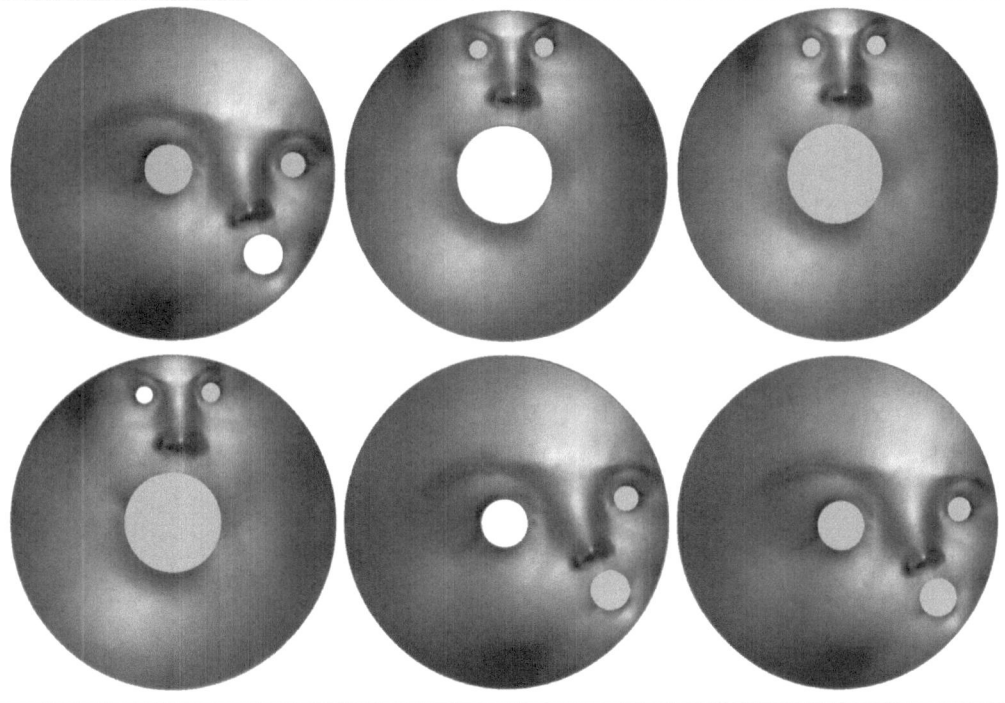

Figure 19.24 Koebe's iteration, first two iterations.

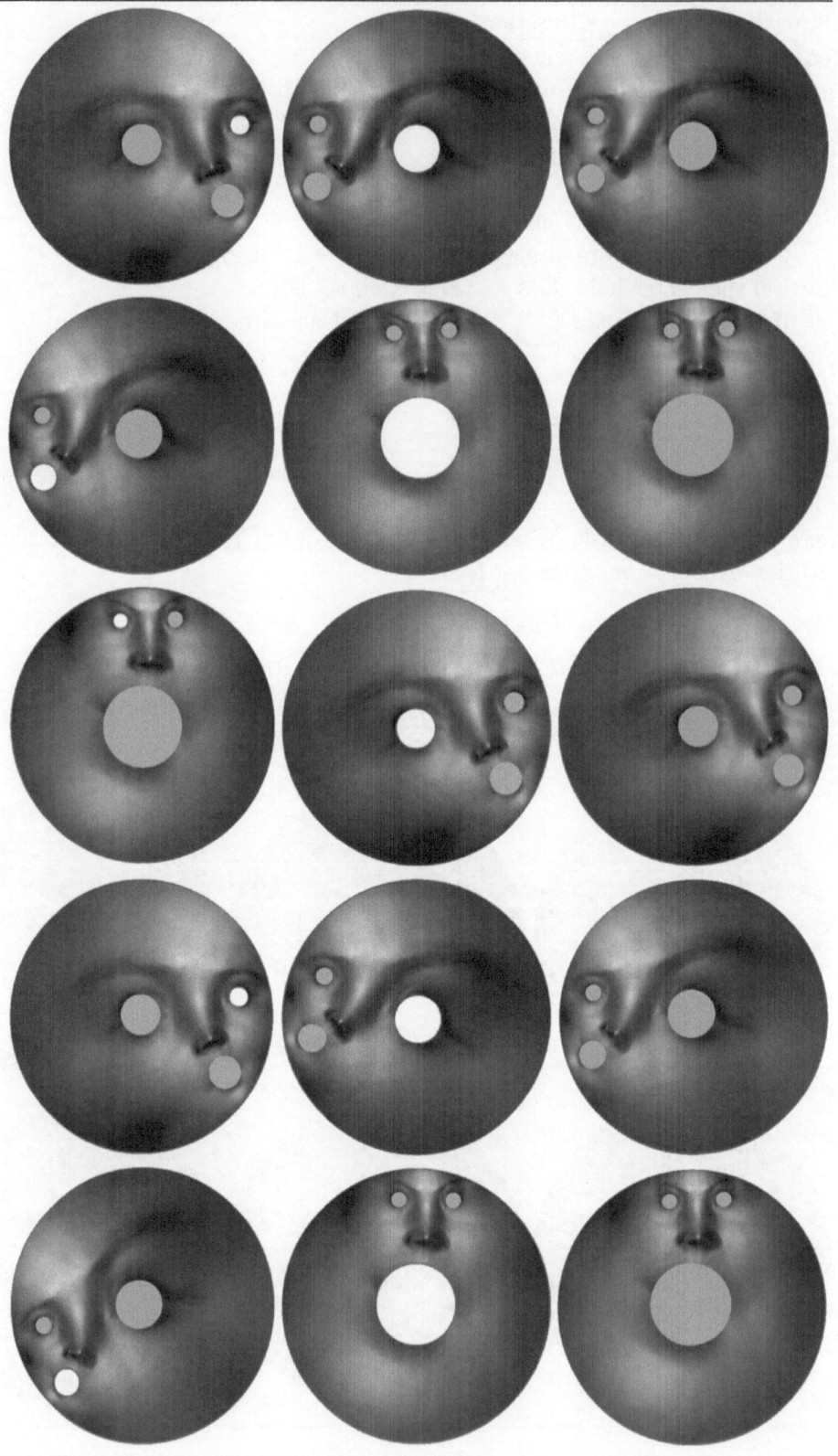

Figure 19.25 Koebe's iteration, iterations 3-7.

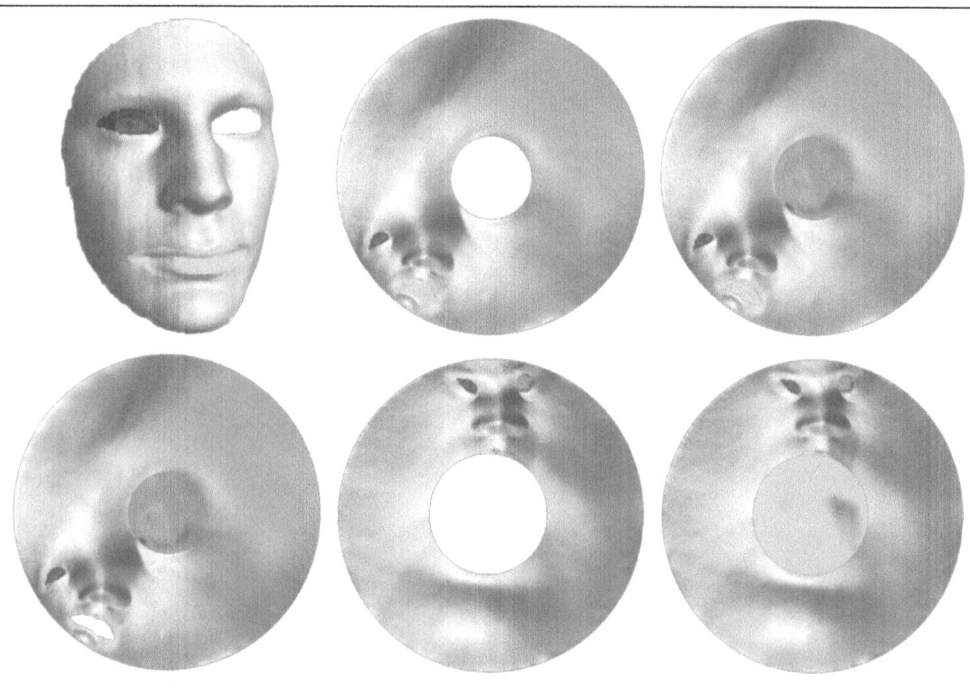

Figure 19.26 Koebe's iteration for a male face, the first two iterations.

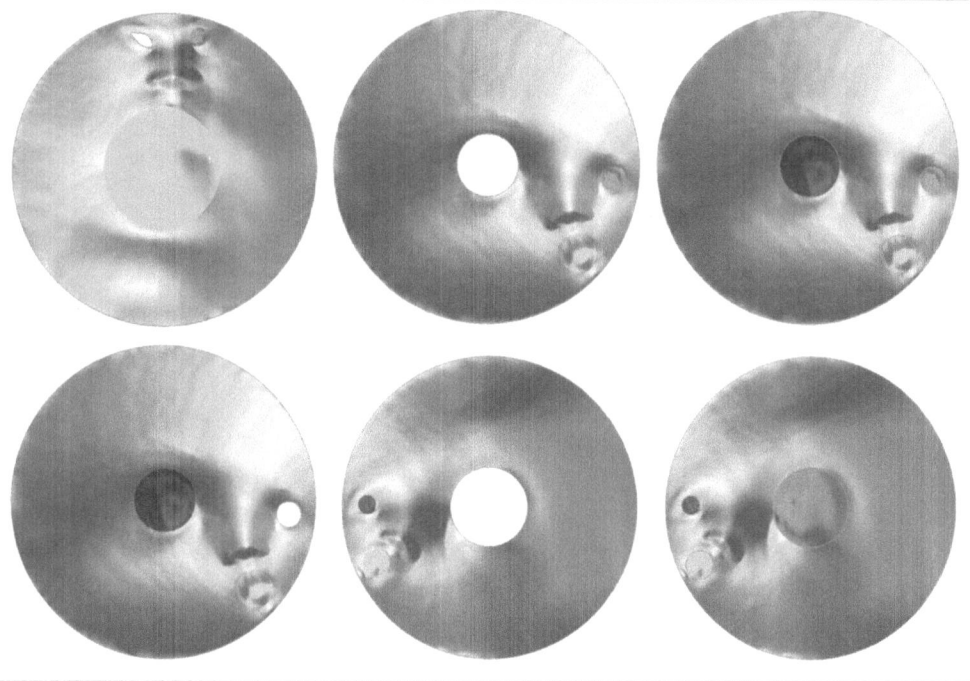

Figure 19.27 Koebe's iteration for a male face, the 3rd and the 4th iterations.

which maps $\Gamma_{k-1,k}$ to the unit circle, C_{k-1} to C_k, $h_k(\infty) = \infty$ and $h'_k(\infty) = 1$. We recursively define the symbols as follows:

$$C_k = h_k(C_{k-1}), \quad \Gamma_{k,i} = h_k(\Gamma_{k-1,i}), \quad D_{k,i} = h_k(D_{k-1,i}), \quad i \neq k,$$

$D_{k,k}$ is the unit disk $\mathbb{D}$, $\Gamma_{k,k}$ the unit circle. We construct a biholomorphic map $f : C_0 \to C_k$,

$$f_k = h_k \circ h_{k-1} \cdots \circ h_1,$$

and the biholomorphic map from the circle domain $\mathcal{C}$ to C_k, $g_k : \mathcal{C} \to C_k$, is defined as

$$g_k := f_k \circ f^{-1},$$

g_k satisfies the normalization condition $g_k(\infty) = \infty$, $g'_k(\infty) = 1$. When k goes to infinity, g_k converges to the identity.

Consider the circle domain $\mathcal{C}$ in the left frame in the top row, we define

$$\delta := \min_{i \neq j} \text{dist}(\Gamma_i, \Gamma_j), \quad i \neq j.$$

For each circle Γ_i with center p_i and r_i, we keep the center and increase the radius to $\mu^{-1} r_i$, and enlarge the circle to $\tilde{\Gamma}_i$. We choose μ as small as possible, such that there exist two $\tilde{\Gamma}_i$ and $\tilde{\Gamma}_j$ that are tangent to each other. We can find a circle Γ_ρ, centered at the origin, with radius ρ big enough to enclose all the $\tilde{\Gamma}_i$'s. Then we have the following estimate:

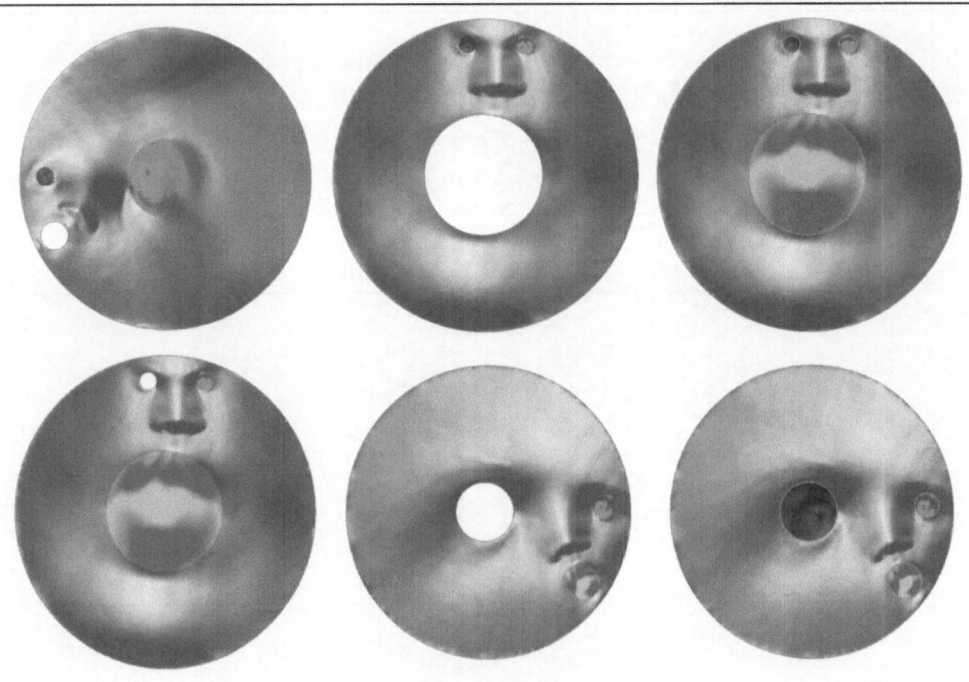

Figure 19.28 Koebe's iteration for a male face, the 5th and the 6th iterations.

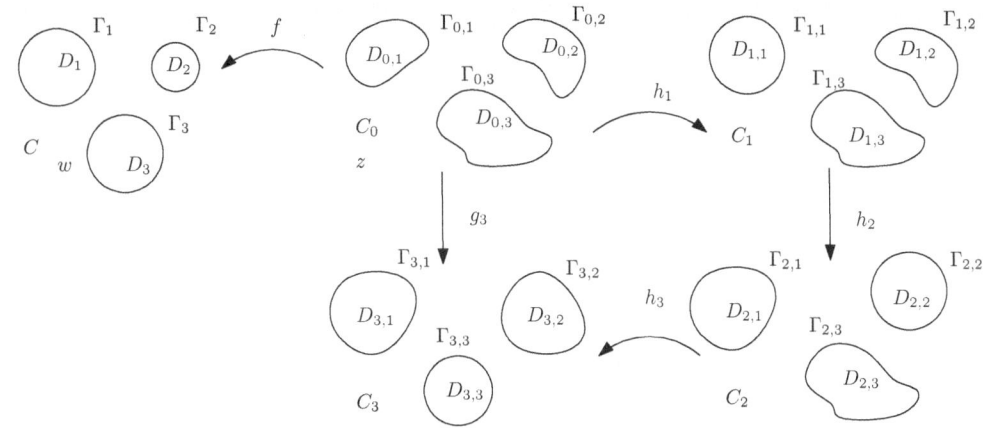

Figure 19.29 Error estimate for Koebe's iteration.

Theorem 19.5.3 *In Koebe's iteration algorithm, when the iteration number $k > mn$,*

$$|g_k(w) - w| \le \frac{1}{4\delta} \left(\frac{-8\pi^2\rho^2}{\log \mu} + 4\rho^2 \right) \mu^{4m}.$$

Detailed proof can be found in [162]. This shows the convergence rate is controlled by the conformal invariance μ.

19.6 TOPOLOGICAL TORUS

Suppose $(S, \mathbf{g})$ is a topological torus, namely a closed genus one oreinatable surface, its universal covering space is $\tilde{S}$, $\pi : \tilde{S} \to S$ is the projection map. The fundamental group of the surface is $\pi_1(S, p) = \langle a, b | aba^{-1}b^{-1} \rangle$, where a, b are canonical basis, and the homology group is $H^1(S, \mathbb{Z}) = \mathbb{Z}^2$. According to Hodge theorem 16.4.15, the holomorphic 1-form group $\Omega^1(S)$ is 2 real dimensional. Suppose $\omega \in \Omega^1(S)$, satisfies the condition

$$\int_a \omega = 1, \quad \int_b \omega = \lambda \in \mathbb{C}, \tag{19.6.1}$$

then we can lift ω to the holomorphic 1-form $\tilde{\omega}$ defined on the universal covering space $\tilde{S}$, by $\tilde{\omega} = \pi^*\omega$, namely for all 1-chain $\tilde{\gamma} \in C_1(\tilde{S}, \mathbb{Z})$, we have

$$\tilde{\omega}(\tilde{\gamma}) = \omega \circ \pi(\tilde{\gamma}).$$

Let $\tilde{q} \in \tilde{S}$ be a pre-image of the base point $p \in S$, $\tilde{p} \in \pi^{-1}(p)$. For every point $\tilde{q} \in \tilde{S}$, we find a path $\tilde{\gamma} : [0, 1] \to \tilde{S}$ connecting $\tilde{p}$ and $\tilde{q}$, $\tilde{\gamma}(0) = \tilde{p}$, $\tilde{\gamma}(1) = \tilde{q}$, then we define the mapping

$$\tilde{\varphi}(\tilde{q}) = \int_{\tilde{\gamma}} \tilde{\omega}. \tag{19.6.2}$$

Because $\tilde{S}$ is simply connected, the integration is independent of the choice of the path $\tilde{\gamma}$, the mapping $\tilde{\varphi}$ is well defined. $\tilde{\varphi}$ conformally maps the universal covering space to the whole complex plane, $\tilde{\varphi} : \tilde{S} \to \mathbb{C}$. The preimages of the base point $\pi^{-1}(p)$ are mapped to a grid Λ on the plane,

$$\Lambda = \{m + n\lambda \mid m, n \in \mathbb{Z}\}.$$

the quotient space $\mathbb{C}/\Lambda$ is a flat torus. Given a point $q \in S$, given two points in its orbit on the universal covering space $\tilde{q}_1, \tilde{q}_2 \in \pi^{-1}(q)$. Suppose a path connecting them is $\tilde{\tau}$, $[\pi(\tilde{\tau})] = m[a] + n[b]$, then

$$\tilde{\varphi}(\tilde{q}_1) - \tilde{\varphi}(\tilde{q}_2) = m + n\lambda,$$

therefore $\tilde{\varphi}$ maps both $\tilde{q}_1$ and $\tilde{q}_2$ to the same point in the flat torus $\mathbb{C}/\Gamma$. This shows $\tilde{\varphi}$ maps every point in the orbit of q to the same point on the flat torus. This induces a conformal map from the original topological torus to the flat torus. Figure 19.30 shows an example of the conformal mapping of a topological torus, the kitten surface model. The red parallelogram shows one fundamental polygon of the surface. The algorithmic details can be found in Alg. 22.

Algorithm 22 Topological Torus Conformal Mapping

Require: A topological torus $(S, \mathbf{g})$;
Ensure: A conformal map $\varphi : S \to \mathbb{C}/\Lambda$;
    Compute the fundamental group $\pi_1(S, p)$ using Alg.4, find generators a, b;
    Compute the holomorphic 1-form basis ω_1, ω_2 using Alg. 13;
    Find real coefficient λ_1, λ_2, $\omega = \lambda_1\omega_1 + \lambda_2\omega_2$, such that Eqn. (19.6.1) holds;
    Construct a finite portion of the universal covering space $\tilde{S}$ using Alg. 5;
    Compute the conformal map $\tilde{\varphi} : \tilde{S} \to \mathbb{C}$ in Eqn. (19.6.2);
    Extract the orbit $\pi^{-1}(p)$ and obtain the Lattice Λ;
    return The conformal mapping $\varphi : S \to \mathbb{C}/\Lambda$;

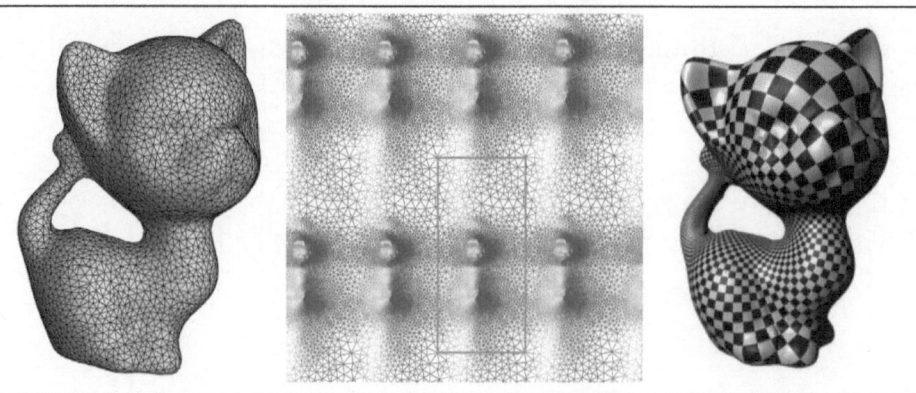

Figure 19.30   Conformal mapping for a topological torus, the kitten surface model.

CHAPTER 20

Discrete Surface Curvature Flows

This chapter introduces the computational algorithms for surface Ricci flow. Ricci flow is a classical method in geometric analysis, proposed by Richard Hamilton [158] for the purpose of proving Poincaré's conjecture . Poincaré conjectured that any simply connected compact 3-manifold is a topological sphere. Hamilton's intuitive idea is as follows: first, we can pick a Riemanian metric on the manifold, then deform the metric, the deformation rate is proportional to the current Ricci curvature, the curvature evolves according to a diffusion-reaction process, eventually the curvature becomes constant everywhere, therefore the manifold is isometric to the sphere. Perelman usued Ricci flow to prove the Poincaré's conjecture [247, 248]. The Ricci flow theory induces a very powerful computational method to design Riemannian metrics by prescribing the target curvatures. This method plays a very important role in engineering and medical fields.

Classical Ricci flow is defined on smooth manifolds, the curvature definition requires the Riemannian metric tensor to be C^2 smooth. However on computers, the polyhedral surfaces are not C^2 smooth, classical curvature cannot be defined directly. Therefore, we need to generalize the Ricci flow theory from smooth manifolds to the discrete settings.

20.1 YAMABE EQUATION

Suppose S is a surface with a Riemannian metric $\mathbf{g}$, for any point $p \in S$, there is a neighborhood $U(p)$ with the isothermal coordinates,

$$\mathbf{g} = e^{2\lambda(u,v)}(du^2 + dv^2).$$

where $\lambda : U(p) \to \mathbb{R}$ is a smooth real function defined on $U(p)$. Figure 20.1 shows the isothermal coordinates on the Stanford bunny surface.

Under the isothermal coordinates, the Gaussian curvature has the simple form

$$K(u,v) = -\Delta_{\mathbf{g}}\lambda(u,v) = -\frac{1}{e^{2\lambda(u,v)}}\Delta\lambda(u,v), \qquad (20.1.1)$$

DOI: 10.1201/9781003350576-20

Figure 20.1 The isoterhmal coordinates on the Stanford bunny model.

where the Laplace operator Δ is defined as

$$\Delta = \frac{\partial^2}{\partial u^2} + \frac{\partial^2}{\partial v^2},$$

$\Delta_{\mathbf{g}}$ is called the *Laplace-Beltrami* operator. The geodesic curvature on boundary point $p \in \partial S$ is given by

$$k_g(u, v) = -\frac{1}{e^{\lambda(u,v)}} \frac{\partial}{\partial \mathbf{n}} \lambda(u, v). \qquad (20.1.2)$$

Suppose the boundary of the surface ∂S are piecewise smooth curves, at the corner points the exterior angles are θ_i's, then the total curvature satisfies the Gauss-Bonnet theorem:

$$\int_S K dA + \int_{\partial S} k_{\mathbf{g}} ds + \sum_i \theta_i = 2\pi \chi(S), \qquad (20.1.3)$$

where $\chi(S)$ is the Euler characteristic number of the surface.

Definition 20.1.1 (Conformal Metric) *Suppose S is a surface with a Riemannian metric tensor* $\mathbf{g}$*,* $\lambda : S \to \mathbb{R}$ *is a function defined on the surface, then* $e^{2\lambda}\mathbf{g}$ *is also a Riemannian metric on Σ and called a* conformal metric. λ *is called the conformal factor. The transformation*

$$\mathbf{g} \to e^{2\lambda}\mathbf{g}$$

is called a conformal metric deformation.

Angles are invariant measured by conformal metrics. Suppose $\bar{\mathbf{g}} = e^{2\lambda}\mathbf{g}$ is a conformal metric on the surface, then by Eqn. (20.1.1) we obtain the Gaussian curvature on interior points are

$$\bar{K} = e^{-2\lambda}(-\Delta_{\mathbf{g}}\lambda + K), \qquad (20.1.4)$$

Similarly, the geodesic curvature on the boundary becomes

$$\bar{k}_g = e^{-\lambda}(-\partial_n \lambda + k_g). \qquad (20.1.5)$$

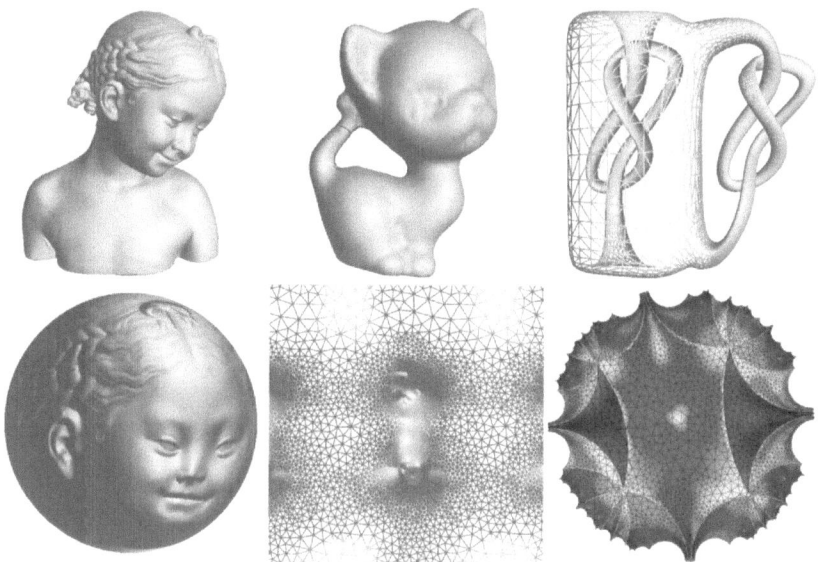

Figure 20.2 Uniformization for closed surfaces.

Eqn. (20.1.4) and Eqn. (20.1.5) are called the *Yamabe equations*. By solving the Yamabe equation, we can find the conformal metric satisfying the prescribed Gaussian curvature.

As shown in Figure 20.2, one of the most fundamental theorem in surface differential geometry is the following uniformization theorem, which states all the metric surfaces can be conformally deformed to one of three canonical shapes, the sphere, the Euclidean plane and the hyperbolic plane.

Theorem 20.1.2 (Koebe-Poincaré Uniformization Theorem) *Let $(\Sigma, \mathbf{g})$ be a compact, orientable, closed 2-dimensional Riemannian manifold. Then there is a metric $\tilde{\mathbf{g}} = e^{2\lambda}\mathbf{g}$ conformal to $\mathbf{g}$ which has constant Gauss curvature. The constant is one of three choices $\{+1, 0, -1\}$, if the genus of the surface S is 0, 1 or greater than 1 respectively.*

One of the most fundamental problems in discrete differential geometry is to compute the uniformizaiton metric, therefore it is crucial to develop algorithms to solve Yamabe equation.

20.2 SURFACE RICCI FLOW

Yamabe equation is highly non-linear, which cannot be solved using conventional methods. The most effective method is the Hamilton's Ricci flow [158]. The Ricci flow on a Riemannian manifold $(M, \mathbf{g})$ is defined as

$$\partial_t \mathbf{g}(t) = -2\mathrm{Ric}(t),$$

where Ric is the Ricci curvature tensor. On a Riemannian surface,

$$\mathrm{Ric} = \frac{1}{2}R\mathbf{g},$$

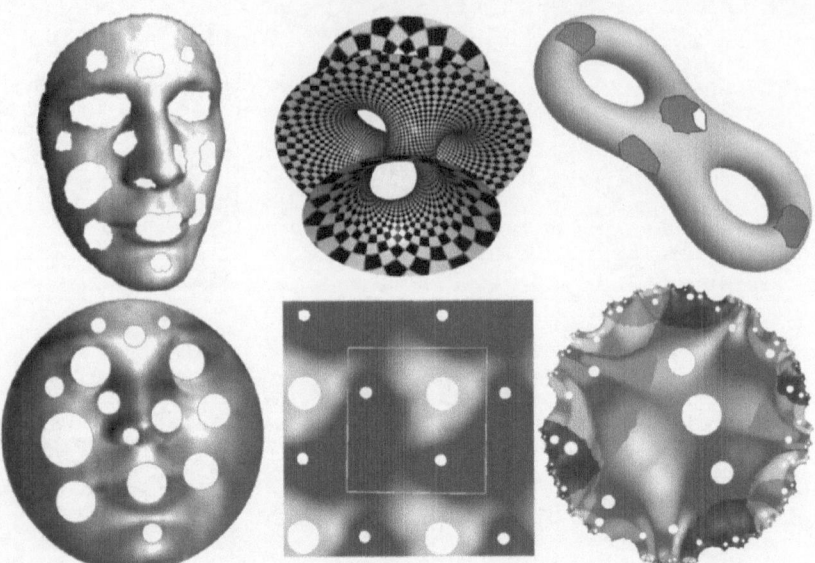

Figure 20.3 Uniformization for open surfaces.

where $R = 2K$ is the scalar curvature, two times of the Gaussian curvature. Therefore, we have

Definition 20.2.1 (Surface Ricci Flow) *Given a Riemannian surface $(S, \mathbf{g})$, the* surface Ricci flow *is defined as*

$$\partial_t \mathbf{g}(t) = -2K(t)\mathbf{g}(t). \tag{20.2.1}$$

The normalized surface Ricci flow *is defined as*

$$\partial_t \mathbf{g}(t) = (\rho - R(t))\mathbf{g}(t) = \left(\frac{4\pi\chi(S)}{A(0)} - 2K(t) \right) \mathbf{g}(t), \tag{20.2.2}$$

where $A(0)$ is the initial total surface area, $\chi(S)$ the Euler characteristic number of the surface.

It can be proven that the surface Ricci flow preserves the conformal class of the metric, $\mathbf{g}(t) = e^{2\lambda(t)}\mathbf{g}(0)$. We plug it into the Ricci flow Eqn. (20.2.2) to obtain the evolution equation for the conformal factor,

$$\partial_t \lambda(t) = \frac{1}{2}(\rho - R(t)). \tag{20.2.3}$$

It is easy to show that the normalized surface Ricci flow preserves the total area of the surface. The scalar curvature evolves according to the following equation:

$$\partial_t R = \Delta_{\mathbf{g}(t)} R + R(R - \rho), \tag{20.2.4}$$

where $R(t) = 2K(t)$ and $\rho = \frac{4\pi\chi(S)}{A(0)}$ is the initial mean scalar curvature.

The following two theorems show that the normalized surface Ricci flow converges to a metric such that the Gaussian curvature is constant $\bar{K} = \frac{2\pi\chi(S)}{A(0)}$ every where.

Figure 20.4   A discrete surface.

Theorem 20.2.2 (Hamilton [158]) *For a closed surface of non-positive Euler characteristic, if the total area of the surface is preserved during the flow, the Ricci flow will converge to a metric such that the Gaussian curvature is constant (equals to $\bar{K}$) every where.*

Theorem 20.2.3 (Chow [86]) *For a closed surface of positive Euler characteristic, if the total area of the surface is preserved during the flow, the Ricci flow will converge to a metric such that the Gaussian curvature is constant (equals to $\bar{K}$) every where.*

20.3   DISCRETE SURFACE

In practice, smooth surfaces are approximated by *discrete surfaces*. In general, a discrete surface is represented by tuples $(S, V, \mathbf{d}, \mathcal{T})$, where S is a topological surface, V a discrete point set on S, $\mathcal{T}$ a triangulation of S with V as the vertex set, $\mathbf{d}$ a Riemannian metric with cone singularities at V. For example, a triangualted polyhedral surface is a discrete surface, V is the vertex set, $\mathbf{d}$ is the polyhedral metric. The metric is flat everywhere with cone singularities at the vertices. Figure 20.4 shows an example of discrete surface, a triangular mesh of Michelangelo's King David sculpture. Each discrete surface is constructed by isometrically gluing Euclidean $\mathbb{E}^2$ triangles. In general situations, the triangular mesh can be assembled by isometrically gluing spherical $\mathbb{S}^2$ or hyperbolic $\mathbb{H}^2$ triangles as well. Accordingly, we say a triangular mesh is with Euclidean, spherical or hyperbolic *background geometry* respectively. The key concepts such as Riemannian metric, curvature and conformal metric deformation need to be generalized to the discrete surfaces.

The discrete Riemannian metric on a discrete can be represented as the edge length function satisfying the triangle inequality. Fix the combinatorial structure, a marked surface with a triangulation $(S, V, \mathcal{T})$ has infinitely many metrics. Figure 20.5 shows two different metrics on the same combinatorial structure.

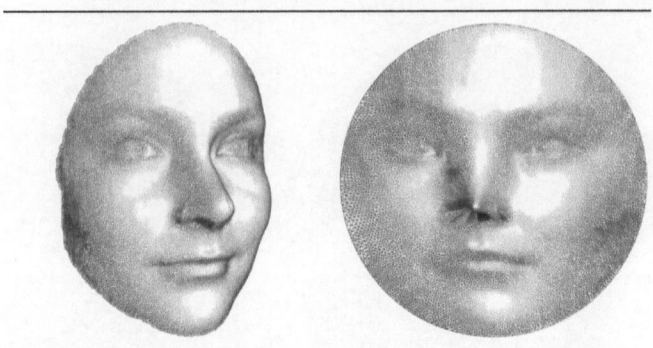

Figure 20.5 Two different discrete metrics on the same combinatorial structure.

The discrete Gaussian curvature is defined as the angle deficit in Eqn. (12.2.2), the total discrete Gaussian curvature satisfies Gauss-Bonnet theorem Eqn. (12.2.3). In smooth cases, the Riemannian metric determines the Gaussian curvature. Similarly, in discrete cases, the discrete metric determines the discrete curvature by the cosine laws Eqn. (12.2.4).

We have defined the concepts of discrete Riemannian metric and discrete Gaussian curvature, the next key concept is discrete conformal metric deformation. On smooth surfaces, the conformal metric deformation can be represented as $\mathbf{g} \mapsto e^{2\lambda}\mathbf{g}$. In analogy to the smooth situation, in the discrete situation, we can deform the edge length by multiplying the exponential of the conformal factors defined on the end vertices.

Definition 20.3.1 (Vertex Scaling Transformation) *Given a discrete surface* $(S, V, \mathbf{d}, \mathcal{T})$, *the* vertex scaling *tranforms the metric* $\mathbf{d}$ *to a new metric, denoted as* $u * \mathbf{d}$, *such that for each edge* $[v_i, v_j] \in E(\mathcal{T})$, *its length is transformed to*

$$u * \mathbf{d} : l_{ij} \mapsto e^{\frac{u_i}{2}} l_{ij} e^{\frac{u_j}{2}}. \tag{20.3.1}$$

The vertex scaling operator transforms the discrete metric, therefore changes the discrete curvature, which is governed by the following differential cosine law Eqn. (12.2.5). From the derivative cosine law, we can obtain the symmetric relation for vertex scaling operator Eqn. (12.2.7). This leads to the symmetry of the partial derivative of the discrete curvature with respect to the discrete conformal factor on the discrete surface.

Lemma 20.3.2 (Symmetry) *Given a discrete surface* $(S, V, \mathbf{d}, \mathcal{T})$ *in the vertex scaling transformation* $u * \mathbf{d}$, *for each pair of adjacent vertices* $[v_i, v_j] \in E(\mathcal{T})$, *we have the equation*

$$\frac{\partial K_i}{\partial u_j} = \frac{\partial K_j}{\partial u_i} = -w_{ij}, \tag{20.3.2}$$

where w_{ij} *is the cotangent edge weight in Eqn. (16.5.1) on the edge* $[v_i, v_j]$.

By the above symmetry lemma, we can define the *entropy energy* for a discrete surface.

Definition 20.3.3 (Entropy Energy) *Given a discrete surface* $(S, V, \mathbf{d}, \mathcal{T})$, *the vertex scaling operator transforms the metric to* $u * \mathbf{d}$, *the discrete entropy energy is defined as*

$$\mathcal{E}(\mathbf{u}) := \int^{\mathbf{u}} K_1 du_1 + K_2 du_2 + \cdots + K_n du_n, \qquad (20.3.3)$$

where $\mathbf{u} = (u_1, u_2, \ldots, u_n)$ *is the discrete conformal factor defined on the vertices.*

By the symmetry lemma 20.3.2, we can easily show the strict convexity of the entropy energy, its gradient maps the discrete conformal factor to the discrete Gaussian curvature diffeomorphically.

Theorem 20.3.4 *Given a discrete surface* $(S, V, \mathbf{d}, \mathcal{T})$, *let* $\mathbf{u} = (u_1, u_2, \ldots, u_n)^T$, *for every face* $f_\alpha \in F(\mathcal{T})$, $f_\alpha = [v_i, v_j, v_k]$, *we define*

$$\Omega_\alpha := \left\{ \mathbf{u} \left| \frac{l_{jk}}{e^{u_i}} + \frac{l_{ki}}{e^{u_j}} > \frac{l_{ij}}{e^{u_k}}, \frac{l_{ki}}{e^{u_j}} + \frac{l_{ij}}{e^{u_k}} > \frac{l_{jk}}{e^{u_i}}, \frac{l_{ij}}{e^{u_k}} + \frac{l_{jk}}{e^{u_i}} > \frac{l_{ki}}{e^{u_j}}, \right. \right\}$$

the domain of the entropy is

$$\Omega = \bigcap_{f_\alpha \in F(\mathcal{T})} \Omega_\alpha \bigcap \left\{ \mathbf{u} \left| \sum_{i=1}^{n} u_i = 0 \right. \right\},$$

the gradient of the entropy maps the discrete conformal factor to the curvature space

$$\mathcal{K} := \left\{ \mathbf{K} \left| \sum_{i=1}^{n} K_i = 2\pi\chi(S) \right. \right\}.$$

The entropy energy Eqn. (20.3.3) $\mathcal{E} : \Omega \to \mathbb{R}$ *is strictly convex, the gradient map* $\nabla\mathcal{E} : \Omega \to \mathcal{K}$, $\mathbf{u} \mapsto \mathbf{K}$ *is a local diffeomorphism.*

20.4 DISCRETE SURFACE YAMABE FLOW

The key concept is the conformal equivalent relation among Riemannian metrics, we can generalize it to discrete surfaces as follows:

Definition 20.4.1 (Discrete Conformality) *Two discrete Riemannian metrics* $\mathbf{d}$ *and* $\mathbf{d}'$ *on* (S, V) *are discrete conformal if there exists a sequence of discrete metrics on* (S, V),

$$\mathbf{d} = \mathbf{d}_1, \mathbf{d}_2, \ldots, \mathbf{d}_m = \mathbf{d}',$$

and a sequence of triangulations of (S, V),

$$\mathcal{T} = \mathcal{T}_1, \mathcal{T}_2, \ldots, \mathcal{T}_m = \mathcal{T}',$$

satisfying the following conditions:

1. each $\mathcal{T}_i$ is Delaunay in the discrete metric $\mathbf{d}_i$;

2. if $\mathcal{T}_i = \mathcal{T}_{i+1}$, there exists a discrete conformal factor $u : V \to \mathbb{R}$, for each edge $e \in E(\mathcal{T}_i)$ with vertices v_1 and v_2, then

$$l_{\mathbf{d}_{i+1}}(e) = l_{\mathbf{d}_i}(e)e^{u(v_1)+u(v_2)},$$

3. if $\mathcal{T}_i \neq \mathcal{T}_{i+1}$, then $(S, \mathbf{d}_i)$ is isometric to $(S, \mathbf{d}_{i+1})$ by an isometry homotopic to the identity in (S, V).

The discrete conformal class of discrete Riemannian metrics on (S, V) is called a *discrete conformal structure* or a *discrete Riemann surface* of (S, V).

Definition 20.4.2 (Discrete Surface Yamabe Flow with Surgery) *Given a discrete surface $(S, V, \mathbf{d}, \mathcal{T})$ and a target discrete curvature $\bar{K} : V \to \mathbb{R}$, satisfying the following conditions:*

1. *for each vertex $v_i \in V$, $\bar{K}(v_i) \in (-\infty, 2\pi)$;*

2. *the total curvature satisfies the Gauss-Bonnet condition:*

$$\sum_{v_i \in V} \bar{K}(v_i) = 2\pi\chi(S).$$

The discrete surface Yamabe flow is defined as: for each vertex $v_i \in V$, the conformal factor $u : V \to \mathbb{R}$ satisfies the ordinary differential equation

$$\frac{du(v_i)}{dt} = \bar{K}(v_i) - K(v_i, t). \tag{20.4.1}$$

The discrete metric evolves with time t, denoted as $\mathbf{d}(t)$, the triangulation also changes, denoted as $\mathcal{T}(t)$: there exists times

$$0 = t_0 < t_1 < \cdots < t_n < \cdots,$$

satisfying

1. *for any time t, $\mathcal{T}(t)$ is Delaunay in the discrete metric $\mathbf{d}(t)$;*

2. *in the time interval $[t_k, t_{k+1}]$, the triangulation keeps unchanged;*

3. *for any time $t \in [t_k, t_{k+1}]$, the metric differs by a vertex scaling transformation:*

$$\mathbf{d}(t) = (u(t) - u(t_k)) * \mathbf{d}(t_k); \tag{20.4.2}$$

4. *at any time t_k, $k = 0, 1, 2, \ldots$, there are finite number of edges in $\mathcal{T}(t_k)$ with 0 cotangent weights, flip these edges to update the triangulation to obtain $\mathcal{T}(t_k + \varepsilon)$, $\varepsilon > 0$ is positive infinitesimal.*

From variational point of view, the discrete surface Yamabe flow is the gradient flow of the discrete entropy energy. The following theorem can be easily proved.

Theorem 20.4.3 *Given a discrete surface* $(S, V, \mathbf{d}, \mathcal{T})$, *discrete Yamabe flow in Eqn. (20.4.1) is the gradient flow of the entropy energy*

$$\mathcal{E}(\mathbf{u}) = \int^{\mathbf{u}} \sum_{i=1}^{n} (\bar{K}(v_i) - K(v_i)) du_i, \tag{20.4.3}$$

the energy is strictly concave in the hyper-plane $\{\mathbf{u} | \sum_{i=1}^{n} u_i = 0\}$.

The following theorem shows the exitence and the uniqueness of the solution to the discrete surface Yamabe flow in Eqn. (20.4.1).

Theorem 20.4.4 (Discrete Surface Uniformization [152]) *Suppose* (S, V) *is a closed connected marked surface and* $\mathbf{d}$ *is any discrete metric on* (S, V). *Then for any discrete Gaussian curvature* $\bar{K} : V \to (-\infty, 2\pi)$ *with* $\sum_{v \in V} \bar{K}(v) = 2\pi \chi(S)$, *there exists a discrete metric* $\mathbf{d}'$, *unique up to scaling on* (S, V), *so that* $\mathbf{d}'$ *is discrete conformal to* $\mathbf{d}$ *and the discrete curvature of* $\mathbf{d}'$ *in* $\bar{K}$. *Furthermore, the discrete Yamabe flow with surgery associated to the curvature* $\bar{K}$ *with initial value* $\mathbf{d}$ *converges to* $\mathbf{d}'$ *exponentially fast.*

The computation of discrete surface Yamabe flow is equivalent to the optimization of the concave entropy energy, the desired conformal factor is the unique maximum of the energy. We can use the Newton's method to directly optimize the entropy to obtain the target discrete metric.

Assume at the k-th step, the discrete conformal factor is $\mathbf{u}_k$, we would like to change by $\delta \mathbf{u}$ to increase the entropy. The Taylor expansion of the entropy $\mathcal{E}(\mathbf{u}_k + \delta \mathbf{u})$ is

$$\mathcal{E}(\mathbf{u}_k + \delta \mathbf{u}) = \mathcal{E}(\mathbf{u}_k) + \langle \nabla \mathcal{E}(\mathbf{u}_k), \delta \mathbf{u} \rangle + \frac{1}{2} \delta \mathbf{u}^T \left(\frac{\partial^2 \mathcal{E}}{\partial u_i \partial u_j} \right) (\mathbf{u}_k) \delta \mathbf{u} + o(|\delta \mathbf{u}|^2).$$

By ignoring the higher order terms, we approximate the entropy to the second order term, and take the derivative with respect to $\delta \mathbf{u}$, then at the maximum the first derivative equals to 0,

$$\nabla \mathcal{E}(\mathbf{u}_k) + \left(\frac{\partial^2 \mathcal{E}}{\partial u_i \partial u_j} \right) (\mathbf{u}_k) \delta \mathbf{u} = 0,$$

we obtain the updating rule,

$$\mathbf{u}_{k+1} = \mathbf{u}_k + \lambda \delta \mathbf{u} = \mathbf{u}_k - \lambda \left(\frac{\partial^2 \mathcal{E}}{\partial u_i \partial u_j} \right) (\mathbf{u}_k)^{-1} \nabla \mathcal{E}(\mathbf{u}_k),$$

where λ is the step length. The Hessian matrix equals to the negative Laplace-Beltrami operator, the gradient of the energy equals to the difference between the target curvature and the current curvature.

Discrete surface Yamabe flow is a powerful tool to compute surface conformal mappings, which is more flexible than the algorithms based on harmonic mapping or holomorphic differentials.

Algorithm 23 Discrete Surface Yamabe Flow

Require: A discrete surface $(S, V, \mathbf{d}, \mathcal{T})$, target curvature $\bar{K} : V \to (-\infty, 2\pi)$ s.t.
$\sum_i \bar{K}(v_i) = 2\pi\chi(S)$, error threshold ε, step length λ;
Ensure: The unique target metric $\mathbf{d}'$ discrete conformal equivalent to $\mathbf{d}$ and induces
the discrete curvature $\bar{K}$;
Initialize the discrete conformal factor $\mathbf{u}_0 \leftarrow \mathbf{0}$, initialize the discrete polyhedral
metric $\mathbf{d}_0 \leftarrow \mathbf{d}$;
while true **do**
  Compute the corner angles using cosline law;
  Compute the discrete Gaussian curvature $K(\mathbf{u}_k)$ using Eqn. 12.2.2;
  if $\max |K(\mathbf{u}_k, v_i) - \bar{K}(v_i)| < \varepsilon$ **then**
    break;
  end if
  Compute the cotangent edge weight using Eqn. (16.5.1) for all edges, update
  the Laplace-Beltrami operator $\Delta(\mathbf{u}_k)$;
  Solve the linear system

$$\Delta(\mathbf{u}_k)\delta\mathbf{u} = \bar{K} - K(\mathbf{u}_k).$$

  Update the conformal factor $\mathbf{u}_{k+1} \leftarrow \mathbf{u}_k + \lambda\delta\mathbf{u}$, update the metric $\mathbf{d}_{k+1} \leftarrow \lambda\delta\mathbf{u} * \mathbf{d}_k$;
  Update the triangulation $\mathcal{T}_k$ to the Delaunay triangulation $\mathcal{T}_{k+1}$ by edge flipping;
end while
return The current discrete metric $\mathbf{d}_k$;

20.5 TOPOLOGICAL QUADRILATERAL

Suppose $(S, \mathbf{g})$ is a topological disk with a Riemannian metric, $\{v_0, v_1, v_2, v_3\}$ are four markers on the boundary, then there is a conformal mapping $\varphi : S \to \mathbb{C}$, such that the image $\varphi(S)$ is a planar rectangle, the four markers are mapped to the four corner points of the rectangle. Figure 20.6 shows one example.

The computational process based on discrete surface Yamabe flow is straight forward. We set the target Gaussian curvature to be zero everywhere in the interior of the surface, and the target geodesic curvature for the four markers to be $\frac{\pi}{2}$ and zero everywhere else for the points on the boundary. Then we run surface Yamabe flow to compute the flat metric $\bar{\mathbf{g}}$. Finally, we isometrically embed $(S, \bar{\mathbf{g}})$ onto the plane to obtain the rectangle.

Isometric Embedding Suppose $(S, V, \mathcal{T}, \mathbf{d})$ is a discrete surface with a flat metric $\mathbf{d}$, such that for each interior vertex $v_i \in V(\mathcal{T})$, $v_i \notin \partial S$, the discrete Gaussian curvature $K(v_i)$ is zero.

Suppose the first triangular face is $\Delta_0 = [v_0, v_1, v_2]$, the edge e_k is against vertex v_k, $k = 0, 1, 2$ with length l_k. The corner angle at v_k is θ_k, by the Euclidean cosine law,

$$\theta_k = \cos^{-1} \frac{l_i^2 + l_j^2 - l_k^2}{2l_i l_j}. \tag{20.5.1}$$

The images of the vertices under the isometric embedding are given by:

$$\varphi(v_0) = 0, \quad \varphi(v_1) = l_2, \quad \varphi(v_2) = l_1 e^{i\theta_0}. \tag{20.5.2}$$

Then we put all the faces adjacent to Δ_0 into a queue Q. When the queue is not empty, we pop the first face Δ_1 from the queue, suppose $\Delta_1 = [v_i, v_j, v_k]$ is a triangle, the vertices v_i and v_j have been embedded and v_k hasn't been embedded yet. We find all the faces adjacent to v_k and also in the queue, denoted as $\{\Delta_1, \Delta_2, \ldots, \Delta_n\} \subset Q$. We remove all of them from the queue. In the triangle Δ_1, $\varphi(v_k)$ is the intersection between two circles $|\varphi(v_k) - \varphi(v_i)| = l_j$ and $|\varphi(v_j) - \varphi(v_j)| = l_i$, and the orientation of the triangle $[\varphi(v_i), \varphi(v_j), \varphi(v_k)]$ is counter-clockwise, namely

$$\begin{cases} |\varphi(v_k) - \varphi(v_j)| &= l_i \\ |\varphi(v_k) - \varphi(v_i)| &= l_j \end{cases} \tag{20.5.3}$$

and

$$(\varphi(v_k) - \varphi(v_j)) \times (\varphi(v_k) - \varphi(v_i)) \cdot \mathbf{n} > 0, \tag{20.5.4}$$

where $\mathbf{n} = (0, 0, 1)^T$ is the normal to the plane. Then by solving the above equations in Δ_1, we can compute the embedding of v_k, denoted as $\varphi_1(v_k)$. For each triangle Δ_l, we can compute the embedding $\varphi_l(v_k)$. The final embedding of v_k is the average of all $\varphi_l(v_k)$'s,

$$\varphi(v_k) = \frac{1}{n} \sum_{l=1}^{n} \varphi(v_k). \tag{20.5.5}$$

Details can be found in Alg. 24.

Doubling   The discrete surface Yamabe flow is carried out on closed discrete surfaces. If the surface is with boundary components, we need to transform it to a symmetric closed surface using the *doubling technique*.

Suppose $(S, V, \mathcal{T})$ is an oriented discrete surface with boundary components ∂S, then we make two copies of S denoted as S^+ and S^-. For each triangle face $[v_i, v_j, v_k] \in F(\mathcal{T})$, we have a corresponding face $[v_i^-, v_j^-, v_k^-]$ in S^-, we reverse its orientation to $[v_j^-, v_i^-, v_k^-]$. This will reverse the orientation of the whole surface S^-. Then we glue S^+ and S^- along the corresponding boundary components. Namely, we define an equivalence relation

$$v_i^+ \sim v_i^-, \quad \forall v_i \in \partial S,$$

the doubled surface is defined as

$$\bar{S} := S^+ \cup S^- / \sim.$$

Algorithm 24 Isometric Embedding

Require: A discrete surface $(S, V, \mathbf{d}, \mathcal{T})$ with a flat metric $\mathbf{d}$;
Ensure: Isometric embedding $\varphi : S \to \mathbb{C}$;
  for each face $f \in F(\mathcal{T})$ **do**
    Compute the corner angles of f using the formula Eqn. (20.5.1);
    Label f as non-processed;
  end for
  for each vertex $v \in V(\mathcal{T})$ **do**
    Label v as non-embedded;
  end for
  Choose the first face Δ_0 and embed it using Eqn.(20.5.2)
  Label Δ_0 as processed, label all vertices of Δ_0 as embedded
  for each face f adjacent to Δ_0 **do**
    Label f as processed
    Enqueue f to the queue, $Q \leftarrow Q \cup \{f\}$
  end for
  while Q is not empty **do**
    Pop the head, $\Delta_1 \leftarrow Q.pop()$;
    Find the un-embedded vertex $v_k \in \Delta_1$;
    Enlist Δ_1 to the list, $L \leftarrow L \cup \{\Delta_1\}$
    for each face Δ_l adjacent to v_k **do**
      if Δ_l is in the queue Q **then**
        Enlist Δ_l to the list , $L \leftarrow L \cup \{\Delta_l\}$;
        Remove Δ_l from the queue, $Q \leftarrow Q \setminus \{\Delta_l\}$;
      end if
    end for
    for each face Δ_l in the list L **do**
      Compute $\varphi_l(v_k)$ using Eqn. (20.5.3) and Eqn. (20.5.4)
    end for;
    Compute $\varphi(v_k)$ as the mean of $\varphi_l(v_k)$'s using Eqn. (20.5.5);
    Label v_k as embedded;
    for each $\Delta_l \in L$ **do**
      for each face f adjacent to Δ_l **do**
        if f is not processed **then**
          Label f as processed
          Enqueue f to the queue, $Q \leftarrow Q \cup \{f\}$
        end if
      end for
    end for
  end while

The doubled surface $\bar{S}$ is a symmetric closed surface. It is obvious that the doubled surface $\bar{S}$ is a two-fold covering space of S, the projection map is given by $v_i^+ \to v_i$ and $v_i^- \to v_i$. The details can be found in Alg. 25.

Algorithm 25 Surface Doubling Operator

Require: A discrete surface $(S, V, \mathcal{T})$ with boundary components;
Ensure: The doubled surface $\bar{S}$;
 for each vertex $v_i \in S$ **do**
  Construct a vertex v_i^+ in $\bar{S}$;
  Assign $w_i^+ \leftarrow v_i^+$ and $w_i^- \leftarrow v_i^+$;
  if $v_i \notin \partial S$ **then**
   Construct a vertex v_i^- in $\bar{S}$;
   Assign $w_i^- \leftarrow v_i^-$
  end if
 end for
 for each face $f \in S$, $f = [v_i, v_j, v_k]$ **do**
  Construct a face $f^+ = [w_i^+, w_j^+, w_k^+]$ in $\bar{S}$;
  Construct a face $f^- = [w_i^-, w_j^-, w_k^-]$ in $\bar{S}$;
 end for

Conformal Mapping Given a topological quadrialateral $(S, V, \mathcal{T}, \mathbf{d})$, we construct the doubled surface $(\bar{S}, \bar{\mathbf{d}})$, the projection map is $p : \bar{S} \to S$, the metric $\bar{\mathbf{d}} = p^*\mathbf{d}$ is the pulled-back metric induced by the projection. Suppose $\{v_0, v_1, v_2, v_3\} \subset \partial S$ are the boundary markers, their preimages are $\bar{v}_k = p^{-1}(v_k)$, $k = 0, 1, 2, 3$. We set the target curvature of $\bar{v}_k$ to be π and zeros for other vertices. By running discrete Yamabe flow, we can find the discrete conformal metric $\bar{\mathbf{d}}'$, push it forward to S to obtain the metric $\mathbf{d}'$. By isometrically embedding $(S, \mathbf{d}')$ onto a planar rectangle, we obtain the conformal mapping $\varphi : S \to \mathbb{C}$. Details can be found in Alg. 26.

The similar algorithm can be applied for computing the conformal map from a topological disk S with three boundary markers $\{v_0, v_1, v_2\}$ onto a planar equilateral triangle. The only difference is to set the target curvature on markers to be $\frac{2\pi}{3}$. Figure 20.7 shows one computational result.

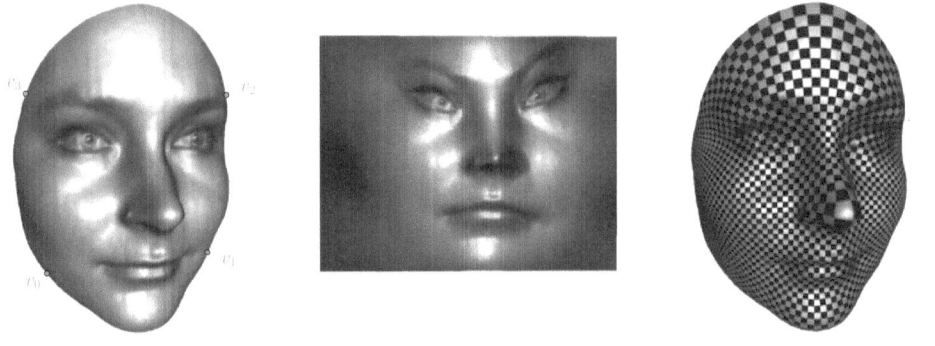

Figure 20.6 Topological quadrilateral conformal mapping using discrete Yamabe flow.

Algorithm 26 Discrere Yamabe Flow for Topological Quadrilateral Conformal Mapping

---

Require: A discrete surface $(S, V, \mathcal{T})$ with a single boundary component and four markers $\{v_0, v_1, v_2, v_3\} \subset \partial S$;

Ensure: A conformal mapping $\varphi : S \to \mathbb{C}$ transformed S to a planar rectangle, the markers are mapped to the corners of the rectangle;

Construct the doubled surface $(\bar{S}, \bar{\mathbf{d}})$ using Alg. 25; The projection map is $p : \bar{S} \to S$;

    for each vertex $\bar{v}_i \in \bar{S}$ **do**

        Set $\bar{K}(\bar{v}_i) \leftarrow 0$;

        if $p(\bar{v}_i) \in \{v_0, v_1, v_2, v_3\}$ **then**

            Set $\bar{K}(\bar{v}_i) \leftarrow \pi$;

        end if

    end for

Run discrete surface Yamabe flow to compute a flat metric $\bar{\mathbf{d}}'$ using Alg. 23;

Push forward the metric from $\bar{S}$ to S,

$$\mathbf{d}' \leftarrow p_* \bar{\mathbf{d}}'.$$

Isometric embed $(S, \mathbf{d}')$ using Alg. 24;

20.6 TOPOLOGICAL ANNULUS

Suppose $(S, \mathbf{d})$ is a topological annulus, $\partial S = \gamma_0 - \gamma_1$, where γ_0 and γ_1 are exterior and interior boundary components. We compute the doubled surface $\bar{S}$, which is a topological torus, $p : \bar{S} \to S$ is the projection map. The pull-back metric on the doubled surface is denoted as $\bar{\mathbf{d}} := p^* \mathbf{d}$. We set the target curvature to be zero everywhere on the doubled surface $\bar{S}$ and use the discrete surface Yamabe flow to compute the flat metric $\bar{\mathbf{d}}'$. The push-forward metric induced by the projection is $\mathbf{d}' = p_* \bar{\mathbf{d}}'$ is also flat on the original surface S.

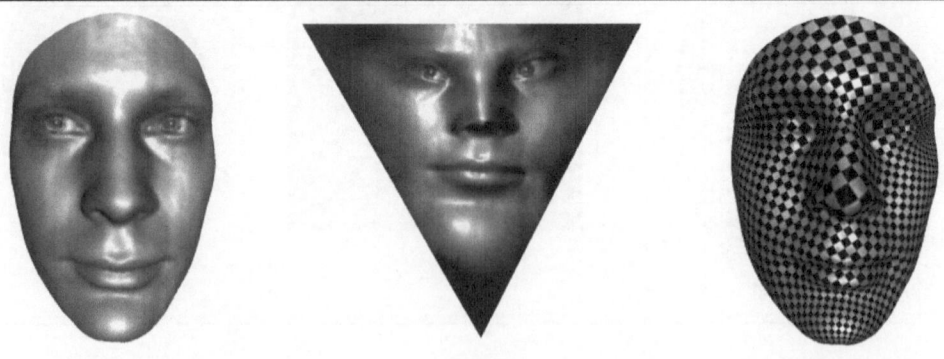

Figure 20.7 Topological triangle conformal mapping using discrete Yamabe flow.

We compute the shortest path τ between γ_0 and γ_1, slice S along τ to form a topological quadrilateral $\hat{S}$ with boundary $\partial\bar{S} = \gamma_0 - \gamma_1 + \tau + +\tau^-$ as shown in Figure 19.5. Then we isometrically embed the topological quadrilateral $(\hat{S}, \bar{\mathbf{d}}')$ onto the plane, $\varphi : \hat{S} \to \mathbb{C}$. By translation, rotation and scaling, we map $\varphi(\gamma_0)$ to the imaginary axis with total height 2π as shown in Figure 19.5. Finally, we compose φ with an exponential map $\exp : z \mapsto e^z$ to conformally map the topological quadrilateral $\hat{S}$ onto a canonical planar annulus $\mathcal{A}$ as shown in Figure 20.8.

The images of the corresponding vertices on τ^- and τ^+ coincide. In fact, this gives the conformal mapping from the original topological annulus $(S, \mathbf{d})$ to the canonical planar annulus $\mathcal{A}$. Details can be found in Alg. 27.

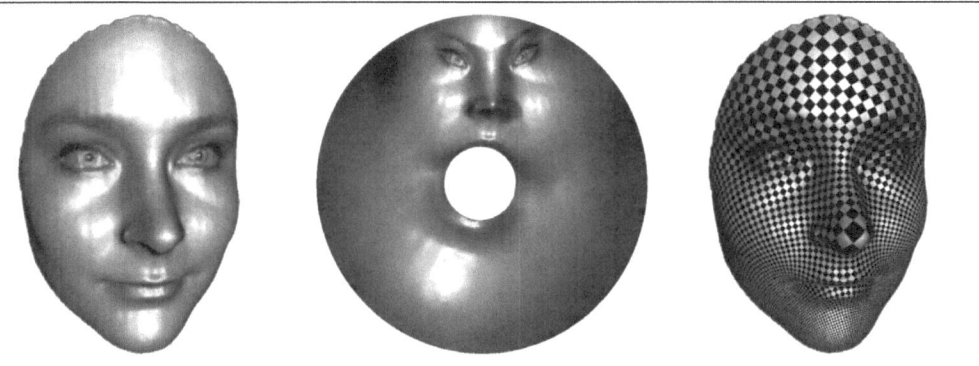

Figure 20.8 Topological annulus conformal mapping using discrete Yamabe flow.

Algorithm 27 Discrere Yamabe Flow for Topological Annulus Conformal Mapping

---

Require: A genus zero discrete surface $(S, V, \mathcal{T})$ with two boundary components $\partial S = \gamma_0 - \gamma_1$;

Ensure: A conformal mapping $\varphi : S \to \mathcal{A}$ from S to a planar annulus $\mathcal{A}$;

Construct the doubled surface $(\bar{S}, \bar{\mathbf{d}})$ using Alg. 25; The projection map is $p : \bar{S} \to S$;

for each vertex $\bar{v}_i \in \bar{S}$ **do**

    Set $\bar{K}(\bar{v}_i) \leftarrow 0$;

end for

Run discrete surface Yamabe flow to compute a flat metric $\bar{\mathbf{d}}'$ on $\bar{S}$ using Alg. 23;

Push forward the metric from $\bar{S}$ to S, $\mathbf{d}' \leftarrow p_*\bar{\mathbf{d}}'$;

Compute the shortest path τ between γ_0 and γ_1;

Slice S along τ to get a topological quadrilateral $\hat{S}$;

Isometric embed $(\hat{S}, \mathbf{d}')$ using Alg. 24, denote the mapping as $\varphi : \hat{S} \to \mathbb{C}$;

Rotate, translate and scale the image of φ, such that $\varphi(\gamma_0)$ is aligned with the imaginary axis with height 2π;

Compose the map with the exponential map $\exp : z \mapsto e^z$ to map $\hat{S}$ to a planar annulus; the images of the corresponding vertices on τ^- and τ^+ coincide;

Copy the image of each vertex from $\hat{S}$ to the input surface S.

20.7   TOPOLOGICAL POLY-ANNULUS

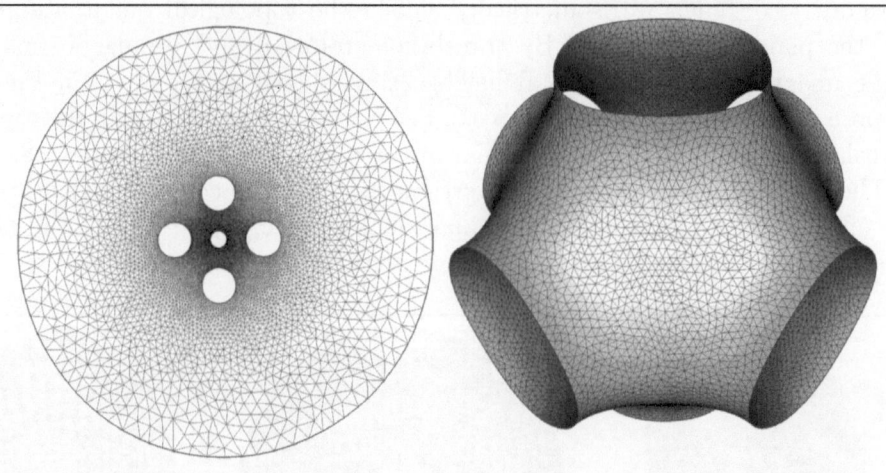

Figure 20.9 Topological poly-annulus conformal mapping using discrete Yamabe flow and Koebe's iteration algorithm.

Discrete surface Yamabe flow can also be applied for topological poly-annulus conformal mapping directly. Suppose $(S, V, \mathcal{T}, \mathbf{d})$ is a discrete surface, which is of genus zero and with multiple boundary components,

$$\partial S = \gamma_0 - \gamma_1 - \gamma_2 \cdots - \gamma_n,$$

as shown in Figure 19.12. We use Koebe's iteration method to compute the conformal map $\varphi : S \to \mathbb{D}^2$, where $\mathbb{D}^2$ is a circle domain on the plane. Figure 20.9 shows the conformal mapping from a minimal surface onto a planar circle domain. Figure 20.10 shows the computational process for a human facial surface with eyes and mouth regions removed.

The hole filling procedure is straight forward: given an inner hole with boundary γ_k, which can be represented as an ordered list of vertices,

$$\gamma_k = v_0, v_1, \ldots, v_{m-1},$$

then we compute the center of all the vertices,

$$c_k := \frac{1}{m} \sum_{i=0}^{m-1} v_i. \tag{20.7.1}$$

Then we construct triangles $[c_k, v_{i-1}, v_i]$ to fill the hole, $i = 1, 2, \ldots, m$, where i is modulo m. The input poly annulus S has n inner holes, after the hole filling procedure, it becomes a topological disk S^0. The original internal boundary γ_k bounds a simply connected domain, denoted as $\partial D_{k-1}^0 = \gamma_k$.

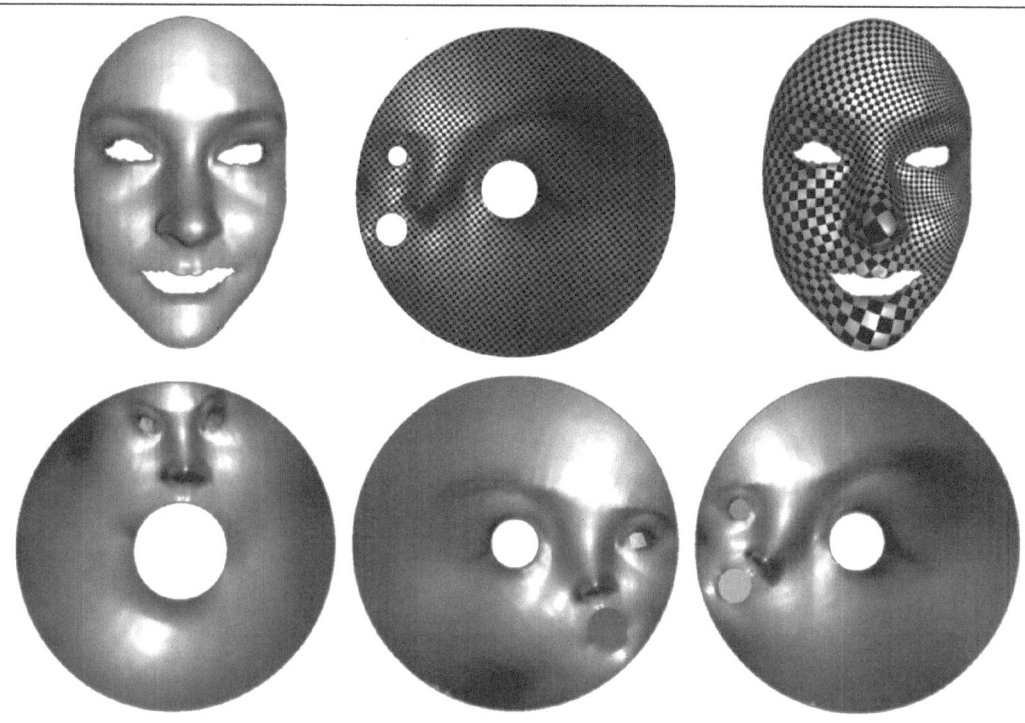

Figure 20.10 Topological poly-annulus conformal mapping using discrete Yamabe flow and Koebe's iteration algorithm.

We remove D_0^0 and map the surface $S^0 \setminus D_0^0$ to the canonical annulus using Alg. 27, $\varphi^0 : S^0 \setminus D_1^0 \to \mathbb{D}^2$, then fill a disk D^0 to the annulus image $\varphi^0(S^0 \setminus D_0^0)$ to obtain S^1,

$$S^1 := \varphi^0(S^0 \setminus D_0^0) \cup D^0.$$

The domains on S^1 corresponding to the original inner inner holes are denoted $D_0^1, D_1^1, \ldots, D_{n-1}^1$,

$$D_k^1 = \varphi^0(D_k^0), k > 0, \quad D_0^1 = D^0.$$

Then we compute $\varphi^1 : S^1 \setminus D_1^1 \to \mathbb{D}^2$, and fill a disk D^1 to the inner circular hole of the annulus $\varphi^1(S^1 \setminus D_1^1)$ and obtain

$$S^2 := \varphi^1(S^1 \setminus D_1^1) \cup D^1,$$

with the interior domains

$$D_i^2 = \varphi^1(D_i^1), i \neq 1, \quad D_1^2 = D^1.$$

We repeat this procedure, at the k-th step, the input surface is S^k, with interior domains $D_0^k, D_1^k, \ldots, D_{n-1}^k$, we puncture a hole D_r^k, $r \equiv k \mod n$ and map it to the annulus by $\varphi^k : S^k \setminus D_r^k \to \mathbb{D}^2$, then we use a disk D^r to fill the inner circular hole of the annulus to obtain

$$S^{k+1} := \varphi^k(S^k \setminus D_r^k) \cup D^r, \quad r \equiv k \mod n,$$

with the interior domains

$$D_i^{k+1} = \varphi^k(D_i^k), \quad i \neq r, \quad D_r^{k+1} = D^r.$$

We repeat this procedure, the inner domains get rounder and rounder, and converge to conical circular disks eventually. The algorithmic details can be found in Alg. 28.

Algorithm 28 Koebe's Iteration Based on Discrete Surface Yamabe Flow

---

Require: A topological poly-annulus $(S, \mathbf{g})$;
Ensure: A conformal mapping $\varphi : S \to \mathbb{D}^2$, φ maps S to a circle domain;
  Initialize $S^0 \leftarrow S$;
  for each interior boundary component γ_k, $k = 1, 2, \ldots, n$ **do**
    Compute the center c_k of γ_k using Eqn. (20.7.1);
    Initialize a domain $D_{k-1}^0 \leftarrow \emptyset$;
    for each each edge $[v_i, v_{i+1}]$, $i = 0, 1, \ldots, m - 1$ **do**
      Construct a triangle face $f_{k,i} := [c_k, v_i, v_{i+1}]$;
      $D_{k-1}^0 \leftarrow D_{k-1}^0 \cup \{f_{k,i}\}$;
    end for
    $S^0 \leftarrow S^0 \cup D_{k-1}^0$;
  end for
  for each $k \in \{0, 1, 2, \ldots, n, \ldots\}$ **do**
    Remove an interior domain D_r^k, $r \equiv k \mod n$;
    Compute a conformal map $\varphi^k : S^k \setminus D_r^k \to \mathbb{D}^2$ using Alg. 27;
    Fill the center hole $S^{k+1} \leftarrow \varphi^k(S^k \setminus D_r^k) \cup D^r$;
    if all the domains D_r^{k+1}, $r = 0, \ldots, n - 1$ are circular enough **then**
      Break;
    end if
  end for
  return S_n;

20.8 TOPOLOGICAL TORUS

The discrete surface Yamabe flow can be applied for computing the conformal module of topological torus directly.

Suppose $(S, V, \mathbf{d}, \mathcal{T})$ is a topological torus, we first compute its fundamental group generators, $\pi_1(S, p) = \langle a, b | aba^{-1}b^{-1} \rangle$, where the base point p is the intersection point of a and b; then slice the surface along a, b to obtain a fundamental domain $\bar{S}$. We run Yamabe flow on S with zero target curvature everywhere to get a flat metric $\mathbf{d}'$, and isometrically embed $(\bar{S}, \mathbf{d}')$ on the plane, $\varphi : \bar{S} \to \mathbb{C}$. The base point p has four images, forming a parallelogram, which is the fundamental polygon of the lattice Λ. This gives the conformal mapping from the surface S to the flat torus $\mathbb{C}/\Lambda$. The algorithmic details can be found in Alg. 29.

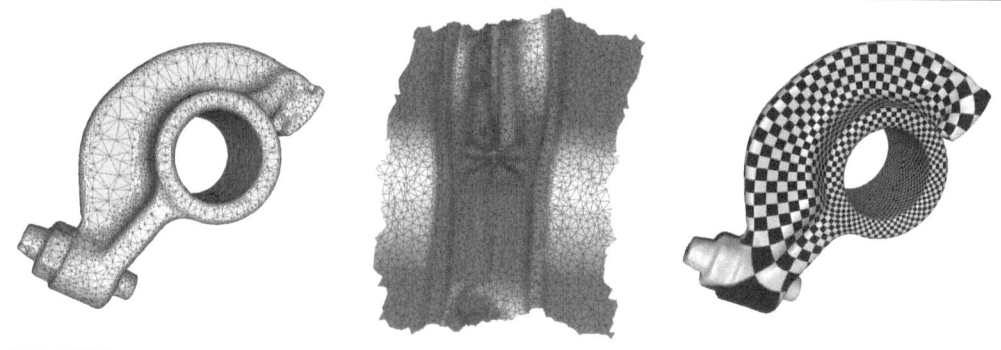

Figure 20.11 Conformal mapping for a topological torus, the kitten surface model.

Figure 20.11 shows one computational result of the genus one surface using the discrete surface Yamabe flow. The left frame shows the input topological torus S; the middle frame displays one fundamental domain $\varphi(\bar{S})$; the right frame demonstrates the conformality of the mapping $\varphi : S \to \mathbb{C}/\Lambda$ by the checkerboard texture mapping.

Algorithm 29 Discrete Yamabe Flow for Topological Torus Conformal Mapping

---

Require: A topological torus $(S, \mathbf{g})$;
Ensure: A conformal map $\varphi : S \to \mathbb{C}/\Lambda$;
  Compute the fundamental group $\pi_1(S, p)$ using Alg.4, find the generators a, b and the intersection point p;
  Compute the holomorphic 1-form basis ω_1, ω_2 using Alg. 13;
  Find the real coefficient λ_1, λ_2, $\omega = \lambda_1 \omega_1 + \lambda_2 \omega_2$, such that Eqn. (19.6.1) holds;
  Construct a finite portion of the universal covering space $\tilde{S}$ using Alg. 5;
  Compute the conformal map $\tilde{\varphi} : \tilde{S} \to \mathbb{C}$ in Eqn. (19.6.2);
  Extract the orbit $\pi^{-1}(p)$ and obtain the Lattice Λ;
  return The conformal mapping $\varphi : S \to \mathbb{C}/\Lambda$;

CHAPTER 21

Mesh Generation Based on Abel-Jacobi Theorem

This chapter introduces the computational algorithms for structured mesh generation based on Abel-Jacobi theorem, which are based on the works in [79, 80, 349].

21.1 QUAD-MESHES AND MEROMORPHIC QUARTIC FORMS

In this section, we show the intrinsic relation between quad-meshes and meromorphic quartic differentials.

Definition 21.1.1 (Quadrilateral Mesh) *Suppose S is a topological surface, $\mathcal{Q}$ is a cell partition of S, if all cells of $\mathcal{Q}$ are topological quadrilaterals, then we say $(S, \mathcal{Q})$ is a* quadrilateral mesh *or a* quadrangulation.

Figure 21.1 shows different quadrilateral meshes for a planar rectangle with two circular holes. The topological valence 4 vertices are called *normal*, otherwise *singular*. The stream lines through the singularities are also labelled, which are called *separatrices*.

On a quad-mesh, the *topological valence* of a vertex is the number of faces adjacent to the vertex.

Definition 21.1.2 (Singularity) *Suppose $(S, \mathcal{Q})$ is a quadrilateral mesh. If the topological valence of an interior vertex is 4, then we call it a* regular vertex, *otherwise a* singularity; *if the topological valence of a boundary vertex is 2, then we call it a* regular boundary vertex, *otherwise a* boundary singularity. *The index of a singularity is defined as follows:*

$$Ind(v_i) = \begin{cases} 4 - Val(v_i) & v_i \notin \partial(S, \mathcal{Q}) \\ 2 - Val(v_i) & v_i \in \partial(S, \mathcal{Q}) \end{cases}$$

where $Ind(v_i)$ and $Val(v_i)$ are the index and the topological valence of v_i.

Theorem 21.1.3 (Qaud-Mesh to Meromrophic Quartic Differential) *Suppose $(S, \mathcal{Q})$ is a closed quadrilateral mesh, then*

DOI: 10.1201/9781003350576-21

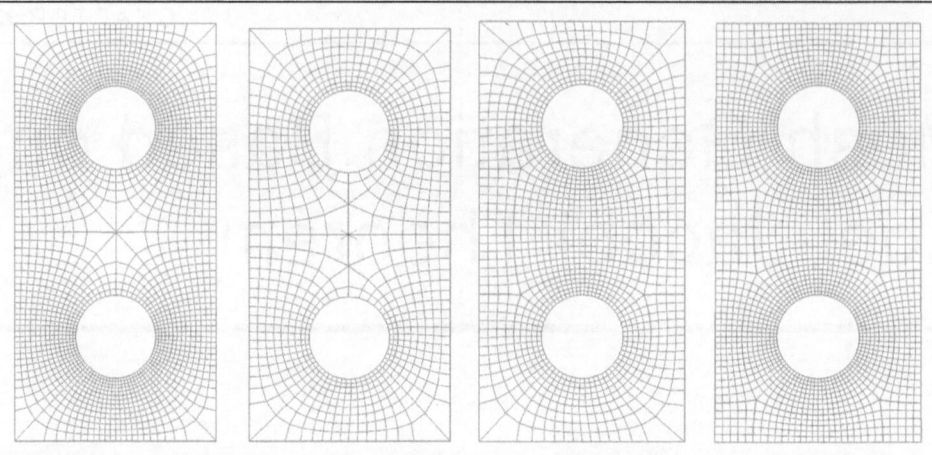

Figure 21.1 Quad-meshes with different number of singularities.

1. *the quad-mesh $\mathcal{Q}$ induces a conformal atlas $\mathcal{A}$, such that $(\Sigma, \mathcal{A})$ form a Riemann surface, denoted as S_Q.*

2. *the quad-mesh $\mathcal{Q}$ induces a meromorphic quartic differential ω_Q on S_Q. The valence-k singular vertices correspond to poles or zeros of order $k - 4$. Furthermore, the trajectories of ω_Q are finite.*

Proof As shown in Figure 21.3, we treat each face as a Euclidean unit planar square, this assigns a flat Riemannian metric to the surface, such that the curvature is zero everywhere except at the singularities. At the singularity with valence k, the cone angle is $k\pi/2$.

First, we construct the open covering of the surface. The interior of each face f is one open set U_f; each edge e is covered by the interior of a rectangle U_e; each vertex v is covered by a disk, U_v. It is obvious that

$$S \subset \bigcup_f U_f \bigcup_e U_e \bigcup_v U_v.$$

Second, we establish the local complex coordinates of each open set. For each face U_f, we choose the center of the face as the origin, the real and imaginary axises are

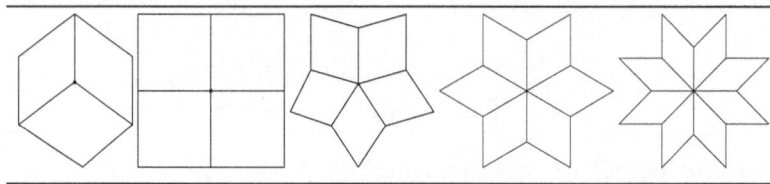

Figure 21.2 Quad-mesh vertices with different topological valences, from left to right, the valences are $3, 4, 5, 6$ and 8 respectively.

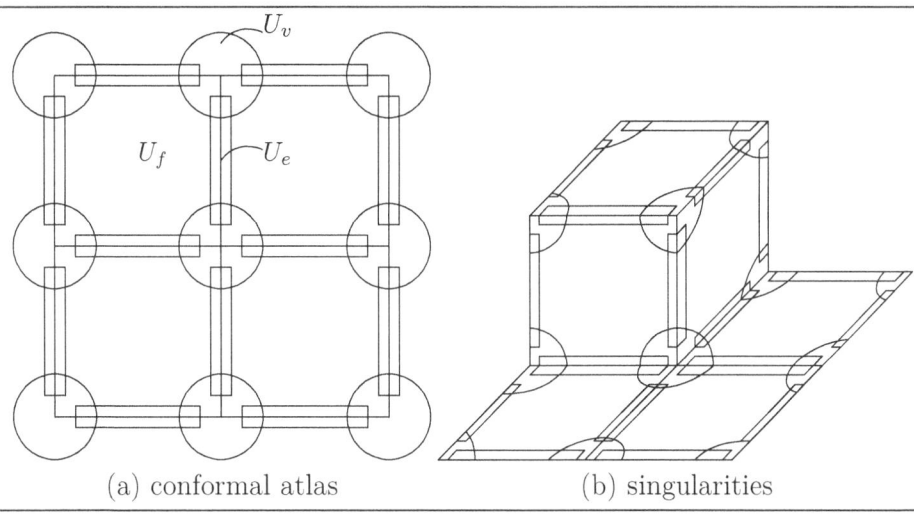

| (a) conformal atlas | (b) singularities |

Figure 21.3 A quad-mesh induces a conformal atlas, such that the surface $(S, \mathcal{Q})$ becomes a Riemann surface.

parallel to the edges of the face, the local parameter is denoted as z_f; for each edge U_e, we choose the center as the origin, the real axis is parallel to the edge direction, the local parameter is z_e; for each regular vertex U_v, we choose the center of the disk as the origin, one edge as the real axis, the local parameter is z_v; for a singular vertex U_v, we first isometrically flatten U_v on the complex plane, to obtain the local parameter w_v, the we use the complex power function to shrink it as

$$w_v^{\frac{4}{k}} \mapsto z_v,$$

where k is valence of the singular vertex.

Now we examine all the coordinate transition maps. We denote a subgroup of the planar rigid motion G generated by

$$z \mapsto e^{i\frac{\pi}{2}} z, \quad z \mapsto z + \frac{1}{2}, \quad z \mapsto z + \frac{i}{2},$$

Then any transition map φ_{fe} between a face and an adjacent edge $z_f \to z_e$ is an element in G, $\varphi_{fe} \in G$; any transition map φ_{ev} between an edge and one of its end vertex $z_e \to z_v$ is an element in G, $\varphi_{ev} \in G$.

The transition between a singulary vertex and its neihboring edge or face is more complicated. For example, the transition from a face to its neighboring singular vertex is given by

$$z_v = \left(e^{i\frac{n\pi}{2}} z_f + \frac{1}{2}(\pm 1 \pm i) \right)^{\frac{4}{k}} \qquad (21.1.1)$$

where $m, n \in \mathbb{Z}$, k is the valence of v.

Hence all the transition maps are biholomorphic. $\mathcal{A} = \{(U_f, z_f), (U_e, z_e), (U_v, z_v)\}$ form a conformal atlas. $(S, \mathcal{A})$ is a Riemann surface, denoted as $S_{\mathcal{Q}}$.

Now, we construct a holomorphic 1-form on each face U_f, dz_f. Because the orientations of all faces are chosen individually, dz_f is not globally defined. Then we define the holomorphic quadric form

$$\omega = (dz_f)^4,$$

then ω is globally defined. Suppose a face f and an edge e are adjacent, then

$$dz_f = e^{i\frac{n\pi}{2}} dz_e, \ n \in \mathbb{Z},$$

then $(dz_f)^4 = (dz_e)^4$. Suppose an edge e has an regular edge vertex v

$$dz_e = e^{i\frac{n\pi}{2}} dz_v, \ n \in \mathbb{Z},$$

therefore $(dz_e)^4 = (dz_v)^4$. Suppose a face f has a regular vertex v, according to Eqn. (21.1.1), where $k = 4$,

$$dz_v = e^{i\frac{n\pi}{2}} dz_f,$$

hence $(dz_v)^4 = (dz_f)^4$.

Suppose v is a singular vertex with valence $k \neq 4$, by Eqn. (21.1.1), we obtain

$$\frac{k}{4} z_v^{\frac{k-4}{4}} dz_v = e^{i\frac{n\pi}{2}} dz_f,$$

therefore

$$\left(\frac{k}{4}\right)^4 z_v^{k-4}(dz_v)^4 = (dz_f)^4 = \omega. \tag{21.1.2}$$

Hence, a valence 3 singular vertex corresponds to a simple pole $\frac{c}{z_v} dz_v^4$, a valence 5 singular vertex becomes a simple zero $c z_v dz_v^4$.

Therefore, the meromorphic quartic differential ω is globally defined, the valence-k singular vertices correspond to poles or zeros with order $k - 4$. The finiteness of the trajectories of ω is obvious.

□

Theorem 21.1.4 (Abel-Jacobi condition) *Suppose $\mathcal{Q}$ is a closed quadrilateral mesh, S_Q is the induced Riemann surface, ω_Q is the induced meromorphic quadric form. Assume ω_0 is an arbitrary holomorphic 1-form on S_Q, then*

$$\mu((\omega_Q) - 4(\omega_0)) = 0 \qquad \text{mod } \Lambda \tag{21.1.3}$$

in the Jacobian $J(S_Q)$.

Proof The ratio $f = \omega_Q/\omega_0^4$ is a meromorphic function, therefore according to Abel theorem $\mu((f)) = 0$ in $J(S_Q)$,

$$\mu((f)) = \mu((\omega_Q/\omega_0^4)) = \mu((\omega_Q) - (\omega_0^4)) = \mu((\omega_Q) - 4(\omega_0)) = 0 \qquad \text{mod } \Lambda.$$

□

By the Abel-Jacobi condition, we can show the non-existences of special types of structured meshes.

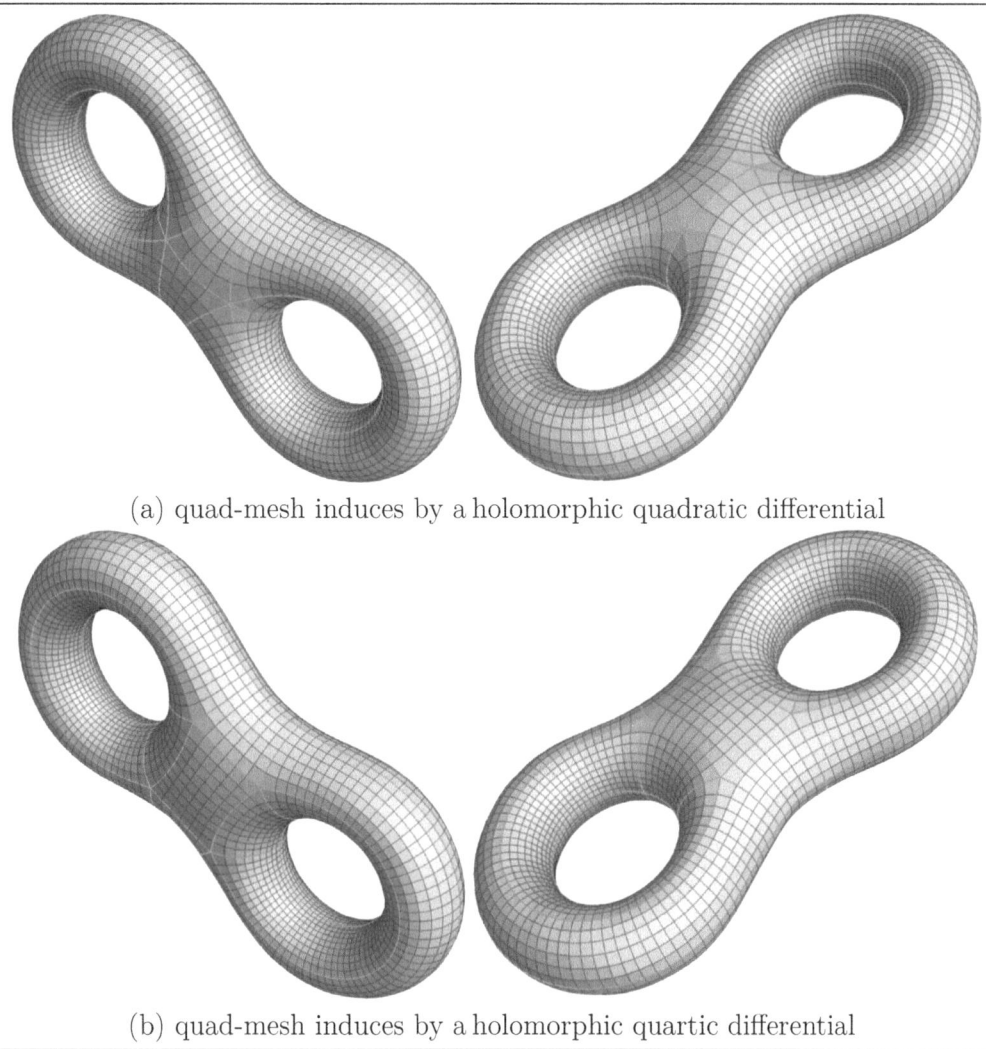

(a) quad-mesh induces by a holomorphic quadratic differential

(b) quad-mesh induces by a holomorphic quartic differential

Figure 21.4   Quadrilateral meshes induced by holomorphic differentials.

Theorem 21.1.5 (Barnette et al [29]) *The torus has no $3, 5$-quadrangulation, that is, no quadrangulation with exactly two exeptional vertices, of degree 3 and 5.*

Proof Suppose there is a $3, 5$-quadrangulation $\mathcal{Q}$, p is the valence-5 vertex, q the valence-3 vertex. $\mathcal{Q}$ induces a Riemann surface $S_\mathcal{Q}$ and a meromorphic quartic differential $\omega_\mathcal{Q}$. We choose a holomorphic 1-form ω_0 without zeros or poles, then $f = \omega_\mathcal{Q}/\omega_0^4$ is a meromorphic function, then

$$\mu((f)) = \mu\left(\frac{\omega_\mathcal{Q}}{\omega_0^4}\right) = \mu((\omega) - 4(\omega_0)) = \mu(p) - \mu(q) = 0.$$

Thus p conincides with q, contradiction to the fact that p is the zero and q the pole of $\omega_\mathcal{Q}$. Therefore, there is no 3-5 quad-mesh on a torus.

□

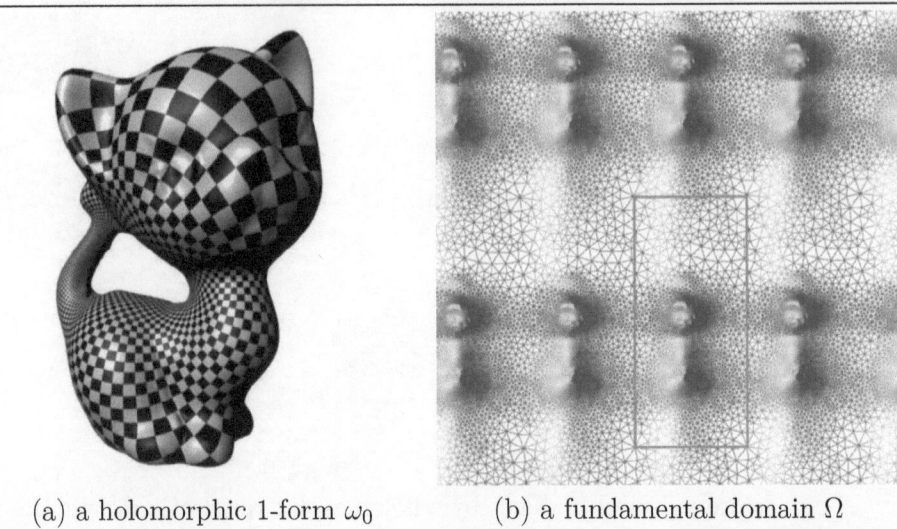

(a) a holomorphic 1-form ω_0 (b) a fundamental domain Ω

Figure 21.5 The proof of Theorem 21.1.5 is based on Theorem 21.1.4 (Abel-Jacobian condition). No genus one closed quad-mesh has only one valence 3 and one valence 5 singular vertices.

Theorem 21.1.6 (Izmestiev et al 3 [172]) *The vertices of a 2, 4-hexangulation of the torus cannot be bicolored (that is, the 1-skeleton is not bipartite).*

Proof Assume there exists a 2, 4-hexangulation of the torus, denoted as $\mathcal{H}$ and all the vertices are colored as black and white. Similar to the theorem 21.1.3, we construct a conformal at atlas $\mathcal{A}$. As shown in Figure 21.6, each edge e has an orientation from the black vertex to the white vertex, the origin of z_e is the mid-point, the real axis points to the positive direction of e. Each face f is treated as a canonical planar

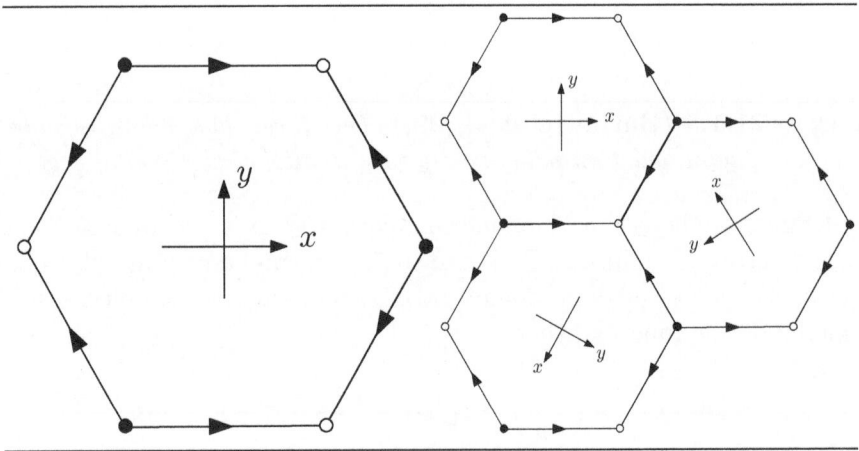

Figure 21.6 A hexangulation induces a conformal atlas, such that the surface becomes a Riemann surface.

hexagon, the center of z_f is at the origin, the real axis is aligned with one edge whose orientation is consistent with the face orientation, hence there are 3 choices. For each vertex v, the origin is the vertex, the real axis is along one edge with consistent orientation. The rotation components of all transition maps among faces, edges and regular vertices are $e^{i\frac{2n\pi}{3}}$, $n \in \mathbb{Z}$. For a singular vertex v with degree k, we have

$$z_v = w_v^{\frac{3}{k}}.$$

Then on faces, edges and regular vertices, the holomorphic differential $\omega = (dz)^3$ is globally defined. On the singular vertex, we have

$$dw_v = \frac{k}{3}(z_v)^{\frac{k-3}{3}}dz_v,$$

then we extend ω to the singular vertex,

$$(dw_v)^3 = \left(\frac{k}{3}\right)^3 (z_v)^{k-3}(dz_v)^3.$$

Then ω is globally well-defined. The meromorphic function is given by $f = \frac{\omega}{(\omega_0)^3}$. Similar to the proof of theorem 21.1.5, by Abel-Jacobi theorem, we obtain a contradiction. □

Theorem 21.1.7 (Izmestiev et al 4 [172]) *The faces of a $4, 8$-triangulation of the torus cannot be 2-colored (that is, the dual graph is not bipartite).*

Proof Suppose there is a $4, 8$-triangulation, and all the faces are 2-colored as shown in the left frame of Figure 21.7. For each edge e, we define an orientation, such that the black face is on the left. The proof is similar to that of theorem 21.1.3, the

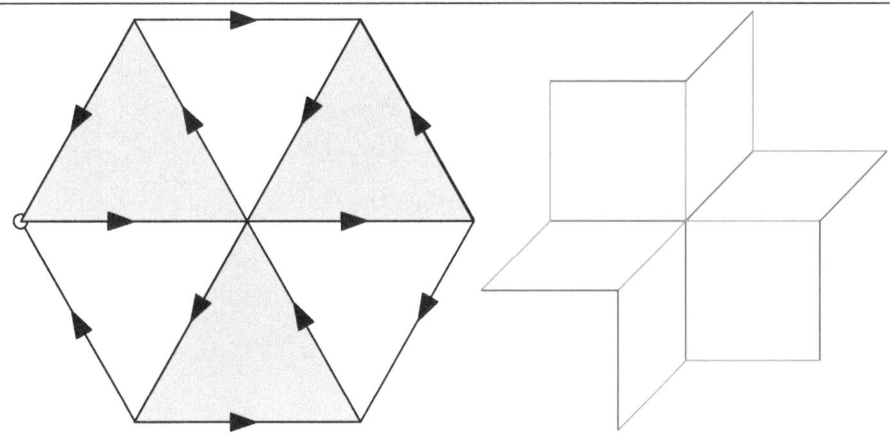

Figure 21.7 A triangle-mesh with face 2-coloring (left), a quad-mesh with edge 2-coloring.

rotational components of the transition maps among faces, edges and regular vertices are $e^{i\frac{2n\pi}{3}}$, $n \in \mathbb{Z}$. The singularity v must have even degree $2k$, then $z_v = w_v^{\frac{3}{k}}$. The global meromorphic differntial $\omega = (dz)^3$ on faces, edges and regular vertices, and for singular vertex v with degree $2k$, $\omega = (\frac{k}{3})^3 (z_v)^{k-3} (dz_v)^3$. Hence the degree 4 and 8 vertices become the pole and the zero of the meromorphic function $\frac{\omega}{(\omega_0)^3}$.

$\square$

Theorem 21.1.8 (Izmestiev et al 5 [172]) *The edges of a 2,6-quadrangulation of the torus cannot be bicolored with colors alternating around each face (or equivalently, around each vertex).*

Proof Suppose there is a 2,6-quadrangulation, and all the edges are 2-colored as shown in the right frame of Figure 21.7.

For each face f, the real axis of z_f is along red edge direction; for each red edge e, the real axis of z_e is along the edge; for each blue edge e, the imaginary axis is along the edge. For each vertex v, the degree must be even, and the real axis of z_v is along one red edge. The rotational components of the transition maps among faces, edges and regular vertices are $e^{in\pi}$, $n \in \mathbb{Z}$. The singularity v must have even degree $2k$, then $z_v = w_v^{\frac{2}{k}}$. The global meromorphic differntial $\omega = (dz)^2$ on faces, edges and regular vertices, and for singular vertex v with degree $2k$, $\omega = (\frac{k}{2})^2 (z_v)^{k-2} (dz_v)^2$. Hence the degree 2 and 6 vertices become the pole and the zero of the meromorphic function $\frac{\omega}{(\omega_0)^2}$.

$\square$

21.2   METRICS WITH SPECIAL HOLONOMIES

Each quad-mesh also induces a flat metric with cone singularities.

Definition 21.2.1 (Quad-Mesh Metric) *Suppose S is a topological surface, $\mathcal{Q}$ is a quadrilateral mesh on S. If each quad-face is treated as a unit planar square, then $\mathcal{Q}$ induces a flat metric with cone singularities, denoted as $\mathbf{g}_\mathcal{Q}$.*

For a vertex $v_i \in Q$, its discrete Gaussian curvature is given by

$$K(v) = \begin{cases} \frac{\pi}{2}(4 - k(v)) & v \notin \partial\mathcal{Q} \\ \frac{\pi}{2}(2 - k(v)) & v \in \partial\mathcal{Q} \end{cases}$$

Then the total Gaussian curvature satisfies the Gauss-Bonnet theorem.

Lemma 21.2.2 *Given a quad-mesh $\mathcal{Q}$ on a surface S, the induced metric is $\mathbf{g}_\mathcal{Q}$, the total Gaussian curvature satisfies*

$$\sum_{v_i \in \partial\mathcal{Q}} K(v_i) + \sum_{v_i \notin \partial\mathcal{Q}} K(v_i) = 2\pi\chi(S). \tag{21.2.1}$$

Proof Suppose S is an oriented closed surface, then each face has 4 edges, each edge is shared by 2 faces, therefore $4F = 2E$. The total curvature

$$\sum_{v_i \in V} K(v_i) = 2\pi V - 2\pi F = 2\pi(V - F) = 2\pi(V + F - E) = 2\pi\chi(S).$$

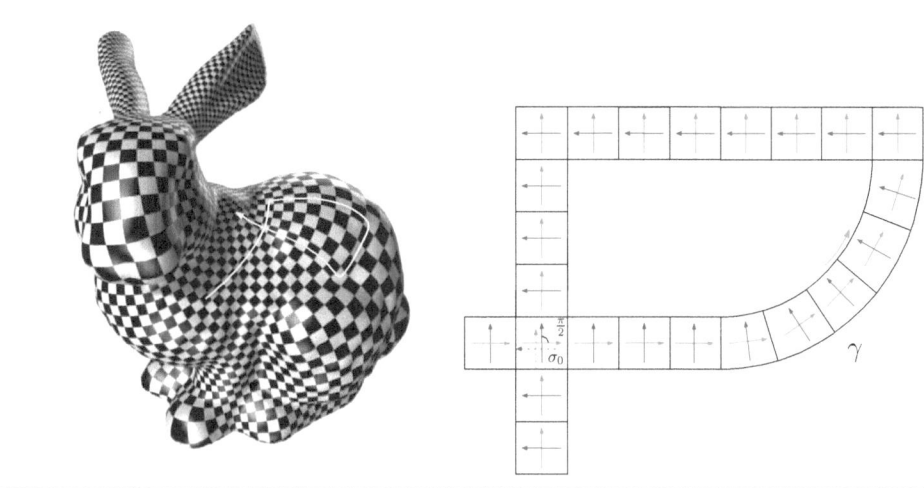

Figure 21.8 Parallel transportation along a face loop.

Namely, Eqn. (21.2.1) still holds. If S has boundaries, then we double the surface to obtain $\bar{S}$. The Euler numbers satisfy $\chi(\bar{S}) = 2\chi(S)$. The total curvature of $\bar{S}$ equals to the double of that of S, therefore Eqn. (21.2.1) still holds.

□

Given a genus one surface S, for any two distinct points $p, q \in S$, we can always find a flat metric $\mathbf{g}$ with cone singularities at p and q, such that the discrete Gauss curvatures satisfy $K(p) = \frac{\pi}{2}$ and $K(q) = -\frac{\pi}{2}$. But according to theorem 21.1.5, there is no 3-5 quad-mesh on the torus. This means the quad-mesh metric satisfies more constraints than Gauss-Bonnet Eqn. (21.2.1). The key is the holonomy condition.

As shown in Figure 21.8, suppose $\mathcal{Q}$ is a quad-mesh, we fix a base face σ_0. Let γ be a face-loop, namely a sequence of faces $\gamma = \sigma_0, \sigma_1, \ldots, \sigma_{n-1}, \sigma_n$, where $\sigma_n = \sigma_0$, each pair of consecutive faces σ_i and σ_{i+1} share a common edge. We select an orthonormal frame in σ_0, whose axis are parallel to the edges of the face, then parallel transport the frame to the next face σ_1, and parallel transport to the 3rd face σ_2 and so on. When we return to the original face, the transported frame and the original frame differ by a rotation. All such kind of rotations form a group, the so-called *holonomy group* of the metric $\mathbf{g}_\mathcal{Q}$.

Definition 21.2.3 (Holonomy Group) *Given a quad-mesh $\mathcal{Q}$, which induces the quad-mesh metric $\mathbf{g}_\mathcal{Q}$. Select a base face σ_0, and for any face loop γ starting from σ_0, the parallel transportation along γ induces a planar rotation $e^{i\theta_\gamma}$. All such kind of rotations form a group, which is called the* holonomy group *of $\mathcal{Q}$, denoted as $\mathcal{H}_\mathcal{Q}$.*

Lemma 21.2.4 (Quad-mesh Holonomy) *Suppose S is an oriented surface, $\mathcal{Q}$ is a quad-mesh of S with induced quad-mesh metric $\mathbf{g}_\mathcal{Q}$. The holonomy group of Q is a subgroup of the rotation group,*

$$\mathcal{H}_\mathcal{Q} \subset \left\{ e^{i\frac{k}{2}\pi}, k = 0, 1, 2, 3 \right\}. \tag{21.2.2}$$

Proof Suppose γ is a face loop, consists of quad-faces $\sigma_0, \sigma_1, \ldots, \sigma_n$, $\sigma_n = \sigma_0$. We set an initial frame on σ_0, such that the axises are parallel to the edge. When we parallel transport the frame from one face to the next, the frame axises are always parallel to the edges. When we reach σ_n, the result frame axises are also parallel to the edges. Since $\sigma_0 = \sigma_n$, the rotation of the frame is $e^{i\frac{k}{2}\pi}$, $k \in \mathbb{Z}$. Since γ is arbitrarily chosen, Eqn. (21.2.2) holds.

□

We can consider a flat cone metric $\mathbf{g}$ on S satisfying the holonomy condition Eqn. (21.2.2), then natually we can define a meromorphic quartic differential $\omega_{\mathbf{g}}$, such that the cone singularities become the zeros or poles of $\omega_{\mathbf{g}}$. For any cone singular point $p \in S$, its cone angle is $\frac{k_p}{2}\pi$, then $\mathrm{ord}_p(\omega_{\mathbf{g}}) = k_p - 4$, the divisor of the differential $\omega_{\mathbf{g}}$ is

$$(\omega_{\mathbf{g}}) = \sum_{p \in S}(k_p - 4)p. \tag{21.2.3}$$

The Abel-Jacobi condition for the metric $\mathbf{g}$ holds.

Theorem 21.2.5 (Quad-mesh Holonomy Group) *Suppose S is a closed surface with genus $g > 0$, $\mathbf{g}$ is a flat metric with cone singularities, such that the holonomy group is a subgroup of $\left\{e^{i\frac{k}{2}\pi}, k = 0, 1, 2, 3\right\}$. Suppose ω_0 is a holomorphic 1-form on $(S, \mathbf{g})$, then*

$$\mu\left(\sum_{p \in S}(k_p - 4)p - 4(\omega_0)\right) = 0. \tag{21.2.4}$$

Proof We select one point $p \in S$ which is not a cone singularity, then choose an orthonormal frame of its tangent plane T_pS and treat it as a cross. We parallel transport the frame on the surface. Because the metric $\mathbf{g}$ satisfies the holonomy condition in Eqn. (21.2.4), the parallel transportation result is independent of the choice of the path. Then we globally define a cross field everywhere on the surface except the cone singularities.

For each point $p \in S$, which is not a cone singularity, we find a neighborhood $U(p)$ without any singularity in it, and isometrically embed $U(p)$ on the complex plane, such that the cross field on $U(p)$ is aligned with the real and imaginary axises, the mapping is denoted as φ_p. Then, for each singularity, we find a cone $V(p)$, isometrically flatten onto a fan shape $\{(r, \theta)|0 \le \theta \le \frac{k_p}{2}2\pi, r \in [0, \varepsilon)\}$, then use an exponential map $z \mapsto z^{\frac{4}{k_p}}$ to transform the fan shape to a disk, the mapping is denoted as ψ_p. Then the union of the local charts $(U(p), \varphi_p)$'s for all normal points and $(V(p), \psi_p)$'s for all singularities form a conformal atlas. On each chart $(U(p), \varphi_p)$, we define a holomorphic quartic differential dz_p^4 and extended to the charts of singularities, this globally defines a meromorphic quartic differential $\omega_{\mathbf{g}}$. The divisor of $\omega_{\mathbf{g}}$ is given by Eqn. (21.2.3).

Suppose ω_0 is a holomorphic 1-form, then $f = \frac{\omega_{\mathbf{g}}}{\omega_0^4}$ is a meromorphic function, according to Abel theorem,

$$\mu((f)) = \mu((\omega_{\mathbf{g}})) - 4\mu((\omega_0)) = 0,$$

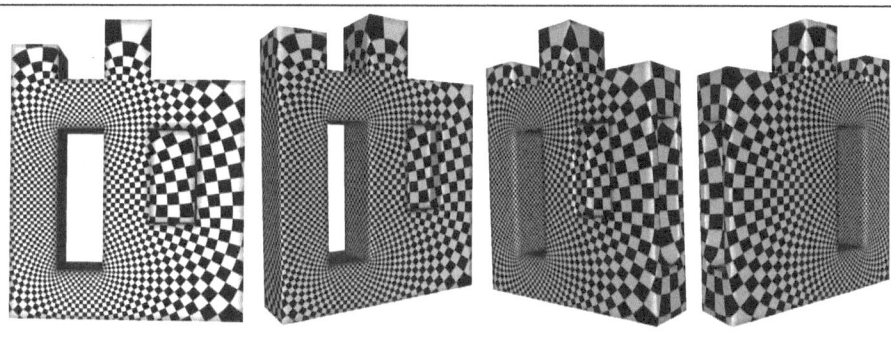

Figure 21.9 A genus one polycube surface.

hence Eqn. (21.2.4) holds.

□

In order to verify the theorem, we show an example in Figure 21.9. The surface S is the union of cuboids, which is called a *polycube* surface and satisfies the holonomy condition in Eqn. 21.2.4. We compute a holomorphic 1-form ω using Alg. 13and the homology basis $\{a, b\}$ using Alg. 4. The base point p is the intersection point between a and b. As shown in Figure 21.10, we slice the surface along $\{a, b\}$ to get a fundamental domain D, $\partial D = abab^{-1}b^{-1}$, and compute the Abel-Jacobi mapping $\mu : S \to J(S)$ as

$$\mu(q) = \int_p^q \omega,$$

where p is the base point and the integration path is arbitrarily chosen inside the fundamental domain D. As shown in Figure 21.11, theorem 21.2.5 is verified. Suppose q_i's are valence 3 singularities (poles), p_j's are valence 5 singularities (zeros), then we have found that the number of poles equals to that of the zeros, furthermore, the mass center of the Abel-Jacobi map images of the poles coincides with that of the

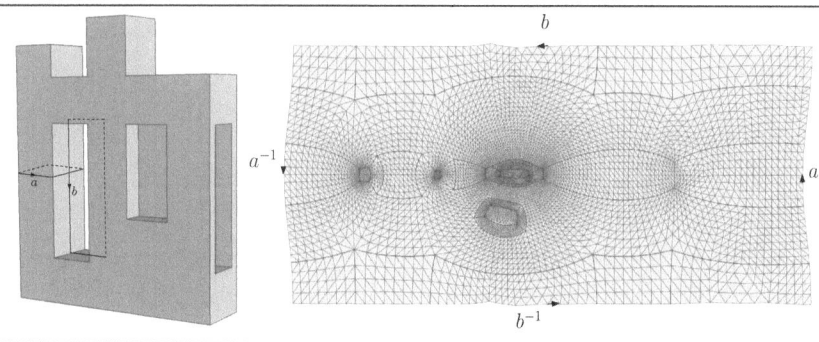

Figure 21.10 Abel-Jacobi map $\mu : S \to J(S)$.

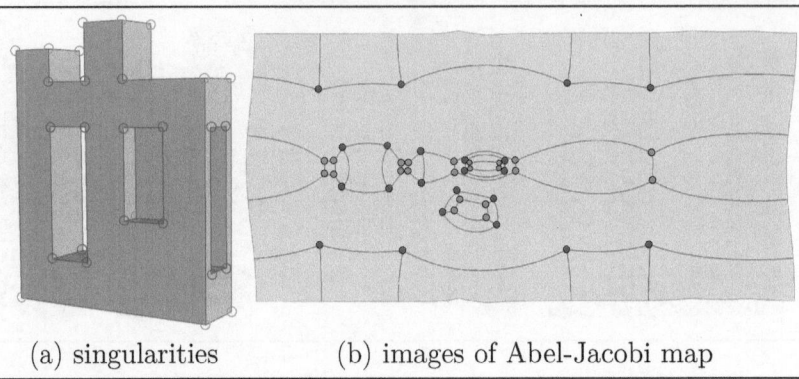

(a) singularities   (b) images of Abel-Jacobi map

Figure 21.11   The mass center of the Abel-Jacobi map images of the zeros (blue) equal to that of the poles (red)

zeros,

$$\sum_{j=1}^{22} \mu(p_j) - \sum_{i=1}^{22} \mu(q_i) = 0.$$

As shown in Figure 21.12, theorem 21.2.5 is verified again by a genus two polycube surface. The cone singularities are labeled, including the valence 3 (poles) and valence 5 (zeros). We compute a holomorphic 1-form ω using the Alg. 13 and visualized using checkerboard texture mapping. The zero points of ω are also marked as the centers of octagons on the textures. There are $2g - 2$ zeros of the holomorphic 1-form ω, one is in the front and the other is in the back.

We compute a set of canonical basis of the fundamental group of the polycube surface $\{a_1, b_1, a_2, b_2\}$ as shown in the left frame of Figure 21.13. There are two handles on the surface, each handle corresponds a pair of bases $\{a_k, b_k\}$, $k = 1, 2$. Furthermore, in order to compute a fundamental domain, we add a path τ connecting two handles. We slice the surface along a_1, b_1, a_2, b_2 and τ to obtain a fundamental domain D, then obtain (one component of) the Abel-Jacobi map $\mu : S \to \mathbb{C}$ by integrating ω:

$$\mu(q) = \int_p^q \omega, \quad \forall q \in S$$

where p is a fixed base point, the integration path is arbitrarily chosen inside D. As shown in the right frame of Figure 21.13, roughly speaking, each handle is conformally mapped to a planar parallelogram with boundaries $\mu(a_k b_k a_k^{-1} b_k^{-1})$. The two parallelograms are attached together through two branching points c_1 and c_2, which are the zeros of ω.

The 16 zeros p_i's and the 8 poles q_j's of the meromorphic quartic differential $\omega_{\mathbf{g}}$ are marked in the right frame of Figure 21.14. We also label the sharp edges of the polycube surface. By direct computation, we can verify the Abel-Jacobi condition in

Eqn. (21.2.4),

$$\sum_{i=1}^{16} \mu(p_i) - \sum_{j=1}^{8} \mu(q_j) = 4 \sum_{k=1}^{2} \mu(c_k),$$

the mass center of zeros minus the mass center of the poles coincides with four times of that of the branching points.

21.3 MESH GENERATION

Mesh generation plays a fundamental role in many engineering and medical fields, especially CAD (Computer Aided Design), CAE (Computer Aided Engineering) fields. In CAD field, most geometric objects are represented as Spline surfaces, such as NURBS (Non-Uniform Rational BSpline Surface), which requires global structured quadrilateral mesh generation. Recently, TSplines becomes more popular in CAD field, which requires T-mesh generation (a flat metric with cone singularities satisfying the holonomy condition Eqn. (21.2.4)). In CAE field, the central task is to simulate physics phenomena on complicated geometric shapes, which is reduced to solve partial differential equations on Riemannian manifolds. This can be achieved using Finite Element Method, Finite Volume Method or Isogeometric analysis method, all of them depend on high quality mesh generation.

Triangle Mesh Generation   Conformal geometry is crucial for mesh generation, because it can convert surface mesh generations to planar mesh generations. Mesh quality is mainly measured by the corner angles of cells and the uniformity of samples. Conformal mappings preserve angles, and the sampling density can be adapted to conformal factors, therefore conformal mappings can reduce surface meshing

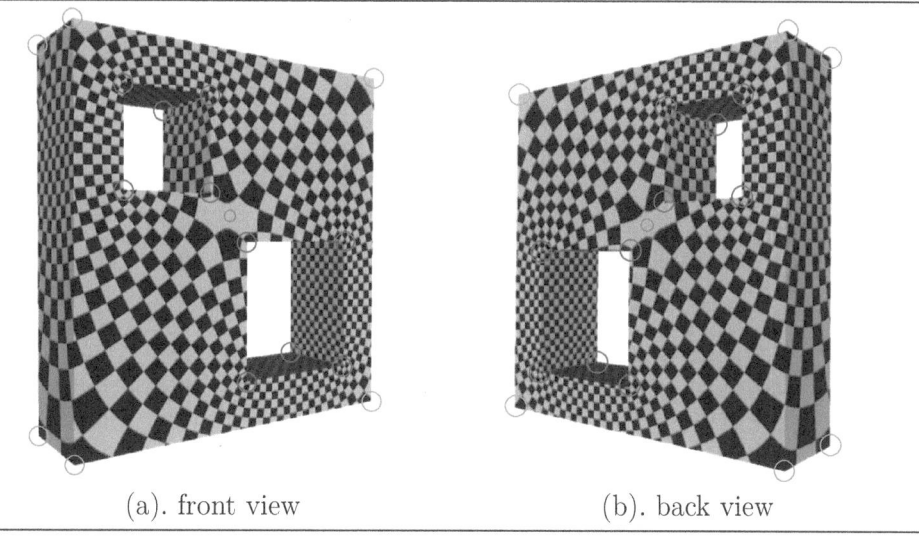

| (a). front view | (b). back view |

Figure 21.12 A genus two polycube surface with a holomorphic 1-form.

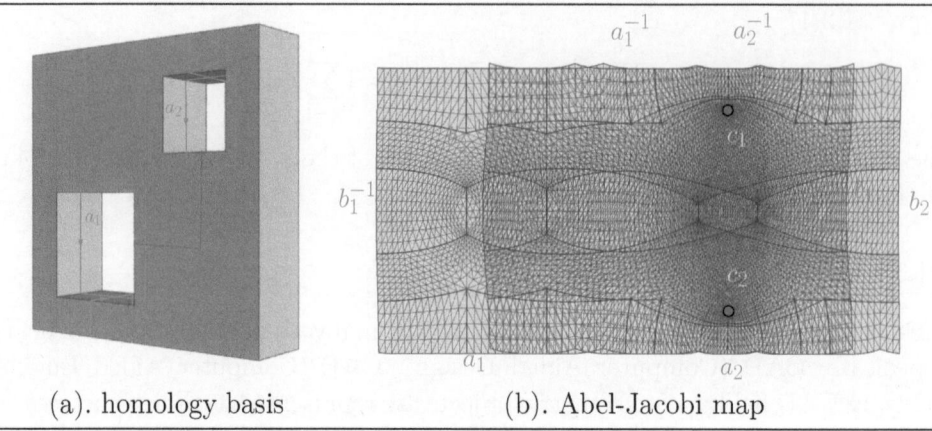

(a). homology basis   (b). Abel-Jacobi map

Figure 21.13 Abel-Jacobi map of the genus two polycube surface.

problems to planar ones. Planar mesh generation algorithms are relatively mature, most of them are based on Delaunay refinement method [115, 301], for examples Chew's second algorithm [85], Ruppert's algorithm [263], Centroidal Voronoi Tessellation algorithm [108] and so on.

Chew's second algorithm [85] is as follows: given a convex planar domain Ω, we first sample the boundary, such that the distance between two consecutive samples is ε, the sample point set is denoted as $\mathcal{S}$. We compute the Delaunay triangulation of the current sample point set $\mathcal{S}$. If there is a triangle with circum-radius great than ε, then we insert its circum-center c to the sample point set, $\mathcal{S} \leftarrow \mathcal{S} \cup \{c\}$, and update the Delauney triangulation. The circum-center c is called a *Steiner point*. We repeat inserting the Steiner points, until all the circum-radii are no greater than ε. First, we show the algorithm terminates with a finite number of steps. Due to the empty circle property of Delaunay triangulation, for each Steiner points c_i in $\mathcal{S}$, the disk

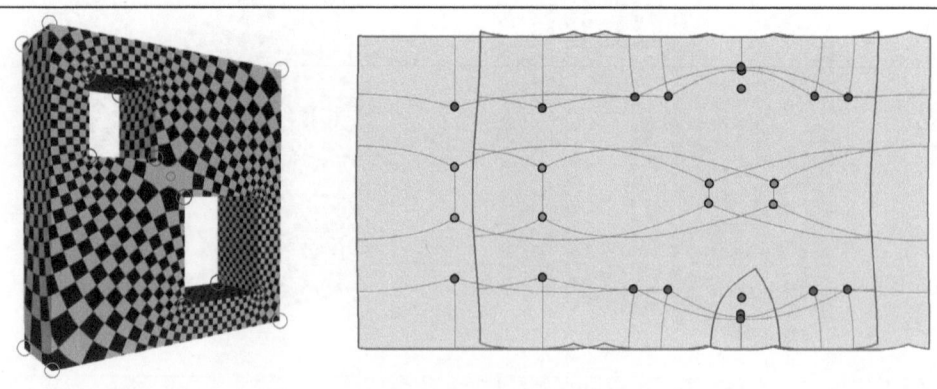

Figure 21.14 The zeros and poles of the meromorphic quartic differential $\omega_{\mathbf{g}}$.

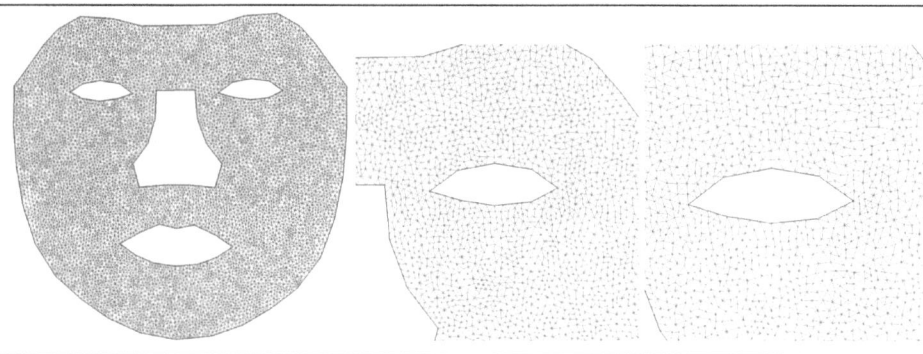

Figure 21.15 Ruppert's Delaunay refinement algorithm.

$B(c_i, \varepsilon/2)$ is disjoint from the disk $B(c_j, \varepsilon/2)$, for any $c_i \in \mathcal{S}$ and $c_j \neq c_i$. Therefore, the total number of possible Steiner points is less than the area of Ω divided by $\pi\varepsilon^2/4$. Furthermore, the Steiner point c_i is inserted at the i-th step, every point in $c_j \in \mathcal{S}$ is outside the disk $B(c_i, \varepsilon)$, the new edges in the updated Delaunay triangulation connect c_i with some c_j's, therefore all the new edge lengths are greater than ε. When the algorithm ends, for each triangle, the circum-radius is no greater than ε and the edge lengths are no less than ε, hence the corner angles are no less than $30°$. This ensures the quality of the triangulation. Rupper's Delaunay Refinement algorithm [263] can handle more general planar domains, and the minimal angle is no less than $20.7°$, as shown in Figure 21.15.

Given a surface $(S, \mathbf{g})$, we can compute a conformal mapping $\varphi : S \to \mathbb{C}$, then the surface metric can be represented as

$$\mathbf{g} = e^{2\lambda(z)} dz d\bar{z},$$

where $\lambda : S \to \mathbb{R}$ is the conformal factor. We use a planar Delaunay refinement algorithm to compute a high quality triangulation on the image $\varphi(S)$, the sampling

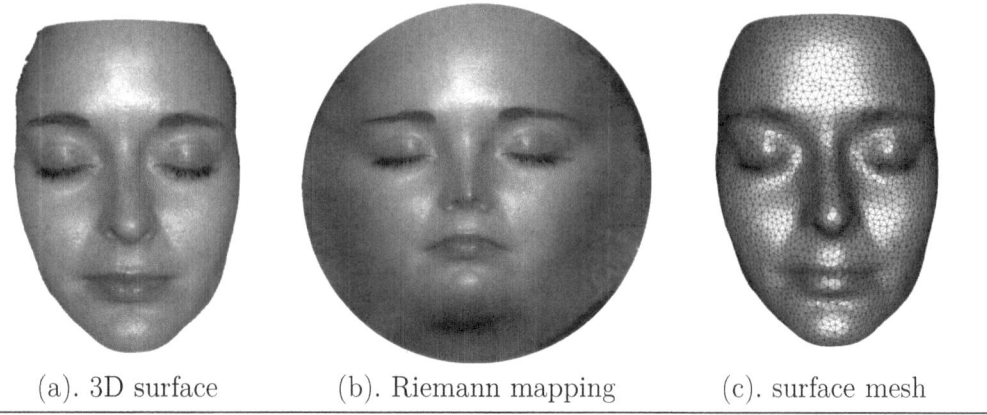

| (a). 3D surface | (b). Riemann mapping | (c). surface mesh |

Figure 21.16 Surface meshing based on conformal mapping and planar Delaunay Refinement algorithm.

(a) Buddha surface   (b). Conformal mapping   (c). Optimal transport map

Figure 21.17   Conformal and optimal transport maps.

density is adapted to the conformal factor $e^{2\lambda}$. Figure 21.16 shows one example of surface triangulation combing a conformal mapping and a planar Delaunay Refinement algorithm.

The sampling density can be adjusted by the so-called optimal transport map. As shown in the left frame of Figure 21.17, the Buddha surface is embedded in $\mathbb{R}^3$, with the induced the Euclidean metric $\mathbf{g}$. The Riemann mapping $\varphi : (S, \mathbf{g}) \rightarrow \mathbb{D}$ is computed using our Alg. 17, the metric is represented as the conformal factor $\mathbf{g} = e^{2\lambda} dz d\bar{z}$. The image of the Riemann mapping is shown in the middle frame. The surface area measure is denoted as $dA_\mathbf{g}$, the Riemann mapping φ pushes it to the disk, the push-forward measure can be represented as

$$\varphi_{\#} dA_\mathbf{g} = e^{2\lambda(x,y)} dx \wedge dy.$$

The optimal transportation map is an automorphism of the disk, $T : \mathbb{D}^2 \rightarrow \mathbb{D}^2$, it transforms the measure induced by the Riemann mapping $\varphi_{\#} dA_\mathbf{g}$ to the Lebesgue measure $\mathcal{L} = dx \wedge dy$ and minimizes the total transportation cost

$$\int_{\mathbb{D}} \|p - T(p)\|^2 e^{2\lambda(p)} dp.$$

The composition $T \circ \varphi : S \rightarrow \mathbb{D}^2$ is an area-preserving mapping. If we uniformly sample on the disk, and pull back the samples on the surface, then we obtain uniform sampling on the surface. We can design different sampling densities and use optimal transport maps to convert them to uniform ones. The computation of optimal transport maps is equivalent to solving Monge-Amperè equations, and can be achieved using a geometric variational approach [97, 153].

Figures 21.18 and 21.19 show an example for multi-resolution remeshing of a facial surface. The facial surface in the left frame of Figure 21.18 is conformally mapped onto the planar unit disk in the middle frame. The optimal transport map transforms the push-forwarded surface area element to the Lebesgue measure as shown in the right frame. We sample on the image of the optimal transport map, and map back the samples to the conformal mapping image. The Delaunay triangulation is computed

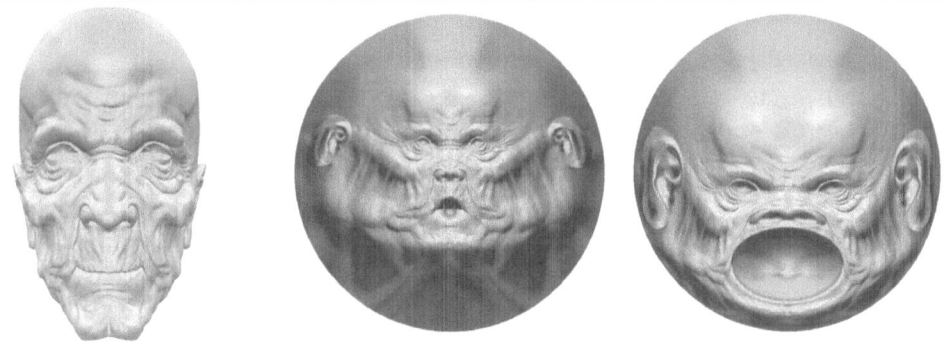

Figure 21.18 Conformal mapping and optimal transport mapping of an old man's facial surface.

on the conformal mapping image and pulled back to the original surface. By changing the total number of samplings, we obtain multi-resolution remeshing results as shown in Figure 21.19.

T-Mesh and Quad-Mesh Generation   In order to represent surfaces using T-Splines [299] and perform analysis using Isogeometric analysis [95], we would like to compute a flat metric with cone singularities satisfying the holonomy condition in Eqn. (21.2.2), this is equivalent to find the cone singularities satisfying the Abel-Jacobi condition in Eqn. (21.2.4).

Given a surface S of genus $g > 0$, we compute a cut graph Γ of the surface using Alg. 3 and slice the surface along the cut graph to obtain a fundamental domain $\bar{S}$. We also calculate a set of canonical basis of the fundamental group using Alg. 4, $\{a_1, b_1, a_2, b_2, \ldots, a_g, b_g\}$. We then compute a basis of the holomorphic 1-form group using Alg. 13, $\{\omega_1, \omega_2, \ldots, \omega_g\}$. We compute the period matrix and the Lattice Λ of

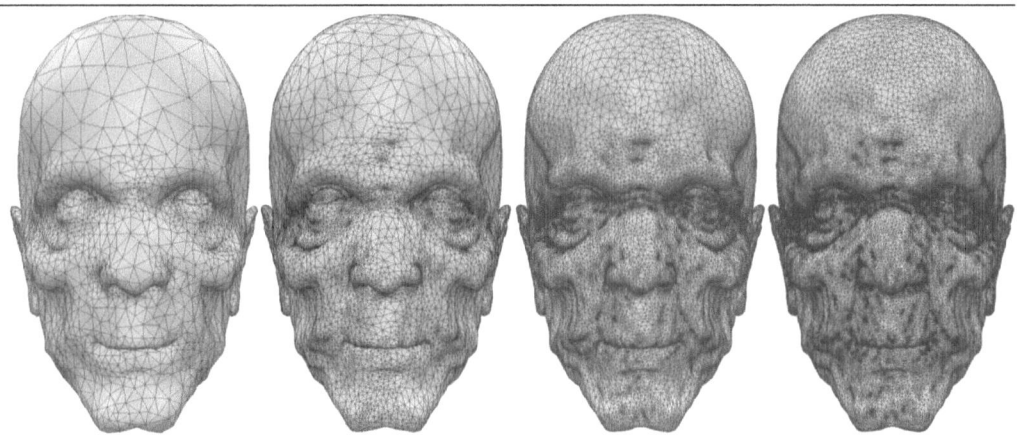

Figure 21.19 Surface multi-resolution remeshing.

Figure 21.20 Step 1. Compute the singularities by optimizing Abel-Jacobi condition.

the surface using Eqn. (18.4.1). The Abel-Jacobi map $\mu : S \to J(S)$ can be carried out using the formula in Eqn. (18.4.2) directly.

As shown in Figure 21.20, we compute the singularities. On the surface, we compute the Gaussian curvature $K : S \to \mathbb{R}$ and find the extremal points as the initial divisor, the local maxima are treated as poles q_i's, the local minima are treated as zeros p_j's, the order of each singularity is determined by its curvature. The goal is to adjust the positions of the zeros and poles, such that $D = \sum_{i=1}^{n} p_i - \sum_{j=1}^{m} q_j - \sum_{k=1}^{4}(\omega_k)$ is principle, where ω_k's are holomorphic 1-forms, obtained by linear combination of the holomorphic 1-form basis. We design a special energy,

$$E(p_1, \ldots, p_n, q_1, \ldots, q_j) = \left| \sum_{i=1}^{n} \mathrm{Ord}(p_i)\mu(p_i) - \sum_{j=1}^{m} \mathrm{Ord}(q_j)\mu(q_j) - \sum_{k=1}^{4} \mu((\omega_k)) \right|^2,$$

where $\mathrm{Ord}(p_i)$ is the order of the singularity p_i. Locally, the Abel-Jacobi map is invertible,

$$d\mu(p) = (\omega_1(p), \omega_2(p), \ldots, \omega_g(p)),$$

therefore the energy can be optimized using gradient descend method.

As shown in Figure 21.21, we compute the flat metric with cone singularities and the motorcycle graph. When the principle divisor is obtained, we compute the flat metric $\bar{\mathbf{g}}$ with cone singularities by setting the target curvature at each pole q_j as $\frac{\pi}{2}$, and at each zero p_j as $-\frac{\pi}{2}$ and using discrete surface Ricci flow in Alg. 23. For each pole q_i, we compute the shortest path τ_i from the pole to the cut graph; similarly, for each zero p_j, we find the shortest path τ_j. We slice the whole surface along $\Gamma \cup_i \tau_i$ to obtain a topological disk D, then isometrically flatten D with the flat metric $\bar{\mathbf{g}}$. We denote the flattening map as $\iota : D \to \mathbb{C}$. All the further computations can be conducted on the image $\iota(D)$. We define the horizontal and vertical directions on $(S, \bar{\mathbf{g}})$ by those directions on $\iota(D)$. Through each zero point p_i, we issue the horizontal and vertical

Figure 21.21 Compute the flat cone metric using surface Ricci flow, and compute the motorcycle graph.

geodesics, there are 5 geodesics in total; for each pole q_j, we issue three horizontal and vertical geodesics. When two geodesic transversely intersect at some point on the surface, the longer geodesic stops and the shorter geodesic continues. These geodesics form the so-called *motorcycle graph*. Figure 21.21 shows the motorcycle graph on the buddha surface.

As shown in Figure 21.22, the motorcycle graph partitions the surface into topological quadrilaterals, each cell can be isometrically mapped onto a planar rectangle by the flat metric $\bar{g}$. This cell decomposition induces the desired T-mesh: each cell is a rectangle, and at the T-junctions, two edges intersect orthogonally as shown in Figure 21.23. We represent the T-mesh as a cell complex $\mathcal{C}$, as shown in the right frame, the vertices of $\mathcal{C}$ are the cone singularities or the T-junctions, the cells of $\mathcal{C}$ are Euclidean rectangles. We treat each rectangular cell as one chart, then the chart

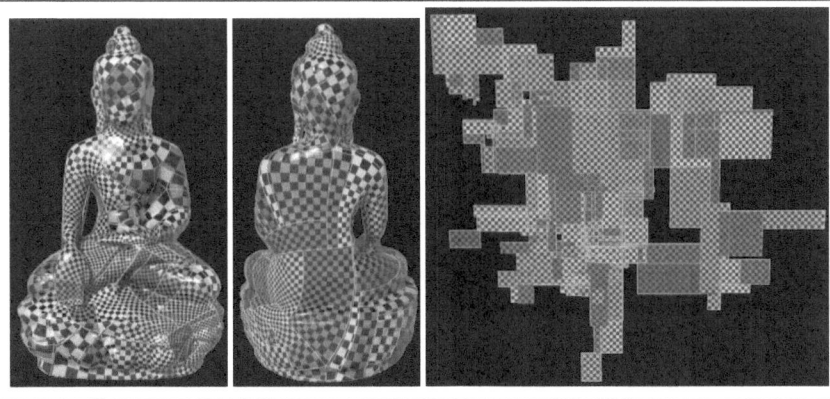

Figure 21.22 The motorcycle graph partitions the surface into patches, each patch is conformally flattened onto a planar rectangle.

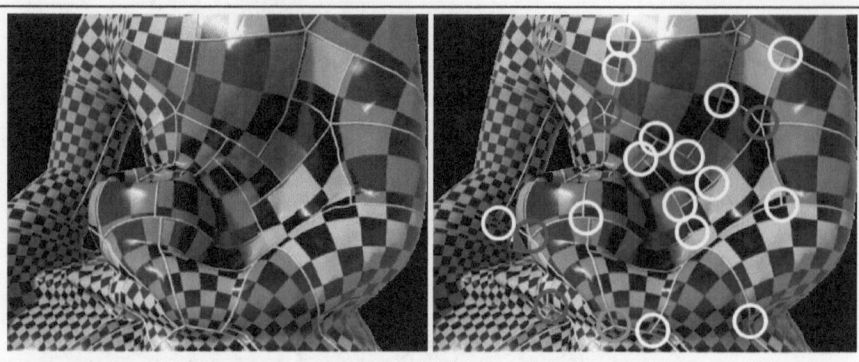

Figure 21.23 The T-mesh cell decomposition of the surface form a cell complex $\mathcal{C}$, whose vertices are cone singularities and T-junctions.

transitions are planar translations and rotations by angle $k\pi/2$, where k is an integer. The T-mesh representation can be applied for constructing T-Spline directly.

Furthermore, in order to generate a quad-mesh, we need to deform the T-mesh. There are various ways for the deformation, edge length deformation is one of them. We compute the generators of the fundamental group of the punctured surface using the cell complex $\mathcal{C}$,

$$\pi_1(S \setminus \{\cup_{i=1}^n p_i \cup_{j=1}^m q_j\}, b),$$

where the base point b is a T-junction vertex. Each fundamental group generator corresponds to a deck transformation group generator. Each edge on the cell complex $\mathcal{C}$ is either horizontal or vertical, the deck transformations can be represented by the edge lengths. We adjust the edge lengths such that all the deck transformations are integral (or equivalently rational) and all the cells are rectangular.

Figures 21.25, 21.26 and 21.24 show the quad-mesh generation process of a genus 3 surface model. In Figure 21.25, the left two frame show the cone singularities, the right two frame display the motocycle graph. In Figure 21.26, the left two frames

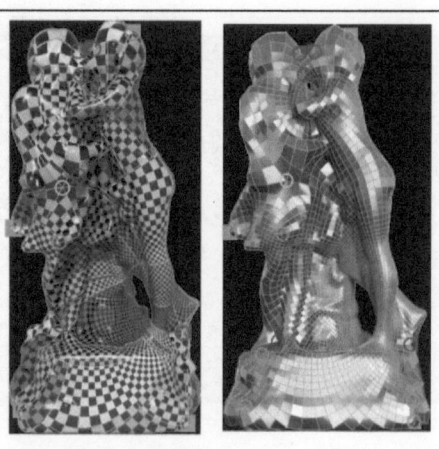

Figure 21.24 The quad-mesh of the kissing sculpture.

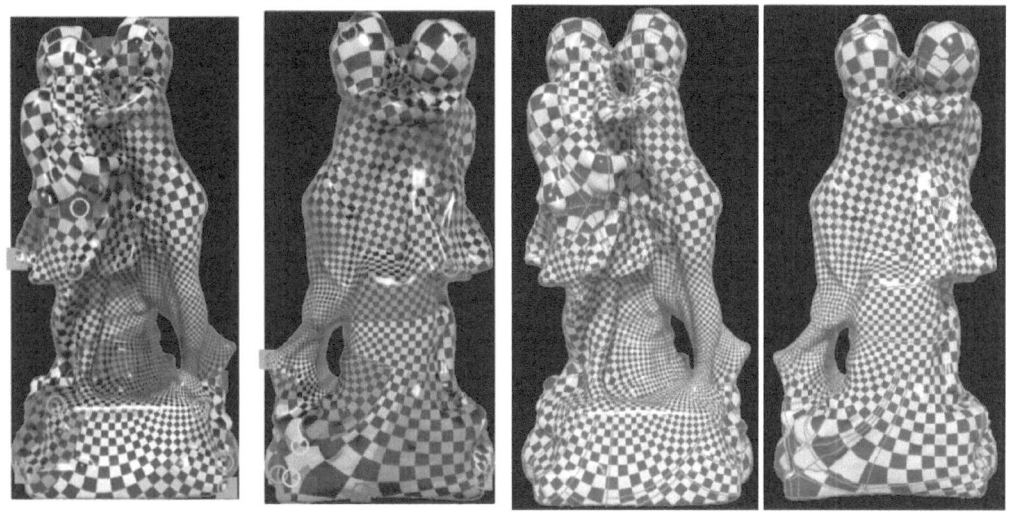

Figure 21.25 The cone singularities and the motocycle graph of a genus three kissing sculpture model.

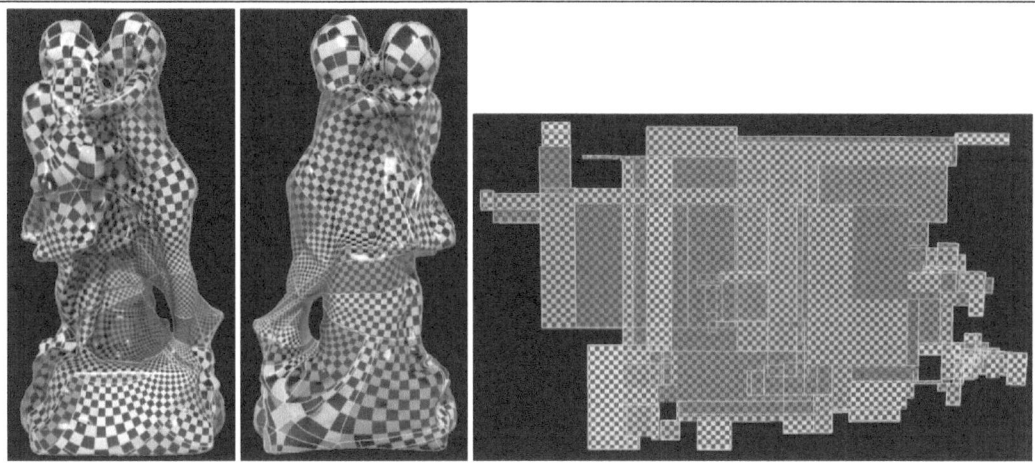

Figure 21.26 The T-mesh cell complex of the kissing sculpture.

show the T-mesh complex on the surface, the right frame shows the rectangular cells on the parameter plane. The quad-mesh of the surface is shown in Figure 21.24. The algorithm pipeline can be found in Alg. 30.

Algorithm 30 Algorithm for T-Mesh and Quad-mesh Generation

Require: A closed triangle mesh M of genus g

Ensure: A quad-mesh Q on M

Compute the singularity configuration by an optimization to satisfy Abel-Jacobi condition Eqn. 21.1.3;

Compute the flat metric $\bar{\mathbf{g}}$ with cone singularities using discrete surface Ricci flow Alg. 23;

Isometrically immerse the punctured surface with $\bar{\mathbf{g}}$, determine the horizontal and the vertical directions;

Compute the motorcycle graph to construct the T-mesh cell complex $\mathcal{C}$;

Adjust the edge lengths of the T-mesh complex to make all the deck transformations to be integral;

Construct the quad-mesh Q on the cell complex $\mathcal{C}$.

Appendices

Alexandrov Curvature

We have introduced in Section 3.2 the Wald metric curvature which is a comparison curvature, as we have seen. While originally the idea of comparison curvature did not gain traction, further developments transformed this type of geometry into a main staple of modern Mathematics, being applied, in particular, for the celebrated resolution [247],[248] of the famous Poincaré conjunction. The moder garb of this type of curvature is the so called *Alexandrov curvature*, which we briefly present below and, furthermore, explore its connection with the older Wald curvature.

A.1 ALEXANDROV CURVATURE

We begin by reminding the reader the definition of Alexandrov comparison curvature. Before bringing the formal definition, let us just specify the main difference between this approach and Wald curvature: In defining Alexandrov curvature, one makes appeal to *comparison triangles* in the model space (i.e. gauge surface $\mathcal{S}_k$), rather than quadrangles, as in the definition of Wald curvature. Before being able to do so, we need to recall the following definition:

Definition A.1.1 *A metric space (X, d) is called and inner metric space if distances are realized as lengths of shortest curves, i.e. if for any $x, y \in X$, $d(x, y) = \inf_{\gamma \in \Gamma} \ell(\gamma)$, where $\ell(\gamma)$ denotes, as usual, the length of γ, i.e. $\ell(\gamma) = \sup_{p \in P} \sum_0^m d(\gamma(t_i), \gamma(t_{i+1}))$, and where Γ and P denote, respectively, the set of continuous curves between x and y and the set of partitions of $[0, 1]$, that is $\Gamma = \{\gamma : [0, 1] \to X \mid \gamma \text{ continuous}; \gamma(0) = x, \gamma(1) = y\}$, $P = \{0 = t_0, t_1 < \cdots t_n = 1\}$.*

A space is said to be of *Alexandrov curvature* $\geq k$ iff any of the following equivalent conditions holds *locally*, and to be of *Alexandrov curvature* $\leq k$ iff any of the conditions below holds with the opposite inequality:

Definition A.1.2 *Let X be an inner metric space, let $T = \triangle(p, q, r)$ be a geodesic triangle, of sides $\overline{pq}, \overline{pr}, \overline{qr}$, and let $\widetilde{T}$ denote its(a) representative triangle in $\mathcal{S}_k$.*
A0 *Given the triangle $T = \triangle(p, q, r)$ and points $x \in \overline{pq}$ and $y \in \overline{pr}$, there exists a representative triangle $\widetilde{T}$ in $\mathcal{S}_k$. Let $\tilde{x}, \tilde{y}$ represent the corresponding points on the sides of $\widetilde{T}$. Then $d(x, y) \geq d(\tilde{x}, \tilde{y})$.*

DOI: 10.1201/9781003350576-A

A01 *Given the triangles $T = \triangle(p, q, r)$ and $\tilde{T}$ as above, and a point $x \in \overline{pq}$,*
$d(x, c) \geq d(\tilde{x}, \tilde{c})$.

A1 *Given the triangle $T = \triangle(p, q, r)$, there exists a representative triangle $\tilde{T}$ in $\mathcal{S}_k$, and $\measuredangle(\overline{pq}, \overline{pr}) \geq \measuredangle(\overline{\tilde{p}\tilde{q}}, \overline{\tilde{p}\tilde{r}})$, where $\measuredangle(\overline{pq}, \overline{pr})$ denotes the angle between $\overline{pq}$ and $\overline{pr}$, etc.*

A2 *For any hinge $H = (\overline{pq}, \overline{pr})$, there exists a representative hinge $\tilde{H} = (\overline{\tilde{p}\tilde{q}}, \overline{\tilde{p}\tilde{r}})$ in $\mathcal{S}_k$ and, moreover, $d(p, q) \leq d(\tilde{p}, \tilde{q})$, where a hinge is a pair of minimal geodesics with a common end point.*

Remark A.1.3   *1. Axiom A0 represents the basic one in defining Alexandrov comparison. It is also the one used by Rinow [260], in his seemingly (semi-) independent development of comparison geometry. – See Section A.3 below. This condition represents nothing more the transformation into an axiom of the following essential geometric fact: Thales Theorem does not hold in Spherical and Hyperbolic Geometry. (In particular – and most spectacular – the line connecting the midpoints of two of the edges of a triangle does not equate half of the third one.[1])*

   2. Condition A01 clearly represents just a particular case of A0, by fixing y to be one of the end points of $\overline{pr}$, say $y = r$. However, it is, actually, equivalent to A0 and it is, in fact, usually easier to check. It is sometimes used as the basic definition of Alexandrov comparison, saying, for instance that "negatively curved spaces have short ties" (the figure of speech being, we believe, self explanatory) – see, e.g. [328].

   The role of sd-quads in such fundamental results as Theorem 3.2.21 becomes now less strange and, in fact, it will become quite clear once the result in the next section is introduced. We anticipate somewhat by adding that now Robinson's method in Theorem 3.2.36 shows itself as it truly is: A method[2] of approximating the relevant k appearing in Axiom A01.

   3. We have used here (for the most part) Plaut's notation in [?]. For an excellent, detailed, clear (and by now already classical) presentation of the various comparison conditions, see [63].

   4. Conditions A1 and A2 show that one can introduce comparison Geometry via angle comparison. However, we prefer a "purely" metric approach, even if it is somewhat illusory. (See Chapter 11 for the application of this approach to a "purely" metric Regge Calculus.)

A.2 ALEXANDROV CURVATURE VS. WALD CURVATURE

Loosely formulated, the important fact regarding the connection between Wald and Alexandrov curvatures is that *(in the presence of sufficiently many minimal geodesics)*

---

[1]This well-known "paradox" of the foundations of Geometry is, unfortunately, generally overlooked in certain applications in Imaging and Graphics, which results in a penalty on the quality of the numerical results.

[2]developed *avant la lettre*

Wald curvature is (essentially) equivalent to Alexandrov curvature or, slightly more precisely, that inner metric spaces with Wald curvature $\geq k$ satisfy the condition of having sufficiently many geodesics. (This fact may be viewed as an extended, weak Hopf-Rinow type theorem.) The formal enouncement of this result requires yet more technical definitions, which we present below for the sake of completeness:

Definition A.2.1 *Let X be an inner metric space and let $\gamma_{pq} \subset X$ be a minimal geodesic connecting the points p and q. γ is called*

1. *extendable beyond q if there exists a geodesic $\tilde{\gamma}$, such that $\gamma = \tilde{\gamma}|_{(p,q)}$ and $q \in \operatorname{int}\tilde{\gamma}$.*

2. *almost extendable beyond q if for any $\varepsilon > 0$, there exists an $r \in X \setminus \{p,q\}$, such that $\sigma(q;p,r) < \varepsilon$, where $\sigma(q;p,r)$ denotes the strong excess*

$$\sigma(q;p,r) = \frac{e(T)}{\min d(p,r), d(r,q)}, \qquad (A.2.1)$$

where $T = \Delta(q,p,r)$ (and where $e(T)$, stands, as above, for its excess).

We shall also need the following

Definition A.2.2 *Let X be as above and let $p \in X$. We denote*

$$J_p = \{q \in X \mid \exists! \, \text{minimal geodesic } \gamma_{pq} \text{ almost extendable beyond } q.\}. \qquad (A.2.2)$$

We have the following result, for the who's enunciation we need a more technical notion and result from the theory of metric spaces. (An excellent reference to these is [229], but the reader might omit the reminder of this if the subject is not of subsection if the subject is not of particular interest to her/him.)

Theorem A.2.3 (Plaut [250], [251]) *Let X be an inner metric space of Wald-Berestovskii curvature $\geq \kappa$. Then, for any $p \in X$, J_p contains a dense G_δ set.*

Since, by the Baire Category Theorem, the intersection of countably many dense G_δ sets is a dense G_δ set, we obtain the following corollary:

Corollary A.2.4 *Let X be an inner metric space of Wald-Berestovskii curvature $\geq \kappa$, and let $p_1, p_2, \ldots \in X$. Then there exist points $p'_1, p'_2, \ldots \in X$ such that*

1. *p_i is arbitrarily close to p_i, for all i;*

2. *There exists a unique minimal geodesic connecting p_i and p'_i.*

Moreover, one can take $p'_1 = p_1$.

In other words, given any three points p, q, r in X, there exist points p_1, q_1, r_1 arbitrarily close to them (respectively) such that p_1, q_1, r_1 represent the vertices of a triangle whose sides are minimal geodesics or, simply put, one can construct (minimal geodesic) triangles "almost everywhere".

From the corollary above and from the Toponogov Comparison Theorem (see Remark 3.2.29) we obtain the announced theorem that establishes the essential equivalence between Alexandrov and Wald curvature, once the existence of "enough geodesics" is assured:

Theorem A.2.5 (Plaut [251]) *Let X be an inner metric space. X is a space of Alexandrov curvature bounded from below if and only if for every $x \in X$, there exists an open set U, $x \in U$, such that for every $y \in U$ the set J_y contains a dense G_δ set of U.*

(For a different formulation of the results above see [252], Corollary 40.)

In view of the result above, we can identify, for all practical purposes, finitely dimensional spaces of *Wald* curvature $\geq k$ with *Alexandrov spaces*, that is metric spaces of bounded (from below) Alexandrov curvature. However, Wald curvature allows us to discard conditions as are usually used when employing the Alexandrov triangle comparison, e.g. local compactness, while still being able to obtain many important theorems, such as the Toponogov Theorem and the Hopf-Rinow Theorem that we have discussed above, as well as fitting variants of such results as the Maximal Radius Theorem and of the Sphere Theorems, which are beyond the scope of this book. (For more information, the reader is invited to consult [251].)

A.3   RINOW CURVATURE

The curvatures introduced before may seem a bit archaic in comparison to the more fashionable approach of *comparison triangles* (see Section A.1), with their far reaching applications. We present here one of these comparison criteria and show its equivalence with the Wald curvature. We start with the following definition:

Definition A.3.1 Let (M, d) be a metric space, together with the intrinsic metric induced by d. Let $R = int(R) \subseteq M$ be a region of M. We say that R is a *region of curvature $\leq \kappa$* $(\kappa \in \mathbb{R})$ iff

1. For any $p, q \in R$ there exists a geodesic segment $pq \subset R$;

2. Any triangle $T(p, q, r) \subset R$ is isometrically embeddable in $\mathcal{S}_\kappa$;

3. If $T(p, q, r) \subset R$ and $x \in pq, y \in pr$ and if the points $p_\kappa, q_\kappa, r_\kappa, x_\kappa, y_\kappa \in \mathcal{S}_\kappa$ satisfy the following conditions:

   (a) $T(p, q, r) \cong T(p_\kappa, q_\kappa, r_\kappa)$;
   (b) $T(p, q, x) \cong T(p_\kappa, q_\kappa, x_\kappa)$;
   (c) $T(p, r, y) \cong T(p_\kappa, r_\kappa, y_\kappa)$;

   then $xy \leq x_\kappa y_\kappa$.

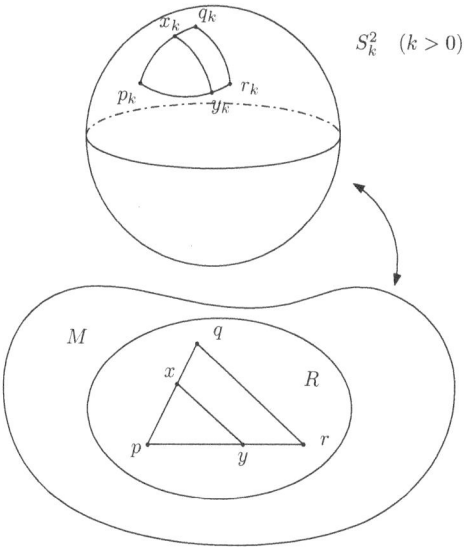

Figure A.1 Positive Rinow Curvature.

By replacing the condition: "$xy \leq x_\kappa y_\kappa$" with: "$xy \geq x_\kappa y_\kappa$", we obtain the definition of a *region of curvature* $\geq \kappa$ (see Figure A.1).

We now pass to the localization of the definition above:

Definition A.3.2 Let (M, d) be a metric space, together with the intrinsic metric induced by d, and let $p \in M$ be an accumulation point. Then M has at p *Rinow curvature* $\kappa_R(p)$ iff

1. There exists a linear neighborhood $N \in \mathcal{N}(p)$;

2. For any $\varepsilon > 0$, there exists $\delta > 0$, such that $B(p; \delta)$ is

    (a) a region of Rinow curvature $\leq \kappa_R(p) + \varepsilon$
        and

    (b) a region of Rinow curvature $\geq \kappa_R(p) - \varepsilon$.

While its greater generality endows the Rinow curvature with more flexibility in applications and makes it easier in generalization, it is even more difficult to compute than Wald curvature. However, this quandary has an almost ideal solution, due to Kirk (see [190]), solution which we briefly expose here:

Theorem A.3.3 ([190]) Let M be a compact, convex metric space, and let $p \in M$. If $\kappa_W(p)$ exists, then $\kappa_R(p)$ exists, and $\kappa_R(p) = \kappa_W(p)$.

Unfortunately, since $\kappa_R(p)$ makes no presumption of dimensionality, the existence of $\kappa_R(p)$ does not imply the existence of $\kappa_W(p)$.

Counterexample A.3.4 *Let $M \equiv \mathbb{R}^3$. Then $\kappa_R(p) \equiv 0$, but $\kappa_W(p)$ does not exist at any point, since every neighborhood contains linear quadruples.*

Kirk's solution of this problem is to consider the *modified Wald curvature κ_{WK}*, defined as follows:

Definition A.3.5 Let (M, d) be a metric space, together with the intrinsic metric induced by d, and let $p \in M$. Then M has *modified Wald curvature $\kappa_{WK}(p)$* at p iff

1. There exists a linear neighborhood $N \in \mathcal{N}(p)$;

2. For any $\varepsilon > 0$, there exists $\delta > 0$, such that if $Q \subset B(p; \delta)$ is a non-degenerate sd-quad, then $\kappa_W(Q)$ exists and $|\kappa_{WK}(p) - \kappa_W(Q)| < \varepsilon$.

Remark A.3.6 If $\kappa_W(p)$ exists, then $\kappa_{WK}(p)$ exists but the existence of $\kappa_{WK}(p)$ does not imply that of $\kappa_W(p)$. Indeed, if $p \in \mathbb{R}^3$, then $\kappa_{WK}(p) = 0$ but $\kappa_W(p)$ does not exist.

This modified curvature indeed represents the wished for solution, as proved by the following two theorems:

Theorem A.3.7 ([190]) Let (M, d) be a metric space. Then: if $\kappa_R(p)$ exists, then $\kappa_{WK}(p)$ also exists and $\kappa_R(p) = \kappa_{WK}(p)$.

Theorem A.3.8 ([190]) Let (M, d) be a metric space together with the associated intrinsic metric, and let $p \in M$. Suppose that

1. $\kappa_{WK}(p)$ exists

    and

2. There exists $B(p; \rho) \in \mathcal{N}(p)$, such that $qr \subset B(p; \rho)$, for all $q, r \in B(p; \rho)$.

Then $\kappa_R(p)$ exists and $\kappa_R(p) = \kappa_{WK}(p)$.

Thick Triangulations Revisited

We reconsider here the definition of thick triangulations and we show that it can be expressed solely in terms of the metric (i.e. in terms of the lengths of the edges of a triangulation or piecewise flat approximation). As we have noted in Chapter 11, the condition of thickness lacks this purely metric aspect. Indeed, while volume and edge lengths of a Euclidean simplex (or, for that matter, in any space form) are closely related, in a general metric space no volume is priorly postulated. (Admittedly, one can attempt a fitting definition for *metric measure spaces* [274], but this has only limited purely mathematical applications and, furthermore, it also departs from the purely metric approach.[1])

Fortunately, this fault is easy to amend, using the generalized (n-dimensional) *Cayley-Menger determinant*, that expresses the volume of the n-dimensional Euclidean simplex $\sigma_n(p_0, p_1, \ldots, p_n)$ of vertices $p_0, p_1, \ldots, p_n$ as a function of its edges $d_{ij}, 0 \leq i < j \leq n$:

$$D(p_0, p_1, \ldots, p_n) = \begin{vmatrix} 0 & 1 & 1 & \cdots & 1 \\ 1 & 0 & d_{01}^2 & \cdots & d_{1n}^2 \\ 1 & d_{10}^2 & 0 & \cdots & d_{1n}^2 \\ \vdots & \vdots & \vdots & \ddots & \vdots \\ 1 & d_{n0}^2 & d_{n1}^2 & \cdots & 0 \end{vmatrix} ; \qquad (\text{B.0.1})$$

namely

$$\text{Vol}^2(\sigma_n(p_0, p_1, \ldots, p_n)) = \frac{(-1)^{n+1}}{2^n (n!)^2} D(p_0, p_1, \ldots, p_n) . \qquad (\text{B.0.2})$$

(Similar expressions for the Hyperbolic and Spherical simplex also exist, see, e.g. [212], [47]. However, we do not bring them here, not least because they are far too technical for this limited exposition; suffice therefore to add that they essentially reproduce the proof given in the Euclidean case, taking into account the fact that,

---

[1]even if such an approach is, in the light of the literature mentioned above, less useful in Physics than previously believed

DOI: 10.1201/9781003350576-B

when performing computations in the spherical (resp. hyperbolic) metric, one has to replace the distances d_{ij} by $\cos d_{ij}$ (resp. $\cosh d_{ij}$) – see [47] for the full details.)

Evidently, the Cayley-Menger determinant makes sense for any *metric* $(n+1)$-*tuple*, i.e. for any metric metric space with $n+1$ points $p_0, p_1, \ldots, p_n$ and mutual distances $d_{ij}, 0 \leq i \leq j \leq n$.

As implicitly mentioned above, the definition of thickness as given by Definition 11.2.9 is only one of the (many) existing ones. An alternative one, given solely in terms of distances, is given in [327]. Tukia's fatness[2] of a simplex $\sigma = \sigma(p_0, \ldots, p_n)$ can be written as

$$\varphi_T(\sigma) = \max_\pi \frac{\delta(p_0, \ldots, p_n)}{\text{diam}\,\sigma} \,, \tag{B.0.3}$$

where $\delta = \delta(p_0, \ldots, p_n) = \text{dist}(p_i, F_i)$, where F_i denotes the $(n-1)$-dimensional face (of σ) opposite to p_i, and where the maximum is taken over all the permutations π of $0, \ldots, n$. In other words, δ represents the maximal height h_{Max} from the vertices of σ to its $(n-1)$-dimensional faces (or *facets*). Note that this definition of thickness can also be used, for instance, for Hyperbolic simplices (as, indeed, Tukia did).

Since the diameter of a simplex is nothing but the longest edge l_{Max}, that is $\text{diam}\,\sigma = \max_{0 \leq i < j \leq n} d_{ij} = \max_{1 \leq m \leq n(n+1)/2} l_m$, Tukia's definition of thickness is, clearly, purely metric. It is true that the expression of φ_T is not given solely in terms of the distances between the vertices (of σ), however it is easy to remedy this by computing h_i – the distance from p_i to the face F_i, using the classical and well-known formula:

$$\text{Vol}(\sigma) = \frac{1}{n} h_i \text{Area}(F_i) \tag{B.0.4}$$

and then expressing the volume of F_i using the fitting Cayley-Menger determinant, where we denoted the $(n-1)$-volume by "Area". (One has to proceed somewhat more carefully for the case, say, of Hyperbolic simplices, but we are not concerned here with this case.)

Peltonen's definition – see [245] – is, perhaps the easiest to express, even not if the simplest for actual computation, as she defines fatness (of a simplex) as:

$$\varphi_P(\sigma) = \frac{r}{R} \,, \tag{B.0.5}$$

where r denotes the radius of the inscribed sphere of σ (*inradius*) and R denotes the radius of the circumscribed sphere of σ (*circumradius*). The problem with this approach is that it is not given – explicitly, that is – in terms of the lengths of the edges of the simplex. Again, this does not represent a serious inconvenience, since, given a simplex $\sigma = \sigma(p_1, \ldots, p_n)$, the circumradius R is given by the following formula:

$$R = -\frac{1}{2} \frac{\Delta(p_1, \ldots, p_n)}{D(p_1, \ldots, p_n)} \,, \tag{B.0.6}$$

---

[2]In fact he defines the reciprocal quantity which he calls *flatness* and denotes by $F(\sigma)$.

where

$$\Delta(p_1, \ldots, p_n) = \begin{vmatrix} 0 & d_{12}^2 & \cdots & d_{1n}^2 \\ d_{21}^2 & 0 & \cdots & d_{1n}^2 \\ \vdots & \vdots & \ddots & \vdots \\ d_{n1}^2 & d_{n2}^2 & \cdots & 0 \end{vmatrix} . \tag{B.0.7}$$

Moreover, the inradius r can be computed quite simply using the fact (akin to formula (B.0.4)) that

$$\mathrm{Vol}\,\sigma_n = \frac{1}{n} \sum_{1}^{n+1} \mathrm{Area}(F_i), \tag{B.0.8}$$

and expressing again $\mathrm{Area}(F_i)$ as a function of its edges.

On the other hand, the equivalence of Tukia's and Peltonen's definitions follows using Peltonen's computation of the h_i-s (and diamσ), which is given – *inter alia* – in [245]. However, the proof requires additional notations and definitions as well as some quite extensive technical details. Therefore, we refer the reader to the original paper of Peltonen where the connection between φ_P and h (hence φ, φ_T) is given (even if not quite explicitly).

Munkres' definition is, for the specific case of Euclidean simplices[3], somewhat of a compromise between Cheeger's and Tukia's ones:

$$\varphi_M = \frac{\mathrm{dist}(b, \partial\sigma)}{\mathrm{diam}\,\sigma}, \tag{B.0.9}$$

where b denotes the *barycenter* of σ and $\partial\sigma$ represents the standard notation for the *boundary* of σ (i.e the union of the $(n-1)$-dimensional faces of σ). From the considerations above (and from [229], Section 9) it follows that for the Euclidean case, this definition of thickness is also equivalent to the previous ones.

Remark B.0.1 *Fu [125] also introduces a definition of fatness that (up to the quite different notation) is identical to that of [75], when restricted to individual simplices. However, for triangulations, his definition exceeds, in general, the one given in [75], due to the fact that Fu discards, in his use of fatness, the requirement that the approximations become arbitrarily fine.*

Remark B.0.2 *Of course, there exists a (well know) duality between the spherical distances and dihedral angles, as expressed by the Gram determinant (see, e.g. [194] and the references therein), but unfortunately this duality does not hold precisely in the case that we are interested in, that is constant sectional curvature $K \equiv 0$, so we do not discuss it here. Let us note only the fact that this type of analysis involves a generalization of the Cayley-Menger determinant.*

To complete the circle – so to say – we should emphasize that, given the dihedral angles and the areas of the $(n-1)$-faces of the Euclidean (piecewise flat) triangulation,

---

[3]But only in this case, as it represents a generalization of the Euclidean case for general simplices (as befitting Differential Topology goals).

one can calculate the edge lengths. For example, one can devise explicit computations for the case important in Ricci calculus, i.e. 4-simplices, minus a number of so called "dangerous configurations" (and implicit ones for the general case) in [103]. (A related formula – explicit for the 3-dimensional case – this time involving the Cayley-Menger determinant, can be found in [47], p. 344.)

The Gromov-Hausdorff Distance

We have encountered a number of times throughout the book the notion of Gromov-Hausdorff distance and, while to some readers this notion might be familiar, is at least equally probable that it is unknown (and even intimidating to many others). We therefore bring here an overview of this notion, as well as that of some related ones, since they all have become quite important and quite common in many applications, not just in Graphics and Imaging but also in Machine Learning and related fields.

We begin by introducing the classical *Hausdorff distance (metric)*:

Definition C.0.1 *Let* (X, d) *be a metric space and let* $A, B \subseteq X$. *We define the* Hausdorff distance *between* A *and* B *as:*

$$d_H(A, B) = \inf\{r > 0 \,|\, A \subset U_r(B), \, B \subset U_r(A)\}, \tag{C.0.1}$$

where $U_r(A)$ *is the* r-*neighborhood of* A, $U_r(A) = \bigcup_{a \in A} B_r(a)$.

Another equivalent way of defining the Hausdorff distance is as follows:

$$d_H(A, B) = \max\{\sup_{a \in A} d(a, B), \, \sup_{b \in B} d(b, A)\}. \tag{C.0.2}$$

Some readers might find such definitions annoyingly technical (and perhaps unsettling), but even taking a brief look at Berger's illustration of the idea[1] is immediately suggestive of applications. For some of these in the fields of Graohics, see, e.g. [31], [187], and the references therein.

In general d_H is only a *semi-metric*,[2] however it becomes a proper metric if we restrict ourselves to an important class of subsets. More precisely, we have the following result:

Proposition C.0.2 *Let* (X, d) *be a metric space. Then* d_H *is a metric on the set* $\mathcal{M}(X)$ *of closed subsets of* X.

---

[1]the so called "Pithecanthropus Geometricus" drawing, to be more precise
[2]i.e. $d_H(X, Y)$ does not necessarily imply that $X = Y$.

DOI: 10.1201/9781003350576-C

Moreover, the space $(\mathcal{M}(X), d_H)$ inherits some important properties of (X, d):

Theorem C.0.3 *(a) If X is complete, then $\mathcal{M}(X)$ complete.*
(b) (Blaschke) If X is compact, then $\mathcal{M}(X)$ compact.

(For proofs, see [63], pp. 253-254.)

It is only natural to extend the Hausdorff metric to non-compact spaces. For this, we proceed along the following basic guide-lines: We want to obtain the *maximum* distance d_{GH} that satisfies the two conditions below:

1. $d_{GH}(A, B) \leq d_H(A, B)$, for any $A, B \subset X$ (i.e. sets that are close as subsets of a given metric space X will still be close as abstract metric spaces);

2. X is isometric to Y iff $d_{GH}(X, Y) = 0$.

The sought definition is then the following one:

Definition C.0.4 *Let X, Y be metric spaces. Then the* Gromov-Hausdorff distance *between X and Y is defined by:*

$$d_{GH}(X, Y) = \inf d_H^Z(f(X), g(Y)); \qquad (C.0.3)$$

where the infimum is taken over all metric spaces Z in which both X and Y can be isometrically embedded and over all such isometric embeddings.

Remark C.0.5 *In fact, it suffices to consider embeddings f into the disjoint union $X \coprod Y$ of the spaces X and Y. $X \coprod Y$ is made into a metric space by defining*

$$d(x, y) = \begin{cases} \inf_{z \in X \cap Y} \{d_X(x, z) + d_Y(z, y)\}, & (x \in X) \text{ and } (y \in Y); \\ \infty, & X \cap Y = \emptyset. \end{cases} \qquad (C.0.4)$$

The following notion is not only important for theoretical ends (see, e.g. [142]), it is also highly relevant to our applicative purposes (in Graphics, Vision, etc.)

Definition C.0.6 *Let (X, d) be a metric space, and let $A \subset X$. A is called an ε-net iff $d(x, A) \leq \varepsilon$, for all $x \in X$.*

It should be evident, from the definition and the preparatory words preceding it, that the approximation of spaces by ε-nets will represent one of the main topics in the sequel. We begin by citing the following fundamental result:

Theorem C.0.7 *d_{GH} is a finite metric on the set of isometry classes of compact metric spaces.*

Remark C.0.8 *It is most alluring (see [3], [4]) to make appeal to the Gromov-Hausdorff distance between subsets of $\mathbb{R}^3$ (usually triangular meshes), for Graphics and Imaging goals (such as shape comparison, registration, etc.). However, this is a fallacy, since the Gromov-Hausdorff distance between two sets is* not *achieved by* *their embeddings in any Euclidean space. This was pointed out by Gromov, in [142], immediately after introducing his modification of the classical Hausdorff metric, via a simple example (see Remark (c), p. 72): Let X be an equilateral triangle of edge 1, and let Y be a point. Then $d_{GH}(X, Y) = 1/2$, although the Hausfdorff distance between their embeddings in $\mathbb{R}^2$, hence between their embeddings in any $\mathbb{R}^n$, must satisfy the inequality $d_{GH}^{\mathbb{R}^n}(X, Y) > \sqrt{3}/3$.*

Since, moreover, computational issues also arise when trying to compute the Gromov-Hausdorff distances, between, say, two large triangular meshes, one is forced, in practice, to compute the simple Hausdorff distance, which it is not, itself, without certain complications (see e.g [99] and the bibliography therein).

In addition, ε-nets in compact metric spaces have the following important property:

Proposition C.0.9 *Let $X, \{X_n\}_{n=1}^\infty$ be compact metric spaces. Then $X_n \underset{GH}{\longrightarrow} X$ iff for all $\varepsilon > 0$, there exist finite ε-nets $S \subset X$ and $S_n \subset X_n$, such that $S_n \underset{GH}{\longrightarrow} S$ and, moreover, $|S_n| = |S|$, for large enough n.*

The importance of the proposition above does not reside solely in the fact that compact metric spaces can be approximated by finite ε-nets (after all, just the existence of *some* approximation by such sets is hardly surprising), but rather in the fact that, as we shall shortly see, it assures the convergence of geometric properties of S_n to those of S, as $X_n \underset{GH}{\longrightarrow} X$. A very important – and extremely significant for us here – consequence is that of the *intrinsic metric* i.e. the metric induced by a *length structure* (i.e. path length) by a metric on a subset of a given metric space. (The motivating example, both in the theoretical setting and for this study is that of surfaces in $\mathbb{R}^3$.)

Moreover, the sequence $\{S_n\}_{n \geq 1}$ of ε-nets corresponds to the situation encountered frequently in practice, where in many instances one has to approximate a smooth object (manifold), having a finite number of fixed ("sampling") point, by a sequence of PL (metric, discrete) approximations, having also a fixed set of "marked" points that also converge to the "samples" chosen on the target manifold.

Proposition C.0.9 can also be reformulated in an equivalent, but less concise and elegant manner that is, on the other hand, far more useful in concrete instances as well as being more familiar for the Applied Mathematics community:

Proposition C.0.10 *Let X, Y be compact metric spaces. Then:*
(a) If Y is a (ε, δ)-approximation of X, then $d_{GH}(X, Y) < 2\varepsilon + \delta$.
(b) If $d_{GH}(X, Y) < \varepsilon$, then Y is a 5ε-approximation of X.

Recall that ε-δ-*approximations* are defined as follows:

Definition C.0.11 *Let X, Y be compact metric spaces, and let $\varepsilon, \delta > 0$. X, Y are called ε-δ-approximations (of each-other) iff: there exist sequences $\{x_i\}_{i=1}^N \subset X$ and $\{y_i\}_{i=1}^N \subset Y$ such that*
(a) $\{x_i\}_{i=1}^N$ is an ε-net in X and $\{y_i\}_{i=1}^N$ is an ε-net in Y;

(b) $|d_X(x_i, x_j) - d_{(y_i, y_j)}| < \delta$ *for all* $i, j \in \{1, ..., N\}$.
An $(\varepsilon, \varepsilon)$-approximation is called, for short an ε-approximation.

Among metric spaces, those whose metric d is intrinsic, are called *length spaces* and are of special interest in Geometry. The following theorem shows that length spaces are closed in the *Gromov-Hausdorff topology*:

Theorem C.0.12 *Let $\{X_n\}_{n=1}^{\infty}$ be length spaces and let X be a complete metric space such that $X_n \underset{GH}{\to} X$. Then X is a length space.*

Using the language of *ε-approximations* one can prove the following theorem and corollary, that are of paramount importance, not only for the goals of this overview, but in a far more extensive and powerful context (see e.g. [142] and [63]):

Theorem C.0.13 (Gromov) *Any compact length space is the GH-limit of a sequence of finite graphs.*

The proof of the theorem above is constructive and thus potentially adaptable for practical applications (especially in Graphics, Imaging and related fields). Because of this reason we bring it below:

Proof C.1 *Let ε, δ $(\delta \ll \varepsilon)$ small enough, and let S be a δ-net in X. Also, let $G = (V, E)$ be the graph with $V = S$ and $E = \{(x,y) \,|\, d(x,y) < \varepsilon\}$. We shall prove that G is an ε-approximation of X, for δ small enough (more precisely, for $\delta < \frac{\varepsilon^2}{4\mathrm{diam}(X)}$).*
But, since S is an ε-net both in X and in G, and since $d_G(x,y) \geq d_X(x,y)$, it is sufficient to prove that:
$$d_G(x,y) \leq d_X(x,y) + \varepsilon.$$
Let γ be the shortest path between x and y, and let $x_1, ..., x_n \in \gamma$, such that $n \leq \mathrm{length}(\gamma)/\varepsilon$ (and $d_X(x_i, x_{i+1}) \leq \varepsilon/2$). Since for any x_i there exists $y_i \in S$, such that $d_X(x_i, y_i) \leq \delta$, it follows that $d_X(y_i, y_{i+1}) \leq d_X(x_i, x_{i+1}) + 2\delta < \varepsilon$.
Therefore, (for $\delta < \varepsilon/4$), there exists an edge $e \in G, e = y_i y_{i+1}$. From this we get the following upper bound for $d_G(x,y)$:

$$d_G(x,y) \leq \sum_{n=0}^{n} d_X(y_i, y_{i+1}) \leq \sum_{n=0}^{n} d_X(x_i, x_{i+1}) + 2\delta n$$

But $n < 2\,\mathrm{length}(\gamma)/\varepsilon \leq 2\mathrm{diam}(X)/\varepsilon$. Moreover: $\delta < \varepsilon^2/4\,\mathrm{diam}(X)$. It follows that:

$$d_G(x,y) \leq d_X(x,y) + \delta \frac{4\mathrm{diam}(X)}{\varepsilon} < d_X(x,y) + \varepsilon.$$

Thus, for any $\varepsilon > 0$, there exists a graph an ε-approximation of X by a graph G, $G = G_\varepsilon$. Hence $G_\varepsilon \underset{\varepsilon}{\to} X$.

In fact, one can infer the more stronger (and useful in applications)

Corollary C.0.14 *Let X be a compact length space. Then X is the Gromov-Hausdorff limit of a sequence $\{G_n\}_{n \geq 1}$ of finite graphs, isometrically embedded in X.*

Remark C.0.15 *Some care still should be paid when using the theorem above. Indeed:*

1. *If $G_n \xrightarrow{\varepsilon} X$, $G_n = (V_n, E_n)$. If there exists $N_0 \in \mathbb{N}$ such that*

   $$(*) \quad |E_n| \leq N_0, \text{ for all } n \in \mathbb{N},$$

   then X is a finite graph.

2. *If condition $(*)$ is replaced by:*

   $$(**) \quad |V_n| \leq N_0, \text{ for all } n \in \mathbb{N},$$

   then X will still be always a graph, but not necessarily finite.

Remark C.0.16 *In fact, one can strengthen the result of Theorem C.0.13. More precisely, it was shown by Cassorla [70] that compact inner metric spaces can be, in fact, Gromov-Hausdorff approximated by smooth surfaces that, moreover, are embedded in $\mathbb{R}^3$. The significance for Graphics and Imaging of this result is evident and it is not lost on Cassorla, either: It shows that one can, in fact, visualize in $\mathbb{R}^3$ (up to some predetermined but arbitrarily small error) any compact inner metric space. This is a truly surprising fact. Unfortunately, it usually happens with "gifts", it comes at a price: The genus of the approximating surfaces cannot remain bounded. (So, in effect, to obtain a good approximation even of a simple space, via the method given in the theorem's proof, one has to increase the topological complexity of the approximating surface.)*

Note that there is no geometric (curvature) restriction on the approximating surfaces. In fact, the author also states – in what represents a seemingly still unpublished result – that one can approximate the given spaces with a series of smooth surfaces having Gaussian curvature bounded from above by -1. However, this comes at the price of loosing the embeddability in $\mathbb{R}^3$ of the approximating surfaces. Recall that we have addressed the problem of Gromov-Hausdorff approximating, under curvature control, of surfaces by surfaces in the next paper of the series, by presenting a result of Brehm and Kühnel, as well as our own extension of their result to the metric curvatures – see Subsection 12.2.2.

Before concluding this remark, we should add a few words regarding Cassorla's proof, especially so since, as we mentioned above, it may prove to be useful for visualization purposes: He begins by constructing an approximation by graphs, following Gromov, then he considers the (smooth) boundaries of canonical tube neighborhoods (or, in other words, he builds the smooth surfaces having as axes (or nerve) the graph constructed previously). Since we shall encounter a very similar construction, but rather more precise, in the proof of Brehm and Kühnel theorem, we do not elaborate here anymore on Cassorla's one.

Remark C.0.17 *For more (and far-reaching) results regarding both the Gromov-Hausdorff approximations of metric spaces graphs, as well as that by surfaces and other "nicer" spaces, see [142],* **3.32.–3.34.**$\frac{1}{2}_{+}$.

The Lipschitz Distance We bring briefly here few facts about another type of distance between metric spaces, that is closely related to the Hausdorff distance but may prove more easy to compute in many situations, namely the *Lipschitz distance*. (We do not elaborate more on this theme because we do make appeal to it in the sequel. For further details see [142], [63] or, for a "digest" somewhat more detailed than this one, [296].)

This definition of the Lipschitz distance is based upon a very simple (that is to say: very intuitive, motivated by routine, every day physical measurements) idea: It measures the relative difference between metrics, more precisely it evaluates their ratio. That is to say that the metric spaces (X, d_X), (Y, d_Y) are close (to each other) iff there exists a homeomorphism $f : X \overset{\sim}{\to} Y$, such that $\frac{d_Y(f(x),f(y))}{d_X(x,y)} \approx 1$, for all $x, y \in X$.

Technically, we give the following

Definition C.0.18 *Let (X, d_X), (Y, d_Y) be metric spaces. Then the* Lipschitz distance *between (X, d_X) and (Y, d_Y) is defined as:*

$$d_L(X, Y) = \inf_{\substack{f:X \overset{\sim}{\to} Y \\ f\ bi-Lip.}} \log \max\left(\mathrm{dil} f, \mathrm{dil} f^{-1}\right) \tag{C.0.5}$$

where the *dilatation* $\mathrm{dil} f$ of f is defined as follows:

Definition C.0.19 *Let $(X, d_X), (Y, d_Y)$ be metric spaces. Given a Lipschitz map $f : X \to Y$, the dilatation $\mathrm{dil} f$ of f is defined as*

$$\mathrm{dil} f = \sup_{x \neq y \in X} \frac{d_Y(f(x), f(y))}{d_X(x, y)}. \tag{C.0.6}$$

We have the following important result (analogous to Proposition 2.2):

Proposition C.0.20 *The Lipschitz distance d_L is a proper distance on the space of isometry classes of compact metric spaces.*

(This means, in particular, that $d_L(X, Y) = 0$ iff X and are isometric (not only homeomorphic) – a fact of clear importance in any applicative setting.)

Remark C.0.21 *The following facts are very important in understanding the Lipschitz distance and its relationship with the Gromov-Hausdorff metric:*

1. *Notice that in the definition of $d_L(X, Y)$ one presumes the existence of homeomorphisms between X and Y, that are supposed, moreover, to be bi-Lipschitz. In the absence of such homeomorphisms one defines the Lipschitz distance between X and Y as $d_L(X, Y) = \infty$. It follows that the Lipschitz distance is not suited for measuring the distance between spaces that are not homeomorphic (a fact that one should keep in mind in applications).*

2. *The Lipschitz topology is stronger than the Gromov-Hausdorff one, therefore convergence in d_L implies the convergence in d_{GH} (but, in general, not the other way around).*

3. *Since the computation of the Gromov-Hausdorff distance between two given metric spaces necessitates the construction of a new, "universal" one, the actual computation can be quite problematic even in simple cases. A way of circumventing this obstruction is by making appeal, as in the definition of the Lipschitz metric, to a (properly defined) notion of distortion. However, in this case, instead of functions one makes appeal to the more general notion of correspondence – for further details see, again, [142], [63] and, for a very brief summary, [296].*

4. *Both of the metrics introduced above have their respective drawbacks: The Lipschitz metric presumes that the spaces are homeomorphic (since otherwise their distance infinite, hence unable to encode any further information); while the Gromov-Hausdorff distance is not, in general, finite if the spaces are not bounded (which is also quite restrictive, not just for geometric ends, but in Imaging also). Therefore, one is induced to combine the two metrics into a single one, such that each of the basic metrics "blurs" the defects of the other one. This is done in a rather standard manner[3] as follows:*

Definition C.0.22 *The Gromov-Hausdorff-Lipschitz distance between two metric spaces X and Y is defined as:*

$$d_{GHL}(X, Y) = d_{GH}(X, X_1) + d_L(X_1, Y_1) + d_{GH}(Y_1, Y) \qquad (C.0.7)$$

where X_1, Y_1 are arbitrary given metric spaces.

*For further details and many beautiful applications of this metric the reader is advised to consult [142], **3.C.**, and we contend ourselves with concluding this "detour" with the following (suggestive, as far as all kinds applications are concerned) definition:*

Definition C.0.23 *Two metric spaces X, Y are said to be quasi-isometric iff $d_{GHL}(X, Y) < \infty$.*

---

[3]For other applications of this method in the construction of new metrics from old ones, see [328].

Remark C.0.24 *As we mentioned above, in many instances it is more easy to measure the Lischitz distance, rather than the Hausdorff (hence also than the Gromov-Hausdorff) one. It is only natural, therefore, that people looked into its applications into Graphics and Imaging – see [61] (an idea we already suggested in [296]).*

Remark C.0.25 *We conclude this appendix dedicated to the Hausdorff distance and some of its derived notions by mentioning that the Lipschitz metric was employed for applied goals by Memoli [119] (an author who dedicated a substantial part of his research to the study of implementations of various extensions and generalizations of the common metrics usualy employed in the what one might call the "visual computer" uses).*

Bibliography

[1] Discrete differential geometry: An applied introduction, SIGGRAPH'06 courses. Boston, MA, USA, July 2006. ACM.

[2] A. Bernig. Curvature bounds of subanalytic spaces, preprint, 1–23, 2003.

[3] M. M. Bronstein, A. M. Bronstein, and R. Kimmel. On isometric embedding of facial surfaces into S3. In *Proceedings of International Conference on Scale Space and PDE Methods in Computer Vision*, pages 622–631, 2005.

[4] M. M. Bronstein, A. M. Bronstein, and R. Kimmel. Three-dimensional face recognition. *International Journal of Computer Vision*, 64(1):5–30, 2005.

[5] S. B. Agard and F. W. Gehring. Angles and quasiconformal mappings. In *Proceedings of the London Mathematical Society*, volume 3, pages 1–21, 1965.

[6] S. B. Alexander and R. L. Bishop. Comparison theorems for curves of bounded geodesic curvature in metric spaces of curvature bounded above. *Differential Geometry and Its Applications*, 6:67–86, 1996.

[7] A. D. Alexandrov and Yu. G. Reshetnyak. *General Theory of Irregular Curves*, volume 29 of *Mathematics and Its Applications*. Kluwer, Dordrecht, 1989.

[8] Paul M. Alsing, Jonathan R. McDonald, and Warner A. Miller. The simplicial Ricci tensor. *Classical and Quantum Gravity*, 28(15):155007, 2011.

[9] P. M. Alsing, W. A. Miller, and S. T. Yau. A realization of Thurston's geometrization: discrete Ricci flow with surgery. *Annals of Mathematical Sciences and Applications*, 3(1):31–45, 2018.

[10] N. Amenta and M. Bern. Surface reconstruction by voronoi filtering. *Discrete and Computational Geometry*, 22:481–504, 1999.

[11] B. Andrews. Gauss curvature flow: the fate of the rolling stones. *Inventiones Mathematicae*, 138:151–161, 1999.

[12] B. Andrews. Positively curved surfaces in three-sphere. In *Proceedings of the ICM*, volume 2, pages 221–230, Beijing, 2002.

[13] B. Andrews, B. Chow, C. Guenther, and M. Langford. *Extrinsic Geometric Flows (Graduate Studies in Mathematics 206)*. American Mathematics Society, Providence, RI, 2020.

[14] S. Angenent, E. Pichon, and A. Tannenbaum. Mathematical Methods in Medical Image Processing. *Bulletin of the AMS*, 43(3):359–396, 2006.

[15] E. Appleboim. From normal surfaces to normal curves to geodesics on surfaces. *Axioms*, 6(3):26, 2017.

[16] E. Appleboim, Y. Hyams, S. Krakovski, C. Sageev, and Saucan E. The Scale-Curvature Connection and its Application to Texture Segmentation. *Theory and Applications of Mathematics & Computer Science*, 3(1):38–54, 2013.

[17] E. Appleboim and E. Saucan. Normal approximations of geodesics on smooth triangulated surfaces. *Scientific Studies and Research, Series Mathematics and Informatics*, 21(1):17–36, 2011.

[18] E. Appleboim, E. Saucan, and Y. Y. Zeevi. Ricci Curvature and Flow for Image Denoising and Superesolution. In *Proceedings of EUSIPCO*, pages 2743–2747, 2012.

[19] A. Ardentov, G. Bor, E. Le Donne, R. Montgomery, and Y. Sachkov. Bicycle paths, elasticae and sub-Riemannian geometry. *Nonlinearity*, 34(7):4661–4683, 2021.

[20] S. Asoodeh, T. Gao, and J. Evans. Curvature of hypergraphs via multi-marginal optimal transport. In *Proceedings of 2018 IEEE Conference on Decision and Control (CDC)*, pages 1180–1185, 2018.

[21] O. Attie. Quasi-isometry classification of some manifolds of bounded geometry. *Mathematische Zeitschrift*, 216:501–527, 1994.

[22] B. Bahr and B. Dittrich. Regge calculus from a new angle. *New Journal of Physics*, 12, 2010.

[23] D. Bakry, Diffusions hypercontractives. In *Séminaire de Probabilités XIX 1983/84*, pages 177–206, Berlin, Heidelberg, 1985. Springer Berlin Heidelberg.

[24] D. Bakry and M. Ledoux. Lévy-Gromov's isoperimetric inequality for an infinite dimensional diffusion generator. *IInventiones Mathematicae*, 123:259–281, 2004.

[25] T. A. Banchoff. Critical points and curvature for embedded polyhedra. *Journal of Differential Geometry*, 1:257–268, 1967.

[26] T. A. Banchoff. Critical points and curvature for embedded polyhedral Surfaces. *American Mathematical Monthly*, 77:475–485, 1970.

[27] V. Barkanass, W. Chen, N. Lei, and E. Saucan. Textures Curvature (Do Stochastic Textures Exist?). In preparation.

[28] Vladislav Barkanass, Jürgen Jost, and Emil Saucan. Geometric Sampling of Networks. *arXiv:2010.15221v1 [math.DG], MPI MIS Preprint*, 2020.

[29] D. Barnette, E. Jucovic, and M. Trenkler. Toroidal maps with prescribed types of vertices and faces. *Mathematika*, 18(1):82–90, 1971.

[30] R. H. Bartels, J. C. Beatty, and B. A. Barsky. *An Introduction to Splines for Use in Computer Graphics and Geometric Modeling*. Morgan Kaufmann, 1987.

[31] M. Bartoň, I. Hanniel, G. Elber, and M. S. Kim. Precise Hausdorff distance computation between polygonal meshes. *Computer Aided Geometric Design*, 27:580–591, 2010.

[32] F. Bauer, B. Hua, J. Jost, S. Liu, and G. Wang. *The Geometric Meaning of Curvature: Local and Nonlocal Aspects of Ricci Curvature*, pages 1–62. Springer International Publishing, 2017.

[33] A. Belyaev. *Curvature Features for Shape Interrogation*. The University of Aizu, 2002.

[34] V. N. Berestovskii. Spaces with bounded curvature and distance geometry. *Siberian Math. J.*, 27(1):8–19, 1986.

[35] M. Berger. *Geometry II (Universitext)*. Spinger-Verlag, Berlin, 1987.

[36] M. Berger. Encounter with a geometer, part ii. *Notices of the AMS*, 47(3):326–340, 2000.

[37] M. Berger. *A Panoramic View of Riemannian Geometry*. Springer-Verlag, Berlin, 2003.

[38] M. Berger. *Geometry I*. Springer-Verlag, Berline, 2009.

[39] M. Berger. *Geometry Revealed: A Jacob's Ladder to Modern Higher Geometry, 1232*. Springer-Verlag, Berlin-Heidelberg, 2010.

[40] M. Berger and B. Gostiaux. *Differential Geometry: Manifols, Curves and Surfaces*. Spinger-Verlag, New York, 1987.

[41] M. Bern, L. P. Chew, D. Eppstein, and J. Ruppert. Dihedral Bounds for Mesh Generation in High Dimensions. In *Proceedings of the 6th ACM-SIAM Symposium on Discrete Algorithms*, pages 189–196, 1995.

[42] A. Bernig. Curvature tensors of singular spaces. *Differential Geometry and its Applications*, 24:191–208, 2006.

[43] W. Blaschke. *Vorlesungen iiber Differentialgeometrie, Band 1, Elementare Dijferentialgeometrie, 4. AuR*. Spinger, Berline.

[44] E. D. Bloch. The angle defect for arbitrary polyhedra. *Beiträge Algebra Geom*, 39:379–393, 1998.

[45] E. D. Bloch. Critical points and the angle defect. *Geometriae Dedicata*, 109:121–137, 2004.

[46] Ethan Bloch. Combinatorial Ricci curvature for polyhedral surfaces and posets, 2014.

[47] L. M. Blumenthal and K. Menger. *Studies in Geometry*. Freeman and Co., San Francisco, 1970.

[48] A. I. Bobenko. A conformal energy for simplicial surfaces. *Combinatorial and Computational Geometry, MSRI Publications*, 52:133–143, 2005.

[49] A. I. Bobenko and P. Schröder. Discrete willmore flow. In *Eurographics Symposium on Geometry Pocessing*, pages 101–110, 2005.

[50] S. Bochner. Curvature and betti numbers. *Annals of Mathematics*, 49(2):379–390, 1948.

[51] A.-I. Bonciocat and K.-T. Sturm. Mass transportation and rough curvature bounds for discrete spaces. *Journal of Functional Analysis*, 256(9):2944–28966, 2009.

[52] V. Borrelli, F. Cazals, and J-M. Morvan. On the angular defect of triangulations and the pointwise approximation of curvatures. *Research Report RR-4590, INRIA*, 2002.

[53] V. Borrelli, F. Cazals, and J.-M. Morvan. On the angular defect of triangulations and the pointwise approximation of Curvatures. *Computer Aided Geometric Design*, 20:319–341, 2003.

[54] B. H. Bowditch. Bilipschitz triangulations of Riemannian manifolds, 2020.

[55] A. Braunmühl. Geodätische linien auf dreiachchsigen flächen zwitein grades. *Mathematische Annalen*, 20:557–586, 1882.

[56] U. Brehm and W. Kühnel. Smooth approximation of polyhedral surfaces regarding curvatures. *Geometriae Dedicata*, 12:435–461, 1982.

[57] S. Brendle. Curvature flows on surfaces with boundary. *Mathematische Annalen*, 324:491–519, 2002.

[58] S. Brendle. A family of curvature flows on surfaces with boundary. *Mathematische Zeitschrift*, 241:829–869, 2002.

[59] William Breslin. Thick triangulations of hyperbolic n-manifolds. *Pacific Journal of Mathematics*, 241(2):215–225, 2009.

[60] A. M. Bronstein, M. M. Bronstein, A. M. Bruckstein, and R. Kimmel. Paretian similarity for partial comparison of non-rigid objects. In *Lecture Notes in Computer Science*, volume 4485, pages 264–275, 2007.

[61] Alexander M. Bronstein, Michael M. Bronstein, Alfred M. Bruckstein, and Ron Kimmel. Paretian similarity for partial comparison of non-rigid objects. In Fiorella Sgallari, Almerico Murli, and Nikos Paragios, editors, *Scale Space and Variational Methods in Computer Vision*, pages 264–275, Berlin, Heidelberg, 2007. Springer Berlin Heidelberg.

[62] S. M. Buckley, J. MacDougall, and D. J. Wraith. On ptolemaic metric simplicial complexes. *Mathematical Proceedings of the Cambridge Philosophical Society*, 149(1):93–104, 2010.

[63] D. Burago, Y. Burago, and S. Ivanov *Course in Metric Geometry*, volume 33 of *Graduate Studies in Mathematics*. AMS, Providence, 2000.

[64] Yu. D. Burago, M. L. Gromov, and G. Ya. Perelman. A. d. aleksandrov spaces with curvatures bounded below. *Russian Math Surveys*, 47(2):1–58, 1992.

[65] H. Buseman. *The Geometry of Geodesics*. Academic Press Inc., New York, 1995.

[66] S. S. Cairns. On the triangulation of regular loci. *Annals of Mathematics*, 35:579–587, 1934.

[67] S. S. Cairns. Polyhedral approximation to regular loci. *Annals of Mathematics*, 37:409–419, 1936.

[68] S. S. Cairns. A simple triangulation method for smooth manifolds. *Bulletin of American Mathematical Society*, 67:380–390, 1961.

[69] L. Cao, J. Zhao, J. Xu, S.-M. Chen, G. Liu, S.-Q. Xin, Y. Zhou, and Y. He. Computing smooth quasi-geodesic distance field (qgdf) with quadratic programming. *Computer-Aided Design*, 127:102879, 2020.

[70] M. Cassorla. Approximating compact inner metric spaces by surfaces. *Indiana University Mathematical Journal*, 41:505–513, 1992.

[71] R. Charney and N. Davis. The euler characteristic of a nonpositively curved, piecewise euclidean manifold. *Pacific Journal of Mathematics*, 171(1):117–137, 1995.

[72] J. Cheeger. Finiteness theorems for Riemannian manifolds. *American Journal of Mathematics*, 92:61–74, 1970.

[73] J. Cheeger and D. G. Ebin. *Comparison Theorems in Riemannian Geometry*. AMS Chelsea Publishing, Providence, 1975.

[74] J. Cheeger, W. Müller, and R. Schrader. Lattice gravity or Riemannian structure on piecewise linear spaces. In *Unified Theories of Elementary Particles, Lecture Notes in Physics 160*, pages 176–188. Springer, Berlin, 1982.

[75] J. Cheeger, W. Müller, and R. Schrader. On the Curvature of Piecewise Flat Spaces. *Communications in Mathematical Physics*, 92:405–454, 1984.

[76] J. Cheeger, W. Müller, and R. Schrader. Kinematic and Tube Formulas for Piecewise Linear Spaces. *Indiana University Mathematical Journal*, 35(4):737–754, 1986.

[77] Jeff Cheeger. A vanishing theorem for piecewise constant curvature spaces. In Katsuhiro Shiohama, Takashi Sakai, and Toshikazu Sunada, editors, *Curvature and Topology of Riemannian Manifolds*, Lecture Notes in Math. 1201, pages 33–40, Berlin, Heidelberg, 1986. Springer Berlin Heidelberg.

[78] Wei Chen, Min Zhang, Na Lei, and David Xianfeng Gu. Dynamic unified surface Ricci flow. *Geometry, Imaging and Computing*, 3(1):31–56, 2016.

[79] Wei Chen, Xiaopeng Zheng, Jingyao Ke, Na Lei, Zhongxuan Luo, and Xianfeng Gu. Quadrilateral mesh generation i: Metric based method. *Computer Methods in Applied Mechanics and Engineering*, 356:652–668, 2019.

[80] Wei Chen, Xiaopeng Zheng, Jingyao Ke, Na Lei, Zhongxuan Luo, and Xianfeng Gu. Quadrilateral mesh generation ii: Meoromorphic quartic differentials and Abel-Jacobi condition. *Computer Methods in Applied Mechanics and Engineering*, 366:112980, 2020.

[81] S. W. Cheng, T. K. Dey, E. A. Ramos, and Tathagata Ray. Sampling and meshing a surface with guaranteed topology and geometry. In *Proceedings of the twentieth annual symposium on Computational geometry*, pages 280–289, 2004.

[82] S. S. Chern. Curves and surfaces in Euclidean space. *Studies in Global Geometry and Analysis, MAA Studies in Mathematics*, 4:1–16, 1967.

[83] S. S. Chern and R. K. Lashof. On the total curvature of immersed manifolds. *American Journal of Mathematics*, 79(2):306–318, 1957.

[84] Shiing-shen Chern. An elementary proof of the existence of isothermal parameters on a surface. *Proceedings of American Mathematical Society, American Mathematical Society*, 6(5):771–782, 1955.

[85] L. Paul Chew. Guaranteed-quality mesh generation for curved surfaces. In *Proceedings of the Ninth Annual Symposium on Computational Geometry*, pages 274–280, 1993.

[86] B. Chow. The Ricci flow on the 2-sphere. *Journal of Differential Geometry*, 33(2):325–334, 1991.

[87] B. Chow and D. Knopf. *The Ricci Flow: An Introduction (Mathematical Surveys and Monographs 110)*. AMS Providence, RI, 2004.

[88] B. Chow and F. Luo. Combinatorial Ricci Flows on Surfaces. *Journal of Differential Geometry*, 63(1):97–129, 2003.

[89] U. Clarenz, G. Dziuk, and M. Rumpf. *On Generalized Mean Curvature Flow in Surface Processing*, pages 217–248. Springer, Berline, Heidelberg, 2003.

[90] H. Cohen   Semantic Gravitation: Predicting Movement of the Stream of Thoughts, PhD Thesis, 2001.

[91] D. Cohen-Steiner and J.-M. Morvan. Approximation of the curvature measures of a smooth surface endowed with a mesh. *Research Report 4867, INRIA*, 2003.

[92] D. Cohen-Steiner and J. M. Morvan. Restricted Delaunay triangulations and normal cycle. In *Proceedings of the 19th Annual ACM Symposium on Computational Geometry*, pages 237–246, 2003.

[93] D. Cohen-Steiner and J-M Morvan. Restricted Delaunay Triangulations and Normal Cycle. In *Proceedings of the Nineteenth Annual Symposium on Computational Geometry*, pages 312–321, June 2003.

[94] D. Cooper and I. Rivin. Combinatorial scalar curvature and rigidity of ball packings. *Mathematical Research Letters*, 3(1):51–60, 1996.

[95] J. Austin Cottrell, Thomas J. Hughes, and Yuri Bazilevs. *Isogeometric Analysis: Toward Integration of CAD and FEA*. Wiley.

[96] H. S. M. Coxeter. *Introduction to Geometry, 2nd ed.* John Wiley & Sons, 1969.

[97] Li Cui, Xin Qi, Chengfeng Wen, Na Lei, Xinyuan Li, Min Zhang, and Xianfeng Gu. Spherical optimal transportation. *Computer-Aided Design*, 115:181–193, 2019.

[98] D. Cohen-Steiner and J. M. Morvan. Approximation of Normal Cycles, 2003.

[99] J. Dai, W. Luo, M. Jin, W. Zeng, Y. He, S.-T. Yau, and X. Gu. Geometric accuracy analysis for discrete surface approximation. *Computer Aided Geometric Design*, 24:323–338, 2007.

[100] Y. Colin de Verdière et Alexis Marin. Triangulations presque équilatérales des surfaces. *Journal of Differential Geometry*, 32(1):199–207, 1990.

[101] Tamal K. Dey, Kuiyu Li, Jian Sun, and David Cohen-Steiner. Computing geometry-aware handle and tunnel loops in 3d models. *ACM Transaction on Graphics*, 27(3):1–9, 2008.

[102] M. Deza and E. Deza. *Encyclopedia of Distances*. Springer, Berlin, 2009.

[103] B. Dittrich and S. Speziale. Area angle variables for general relativity. *New Journal of Physics*, 10(8):083006, 2008.

[104] M. P. do Carmo. *Riemannian Geometry (Mathematics: Theory & Applications)*. Birkhäuser, Boston, Mass.

[105] M. P. do Carmo. *Differential Geometry of Curves and Surfaces*. Prentice-Hall, Englewood Cliffs, N.J., 1976.

[106] P. Dombrowski. 150 years after Gauss' "disquisitiones generales circa superficies curvas". *Asterisque*, (62):1–157, 1979.

[107] Simon Donaldson. *Riemann Surfaces*, volume 22 of *Oxford Graduate Texts in Mathematics*. Oxford University Press, 2011.

[108] Qiang Du, Vance Faber, and Max Gunzburger. Centroidal voronoi tessellations: Applications and algorithms. *SIAM Review*, 41(4):637–676, 1999.

[109] E. Saucan, C. Sagiv, and E. Appleboim. Geometric Wavelets for Image Processing: Metric Curvature of Wavelets. In *Proceedings of SampTA*, pages 85–89, 2009.

[110] E. Appleboim. Quasi Normality of least Area Incompressible Surfaces in 3-Manifolds, 2014.

[111] A. Samal, E. Saucan, and J. Jost. A simple differential geometry for networks and its generalizations. In *Proceedings of Complex Networks and Their Applications VIII, Complex Networks 2019 (SCI)*, volume 881, pages 943–954. Springer, 2020.

[112] E. Sonn. Ricci Flow for Image Processing and Computer Vision, MSc Thesis, Technion.

[113] E. Sonn. Ricci flow for image processing and computer vision, June 2015.

[114] H. Edelsbrunner. *Geometry and Topology for Mesh Generation*. Cambridge University Press, Cambridge, 2001.

[115] Herbert Edelsbrunner. *Geometry and Topology for Mesh Generation*. Number 7 in Cambridge Monographs on Applied and Computational Mathematics. Cambridge University Press, 1 edition, 2001.

[116] P. Elumalai, Y. Yadav, N. Williams, E. Saucan, J. Jost, and A. Samal. Graph Ricci curvatures reveal disease-related changes in autism spectrum disorder. Scientific Reports, 10, 10819, 2022.

[117] D. B. A. Epstein et al. (editor). *Word Processing in Groups*. Jones and Bartlett, Boston, MA, 1992.

[118] F.-E. Wolter. Cut Loci in Bordered and Unbordered Riemannian Manifolds, 1985.

[119] G. Sapiro, F. Mèmoli, and P. Thomson. Geometric surface and brain warping via geodesic minimizing Lipschitz extension. In *Proceedings of 1st MICCAI Workshop on Mathematical Foundations of Computational Anatomy: Geometrical, Statistical and Registration Methods for Modeling Biological Shape*, pages 58–67, 2006.

[120] H. Federer. Curvature measures. *Trans. Amer. Math. Soc.*, 93:418–491, 1959.

[121] W. Firey. On the shapes of worn stones. *Mathematika*, 21:1–11, 1974.

[122] F. J. Flaherty. Curvature measures for piecewise linear manifolds. *Bulletin of American Mathematical Society*, 79(1):100–102, 1973.

[123] F. Forman. Bochner's method for cell complexes and combinatorial Ricci curvature. *Discrete and Computational Geometry*, 29(3):323–374, 2003.

[124] M. Fréchet. Sur la distance de deux lois de probabilité. *Comptes Rendus Hebdomadaires des Seances de L Academie des Sciences*, 244(6):689–692, 1957.

[125] J. H. G. Fu. Convergence of Curvatures in Secant Approximation. *Journal of Differential Geometry*, 37:177–190, 1993.

[126] G. Alexits. La torsion des espaces distanciés. *Compositio Mathematica*, 6:471–477, 1938-1939.

[127] G. Gilboa and S. Osher Nonlocal linear image regularization and supervised segmentation. *SIAM Multiscale Modeling and Simulation*, 6(2):595–630, 2007.

[128] G. Gilboa and S. Osher Nonlocal operators with applications to image processing. *SIAM Multiscale Modeling and Simulation*, 7(3):1005–1028, 2008.

[129] M. Gage and R. S. Hamilton. The heat equation shrinking convex plane curves. *Journal of Differential Geometry*, 23(1):69–96, 1986.

[130] K. F. Gauss. Disquisitiones generales circa superficies curvas, "General Investigations of Curved Surfaces" (published 1965), (J. C. Morehead and A. M. Hiltebeitel, trans.). Raven Press, New York. *Commentationes Societatis Regiae Scientiarum Gottingesis Recentiores*, VI:99–146, 1823-1827.

[131] W. F. Gehring and J. Väisälä. The coefficients of quasiconformality. *Acta Math.*, 114:1–70, 1965.

[132] A. L. Gibbs and F. E. Su. On Choosing and Bounding Probability Metrics. *International Statistical Review*, 70.

[133] G. Gilboa, E. Appleboim, E. Saucan, and Y. Y. Zeevi. On the Role of Nonlocal Menger Curvature in Image Processing. In *Proceedings of ICIP*, pages 4337–4341, 2015.

[134] D. Glickenstein. A combinatorial Yamabe flow in three dimensions. *Topology*, 44(4):791–808, 2005.

[135] D. Glickenstein. A maximum principle for combinatorial Yamabe flow. *Topology*, 44(4):809–825, 2005.

[136] D. Glickenstein. Discrete conformal variations and scalar curvature on piecewise at two and three dimensional manifolds. *Journal of Differential Geometry*, 87(2):201–238, 2011.

[137] A. Gray. *Tubes*. Addison-Wesley, Redwood City, CA, 1990.

[138] A. Gray and L. Vanhecke. Riemannian geometry as determined by the volume of small geodesic balls. *Acta Mathematica*, 142:157–198, 1979.

[139] M. A. Grayson. The heat equation shrinks embedded plane curves to round points. *Journal of Differential Geometry*, 26(2):285–314, 1987.

[140] M. A. Grayson. Shortening embedded curves. *Annals of Mathematics. Second Series*, 129(1):71–111, 1989.

[141] M. J. Greenberg. Euclidean and non-euclidean geometries: Development and history 3d edn. 1993.

[142] M. Gromov. *Metric structures for Riemannian and non-Riemannian spaces (Progress in Mathematics Volume 152)*. Birkhäuser, Boston, 1999.

[143] M. Gromov. Isoperimetry of waists and concentration of maps. *Geometric and Functional Analysis*, 13:178–215, 2003.

[144] K. Grove and S. Markvorsen. Curvature, Triameter and Beyond. *Bulletin of American Mathematical Society*, 27:261–265, 1995.

[145] K. Grove and S. Markvorsen. New extremal problems for the Riemannian recognition program via Alexandrov geometry. *Journal of American Mathematical Society*, 8(1):1—-28, 1995.

[146] K. Grove and P. Petersen. Bounding homotopy types by geometry. *Annals of Mathematics*, 128:195–206, 1988.

[147] D. X. Gu and E. Saucan. Metric Ricci curvature for PL manifolds. *Geometry*, 2013(Article ID 694169, 12 pages), 2013.

[148] X. Gu. Parametrization for surfaces with arbitrary topologies. *PhD Thesis, Computer Science*, 2012.

[149] X. Gu and S.-T. Yau. Computing Conformal Structures of Surfaces. *Communications in Information and Systems*, 2(2):121–146, 2002.

[150] X. D. Gu and S.-T. Yau. *Computational Conformal Geometry*, volume 3 of *Advanced Lectures in Mathematics*. International Press, Somerville, MA, 2008.

[151] Xianfeng Gu, Ren Guo, Feng Luo, Jian Sun, and Tianqi Wu. A discrete uniformization theorem for polyhedral surfaces i. *Journal of Differential Geometry (JDG)*, 109(3):431–466, 2018.

[152] Xianfeng Gu, Feng Luo, Jian Sun, and Tianqi Wu. A discrete uniformization theorem for polyhedral surfaces i. *Journal of Differential Geometry (JDG)*, 109(2):223–256, 2018.

[153] Xianfeng Gu, Feng Luo, Jian Sun, and S-T Yau. Variational principles for minkowski type problems, discrete optimal transport, and discrete monge-ampere equations. *Asian Journal of Mathematics (AJM)*, 20(2):383–398, April 2016.

[154] H. Cohen, A. Maril, and E. Saucan. Comparative analysis of four discrete curvatures for semantic graphs, in preparation, 2021.

[155] Y.-L. Yang, Y.-K. Lai, H. Pottmann, J. Wallner, and S.-M. Hu. Principal curvatures from the integral invariant viewpoint. *Computer Aided Geometric Design*, 24:428–442, 2007.

[156] S. Haker, L. Zhu, A. Tannenbaum, and S. Angenent. Optimal Mass Transport for Registration and Warping. *International Journal of Computer Vision*, 60.

[157] B. Hamann. Curvature approximation for triangulated surfaces. *Computing Suppl.*, (8):1–14, 1993.

[158] R. Hamilton. The Ricci Flow on Surfaces. *AMS Contemporary Mathematics*, 71.

[159] T. Har'el. Curvature of curves and surfaces – A parabolic approach. *preprint*, pages 1–14, 1995.

[160] A. Hatcher. *Algebraic Topology*. Cambridge University Press, 2001.

[161] E. Heintze and H. Karcher. A general comparison theorem with applications to volume estimates for submanifolds. *Annales scientifiques de l'Ècole Normale Supèrieure*, 11:451–470, 1978.

[162] Peter Henrici. *Applied and Computational Complex Analysis Vol. 1: Power Series Integration, Conformal Mapping, Location of Zeros*. Wiley & Sons, Incorporated, John, 1988.

[163] D. Hilbert and S. Cohn-Vossen. *Geometry and the Imagination*. American Mathematics Society, (German original: Springer, Berlin,1932), Chelsea, Providence, RI 195, 1999.

[164] T. P. Hill. On the oval shapes of beach stones. *AppliedMath*, 2(1):16–38, 2022.

[165] H. Hopf. *Differential Geometry in the Large*. Lecture Notes in Mathematics (LNM, volume 1000). Springer, Berlin, 1983.

[166] D. Horak and J. Jost. Spectra of combinatorial Laplace operators on simplicial complexes. *Advances in Mathematics*, 244:303–336, 2013.

[167] J. Howland and J. H. G. Fu. Tubular neighbourhoods in Euclidean spaces. *Duke Mathematical Journal*, 52(4):1025–1045, 1985.

[168] C.-C. Hsiung. *A First Course in Differential Geometry*. John Wiley & Sons, 1981.

[169] J. F. Hudson. *Piecewise Linear Topology (Mathematics Lecture Notes Series)*. Benjamin, New York, 1969.

[170] D. Hug, G. Last, and W. Weil. A local Steiner-type formula for general closed sets and applications. *Mathematische Zeitschrift*, 246(1-2):237–272, 2004.

[171] S. Hurder V. Šešum I. Corwin, N. Hoffman, and Y. Xu. Differential Geometry of Manifolds with Density. *Rose Hulman Undergraduate Journal of Mathematics*, 7(1):15, 2006.

[172] Ivan Izmestiev, Robert B Kusner, Günter Rote, Boris Springborn, and John M Sullivan. There is no triangulation of the torus with vertex degrees 5, 6,..., 6, 7 and related results: Geometric proofs for combinatorial theorems. *Geometriae Dedicata*, 166(1):15–29, 2013.

[173] J. Gaddum. Metric methods in integral and differential geometry. *American Journal of Mathematics*, 75(1):30–42, 1953.

[174] J. Giesen. Curve Reconstruction, the Traveling Salesman Problem and Menger's Theorem on Length. In *Proceedings of the 15th ACM Symposium on Computational Geometry (SoCG)*, pages 207–216, 1999.

[175] J. Haantjes. Distance geometry. Curvature in abstract metric spaces. In *Proc. Kon. Ned. Akad. v. Wetenseh., Amsterdam 50*, pages 496–508, 1947.

[176] J.-L. Maltret and M. Daniel. Discrete curvatures and applications: a survey, 2002.

[177] Svante Janson. Lecture notes on Euclidean, spherical and hyperbolic trigonometry. 2015.

[178] J. Jost. *Riemannian Geometry and Geometric Analysis*. 7 edn.

[179] Jügen Jost. *Harmonic Maps Between Surfaces*, volume 1062 of *Lecture Notes in Mathematics*. Springer Berlin, Heidelberg, 1984.

[180] Jürgen Jost and Florentin Münch. Characterizations of forman curvature, 2021.

[181] Frederic J Almgren Jr and Igor Rivin. The mean curvature integral is invariant under bending. *Geometry & Topology Monographs*, 1:1–21, 1998.

[182] K. Watanabe and A. G. Belyaev. Detection of salient curvature features on polygonal surfaces. In *Estimating the Tensor of Curvature of a Surface from a Polyhedral Approximation*, volume 20, 2001.

[183] K. A. Murgas, E. Saucan, and R. Sandhu. Beyond Pairwise Interactions: Higher-Dimensional Geometry in Protein Interaction Networks, preprint, 2022. https://doi.org/10.1101/2022.05.03.490479.

[184] D.C. Kay. Arc curvature in metric spaces. *Geometriae Dedicata*, 9(1):91–105, 1980.

[185] M. Keller and F. Münch. A new discrete Hopf–Rinow theorem. *Discrete Mathematics*, 342(9):2751–2757, 2019.

[186] Matthias Keller. Intrinsic metrics on graphs: a survey. In Mugnolo D., editor, *Mathematical Technology of Networks*, volume 128 of *Springer Proceedings in Mathematics & Statistics*, pages 81–119. Springer, Cham, 2015.

[187] Y.-J. Kim, Y.-T. Oh, S.-H. Yoon, M.-S. Kim, and G. Elber. Precise Hausdorff distance computation for planar freeform curves using biarcs and depth buffer. *The Visual Computer*, 26:1007–1116, 2010.

[188] R. Kimmel, R. Malladi, and N. Sochen. Images as embedded maps and minimal surfaces: Movies, color, texture, and volumetric medical images. *International Journal of Computer Vision*, 39(2):111–129, 2000.

[189] R. Kimmel and J. Sethian. Computing geodesic paths on manifolds. In *Procedings of National Academy of Sciences*, volume 95, pages 8431–8435, 1998.

[190] W. A. Kirk. On Curvature of a Metric Space at a Point, year = 1964. *Pacific Journal of Mathematics*, 14:195–198.

[191] O. Knill. A discrete Gauss-Bonnet type theorem. *Elemente der Mathematik*, 67(1):1–17, 2012.

[192] Oliver Knill. A graph theoretical gauss-bonnet-chern theorem, 2011.

[193] Oliver Knill. A graph theoretical Poincare-Hopf theorem, 2012.

[194] J. J. Koenderink. *Solid Shape*. Cambridge, Massachusetts: MIT Press, 1990.

[195] J. Kunegis. Les miserables, KONECT: the Koblenz Network Collection. In *Proceedings of the 22nd International Conference on World Wide Web*, pages 1343–1350, May 2013.

[196] Xin Lai, Shuliang Bai, and Yong Lin. Normalized discrete Ricci flow used in community detection. *Physica A: Statistical Mechanics and Its Applications*, 597:127251, 2022.

[197] S. Lang. *Linear Algebra, (third ed.)*. Springer, New York, 1987.

[198] D. Larman and C. Rogers (Eds.). Proceedings of the durham symposium on the relations between infinite-dimensional and finite-dimensional convexity. *The Bulletin of the London Mathematical Society*, 8:1–33, 1976.

[199] N. Lei, D. An, Y. Guo, K. Su, S. Liu, Z. Luo, S.-T. Yau, and X. Gu. A geometric understanding of deep learning. *Journal of Engineering*, 6(3):361–374, 2020.

[200] N. Lei, K. Su, L. Cui S.-T. Yau, and X. Gu. A geometric view of optimal transportation and generative model. *Computer Aided Geometric Design*, 68:1–21, 2019.

[201] Na Lei, Jisui Huang, Yuxue Ren, Emil Saucan, and Zhenchang Wang. Ricci curvature based volumetric segmentation of the auditory ossicles, 2020.

[202] R. Lev, E. Saucan, and G. Elber. Curvature based clustering for dna microarray data analysis. In *Lecture Notes in Computer Science, Mathematics of Surfaces*, volume 4647, pages 275–289. Springer-Verlag, 2007.

[203] Ronen Lev, Emil Saucan, and Gershon Elber. Curvature estimation over smooth polygonal meshes using the half tube formula. In *Lecture Notes in Computer Science, Mathematics of Surfaces: 12th IMA International Conference*, volume 4647, pages 275–289. Springer-Verlag, 2007.

[204] H. Li, W. Zeng, J-M Morvan, L. Chen, and X. Gu. Surface meshing with curvature convergence. *IEEE Transactions on Visualization and Computer Graphics*, 20(6):919–934, 2014.

[205] X.-Y. Li and S.-H. Teng. Generating Well-Shaped Delaunay Meshes in 3D. In *Proceedings of SODA*, pages 28–37, 2001.

[206] S. Lin, Z. Luo, J. Zhang, and E. Saucan. Generalized Ricci Curvature Based Sampling and Reconstruction of Images. In *Proceedings of EUSIPCO*, pages 604–608, 2015.

[207] A. Linnér. Curve-straightening in closed euclidean submanifolds. *Commun. Math. Phys.*, 138:33–49, 1991.

[208] Gabriele Lohmann, Eric Lacosse, Thomas Ethofer, Vinod J. Kumar, Klaus Scheffler, and Jürgen Jost. Predicting intelligence from fmri data of the human brain in a few minutes of scan time. *bioRxiv*, 2021.

[209] J. Lott and C. Villani. Ricci curvature for metric-measure spaces via optimal transport. *Annals of Mathematic*, 169(3):903–991, 2009.

[210] Feng Luo. Combinatorial Yamabe flow on surfaces. *Communications in Contemporary Mathematics*, 6(5):765–780, 2004.

[211] R. C. Lyndon and P. E. Schupp. *Combinatorial Group Theory*. Ergebnisse der Mathematik und ihrer Grengebiete 89. Springer-Verlag, Berlin-Heidelberg-New York, 1977.

[212] L. M. Blumenthal. *Distance Geometry – Theory and Applications*. Claredon Press, Oxford, 1953.

[213] M. Gelbrich. On a formula for the L^2 Wasserstein metric between measures on Euclidean and Hilbert spaces. *Mathematische Nachrichten*, 147.

[214] M. Zähle. Curvature Theory for Singular Sets in Euclidean Spaces, 2005.

[215] Y. Machigashira. The Gaussian curvature of Alexandrov surfaces. *Journal of the Mathematical Society of Japan*, 50(4):859–878, 1998.

[216] T. Maekawa, F.-E. Wolter, and N. M. Patrikalakis. Umbilics and lines of curvature for shape interrogation. *Computer Aided Geometric Design*, 13(2):133–161, 1996.

[217] R. Martin. Estimation of principal curvatures from range data. *International Journal of Shape Modeling*, 13(2):99–111, 1998.

[218] J. C. Mathews, S. Nadeem, M. Pouryahya, Z. Belkhatir, J. O. Deasy, A. J. Levine, and A. R. Tannenbaum. Functional network analysis reveals an immune tolerance mechanism in cancer. In *Proceedings of the National Academy of Sciences*, volume 117, pages 16339–16345, 2020.

[219] H. Matsuda and S. Yorozu. Notes on Bertrand curves. *Yokohama Mathematical Journal*, 50:41–58, 2003.

[220] D. Meek and D. Walton. On surface normal and Gaussian curvature approximations given data sampled from a smooth surface. *Computer Aided Geometric Design*, 17(6):521–543, 2000.

[221] K. Menger. Untersuchungen über allgemeine Metrik. *Mathematische Annalen*, 103:466–501, 1930.

[222] K. Menger. La geométrie des distances et ses relations avec les autres branches des mathématiques. *L'Enseignement math.*, 105:348–372, 1935.

[223] K. Menger. *Calculus - A Modern Approach*. Ginn and Company, Boston, 1955.

[224] W. A. Miller, J. R. McDonald, P. M. Alsing, D. X. Gu, and S.-T. Yau. Simplicial Ricci flow. *Communications in Mathematical Physics*, 329(2):579–0608, 2014.

[225] J. Milnor. *Morse Theory*. Princeton University Press, 1963.

[226] John Milnor and James D. Stasheff. *Characteristic Classes, Annals of Mathematics Studies 76*. Princeton University Press, 1974.

[227] R. Milo, S. Shen-Orr, S. Itzkovitz, N. Kashtan, D. Chklovskii, and U. Alon. Network motifs: simple building blocks of complex networks. *Science*, 298(5594):824–827, 2002.

[228] F. Morgan. Manifolds with density. *Notices of the American Mathematical Society*, 52:853–858, 2005.

[229] J. R. Munkres. *Elementary Differential Topology, (rev. ed.)*. Princeton University Press, Princeton, NJ, 1966.

[230] J. R. Munkres. *Topology, A First Course*. Prentice-Hall, 1974.

[231] A. Naitsat, S. Cheng, X. Qu, Y. Y Zeevi, X. Fan, and E. Saucan. An elementary analogue of the Gauss-Bonnet theorem. *American Mathematical Monthly*, 61:601–603, 1954.

[232] A. Naitsat, E. Saucan, and Y. Y. Zeevi. Volumetric quasi-conformal mappings. In *Proceedings of GRAPP/VISIGRAPP*, pages 46–57, 2015.

[233] Chien-Chun Ni, Yu-Yao Lin, Jie Guo, and Xianfeng Gu. Network alignment by discrete Oliivier-Ricci flow. In *International Symposium on Graph Drawing and Network Visualization*, pages 447–462. Springer, Cham, 2018.

[234] Chien-Chun Ni, Yu-Yao Lin, Jie Guo, Xianfeng Gu, and Emil Saucan. Ricci curvature of the internet topology. In *Proceedings of INFOCOM*, pages 2758–2766. IEEE, 2015.

[235] Chien-Chun Ni, Yu-Yao Lin, Feng Luo, and Jie Gao. Community detection on networks with Ricci flow. *Scientific Reports*, 9(1):1–12, 2019.

[236] G. Elber O. Soldea, and E. Rivlin. Global segmentation and curvature analysis of volumetric data sets using trivariate b-spline functions. *IEEE Transactions on Pattern Analysis and Machine Intelligence*, 28(2):265–278, 2006.

[237] Jung Hun Oh, Maryam Pouryahya, Aditi Iyer, Aditya P. Apte, Joseph O. Deasy, and Allen Tannenbaum. A novel kernel wasserstein distance on gaussian measures: An application of identifying dental artifacts in head and neck computed tomography. *Computers in Biology and Medicine*, 120:103731, 2020.

[238] Y. Ollivier. Ricci curvature of markov chains on metric spaces. *Journal of Functional Analysis*, 256(3):81–864, 2009.

[239] Y. Ollivier. A visual introduction to riemannian curvatures and some discrete generalizations. In Galia Dafni, Robert McCann, and Alina Stancu, editors, *Analysis and Geometry of Metric Measure Spaces: Lecture Notes of the 50th Séminaire de Mathématiques Supérieures (SMS), Montréal, 2011*. AMS, 2013.

[240] B. O'Neill. *Elementary Differential Geometry*. Academic Press, N.Y., 1966.

[241] S. Osher and J. A. Sethian. Fronts propagating with curvature-dependent speed: Algorithms based on hamilton–jacobi formulation. *Journal of Computational Physics*, 79:12–49, 1988.

[242] J. Pach and P. K. Agarwal. *Combinatorial Geometry*. Wiley-Interscience, 1995.

[243] H. Pajot. *Analytic Capacity, Rectificabilility, Menger Curvature and the Cauchy Integral.* Lecture Notes in Mathematics (LNM, volume 1799). Springer-Verlag, Berlin, 2002.

[244] N. M. Patrikalakis and T. Maekawa. *Shape Interrogation for Computer Aided Design and Manufacturing.* Springer Verlag, Berlin, Heidelberg, 2002.

[245] K. Peltonen. On the existence of quasiregular mappings. *Annales Academiae Scientiarum Fennicae Mathematica, Series I Math. Dissertationes*, 1992.

[246] G. Perelman. Alexandrov's spaces with curvature bounded from below II. *preprint*, 1991.

[247] G. Perelman. The entropy formula for the Ricci flow and its geometric applications, 2002.

[248] G. Perelman. Ricci flow with surgery on three-manifolds, 2003.

[249] P. Petersen. *Riemannian Geometry.* Springer-Verlag, New York, 1998.

[250] C. Plaut. Almost Riemannian Spaces. *Journal of Differential Geometry*, 34:515–537, 1991.

[251] C. Plaut. Spaces of Wald-Berestowskii Curvature Bounded Below. *Journal of Geometric Analysis*, 6(1):113–134, 1996.

[252] C. Plaut. *Metric Spaces of Curvature $\geq k$*, pages 819–898. Elsevier, Amsterdam., 2017.

[253] A. V Pogorelov. Quasigeodesic lines on a convex surface. *American Mathematical Society Translation*, 6(72):430–472, 1952.

[254] W. K. Pratt. *Digital Image Processing.* Joh Whiley & Sons, Inc., N.Y., 2001.

[255] J. R. Weeks. *The Shape of Space.* Marcel Dekker, Inc., New York - Basel, 2002.

[256] H. E. Rauch. Geodesics, symmetric spaces, and differential geometry in the large. *Commentarii Mathematici Helvetici*, 27:38–55, 1951.

[257] T. Regge. General relativity without coordinates. *Nuovo Cimento*, 19:558–571, 1961.

[258] T. Regge and R. M. Williams. Discrete structures in gravity. *Journal of Mathematical Physics*, 41(6):3964–3984, 2000.

[259] Thomas Richard. Canonical smoothing of compact alexandrov surfaces via Ricci flow, 2012.

[260] W. Rinow. *Die innere Geometrie der metrischen Räume.* Springer, Berlin, 1961.

[261] Igor Rivin. An extended correction to "combinatorial scalar curvature and rigidity of ball packings," (by D. Cooper and I. rivin), 2003.

[262] C. V. Robinson. A Simple Way of Computing the Gauss Curvature of a Surface. *Reports of a Mathematical Colloquium, Second Series*, 5-6:16–24, 1944.

[263] Jim Ruppert. A delaunay refinement algorithm for quality 2-dimensional mesh generation. *Journal of Algorithms*, 18(3):548–585, 1995.

[264] A. Samal, H. K. Pharasi, S. J. Ramaia, E. Saucan, J. Jost, and A. Chakraborti. Network geometry and market instability. *Royal Society Open Science*, 8:201734, 2021.

[265] A. Samal, R. P. Sreejith, J. Gu, S. Liu, E. Saucan, and J. Jost. Comparative analysis of two discretizations of Ricci curvature for complex networks. *Scientific Reports*, 8(1):8650, 2018.

[266] R. Sandhu, T. Georgiou, E. Reznik, L. Zhu, I. Kolesov, Y. Senbabaoglu, and A. Tannenbaum. Graph curvature for differentiating cancer networks. *Scientific Reports*, 5:1–13, 2015.

[267] G. Sapiro. Vector (self) snakes: a geometric framework for color, texture, and multiscale image segmentation. In *Proceedings of 3rd IEEE International Conference on Image Processing*, volume 1, pages 817–820, 1996.

[268] E. Saucan. Discrete Morse Theory, Persistent Homology and Forman-Ricci Curvature. *Mathematics, Computation and Geometry of Data*, 2, 2002.

[269] E. Saucan. Note on a theorem of Munkres. *Mediterranean Journal of Mathematics*, 2(2):215–229, 2005.

[270] E. Saucan. A place for differential geometry? In *Proceedings of The 4th International Colloquium on the Didactics of Mathematics*, volume II, pages 267—276, 2005.

[271] E. Saucan. *Curvature – Smooth, Piecewise-Linear and Metric*, pages 237–268. Advanced Studies in Mathematics and Logic. Polimetrica, Milano, 2006.

[272] E. Saucan. The existence of quasimeromorphic mappings in dimension 3. *Conformal Geometry and Dynamics*, 10:21–40, 2006.

[273] E. Saucan. Remarks on the existence of quasimeromorphic mappings. *Contemporary Mathematics*, 455:325–331, 2008.

[274] E. Saucan. Curvature based triangulation of metric measure spaces. *Contemporary Mathematics*, 554:207–227, 2011.

[275] E. Saucan. Geometric sampling of infinite dimensional signals. *Sampling Theory in Signal and Image Processing*, 10(1-2):59–76, 2011.

[276] E. Saucan. Isometric Embeddings in Imaging and Vision: Facts and Fiction. *Journal of Mathematical Imaging and Vision*, 43(2):143–155, 2012.

[277] E. Saucan. On a construction of Burago and Zalgaller. *The Asian Journal of Mathematics*, 16(4):587–606, 2012.

[278] E. Saucan. Metric Ricci curvature and flow for *pl* manifolds. *Actes des rencontres du C.I.R.M.*, 3(1):119–129, 2013.

[279] E. Saucan. A Metric Ricci Flow for Surfaces and its Applications. *Geometry, Imaging and Computing*, 1(2):259–301, 2014.

[280] E. Saucan. *Metric Curvatures Revisited – A Brief Overview*, volume 2184, pages 63–114. Springer International Publishing, 2017.

[281] E. Saucan. *Metric Curvatures Revisited – A Brief Overview*, pages 63–114. Springer International Publishing, Cham, 2017.

[282] E. Saucan. Can i keep the shape of my network? *preprint*, 2019.

[283] E. Saucan. Metric Curvatures and their Applications 2: Metric Ricci Curvature and Flow. *Mathematics, Computation and Geometry of Data*, 1(1):59–97, 2020.

[284] E. Saucan, Samal A., and Jost J. A Simple Differential Geometry for Complex Networks. *Network Science*, 9, Special Issue S1: Complex Networks: S106–S133, October, 2021.

[285] E. Saucan and E. Appleboim. Metric methods in surface triangulation. In *Proceedings of IMA Conference "Mathematics of Surfaces XIII", Lecture Notes in Computer Science*, volume 5654, pages 335–355. Springer-Verlag, 2000.

[286] E. Saucan and E. Appleboim. Curvature based clustering for dna microarray data analysis. In *Pattern Recognition and Image Analysis, Lecture Notes in Computer Science*, volume 3523, pages 405–412. Springer Berlin Heidelberg, 2005.

[287] E. Saucan, E. Appleboim, E. Barak-Shimron, R. Lev, and Y. Y. Zeevi. Local versus global in quasiconformal mapping for medical imaging. *Journal of Mathematical Imaging and Vision*, 32(3):293–311, 2008.

[288] E. Saucan, E. Appleboim, G. Wolansky, and Y. Zeevi. Combinatorial Ricci curvature for image processing. *MICCAI 2008 Workshop "Manifolds in Medical Imaging: Metrics, Learning and Beyond"*, 2008.

[289] E. Saucan, E. Appleboim, and Y. Y. Zeevi. Sampling and reconstruction of surfaces and higher dimensional manifolds. *Journal of Mathematical Imaging and Vision*, 30(1):105–123, 2008.

[290] E. Saucan, E. Appleboim, and Yehoshua Y. Zeevi. Sampling and reconstruction of surfaces and higher dimensional manifolds. *Journal of Mathematical Imaging and Vision*, 30(1):105–123, 2008.

[291] E. Saucan, E. Appleboim, G. Wolansky, and Y. Y. Zeevi. Combinatorial Ricci curvature and laplacians for image processing. In *Proceedings of 2nd International Congress on Image and Signal Processing (CISP'09)*, volume 2, pages 992–997, 2009.

[292] E. Saucan, A. Samal, and J. Jost. A simple differential geometry for complex networks. *Network Science*, 9(S1):S106–S133, 2021.

[293] E. Saucan, A. Samal, M. Weber, and J. Jost. Discrete curvatures and network analysis. *MATCH Communications in Mathematical and in Computer Chemistry*, 20(3):605–622, 2018.

[294] E. Saucan, R. P. Sreejith, R. P. Vivek-Ananth, J. Jost, and A. Samal. Discrete Ricci curvatures for directed networks. *Chaos, Solitons & Fractals*, 118:347 – 360, 2019.

[295] E. Saucan and M. Weber. Forman's Ricci curvature - From networks to hypernetworks. In *Proceedings of Complex Networks and Their Applications II, Studies in Computational Intelligence (SCI)*, pages 706–717. Springer, 2019.

[296] Emil Saucan. Surface triangulation - the metric approach. *CoRR*, cs.GR/0401023, 2004.

[297] Emil Saucan and Jurgen Jost. Network topology vs. geometry: From persistent homology to curvature. In *Proceedings of Learning in High Dimensions with Structure Workshop of the Conference on Neural Information Processing Systems (NIPS LHDS 2016)*, 2017.

[298] Richard Schoen and Shing-Tung Yau. *Lectures on Harmonic Maps*. International Proess of Boston, 2013.

[299] Thomas W. Sederberg, Jianmin Zheng, Almaz Bakenov, and Ahmad Nasri. T-splines and t-nurccs. *ACM Transactions on Graphics*, 22:477–484, 2003.

[300] J. Sethian. A fast marching level set method for monotonically advancing fronts. In *Proceedings of National Academy of Sciences*, volume 93, pages 1591–1595. SIAM Press, 1996.

[301] Jonathan Richard Shewchuk. Delaunay refinement algorithms for triangular mesh generation. *Computational Geometry*, 22(1):21–74, 2002. 16th ACM Symposium on Computational Geometry.

[302] P. Shvartsman. Sobolev L_p^2-functions on closed subsets of $\mathbb{R}^2$. *Advances in Mathematics*, 252:22–113, 2014.

[303] J. Sia, E. Jonckheere, and P. Bogdan. Ollivier-Ricci curvature-based method to community detection in complex networks. *Scientific Reports*, 9(1):1–12, 2019.

[304] N. Sochen, R. Kimmel, and R. Malladi. A general framework for low level vision. *IEEE Transactions on Image Processing*, 7(3):310–318, 1998.

[305] D.M.Y. Sommerville. *An Introduction to the Geometry of N Dimensions*. Dover Publications, New York, 1958.

[306] E. Sonn, E. Saucan, E. Appelboim, and Y. Y. Zeevi. Ricci flow for image processing. In *Proceedings of IEEEI*, 2014.

[307] R. Sorkin. The electromagnetic field on a simplicial net. *Journal of Mathematical Physics*, 16(12):2432–2440, 1975.

[308] M. Spivak. *A Comprehensive Introduction to Differential Geometry, Vol. 1, 3d ed.* Publish or Perish, Houston, TX, 1970.

[309] M. Spivak. *A Comprehensive Introduction to Differential Geometry, Vol. 2, 3d ed.* Publish or Perish, Houston, TX, 1970.

[310] M. Spivak. *A Comprehensive Introduction to Differential Geometry, Vol. 3, 3d ed.* Publish or Perish, Houston, TX, 1970.

[311] E. Stokely and Wu. S. Y. Surface parameterization and curvature measurement of arbitrary 3d-objects: Five practical methods. *IEEE Transactions on Pattern Analysis and Machine Intelligence*, 14(8):833–840, August 1992.

[312] D. Stone. Geodesics in piecewise linear manifolds. *Transactions of American Mathematical Society*, 215:1–44, 1976.

[313] D. A. Stone. Sectional curvatures in piecewise linear manifolds. *Bulletin of American Mathematical Society*, 79(5):1060–1063, 1973.

[314] D. A. Stone. A combinatorial analogue of a theorem of myers. *Illinois Journal of Mathematics*, 20(1):12–21, 1976.

[315] D. A. Stone. Correction to my paper: "a combinatorial analogue of a theorem of myers". *Illinois Journal of Mathematics*, 20:551–554, 1976.

[316] D. A. Stone. Geodesics in piecewise linear manifolds. *Transactions of American Mathematical Society*, 215:1–44, 1976.

[317] D. Struik. *Lectures in Classical Differential Geometry, 2nd ed.* Addison-Wesley, 1961.

[318] T. Surazhsky, E. Magid, O. Soldea, G. Elber, and E. Rivlin. A Comparison of Gaussian and Mean Curvatures Estimation Methods on Triangular Meshes. In *Proceedings of the IEEE International Conference on Robotics and Automation Taipei, Taiwan*, pages 1021–1026, 2003.

[319] T. Surazhsky, E. Magid, O. Soldea, G. Elber, and E. Rivlin. A Comparison of Gaussian and Mean Curvatures Estimation Methods on Triangular Meshes. In *Proceedings of the IEEE International Conference on Robotics and Automation. Taipei, Taiwan*, pages 1021–1026, September 2003.

[320] G. Taubin. Estimating the tensor of curvature of a surface from a polyhedral approximation. In *Estimating the Tensor of Curvature of a Surface from a Polyhedral Approximation*, pages 902–907, 1995.

[321] Chan T.F., Sandberg B.Y., and Vese L.A. Active contours without edges for vector-valued images. *Journal of Visual Communication and Image Representation*, 11(2):130–141, 2000.

[322] W. Thurston. *Three-Dimensional Geometry and Topology (Levy S, Ed.)*. Princeton University Press, Princeton, 1997.

[323] P. M. Topping. *Lectures on the Ricci Flow (London Mathematical Society Lecture Note Series 325)*. Cambridge University Press, Cambridge, 2006.

[324] T. Toro. Surfaces with generalized second fundamental form in L^2 are Lipschitz manifolds. *J. Differential Geom.*, 39(1):65–101, 1994.

[325] William F. Trench. *Elementary Differential Equations*. Brooks/Cole, 2001.

[326] Aaron Trout. Positively curved combinatorial 3-manifolds. *The Electronic Journal of Combinatorics*, pages R49–R49, 2010.

[327] P. Tukia. Automorphic Quasimeromorphic Mappings for Torsionless Hyperbolic Groups. *Annales Academiæ Scientiarum Fennicæ Mathematica*, 10:545–560, 1985.

[328] C. Villani. *Optimal Transport, Old and New (Grundlehren der mathematischen Wissenschaften 338)*. Springer, Berlin-Heidelberg, 2009.

[329] Klingenberg W. *A Course in Differential Geometry*. Springer, Berlin, 1977.

[330] A. Wald. Sur la courbure des surfaces. *Comptes Rendus de l'Acadèmie des Sciences*, 201:918–920, 1935.

[331] A. Wald. Begreudeung einer koordinatenlosen Differentialgeometrie der Flächen. *Ergebnisse e. Mathem. Kolloquims, First Series*, 7:24–46, 1936.

[332] H. Wang, D. Pellis, F. Rist, H. Pottmann, and C. Muller. Discrete geodesic parallel coordinates. *ACM Transactions on Graphics (TOG)*, 38(6):1–13, 2019.

[333] Frank W. Warner. *Foundations of Differential Manifolds and Lie Groups GTM 94*. Springer-Verlag Berlin Heidelberg, 1983.

[334] M. Weber, E. Saucan, and J. Jost. Characterizing complex networks with Forman-Ricci curvature and associated geometric flows. *Journal of Complex Networks*, 5(4):527–550, 2017.

[335] M. Weber, E. Saucan, and J. Jost. Coarse geometry of evolving networks. *J. Complex Netwowrks*, 6(5):706–732, 2018.

[336] Melanie Weber, Jürgen Jost, and Emil Saucan. Forman-Ricci flow for change detection in large dynamic data sets. *Axioms*, 5(4):26, 2016.

[337] Melanie Weber, Jürgen Jost, and Emil Saucan. Detecting the coarse geometry in networks. In *International Workshop on Complex Networks and Their Applications (NIPS 2018)*, pages 706–717. Springer, Cham, 2018.

[338] J. Weickert. Anisotropic diffusion in image processing. *Stuttgart: Teubner*, 1, 1998.

[339] R. Weizenböck. *Invariantentheory*. P. Noordhoff, Groningen, 1923.

[340] T. J. Willmore. *Riemannian Geometry*. Clarendon Press, Oxford, 1993.

[341] S.-Q. Xin, Y. He, and C.-W. Fu. Efficiently computing exact geodesic loops within finite steps. *IEEE Transactions on Visualization and Computer Graphics*, 18(6):879–889, 2012.

[342] Otsu Y. On manifolds with small excess. *American Journal of Mathematics*, 115:1229–1280, 1993.

[343] Y. I. Perelman. Geometry for Entertainment, 1929.

[344] Y. Yadav, A. Samal, and E. Saucan. A Poset-Based Approach to Curvature of Hypergraphs. *Symmetry*, 14(2):420, 2022.

[345] N. Yakoya and M. Levine. Range image segmentation based on differential geometry: A hybrid approach. *IEEE Transactions on Pattern Analysis and Machine Intelligence*, 16(6):643–649, June 1989.

[346] X. Ying, X. Wang, and Y. He. Saddle vertex graph (svg): A novel solution to the discrete geodesic problem. *ACM Transactions on Graphics (SIGGRAPH Asia '13)*, 32(6):1–12, 2013.

[347] S. Yoshizawa, A. Belyaev, and H. Yokota. Shape and Image Interrogation with Curvature Extremalities. *Journal for Geometry and Graphics*, 16(1):81–95, June 2001.

[348] Min Zhang, Ren Guo, Wei Zeng, Feng Luo, Shing-Tung Yau, and Xianfeng Gu. The unified discrete surface Ricci flow. *Graphical Models*, 76(5):321–339, 2014.

[349] Xiaopeng Zheng, Yiming Zhu, Wei Chen, Na Lei, and Zhongxuan Luo. Quadrilateral mesh generation III : Optimizing singularity configuration based on Abel-Jacobi theory. *Computer Methods in Applied Mechanics and Engineering*, 387:114146, 2021.

[350] A. Cayley. On a theorem in the geometry of position. *Cambridge Mathematical Journal*, Vol. II:267–271, 1941.

[351] P. M. Topping. Uniqueness and nonuniqueness for Ricci flow on surfaces: Reverse cusp singularities. *International Mathematics Research Notices*, 2012:2356–2376, 2012.

[352] M. Jin, J. Kim, and X. D. Gu. Discrete surface Ricci flow: Theory and applications. LNCS, 4647:209–232, 2007.

[353] E. Appleboim, E. Saucan, and Yehoshua Y. Zeevi. Ricci curvature and flow for image denoising and super-resolution. *Proceedings of EUSIPCO*, 2012:2743–2747, 2012.

[354] G. E. P. Box, D. R. Cox. An analysis of transformations. *Journal of the Royal Statistical Society*, Series B. 26(2):211–252, 1964.

[355] Y. Messerli, S. Bouaziz, E. Appleboim, and E. Saucan. Constrained Local Model with the Discrete 2D, Bochner Laplacian. preprint, 2012.

[356] Kevin A. Murgas, Emil Saucan, and Romeil Sandhu. Quantifying cellular pluripotency and pathway robustness through Forman-Ricci curvature. *Proceedings of COMPLEX NETWORKS AND THEIR APPLICATIONS* 2021, SCI 1016:616–628.

[357] Jakub Bober, Anthea Monod, Emil Saucan, and Kevin Webster, Rewiring Networks for Graph Neural Network Training Using Discrete Geometry. *arXiv:2207.08026v1 [stat.ML]*, 2022.

Index